D1190673

Handbook of
Toxic Properties of Monomers and Additives

Victor O. Sheftel

LEWIS PUBLISHERS

Boca Raton   New York   London   Tokyo

Library of Congress Cataloging-in-Publication Data

Sheftel, Victor O.
  Handbook of toxic properties of monomers and additives.
    p.   cm.
  Includes bibliographical references and index.
  ISBN 1-56670-075-2
   1. Plastics—Additives—Toxicology. 2. Monomers—Toxicology. I. Title.
  RA1242.P66S54 1994
  615.9′51—dc20
                                                   93-47296
                                                     CIP

© 1995 by CRC Press, Inc.
Lewis Publishers is an imprint of CRC Press

No claim to original U.S. Government works
International Standard Book Number 1-56670-075-2
Library of Congress Card Number 93-47296
Printed in the United States of America 1 2 3 4 5 6 7 8 9 0
Printed on acid-free paper

THE AUTHOR

Dr. Victor O. Sheftel is Chief Environmental Toxicologist of the Ministry of Health, State of Israel, Jerusalem.

Dr. Sheftel received his higher education (M.D.) at the Kiev Medical Institute, Ukraine. After having obtained a Ph.D. in 1966, he was a senior researcher and team manager at the All-Union Research Institute of Hygiene and Toxicology of Pesticides, Polymers and Plastic Materials (Kiev, Ukraine). He performed research work aimed at the development of the modern toxicology of plastics as a new branch of applied toxicology.

By 1978 he earned his third degree (D.Sc.) at the Sysin Research Institute of Environmental Hygiene (Moscow) in the Academy of Medical Sciences of the USSR.

Dr. Sheftel has published 12 monographs and over 50 scientific papers, mainly related to the toxicological evaluation of food- and water-contact materials and contaminants, the methodology of regulatory process in this field, and other actual problems of general and applied toxicology. In 1990 Dr. Sheftel emigrated to Israel. At present his work in the Ministry of Health is concerned mainly with the problems of drinking water safety. Simultaneously he is involved in research work targeting the development of a toxicity prediction system for carcinogens, developmental toxicants, etc., on the basis of advanced software for building quantitative structure–activity relationship models.

PREFACE

Plastics or *polymeric materials* (PM) appear to be an important source of chemical contamination of food and environment, along with industrial wastes and pesticides. For the last several decades, the safety assessment of plastics intended for use in contact with foodstuffs or drinking water continues to present a serious challenge for industry and regulatory agencies.

Toxicology of plastics studies a potential hazard of polymeric materials and their ingredients for human health and develops recommendations for production and safe use of such materials.

Since it is a comparatively new and insufficiently investigated branch of applied toxicology, the experimental data obtained in this field hitherto have not been collected and generalized. Incomplete and fragmentary information on the subject could be found in the following sources: Patty's *Industrial Hygiene and Toxicology,* 3rd ed. (1982); *Practical Toxicology of Plastics,* CRC Press, 1968, and *Les Matieres Plastiques dans l'Industrie Alimentaire,* Paris, 1972 (in French), by R. Lefaux; *Industrial Hazards of Plastics and Synthetic Elastomers,* by J. Jarvisalo, P. Pkaffli, and H. Vainio, Eds., Alan R. Liss, Inc. (1984). Unfortunately all of these books are already out of date.

This handbook is an attempt to provide comprehensive information on the toxic effects of plastics ingredients that enter the body mainly by the oral route and thus it may serve as a sort of encyclopedia for specialists and practitioners in their field. Basic toxicological and other scientific data necessary to identify, characterize, measure, and predict hazards of plastic-like materials use have been assembled from the scientific literature and from regulatory national and international documents.

The contents of this handbook factually overstep the limits of toxicology of plastics because they comprise information concerning many of the widespread food and water contaminants: heavy metals, solvents, monomers, plasticizers, etc. Because toxic properties of PM depend on toxic properties of the substances released by them, this book will be of use when assessing toxic properties not only of the existing plastics but also of future materials containing ingredients that have already undergone toxicological evaluation.

The handbook includes the thoroughly reviewed American and European toxicology literature, as well as the screened Russian toxicology data, unknown in the West but reasonably fit to interpretation. Since the toxic properties of PM are determined by the toxic properties of the substances released by them, toxic hazard assessment of ingredients migrating into food and water is the essential part either of new material development or of the regulatory decision-making process.

It should be borne in mind that assessment of toxic potential is an exclusively complicated task even for those who have many years of experience in toxicology testing. Such assessment requires examining and scrutinizing of complex data that describe the toxicology of a substance, selecting the appropriate valid information from often conflicting or incomplete data, and arriving at a conclusion on the relevance of the information to human health risk.

For these and other reasons, it has not been possible to assess exactly a relative validity either of data obtained before GLP implementation or of ''Russian'' toxicology data. On the other hand, a toxicology profile will be incomplete today if only findings of the last decade are taken into account. Russian toxicology developed separately from that of the West. Since much data collected in Russia is unique, it cannot be ignored, even if it does not completely conform to GLP requirements. In every case, references should enable easy identification of the place and year of each research cited.

The following remarks concerning presentation of the data in this handbook are to be made:

1. For each chemical, the following data are provided when available:

   - Substance Prime Name
   - Synonyms
   - CAS Number
   - Properties (sometimes Composition)
   - Applications and Exposure (sometimes Migration Levels)
   - Acute Toxicity
   - Repeated Exposure
   - Short-term Toxicity

- Long-term Toxicity
- Immunotoxicity or Allergenic Effect
- Reproductive Toxicity (Embryotoxicity, Teratogenicity, and Gonadotoxicity)
- Genotoxicity
- Carcinogenicity (including IARC, USEPA and NTP Cancer Classification)
- Chemobiokinetics
- Standards, Guidelines, Regulations, Recommendations
- References

2. The oral route of administration is intended throughout the book unless otherwise specified. In the absence of oral toxicity data, information from inhalation or dermal toxicity studies, as well as from administration via i/p, i/v, or other routes, is presented.

3. Description of the common toxic effects is subdivided into acute, repeated, short-term, and long-term (chronic) toxicity: **acute toxicity** refers to the result of a single oral exposure; **repeated exposure** refers to a length of exposure being about 2 weeks to 2 months; **short-term toxicity** refers to a period of treatment not less than 3 to 4 months; **long-term toxicity** refers to a length of exposure not less than 6 months.

4. A quantitative assessment of functional accumulation is given. **Coefficient of accumulation** (on the lethal level) has been determined by one of the following three methods: the method of Yu. S. Kagan and V. V. Stankevich (designated as ''by Kagan'') stipulates administration of the agent to experimental animals at equal daily doses of $\frac{1}{5}$ to $\frac{1}{50}$ LD_{50} for 2 to 4 months; the method of R. K. Lim, et al. (designated as ''by Lim'') stipulates administration of the substance at gradually increasing doses, beginning with $\frac{1}{10}$ LD_{50} for no more than 4 weeks; the method of S. N. Cherkinsky, et al. (designated as ''by Cherkinsky'') stipulates administration of $\frac{1}{5}$ LD_{50} for 20 days.

5. Data on certain long-term or ''delayed'' effects, in particular carcinogenicity, may not present appropriate information about the safe levels. These data are usually obtained at high dose-levels, and recommendations for carcinogenicity evaluation have been computed from hypothetical mathematical models that cannot be verified experimentally.

6. There are three most acknowledged **carcinogenicity classifications:** IARC, USEPA and NTP. All substances tested could be subdivided according to **IARC weight of evidence** for carcinogenicity: **1**—*Human carcinogens;* **2A**—*Probable carcinogens;* **2B**—*Possible carcinogens;* **3**—*Not classified;* **4**—*Probably not carcinogenic to humans.* According to **USEPA weight of evidence** for carcinogenicity the classification is as follows: **A**—*Human carcinogens;* **B1** and **B2**—*Probable human carcinogens;* **C**—*Possible human carcinogens;* **D**—*Not classified;* **E**— *No evidence of carcinogenicity in humans.* **NTP categorization** according to the weight of the experimental evidence is presented by the following groups: **CE**—*Clear evidence of carcinogenic activity;* **SE**—*Some evidence of carcinogenic activity;* **EE**—*Equivocal evidence of carcinogenic activity;* **NE**—*No evidence of carcinogenic activity;* **IS**—*Inadequate study of carcinogenic activity.* Earlier NTP designations are as follows: **P**—*Positive;* **E**—*Equivocal;* **N**— *Negative.* Designations of this categorization are displayed in the following order: *Male Rats— Female Rats—Male Mice—Female Mice.* Absence of the data is designated as **XX.** Rodent carcinogens with significantly elevated tumor rate at some dose(s) below MTD are marked with * (according to J. K. Haseman and A. Lockhart, 1993).

7. Within chapters, ingredients are placed in the English alphabetical order of their prime names, ignoring special characters such as Greek letters or numerals. Toxicity data are transformed into a special, newly developed format to facilitate the use of available toxicology information to evaluate potential migration levels of plastic ingredients into food or drinking water.

8. WHO, EEC, U.S., and some other available national standards, guidelines, and recommendations, taken from the following sources, are presented: ''WHO Guidelines for Drinking-Water Quality,'' November, 1992; ''Council Directive of 15 July 1980,'' *Official Journal of European Communities,* 30, 8, August 1980; ''Drinking Water Regulations and Health Advisories'' by USEPA, 1991; Commission Directive 90/128/EEC of 23 February 1990 relating to plastics materials and articles intended to come in contact with foodstuffs,'' *Official Journal of European Communities,* 33(L75), 19, 1990; Code of Federal Regulations, Food and Drugs, 21 CFR Part 175–179, 1993; List of Maximum Allowable Concentrations set by Ministry of Health, the

USSR, Appendix 2, Sanitary Rules and Standards, No. 4630–88, 1988. References to USFDA are cited according to CFR.

9. Definitions of abbreviations:

ADI—Acceptable Daily Intake
CFR—U.S. Code of Federal Regulations
DWEL—Drinking Water Equivalent Level
ET$_{50}$—Median time from ingestion up to death of animals after LD$_{50}$ administration
GLP—Good Laboratory Practice
GRAS—Generally Recognized As Safe
K$_{acc}$—Coefficient of Accumulation
LC$_i$—Lethal Concentration (subscript indicates percentage of mortality)
LD$_i$—Lethal Dose (subscript indicates percentage of mortality)
LOAEL—Lowest-Observed-Adverse-Effect-Level
LOEL—Lowest-Observed-Effect-Level
MAC—Maximum Allowable Concentration in water bodies
MCL—Maximum Contaminant Level in drinking water
MCLG—Maximum Contaminant Level Goal in drinking water
MPC—Maximum Permissible Concentration in food
MTD—Maximum Tolerable Dose
n/m—not monitored
NOAEL—No-Observed-Adverse-Effect-Level
NOEL—No-Observed-Effect-Level
NTP—U.S. National Toxicology Program
organolept.—organoleptic criterion
PML—Permissible Migration Level to food or water
PTWI—Provisional Tolerable Weekly Intake
QM—Maximum Permitted Quantity of the residual substance in the material or article
RfD—Reference Dose
RTECS—Registry of Toxic Effects of Chemical Substances
SML—Specific Migration Limit in food or food simulant
TDI—Tolerable Daily Intake
UF—Uncertainty factor
ALT—alanine aminotransferase
AST—aspartate aminotransferase
BW—body weight
CA—chromosome aberrations
CNS—central nervous system
DLM—dominant lethal mutations
DNA—deoxyribonucleic acid
ECG—electrocardiogram
EEG—electroencephalogram
GI—gastrointestinal
Hb—hemoglobin
LDH—lactate dehydrogenase
MetHb—methemoglobin
NS—nervous system
RNA—ribonucleic acid
STI—summation threshold index
SCE—sister chromatid exchanges
i/g—intragastric administration
i/m—intramuscular injection
i/p—intraperitoneal injection
i/v—intravenous injection
s/c—subcutaneous injection
ppm—parts per million
ppb—parts per billion

10. Reference numbers related to the most repeatedly used literature sources:

- 01—Alexander, H. C., McCarty, W. A., and Bartlett, E. A., Aqueous odour and taste threshold values of industrial chemicals, *J. Am. Water Works Assoc.,* 74, 595, 1982.
- 02—Amoore, J. E. and Hautala, E., Odor as an aid to chemical safety: odor threshold compared with threshold limit values and volatilities for 214 industrial chemicals in air and water dilution, *J. Appl. Toxicol.,* 3, 272, 1983.
- 03—Patty, F. A., Ed., *Industrial Hygiene and Toxicology,* 3rd ed., John Wiley & Sons, New York, 1982.
- 04—Lefaux, R., *Chimie et Toxicologie des Matieres Plastiques,* Paris, 1964 (in French).
- 05—Lefaux, R., *Les Matieres Plastiques dans l'Industrie Alimentaire,* Paris, 1972 (in French).
- 06—*Registry of Toxic Effects of Chemical Substances,* 1985–86 ed., User's Guide, D. V. Sweet, Ed., U.S. Dept. Health and Human Services, Washington, D.C., 1987.
- 07—Sheftel, V. O., *Toxic Properties of Monomers and Additives,* E. Inglis, and S. Dunstall, Eds., Rapra Technology Ltd., U.K., 1990.
- 08—Broitman, A. Ya., Basic Problems of the Toxicology of Synthetic Antioxidants Intended for Stabilizing Plastics, Author's abstract of thesis, Leningrad, 1972, p. 48 (in Russian).
- 09—Gold, L. S., Manley, N. B., Slone, T H., Garfinkel, G. B., Rohrbach, L., and Ames, B. N., The fifth plot of the Carcinogenic Potency Database: results of animal bioassays published in the general literature through 1988 and by the National Toxicology Program through 1989, *Environ. Health Persp.,* 100, 65, 1993.
- 010—*Identification and Treatment of Tastes and Odors in Drinking Water,* Mallevialle, J. and Suffet, I. H., Eds., AWWA Research Foundation, Lyonnaise des Eaux-Dumez, 1987, p. 292.

This handbook is intended for specialists in industry as well as for health care providers, legislators, regulators, scientists, and practitioners of occupational and environmental medicine, various national and international agencies and organizations, national and local governmental authorities and consumer associations, movements such as Greenpeace, etc.

Application of the appropriate data base should present complete and necessary information for many of those concerned with research, application, and legislation relevant to toxic hazards of packaging materials and food-contact coatings and articles. The author has made every effort to ensure that the information presented in this handbook is accurate and up-to-date. Nevertheless, despite reasonable screening and evaluation of presented data, inclusion herein does not imply endorsement of the cited literature. No claims of assurance or liability for the misaccuracy of information presented are assumed, either by the author or the publisher. Final evaluation of the references included is the responsibility of the readers.

Victor O. Sheftel, M.D., Ph.D., D.Sc.
Jerusalem, 1994

CONTENTS

HARMFUL SUBSTANCES IN PLASTICS

As it stands today, not only at the production stage but also in everyday living, it is unlikely that the population could avoid coming into contact with plastics. Approximately 70 to 80% of food is packaged in various *polymeric materials* (PM).

Unfortunately, PM appear to be a potential source of the release of chemicals into the environment; they may have a variety of effects on human health through water, air, or skin contamination. The principal harmful factor associated with the use of PM remains the possible contamination of food.

The absence of acute poisonings with fatal outcomes does not prove the safety of synthetic packaging materials. Nevertheless, it must be remembered that we do not completely realize the real contribution of PM to the actual contamination of foodstuffs.

It is true that PM ingredients do not act like pesticides (or a variety of other highly bioactive substances), and one can hardly expect immediate and pronounced clinical manifestations of their toxic action. The occurrence of acute toxicity due to plastics materials used in contact with food and drinking water is most unlikely, since only trace quantities of toxic substances are likely to migrate. PM may produce chronic effects as the result of repeated ingestion of a number of small doses, each in itself insufficient to cause an immediate acute reaction but in the long term having a cumulative toxic effect. Thus, PM and other widely used chemicals have brought to light the problem of the protracted action of low concentrations of chemicals upon human health.

PM are likely to be a *depot* of organic (sometimes also inorganic) compounds which, during the ''life-span'', are discharged into the environment, polluting various contact media, such as food, water, air, skin surface, etc. (Whenever food or drinks contact a solid surface, the resulting food contaminants could be called *migrants*). Since PM ingredients have the potential to migrate from the plastics packaging or wrapping materials into the food in measurable amounts and thereby become the *indirect food additives,* these migrants have to be appropriately regulated (Title 21 of the *Code of Federal Regulations,* part 175–179). U.S. CFR lays out general safety requirements for all *indirect food additives* covering the safe use of food-contact PM.

Under the *Food, Drug, and Cosmetic Act,* an industry must show that a new PM having indirect contact with food such as packaging material or can coating material, is safe for the intended use. Appropriate toxicology information, along with the results of animal toxicity tests, is submitted to the *Food and Drug Administration* (FDA) for review as a part of a *Food Additive Petition.*

Thus, a toxicological evaluation of extractable contaminants is essential in selecting materials for use in contact with food and/or drinking water for verifying that even if migration occurs, no known toxic hazard will exist to the consumer.

It is known that food contact applications are numerous and include the use of plastics, cellulose, paper, aluminum foil, glass, rubber, printing inks, and coatings. PM are widely used in particular in contact with foodstuffs, namely, in food processing equipment, food utensils, and as food packaging.

PM are manufactured by polymerization or polycondensation of one or more monomers and/or other starting substances. As basic polymers the following compounds are most widely used (21 CFR):

- Vinyl resinous substances (polyvinyl acetate, polyvinyl alcohol, polyvinyl butyral, polyvinyl chloride, polyvinyl formal, polyvinylidene chloride, polyvinyl pyrrolidone, polyvinyl stearate, a number of polyvinyl chloride copolymers);
- Styrene polymers (polystyrene α-methyl styrene polymer, styrene copolymers with acrylonitrile and α-methyl styrene);
- Polyethylene and its copolymers;
- Polypropylene and its copolymers;
- Acrylics and their copolymers;
- Elastomers (butadiene-acrylonitrile copolymer, butadiene-acrylonitrile-styrene copolymer, butadiene-styrene copolymer, butyl rubber, chlorinated rubber, 2-chloro-1,3-butadiene/neoprene, natural rubber, polyisobutylene, rubber hydrochloride, styrene-isobutylene copolymer).

In the preparation of PM, numerous additives are used, and the nature of these is dependent on the type of polymer being produced. Examples of the additives that may be used are *plasticizers, antioxidants, catalysts, suspension and emulsifying agents, stabilizers, and polymerization inhibitors, pigments, fillers,* etc. These additives are bound either chemically or physically into the polymer and may be present in their original or altered form. In addition, the polymerization process may leave trace quantities of residual monomer or low-molecular-mass polymer in the PM. It is therefore necessary to specify the purity of the polymer to be used in the preparation of PM intended for food and/or drinking water contact use.

Subsequently, PM contain a spectrum of polymers of different molecular mass, products of side reactions, and residues of all the auxiliary chemicals.

The migration potency of PM is predominantly determined by the availability of unpolymerized monomers, residues of catalysts, transformation products of other starting ingredients, and destruction products (about 3000 PM components are listed by the EC Commission as conceivable migrants). The structure of PM could be changed through time as a result of destruction and ageing processes and of leaching or evaporation of PM ingredients or of their interaction product.

All additives are liable to break down during processing; some, such as antioxidants, are intended to do so to fulfill their function. A number of PM release not only additives into the environment but also monomers that could be present in PM as residues or have appeared as a result of destructive processes. As a matter of fact, PM appears to be a complicated and mobile system that is more or less stable, depending on its age, production technology, and conditions of actual use.

Potential migrants encompass a large group of substances of differing molecular mass and physical properties. In some cases it is impossible to determine accurate amounts of ingredients migrating from PM into contact media. Migration levels can be affected considerably by destruction processes and ageing of plastics, by the presence of unbound low-molecular compounds. The extent to which migration occurs will depend upon such factors as the contact area, the rate of transfer, the type of PM, the temperature, and the contact time. The migration of substances from PM into food is also related to the type of food packaged in PM. Alcoholic beverages and edible fats and oils will extract substances more readily than dry foods such as cereals.

The *high-molecular mass polymer* itself does not pose a toxic hazard, being inert and essentially insoluble in food. *Monomers* are very reactive and biologically aggressive. Some of them have been shown to cause allergic effects, to damage the liver and reproductive functions, and to induce carcinogenicity.

Plasticizers are used to assist processing and impart flexibility to plastics. Plasticizers can be present in food packaging materials in significant amounts and have the potential to migrate into food.

In the food packaging and food processing industries, the major plasticizers used are *di(2-ethylhexyl) adipate* [DEHA], polymeric species, *epoxidized soybean oil* [ESBO], and *acetyl tributyl citrate* [ATBC] in packaging films and *di(-2ethylhexyl) phthalate* [DEHP], *diisodecyl phthalate,* and *diisooctyl phthalate* in closure seals for containers.

Those currently used in polyvinyl chloride [PVC] include *phthalates, phosphates, aliphatic dibasic acid esters,* and *polyesters.* The plasticizers most widely used in PVC for food contact applications are DEHA, DEHP, and polymeric species. Different plastics may contain markedly different levels of phthalate plasticizers. ATBC is commonly used in vinylidene chloride copolymers, ESBO is also used in polyvinyl chloride and vinylidene chloride copolymer films.

The main function of *stabilizers* is to prevent destruction of PM. *Antioxidants* are introduced to avoid undesirable oxidation. Stabilizers and antioxidants are not bound to polymeric macromolecules and could be leached easily into contact liquid media. Thermal stabilizers contribute to food contamination with their own residues.

Catalysts and *hardeners* usually occur in the finished product. Catalysts of polycondensation (e.g., alkali and acids) seem to be less aggressive in comparison with catalysts used in polymerization.

A number of PM ingredients listed for use in the U.S. and EEC regulations have been shown to migrate into the foodstuffs and simulant media under ordinary conditions at significant concentrations. Modern chromatography techniques, including capillary gas chromatography and high-performance liquid chromatography (HPLC), together with other highly-selective detectors (FID, electron capture, nitrogen-phosphorus, MS for GC, UV, electrochemical, fluorescent, mass spectrometry for HPLC) allow the separation, identification, and precise determination of the majority of toxic substances migrating from plastics to contact media at the levels required for safety evaluation.

Advances in analytical chemistry have made it possible in many cases to decode the complex set of chemicals released from PM. Analytical chemistry has gradually become a method of routinely monitoring the safety properties of plastics that can estimate or measure directly PM contamination.

In order to prevent or eliminate the risk of a health hazard to a population exposed to PM, U.S. CFR lays out general safety requirements for all indirect food additives covering the safe use of food- and water-contact PM. It regulates the use of such materials and articles in contact with food in accordance with prescribed conditions. EEC regulations of plastic materials and articles intended to come in contact with foodstuffs in some cases prescribe an overall migration limit in food and food simulants and maximum permitted quantities of the 'residual' substance in the plastic materials and articles. These regulations include provisions applicable when checking migration limits and a positive list of monomers and other starting substances that may be used in the manufacture of materials and articles intended to come into contact with foodstuffs.

Neither the CFR nor the EEC regulations explain why they are publishing the long records of PM ingredients. The positive lists produce a certain unfavorable effect, creating the illusion that the PM composition is safe. This opinion, shared by many, is erroneous. Sometimes a legislator must admit reluctantly that "such a list *would offer no tangible benefit* (bolded by the author—V. S.) in terms of safeguarding human health" (Commission Directive 90/128/EEC). Then what benefit can they offer? Positive lists neither contribute to the problem nor help with the solution.

Some approaches, based on an artificial hypothesis aimed to accept a level of toxicological insignificance, have been suggested in order to avoid toxicity testing of all migrants that might be present (the *overall migration test* and the so-called *threshold of regulation*). Unfortunately, so far there is little scope for such approaches since each such recommendation needs to have a firm experimental base.

Successful regulation of PM seems to be possible with the help of the newest analytical methods and achievements of modern experimental and regulatory toxicology. A correct strategy in toxicology of plastics comprises precise chemical analysis of potential contamination of food, water, or simulant media under specified conditions. Obtained in this way, analytical results must be compared with available toxicology data and safety standards.

In the author's opinion (and those who have labored long in the area will notice this), this handbook provides an opportunity to make an advancement in the regulatory toxicology of plastics, specifically to implement the modern toxicology approach instead of the application of an out-of-date limit of total extractives and positive lists.

Chapter 1

MONOMERS

ACETALDEHYDE (CAS No 75–07–0)

Synonyms. Acetic aldehyde; Ethanal; Ethylaldehyde.

Properties. Colorless liquid with a pungent odor of rotten apples. Readily miscible with water and alcohol. Odor perception threshold is reported to be 0.034 mg/l,[02] taste perception threshold is 0.21 mg/l.

Applications and Exposure. Used in the production of synthetic rubbers, alkyd resins, and epoxy compounds. A. occurs in common dietary components such as vegetables, fruits, and beverages, and in tobacco smoke. It is a food additive.

Acute Toxicity. LD_{50} is 1.93 g/kg BW for rats,[1] and 1.2 g/kg BW for mice. The treated animals displayed adynamia and labored respiration, followed by convulsions. Death occurs within 3 to 10 min after administration.

Repeated Exposure revealed very low accumulation of A. because of its rapid decomposition in the body (7 to 8 mg/min in rabbits).[2] Rats received the doses of 25 mg/kg, 125 mg/kg, and 675 mg/kg BW via their drinking water over a period of 4 weeks. Food and liquid intake was decreased in the top-dose group. Hyperkeratosis of the forestomach in the top-dosed rats was the only adverse effect observed. The NOAEL of 125 mg/kg BW was identified in this study.[3]

Long-term Toxicity. Rats were dosed by gavage with 10 mg/kg and 100 mg/kg BW for 6 months. The treatment affected CNS functions and increased arterial pressure.

Reproductive Toxicity. Rats were given A. on days 10 to 12 of gestation. *Embryotoxic* effects comprised a great number of fetal resorptions, edema, microcephaly, hemorrhaging, retardation of fetal growth, and other lesions, including skeletal abnormalities. The treatment resulted in reduced placenta weights and umbilical cord length.[4] Inhalation exposure to 5 mg A./m³ produced embryotoxic effect in rats. Morphology changes in the placenta have also been reported.[5]

Genotoxicity. No data are reported on the genetic and related effects of A. in humans. However, it is capable of inducing gene mutations at the *hprt* locus in human cells.[6] A. is a well-known clastogen and SCE inducer in cultured human and hamster cells.[7] It caused DNA cross-links and CA in human cells *in vitro*. It increased the incidence of SCE in bone marrow cells of mice and hamsters *in vivo*, induced CA in rat embryos exposed *in vivo* and micronuclei in cultured rodent cells (IARC 36–101).

Carcinogenicity. Inhalation exposure increased the incidence of GI tract tumors and carcinomas of the nasal cavity in rats and of the larynx in hamsters.[8] *Carcinogenicity classification.* IARC: Group 2B.

Chemobiokinetics. A. is absorbed and metabolized to **acetic acid** by NAD-dependent aldehyde dehydrogenase.[9] According to Casier and Polet (1959), A. is likely to play a role in the acetylation of coenzyme A and in subsequent synthesis of cholesterol and fatty acids. Urinary excretion appeared to be nonexistent.

Regulations. USFDA (1993) regulates A. as a direct food additive. It is considered to be GRAS for its intended use as a synthetic flavoring substance and adjuvant. A. is also approved for use as a component of phenolic resins in molded articles intended for repeated use in contact with nonacid food (pH above 5).

Standards. Russia (1988). MAC: 0.2 mg/l (organolept., taste).

References:
1. Smyth, H. F., Carpenter, C. P., and Weil, C. S., Range-finding toxicity data: list IV, *AMA Arch. Ind. Health Occup. Med.,* 4, 119, 1951.
2. Tsai, L. M., On the problem of acetaldehyde metabolism in the body, *Gig. Truda Prof. Zabol.,* 12, 33, 1962 (in Russian).
3. Til, H. P., Woutersen, R. A., and Feron, V. J., Evaluation of the oral toxicity of acetaldehyde and formaldehyde in a 4-week drinking water study in rats, *Food Chem. Toxicol.,* 26, 447, 1988.
4. Sreenathan, R. N., Padmanabhan, R., and Singh, S., Teratogenic effects of acetaldehyde in the rat, *Drug Alcohol. Depend.,* 9, 339, 1982.
5. *Structural Principles and Regulation of Compensation Adaptation Reactions,* Omsk Medical Institute, 1986, 72 (in Russian).

6. Sai-Mei He and Bdanbert, Acetaldehyde-induced mutations at the *hprt* locus in human lymphocytes *in vitro, Environ. Mutat.,* 16, 57, 1990.
7. Bohlke, J. U., Singh, S., and Goedde, H. W., Cytogenetic effects of acetaldehyde in lymphocytes of Germans and Japanese: SCE, clastogenic activity, and cell cycle delay, *Hum. Genet.,* 63, 285, 1983.
8. Woutersen, R. A., Appelman, L. M., van Garderen-Hoetmer, A., et al., Inhalation toxicity of acetaldehyde in rats. III. Carcinogenicity study, *Toxicology,* 41, 213, 1986.
9. Brien, J. F. and Loomis, C. W., Pharmacology of acetaldehyde, *Can. J. Physiol. Pharmacol.,* 61, 1, 1983.

ACETONITRILE (CAS No 75–05–8)

Synonyms. Acetic acid nitrile; Cyanomethane; Ethanenitrile; Ethylnitrile; Methanecarbonitrile; Methylcyanide.

Properties. Colorless liquid with an ether-like odor. In water solution, A. is hydrolyzed to form **acetamide** and **acetic acid.** May be reduced to **ethyl amine.** Readily miscible with water and alcohol. Odor perception threshold is 2.4 mg/l (Rubinsky, 1969). According to other data, odor perception threshold is 0.75 mg/l at 60°C. However, odor perception threshold of 300 mg/l is also reported.[02]

Applications. A monomer and a high-polarity organic solvent. Used also in consumer goods such as cosmetics and in various chemical industries and laboratories.

Acute Toxicity. LD_{50} is 3.8 to 3.9 g/kg BW in rats, 0.2 to 0.33 g/kg BW in mice, and 0.14 to 0.35 g/kg BW in guinea pigs (Rubinsky). According to Hashimoto, LD_{50} is 170 to 520 mg/kg BW in mice, which are shown to be among the most susceptible animals to A.[1] Rats exhibited different age sensitivity to acute poisoning; in young animals (BW up to 50 g), LD_{50} was 200 mg/kg BW, while in adult rats (80 to 100 g), it was 3900 mg/kg BW, and in old animals (300 to 400 g), it was 4400 mg/kg BW.[2] Poisoning was accompanied by adynamia or agitation, coordination disorder, convulsions, dyspnea, depression of reflexes, hypothermia, etc. Lung emphysema was found to develop in 3 h after administration of A. Death occurs as a result of respiratory arrest.

Repeated Exposure failed to reveal cumulative properties. Guinea pigs were dosed by gavage with 1/5 and 1/20 LD_{50} for 75 d. The treatment resulted in decreased CO_2 production, reticulocytosis, leukocytosis, and increased content of ascorbic acid in the liver, kidneys, and adrenal glands.[1] Gross pathology examination revealed dystrophic changes in the viscera and reduced spleen relative weights.

Long-term Toxicity. In a 6-month study, guinea pigs were exposed to the oral doses of 0.7 and 3.5 mg/kg BW. The treatment with the higher dose caused reduced catalase activity, increased relative weights of the adrenals, and elevated ascorbic acid levels in the liver and spleen. Histological examination revealed moderate dystrophic changes in the visceral organs (Rubinsky).

Reproductive Toxicity. *Embryotoxicity.* Sprague-Dawley rats were administered the doses of 125 to 175 mg/kg BW on days 6 to 19 of gestation. Maternal toxicity and embryotoxicity effects were observed at the high dose-level.[3] *Teratogenicity.* Inhalation by pregnant animals may produce malformations in the offspring (axial skeletal disorders) at maternal toxic levels. Significant teratogenic effect (exencephalia, medullary hernias, fusion of the ribs) and increased embryolethality were found after ingestion of 100 to 400 mg A./kg BW on day 8 of gestation. The treatment produced a reduction in BW of fetuses. According to Willhite,[4] these malformations were likely to occur because of the release of CN^- during A. metabolism. However, no teratogenic effect had been observed in the study.[3]

Genotoxicity. A. is not mutagenic in the standard test using *Salmonella typhimurium.*

Chemobiokinetics. A. is readily absorbed after ingestion. Cytochrome P-450 IIE1 is a probable catalyst in oxidation of AN to **cyanide** by microsomes.[5] Freeman and Hayes[6] believe the metabolism of A. occurs by cytochrome P-450-dependent pathway and not by a nucleophilic substitution reaction with glutathione. The toxic effects are attributable to the metabolic release of cyanide, but the symptoms of poisoning may be delayed due to slow hepatic metabolism.

Standards. Russia (1988). MAC: 0.7 mg/l (organolept., odor).

References:

1. Hashimoto, K., Toxicology of acetonitrile, *Sangyo-Igaku* 33, 463, 1991 (in Japanese).
2. Kimura, E. T., Ebert, D. M., and Dodge, P. W., Acute toxicity and limits of solvent residue for sixteen organic solvents, *Toxicol. Appl. Pharmacol.,* 19, 699, 1971.
3. Johannsen, F. R., Levinskas, G. J., Berteau, P. E., et al., Evaluation of the teratogenic potential of three aliphatic nitriles in the rat, *Fundam. Appl. Toxicol.,* 7, 33, 1986.
4. Willhite, C. C., Developmental toxicology of acetonitrile in the Syrian golden hamster, *Teratology,* 27, 313, 1983.
5. Feierman, D. E. and Cederbaum, A. I., Role of cytochrome P-450 IIE1 and catalase in the oxidation of acetonitrile to cyanide, *Che. Res. Toxicol.,* 2, 359, 1989.
6. Freeman, J. J. and Hayes, E. P., The metabolism of acetonitrile to cyanide by isolated rat hepatocytes, *Fundam. Appl. Toxicol.,* 8, 263, 1987.

ACROLEIN (CAS No 107–02–8)

Synonyms. Acraldehyde; Acrylaldehyde; Allyl aldehyde; Aqualin; 2-Propenal; 2-Propen-1-on.

Properties. Colorless, volatile, transparent liquid with a pungent odor. A. is fairly soluble in water: 200 g/l at 25°C,[02] and in organic solvents (ethanol, diethyl ether, etc.). Odor perception threshold is reported to be 0.2 mg/l[1] or 0.11 mg/l.[02]

Applications and Exposure. A. is an intermediate in the production of polymers and copolymers of acrylic acid and acrylonitrile, and cycloaliphatic epoxy resins. The product of degradation of synthetic polymers. A. could be detected in cigarette smoke and in volatile components of some foods.

Acute Toxicity. LD_{50} values in Wistar rats, mice, and rabbits are 39 to 56 mg/kg, 28 mg/kg, and 7 mg/kg BW, respectively (IARC 36–143).[2,3] However, LD_{95} for Charles River rats is 11.2 mg/kg BW, and five out of ten rats died following administration of a single oral dose of 10 mg/kg by gavage.[4]

Repeated Exposure. Mice were dosed with 1.5 mg/kg BW for a month. The treatment caused decrease in the food consumption and morphological changes in the liver and kidneys. The NOEL appeared to be 0.17 mg/kg BW.[1]

Long-term Toxicity. Chronic human exposure is unlikely due to severe irritating properties of A.[5] In a 6-month study,[1] disorders of the kidney (protein in the urine) and liver functions (shortening of the prothrombin time) were found in rats. Histological examination revealed pneumonia and reduced relative liver weights. In a 1-year study, beagle dogs received 0.1, 0.5, and 1.5 mg A./kg BW (as 0.1% aqueous solution in gelatin capsules). After 4 weeks the highest dose was increased to 2 mg A./kg. The major test effect was frequent vomiting after dosing. This was considered to be an adaptive effect. Serum albumin, calcium, and total protein values were found to be depressed only in high-dosed animals.[6]

Reproductive Toxicity. A. has not been found to be a selective reproductive toxin in the rat. It produced toxic effects down to a dose-level of 3 mg/kg BW.[7] *Embryotoxicity.* A. is a metabolite of **cyclophosphamide,** which is a known embryotoxic agent. A. produces the embryotoxic effect in rats, inhibiting fetal growth[8] but only at maternal toxicity level. The data on the toxic effect of A. on chick embryos are contradictory.[9,10] *Teratogenicity.* There is no solid evidence that A. produces fetal malformations. Schmidt et al. found no teratogenic activity in Sprague-Dawley rats.[11] A. was not found to be a developmental toxicant or teratogen in New Zealand white rabbits at doses not toxic to the dams (up to 2 mg/kg BW). The bigger doses (4 and 6 mg/kg BW) produced a high incidence of maternal mortality, spontaneous abortions, resorptions, and gastric ulcerations.[12]

Genotoxicity. *Humans.* There are no data available on the genetic and related effects. *Animals.* Assays on mutagenic potential revealed conflicting results. A. is unlikely to be a direct-acting mutagen.[5] It is negative in *Salmonella* mutagenicity bioassay (NTP–92) and does not induce DLM in mice but causes SCE in Chinese hamster ovary cells *in vitro* and is found to be mutagenic to bacteria (IARC 19–479; IARC 36–133).

8

Carcinogenicity. Humans. There is no evidence that A. is a human carcinogen. *Animals.* In a 104-week study, Fisher 344 rats were exposed to A. in their drinking water at a concentration of 625 ppm. No decrease in survival or increase in tumor incidence was reported. Out of 25 female rats, 5 developed adrenal cortical adenomas; 2 out of 20 rats had neoplastic nodules in the adrenal cortex. The authors did not consider the study to be a definitive carcinogenicity bioassay.[3] Parent et al. have revised the tissues from this study and found no proof of A. carcinogenicity.[14] Sprague-Dawley rats received 10 mg A./l in their drinking water for 102 weeks (equivalent to 0.05 to 2.5 mg/kg BW). The only effect was consistent depression of creatinine phosphokinase levels. No microscopic lesions in the treated rats, whether neoplastic or nonneoplastic, were noted.[14] In other chronic studies,[14] A. was given by gavage to Sprague-Dawley rats and CD-1 mice. The treatment did not produce carcinogenic response. In addition, the authors failed to observe any significant systemic effect other than increased mortality and retardation of BW gain. ***Carcinogenicity classification.*** IARC: Group 3.

Chemobiokinetics. After ingestion, A. is found to be readily absorbed in the GI tract of experimental animals. Its transport throughout the body is very low. Being a highly reactive compound, it reacts with the substances present in the tissues at the site of the contact. A. can combine with glutathione. It inhibits metabolism of xenobiotics. A. is likely to be converted into **acrylic acid.**

Standards. Russia (1988). MAC: 0.01 mg/l.

Regulations. USFDA (1993) approved the use of A. (1) as an ingredient of resinous and polymeric coatings for polyolefin films to be safely used as a food-contact surface and subject to certain limitations, (2) as a slimicide in the manufacture of paper and paperboard products to contact with food and (3) as an etherifying agent in the manufacture of food additives or of "modified food starch" in an amount not exceeding that reasonably required to reach the intended effect (GMP) or not exceeding 4% when added alone or 0.6% when added with vinyl acetate.

References:

1. *Proc. Sanitary-Hygiene Medical Institute,* Leningrad, 81, 1965, p. 58 (in Russian).
2. Newell, G. W., Acute and Subacute Toxicity Study of Acrolein at Standard Research Institute, SRI Prg N S–568–2, prepared for Shell Development Corp., Modesto, CA, 1958.
3. Lijinsky, W. and Reuber, M. D., Chronic carcinogenesis studies of acrolein and related compounds, *Toxicol. Ind. Health,* 3, 337, 1987.
4. Draminski, W., Eder, E., and Henschler, D., A new pathway of acrolein metabolism in rats, *Arch. Toxicol.,* 52, 243, 1983.
5. Beauchamp, R.O., Andjelkowich, D. A., Kleigerman, A. D., et al., A critical review of literature on acrolein toxicity, *Crit. Rev. Toxicol.,* 14, 309, 1985.
6. Parent, R. A., Caravello, H. E., Balmer, M. F., et al., One-year toxicity of orally administered acrolein to the beagle dogs, *J. Appl. Toxicol.,* 12, 311, 1992.
7. Parent, R. A., Caravello, H. E., and Hoberman, A. M., Reproductive study of acrolein in two generations of rats, *Fundam. Appl. Toxicol.,* 19, 228, 1992.
8. Mirkes, P. E., Fantel, A. G., Greenaway, R. N., et al., Teratogenicity of cyclophosphamide metabolites: phosphoramide mustard, acrolein, and 4-ketocyclophosphamide in rat embryos cultured *in vitro, Toxicol. Appl. Pharmacol.,* 58, 322, 1981.
9. Chibber, G. and Gilani, S. H., Acrolein and embryogenesis: an experimental study, *Environ. Res.,* 39, 44, 1986.
10. Kankaanpaa, E. E., Hemminki, K., and Vainio, H., Embryotoxicity of acrolein, acrylonitrile and acrylic acid in developing chick embryos, *Toxicol. Lett.,* 4, 93, 1979.
11. Schmidt, B. P. and Schon, H., The post-implantation rodent embryo culture system: a potential prescreen in teratology, *Experientia,* 37, 675, 1981.
12. Parent, R. A., Caravello, H. E., Christian, M. S., et al., Developmental toxicity of acrolein in New Zealand white rabbits, *Fundam. Appl. Toxicol.,* 20, 248, 1993.
13. Parent, R. A., Caravello, H. E., and Long, J. E., Two-year toxicity and carcinogenicity study of acrolein in rats, *J. Appl. Toxicol.,* 12, 131, 1992.
14. Parent, R. A., Caravello, H. E., and Long, J. E., Eighteen month oncogenicity study of acrolein in mice, *J. Am. Coll. Toxicol.,* 10, 647, 1992.

ACRYLAMIDE (CAS No 79–06–1)

Synonyms. Acrylic acid, amide; Ethylenecarboxamide; 2-Propenamide.

Properties. Colorless, odorless crystals. Water solubility is 204 g/100 ml, soluble in alcohol. When hydrolyzed, A. forms **acrylic acid.** Concentrations of 0.01 to 0.5 mg/l give water neither color or opalescence nor any unrequired odor. A concentration of 0.001 to 0.01 mg/l gives water no unrequired taste.[1]

Applications and Exposure. A. is predominantly used as a monomer in the production of poly-acrylamide (PAA), a number of copolymers and synthetic rubber. Polymers of A. are used in the production of lacquers, paints, adhesives, and stomatological compositions. Residual A. monomer occurs in PAA coagulants used in the treatment of drinking water. Generally, the maximum authorized dose of polymer is 1 mg/l, corresponding to a maximum theoretical concentration of 0.0005 mg monomer/l in water. Practical concentration may be lower by a factor of 2 to 3. This applies to the anionic and nonanionic PAA, but residual levels from cationic PAA may be higher. PAA are also used as grouting agents in the construction of drinking water reservoirs and wells. Additional human exposure might result from food owing to the use of PAA in food processing.

Acute Toxicity. LD_{50} is reported to be 120 to 180 mg/kg BW in rats, 100 to 170 mg/kg BW in mice, 150 to 170 mg/kg BW in guinea pigs, and 150 to 280 mg/kg BW in rabbits.[1–3,05] In 12 h after administration of 50 to 200 mg/kg BW, poisoned animals displayed impaired hind limb functioning, convulsions, and diffuse damage to different sections of the NS. Gross pathological examination revealed circulatory disturbances in the parenchymatous organs and brain matter as well as signs of parenchymatous dystrophy of the liver and kidneys. The treatment caused marked lysis of tigroidal matter in the cerebellum neurons.[2] The acute effect threshold for the change in STI value appeared to be 40 mg/kg BW.

Cats seem to be more sensitive to A. than rats and monkeys.

Repeated Exposure. *Humans.* Five individuals were exposed through ingestion and external use of well water contaminated with A.[4] Signs of intoxication included confusion, disorientation, memory disturbances, hallucinations, and truncal ataxia. All exposed persons recovered completely within 4 months. *Animals.* A. exhibited pronounced cumulative properties: administration of $\frac{1}{5}$ and $\frac{1}{20}$ LD_{50} resulted in K_{acc} of 2.2 and 2.4 respectively (by Kagan). According to Lim method, K_{acc} appeared to be 1.07 in mice and 1.56 in rats.[1] A. is shown to reduce the amount of large-diameter fibers in the peripheral nerves and to cause their degeneration. Doses of 50 to 100 mg/kg BW impair functioning of the hind legs, but the front legs remain unimpaired, even with 400 mg/kg BW dose.[3] Repeated administration of the doses up to 50 mg/kg BW by any route produced neuropathy of the peripheral nerves in a majority of laboratory animals, including nonantropoidal monkeys. Changes observed seem to be reversible, but secondary effects include atrophy of the skeletal muscles. Administration of 300 mg A./kg BW produced muscular weakness in the limbs and the loss of reflexes. A dose of 100 mg A./kg BW in the feed caused retardation of BW gain. An addition of 10 to 50 mg A./kg BW to the diet of young rats had no clear toxic effect (IARC 19–93).

Short-term Toxicity. Young rats were given a dose of 31 mg/kg BW for 3 months. The treatment caused retardation of BW gain and anemia. Gross pathological examination revealed edema of the mucosa and submucosa of the stomach and small intestine, moderate epithelial sloughing in the large intestine, dystrophic changes in the brain, parenchymatous dystrophy in the liver and kidneys, and increased relative weights of the liver.[5,6] Rats exposed to A. solution at dose levels of 5 to 20 mg/kg BW for 3 months became weak and dragged their hind legs. The treatment caused retardation of BW gain. Histological examination revealed the loss of axones and degeneration and demyelination of the peripheral nerves.[7] Administration of 10 mg/kg BW to male rats three times a week for 13 weeks resulted in retardation of BW gain observed 7 weeks after the onset of treatment.[3] In one study,[7] macaques received an overall dose of 320 to 450 mg/kg BW until toxic effect developed. Animals displayed reduced BW gain, preceded by neurological disturbances, including loss of equilibrium, tremors, reduction in activity, and coordination disorder. The vibration sensitivity threshold was increased, but there was no change in sensitivity to electrostimulation. A dose of 0.2 mg/kg BW appeared to be ineffective in a 3-month experiment in rats.[8]

Long-term Toxicity. Rats were fed 100 to 400 mg A./kg BW for 6 to 48 weeks. The treatment disrupted nervous conduction in the hind legs and caused degenerative changes in the peripheral NS. Recovery began a long time after administration had ceased (IARC 19–99). Baboons received 10 to 20 mg A./kg BW with their drinking water for 29 to 192 d. The treated animals displayed weakness and ataxia of their limbs, weakness of the eye muscles, and later, development of tetraplegia. Histological examination revealed an extensive effect on the peripheral nerves.[9] The NOAEL in rat chronic toxicity studies was reported to be 0.2 mg/kg[7] or 0.3 mg/kg BW.[05]

Immunotoxicity. Novikov and Ostapova[6,10] found no sensitizing effect during acute or chronic administration of A. With repeated administration of 12 mg A./kg BW, there were changes in immune reactivity. A dose of 0.1 mg/kg BW caused an increase in the amount of plaque-forming cells, but a dose of 0.01 mg/kg BW had no such effect.

Reproductive Toxicity. *Gonadotoxicity.* A. affects germ cells and impairs reproductive function. The dose of 500 mg/kg led to a reduction in testicle weights.[11] A marked degeneration of the seminiferous tubules was observed.[11,12] The NOEL of 1.5 mg/kg BW was identified based on the absence of DLM in male sperm. Rumyantzev et al.[2] did not observe gonadotoxic effect at dose-levels up to 0.05 mg A./kg BW. *Embryotoxicity.* No adverse effect of A. on offspring development was found in the study where Porton rats were given A. with their feed during gestation at doses of 0.2 to 0.4 g A./kg BW with the simultaneous administration of an i/v dose of 0.1 g/kg on days 9, 14, or 21 of gestation.[13] Maternal toxicity at the highest dose-level was demonstrated in mice gavaged with 3 to 45 mg/kg BW on days 6 to 17 of gestation, and in rats given 2.5 to 15 mg/kg BW on days 6 to 20 of gestation. BW was decreased and hind limb splaying occurred in mice only.[14] In another study, the NOEL for embryotoxic effect was found to be 2 mg/kg BW.[1] *Teratogenicity.* No fetal abnormalities were shown after oral administration of high doses.[15,16]

Genotoxicity. Although A. is structurally similar to the mutagens and carcinogens *acrylonitrile* and *vinyl chloride* and is capable of reacting with nucleophilic sites or compounds, published data on its mutagenicity are contradictory. A. is reported to produce a mutagenic effect in the germ cells of mice given it at a dose-level of 500 mg/kg in the diet for 3 weeks.[11] A. induced gene mutations in mammalian cells and CA *in vitro* and *in vivo*.[13] It appeared to be a clastogenic agent, inducing CA, DLM, SCE, and unscheduled DNA synthesis. There was no increase of micronuclei in CD-1 mice.[17] Shelby et al. found that A. produced DLM and translocations in early spermatozoa and late spermatides of the mouse, but not at earlier stages.[18,19] According to Hurt et al.,[20] A. caused unscheduled DNA synthesis in pachytene spermatocytes of rats. According to Knaap et al.,[21] A. exhibits no mutagenic activity in different test-systems, including *Dr. melanogaster, Salmonella,* and a test on gene mutations in murine lymphoma cells.

Carcinogenicity. There was an increased incidence of lung adenomas in mice exposed to average daily doses of 2.7 to 10.7 mg/kg BW.[22] In a 2-year study, rats were administered 0.5 to 2.0 mg/kg BW via drinking water. Increased rates of scrotal mesotheliomas, mammary gland tumors, thyroid adenomas, uterine adenocarcinomas, clitorial gland adenomas, and oral papillomas were found.[23] A. is considered to be a genotoxic carcinogen (WHO, 1991). *Carcinogenicity classification.* IARC: Group 2A; USEPA: Group B2.

Chemobiokinetics. Following ingestion, A. is readily absorbed from the GI tract and widely distributed in the body fluids. Accumulation in the liver and kidneys as well as in the male reproductive system has been demonstrated. A. is likely to cross the placenta. Less than 1% of the administered dose accumulates in the NS. A. is found mainly in the muscles and skin. A. is electrophilic and, subsequently, undergoes reactions with nucleophiles. In rats, biotransformation of A. occurs through glutathione conjugation and decarboxylation. The major metabolite is *N*-acetyl-*S*-(3-amino-3-oxypropyl)cysteine, which accounted for 48% of the oral dose.[24] Despite much research, it is still unclear whether the parent compound or a metabolite is responsible for the observed toxic effects.[25] A. is mainly excreted as metabolites in the urine and bile.

Guidelines. WHO (1992). Guideline value for drinking water: 0.0005 mg/l. Generally A. cannot be detected at concentrations of 0.001 mg/l or less but can be controlled by product specification.

Standards. USEPA (1991). MCL: TT; MCLG: zero. **Russia** (1988). MAC and PML: 0.01 mg/l.

Regulations. USFDA (1993) regulates A. as an indirect food additive, a component of food-contact surfaces for single and repeated use. Polyacrylamide (PAA) solutions food-contact MPC—0.02% (1990); PAA food additive MPC—0.05 to 0.2% (1983), various applications. FDA has established regulations for use of A.: (1) it can be used in washing or to assist in the peeling of fruits and vegetables using lye if the concentration does not exceed 10 mg/l in the wash water and if no more than 0.2% A. is present; (2) **A.-sodium acrylate resins** can be used as boiler water additives in the preparations that will be in contact with food, if the water contains not more than 0.05% by weight of A.; (3) **PAA** can be used as a film former in the imprinting of soft-shell gelatin capsules if not more than 0.2% the monomer is present; (4) **homopolymers and copolymers of A.** may be safely used as food packaging adhesives, as long as the amount used does not exceed that "reasonably required to accomplish the intended effect"; (5) **A.-acrylic acid resins** may be safely used as components in the production of paper or paperboard used for packaging food, as long as the resin contains less than 0.2% residual monomer and the resin does not exceed 2% by weight of the paper or paperboard. In **Germany,** PAA flocculants must not contain more than 0.05% of the monomer.[11] The addition of PAA is limited to 1 mg/l in treated water. The concentration of the monomer will therefore not exceed 0.0005 mg/l in finished water.

Recommendations. The U.K. Committee on Carcinogenicity[26] recommended that the level of A. monomer in polyacrylamide used for drinking water treatment should be reduced to the lowest practicable level.

References:

1. Kozeyeva, E. E., Production and Experimental Investigations on Hygienic Assessment of Acrylamide, Author's abstract of thesis, Moscow, 1980, p. 23 (in Russian).
2. Rumyantsev, G. I., Novikov, S. M., Kozeeva, E. E., et al., Experimental study of biological effects of acrylamide, *Gig. Sanit.,* 9, 38, 1980 (in Russian).
3. Tilson, H. and Cabe, P., The effects of acrylamide given acutely or in repeated doses on fore- and hindlimb function of rats, *Toxicol. Appl. Pharmacol.,* 47, 253, 1979.
4. Igusu, H., Goto, I., Kawamura, Y., et al., Acrylamide encephalo-neuropathy due to well water pollution, *J. Neurol. Neurosurg. Psychiatr.,* 38, 581, 1975.
5. *Physiology of Autonomic Nervous System,* Kuibyshev, 1, 1979, p. 321 (in Russian).
6. Ostapova, I. F., Pharmaceutical Kinetics, Toxicology, and Pharmaceutical Dynamics of Acrylamide, Methacrylamide, and Diacetone Acrylamide, Author's abstract of thesis, Yaroslavl, 1981, 333 (in Russian).
7. Burek, J. D., Albee, R. R., Beyer, J. E., et al., Subchronic toxicity of acrylamide administered to rats in the drinking water followed by up to 144 days of recovery, *J. Environ. Pathol. Toxicol.,* 4, 157, 1980.
8. Maurissen, J. P., Weiss, B., Davis, H. T., et al., Somatosensory thresholds in monkeys exposed to acrylamide, *Toxicol. Appl. Pharmacol.,* 71, 266, 1983.
9. Hopkins, A., The effect of acrylamide on the peripheral nervous system of the baboon, *J. Neurol. Neurosurg. Psychiatr.,* 33, 805, 1970.
10. Novikov, S. M., Problems of Occupational Safety and Industrial Toxicology in Production and Use of Acrylamide, Author's abstract of thesis, Moscow, 1974, 23 (in Russian).
11. Shiraishi, Y., Chromosome aberrations induced by monomeric acrylamide in bone marrow and germ cells of mice, *Mutat. Res.,* 57, 313, 1978.
12. McCollister, D. D., Oyen, F., and Rowe, V. K., Toxicology of acrylamide, *Toxicol. Appl. Pharmacol.,* 6, 172, 1964.
13. Zenick, H., Hope, E., and Smith, M. K., Reproductive toxicity associated with acrylamide treatment in male and female rats, *J. Toxicol. Environ. Health,* 17, 457, 1986.
14. Field, E. A., Price, C. J., Sleet, R. B., et al., Developmental toxicity evaluation of acrylamide in rats and mice, *Fundam. Appl. Toxicol.,* 14, 502, 1990.
15. Edwards, P. M., The insensitivity of the developing rat fetus to the toxic effects of acrylamide, *Chem. Biol. Interact.,* 12, 13, 1976.
16. See **ACROLEIN,** #10.
17. Dearfield, K. L., Abernathy, C. O., Ottley, M. S., et al., Acrylamide: its metabolism, developmental and reproductive effects, genotoxicity, and carcinogenicity, *Mutat. Res.,* 195, 45, 1988.
18. Shelby, M. D., Cain, K. T., Hughes, L. A., et al., Dominant lethal effects of acrylamide in male mice, *Mutat. Res.,* 173, 35, 1986.
19. Shelby, M. D., Cain, K. T., Cornett, C. V., et al., Acrylamide: induction of heritable translocations in male mice, *Environ. Mutat.,* 9, 363, 1987.

20. Hurt, M. E., Bentloy, K., Working, P. K., Effects of acrylamide and acrylonitrile on unscheduled DNA synthesis in rat spermatocytes, *Environ. Mutat.*, 9, 49, 1987.
21. The 14th Annual Conf. Europ. Soc. Environ. Mutagens, Abstracts, Moscow, 1984, 247 (in Russian).
22. Bull, R. J., Robinson, M., Laurie, R. D., et al., Carcinogenic effects of acrylamide in SENCAR and A/J mice, *Cancer Res.*, 44, 107, 1987.
23. Johnson, K. A., Gorzinski, S. J., Bodner, K. M., et al., Chronic toxicity and oncogenicity study on acrylamide incorporated in the drinking water of Fisher 344 rats, *Toxicol. Appl. Pharmacol.*, 85, 154, 1986.
24. Miller, M. J., Carter, D. E., and Sipes, I. G., Pharmacokinetics of acrylamide in Fisher 344 rats, *Toxicol. Appl. Pharmacol.*, 63, 36, 1982.
25. Calleman, C. J., Bergmark, E., and Costa, L. G., Acrylamide is metabolized to glycidamide in the rat: evidence from hemoglobin adduct formation, *Che. Res. Toxicol.*, 3, 406, 1990.
26. 1992 Annual Report of the Committee on Toxicity, Mutagenicity, Carcinogenicity of Chemicals in Food, Consumer Products and the Environment, Dept. of Health, HMSO, London, 1993, 72.

ACRYLIC ACID (AA) (CAS No 79–10–7)

Synonyms. Acroleic acid; Ethylenecarboxylic acid; Propenoic acid; Vinylformic acid.

Properties. Colorless, fuming, corrosive liquid with an acrid odor. Readily soluble in alcohol and water, insoluble in ether. Odor perception threshold is 0.57 or 0.094 mg/l.[010] Taste (practical threshold of 50 mg/l) and odor disappeared entirely after 24 h.[1]

Applications and Exposure. Used in the production of various acrylic esters, raw materials for a wide variety of polymers; human exposure may be expected to occur.

Acute Toxicity. LD_{50} is found to be 830 and 250 mg/kg BW for mice and rabbits, respectively. The LD_{50} values of crystalized AA in rats have been reported to range from 193[2] to 350 mg/kg BW,[3,4,03] with estimated LD_{50} to lie within the range of 2100 to 3200 mg/kg BW. Administration of 0.5 to 1 mmol AA per kilogram BW can cause hyperglycemia development in 4 h after poisoning (Vodicka et al., 1985). Poisoned animals displayed CNS inhibition and convulsions and death within a few days. Histopathological lesions consisted of liver congestion and spleen enlargement.[1]

Repeated Exposure revealed cumulative properties.

Short-term Toxicity. High mortality was noted in Wistar rats dosed by gavage with 150 or 375 mg AA per kilogram BW for 3 months. Gross pathological examination revealed a dose-dependent pronounced irritation in the forestomach and glandular stomach with ulcerations and necroses.[5] In a 3-month drinking water study, the NOEL of 83 mg/kg BW was established in Fisher 344 rats.[6]

Long-term Toxicity. In a 12-month study, Wistar rats were given drinking water containing 120 to 5000 ppm AA (providing doses 9 to 331 mg/kg BW). Feed and water consumption was reduced in the high-dosed groups. There was no indication on systemic toxicity and/or any carcinogenic potential.[5]

Reproductive Toxicity. *Embryotoxicity* and *Teratogenicity*. Doses of 4.7 and 8.0 mg/kg, but not 2.5 mg/kg BW, caused an increase in embryolethality and produced malformations in the embryos. The 8 g/kg BW dose caused a significant increase in the incidence of skeletal abnormalities. The same dose produced a marked embryotoxic effect.[7]

Genotoxicity. Negative results were found in the *Salmonella* mutagenicity assay (NTP-90). AA did not induce micronuclei, unscheduled DNA synthesis, and morphological transformations in cultured Syrian hamster embryo fibroblast cells.[8,9] The rapid clearance of AA in animals and the weight of evidence of genetic toxicity tests in both somatic and germ cells *in vivo* indicate that AA lacks mutagenic potential.[10]

Carcinogenicity. Wistar rats were given AA in their drinking water at concentrations of 0, 120, 400, or 1200 ppm (8, 27, or 78 mg/kg BW) over 26 (males) or 28 (females) months. The study did not reveal any toxic changes or indications of a carcinogenic potential.[5] *Carcinogenicity classification.* IARC: Group 3.

Chemobiokinetics. AA metabolism involves β and ω oxidation in the body; it takes part in β-alanine synthesis. AA alters the processes of phosphorylation. The principal route of detoxication of AA in mammals comprises rapid incorporation into a mitochondrial pathway for propionic acid catabolism that results in the release of CO_2 and possible bioincorporation as **acetate.** Administered dose is excreted as CO_2 (60 to 80%) within 2 to 8 h of oral dosing in rats.[10]

Standards. Russia (1988). MAC and PML: 0.5 mg/l.

Regulations. USFDA (1993) listed AA as an ingredient (1) in resinous and polymeric coatings for polyolefin films to be safely used in the food-contact surface, (2) in homopolymers and copolymers in production of paper and paperboard intended for contact with dry food, and (3) in polyethylene phthalate polymers that may be safely used as articles, or components of plastics intended for use in contact with food.

References:

1. *Industrial Pollution of Water Bodies,* Meditsina, Moscow, 9, 1969, p. 171 (in Russian).
2. Union Carbide Corp., *Toxicology Studies—Acrylic Acid,* Ind. Med. Toxicol. Dept., Glacial, New York, 1977.
3. Carpenter, C. P., Weil, C. S., and Smyth, H. F., Jr., Range-finding toxicity data: list VIII, *Toxicol. Appl. Pharmacol.,* 28, 313, 1974.
4. Plastics, resins, rubbers, fibers, in *Encyclopedia of Polymer Science and Technology,* Vol. 1, N. M. Bikales, Ed., Interscience, New York, 1964, p. 197.
5. Helwig, J., Deckardt, K., and Freisberg, K. O., Subchronic and chronic studies of the effects of oral administration of acrylic acid to rats, *Food Chem. Toxicol.,* 31, 1, 1993.
6. DePass, L. R., Weil, C. S., and Frank, F. R., Acrylic acid $CH_{2=CH-COOH}$, Subchronic and reproductive toxicity studies on acrylic acid in the drinking water of the rat, *Drug Chem. Toxicol.,* 6, 1, 1983.
7. Singh, A. R., Lawrence, W. H., and Autian, J., Embryonic-fetal toxicity and teratogenic effects of a group of methacrylic esters in rats, *J. Dent. Res.,* 51, 1632, 1972.
8. Wiegand, H. J., Schiffmann, D., and Henschler, D., Non-genotoxicity of acrylic acid and *n*-butyl acrylate in a mammalian cell system (SHE cells), *Arch. Toxicol.,* 63, 250, 1989.
9. McCartny, K. I., Thomas, W. C., Aardema, M. J., et al., Genetic toxicology of acrylic acid, *Food Chem. Toxicol.,* 30, 505, 1992.
10. Finch, L. and Frederick, C. B., Rate and route of oxidation of acrylic acid to carbon dioxide in rat liver, *Fundam. Appl. Toxicol.,* 19, 498, 1992.

ACRYLONITRILE (AN) (CAS No 107–13–11)

Synonyms. Cyanoethylene; 2-Propenenitrile; Vinylcyanide.

Properties. Clear, colorless liquid with a slight odor. Forms **acrylamide** when added to water. Water solubility is 73 g/l at 25°C.[02] The threshold of the change of water organoleptic properties is 0.01 to 0.05 mg/l (does not affect the color of water).[1] Odor perception threshold is 9.1 mg/l.[02]

Applications and Exposure. AN is used in the production of acrylic and modacrylic resins and rubbers. Main human exposure occurs in chemical industry. Free AN can migrate into food from plastics when they are used as packaging materials in the food industry (migration levels up to 0.15 mg/kg). Migration of AN from the ABC-polymers (residual AN level 12 mg/kg) was found to be 0.109 mg/l into 8% ethanol solution (exposure 24 h, 49°C) and 0.306 mg/l into water (exposure 10 d, 49°C). From the materials containing 49 mg residual AN/kg, migration was 0.65 and 2.6 mg/l respectively.[2]

Acute Toxicity. The LD_{50} values are reported to be 78 to 150 mg/kg BW in rats, 20 to 35 mg/kg BW in mice, 50 to 90 mg/kg BW in guinea pigs, and 93 to 100 mg/kg BW in rabbits.[1,05] Toxic symptoms included agitation immediately after exposure followed by depression. Death occurred within 5 h from respiratory arrest. Administration of single oral doses of 20 to 80 mg/kg BW to male Sprague-Dawley rats caused dose-dependent acute neurotoxic effects cholinemetic in nature.[3] A single administration of 7.5 to 15 mg AN/kg BW produced CNS inhibition, a reduction in blood serum content of SH-groups and in body temperature.[4] Gross pathological examination revealed integumental hyperemia of the visceral organs, especially of the spleen, and brain degenerative changes. Hemorrhages in the GI tract of animals were caused by immediate irritating effect of AN.

Repeated Exposure failed to reveal cumulative properties. Exposure to high doses of AN caused unspecific changes, which are characteristic of the general adaptation syndrome: neutrophilic leucocytosis with relative lymphopenia. A rapid increase in the content of hormones, enzymes, and biologically active substances was observed. Rats received AN via drinking water. Consumption of AN at a concentration of 0.5 mg/l for 21 d caused an increase in the glutathione content of the liver. Concentrations of 0.01 to 0.2% AN in water or equivalent *i/g* doses given for 7 to 60 d reduced the concentration of plasma corticosterone and caused cortical atrophy in the adrenal glands. Gross pathological examination revealed mucosal hyperplasia in the stomach and duodenum.[5] Administration of AN threshold doses produced hematological changes such as hypochromic anemia, granulocytopenia, and lymphocytosis. A dose dependence was observed.[6]

Short-term Toxicity. Rats received 0.1% aqueous solution of AN (1 mg/l) for 13 weeks. The treated animals displayed emaciation (adult rats) and reduced BW gain (young animals).[05]

Long-term Toxicity. In a 6-month study, pneumonia, chronic gastritis, and gastric polyps were observed in rabbits and rats.[7] Negligible (if any) effect was noted in rats exposed to 0.5 mg/l in drinking water for 2 years.[05]

Immunotoxicity. The sensitizing effect of AN administration was found. *I/a* injections exerted dose-dependent immunodepression.[8]

Reproductive Toxicity. Embryotoxic and teratogenic effects were found in rats gavaged with 65 mg AN/kg BW during gestation.[9] Mehrota et al. reported no postnatal ill-effects in rats given 5 mg AN/kg BW.[10]

Genotoxicity. *Humans.* AN did not enhance the frequency of CA in lymphocytes of the exposed workers. It induced SCE, mutation, and unscheduled DNA synthesis but not CA in human cells *in vitro* (IARC 19–73). *Animals.* AN did not induce DLM or micronuclei in mice or CA in rat bone marrow cells. It bound covalently to rat liver DNA *in vivo* and induced unscheduled DNA synthesis in rat liver but not brain. AN induced cell transformation in several test systems and inhibited intercellular communication in Chinese hamster V79 cells.[11] (IARC 19–73). Zhurkov[12] believes there are no data indicating mutagenic activity of AN for mammals.

Carcinogenicity. AN can produce a carcinogenic effect through alkylation of DNA in the extrahepatic target tissues, stomach, and brain.[13] Rats were exposed to 5 mg AN/kg BW three times for a year. After 131 weeks from the start of the experiment, there was a certain increase in the rate of fibroadenomas and carcinomas of the mammary gland in the test animals as compared to the controls.[14] *I/g* administration caused tumors of the brain, squamous cell carcinoma of the stomach and Zymbal gland, and cancer of the tongue, small intestine and mammary gland to develop.[15] ***Carcinogenicity classification.*** IARC: Group 2A; USEPA: Group B1.

Chemobiokinetics. There is little evidence of AN accumulation in animal tissues following prolonged exposure. The radioisotope method showed the accumulation of AN in the erythrocytes and the liver.[16,17] Distribution in the body is studied by whole-body autoradiography after AN is administered orally. Uptake of radioactivity was seen in the blood, liver, kidney, lung, and adrenal cortex.[18] Metabolism of organic cyanides in the body culminates in their transformation into thiocyanates, and their concentration in the urine of animals increases after poisoning.[19] ***S*-(2-cyanoethyl)mercapturic acid** is likely to be the main metabolite of AN in rats. AN combines with erythrocytes to form an integral molecule. Thiocyanate (SCN^-) and **hydroxyethyl-mercapturic acid** are also found to be the major metabolites of A. *in vivo*. The character of AN excretion depends on the route of administration and species. As much as 90% AN is removed with the urine, and only traces of AN are found in feces. Excretion via expired air is negligible.[20,21]

Standards. EEC (1989). SML: 'not detectable' (detection limit 0.02 mg/kg). **Russia** (1986). MAC: 2 mg/l, PML 0.02 mg/l.

Regulations: USFDA (1993) has banned the use of AN copolymers for beverage containers and proposed to limit its migration from other food-contact materials:

—0.003 mg/in.^{2} in the case of single-use articles having a volume to surface ratio of >10 ml/in.^{2} of food contact surface when extracted to equilibrium at 120°F, and in the case of repeated-use articles when extracted at a time equivalent to initial batch usage;

—0.3 ppm in the case of single-use articles having a volume to surface ratio of <10 ml/in.^{2} calculated on the basis of the volume of the container when extracted to equilibrium at 120°F.

FDA is reviewing all AN regulations to align them with current carcinogenicity data. The review includes an advisory opinion and a subsequent proposal for use of a new type of container that limits migration of AN.

References:
1. *Protection of Water Reservoirs Against Pollution by Industrial Liquid Effluents,* Medgiz, Moscow, 4, 1960, p. 147 (in Russian).
2. Lickly, T. D., Markham, D. A., and Rainey, M. L., The migration of acrylonitrile from acrylonitrile/butadiene/styrene polymers into food-simulating liquids, *Food Chem. Toxicol.,* 29, 25, 1991.
3. Chanayem, B. I., Farooqui, M. Y., Elshabrawy, O., et al., Assessment of the acute acrylonitrile-induced neurotoxicity in rats, *Neurotoxicol. Teratol.,* 13, 499, 988.
4. Stasenkova, K. P., Bondarev, G. I., and Murav'iova, S. I., Evaluation of toxicity of acrylonitrile, *Kauchuk i Resina,* 3, 29, 1978 (in Russian).
5. Szabo, S., Bailey, K. A., Boor, P. J., et al., Acrylonitrile and tissue glutathione—differential effect of acute and chronic interaction, *Biochem. Biophys. Res. Commun.,* 79, 32, 1977.
6. Kuznetzov, P. P. and Shustov, V. Ia., Influence of acrylonitrile on blood indices, *Gig. Truda Prof. Zabol.,* 7, 41, 1986 (in Russian).
7. Grushko, Ya. M., *Harmful Organic Compounds in the Liquid Industrial Effluents,* Khimiya, Leningrad, 1976, p. 23 (in Russian).
8. *Metabolic Aspects of the Effect of Industrial Chemical Compounds on the Body,* Krasnoyarsk, 1982, p. 56 (in Russian).
9. Murray, F. J., Schwetz, B. A., Nitsche, K. D., et al., Teratogenicity of acrylonitrile given to rats by gavage or by inhalation, *Food Cosmet. Toxicol.,* 16, 547, 1978.
10. Mehrota, J., Khannal, V. K., Hussain, R., et al., Biochemical and developmental effects in rats following *in utero* exposure to acrylonitrile: a preliminary report, *Ind. Health,* 26, 251, 1988.
11. Leonard, A., Garny, V., Poncelet, F., et al., Mutagenicity of acrylonitrile in mouse, *Toxicol. Lett.,* 7, 329, 1981.
12. Zhurkov, V. S., Shram, R. L., and Dugan, A. M., Analysis of mutagenic activity of acrylonitrile, *Gig. Sanit.,* 1, 71, 1983 (in Russian).
13. Farooqui, M. Y. G. and Ahmed, A. E., Molecular interaction of acrylonitrile and potassium cyanide with rat blood, *Chem. Biol. Interact.,* 38, 145, 1982.
14. Maltoni, C., Ciliberti, A., and Carretti, D., Experimental contributions in identifying brain potential carcinogens in the petrochemical industry, *Ann. N.Y. Acad. Sci.,* 381, 216, 1982.
15. Bigner, D. D., Bigner, S. H., Burger, P. C., et al., Primary brain tumours in Fisher 344 rats chronically exposed to acrylonitrile in their drinking water, *Food Chem. Toxicol.,* 24, 129, 1986.
16. Sokal, J. A. and Klyszejko-Stefanovicz, L., Nicotinamide adenine dinucleotides in acute poisoning with some toxic agents, *Lodz. Tow. Nauk. Prac. Wydz.,* 3, 104, 1972 (in Polish).
17. Sapota, A., The disposition of ^{14}C-acrylonitrile in rats, *Xenobiotica,* 12, 259, 1982.
18. Sondberg, E. Ch. and Salnina, P., Distribution of 1–[14]C-acrylonitrile in rat and monkey, *Toxicol. Lett.,* 6, 187, 1980.
19. Ahmed, A. E. and Petel, K., Acrylonitrile: *in vivo* metabolism in rats and mice, *Drug Metab. Dispos.,* 9, 219, 1981.
20. Lambotte-Vandepaer, M. and Duverger-van-Bogaert, M., Genotoxic properties of acrylonitrile, *Mutat. Res.,* 134, 49, 1984.
21. Kopecky, J., Zachardova, D., Gut, I., et al., Metabolism of acrylonitrile in rats *in vivo,* *Prac. Lek.,* 31, 203, 1979 (in Czech).

ADIPIC ACID (AA) (CAS No 124–04–9)

Synonyms. Adipinic acid; **1,4-**Butanedicarboxylic acid; **1,6-**Hexanedioic acid.

Properties. White to yellowish prisms with a bone black odor. Water solubility is 1.4% at 15°C, readily soluble in alcohol. AA has no effect on water color, odor, or foaming. Organoleptic perception threshold is 200 mg/l.[1]

Applications and Exposure. Monomer in the production of polyamides, a curing agent for epoxy resins, etc.

Acute Toxicity. In mice, LD_{50} is 4.2 g/kg BW.[2] According to Kropotkin et al. (1981), rats tolerate this dose. AA is a polytropic toxin affecting primarily the parenchymatous organs, NS, and enzyme systems, and disrupting exchange processes.

Repeated Exposure failed to reveal cumulative properties. Rats tolerated administration of 10 g/kg BW. The treatment with 42 to 420 mg/kg BW resulted in retardation of BW gain, a reduction in the STI value and in the total blood serum protein, and changes in the enzyme activity. Gross pathological examination revealed changes in the liver, kidneys, and GI tract.[3]

Long-term Toxicity. In a 6-month study, rabbits were dosed by gavage with 5.0 mg **sodium adipate**/kg BW. The increased catalase activity and reduced cholinesterase activity were noted in the blood serum.[4] AA was found to produce an irritating effect on the kidneys.

Chemobiokinetics. Following ingestion, AA and its polyesters are poorly absorbed. AA metabolism involves its oxidation. AA metabolites are discovered in the urine: urea, glutamic, lactic, β-**ketoadipic,** and **citric acids,** etc.[5]

Standards. Russia (1988). MAC and PML: 2 mg/l.

Regulations. USFDA (1993) approved the use of AA (1) in cross-linked polyester resins that may be safely used as articles or components of articles intended for repeated use in contact with food, (2) in polyurethane resins safely used as food-contact surfaces of articles intended for use in contact with dry food, and (3) in resinous and polymeric coatings of food-contact surfaces of articles intended for use in producing, manufacturing, packing, transporting, or holding food.

References:
1. Andreyev, I. A., The method for rapid elaboration of hygienic norms of the content of chemical compounds (as exemplified by adipinic and sebacic acids) in water bodies, *Gig. Sanit.,* 7, 10, 1985 (in Russian).
2. *Hygiene and Toxicology of High Molecular Weight Compounds and of the Chemical Raw Material Used for Their Synthesis,* Proc. 6th All-Union Conf., Kalinin, B. Yu., Ed., Leningrad, 1979, p. 224 (in Russian).
3. Novikov, Yu. V., Andreyeva, I. A., Ivanov, Yu. V., et al., Hygienic standardization of sebacic and adipic acids in water bodies, *Gig. Sanit.,* 9, 72, 1983 (in Russian).
4. *Protection of Water Reservoirs Against Pollution by Industrial Liquid Effluents,* Meditsina, Moscow, 6, 1964, p. 118. (in Russian).
5. Rusoff, J. J., Baldwin, R. R., Domingues, F. J., et al., Intermediary metabolism of adipic acid, *Toxicol. Appl. Pharmacol.,* 2, 316, 1960.

11-AMINOUNDECANOIC ACID (CAS No 2432–99–7)

Properties. White, crystalline substance.

Applications. Used in undecane (or nylon-11) production.

Acute Toxicity. Male rats tolerate a dose of 21.5 g/kg BW for 14 d. One female out of five died from a dose of 14.7 g/kg BW, and all females died following exposure to 21 g/kg BW.[1]

Long-term Toxicity. Mice and rats were administered 7.5 and 15 g/kg BW for 2 years. The treated animals displayed dose-dependent changes consisting of transitional epithelium hyperplasia in the renal pelvis and bladder in rats of both sexes, changes in the cortical and medullary layers of the kidneys in female rats, and mineralization of the kidneys and vacuolization of the hepatocytes in mice.

Genotoxicity. A. appeared to be negative in the sex-linked recessive mutation test.[2]

Carcinogenicity. B6C3F$_1$ mice were exposed to dietary levels of 7.5 and 15 g/kg BW for 2 years. Only males developed malignant lymphomas. In male rats given the same doses, transitional-cell bladder carcinomas were found. A dose-dependent hyperplasia of the bladder and renal pelvis cells was observed in rats. Nodular neoplasms were found in the liver of males.[1,3] *Carcinogenicity classification.* NTP: P—N—E—P.

Regulations. USFDA (1993) permits the use of nylon 11 resins in articles intended for use in contact with food. They are also approved for use as components of sideseam cements intended for a single use in contact with food.

References:

1. NTP, *Carcinogenesis Bioassay of 11-Aminoundecanoic Acid,* NTP Technical Report, Series No. 216, Research Triangle Park, NC, 1982.
2. Yoon, J. S., Mason, J. M., Valencia, R., et al., Chemical mutagenesis testing in Drosophila. IV. Results of 45 coded compounds tested for the NTP, *Environ. Mutat.,* 7, 349, 1985.
3. Dunnick, J. K., Huff, J. E., Haseman, J. K., et al., Lesions of the urinary tract produced in Fisher 344 rats and B6C3F₁ mice after chronic administration of 11-aminoundecanoic acid, *Fundam. Appl. Toxicol.,* 3, 614, 1983.

ANILINE (CAS No 62–53–3)

Synonyms. Aminobenzene; Aminophen; Anyvim; Benzenamine; Phenylamine.

Properties. Colorless, oily liquid, darkening rapidly in air and light, with an aromatic odor. Water solubility is 34 g/l or 37 g/l,[02] readily soluble in alcohol and fats. Odor threshold concentration is reported to be 50 mg/l,[1] 70 mg/l,[2] or 2 mg/l.[3] A concentration of 5 mg/l gives water a yellow color (Obukhov).

Applications and **Exposure.** Used in the manufacture of plastics, vulcanizates, and dyes.

Acute Toxicity. LD_{50} is found to be 300 to 750 mg/kg BW in rats and 1075 mg/kg BW in mice. According to other data, LD_{50} is 460 mg/kg for mice and 250 mg/kg BW for cats.[4] There was no correlation between the toxic effect severity and methemoglobin formation. Even lethal doses did not produce metHb in mice and rabbits, but it was found in the blood of guinea pigs, cats, and dogs. Poisoning was followed by body tremor. In 20 to 30 min, animals became irresponsive to the external stimuli and assumed side position. In 2 to 3 h, the ears, tails, and legs turned greyish-blue. Persistent anemia developed, followed by reduction in blood viscosity and disturbance in protein exchange. Gross pathological examination revealed a brown color of all organs and tissues.

Repeated Exposure failed to reveal cumulative properties on administration of ⅕ LD_{50} for 5 weeks. Rats were orally exposed to 50 mg/kg BW for 14 d. The treatment resulted in elevated relative weights of the visceral organs, dystrophic and necrobiotic changes in hepatocytes and in the spleen.[5]

Short-term Toxicity. Male Sprague-Dawley rats were given A. at a concentration of 600 ppm in their drinking water for 3 months. The main toxic effect was the formation of MetHb. Toxicity to the hemopoietic system comprised a 65% increase in leukocyte count in 30 d, whereas no changes were recorded at later time points. Erythrocyte count appeared to be significantly decreased. Histopathological examination revealed striking changes in the spleen, including marked red pulp expansion and light-brown pigment of heme origin. Testes size was slightly decreased in 60 d. There was no evidence of neoplasia.[6]

Reproductive Toxicity. Pregnant rats were gavaged with 100 mg/kg BW on days 7 to 20 of gestation.[7] The treatment caused no teratogenic effect.

Genotoxicity. No data are available on the genetic and related effects in humans. Negative results were reported for SCE assay in human cells *in vitro*. A. induced SCE but not micronuclei in the bone marrow cells of mice treated *in vivo;* DNA strand breakage was induced in the liver and kidney of rats *in vivo*. Syrian hamster embryo cells and virus-infected Fisher rat embryo cells were not transformed, but BALB/c 3T3 cells were. A. induced SCE and CA but not DNA strandbreaks or unscheduled DNA synthesis in mammalian cells *in vitro*[8] (IARC 4–27; IARC 27–39).

Carcinogenicity. *Humans.* Increased risk of bladder cancer was strongly associated (p <0.001) with increased occupational exposure to aniline. Simultaneously, *o*-toluidine, a well-known carcinogen, was detected in the air in this study.[9] *Animals.* An oral study in mice did not show any increase in tumor incidence in males and females. In rats, A. produced fibrosarcomas, sarcomas, and hemoangiosarcomas of the spleen and peritoneal cavity (IARC 27–39). *Carcinogenicity classification.* IARC: Group 3; NTP: P*—P—N—N.

Chemobiokinetics. Formation of MetHb could result from the oxidation of Hb by A. metabolites such as **phenylhydroxyaniline,** 2-**aminophenol,** and 4-**aminophenol,** of which phenylhydroxyaniline has been demonstrated to be the principal mediator of A.-induced methemoglobinemia in rats.[10] A. is metabolized via hydroxylation of the aromatic ring. Metabolites (predominantly, **aminophenols, phenylsulphamic acid,** and **acetanimide**) are excreted with the urine in the form of conjugates. See also Parke.[11]

Standards. Russia (1988). MAC and PML: 0.1 mg/l.

References:

1. *Industrial Pollution of Water Reservoirs,* Meditsina, Moscow, 8, 1967, p. 156. (in Russian).
2. Zoetman, B., *Organoleptic Assessment of Water Quality,* Stroyizdat, Moscow, 1984, p. 160 (Russian translation from English).
3. See **ACRYLONITRILE,** #7, 35.
4. Jacobson, K. H., Acute oral toxicity of mono- and dialkyl-ring-substituted derivatives of aniline, *Toxicol. Appl. Pharmacol.,* 22, 153, 1972.
5. Agranovsky, M. Z., Experimental study of lipamide as a means of prevention of aniline methemoglobinemia, *Gig. Sanit.,* 10, 96, 1973 (in Russian).
6. Khan, M. F., Kaphalia, B. S., Boor, P. J., et al., Subchronic toxicity of aniline hydrochloride in rats, *Arch. Environ. Contam. Toxicol.,* 24, 368, 1993.
7. Price, C. G., Tyl, R. W., Marks, T. A., et al., Teratologic and postnatal evaluation of aniline hydrochloride in the Fischer 344 rat, *Toxicol. Appl. Pharmacol.,* 77, 465, 1985.
8. *Proc. 4th Conf. Geneticist Breeders,* Naukova Dumka, Kiev, 1981, p. 143 (in Russian).
9. Ward, E., Carpenter, A., Markowitz, S., et al., Excess number of bladder cancer in workers exposed to *o*-toluidine and aniline, *J. Natl. Cancer Inst.,* 83, 501, 1991.
10. Harrison, J. H. and Jollow, D. J., Role of aniline metabolites in aniline-produced hemolitic anemia, *J. Pharmacol. Exp. Ther.,* 238, 1045, 1986.
11. Parke, D. B., *Biochemistry of Foreign Compounds,* Meditsina, Moscow, 1973, p. 269 (Russian translation from English).

BISPHENOL A (BPA) (CAS No 80–05–7)

Synonyms. Bis(*p*-hydroxyphenyl)propane; Diphenylolpropane; Diane; **4,4'**-Isopropylidenediphenol.

Properties. Greyish or colorless, crystalline powder. Water solubility is 0.04%. Readily soluble in alcohol, insoluble in oils. Odor perception threshold is 50 mg/l; taste perception threshold is 0.25 mg/l (astringent).[1] BPA forms chlorophenol odors (theshold concentration 0.01 mg/l).

Applications and **Exposure.** Monomer in the production of polycarbonates and epoxy, phenolic, ethoxylene, ion-exchange resins, corrosion-resistant unsaturated polyester-styrene resins used in inferior coatings of cans and drums, reinforced pipes, food packaging materials, and vulcanizates. A thermal stabilizer of PVC. Migration into water from epoxy coatings (exposure 7 d, 37°C) was determined to be 0.004 mg/l (Krat et al., 1986).

Acute Toxicity. Oral LD_{50} values were reported to range from 4.24 to 12 g/kg BW in rats, and from 2.4 to 2.5 g/kg BW in mice. In rabbits and guinea pigs, LD_{50} is 4 g/kg BW.[1] Alcohol enhances toxicity of BPA. Poisoning is characterized by a transient agitation followed by CNS inhibition and labored respiration, coordination disorder, and convulsions;[2] profuse diarrhea was observed.[05] Death within 2 h after administration. Gross pathological examination revealed congestion in the visceral organs and fatty dystrophy in the liver and kidneys. BPA has a diuretic effect five times greater than that of urea (Bornmann and Loeser, 1959).

Repeated Exposure failed to reveal cumulative properties. Two rats out of ten died after oral exposure to 1 g/kg BW for 10 d. Histological examination revealed fatty dystrophy of the liver, parenchymatous dystrophy of the renal tubular epithelium, and irritation of the spleen pulp.[2] Rats and rabbits were given a dose of 0.5 g BPA/kg BW for 2 months. The treatment led to a reduction in BW gain and to an increase in the acid resistance of erythrocytes and in the content of oxidized glutathione. The level of **free phenols** in the urine was increased. Rabbits displayed erythropenia and elevated liver and spleen weights.[1]

Long-term Toxicity. A 6-month study in rats and rabbits revealed signs of anemia. The treatment affected the NS, the activity of SH-groups of tissue proteins, and the detoxication system of the body.[1]

Reproductive Toxicity. *Embryotoxicity.* CD-1 mice received 0.25 to 1% BPA in the diet for 18 weeks. The treatment caused a decrease in mean number of litters per pair and in mean number of live pups per litter in 0.5 and 1% treated groups.[3] Sprague-Dawley rats and CD-1 mice were dosed by gastric intubation with 160 to 640 mg/kg and 500 to 1250 mg/kg BW, respectively, on days 6 to 15 of gestation. Higher doses produced fetal toxicity in mice but not in rats and did not alter fetal morphological development in either species.[4] *Teratogenic* effect has been reported in rats but not in mice.[5] *Gonadotoxicity.* BPA has an estrogenic effect.[05]

Genotoxicity. BPA is reported to be negative in *Salmonella* mutagenicity bioassay and mouse lymphoma assay; it did not cause CA in cultured mammalian cells.[6]

Carcinogenicity. In a long-term carcinogenicity bioassay, rats were given 1000 and 2000 ppm in their diet, mice were given 1000 or 5000 ppm (males) and 5000 or 10,000 ppm (females). BPA did not appear to be carcinogenic; however, increased incidence of leukemia in male rats and lymphoma in male mice was associated with the test chemical.[7] *Carcinogenicity classification.* NTP: E—N—N—N.

Chemobiokinetics. BPA metabolism occurs through partial conversion into **phenols,** increasing their urinary content in a free and bound form. BPA is passed from the body unaltered in the urine and feces in the form of **glucuronides.**[8]

Standards. EEC (1990). SML: 3 mg/kg. **Russia** (1988). MAC and PML in food and drinking water: 0.01 mg/l (organolept.).

References:

1. Fedyanina, V. N., Study of the Effect of Epichlorohydrin and Diphenylolpropane Upon the Body, Author's abstract of the thesis, Novosibirsk, 1970, p. 22 (in Russian).
2. *Hygiene and Toxicology of High Molecular Weight Compounds and of the Chemical Row Material Used in Their Synthesis,* Proc. 4th All-Union Conf., S. L. Danishevsky, Ed., Khimiya, Leningrad, 1969, p. 180 (in Russian).
3. Reel, J. R., George, J. D., Lawton, A. D., et al., Bisphenol A: Reproduction and Fertility Assessment in CD-1 Mice When Administered in the Feed, Final study report, NTP/NIENS Contract No. ES-2–504, NTIS Accession N PB86103207, 1985.
4. Morrissey, R. E., George, J. D., Price, C. J., et al., The developmental toxicity of bisphenol A in rats and mice, *Fundam. Appl. Toxicol.,* 8, 571, 1987.
5. Schardein, J. L., *Chemically Induced Birth Defects,* 2nd ed., Marcel Dekker, Inc., 1993, p. 790.
6. Zeiger, E., Haseman, J. K., Shelby, M. D., et al., Evaluation of four *in vitro* genetic toxicity tests for predicting rodent carcinogenicity: confirmation of earlier results with 41 additional chemicals, *Environ. Mutat.,* 16 (Suppl. 18), 1, 1990.
7. NTP Technical Report Series No 215, Litton Bionetics, Inc., 1980.
8. Knaak, J. and Sullivan, L., Metabolism of bisphenol A in the rat, *Toxicol. Appl. Pharmacol.,* 8, 175, 1966.

BISPHENOL A, DIGLYCIDYL ETHER (CAS No 1675–54–3)

Synonyms. 2,2'-[(1-Methylethylidene)bis(4,1-phenyleneoxy-methylene)]bis(oxirane); oligomer 340.

Composition and **Properties.** A mixture of monomer, dimer, trimer, and tetramer. A medium viscosity, unmodified liquid epoxy resin. Poorly soluble in water. Soluble in acetone and toluene.

Applications and **Exposure.** The most common active component of epoxy resins, which are used in protective coatings, in reinforced plastic laminates and composites and adhesives. When used in acrylic-epoxy adhesive, it was found to migrate into meat-and-vegetable-filled pastry at the level of 1.26 µg/cm^2, and into food-simulating liquid (Miglyol 812) at the level of 5.67 µg/cm^2.[1]

Acute Toxicity. LD$_{50}$ is 19.6 ml/kg BW for rats.[2]

Allergenic Effects. *Humans.* B. has been shown to cause allergenic contact dermatitis in humans. The workers exposed to epoxy resins in different factories all gave positive reactions in a patch test.[3] *Animals.* Skin sensitization in experimental animals was established. B. sensitized all animals in a guinea pigs skin maximization test and was classified as an extreme allergen.[4]

Reproductive Toxicity. B. induced prenatal toxicity in New Zealand white rabbits following dermal exposure. Teratogenic effect was not observed.[5]

Genotoxicity. CA were found in the peripheral lymphocytes of workers exposed to B.

Carcinogenicity. B. was tested only by skin applications in mice. It caused increased incidence of epidermal tumors.[6] (IARC 47–245)

Chemobiokinetics. B. is rapidly metabolized in mice via the epoxide groups to form the corresponding **bis-diol**. Metabolites including conjugates are excreted in feces and urine.[7]

Regulations. USFDA (1993) permits the use of epoxy resins as components of coatings that may come into contact with food.

References:

1. Begley, T. H., Biles, J. E., and Hollifield, H. C., Migration of an epoxy adhesive compound into a food-simulating liquid and food from microwave susceptor packaging, *J. Agric. Food Chem.*, 39, 1944, 1991.
2. Weil, C. S., Condra, N., Haun, C., et al., Experimental carcinogenicity of representative epoxides, *Am. Ind. Hyg. Assoc. J.*, 24, 305, 1963.
3. Fregert, S. and Thorgeirsson, A., Patch testing with low molecular oligomers of epoxy resins in humans, *Contact Dermat.*, 3, 301, 1977.
4. Zakova, N., Froehlich, E., and Hess, R., Evaluation of skin carcinogenicity of technical **2,2-**bis(*p*-glycidyloxyphenyl) propane in CF₁ mice, *Food Chem. Toxicol.*, 23, 1081, 1985.
5. Breslin, W. J., Kirk, H. D., and Johnson, K. A., Teratogenic evaluation of diglycidyl ether of bisphenol A (DGEBPA) in New Zealand white rabbits following dermal exposure, *Fundam. Appl. Toxicol.*, 10, 736, 1988.
6. Holland, J. M., Gosslee, D. G., and Williams, H. J., Epidermal carcinogenicity of bis(**2,3**-epoxycyclopentyl) ether, **2,2**-bis(*p*-glycidyloxyphenyl)propane and *m*-phenylenediamine in male and female C3H and C57BL/6 mice, *Cancer Res.*, 39, 1718, 1979.
7. Climie, I. J. G., Hutson, D. H., and Stydin, G., Metabolism of the epoxy resin component **2,2-**bis(*p*-glycidyloxyphenyl)propane, the diglycidyl ether of bisphenol A in the mouse, *Xenobiotica*, 11, 401, 1981.

1, 3-BUTADIENE (CAS No 106–99–0)

Synonyms. Biethylene; Bivinyl; Divinyl; Erythrene; Pyrrolylene; Vinylethylene.

Properties. A colorless gas with a garlic and horseradish odor. Solubility in water is 1.3 g/l at 15°C or 735 mg/lg at 20°C, soluble in alcohol. Odor perception threshold is reported to be 0.45 mg/l[010] or 0.0014 mg/l.[02]

Applications and **Exposure.** A chemical intermediate and a polymer component in the manufacture of synthethic rubbers. Used as comonomer for ABC resins and styrene-butadiene latexes, neoprene elastomers, and other polymers or copolymers. Monomer in the production of a wide range of polymers and copolymers: styrene-butadiene, polybutadiene, butadiene-acrylonitrile, acrylonitrile-butadiene-styrene.

Acute Toxicity. LD_{50} is 5.48 g/kg BW in rats and 3.2 g/kg BW in mice.[1,2]

Short-term Toxicity. Rats were administered the dose of 100 mg/kg BW for 3 months. Manifestations of toxic action included adynamia, decreased BW gain, and depressed erythropoiesis and activity of acetylcholinesterase and cholinesterase. Inhalation of 1250 ppm over a period of 6 to 24 weeks led to changes in the spleen and thymus.[2] **1,3**-B. is found to affect the CNS.

Immunotoxicity. Inhalation exposure to **1,3**-B. resulted in transient changes in the immune system function.[2]

Reproductive Toxicity. Inhalation exposure to **1,3**-B. caused no reduction in fertility of rats, guinea pigs, and rabbits. **1,3**-B. might produce tumors and atrophy of the gonads of mice.[3] Effect on reproduction was observed in Swiss mice but not in Sprague-Dawley rats.[4,5]

Genotoxicity. 1,3-B. is positive in *Salmonella* mutagenicity assay and is assumed to be mutagenic for human embryo cells.[3] It caused sperm head abnormalities in mice but not in rats, and produced micronuclei and SCE.[6]

Carcinogenicity. *Humans.* Workers displayed increased risk for leukemia (IARC 54–273). Inhalation exposure induced a dose-related increase in the incidence of tumor development.[1,7] *Animals.* As a result of inhalation exposure of mice to 625 and 1250 ppm **1,3**-B., the increased incidence of myocardial hemoangiosarcomas, malignant lymphomas, and tumors of the lungs and stomach was reported.[8] Mice are about 1000 times more sensitive than rats. An increased rate of thyroid, pancreas, and mammary gland tumors was found in rats.[9] Carcinogenicity of **1,3**-B. is likely to be the result of its biotransformation to epoxide (see **Chemobiokinetics**). *Carcinogenicity classification.* IARC: Group 2A.

    Chemobiokinetics. In rats, mice, and monkeys, **1,3-B.** undergoes transformation into epoxides. It is metabolized by liver microsomes to **butadiene oxide,** with subsequent conversion into 3-**butane**-1,2-**diol** by microsome epoxyhydrolases. Four other metabolites were found as a result of incubation of butadiene oxide with NADP and microsomes.[10] Human and rat liver tissues produce much less of **1,3-B.** monoepoxide than do those of B6C3F$_1$ mice. The human liver tissue rapidly detoxifies the metabolite, while the mouse liver tissue very slowly degrades **1,3-B.** monoepoxide to nontoxic products.[11]

    Standards. EEC (1990). Maximum Permitted Quantity in materials and aricles is 1 mg/kg, or SML not detectable (detection limit—0.02 mg/kg). **Russia** (1988). MAC: 0.05 mg/l (organolept., odor).

    Regulations. USFDA (1993) regulates **1,3-B.** as an indirect food additive. Polymers or copolymers of **1,3**-butadiene are allowed to be used (1) as components of articles coming in contact with food, (2) in pressure-sensitive adhesives, (3) in resinous or polymeric coatings and rubber articles intended for repeated use, (4) in can-end cements, (5) as a coating or component of coating, limited to a level not to exceed 1.0% by weight of paper and paperboard, (6) in semirigid and rigid acryl polymers in repeated use articles, (7) in acrylonitrile-styrene-butadiene copolymers used in closures with sealing gaskets that may be safely used on containers, and (8) in textiles and textile fibers intended for use in contact with food.

References:

1. Toxicology and Carcinogenesis Studies of **1,3**-Butadiene in B6C3F$_1$ Mice (Inhalation Studies), NTP Technical Report Series 288, Research Triangle Park, NC, 1984.
2. Thurmond, L. M., Lauer, L. D., House, R. V., et al., Effect of short-term inhalation exposure to **1,3**-butadiene on murine immune functions, *Toxicol. Appl. Pharmacol.*, 86, 170, 1986.
3. Rosenthal, S. L., The reproductive effects assessment group's report on the mutagenicity of **1,3**-butadiene and its reactive metabolites, *Environ. Mutat.*, 7, 933, 1985.
4. Hackett, P. L., Sikov, M. R., Mast, T. J., et al., Inhalation Developmental Toxicology Studies of **1,3**-Butadiene in the Rat, Final Report No. NIH-401-ES-40131, Pacific Northwest Laboratory, Richland, WA, 1987.
5. Morrissey, R. E., Schwetz, B. A., Hackett, P. L., et al., Overview of reproductive and developmental toxicity studies of **1,3**-butadiene in rodents, *Environ. Health Perspect.*, 86, 79, 1990.
6. de Meester, C., Genotoxic properties of **1,3**-butadiene, *Mutat. Res.*, 195, 273, 1988.
7. Melnick, R. L., Huff, J., Chou, B. J., et al., Carcinogenicity of **1,3**-butadiene in C57Bl/6 × C3HF$_1$ mice at low exposure concentrations, *Cancer Res.*, 50, 6592, 1990.
8. Huff, J. E., Melnik, R. L., Solleveld, H. R., et al., Multiple organ carcinogenicity of **1,3**-butadiene in B6C3F$_1$ mice after 60 weeks of inhalation exposure, *Science,* 227, 548, 1985.
9. Owen, P. E., Pullinger, D. H., Glaister, J. R., et al., **1,3**-Butadiene: two-year inhalation toxicity/carcinogenicity study in the rat (abstract No. P34), in *26th Congr. Eur. Soc. Toxicol.*, H. Hanhijarvi, Ed., 16–19 June, 1985, University of Kuopio, Kuopio, Finland, 1985, p. 69.
10. Malvoisin, E. and Roberfroid, M., Hepatic microsomal metabolism of **1,3**-butadiene, *Xenobiotica,* 12, 137, 1982.
11. Csanady, G. A., Guengerich, F. P., and Bond, J. A., Comparison of the biotransformation of **1,3**-butadiene and its metabolite, butadiene monoepoxide, by hepatic and pulmonary tissues from humans, rats and mice, *Carcinogenesis,* 13, 1143, 1992.

1,3-BUTANEDIOL (CAS No 107–88–0)

    Synonyms. Butane-1,3-diol; **1,3**-Butylene glycol; β-Butyleneglycol; **1,3**-Dihydroxybutane; Methyl-3-methyleneglycol.

    Properties. Colorless, viscous, hygroscopic liquid. Soluble in water, acetone, and ethanol.

    Applications. Used in the production of polyester plasticizers; a humictant for cellophane.

    Acute Toxicity. The LD$_{50}$ values are reported to be 18.6 to 22.8 g/kg BW for rats, 13 g/kg BW for mice, and 11 g/kg BW for guinea pigs.

Repeated Exposure. Exposure to high dietary levels (about 200 mg/kg BW) produces ketosis in man and rats.[1]

Long-term Toxicity. In a 2-year feeding study in dogs in which they were exposed to the levels of up to 30 g/kg BW, no adverse effects were reported.[1]

Reproductive Toxicity. In a five-generation study, Wistar rats were fed the diet containing 5, 10, and 24% B. Embryolethal effect was noted: no pups were obtained in the high-dose level group of the fifth series of litters. Slight teratogenic action was observed.[2]

Genotoxicity. Mutagenic potential was not demonstrated in the DLM assay.[2]

Carcinogenicity. No tumors were found in the described studies.[2]

Recommendations. JECFA (1980). ADI for man: 0 to 4 mg/kg.

Regulations. USFDA (1993) regulates the use of **1,3**-B. as a secondary direct food additive permitted in food for human consumption. It may also be used (1) as a component of adhesives for food-contact surface, (2) in cellophane for packaging food, (3) in the manufacture of closures with sealing gaskets for food containers, (4) in cross-linked polyester resins for repeated use in contact with food, (5) in polyurethane resins for contact with dry food, (6) in resinous and polymeric coatings for polyolefin films to be safely used as a food-contact surface, (7) as a stabilizer for polymers, and (8) as a defoaming agent that may be safely used in the manufacture of paper and paperboard intended for use in producing, manufacturing, packing, transporting, or holding food.

References:

1. The 23rd Report of the JECFA, Technical Report Series 648, WHO, Geneva, 1980, 15.
2. Hess, F. G., Cox, G. E., Bailey, D. E., et al., Reproduction and teratology study of **1,3**-butanediol in rats, *J. Appl. Toxicol.*, 1, 202, 1981.

1,4-BUTANEDIOL (CAS No 110–63–4)

Synonyms. **1,4**-Butylene glycol; **1,4**-Dihydroxybutane; Tetramethylene glycol.

Properties. Glycerine-like, sweetish liquid with a faint odor. Readily miscible with water. Odor perception threshold is 15 mg/l.[1]

Applications. Used as a monomer in the production of polyurethanes; an intermediate product of the synthesis of the blood substitute *poly-N-vinyl-2-pyrrolidone.*

Acute Toxicity. LD_{50} is reported to be 2060 mg/kg BW for mice, 1525 mg/kg BW for rats, 1200 mg/kg BW for guinea pigs, and 2530 mg/kg BW for rabbits.[1] All rats and mice died within 1 to 2 d after exposure to the doses of 5 and 1 g/kg BW respectively.[2] Poisoning is accompanied by a transient agitation with subsequent depression and respiratory disorders. Gross pathological examination revealed hemorrhages and congestion of the visceral organs and dystrophic changes in the renal tubular epithelium.

Repeated Exposure failed to reveal evidence of accumulation in rats and guinea pigs.

Long-term Toxicity. Mice were dosed by gavage with 30.3 mg/kg BW for 6 months. The treatment caused a reduction in the blood serum cholinesterase activity and dysbalance in serum proteins. Decrease in the blood SH-group content was also noted. There was a reduction in the activity and content of cholinesterase in the organs and increased activity of transaminase of the blood serum.[1]

Standards. Russia (1986). MAC and PML: 0.5 mg/l.

Regulations. USFDA (1993) regulates the use of **1,4**-B. (1) in adhesives as a component of articles intended for use in packaging, transporting, or holding food, (2) as an ingredient of polyurethane resins for contact with dry food, (3) as a defoaming agent that may be used safely in the manufacture of paper and paperboard intended for use in producing, manufacturing, packing, transporting, or holding food, and (4) in cross-linked polyester resins to be used safely as articles or components of articles intended for repeated use in contact with food.

References:

1. Knyshova, S. P., Biological effect and hygienic standardization of **1,4**-butynediol and **1,4**-butanediol, *Gig. Sanit.*, 1, 37, 1968 (in Russian).
2. *Toxicology of New Industrial Chemical Substances,* Vol. 7, Meditsina, Moscow, 1965, p. 5 (in Russian).

n-BUTYL ACRYLATE (CAS No 141–32–2)

Synonyms. Acrylic acid, butyl ester; Butyl-2-propeonate; 2-Propenoic acid, butyl ester.

Properties. Colorless liquid with an unpleasant odor. Water solubility is 1.4 g/l at 20°C, soluble in ethanol, diethyl ether, and acetone. Odor perception threshold is reported to be 0.17 mg/l[1] or 0.0078 mg/l.[02] Taste perception threshold is 0.005 mg/l.[1]

Acute Toxicity. LD_{50} is 4.8 g/kg BW in rats and 5.4 g/kg BW in mice.[2–4] A single administration of 520 mg/kg dose in oil solution to rats produced no edema of the gastric mucosa (unlike methyl- or ethylacrylates) but did cause edema when the same dose was administred in an aqueous solution.[5]

Repeated Exposure revealed no pronounced cumulative properties.[4]

Long-term Toxicity. Rabbits were administered a daily dose of BA for 1 to 16 months. The treatment caused changes in the blood and bone marrow, changes in the ratio of blood serum protein fractions, and dystrophic changes in the liver.[6]

Allergenic Effects. BA was shown to be a sensitizing agent in the guinea pig maximization test.[7]

Reproductive Toxicity. No fetotoxic or teratogenic effects were observed in the inhalation study (700 and 1310 mg/m³) in Sprague-Dawley rats. An increase in postimplantation embryolethality was noted.[8]

Genotoxicity. BA was negative in *Salmonella* mutagenicity assay; it did not produce CA in bone marrow cells or Chinese hamster ovary cells.[9]

Carcinogenicity. BA appeared to be negative in dermal carcinogenicity study in mice.[10] *Carcinogenicity classification.* IARC: Group 3.

Chemobiokinetics. In *in vitro* studies, BA was found to be rapidly eliminated from rat blood and hepatic homogenate due to hydrolysis under the action of unspecific enzymes. The disappearance of BA from the blood may also occur because of its combination with erythrocytes. After *i/p* administration to rats, 6% BA was passed with the urine over the course of 24 h in the form of **mercapturic acid.**[11]

Standards. Russia (1988). MAC and PML: 0.01 mg/l (organolept., odor).

Regulations. USFDA (1993) permits the use of BA (1) as a monomer or comonomer in acrylate polymers or copolymers, (2) in semirigid and modified rigid acrylic and vinyl chloride plastics for food-contact articles, (3) as a component of adhesives for food-contact surface, (4) as an ingredient of resinous and polymeric food-contact coatings, (5) in paper and paperboard intended for use in contact with dry food, (6) in cross-linked polyester resins for repeated use, and (7) in polyethylene phthalate polymers to be safely used as articles or components of articles intended for use in contact with food.

References:

1. *Reports on the Methods for Removing Harmful Substances from Gaseous Emissions and Industrial Effluents,* Abstracts, Dserzhinsk, 1967, p. 37 (in Russian).
2. *Chemical Industry Series: Toxicology and Sanitary Chemistry of Plastics,* 2, 1979, p. 22 (in Russian).
3. See **ACRYLIC ACID, #3.**
4. *Toxicology and Sanitary Chemistry of Plastics,* Abstracts, NIITEKHIM, Moscow, 2, 1979, p. 22 (in Russian).
5. Ghanayem, B. J., Maronpot, R. R., and Mattehews, H. B., Ethyl acrylate-induced gastric toxicity. II. Structure-toxicity relationship and mechanisms, *Toxicol. Appl. Pharmacol.,* 80, 323, 1985.
6. *Problems of Toxicology and Hygiene of Synthetic Rubber Production,* Voronezh, 1968, p. 67 (in Russian).
7. van der Walle, H. B. and Bensink, T., Cross reaction pattern of 26 acrylic monomers on guinea pig skin, *Contact Dermat.,* 8, 376, 1982.
8. Merkle, J. and Klimish, H.-J., *n-* Butyl acrylate: prenatal inhalation toxicity in the rat, *Fundam. Appl. Toxicol.,* 3, 443, 1983.
9. Engelhardt, G. and Klimish, H.-J., *n*-Butyl acrylate cytogenic investigation in the bone marrow of Chinese hamsters and rats after 4-day inhalation, *Fundam. Appl. Toxicol.,* 3, 640, 1983.
10. de Pass, L. W., Fowler, E. N., Meckley, D. R., et al., Dermal oncogenicity bioassays of acrylic acid, ethyl acrylate, and butyl acrylate, *J. Toxicol. Environ. Health,* 14, 115, 1986.

11. Miller, R. R., Young, J. T., Kociba, R. J., et al., Chronic toxicity and oncogenicity bioassay of inhaled ethyl acrylate in Fisher 344 rats and B6C3F₁ mice, *Drug Chem. Toxicol.*, 8, 1, 1985.

BUTYLENE (CAS No 25167–67–3)

Synonyms. 1-Butene; α-Butylene.

Properties. Colorless gas. Water solubility is 425 mg/l. Rapidly volatilizes from the open surface of water. Odor perception threshold is 0.2 mg/l. Has no effect on the color and transparency of water.

Acute Toxicity. Rats tolerate administration of 0.5 ml of 350 mg B. per liter solution without signs of toxicity.

Repeated Exposure. Accumulation of B. is impossible because of its rapid removal from the body. Mice were dosed by gavage with 3.75 mg/kg BW for 4 months. The treatment caused no abnormalities in behavior, BW gain, or oxygen consumption. Gross pathological examination revealed no changes in the visceral organs and their relative weights.

Long-term Toxicity. Rats were fed a dose of 0.05 mg/kg BW for 6 months. In the course of this experiment, no changes were observed in general condition or BW gain of the treated animals. No changes were found in the phagocytic activity of leukocytes, in the activity of cholinesterase in the blood serum, or in conditioned reflex activity.

Chemobiokinetics. Direct chemical interaction between B. and biological media in the body is unlikely to occur. B. is rapidly removed unchanged via the lungs.

Standards. Russia (1988). MAC and PML: 0.2 mg/l.

Regulations. USFDA (1992) regulates B. as a component of adhesives to be safely used in food-contact surfaces.

Reference:

Protection of Water Reservoirs Against Pollution by Industrial Liquid Effluents, Meditsina, Moscow, 7, 1965, p. 28 (in Russian).

BUTYL METHACRYLATE (BMA) (CAS No 97–88–1)

Synonyms. Butyl-2-methylpropenoate; Methacrylic acid, butyl ester.

Properties. Colorless, transparent liquid with a characteristic unpleasant odor. Water solubility is 10.2% (20°C). Odor perception threshold is 0.022 mg/l.[1]

Acute Toxicity. LD_{50} is reported to be 16 to 25 g/kg BW for rats, 12.9 to 15.8 g/kg BW for mice, and 25 g/kg BW for rabbits.[1,2] Poisoning was accompanied with slower movements, depression, reddening of the nose, ears, and paws, dyspnea, and irritation of the GI tract. Death is preceded by loss of coordination and reflexes, side position, and labored respiration.

Repeated Exposure revealed mild cumulative properties of BMA during administration of $^1/_{10}$ and $^1/_5$ LD_{50} to mice for 30 d.[1]

Short-term Toxicity. Rats tolerated administration of 0.9 g/kg BW for 4 months. The animals developed retardation of BW gain from the seventh week of the experiment and increased liver weights.[2]

Long-term Toxicity. Rats and rabbits were dosed by gavage with up to 5 mg BMA/kg BW for 8 to 9 months.[1] The treatment predominantly affected the liver, oxidizing enzymes, and erythrocytes. Changes in CNS function were also noted.

Reproductive Toxicity. Similar to other acrylates, BMA produced embryotoxicity and teratogenicity effects in rats.[3]

Chemobiokinetics. BMA metabolism occurs through its hydrolysis to acid.

Standards. Russia (1988). MAC and PML: 0.02 mg/l (organolept., odor).

Regulations. USFDA (1993) approved the use of BMA as a monomer (1) in resinous and polymeric coatings in a food-contact surface, (2) in semirigid and rigid acrylic and modified acrylic plastics for food-contact surfaces, (3) in cross-linked polyester resins to be safely used as articles or components of articles intended for repeated use in contact with food, and (4) as a solvent in polyester resins.

References:
1. Klimkina, N. V., Ekhina, R. S., and Sergeev, A. N., Experimental data on the MAC of methyl and butyl ether of methacrylic acid, *Gig. Sanit.*, 4, 6, 1976 (in Russian).
2. Shepelskaya, N. R., Hygienic comparative study of some methylmetacrylate-based materials, *Gig. Sanit.*, 1, 93, 1976 (in Russian).
3. See **ACRYLIC ACID, #7.**

Σ-CAPROLACTAM (CAS No 105–60–2)

Synonyms. Aminocaproic lactame; 2-Azacycloheptanone; Cyclohexanoneisooxyme; Hexa-hydro-**2H**-azepin-2-one; **6**-Hexanelactam; **2**-Perhydroazepinone.

Properties. White, hygroscopic, crystalline solid. Water solubility is 525 g/100 ml at 25°C, highly soluble in ethanol and chloroform. Organoleptic perception threshold is reported to be 360 mg/l.[1]

Applications and **Exposure.** Monomer in the production of polycaprolactam commonly known as **nylon 6, perlon, capron,** etc. Readily migrates to water and food simulants due to its high water solubility: concentrations detected up to 5 to 10 mg/l.[2]

Acute Toxicity. LD_{50} values are reported to be 0.93 g/kg BW in mice and 1 g/kg BW in rabbits.[3] LD_{50} is 2.1 g/kg BW in male and 2.5 g/kg BW in female $B6C3F_1$ mice, and 1.6 and 1.2 g/kg BW, respectively, in male and female Fisher 344 rats.[4] According to Polushkin, 1974, LD_{50} appeared to be 0.58 g/kg BW in rats.[5] Poisoning is accompanied by severe convulsions and increased diuresis. Sensitivity to C. decreases in the following order: mice, rabbits, rats, guinea pigs.[6]

Short-term Toxicity. A target organ or tissue for C. toxic action was not found. Administration of 1 g/kg BW to rats resulted in a decrease of their body temperature to 34 to 35°C and a 10 to 20% BW loss. The treatment caused 50% mortality in rats within 4 to 11 d. Rats tolerate administration of 10 mg/kg BW for 30 d without evident toxic effect.[3] Ingestion of C. in the form of 1% paste for 8 weeks caused no changes in behavior, hematological analysis, or skeletal tissues.[7]

Long-term Toxicity. Rats were given drinking water containing 0.5% C. for 12 months. There were no changes in their general condition, hematological indices, or skeletal tissues but a slight decrease in BW gain.[7] In a 6-month study, the dose of 500 mg/kg caused decrease of BW gain, anemia, and reticulocytosis in rabbits. Histological examination revealed changes in the gastric mucosa, brain congestion, and perivascular and pericellular edema.[1]

Allergenic Effects. Caused dermal irritation and sensitization. Guinea pigs developed allergy after inhalation of C. In rats, the same effect was noted after oral treatment.[8]

Reproductive Toxicity. Controversial data are reported. *Humans.* C. is shown to produce an increased rate of various complications during pregnancy in women exposed occupationally.[9] *Animals.* *Embryotoxicity.* C. was added to the diet of rats at doses of up to 10 g/kg. No adverse effect in the parent animals was noted. Decreased BW, food consumption, and nephropathy were the only signs of C. toxicity.[10] There was no effect on reproduction in rats treated with 0.5% aqueous solution of C. for a year.[7] However, in other studies, C. has been shown to modify reproductive function and produce a **fetotoxic** effect.[11] Mice were treated by oral intubation with 6.5 to 6.7 mg/kg BW dose of ^{14}C-caprolactam and displayed rapid transfer of the radioactivity across the placenta with near complete elimination from the fetal and maternal compartments in 24 h.[12] *Teratogenic* effect was found neither in rats nor in rabbits.[13]

Genotoxicity. C. showed contradictory results in the cultured human lymphocytes[4,14] and no activity in a large number of *in vitro* tests. However, it induced sex-linked recessive lethals in *Dr. melanogaster*. Significant increase in the incidence of CA in cultured Chinese hamster lung cells was observed only with the dose of 12 mg/ml. C. appeared to be negative in mouse lymphoma and *Salmonella* mutagenicity assays and in micronuclear tests. It did not cause SCE.[14,15]

Carcinogenicity. No carcinogenic effect was observed in mice and rats after dietary exposure. $B6C3F_1$ mice were fed a ration with 7.5 and 15 g C./kg BW doses for 103 weeks. Fisher 344 rats were exposed to 3.75 and 7.5 g/kg BW for 103 weeks. Slight reduction in BW gain in animals of both sexes was noted. No treatment-related tumors were observed.[4] *Carcinogenicity classification.* IARC: Group 4; NTP: N—N—N—N.

Chemobiokinetics. After oral administration, ^{14}C-caprolactam was readily absorbed from the stomach and distributed throughout the body, including the fetuses.[12] Efficient elimination by the kidneys and liver was observed. When a dose of 300 mg/kg BW was administered to rats for 2 months, C. was found in the liver (0.46 to 0.77 ppm), kidneys (0.66 to 1.43 ppm), adipose tissue (1.13 to 2.53 ppm), and blood (0.7 ppm).[16] Rabbits metabolize C. almost entirely. High oral doses of C. may serve as a nonspecific nitrogen source that can modify patterns of hepatic amino acid metabolism. After *i/p* administration to rats, C. is removed partially unchanged, partially in the form of Σ-aminocapronic acid.[17] When 0.18 mg/kg of [14C]-tagged C. is administered to rats, 80% is removed with the urine during 24 h, 3.5% with the feces, 1.5% with exhaled air. C. and two of its metabolites were found in the urine.

Standards. EEC (1990). SML: 15 mg/kg. **Russia** (1988). PML in food: 0.5 mg/l, PML in drinking water: 0.3 mg/l.

Regulations. USFDA (1993) permits the use of C. (ethylene-ethyl acrylate) graft polymers as a component of side-seam cements intended for use in contact with food. Nylon 6 may be used for processing, handling, and packaging food.

References:

1. See **ACRYLONITRILE,** #1, 156.
2. Sheftel, V. O. and Katayeva, S. Ye., *Migration of Harmful Substances from Polymeric Materials,* Khimiya, Moscow, 1978, p. 168 (in Russian).
3. Lomonova, G. V., Toxicity of caprolactam, *Gig. Truda Prof. Zabol.,* 10, 54, 1966 (in Russian).
4. Carcinogenesis Bioassay of Caprolactam in F344 Rats and B6C3F₁ Mice (Feed Study), NTP Technical Report Series No. 214, Research Triangle Park, NC, 1982.
5. Polushkin, B. V., Toxic properties of caprolactam monomer, *Pharmacol. Toxicol., 27, 234, 1974 (in Russian).*
6. Savelova, V. A., et al., Substantiation of the maximum allowable concentration for caprolactam in water, *Gig. Sanit.,* 1, 80, 1962 (in Russian).
7. Bornmann, G. and Loeser, A., Monomer-Polymer: Studie uber Kunststoffe aus E-Caprolactam oder Bisphenol, *Arzneimittel-Forsch.,* 9, 9, 1959.
8. Baida, N. A. and Khomak, S. A., Allergenic effect of caprolactam in mammals, *Vrachebnoye delo,* 2, 104, 1988 (in Russian).
9. Martynova, A. P., Lotis, V. N., Khadzhiyeva, E. D., et al., Occupational hygiene of women engaged in the production of capron (**6**-handecanone) fiber, *Gig. Truda Prof. Zabol.,* 11, 9, 1972 (in Russian).
10. Serota, D. A., Hoberman, A. M., Friedman, M. A., et al., Three-generation reproduction study with caprolactam on rats, *J. Appl. Toxicol.,* 8, 285, 1988.
11. Khadzhieva, E. D., Effects of caprolactam on female reproductive functions, in *Problem of Hygienic Standardization and Examination of the Delayed Effects,* 1972, p. 68 (in Russian).
12. Waddell, W. J., Marlowe, C., and Friedman, M. A., The distribution of [14C]caprolactam in male, female and pregnant mice, *Food Chem. Toxicol.,* 22, 293, 1984.
13. Gad, S. C., Robinson, K., Serota, D. G., et al., Developmental toxicity in the rat and rabbits, *J. Appl. Toxicol.,* 7, 317, 1987.
14. Krassov, S. V., Ivanov, V. P., Zhurkov, V. S., et al., Mutagenic risk of caprolactam, *Gig. Sanit.,* 7, 64, 1992 (in Russian).
15. See **BISPHENOL A,** #6.
16. Sheftel, V. O., Batuyeva, L. N., and Sova, R. Ye., Comparative investigation of toxic action under constant and decreasing dose regimen, *Gig. Sanit.,* 8, 97, 1978 (in Russian).
17. Goldblatt, M. W., Farguharson, M. E., Bennett, G., et al., Σ-Caprolactam, *Br. J. Ind. Med.,* 11, 1, 1954.

CHLOROPRENE (CAS No 126–99–8)

Synonym. 2-Chloro-1,3-butadiene.

Properties. Colorless, inflammable, volatile, rapidly polymerizing liquid with a strong odor. Solubility in water is 0.5 mg/l, readily soluble in alcohol. Odor perception threshold is 0.1 mg/l at 20°C and 0.05 mg/l at 60°C.[1] C. is very unstable in water. When 0.1 mg/l is introduced, within as little as 15 to 16 s, its specific odor is undetectable due to its volatility.

Applications. A monomer in polychloroprene synthetic rubber manufacturing. Widely used in the production of elastomers; on copolymerization with acrylonitrile and metacrylonitrile, fat-resistant plastics are formed.

Acute Toxicity. LD_{50} is found to be 250 mg/kg BW in rats, and 260 mg/kg BW in mice. Doses of 400 mg/kg and 500 mg/kg BW caused total mortality in rats and mice. Poisoning was accompanied by CNS inhibition.[2] Gross pathological examination revealed congestion, hemorrhages, and dystrophic changes in the CNS and visceral organs.

Repeated Exposure produced little evidence of accumulation.

Long-term Toxicity. In a 6-month study, rats were gavaged with 15 mg/kg BW dose. Gross pathological examination revealed congestion in all the visceral organs and cloudy swelling in the renal tubular epithelium.[1] Male rats were exposed to C. via their drinking water for 6 months.[3] The treatment caused an increase in the activity of β-galactosidase in the liver, blood serum, and seminal secretions, and of inosindiphosphatases in the gonads.

Reproductive Toxicity. *Gonadotoxicity.* Long-term exposure to C. affected the estrus cycle and ovary structure of rats.[4] The activity of maleate dehydrogenase was reduced in the liver and gonads; sperm mobility time was decreased.[3] *Embryotoxicity* and *teratogenicity.* A significant embryolethal effect was observed in rats. Fetal malformations comprised hydrocephalies and hemorrhages into the thoracic and abdominal cavities.[5]

Genotoxicity. *Humans.* CA found in the lymphocytes of workers exposed to C. (IARC Suppl. 6–164). *Animals.* In *in vitro* studies, C. did not cause CA, SCE, or micronucleus induction (NTP–92). Mutagenic activity was demonstrated in DLM assay in rats; CA were noted in the bone marrow cells of mice treated *in vivo* (IARC Suppl. 7–160).

Carcinogenicity. C. was given orally to pregnant rats, and their offspring were treated by gavage for the life-span. No difference in tumor incidence was found in this study.[6] *Carcinogenicity classification.* IARC: Group 2B.

Chemobiokinetics. Mechanism of C. toxic action is linked to the formation of peroxides, which reinforce the lipid peroxidation process.[2] Under C. intoxication, the quantity of lipid peroxides in the tissues (particulary in the liver) is increased. Evidently, C. is rendered effectively harmless by conjugation with SH-glutathione, which does not react directly with C. but with its **epoxy metabolites,** formed with the help of microsomal enzymes.[7] After a single administration of 20 and 80 mg C./kg BW to rats, the additional excretion of **thioesters** in the urine was increased.

Standards. Russia (1988). MAC and PML: 0.01 mg/l (organolept.).

Regulations. USFDA (1993) approved the use of C. in adhesives as a component of articles intended for use in packaging, transporting, or holding food.

References:

1. See **ACRYLONITRILE,** #1, 169.
2. Semerdzhyan, L. V. and Mkhitaryan, V. G., Toxic effects of chloroprene, *J. Exp. Clin. Med.,* 16, 3, 1976.
3. *Proc. Sanitary-Hygiene Research Institute,* Georgian SSR, No. 14, 1978, p. 65 (in Russian).
4. Melik-Alaverdyan, N. O., Studies on chloroprene toxicity, *Bull. Exp. Biol. (Yerevan),* 6, 107, 1965 (in Russian).
5. Salnikova, L. S. and Fomenko, V. N., Comparative characteristics of chloroprene embryotoxicity depending on exposure regimen at different routes of administration in the body, *Gig. Truda Prof. Zabol.,* 7, 30, 1975 (in Russian).
6. Ponomarkov, V. and Tomatis, L., Long-term testing of vinylidene chloride and chloroprene for carcinogenicity in rats, *Toxicology,* 37, 136, 1980.
7. Summer, K. H. and Greim, H., Detoxification of chloroprene (**2**-chloro-**1,2**-butadiene) with glutathione in rats, *Biochem. Biophys. Res. Commun.,* 96, 566, 1980.

CRESOLS (CAS No 1319–77–3)

Synonyms. Cresylic acids; Hydroxytoluenes; Methylphenols.

Properties. *o*-Cresol occurs as crystals (or liquid) darkening in air; *m*-cresol is a colorless liquid with a phenol odor; *p*-cresols are crystals with a phenol odor. Water solubility (g/100 ml) is 3.1

(40°C), 2.35 (20°C), and 2.4 (40°C), respectively. Dicresol is a mixture of 65% *m*-cresol and 35% *p*-cresol. Tricresol is a technical grade product containing a mixture of the isomers. Cresols are soluble in alcohol. Odor perception threshold is 0.05 mg *o*-cresol/l, 0.68 mg[010] or 0.001 mg *m*-cresol/l, and 0.002 mg *p*-cresol/l (Budeev et al., 1969). According to other data,[02] odor perception threshold of *m*-cresol is 0.037 mg/l.

Applications and **Exposure.** Mainly used in the production of cresol-aldehyde and phenolic resins. Tricresol is also used in the production of heat-resistant lacquers and molding powders. Dicresol is used in the synthesis of phosphate plasticizers (tricresyl phosphate, cresyldiphenyl phosphate), phenol-formaldehyde resins, and stabilizers. *p*-Cresol is used in large quantities in the production of ionol and antioxidant 2246, which are used to stabilize rubbers, plastics, vulcanizates, and foodstuffs.

Acute Toxicity. *Humans.* Symptoms of acute toxicity following ingestion of 1 to 60 ml of cresol include involuntary muscle movements followed by paresis, GI tract disturbances, renal toxicity, initial CNS stimulation followed by depression, etc.[1,2] *Animals.* The LD_{50} values for *o*-, *m*-, and *p*-cresol administered as 10% oil solution are 344, 828, and 344 mg/kg BW (mice) and 1470, 2010, and 1460 mg/kg BW (rats). The LD_{50} of dicresol (10% oil solution) is 1625 mg/kg BW for rats;[3] that of *o*-cresol is 940 mg/kg BW for rabbits, 100 to 500 mg/kg BW for minks, and 300 to 500 mg/kg BW for polecats.[07] Symptoms of poisoning are similar to those produced by phenol and comprise hematuria, agitation, and convulsions. The irritating effect of cresols can be increased by an alkaline medium. Gross pathological examination of mice exposed to *p*-cresol revealed dystrophic changes in the myocardium, liver, and kidney cells, as well as parenchymatous dystrophy of the renal tubular epithelium.[4]

Repeated Exposure. Exposure to cresols has been associated with hemolysis, methemoglobinemia, and acute Heinz-body anemia.[5] Fisher 344 rats and B6C3F₁ mice were given *o*-, *m*-, and *p*-cresol or *m/p*-cresol (60:40) at concentrations of 300 ppm to 30,000 ppm in the diet for 28 d. Some mice given *o*-cresol at a concentration of 30,000 ppm or *m*-cresol and *p*-cresol at concentrations of 10,000 or 30,000 ppm, respectively, died before the end of the studies.[6] Concentration of 3000 ppm increased relative liver and kidney weights without microscopic changes. Atrophic changes in the nasal and forestomach epithelium were presumably considered to be a specific effect of *p*- and *m/p*-cresol and a direct result of the irritant effects.[6]

Short-term Toxicity. Sprague-Dawley rats were given 0 to 600 mg *o*-cresol, 0 to 450 mg *m*-cresol, and 0 to 600 mg *p*-cresol/kg BW by gavage over a period of 13 weeks. Treatment-related mortality was primarily restricted to 600 mg/kg of *o*-cresol and *p*-cresol groups. Depressed BW gain in males receiving doses of 450 mg *m*-cresol and 600 mg *p*-cresol/kg BW was noted.[1,7] In a 13-week study, *o*-cresol or *m/p*-cresol (60:40) were added to the diet at concentrations of 30,000 ppm to rats, 20,000 ppm (*o*-cresol) or 10,000 ppm (*m/p*-cresol) to mice. There were no serious changes in hematology, clinical chemistry, and urinalysis. A deficient hepatocellular function and forestomach hyperplasia were found in mice given diets containing a higher concentration of *o*-cresol.[6] Rats receiving *o*-cresol in their drinking water (0.3 g/l) were sacrificed after 5 to 20 weeks of administration. No adverse effects were observed, but homogenates of the cerebrum had increased RNA content, decreased glutathione, and lower azoreductase activity.[8]

Long-term Toxicity. Rats exposed to 1 mg *p*-cresol/kg BW for 7 months developed decreased BW gain, reduced oxygen consumption and uropepsin level in the urine, and a change in conditioned reflex activity.[4] *o*-Cresol added to the diet of minks (240 to 2520 mg/kg BW) and of polecats (432 to 4536 mg/kg BW) resulted in reduced BW gain and food consumption. Erythrocyte count and Hb level in the blood were decreased while relative liver weights were increased.[4]

Reproductive Toxicity. *Humans.* Women exposed occupationally to varnishes containing tricresol are reported to have gynecological problems.[9] *Animals.* There were no adverse effects on male reproductive function, but the estrus cycle was lengthened in rats and mice receiving a higher concentration of *o*-cresol, and in rats receiving *m/p*-cresol in a 13-week study.[6] In a 6-month study in minks with *o*-cresol at doses of 100 to 1600 mg/kg BW, reproductive function remained unaltered.[4]

Genotoxicity. Cresol isomers are not mutagens in bacteria (NTP-85). None of the isomers increased SCE in mouse bone marrow, lung, or liver cells *in vivo.*[10] *p*- and *m*-Cresol produced CA in Wistar rats given 0.001 to 1 LD_{50} (Ekshtat and Isakova).

Chemobiokinetics. Following oral administration, cresols are readily absorbed from the GI tract of experimental animals. *p*-Cresol is removed with the urine of rabbits in the form of **glucuronide** (60%) and **sulfate** (15%) **conjugates.** About 10% are oxidized to *p*-**hydroxybenzoic acid,** while traces of *p*-cresol are hydroxylated to 4-**methylpyrocatechole** (3,4-dihydroxy-toluene).[11] In addition to urinary excretion, cresols undergo enterohepatic circulation.[1] *p*-cresol is a normal constituent of human urine, with levels of excretion ranging from 16 to 74 mg/d.[1,6]

Standards. Russia (1988). MAC (*m*- and *p*-cresols): 0.004 mg/l.

Regulations. USFDA has established allowable levels of *p*-cresol in food products. Cresols may be safely used (1) in resinous and polymeric coatings for food-contact surfaces, and (2) in phenolic resins used as the food-contact surface of molded articles intended for repeated use in contact with nonacid food (pH above 5).

References:

1. Patty's *Industrial Hygiene and Toxicology,* 3rd ed., Clayton, G. D. and Clayton, F. E., Eds., John Wiley & Sons, New York, 1981, 2A, 2567.
2. Gosselin, R. E., Smith, R. P., and Hodge, H. C., *Clinical Toxicology of Commercial Products,* 5th ed., Williams & Wilkins, Baltimore, 1984.
3. *Proc. Research Institute Hygiene Occup. Diseases,* Ufa, 7, 115, 1972 (in Russian).
4. *Problems of Hygiene,* Sci. Research Sanitary-Hygiene Institute, Coll. works, Novosibirsk, 15, 1965, p. 71 (in Russian).
5. Cote, M. A., Lyonnais, J., and Leblond, P. F., Acute Heinz-body anemia due to severe cresol poisoning: successful treatment with erythrocytophoresis, *Can. Med. Assoc. J.,* 130, 1319, 1984.
6. Toxicity Studies of Cresols in F344/N Rats and B6C3F₁ Mice (Feed Studies), NTP Tox 9, U.S. Department of Health and Human Services, Washington, D.C., p. 128.
7. Henck, J. W., Traxler, D. J., Dietz, D. D., et al., Neurotoxic potential of *ortho-, meta-,* and *para*-cresol, *Toxicologist,* 7, 246, 1987.
8. Savolainen, H., Toxic effects of peroral *o*-cresol intake on the rat brain, *Res. Commun. Chem. Pathol. Pharmacol.,* 25, 357, 1979.
9. Syrovadko, O. N. and Malysheva, Z. V., Working conditions and their effect on some specific functions of women engaged in the manufacture of enamel-insulated wires, *Gig. Truda Prof. Zabol.,* 4, 25, 1977 (in Russian).
10. Cheng, M. and Kligerman, A. D., Evaluation of genotoxicity of cresols using sister-chromatid exchange (SCE), *Mutat. Res.,* 137, 51, 1984.
11. See **ANILINE,** #10, 175.

CYCLOHEXANOL (CAS No 108–93–0)

Synonyms. Anol; Cyclohexyl alcohol; Hexahydrophenol; Hexaline; Hydroxycyclohexane.

Properties. Colorless, slightly oily liquid with an odor of camphor and amyl alcohol. Solubility in water is 3.6% (20°C), soluble in alcohol. Odor perception threshold is 3.5 or 2.8 mg/l.[02]

Applications. Used in the production of caprolactam and polyamide fibres.

Acute Toxicity. For rats LD₁₀₀ appeared to be 1.5 g/kg BW. LD₅₀ for mice is 1.24 g/kg BW. Poisoning is accompanied by ataxia and muscular weakness. Animals become irresponsive to painful stimuli and assume side position. Death is preceded by deep narcosis.

Repeated Exposure failed to reveal cumulative properties. An increase in the liver relative weights and stimulation of microsomal enzyme and **diphenyl-4-hydroxylase** activity were found in animals given 455 to 1500 mg C./kg BW for 7 d (Lake et al., 1982). Rats were dosed by gavage with a 400 mg/kg BW dose. The treatment had no effect on BW gain, general condition, and behavior. There were changes in the glycogen-forming function of the liver, oxidation-reduction processes in the tissues, and in hematological indices.

Long-term Toxicity. In a 6-month study, rabbits were exposed to 2 and 20 mg/kg BW doses. The treatment resulted in decreased BW gain, impaired liver function, and reduced catalase activity. Gross pathological examination revealed changes in the viscera.

Chemobiokinetics. Formation of **glucuronides** of C. is shown in Gunn rats.

Standards. Russia (1988). MAC and PML: 0.5 mg/l.

Regulations. USFDA (1993) listed C. for use (1) as a component of adhesives for food-contact surfaces, (2) as a component of the uncoated or coated food-contact surface of paper and paperboard that may be safely used for producing, manufacturing, packing, transporting, or holding dry food, and (3) as a defoaming agent that may be safely used in the manufacture of paper and paperboard intended for use in contact with food.

Reference:

  Protection of Water Reservoirs Against Pollution by Industrial Liquid Effluents, Medgiz, Moscow, 5, 1962, p. 78 (in Russian).

CYCLOHEXANONE (CAS No 108–94–1)

Synonyms. Cyclohexyl ketone; Ketohexamethylene; Pimelic ketone

Properties. Colorless, oily liquid with a sweet, sharp odor. Miscible with water (50 g/l at 30°C) and most organic solvents: ethanol, diethyl ether, chloroform. Odor perception threshold is reported to be 0.12 mg/l,[010] 1 to 2 mg/l,[1] or 8.3 mg/l.[02]

Applications and **Exposure.** C. is primarily used as an intermediate in the production of nylon and as an additive and an excellent solvent for a variety of products: rubbers, some biomedical polymers, polyurethane lacquers, natural and synthetic resins and gums; C. is also used in the production of adipic acid.

Acute Toxicity. The LD_{50} values are 2.07 g/kg BW for male and 2.1 g/kg BW for female mice, and 1.8 g/kg BW for female rats.[2] Rabbits seem to be less sensitive. Death occurs with signs of narcosis and without an excitation stage.[3] Autopsy revealed edema with hemorrhages in the lung parenchyma, peritoneal and intestinal congestion in mice, suggesting an irritant effect.

Repeated Exposure. C. induced diverse reactions by virtue of a general vascular or tissue reaction and CNS inhibition.[2] Mice were exposed to 280 mg/kg BW for 25 d. The treatment had no effect on their general condition and BW gain.

Short-term Toxicity. Mice were given 0.4 to 47 g c./l of their drinking water for 13 weeks. One- to two-thirds of the animals in the highest dose group died during experiment. Other animals exhibited retardation of BW gain; C. concentration of 47 g/l produced focal liver necrosis and thymus hyperplasia.[4]

Allergenic Effect. C. did not induce skin allergy in the guinea pig maximization test.[5]

Reproductive Toxicity. No significant prenatal toxicity was observed in mice (IARC 47–165). In mice, 1%-dietary administration for several generations was reported to affect the viability and growth of the first-generation males and females. There were no such effects in the second generation.[6] CD-1 mice were exposed by oral intubation to 800 mg/kg BW on days 8 to 12 of gestation. Reproductive function was not affected.[7]

Genotoxicity. C. is positive in *Salmonella* mutagenicity assay. It exhibited a cytogenetic effect in human cultured lymphocytes at concentrations of 0.0005 to 0.01 mg per 100 ml of the culture medium.[8]

Carcinogenicity. B6C3F₁ mice received 6.5 or 13 g C./l in their drinking water for 104 weeks. A slight increase in the incidence of tumors that occur commonly in this strain was found only in animals given the lower dose. In Fisher 344 rats given 3.3 or 6.5 g C./l in their drinking water for 104 weeks, a slight increase in the incidence of adrenal cortical adenomas occurred in males treated with the lower dose.[4] *Carcinogenicity classification.* IARC: Group 3.

Chemobiokinetics. Data on the ability of C. to accumulate in the body appear to be contradictory. Repeated *i/v* administration (overall dose of 284 mg/kg) provided no evidence of enzyme induction.[9] C. is likely to be metabolized to **cyclohexanol,** which is conjugated with glucuronic acid and excreted mainly in the urine (almost 60%); very little C. or free cyclohexanol is found in the urine.[10,11]

  Standards. Russia (1988). MAC and PML: 0.2 mg/l.

References:

1. See **ACRYLONITRILE,** #1, 76.
2. Gupta, P. K., Lawrence, W. H., Turner, J. E., et al., Toxicological aspects of cyclohexanone, *Toxicol. Appl. Pharmacol.,* 49, 525, 1979.
3. See **ANILINE,** #1, 111.
4. Lijinsky, W. and Kovatch, R. M., Chronic toxicity study of cyclohexanone in rats and mice, *J. Natl. Cancer Inst.,* 77, 941, 1986.

5. Bruze, M., Bomon, A., Bergqvist-Karlson, A., et al., Contact allergy to cyclohexanone resin in humans and guinea pigs, *Contact Dermat.,* 18, 46, 1988.
6. Gondry, E., Studies on the toxicity of cyclohexylamine, cyclohexanone and cyclohexanol, metabolites of cyclomate, *J. Exp. Toxicol.,* 5, 227, 1972.
7. *Short-term Bioassay in the Analysis of Complex Environmental Mixtures,* M. Wasters, S. Sandhy, J. Lewtas, et al., Eds., Plenum Press, New York, 1983, p. 417.
8. Dyshlovoi, V. D., Boiko, N. L., Shemetun, A. M., et al., Cytogenetic effect of cyclohexanone, *Gig. Sanit.,* 5, 76, 1981 (in Russian).
9. Martis, L., Tolhurst, T., Koeferl. M. T., et al., Disposition kinetics of cyclohexanone in beagle dogs, *Toxicol. Appl. Pharmacol.,* 55, 545, 1980.
10. Greener, Y., Martis, L., and Indacochea-Redmond, N., Assessment of the toxicity of cyclohexanone administered intravenously to Wistar and Gunn rats, *J. Toxicol. Environ. Health.,* 10, 385, 1982.
11. See **ANILINE,** #10, 267.

4,4′-DIAMINODIPHENYL ETHER (CAS No 101–80–4)

Synonyms. 4-Aminophenyl ether; **4,4′**-Oxybisbenzenamine; **4,4′**-Oxydianiline.
Properties. Colorless, odorless powder. Poorly soluble in water, readily soluble in alcohol and acetone.
Applications. Used in the production of plasticizers and straight poliamide resins, submits outstanding high-temperature resistance. Antioxidant.
Acute Toxicity. LD_{50} ranges from 570 to 725 mg/kg BW in rats; it is 685 mg/kg BW in mice.[1,2] Produces a narcotic effect; causes formation of metHb. The animals die within 5 d.
Repeated Exposure revealed pronounced cumulative properties.[2] Rats and mice were exposed to 3 g/kg BW dose in their diet. Of rats 0 to 20% and 40 to 80% of mice were still alive on day 14 of the treatment.[3] Administration of 72.5 mg/kg BW for 15 d caused decreased blood Hb level and increased spleen and adrenal weights in rats.[1] Liver and kidney injuries were noted in mice and rats.
Short-term Toxicity. In a 90-d study, Fisher 344 rats and B6C3F₁ mice received dietary concentrations up to 2 g/kg BW. The dose of 0.6 g/kg caused decrease in BW gain. Animals fed dietary concentrations of 0.6 g/kg (rats) and 1 g/kg (mice) and the above displayed diffuse parenchymatous dystrophy, pituitary hyperplasia, seminiferous tubular degeneration and atrophy of the prostate and seminal vesicles.[5]
Long-term Toxicity. In a 2-year feeding study, Fisher 344 rats were given 0.2 to 0.5 g/kg BW, while B6C3F₁ mice received 0.15 to 0.8 g/kg BW. Decrease in BW gain and thyroid follicular-cell and pituitary hyperplasia were observed.[3] In a 6-month rat study, a toxic effect on the cardiovascular system was found; dose-dependence was noted.[6]
Genotoxicity. Mutagenic effect in *Salmonella* was found in the absence and presence of a metabolic activation system (IARC 29–209).
Carcinogenicity. Treatment with ¼ LD_{50} given for 14 d inhibited the growth of spontaneous mammary tumors and of transplanted tumors in mice.[7] Administration of 0.15 to 0.8 g/kg BW to mice and of 0.2 to 0.5 g/kg BW to rats for 103 weeks produced liver cell tumors and malignant follicular cell tumors of the thyroid.[3] Administration to rats (a total dose given 14.4 g/kg BW) caused cirrhosis and necrosis in the liver. Incidence of death from malignant hepatomas was increased.[8]
Carcinogenicity classification. NTP: P*—P*—P*—P*.
References:
1. Lapik, A. S., Makarenko, A. A., and Zimina, L. N., Toxicological characteristics of **4,4′**-oxydianiline, *Gig. Sanit.,* 10, 110, 1968 (in Russian).
2. *Toxicology of New Industrial Chemical Substances,* Meditsina, Moscow, 14, 1975, p. 118 (in Russian).
3. Bioassay of **4,4′**-Oxydianiline for Possible Carcinogenicity (Technical Report Series No. 205), DHEW Publ. No. NIH 80–1761, National Cancer Institute, U.S. Department of Health, Education and Welfare, Washington, D.C., 1980.
4. Dzhigoev, F. K., On carcinogenic activity of **4,4′**-oxydianiline, *Probl. Oncol.,* 1, 69, 1975 (in Russian).

32

32

5. Hayden, D. W., Wage, G. G., and Handler, A. H., The goitrogenic effect of **4,4'**-oxydianiline in rats and mice, *Vet. Pathol.*, 15, 649, 1978.
6. Kondratyuk, V. A., Gnat'yuk, M. S., and Volkov, K. S., On the cardiotoxic effect of **4,4'**- diaminodiphenyl ether, *Gig. Sanit.*, 6, 31, 1986 (in Russian).
7. Boyland, E., Experiments on the chemotherapy of cancer. VI. The effect of aromatic bases, *Biochem. J.*, 40, 55, 1946.
8. Steinholf, D., Kancerogene Wirkung von **4,4'**-Diamino-diphenylather bei Ratten, *Naturwissenschaften*, 64, 344, 1977 (in German).

1,1-DICHLOROETHENE (1,1-DCE) (CAS No 75–35–4)

Synonyms. 1,1-Dichloroethylene; Vinylidene chloride (VDC).

Properties. Clear liquid with a sweet odor. Solubility in water is 2.5 g/l at 25°C. **1,1**-DCE is readily soluble in alcohol, miscible with the majority of organic solvents. Odor perception threshold is 1.5 mg/l.[02]

Applications and **Exposure.** A comonomer in the production of a number of polymers used to manufacture food containers and packaging (**Sarran** film). Copolymerizes with other vinyl monomers such as acrylonitrile, alkylacrylates, methacrylates, vinyl acetate, and vinyl chloride. A solvent. In the U.S., a daily drinking water exposure has been estimated, the maximum being as high as 0.001 mg/l. Migration from food packaging materials is expected to be very low due to limited release and rapid degradation.

Acute Toxicity. In adult rats, LD_{50} is 200 to 1800 mg/kg BW;[1] in young and fasted rats, it is about 50 mg/kg BW.[2] In mice and dogs, these values are 200 and 5750 mg/kg BW, respectively.[3]

Short-term Toxicity. Liver and kidney damage is generally observed in animals after a high dose-exposure. Other changes noted are CNS depression and sensitization of the heart. Vacuolization of hepatocytes was observed in rats receiving 5 to 40 mg/kg BW with their drinking water for 3 months.[4]

Long-term Toxicity. Sprague-Dawley rats were given 9, 14, and 30 mg/kg BW (females), or 7, 10, and 20 mg/kg BW (males) with their drinking water for 2 years. The most sensitive endpoint was liver damage. No consistent treatment-related biochemical changes were observed in any parameter measured. The only abnormal histopathology reported was midzonal fatty accumulation in the livers of both sexes receiving the highest dose. No liver degeneration was observed. A NOAEL of 100 ppm (9 mg/kg BW) in females was identified, based upon a trend toward increased fatty deposition in the liver. Changes and hypertrophy of the liver cells in females at all doses and males at the highest dose were reported.[5,6] After doses of 0.5 to 20 mg/kg BW given by gavage in corn oil for 1 year, no adverse effects were observed on autopsy in Sprague-Dawley rats.[7] The dose of 5 mg/kg BW given in corn oil (but not 1 mg/kg BW) for 2 years caused renal inflammation in Fisher 344/N rats. Liver necrosis was found at the dose-level of 10 mg/kg BW in male mice and of 2 mg/kg BW in female mice.[8]

Reproductive Toxicity. Concentrations of 50 to 200 ppm in drinking water did not affect reproductive capacity in a three-generation study using Sprague-Dawley rats.[9] *Teratogenic* effects were not shown in rats following ingestion of **1,1**-DCE in their drinking water at the concentration of 200 mg/l.[10]

Genotoxicity. *Humans.* There are no data available on the genetic and related effects of **1,1**-DCE in humans (IARC 39–211). *Animals.* Adverse effects were not shown in the DLM assay in mice or rats,[2,13] and using V79 Chinese hamster ovary cells.[14] **1,1**-DCE did not induce CA in the bone marrow cells of rats treated *in vivo*. It induced unscheduled DNA synthesis in mice but not in rat hepatocytes. In the short-term tests, positive responses for DNA alkylation, repair, and synthesis were observed. **1,1**-DCE showed mutagenic effects in *Salmonella* and in *Escherichia coli*.[11,12]

Carcinogenicity. The compound is structurally similar to **vinyl chloride.** Oral administration to mice (2 to 10 mg/kg BW) and rats (50 mg/kg BW) every week caused no statistically significant tumor growth, though tumors that were not seen in the controls appeared at a variety of sites in the treated rats.[8] Malignant neoplasms were found in the kidneys, lungs, and mammary glands of mice inhaling **1,1**-DCE.[7] *Carcinogenicity classification.* USEPA: Group C; NTP: N—N—N—N.

Chemobiokinetics. 1,1-DCE is completely absorbed after ingestion and excreted within 72 h. The highest concentrations were discovered in the liver and kidneys. **1,1**-DCE is transformed into 1,1-**dichloroethylene oxide** and **chloroacetyl chloride,** which are probably responsible for the carcinogenic effect in mice.[3] **Monochloroacetic acid** was also found to be a metabolite, and it can be conjugated with glutathione.[3] A greater proportion is removed unaltered through the lungs.[4]

Guidelines. WHO (1992). Guideline value for drinking water: 0.02 mg/l.

Standards. EEC (1989). Proposed limits in food: 0.05 mg/kg, Maximum Permitted Quantity in materials and articles: 5 mg/kg.[15] **USEPA** (1991). MCL and MCLG: 0.007 mg/l. **Russia** (1988). MAC and PML: 0.0006 mg/l.

Regulations. USFDA (1993) permits the use of **VDC copolymers** as components of articles intended for use in contact with food, including (1) adhesives, (2) resinous and polymeric coatings for polyolefin films, (3) coatings for nylon or polycarbonate films, (4) paper or paperboard (providing the finished copolymers contain at least 50 wt% polymer units derived from vinylidene chloride), (5) cellophane to be safely used for packaging food, (6) semirigid and rigid acrylic and modified acrylic plastics, (7) polyethylene-phthalate polymers, (8) packaging materials for use during the irradiation of prepackaged food (providing the film contains not less than 70% by weight of VDC and has a viscosity of 0.5 to 1.5 cP), and (9) polymer modifiers in semirigid and rigid vinyl chloride plastics. **VDC-methyl acrylate copolymers** may be used in contact with food providing that (1) less than 15% by weight of the polymer units is derived from methyl acrylate, (2) the average molecular weight of the copolymer is not less than 50,000, and (3) the residual VDC will not exceed 10 mg/kg.

References:

1. See **CHLOROPRENE,** #6.
2. Andersen, M. E. and Jenkins, L. R., Oral toxicity of **1,1**-dichloroethylene in the rat: effects of sex, age and fasting, *Environ. Health Perspect.,* 21, 157, 1977.
3. Jones, B. D. and Hathway, D. E., Differences in metabolism of vinylidene chloride between mice and rats, *Br. J. Cancer,* 37, 411, 1978.
4. Norris, J. M., Toxicologic and pharmacokinetic studies on inhaled and ingested vinylidene chloride in laboratory animals, in Proc. Tech. Assoc. Paper. Synth. Conf., Chicago, IL, Atlanta, GA, TAPPI, 1977.
5. Rampy, L. W., Quast, J. F., Humiston, C. G., et al., Interim results of two-year toxicological studies in rats of vinylidene chloride incorporated in the drinking water or administered by repeated inhalation., *Environ. Health Perspect.,* 21, 33, 1977.
6. Quast, J. F., Humiston, C. G., Wade, C. E., et al., A chronic toxicity and oncogenicity study in rats and subchronic toxicity study in dogs on ingested vinylidene chloride, *Fundam. Appl. Toxicol.,* 3, 55, 1983.
7. Maltoni, C., Cotti, G., and Chieco, P., Chronic toxicity and carcinogenicity bioassays of vinylidene chloride, *Acta Oncol.,* 5, 91, 1984.
8. *Carcinogenesis Bioassay of Vinylidene Chloride in F344 Rats and B6C3F₁ Mice (Gavage Study),* NTP, U.S. PHS, NTP–80–2, NIH Publ. No. 82–1784, 1982.
9. Nitschke, K. D., Smith, F. A., Quast, J. F., et al., A three-generation rat reproductive toxicity study of vinylidene chloride in the drinking water., *Fundam. Appl. Toxicol.,* 3, 75, 1983.
10. Murray, F. J., Nitschke, K. D., Rampy, L. W., et al., Embryotoxicity and fetotoxicity of inhaled or ingested vinylidene chloride in rats and rabbits, *Toxicol. Appl. Pharmacol.,* 49, 189, 1979.
11. Bartsch, H., Malaveille, C., Barbin, A., et al., Tissue-mediated mutagenicity of vinylidene chloride and 2-chloro-butadiene in *Salmonella typhimurium, Nature,* 155, 641, 1975.
12. Simmon, V. F., Kauhanen, K., and Tardiff, R. G., Mutagenic activity of chemicals identified in drinking water, *Dev. Toxicol. Environ. Sci.,* 2, 249, 1977.
13. Short, R. D., Minor, J. L., Winston, J. M., et al., A dominant lethal study in male rats after repeated exposure to vinyl chloride or vinylidene chloride, *J. Toxicol. Environ. Health,* 3, 965, 1977.
14. Drevon, C. and Kuroki, T., Mutagenicity of vinyl chloride, vinylidene chloride and chloroprene in V79 Chinese hamster cells, *Mutat. Res.,* 67, 173, 1979.
15. IPCS Safety Guide No. 36, Vinylidene Chloride, WHO, Geneva, 1989, p. 25.

DICYCLOPENTADIENE (DCPD) (CAS No 77–73–6)

Synonyms. Cyclopentadiene; **1,3**-Cyclopentadiene.

Properties. Colorless, crystalline powder with a pungent, nauseous odor. Water solubility is 26 g/l. Miscible with alcohol and ether. Odor perception threshold in water is 0.006 mg/l[1] or 0.011 mg/l.[010]

Applications. Used as a monomer in the production of synthetic rubber.

Acute Toxicity. In rats, LD_{50} of pure DCPD is 670 mg/kg BW[1] or 480 mg/kg BW;[2] LD_{50} of the technical-grade product is 1000 mg/kg BW.[3] Death within 1 to 2 d preceded by clonic convulsions.

Repeated Exposure failed to reveal cumulative properties. Rats tolerated administration of $^1/_{10}$ to $^1/_{50}$ LD_{50} for 30 d. DCPD has a polytropic toxic action (See **Norborndiene**). The LOAEL was estimated to be 5 mg/kg BW.[2] No signs of intoxication were noted in mice given 0.2 and 2 mg/kg for 14.[4]

Short-term Toxicity. There were no signs of intoxication in rats given 0.4 and 4 mg DCPD/kg BW or in mice given 0.15 and 1.5 mg DCPD/kg BW for 90 d.[4]

Reproductive Toxicity. No toxic effect on embryos was observed. A dose of 0.1 mg/kg BW had no gonadotoxic effect.[2]

Chemobiokinetics. Following oral administration, labeled ^{14}C-DCPD was found in the bladder, bile, and fatty tissue. Unchanged DCPD is exhaled or excreted via the kidneys. Metabolites were also detected in the urine (Ross and Dacre, 1977).

Standards. Russia (1988). MAC: 0.006 mg/l (organolept.), PML: 0.002 mg/l. Recommended MAC: 0.015 mg/l (organolept., odor).[4]

References:

1. *Toxicology and Hygiene of Petroleum Chemistry Products,* Yaroslavl, 1972, p. 197 (in Russian).
2. Zholdakova, Z. I., Sil'vestrov, A. E., and Mikhailovsky, N. I., Substantiation of maximum allowable concentration for norbornene, norbornadiene and dicyclopentadiene in water bodies, *Gig. Sanit.,* 2, 77, 1986 (in Russian).
3. *Hygiene of Application and Toxicology of Pesticides and Polymeric Materials,* Kiev, 13, 1983, p. 98 (in Russian).

4,4′-DIISOCYANATE-3,3′-DIMETHOXY-1,1′-BIPHENYL (DADI) (CAS No 91–93–0)

Synonyms. **3,3′**-Dimetoxybenzidine-**4,4′**-diisocyanate; Dianizidine diisocyanate.

Properties. Grey to brown powder soluble in esters.

Applications DADI can be used in isocyanate-based adhesive systems and as a component of the polyurethane elastomers.

Genotoxicity. DADI was mutagenic to *Salmonella typhimurium* TA98 in the presence but not in the absence of an exogenous metabolic system; it was not mutagenic to strains TA1535, TA1537 or TA100.[1]

Carcinogenicity. No treatment-related tumors were observed in $B6C3F_1$ mice gavaged initially with 1.5 and 3 g/kg BW for 22 weeks and subsequently given dietary concentrations of 22 or 44 g/kg BW for 56 weeks. A statistically significant increase in the combined incidence of leukemia and malignant lymphomas was observed in rats together with a treatment-related increase in the incidence of tumors of the skin and Zymbal gland.[2]

References:

1. Haworth, S., Lawlor, T., Mortelmans, K., et al., Salmonella mutagenicity test results for 250 chemicals, *Environ. Mutat.,* Suppl. 1, 3, 1983.
2. Bioassay of **3,3′**-Dimethoxybenzidine **4,4′**-diisocyanate for Possible Carcinogenicity, Technical Report No. 128, National Cancer Institute, Bethesda, MD, 1979.

DIMETHYL TEREPHTHALATE (CAS No 120–61–6)

Synonyms. **1,4**-Benzenedicarboxylic acid, dimethyl ester; Dimethyl **1,4**-benzenedicarboxylate; Dimethyl *p*-phthalate; Methyl **4**-carbomethoxybenzoate; Terephthalic acid, dimethyl ester.

Properties. White, crystalline powder. Poorly soluble in cold water. Solubility in hot water is 3.32 g/l. Soluble in alcohol. Odor perception thresholds are 13.4 mg/l (20°C) and 1.8 mg/l (60°C).[1]

Applications and **Exposure.** A monomer in the production of polyesters, mainly of polyethylene terephthalate (PET). The migration of total levels of PET oligomers was detected by using an analytical approach that involves hydrolysis of oligomers to terephthalic acid, methylation, and detection as DMTP. Total levels of migration of PET oligomers were 0.02 to 2.73 mg/kg depending on the food and temperature attained during cooking.[2]

Acute Toxicity. Male rats tolerate a 6.6 g/kg BW dose, and mice did not die from a dose of 10 g/kg BW. Poisoning is accompanied with tremor and increased motor activity (followed by depression in mice). Gross pathological examination revealed no histopathological lesions.[1]

Short-term Toxicity. Oral exposure of rats to 1% emulsion of DMTP in the feed for 96 d, and to 50 and 200 mg/kg BW for 115 d resulted in decreased BW gain.[1] Ten administrations of 2 to 4 g/kg BW doses did not alter cholinesterase activity in the blood serum and produced no histopathological lesions in the visceral organs of rats. A 500 mg/kg dose administered for 35 to 39 d caused no toxic effect (Prusakov). 28-d-old male rats were fed 3% D. for 2 weeks. All the animals developed bladder stones. DMTP concentrations less than 1.5% produced no such effect. Histological examination revealed epithelial hyperplasia.[3]

Long-term Toxicity. 0.5 and 7.5 mg/kg BW doses caused no histopathological changes in the visceral organs of rats and did not affect their conditioned reflex activity.[1]

Allergenic Effect. DMTP produced a sensitizing effect. 0.075 mg/kg BW and higher doses administered *i/g* for 30 d caused the development of allergic and autoallergic changes manifested by specifically increased degranulation of peripheral blood basophils. There was an increase in the number of autoimmune hemolysis plaques and in the formation of antibodies against both hapten and tissue antigen.[4]

Reproductive Toxicity. Inhalation study in pregnant rats revealed no embryotoxic or teratogenic effect.[5]

Genotoxicity. DMTP is not found to produce CA and SCE in Chinese hamster ovary cells (NTP-85). It was positive in *Dr. melanogaster* and in micronuclear tests in mice following a single *i/p* administration of 65 and 195 mg DMTP/kg BW.[6]

Carcinogenicity. DMTP showed no carcinogenic potential in the rat and mouse feeding study at doses of 5 g/kg BW.[7] *Carcinogenicity classification.* NTP: N—N—E—N.

Chemobiokinetics. DMTP is readily absorbed and excreted from the body.

Standards. Russia (1988). MAC: 1.5 mg/l (organolept., odor).

References:

1. Krassavage, W. I., Yanno, F. J., and Terhaar, C. J., Dimethylterephthalate: acute toxicity, subacute feeding and inhalation studies in male rats, *Am. Ind. Hyg. Assoc. J.*, 34, 455, 1973.
2. Castle. L., Mayo, A., Crews, C., et al., Migration of poly(ethylene terephthalate) (PET) oligomers from PET plastics into foods during microwave and conventional cooking and into bottled beverages, *J. Food Prot.*, 52, 337, 1989.
3. Chin, T. Y., Tyl, R. W., Papp, J. A., et al., Chemical urolithiasis. I. Characteristics of bladder stone induction by terephthalic acid and dimethylterephthalate in weanling Fisher 344 rats, *Toxicol. Appl. Pharmacol.*, 58, 307, 1981.
4. Vinogradov, H. I., Vinarskaya, E. I., and Antomonov, M. Yu., Hygienic assessment of dimethylterephthalate allergenic activity at inhalation and oral intake, *Gig. Sanit.*, 5, 7, 1986 (in Russian).
5. Krotov, Ya. A. and Chebotar', N. A., Studies of embryotoxic and teratogenic effects of some industrial chemicals produced in dimethylterephthalate manufacture, *Gig. Truda Prof. Zabol.*, 6, 40, 1972 (in Russian).
6. See **ACRYLAMIDE,** #21, 199.
7. Haseman, J. K. and Clark, A. M., Carcinogenicity results for 114 laboratory animal studies used to assess the predictivity of four *in vitro* genetic activity assays for rodent carcinogenicity, *Environ. Mutat.*, 16 (Suppl. 18), 15, 1990.

EPICHLOROHYDRIN (ECH) (CAS No 106–89–8)

Synonyms. 1-Chloro-2,3-epoxypropane; 3-Chloro-1,2-epoxy-propane; (Chloromethyl)oxyrane; α-Chloropropylene oxide; Glycerol epichlorohydrin; Glycidyl chloride.

Properties. Colorless, mobile liquid with an irritating chloroform-like odor. Solubility in water is 66 g/l at 20°C. ECH is hydrolyzed in aqueous media, causing difficulties for drinking water monitoring. Odor perception threshold is reported to be 1 mg/l[1] or 3 mg/l.[02] Irritancy threshold is 0.1 mg/l.

Applications and **Exposure.** ECH is widely used in the manufacture of unmodified epoxy resins, elastomers, water treatment resins, ion exchange resins, plasticizers, stabilizers, solvents. Migration into food and drinking water from plastics is possible but is expected to be low. Migration into water from epoxy coatings (exposure 7 d, 37°C) was determined at the level of 0.05 mg/l.[2]

Acute Toxicity. *Humans.* Ingestion of ECH may cause death in man due to respiratory insufficiency. *Animals.* Oral doses of 325 and 500 mg/kg BW caused renal damage, vacuolization, and fatty degeneration in the liver of the test animals. Necrotic foci were found in the GI tract.[3] LD_{50} values are found to be 140 to 260 mg/kg BW in rats and mice, 345 mg/kg BW in rabbits, and 280 mg/kg BW in guinea pigs.[1,4] According to other data,[3] ECH exhibits lower acute toxicity: rats and mice tolerate 250 mg/kg BW without visible symptoms, but all of them die from the dose of 325 mg/kg BW. Poisoned animals displayed adynamia, labored respiration, marked cutaneous hyperemia, ataxia, tremor, and abdominal swelling. Histopatological changes in the lungs, liver, kidneys, suprarenals, and thyroid were found.

Repeated Exposure failed to reveal signs of accumulation. Administration of $^1/_{10}$ LD_{50} for 30 d did not cause mortality. However, according to other data,[5] a nephrotoxic effect with a tendency to accumulation, as well as a change in enzyme activity in the kidneys, was found to develop. It exerts a narcotic action and irritates the skin and mucous membranes. Rabbits were given a dose of 80 mg/kg BW for 2 months. The treatment caused retardation of BW gain, reduction in leukocyte count and glutathione content in the blood, hypercholesterinemia, and an increase in the relative weights of the kidneys and suprarenals.[1] Damage to the kidney was demonstrated after administration of 40 and 80 mg/kg BW.[6,7]

Long-term Toxicity. Biological activity of ECH is likely to be determined by presence of Cl in its molecule. ECH affects the CNS, liver and kidney functions, oxidation processes, neuroendocrine control, and the blood system. In a 2-year study, weanling Wistar rats were given ECH in drinking water by gavage (doses of 2 and 10 mg/kg BW). Gradual increase in mortality in males with clinical symptoms, including dyspnea and weight loss, and decrease in leukocyte count at dose-levels of 1.43 and 7.14 mg/kg BW, was observed. The LOAEL of 1.4 mg/kg BW for forestomach hyperplasia was identified.[8]

Reproductive Toxicity. *Gonadotoxicity.* Five daily doses of 50 mg/kg BW or one single dose of 100 mg/kg BW caused permanent sterility in rats.[9] Antifertility effects in male rats after a single oral or *i/p* dose of 50 mg/kg BW was reported.[10] Administration of 10 mg/kg BW for 3 months or 15 mg/kg BW for 12 d also reduced the fertilizing ability of male rats, but a dose of 2 mg/kg BW had no effect.[7,11] The sperm of rats that had received 25 or 50 mg ECH/kg BW showed an increased percentage of abnormal sperm heads at the higher dose and a reduced number of sperm heads at the lower dose, while no changes were observed in the weights and microscopic picture of the testes.[12] *Embryotoxicity* and *teratogenicity.* Administration of ECH to Sprague-Dawley rats and CD-1 mice on days 6 to 15 of pregnancy produced no embryotoxic and teratogenic effect, even at doses causing 10% maternal mortality (up to 160 mg/kg BW); BW was reduced and liver weight increased in pregnant mice.[13]

Genotoxicity. ECH has been shown to be genotoxic *in vitro* and *in vivo. Humans.* ECH causes gene mutations in all cell systems, as well as CA in eukaryote cells and should be considered as potentially dangerous to man. CA have been observed in workers exposed to this compound, although the studies are difficult to interpret (IARC, Suppl. 6–286). ECH induced CA, SCE, and unscheduled DNA synthesis in human cells *in vitro. Animals.* ECH induced SCE in bone marrow cells but not micronuclei or DLM in mice treated *in vivo;* equivocal findings were found for CA. Weakly positive results were obtained in a cell transformation assay in C3H 10T1/2 cells. It induced CA, SCE, mutation, and DNA strand breaks in rodent cells *in vitro* (IARC 11–131).

Carcinogenicity. Positive results were obtained in several carcinogenicity bioassays with rats exposed via multiple routes of administration. Exposure by ingestion (by gavage or via drinking water) caused forestomach tumors, and exposure via inhalation resulted in tumors of the nasal cavity. Tumors were shown to be present only at the site of administration where ECH was highly irritant. After administration of 80 mg/kg BW for 3 months, papillomata and squamous cell carcinoma were found in two out of five rats. Administration of 20 to 100 mg/kg BW in distilled water to Wistar rats for 2 years caused stomach tumors to develop.[7] Eighteen male Wistar rats were given 18, 39, and 89 mg/kg BW with their drinking water for 81 weeks: forestomach tumors at two higher doses and prestomach hyperplasia at all three doses were found. Mortality in animals increased up to 33 to 45%.[14] No carcinogenic effect was found using a rapid method of organotypical culture of pulmonary and renal embryonic tissue.[15] *Carcinogenicity classification.* IARC: Group 2A.

Chemobiokinetics. ECH is a bifunctional alkylating agent. In the gastric juice, it is converted to **chlorohydrin.** The rate of conversion declines with ECH concentration. ECH is rapidly and extensively absorbed following oral, inhalation, or dermal exposure. It is rapidly removed from the blood and is, therefore, not likely to be accumulated during chronic exposure. ECH may bind to the cellular nucleophiles such as glutathione. The major metabolites in the urine were identified as *N*-**acetyl**-*S*-(3-**chloro**-2-**hydroxypropyl**)-L-**cysteine** formed by conjugation with glutathione, and α-chlorohydrin.[16] **Oxalic acid,** which is probably a metabolite of ECH, is responsible for its renal toxicity. Up to 40% of the dose is removed with the urine in 72 h and less than 4% in the feces. Up to 20% of the labeled material is excreted as CO_2 in the expired air in 4 h, indicating the rapid conversion of ECH in the body.[17]

Guidelines. WHO (1992). Provisional Guideline value for drinking water: 0.0004 mg/l.

Standards. Russia (1988). MAC and PML: 0.01 mg/l.

Regulations. USEPA (1989): MCL TT (treatment technique), MCLG zero; the maximum residual content in flocculating agent shall not exceed 0.01% at the maximum usage rate of 20 ppm of polymer (level of ECH would not exceed 0.0022 mg/l). In **Japan,** JWWA standardized epoxide resin coating condition of temperature, humidity, times, etc., according to the resin components so as to be less than 0.001 mg/l of resin ingredients released from the coats into water (JWWA K135, 1989).

References:
1. See **BISPHENOL A,** #1.
2. Krat, A. V., Keselman, I. M., and Sheftel, V. O., Sanitary-chemical evaluation of polymeric goods in water-supply constructions, *Gig. Sanit.,* 10, 18, 1986 (in Russian).
3. *Toxicology of New Industrial Chemical Substances,* Medgiz, Moscow, 2, 1961, p. 28, (in Russian).
4. John, J. A., Quast, J. F., Murray, F. J., et al., Inhalation toxicity of epichlorohydrin: effects on fertility in rats and rabbits, *Toxicol. Appl. Pharmacol.,* 68, 415, 1983.
5. Pallade, S., Dorobantu, M., Bernsten, I., et al., De quelques modifications de l'active enzymatique dans l'intoxication par l'epichlorhydrine, *Arch. Mal. Prof. Med. Trav. Secur. Soc.,* 31, 365, 1970 (in French).
6. Lawrence, W. H., Malik, M., and Autain, J., Toxicity profile of epichlorohydrin, *J. Pharmacol. Sci.,* 61, 1712, 1972.
7. van Ecsh, G. J. and Wester, P. W., Epichlorohydrin, 1982, 25.
8. Wester, P. M., van der Heijden, C. A., Bisschop, A., et al., Carcinogenicity study with epichlorohydrin by gavage in rats, *Toxicology,* 36, 325, 1985.
9. Cooper, E. R. A., Jones, A. R., and Jackson, H., Effects of α-chlorohydrin and related compounds on the reproductive organs and fertility of the male rat, *J. Reprod. Fertil.,* 38, 379, 1974.
10. Jones, A. R., Davies, P., Edwards, K., et al., Antifertility effects and metabolism of α-epichlorohydrin in rat, *Nature (London),* 224, 83, 1969.
11. Hahn, J. D., Post-testicular antifertility effects of epichlorhydrin and **2,3**-epoxypropanol, *Nature (London)* 226, 87, 1970.
12. Cassidy, E. R., Jones, A. R., and Jackson, H., Evaluation of testicular sperm head counting technique using rats exposed to dimetoxyethyl phthalate, glycerol α-monochlorohydrin, epychlorohydrin, formaldehyde, or methylmethanesulphonate, *Arch. Toxicol.,* 53, 71, 1983.
13. Marks, T., Gerling, F. S., and Staples, R. E., Teratogenic evaluation of epichlorohydrin in the mouse and rat and glycidol in the mouse, *J. Toxicol. Environ. Health,* 9, 87, 1982.
14. Konishi, T., Kawabata, A., and Denda, A., Forestomach tumors induced by orally administered epichlorohydrin in male Wistar rats, *Gann,* 71, 922, 1980.
15. Bokaneva, S. A., Epichlorohydrin, Its Toxico-medical Characteristics and Significance in the Hygiene Regulations for New Epoxy Resins, Authors' abstract of thesis, Moscow, 1980, p. 18 (in Russian).

38

16. Gingell, R., Beatty, P. W., Mitschke, H. R., et al., Evidence that epychlorohydrin is not a toxic metabolite of **1,2**-dibromo-**3**-chloropropane, *Xenobiotica,* 17, 229, 1987.
17. Weigel, W. W., Plotnick, H. B., and Conner, W. L., Tissue distribution and excretion of ^{14}C-epichlorhydrin in male and female rats, *Res. Commun. Chem. Pathol. Pharmacol.,* 20, 275, 1978.

ETHYL ACRYLATE (EA) (CAS No 140–88–5)

Synonyms. Acrylic acid, ethyl ether; Ethyl propenoate; 2-Propeonic acid, ethyl ester.

Properties. Colorless liquid with an acrid odor. Solubility in water is 2% at 20°C, soluble in ethanol. Odor perception threshold is reported to be 0.0002 to 0.0067 mg/l.[02,07,010]

Applications. EA is used mainly in surface coatings, as a co-monomer in the production of several polymers, and in the preparation of latexes, paints, textiles, and paper coatings. EA has been found as a residual monomer in polyethylacrylate and in aqueous polymer latexes.

Acute Toxicity. LD_{50} is reported to be 1 to 2.8 g/kg BW in rats,[1,2] 1.8 g/kg BW in mice,[3] and 280 to 420 mg/kg BW in rabbits.[4] Poisoning affected the CNS functions and vessel permeability. Gross pathological examination revealed irritation of the intestinal walls as well as lesions of the liver and kidneys. Administration of 200 to 400 mg/kg BW to rats led to ataxia as well as hypodynamia and bradycardia, indicating a fall in blood catecholamine content.[5]

Repeated Exposure revealed moderate cumulative properties in mice. K_{acc} is 2.76 (by Lim). Administration of 31.5 mg/kg BW to rabbits for 35 d caused no toxic effect.[4] In a 14-d study, rats and mice were given 100 to 800 mg EA/kg BW. Rats developed abdominal adhesions in response to administration of 600 and 800 mg EA/kg BW in vegetable oil. Stomach lesions were found at the 400 mg/kg dose-level, gastritis was noted at 400 to 600 mg/kg in mice. Doses up to 200 mg/kg BW did not produce such an effect.[6,7]

Short-term Toxicity. Fisher 344 rats were given EA in their drinking water over a period of 3 months. A reduction in the consumption of drinking water and feed was noted at a dose of 83 mg/kg. No definite toxicity or any clear target organ was determined in this study (DePass et al., 1980 and 1983).

Long-term Toxicity. No treatment-related lesions were found in female but not in male Wistar rats that received 2000 mg EA/l in their drinking water for 2 years. Retardation of BW gain was the only sign noted. No effect was seen in dogs that were given 1 g/kg in the diet.[8] However, details on survival and results of the pathological examination in this study seem to be insufficient for evaluation (IARC 19–61). Rats and mice exposed to 100 to 200 mg/kg BW for 2 years exhibited inflammation and epithelial hyperplasia of the stomach walls.[6.] Rats were dosed by gavage for 6 months. The treatment resulted in liver function impairment and decreased histamine level in the blood and liver homogenate.[1]

Allergenic Effect was noted.

Reproductive Toxicity. *Embryotoxicity.* Rats were administered *i/p* $^1/_{10}$ to $^1/_3$ LD_{50} on day 5, 10, and 15 of pregnancy. The treatment increased embryomortality and reduced the fetal weights. A dose-effect relationship was noted.[9] Exposure to 25 to 400 mg/kg BW on days 7 to 16 of pregnancy led to an increase in maternal BW. Although 100 to 400 mg/kg BW doses significantly increased embryomortality, the litter size was only slightly affected. Skeletal ossification was retarded.[10] *Teratogenic* effect was noted on inhalation exposure.[11]

Genotoxicity. EA was negative in *Salmonella* and in *Dr. melanogaster* but produced mutations in mouse lymphoma cells, and CA and SCE in Chinese hamster ovary cells (NTP-85). Doses of 0.0075 and 0.015 mg/ml caused a dose-dependent increase of micronuclei and CA in the cultured Chinese hamster lung cells.[12]

Carcinogenicity. Animals were gavaged with 100 and 200 mg EA BW bolus doses in vegetable oil for 103 weeks. Squamous cell papillomas and forestomach carcinomas were revealed in mice B6C3F$_1$ and Fisher 344 rats of both sexes. A dose-dependent relationship was established. There was no apparent effect on other tissues.[6] The lesions developed only at the dosing site, without systemic toxicity. Recent studies have not confirmed these results. A 110 to 160 mg/kg BW dose given in drinking water did not evoke oncogenic response.[13] *Carcinogenicity classification.* IARC: Group 2B; NTP: P*—P*—P—P.

Chemobiokinetics. EA is readily absorbed and rapidly hydrolized into **acrylic acid** and **ethanol** in the blood and liver, and does not circulate in the body.[6] Although toxic at high concentrations,

EA is metabolized and detoxified rapidly at low concentrations. It may bind with nonprotein SH-groups in the erythrocytes.[14] According to other data,[15] ^{14}C-EA could react with both glutathione and protein. A single gavage dose of 200 mg/kg has been shown to cause severe glutathione depletion in the forestomach, with less depletion observed in the glandular stomach and duodenum.[16] EA is metabolized by cellular carboxylesterases and by conjugation with glutathione. **Mercapturic acid** has been shown to be a minor urinary metabolite.[17]

Standards. The Council of Europe (1981) included EA in a list of artificial flavoring substances that may be added to foodstuffs without hazard to public health at a level of 1 mg/kg in food and 0.2 mg/l in beverages. **Russia** (1988). MAC and PML: 0.005 mg/l (organolept.).

Regulations. USFDA (1993) considered EA to be a GRAS synthetic flavoring substance or food adjuvant. FDA regulates the use of EA (1) in polyethylene phthalate polymers to be safely used as articles or components of articles intended for use in contact with food, and (2) as a component of resinous and polymeric coatings to be safely used in contact with food. **EA polymers or copolymers** are permitted for use (1) as components of adhesives for a food-contact surface, (2) in resinous and polymeric coatings in a food-contact surface of articles intended for producing, manufacturing, packing, transporting, or holding food, (3) in paper and paperboard in contact with dry, aqueous, or fatty foods, (4) in semirigid and rigid acrylic and modified acrylic plastics, and (5) in cross-linked polyester resins for repeated use in articles or components of articles coming in contact with food.

References:
1. See **BISPHENOL A,** #2, 224.
2. Pozzani, U. C., Weil, C. S, and Carpenter, C. P., Subacute vapour toxicity and range-finding data for ethyl acrylate, *J. Ind. Hyg. Toxicol.,* 31, 311, 1949.
3. Tanii, H. and Hashimoto, K., Structure-toxicity relationship of acrylates and methacrylates, *Toxicol. Lett.,* 11, 125, 1982.
4. Treon, J. F., Sigmon, H., Wright, H., et al., The toxicity of methyl and ethyl acrylate, *J. Ind. Hyg. Toxicol.,* 31, 317, 1949.
5. Sobezak, Z. and Baransky, B., *Bromatol. i Chem. toksykol.,* 12, 405, 1979 (in Polish).
6. NTP, *Ethyl Acrylate,* NTP Technical Report Series No. 259, Research Triangle Park, NC, 1983.
7. Ghanayem, B. J., Maronpot, R. R., and Mattehews, H. B., Ethyl acrylate-induced gastric toxicity. II. Structure-toxicity relationship and mechanisms, *Toxicol. Appl. Pharmacol.,* 80, 323, 1985.
8. Borzelleca, J. F., Larson, P. C., and Hennigar, G. R., Studies on the chronic oral toxicity of monomeric ethyl acrylate and methyl metacrylate, *Toxicol. Appl. Pharmacol.,* 6, 29, 1964.
9. See **ACRYLIC ACID,** #7.
10. Pietrowicz, D., Owecka, A., and Baranski, B., Disturbances in rat embryonal development due to ethyl acrylate, *Zwierzeta. Lab.,* 17, 67, 1980 (in Polish).
11. John, J. A., Deacon, M. M., Murray, J. S., et al., Evaluation of inhaled allyl chloride and ethyl acrylate for embryotoxic and teratogenic potential in animals, *Toxicologist,* 1, 147, 1981.
12. Ishidate, M., The Databook of Chromosomal Aberration Tests *in vitro* on 587 Chemical Substances using a Chinese Hamster Fibroblast Cell Line, The Realize Inc., Tokyo, 1988, p. 197.
13. Miller, R. R., Young, J. T., Kociba, R. J., et al., Chronic toxicity and oncogenicity bioassay of inhaled ethyl acrylate in Fisher 344 rats and B6C3F$_1$ mice, *Drug Chem. Toxicol.,* 8, 1, 1985.
14. Potter, D. W. and Thu-Ba Tran, Rates of ethylacrylate binding to glutathione and protein, *Toxicol. Lett.,* 62, 275, 1992.
15. Frederick, C. B., Potter, D. W., Chang-Mateu, I. M., et al., A biologically-based pharmacokinetic model for oral dosing of ethyl acrylate, *Toxicologist,* 9, 237, 1989.
16. Stott, W. T. and McKenna, M. J. Hydrolysis of several glycol ether acetates and acrylates by nasal mucosal carboxylesterase *in vitro, Fundam. Appl. Toxicol.,* 5, 399, 1985.
17. De Bethizy, J. D., Udinsky, J. R., Scribner, H. E., et al., The disposition and metabolism of acrylic acid and ethyl acrylate in male Sprague-Dawley rats, *Fundam. Appl. Toxicol.,* 8, 549,1987.

ETHYLENE (CAS No 74–85–1)

Synonym. Acetene; Elayl; Ethene.

Properties. Colorless gas with a faint odor of ether. Solubility in water is 20 mg/l (20°C) and 250 mg/l (0°C). Rapidly volatilizes from the open surface of water. Odor perception threshold is reported to be 0.039 mg/l,[02] 0.5 mg/l,[1] or even 260 mg/l.[010] Does not affect the color or clarity of water.

Acute Toxicity. Mice tolerate administration of 0.5 ml of a solution with a concentration of 150 mg E./l without changes in their behavior.

Repeated Exposure. Accumulation is impossible because of the rapid excretion of E. from the body.

Short-term Toxicity. Mice were dosed by gavage with 3.75 mg/kg BW for 4 months. The treated animals displayed no changes in behavior or in BW gain and oxygen consumption. Gross patholog- ical examination revealed no changes in the relative weights or in the histological structure of the visceral organs.[1]

Long-term Toxicity. Rats were given 0.05 mg/kg BW for 6 months. The treatment produced no abnormalities in behavior, BW gain, leukocyte phagocytic activity, or in cholinesterase and condi- tioned reflex activity.[1]

Genotoxicity and **Carcinogenicity.** Experiments proved E. to be converted in certain species, notably mice and rats, into the carcinogenic and mutagenic **ethylene oxide.**[2] Carcinogenic effect of E. of endogenous origin is suggested.[3] Whether such an effect is possible with oral administration of E. is not clear. No toxic or carcinogenic effects were found after inhalation of 300 to 3000 ppm.[4]

Carcinogenicity Classification. IARC: Group 3.

Chemobiokinetics. It is unlikely that there is a direct chemical interaction between E. and bio- logical media. E. is not broken down in the body. It seems to be rapidly excreted via the lungs.

Standards. Russia (1988). MAC and PML: 0.5 mg/l (organolept., odor).

References:

1. See **BUTYLENE,** 28.
2. Filser, J. G. and Bold, H. M., Exhalation of ethylene oxide by rats on exposure to ethylene, *Mutat. Res.,* 120, 57, 1983.
3. Kokonov, M. T., Ethylene—an endogenous substance in tumor-carrier, in *Problems of Medical Chemistry,* Vol. 58, 1960, p. 158 (in Russian).
4. Rostron, C., Ethylene metabolism and carcinogenicity, *Food Chem. Toxicol.,* 24, 70, 1987.

ETHYLENEIMINE (EI) (CAS No 151–56–4)

Synonyms. Aziridine; Dimethyleneimine.

Properties. Colorless liquid with an amine odor. Readily soluble in ether and acetone, soluble in water and ethyl alcohol and other organic solvents. Odor perception threshold is 170 mg/l.[02]

Applications. EI has industrial and laboratory applications. It is a monomer in the production of polyethyleneimine and copolymers. An ingredient of different coatings and adhesives, ion-exchange resins, polymer stabilizers and surfactants. It is also used in cosmetics. On polymerization, EI forms a thermoplastic water-soluble mass used for clarification of water (polyethyleneimine).

Acute Toxicity. LD_{50} is reported to be 15 mg/kg BW in rats and to range from 4.2 to 8.4 mg/kg BW in mice, guinea pigs, and dogs. Poisoning is accompanied by vomiting and atony. Death occurs from respiratory arrest. Administration of 10 mg/kg increases the urea nitrogen content in the blood.[1,05]

Repeated Exposure usually results in general weakness and renal damage. The NS can remain unaltered.[2]

Reproductive Toxicity. *Gonadotoxic* effects are likely to be specific for EI. At doses that cause no toxic effect, it produces and maintains sterility in males and females and interrupts pregnancy. EI is shown to interfere with the process of RNA synthesis in the testes.[3] *Embryotoxicity* and *teratogenicity.* Inhalation exposure to EI affected reproduction. The LOAEL for embryotoxicity and teratogenicity effects is reported to be 1 mg/kg BW in rats.[4]

Genotoxicity. EI seems to be a most powerful chemical mutagen. It is exclusively genotoxic in all test-systems investigated: in microorganisms, insects, and cultured cells of mammalian bone marrow by DLM method and in human cell culture.[5] The mechanism by which EI exerts a mutagenic effect involves its interference with the metabolism of a pyrimidine precursor of DNA.

Carcinogenic effect is reported as a result of inhalation or skin application.[6] OSHA considered EI to be an 'occupational carcinogen.'

Chemobiokinetics. ^{14}C-labeled EI administered *i/p* can be detected in the urine, both in a free form and as an unidentified metabolite. Accumulation occurs in the liver, kidneys, and spleen. EI content in the fatty tissues appeared to be negligible.[05]

Standards. EEC (1990). SML: not detectable (detection limit 0.01 mg/kg). **Russia** (1988). MAC and PML: 0.0002 mg/l.

References:

1. Carpenter, C. P., Smith, H. F., and Shaffer, C. B., Acute toxicity of ethyleneimine to small animals, *J. Ind. Hyg. Toxicol.,* 30, 2, 1948.
2. Gelembitsky, P. A., et al., *Chemistry of Ethyleneimine,* Nauka, Moscow, 1966, p. 191 (in Russian).
3. See 4,4'-**DIAMINODIPHENYL ETHER,** #2, 16.
4. Bespamyatnova, A. V., Zaugol'nikov, S. D., and Sukhov, Iu. Z., Embryotoxic and teratogenic effects of ethyleneimine, *Pharmacol. Toxicol.,* 33, 347, 1970 (in Russian).
5. Bochkov, N. P., Sram, R. J., Kuleshov, N. P., et al., System for the evaluation of the risk from chemical mutagens for man: basic principles and practical recommendations, *Mutat. Res.,* 38, 191, 1976.
6. *Basic Problems of Long-term Consequences of Exposure to Industrial Poisons,* Proc. Research Institute Occup. Diseases, Plyasunov, A. K. and Pashkova, G. M., Eds., Moscow, 1976, p. 140 (in Russian).

ETHYLENE OXIDE (EtO) (CAS No 75–21–8)

Synonyms. Dimethylene oxide; **1,2**-Epoxyethane; Ethene oxide; Oxacyclopropane; Oxirane.

Properties. Colorless, incombustible gas. EtO has a high solubility in water (270 g/l at 25°C) but will evaporate to a great extent. Dissolves in water and alcohol. Odor perception threshold is 140 mg/l[02] or 260 mg/l.[010]

Applications and **Exposure.** EtO is used in the production of polyethylene oxide, polyethylene terephthalate, nonionogenic surfactants, di-and triethylene glycols, etc. A major source of exposure is its use as a sterilant.

Acute Toxicity. LD$_{50}$ is reported to be 330 mg/kg BW for male rats, 280 mg/kg BW for female rats, and 365 mg/kg BW for mice. When 200 mg/kg BW was administered in olive oil, all five rats died.[1,2] According to other data, LD$_{50}$ of the aqueous solution is 270 mg/kg BW in guinea pigs.[2] Oral administration produced ataxia, prostration, labored breathing, and convulsions in rats and mice.[3]

Repeated Exposure. Wistar rats received by gavage 22 doses of 3 to 30 mg EtO/kg BW and 15 doses of 100 mg EtO/kg BW during 21 d in olive oil. No effects have been reported on mortality rate, growth, hematology, blood-urea nitrogen, organ weights, and gross and histopathological findings. With the 100 mg/kg BW dose, there was a marked BW loss, gastric irritation, and a slight liver damage.[1]

Reproductive Toxicity. *Gonadotoxicity.* EtO caused gonadotoxic effects in males: reduced sperm number and sperm motility and an increased time to traverse a linear path. *Embryotoxicity.* EtO produced depression of fetal weight gain, as well as fetal death and fetal malformation. *Teratogenicity.* The levels needed to cause fetal effects approach the doses needed to produce maternal toxicity.[4]

Genotoxicity. *Humans.* EtO is a potential human mutagen for both somatic and germ cells. A significant increase in Hb alkylation, in the rate of CA and SCE in peripheral lymphocytes and, in a single study, of micronuclei in erythrocytes has been found in workers exposed occupationally to EtO. No oral data are available. *Animals.* EtO appeared to be an effective mutagen in a variety of organisms, namely microorganisms, insects, and mammalian cells, inducting both cell mutations and carcinogenicity. SCE and CA were observed in blood lymphocytes of monkeys and CA in the bone marrow cells of rats exposed by inhalation.[5,6] The dose of 150 mg/kg is positive in DLM assay.[7]

Carcinogenicity. *Humans.* Epidemiological studies revealed an association between EtO inhalation and excessive risk of carcinogenicity, but both studies have limitations. There are no epidemiological reports on effects of consuming EtO-treated foods. *Animals.* Sprague-Dawley rats were placed on regimens of 7.5 and 30 mg EtO/kg BW in salad oil by gavage twice a week over 110 weeks. Observation covered the life-span of these animals. An increased rate of tumor development was observed only in the forestomach. Metastases in ten rats and two fibrosarcomas were found with the dose of 30 mg/kg.[8] In another study, 50 rats were fed EtO-fumigated feed for 2 years. EtO residues in the feed were 50 to 1400 mg/kg.[9] After pathological examination, there was no evidence of excess tumors. Both these studies have flaws, and neither is definitive; they appear to indicate a lack of carcinogenic potential of EtO by the dietary route of exposure. **Carcinogenicity classification.** IARC: Group 1; NTP: XX—XX—CE*—CE*.

Chemobiokinetics. After absorption, equivalent doses of EtO are distributed throughout the body. The degree of alkylation of proteins and DNA varies slightly between the different organs and blood. In man and rodents, the half-life of EtO has been estimated to be 9 to 10 min. EtO is metabolized, including hydrolysis to 1,2-**ethandiol** and conjugation with glutathione, and is excreted primarily via the urine.[10]

Standards. EEC (1990). Maximum Permitted Quantity in the material or article: 1 mg/kg. ECC legislation prohibits presence of EtO in cosmetics. **Canada** (1980). Residues in food must be below 0.1 ppm.

Regulations. USFDA (1993) regulates EtO as a direct and indirect food additive under FD&CA and has proposed maximum residue limits for the compound in drug products and medical devices. The FDA set a 50 ppm tolerance level for EtO in food; the tolerance limitations for EtO are set only for residues that remain after treatment. The FDA is reevaluating its established regulations governing EtO residues, considering recent toxicity data and information concerning the formation of **1,4**- dioxane. The FDA approved the use of EtO for the following purposes: (1) as a fumigant in sizing used as a component of paper and paperboard in contact with dry foods, (2) as an etherifying agent in the production of modified industrial starch, provided the level of reacted EtO in the finished product does not exceed 3%, and (3) as a fumigant for species and other processed natural seasoning materials, except mixtures to which salt has been added, in accordance with prescribed conditions. **JMPR FAO/WHO** (1971). Data available are insufficient to establish ADI.

References:

1. Hollingsworth, R. L., Rowe, V. K., Oyen, F., et al., Toxicity of ethylene oxide determined in experimental animals, *Arch. Ind. Health,* 13, 217, 1956.
2. Smyth, H. F., Seaton, J., and Fisher, L., Single dose toxicity of some glycols and derivatives, *J. Ind. Hyg. Toxicol.,* 23, 259, 1941.
3. Woodard, G. and Woodart, M. Toxicity of residues from ethylene oxide gas sterilization, in *Proc. HIA Technical Symposium,* Health Industry Association, Washington, D.C., 1971.
4. LaBorde, J. B. and Kimmel, C. A., The teratogenicity of ethylene oxide administered intravenously to mice, *Toxicol. Appl. Pharmacol.,* 56, 16, 1980.
5. Dellarco, V. L., Generoso, W. M., Sega, G. A., et al., *Environ. Mutat.,* 16, 85, 1990.
6. See 4,4'-**DIAMINODIPHENYL ETHER,** #2, 11.
7. Generoso, W. M., Cain, K. T., Crishna, M., et al., Heritable translocation and dominant-lethal mutation induction with ethylene oxide in mice, *Mutat. Res.,* 73, 133, 1980.
8. Dunkelberg, H., Carcinogenicity of ethylene oxide and **1,2**-propylene oxide upon intragastric administration to rats, *Br. J. Cancer.,* 46, 924, 1982.
9. Bar, F. und Griepentrog, F., Langzeitfutterungsversuch an Ratten mit Athylenoxidbegastem futter, *Bundesgesundheitblatt,* 11, 105, 1969 (in German).
10. *Ethylene Oxide,* Environmental Health Criteria No. 55, WHO, Geneva, 1985, p. 80.

FORMALDEHYDE (FA) (CAS No 50—00—0)

Synonyms. Methanal; Formalin; Formic aldehyde; Formol; Methylaldehyde; Methylene oxide; Oxymethylene.

Properties. Colorless gas with a pungent, suffocating odor. Freely miscible with water (550 g/l at 25°C). Solution of 35 to 40% is called **formaline.** Extremely active chemically. Odor perception threshold is reported to be 0.6 mg/l;[02] taste perception threshold is 50 mg/l.[1]

Applications and **Exposure.** Of FA production, 80% is used for plastic and resin manufacture: urea-formaldehyde resins, phenolic resins, pentaerythritol and polyacetal resins, and melamine resins. Migration in drinking water from polyacetal plastic fittings is reported.

FA NOELs in Short-term and Long-term Toxicity Studies

Species	Length of Exposure	Medium Tested	NOEL (mg/kg BW)	Ref.
Rat	90 d	Drinking water	100	6
Dog	90 d	Diet	75	2
Rat	24 months	Drinking water	15–21	9
Rat	24 months	Drinking water	10	8

From Bartone, N. F., Grieco, R. V., and Herr, B. S., *JAMA*, 203, 50, 1968.

FA is present as a natural component in fresh and preserved fish, seafood, honey, roasted foods, and many fruits and vegetables: tomatoes (5.7 to 7.7 ppm), apples (17.3 to 22.3 ppm), spinach (3.3 to 7.3 ppm), and carrots (6.7 to 10 ppm). In addition, there is an exposure to FA deriving from food containers or food-packaging materials and to FA used as preservative in the food industry.[2] Concentrations of FA ranging from 3 to 23 mg/kg have been reported in a variety of foodstuffs. FA is present in tobacco smoke, and it is a widespread contaminant of the air of dwellings and public buildings and of the atmosphere.

Acute Toxicity. *Humans.* Of formalin, 100 to 200 ml is likely to cause a fatal poisoning. LD_{50} (37% solution) appears to be 523 mg/kg BW.[3] Since FA is rapidly metabolized into **formic acid,** a severe acidosis may develop, producing a local corrosive action on the upper GI tract. Thus, ulceration, necrosis, perforation, and hemorrhage occur, frequently resulting in death some days later.[4] *Animals.* LD_{50} is 400 to 800 mg/kg BW in mice and rats and 260 mg/kg BW in guinea pigs.[5] Poisoning of animals is accompanied by immediate marked excitation, with a subsequent narcotic effect. Death occurs in the first 2 to 3 h. The survivors remain apathetic and drowsy for 2 weeks, with no appetite.[6]

When **paraformaldehyde** is administered, the LD_{50} is 0.5 g/kg BW in mice and 5 g/kg BW in rats.[07]

Repeated Exposure failed to reveal cumulative properties. Habituation developed. Administration of ⅕ LD_{50} to rats resulted in 20% animal mortality on day 14 to 19 of administration. The survivors suffered from exhaustion. Gross pathological examination revealed parenchymatous dystrophy of the liver and kidneys and circulatory disturbances (Pomerantzeva, 1971). Rats were given doses of 5 to 125 mg/kg BW via their drinking water for 4 weeks. Food and liquid intake were decreased in the top dose group. Yellow discolorations of the fur, decreased protein and albumin levels in the blood plasma, thickening of the limiting ridge and hyperkeratosis in the forestomach, and focal gastritis in the glandular stomach were also noted at this dose level. No treatment-related effects were observed in animals given 5 and 25 mg FA/kg BW.[6]

Short-term Toxicity. *Humans.* The NOAEL in humans that ingested FA daily over a period of 13 weeks appears to be 200 mg.[7] *Animals.* Rats and dogs received oral doses of FA over a period of 3 months. The only adverse effect was depression of BW gain when doses of 100 mg/kg BW and higher were administered.[8]

Long-term Toxicity. *Humans.* Epidemiological investigations carried out hitherto have not made it possible to obtain reliable data about carcinogenic, mutagenic, teratogenic, embryotoxic, and neurotoxic danger of FA. *Animals.* Aldehydes, due to their instability in water, are highly unlikely to have a chronic effect. Histopathological changes in the GI tract were observed in a study where adverse effects were exhibited only with the 300 mg/kg BW dose.[9] Wistar rats were given FA in their drinking water over a period of 2 years at doses of 1.2 to 82 mg/kg BW (males) and 1.8 to 109 mg/kg (females). Adverse effects were demonstrated only in the highest dose groups. The principal target organs appeared to be the stomach (thickening of the mucosal wall) and kidney (renal papillary necrosis in both sexes).[10] A concentration of 0.5 mg/l produced no effect on conditioned reflex activity in rats.[1] NOELs claimed by the authors are reported in the above table.

TDI/ADI appears to be 0.15 mg/kg BW. Such ADI is in agreement with a 20 mg figure calculated on the basis of the NOAEL equal to 200 mg/kg BW demonstrated in humans.[7] Considering that FA daily intake calculated by Owen et al.[11] (11 mg, approximately) is in this range and that the liver may convert 22 mg FA to CO_2 in one min, it can be inferred that occasional ingestion of food containing free FA or its precursor, hexamethylenetetramine, in the order of few ppm, would not cause any harmful effect.

It is likely that at low concentrations exogenous FA is handled in the body in a way that is not significantly different from that of endogenous FA.

Immunotoxicity. Inhalation exposure and oral administration of FA are found to produce an allergenic effect.[12] The threshold sensitizing concentration of 0.5 mg/m^3 is not accompanied by any change in the number and functional activity of lymphocytes, which allows it to be assessed as nonallergenic.[13]

Reproductive Toxicity. *Gonadotoxicity.* A significant increase in sperm abnormalities was found in rats after a single oral administration of 200 mg/kg BW.[1] No effect on sperm was shown in male mice given FA by gavage at a dose of 100 mg/kg BW for 5 d.[14] *Teratogenicity.* A group of about 30 pregnant CD–1 mice was given oral doses of 74 to 185 mg/kg BW on days 6 to 15 of gestation. No teratogenic effects were shown.[15] *Embryotoxicity.* The dose of 540 mg/kg BW was administered to pregnant mice on days 8 to 12 of gestation. No embryotoxic effects were demonstrated.[16] Pregnant beagle dogs were fed FA at dose levels of 3.1 to 9.4 mg/kg BW from day 4 to day 56 after mating. There were no effects on reproductive performance or on health of the offspring.[17] A noticeable decrease in weight was found in the offspring of Sprague-Dawley female rats given FA in drinking water at a concentration of 2500 mg/l since day 15 of pregnancy.[18] Male rats treated with FA at a level of 0.1 mg/l in drinking water exhibited no changes in reproductive function.[19]

Genotoxicity. FA binds readily to proteins, RNA, and single-stranded DNA, including DNA–protein cross-links and breaks in single-stranded DNA.[20] It affects prokaryotic and eukaryotic cells *in vitro*. It would appear that FA reacts readily with macromolecules in cells primarily at the point of exposure and does not reach other sites in the body at sufficient concentration to produce detectable effects. *Humans.* FA caused an increase in the frequency of CA and SCE in peripheral lymphocytes, but negative results also have been reported. FA induced DNA–protein cross-links, unscheduled DNA synthesis, and mutation in the human cells *in vitro*. *Animals.* As Basler et al. indicate, no mutagenic effect is found in mammals *in vivo*.[21] The level of CA and SCE in a culture of SNO cells is increased significantly, but FA does not induce micronuclei and CA in bone marrow cells (dose of 25 mg/kg) or CA in spleen cells of mice.[22] It also induced transformation of mouse C3H 10TI/2 cells and DNA strand breaks and DNA–protein cross-links in rodent cells *in vitro*. It has been shown to be genotoxic in *Dr. melanogaster* (IARC 29–345). FA increases DNA synthesis and the number of micronuclei and nuclear abnormalities in rat epithelial cells.[23]

Carcinogenicity. The mechanism of FA carcinogenicity is not completely understood, although its reactivity with macromolecules is well known. It is thought that the mutagenic and carcinogenic effects of FA appear both as a result of direct damage to the DNA molecule and as a consequence of the inhibition of its repair. In a 2-year study, a dose-dependent increase in the incidence of leukemia at dose-levels of 5 mg/kg BW or greater was noted in Sprague-Dawley rats receiving FA in their drinking water.[18]

FA has been carcinogenic in rats and mice by inhalation at concentrations that caused irritation of the nasal epithelium. Oral and dermal studies have given little indication of the carcinogenic potential of FA. Positive results were demonstrated in rodents following inhalation exposure: after a 24-month inhalation exposure (5.6 up to 14.3 ppm), rats developed an unusual incidence of nasal mucosa tumors, although in mice the same concentration of FA induced a nonsignificant increase of this pathology.[24,25]

In spite of FA being cytotoxic to the mucosa of the glandular and nonglandular part of the stomach, it is unlikely that FA induces a significant increase of gastric tumors or tumors at other sites at the dose of 100 mg/kg BW, according to the findings of several long-term carcinogenicity studies.[9,26] Only one study reports the occurrence of gastrointestinal tumors in the long-term bioassay: Sprague-Dawley rats received FA in drinking water at concentrations of 10 to 1500 mg/l.[18] However, since the doses administered were higher than 100 mg/kg BW, the results cannot be considered to be of physiological significance. It was concluded that the glandular stomach mucosa in rats is more resistant to FA than forestomach mucosa.[3] This is particularly relevant because the target tissue in the rat has no human counterpart. Humans seem to be less sensitive to the carcinogenic effect of inhaled FA than rodents.[27] The weight of evidence indicates that FA is not carcinogenic by the oral route. *Carcinogenicity classification* (on the basis of inhalation studies). IARC: Group 2A; USEPA: Group B1.

Chemobiokinetics. FA is a normal cell metabolite in mammalian systems. Under normal conditions, the level of FA is very low in animal and human tissues because of its rapid oxidation to CO_2 (70 to 80%). FA is recycled in a single-carbon biosynthetic pathway, and hence a small proportion of a FA may be retained in the body. Under normal conditions, FA absorbed by the GI tract mucosa would remain as the aldehyde for only a very short period of time, either binding to macromolecules at the site of absorption or being metabolized by oxidative enzymes to **formic acid** and subsequently **carbon dioxide** and **water.** In rats and mice given a single oral dose, about 65% of radiolabel was excreted in the urine and feces, and 26% was eliminated in the expired air.[28]

Guidelines. WHO (1992). Guideline value for drinking water: 0.9 mg/l.

Standards. Russia (1988). MAC and PML: 0.05 mg/l; PML in food: 5 mg/l.

Regulations: USFDA (1993) regulates FA as an indirect food additive under FD&CA. It is listed in the CFR (1) as a component of adhesives for food-contact surface, (2) as a component of paper and paperboard for contact with dry, aqueous, and fatty food (only as a preservative for coating formulations), (3) in an acrylate ester copolymer coating to be safely used as a food-contact surface, (4) as a defoaming agent that may be used safely as a component of food-contact articles, (5) in phenolic resins for the surface of molded articles intended for repeated use in contact with nonacid food (pH above 5), and (6) as a substance employed in the production of or added to textiles and textile fibers intended for use in contact with food. It is approved for use (7) as a preservative in various aspects of food production. Most of these relate to its presence in a number of food packaging products, but it is also approved for inclusion (8) in defoaming agents containing dimethylpolysiloxane, which are used in food processing; the level of formaldehyde is limited to a maximum of 1.0% of the dimethylpolysiloxane content. FDA also approved the use of FA (9) as an additive in the manufacture of animal feeds based on animal fats and oilseed meals in order to improve the handling characteristics of the feed. The dried feed may contain a maximum of 1% FA. FA is permitted by Italian law in the production of two Italian cheeses, Grana Podano and Provolone (the level of FA in the final product must not exceed 0.5 mg/kg). The Italian law was passed by the **EEC.**[22] FA is permitted for use in a limited number of cosmetics in the european communities, with the following authorized concentrations in the finished cosmetic product: 5% in nail hardeners; 0.2% as a preservative, and 0.1% in mouth hygiene products (EEC, 1976).

References:

1. See **ACRYLONITRILE,** #1, 76.
2. See **EPICHLOROHYDRIN,** #12.
3. Bartone, N. F., Grieco, R. V., and Herr, B. S., Corrosive gastritis due to ingestion of formaldehyde without esophageal impairment, *JAMA,* 203, 50, 1968.
4. Restani, P. and Corrado, L. G., Oral toxicity of formaldehyde and its derivatives, *Crit. Rev. Toxicol.,* 21, 315, 1991.
5. See **ETHYLENE OXIDE,** #2.
6. See **ACETALDEHYDE,** #3.
7. *Occupational Health and Safety,* Vol. 1, McGraw-Hill, San Francisco, 1971, p. 574.
8. Johannsen, F. R., Levinskas, G. V., and Tegeris, A. S., Effects of formaldehyde in the rat and dog following oral exposure, *Toxicol. Lett.,* 30, 1, 1986.
9. Tobe, M., Naito, K., Caldwell, W. M., et al., Chronic toxicity study of formaldehyde administered orally to rats, *Toxicology,* 56, 79, 1988.
10. Til, H. P., Woutersen, R. A., Feron, V. J., et al., Two-year drinking-water study of formaldehyde in rats, *Food Chem. Toxicol.,* 27, 77, 1989.
11. Owen, B. A., Dudney, C. S., Tan, E. L., et al., Formaldehyde in drinking water: comparative hazard evaluation and an approach to regulation, *Regul. Toxicol. Pharmacol.,* 11, 220, 1990.
12. Vinogradov, H. I., Chemical allergens in the environment and their effects on human health, in Review Information, Moscow, 1985, p. 58 (in Russian).
13. The First All-Union Toxicol. Conf., Abstracts of Reports, Rostov-na-Donu, 1986, p. 227 (in Russian).
14. Ward, J. B., Jr., Hokanson, J. A., Smith, E. R., et al., Sperm count morphology and fluorescent body frequency in autopsy service workers exposed to formaldehyde, *Mutat. Res.,* 130, 417, 1984.
15. Marks, T. A., Worthy, W. C., and Staples, R. E., Influence of formaldehyde and Sonacide (potentiated acid gluturaldehyde) on embryo and fetal development of mice, *Teratology,* 22, 51, 1980.
16. Seidenberg, J. M., Anderson, D. G., and Becker, R. A., Validation of an *in vivo* developmental toxicity screen in the mouse, *Teratogen. Carcinogen. Mutagen.,* 6, 361, 1987.

17. Hurni, H. and Ohder, H., Reproduction study with formaldehyde and hexamethylenetetramine in beagle dogs, *Food Cosmet. Toxicol.,* 11, 459, 1977.
18. Soffritti, M., Maltoni, C., Maffei, F., et al., Formaldehyde: an experimental multipotential carcinogen, *Toxicol. Ind. Health,* 5, 699, 1989.
19. Commission of EC, Council Directive of 30 January 1978, Amend. for 13th Dir. 64/54/EEC.
20. Ma, T. M. and Harris, M. M., Review of genotoxicity of formaldehyde, *Mutat. Res.,* 196, 37, 1988.
21. Basler, A. and Hude, W. V. D., Scheutwinkel-Reiche, M., Formaldehyde-induced SCE *in vitro* and the influence of the exogenous metabolizing systems S9 mix and primary rat hepatocytes, *Arch. Toxicol.,* 58, 10, 1985.
22. Natarajan, A. T., Darroudi, F., Bussmann, C. J. M., et al., Evaluation of the mutagenicity of formaldehyde in mammalian cytogenic assays *in vivo* and *in vitro, Mutat. Res.,* 122, 355, 1983.
23. Migliore, L. et al., Micronuclei and nuclear anomalies induced in the gastrointestinal epithelium of rats treated with formaldehyde, *Mutagenesis,* 4, 327, 1989.
24. Svenberg, J. A., Kerns, W. D., Mitchell, R. I., et al., Induction of squamous cell carcinomas of the rat nasal cavity by inhalation exposure to formaldehyde vapor, *Cancer Res.,* 40, 3398, 1980.
25. Kerns, W. D., Pavkov, K. L., Donofrio, D. J., et al. Carcinogenicity of formaldehyde in rats and mice after long-term inhalation exposure, *Cancer Res.,* 43, 4382, 1983.
26. Takahashi, M., Hasegawa, R., Fyrukawa, F., et al., Effects of ethanol, potassium metabisulfite, formaldehyde and hydrogen peroxide on gastric carcinogenesis in rats after initiation with N-methyl-N'- nitro-N-nitrosoguanidine, *Jpn J. Cancer Res. (Gann),* 77, 118, 1986.
27. Squire, R. A. and Cameron, L. I., An analysis of potential carcinogenic risk from formaldehyde, *Regul. Toxicol. Pharmacol.,* 4, 107, 1984.
28. Galli, C. L., Ragusa, C., Resmini, P., et al., Toxicological evaluation in rats and mice of the ingestion of the cheese made from milk with added formaldehyde, *Food Chem. Toxicol.,* 21, 313, 1983.

FURAN (CAS No 110–00–9)

Synonyms. Divinylene oxide; Furfuran; Oxacyclopentadiene; Tetrole.

Properties. Colorless liquid with a specific odor. Solubility in water is 2.3% (20°C), readily soluble in alcohol. Odor perception threshold is 50 to 75 mg/l. F. sweetish taste in water is detected at higher concentrations.[1]

Applications. Used in the production of furan resins for corrosion-resistant polymeric materials. A solvent.

Acute Toxicity. After administration of 2 g/kg BW, death occurs in 30 min. A 0.5 g/kg BW dose causes death of rats in 3 to 8 d. Poisoned animals displayed excitation, with subsequent CNS inhibition. Death occurs from respiratory arrest. Gross pathological examination failed to reveal evident damage but congestion was apparent in the vessels of the visceral organs and in the bone marrow. Visceral organ degeneration develops as a result of F. doses that do not lead to rapid death.[1]

Repeated Exposure revealed cumulative properties. Administration of 100 mg/kg BW causes death in rats in 3 to 10 d, with signs of emaciation and jaundice. Liver changes are of the parenchymatous hepatitis type. Necrosis in the proximal renal tubules and reduction in N-acetylglucosoaminidase in the urine were found to develop in mice (Wiley et al.).

Short-term Toxicity. Rats were given 50 mg/kg BW for 3 months. The treatment caused atrophic liver cirrhosis with typical pseudopathic lobules and a reduction in the vascular flow. In a 4-month study, dystrophic changes in the liver were noted. A dose of 0.1 mg/kg BW was reported to be ineffective for peripheral blood formula, prothrombine time, serum cholinesterase activity, and conditioned reflex activity in rats.[1]

Allergenic Effect (slight) was observed in experiments on guinea pigs.[2]

Genotoxicity. F. was shown to be negative in mouse lymphoma cells (NTP-86) but positive in CA and SCE in Chinese hamster ovary cells (NTP-88). F. does not elicit a hepatic DNA repair response in rodents *in vivo.*[3]

Carcinogenicity. F. is found to be a potent carcinogen, more hepatotoxic than **furfural** or **furfuryl alcohol,** causing extensive distortion of gross liver structure and substantial microscopic pathology. Administration of 2, 4, and 8 mg/kg BW by gavage caused high incidence of cholangiocarcinomas at all doses in male rats.[4] *Carcinogenicity classification.* NTP: CE*—CE*—CE*—CE*.

Chemobiokinetics. Oral administration of F. decreased hepatic P-450 content and the activity of P-450-dependent enzymes. Covalent binding of F. to microsomal protein indicates that P-450 llEl preferentially catalyzes the metabolic activation of F. CO_2 is found to be the major metabolite. Ten metabolites detected in the urine indicate opening of the F. ring, followed by extensive carbon oxidation.[5,6]

Standards. Russia (1988). MAC and PML: 0.2 mg/l.

References:

1. See **BUTYLENE,** 219.
2. Chernousov, A. D., On allergenic properties of furan compounds, *Gig. Sanit.,* 6, 28, 1974 (in Russian).
3. Wilson, D. M., Goldsworthy, T. L., Popp, J., et al., Evaluation of genotoxicity, pathological lesions and cell proliferation in livers of rats and mice treated with furan, *Environ. Mol. Mutagen.,* 19, 209, 1992.
4. *Toxicology and Carcinogenesis of Furan in Fisher 344 Rats and B6C3F₁ Mice,* NTP Technical Report, Research Triangle Park, NC, 1991.
5. Burka, L. T., Washborn, K. D., and Irwin, R. D., Disposition of ^{14}C-furan in the male Fisher 344 rats, *J. Toxicol. Environ. Health,* 34, 245, 1991.
6. Kedderis, G. L., Carfagna, M. A., Held, S. D., et al., Kinetic analysis of furan biotransformation by F-344 rats *in vivo* and *in vitro, Toxicol. Appl. Pharmacol.,* 123, 274 1993.

FURFURAL (CAS No 98–01–1)

Synonyms. 2-Formylfuran; Fural; 2-Furaldehyde; 2-Furancarboxaldehyde; Furfuraldehyde; Furfurole; Pyromucic aldehyde.

Properties. Colorless, clear, oily liquid with a characteristic odor of bitter almond, rapidly turning brown in air. Solubility in water is 8.3% (20°C), readily soluble in alcohol. Odor perception threshold is 1.86 mg/l[1] or 3.5 mg/l.[02]

Applications. Used in the manufacturing of plastics and rubbers, in the production of furfuryl alcohol, tetrahydrofuran and furan-phenolic corrosion-resistant polymers; a solvent in refining nitrocellulose, cellulose acetate, etc.

Acute Toxicity. The LD_{50} values are reported to range from 65 to 127 mg/kg BW in rats, 400 to 425 mg/kg BW in mice, 541 mg/kg BW in guinea pigs, 800 mg/kg BW in rabbits, and 950 mg/kg in dogs. Poisoning is accompanied by CNS depression and changes in behavior and GI tract functioning. Death occurs in the first 5 to 6 h. Gross pathological examination showed circulatory disturbances, liver and spleen enlargement, and hyperemia of the gastric mucosa.[1]

Repeated Exposure failed to reveal cumulative properties. K_{acc} appeared to be ~20. Administration of 2.5 and 25 mg/kg BW to rats for 5 weeks caused enhanced BW gain.

Long-term Toxicity. Rats were dosed by gavage for 6 months. Administration of 75 mg/kg BW caused reduction in the blood chlorides and Hb levels to the end of the experiment.[1]

Genotoxicity. F. was positive in *Salmonella typhimurium TA100.* Produced a doubling of the SCE frequency, and a three-point monotonic increase with at least the highest dose at the $p < 0.001$ significance level.[5] Mutagenic effect does not depend on presence of an activation system or of *benzo[a]pyrene.*[2]

Carcinogenicity. In a 2-year study, Fisher 344 rats were given 30 and 60 mg/kg BW doses. The treatment caused dysplasia with fibrosis, a precursor lesion to cholangiocarcinomas in males.[3] *Carcinogenicity classification.* NTP: SE—NE—CE—CE.

Chemobiokinetics. F. is rapidly converted to **furoic acid, furanacrylic acid,** and **furoylglycine,** most likely by nonmixed-function oxidase-dependent pathways.[4] F. is eliminated in the urine mainly as a **furoic acid.**

Regulations. USFDA (1993) regulates F. as a component of adhesives intended for use in packaging, transporting, or holding food.

Standards. Russia (1988). MAC: 1 mg/l (organolept.).

References:

1. *Proc. Omsk Med. Inst.,* 69, 38, 1967; *Gig. Sanit.,* 5, 7, 1966 (in Russian).
2. Zdzienicka, M., Tudek, B., Zielenska, M., et al., Mutagenic activity of furfural in *Salmonella typhimurium TA 100, Mutat. Res.,* 58, 205, 1978.
3. *Toxicology and Carcinogenesis of Furfural in Fisher 344 Rats and B6C3F₁ Mice,* NTP Technical Report No. 382, Research Triangle Park, NC, 1989.
4. Irwin, R. D., Enke, S. B., and Prejean, J. D., Urinary metabolites of furfural and furfuryl alcohol in Fisher 344/N rats, *Toxicologist,* 5, (Abstr. 960), 240, 1985.
5. Tucker, J. D., Auletta, A., Cimino, M. C., et al., Sister-chromatid exchange: second report of the Gene-Tox Program, *Mutat. Res.,* 297, 101, 1993.

HEXACHLORO-1,3-BUTADIENE (HCBD) (CAS No 87–68–3)

Synonyms. 1,1,2,3,4,4-Hexachloro-1,3-butadiene; Perchlorobutadiene.

Properties. Clear, colorless, oily liquid. Solubility in water is 3.2 mg/l at 20°C, readily soluble in organic solvents. Odor perception threshold is 0.007 to 0.03 mg/l; taste perception threshold is 0.048 mg/l.[1]

Applications and **Exposure.** HCBD is used in the production of synthetic rubbers and as a solvent for elastomers. The estimated daily intake by man seems to be adequate to avoid adverse noncarcinogenic effects on the kidneys (IPCS-91).

Acute Toxicity. *Humans.* Ingestion of high doses led to irritation or corrosion of the mouth, throat, and GI tract tissues. *Animals.* LD_{50} is 200 to 400 mg/kg BW in female rats, 500 to 670 mg/kg BW in male rats,[1] and 50 to 200 mg/kg BW in mice.[2] Poisoning is accompanied with nephrotoxic changes: increased relative kidney weights and blood urea nitrogen as well as renal tubular cell necrosis.[3] Death may occur in several days or even months after exposure.

Repeated Exposure. B6C3F₁ mice were given HCBD at the dose levels of 30 to 3000 mg/kg BW in the diet for 15 d. All mice in the two highest dose groups died by day 7. Toxic responses (primarily in the highest dose groups) included abnormal clinical signs (lethargy, hunched posture, rough hair coats, light sensitivity, and/or coordination disturbances), gross and histopathological changes in the liver, kidneys, and testes.[4] Wistar rats were exposed for 14 days to 3 to 27 mg HCBD/kg BW.[5] The only lesion observed was degeneration of the renal tubular epithelial cells.

Short-term Toxicity. The kidney was found to be a target organ in short-term studies. Rats given HCBD by gavage for 13 weeks exhibited increased liver and kidney weights, some renal function disorders, and degeneration and hyperplasia of the proximal tubular epithelial cells.[4,6] Japanese quails were fed diets containing 0.3 to 30 mg HCBD/kg BW for 90 d. No adverse effects were noted.[1] In a 13-week study, the NOEL was identified to be 1 and 2.5 mg/kg BW for female and male rats, respectively,[5] and 4.9 mg/kg BW for B6C3F₁ mice.[4]

Long-term Toxicity. In a 2-year feeding study, the NOAEL of 0.2 mg/kg BW was identified for renal toxicity in rats.[4,7] TDI was calculated to be 0.0002 mg/kg BW.

Reproductive Toxicity. *Gonadotoxicity.* Decrease in sperm motility was observed in mice dosed with 0.1 to 16.8 mg/kg BW for 13 weeks.[4] *Embryotoxicity.* Rats were fed doses of 0.2 to 20 mg/kg BW for 90 d prior to mating, 125 d during mating, and subsequently throughout gestation (22 d) and lactation. There were no treatment-related effects on pregnancy or neonatal survival. No toxic effects were observed in neonates at given doses of 0.2 to 2 mg/kg BW; at higher doses histological kidney changes were reported.[1] *Teratogenic* effects were found at a dose-level of 8.1 mg/kg BW given to pregnant rats throughout gestation.[5,6]

Genotoxicity. Contradictory results have been obtained in bacterial assays for point mutations; however, several metabolites gave positive results. The percentage of unscheduled DNA synthesis caused by HCBD was increased in Syrian hamster embryo fibroblasts, both with and without metabolic activation.[8] HCBD is considered to be a genotoxic compound.

Carcinogenicity. Administration in the diet for 2 years caused renal tubular adenomas and adenocarcinomas in rats at doses causing renal injury (20 mg/kg BW).[8] HCBD seems to be a possible carcinogen for humans. *Carcinogenicity classification.* IARC: Group 3; USEPA: Group C.

Chemobiokinetics. Following a single oral administration and absorption, HCBD is distributed mainly in the adipose tissue and liver. In rats and mice, HCBD is easily absorbed and metabolized via conjugation with glutathione followed by biliary excretion of *S*-(1,2,3,4,4-**pentachloro**-1,3-**butadienyl)-glytathione**.[9] This conjugate can be further metabolized to nephrotoxic derivatives. The principal route of excretion is in the bile.[10] According to Payan et al., these glutathione conjugates undergo biliary recycling, being excreted in the urine as the corresponding **mercapturates**.[11]

Guidelines. WHO (1991). Guideline value for drinking water: 0.0006 mg/l.

Standards. Russia (1988). MAC in water: 0.01 mg/l (organolept.), in food it is 0.01 mg/kg (1983).

References:

1. Murzakayev, F. G., Toxicity data for hexachlorobutadiene and its intermediates, *Pharmacol. Toxicol.*, 26, 750, 1963 (in Russian).
2. Reports of the Ukrainian Academy of Science, Kiev, 1983, p. 55 (in Ukrainian).
3. Hook, J. B., Ishmael, J., and Lock, E. A., Nephrotoxicity of hexachloro-**1,3**-butadiene in the rat: the effect of monooxygenase inducers, *Toxicol. Appl. Pharmacol.*, 65, 373, 1984.
4. *The Toxicity Studies of Hexacloro-1,3-butadiene in B6C3F₁ Mice (Feed Studies)*, NTP Report, NIH Publ. No. 91–3120, 1991, 22.
5. Harleman, J. H. and Seinen, W., Short-term toxicity and reproduction studies in rats with hexachloro-(**1,3**)-butadiene, *Toxicol. Appl. Pharmacol.*, 47, 1, 1979.
6. Gehring, P. J. and MacDougall, D., Review of the Toxicity of Hexachlorobenzene and Hexachlorobutadiene, Dow Chemical Co., Midland, MI, 1971.
7. Kociba, R. J., Keyes, D. G., Jersy, G. C., et al., Results of a two-year chronic toxicity study with hexachlorobutadiene in rats, *J. Am. Ind. Hyg. Assoc.*, 38, 589, 1977.
8. Schiffman, D., Reichert, D., and Henschler, D., Induction of morphological transformation and unscheduled DNA synthesis in Chinese hamster embryo fibroblasts by hexachlorobutadiene and its putative metabolite pentachloro-butenoic acid, *Cancer Lett.*, 23, 297, 1984.
9. Dekant, W., Schrenk, D., Vamvakas, S., et al., Metabolism of hexachloro-**1,3**- butadiene in mice: *in vivo* and *in vitro* evidence for activation by glutathione conjugation, *Xenobiotica*, 18, 803, 1988.
10. Nash, J. A., King, L. D., Lock, E. A., et al., The metabolism and disposition of hexachloro-**1,3**-butadiene in the rat and its relevance to nephrotoxicity, *Toxicol. Appl. Pharmacol.*, 73, 124, 1984.
11. Payan, J. P., Fabry, J. P., Beydon, D., et al., Biliary excretion of hexachlorobutadiene and its relevance to tissue uptake and renal excretion in male rats, *J. Appl. Toxicol.*, 11, 437, 1991.

HEXAMETHYLENEDIAMINE (HMDA or HDA) (CAS No 124–09–4)

1,6-HEXANEDIAMINE DIHYDROCHLORIDE (HDDC) (CAS No 6055–52–3)

Synonyms. Of HDA: **1,6**-Diaminohexane. Of HDDC: Hexamethylenediamine dichloride; **1,6**-Diamino-*N*-hexane dichloride.

Note. The exact form of the compound is not specified in the majority of literature reports on HDA (HMDA) studies (see **Properties**).

Properties. Colorless leaflets (or lustrous crystals) with a pungent ammonia odor. HDA (HMDA) absorbs water and CO_2 from the atmosphere. Readily soluble in water and alcohol. In its solid form, HDDC crystallizes as needles from water or ethanol. The technical product is supplied as a 70% aqueous solution. Aqueous solutions of HDA are highly basic. 1,6-HDA dihydrochloride (HDDC) is formed by the neutralization of HDA with hydrochloric acid. Odor perception threshold is 500 mg/l. Aqueous solutions are colorless and transparent.[1]

Applications and **Exposure.** Used in the production of nylon-type polyamide resins (especially nylon 66), in the synthesis of polyurethane coatings, wet-strength resins, and polyamide adhesives. Curing agent for epoxy resins. HDA has not been found in U.S. or European drinking water supplies (NAS, 1977).

Acute Toxicity. LD_{50} is 750 to 980 mg/kg BW in rats and 1110 mg/kg BW in rabbits.[2,3,03] Clinical signs of intoxication include weakness, malaise, salivation, diarrhea, tremors, and BW loss. Five guinea pigs given 0.02 g HDA orally each day died within 20 to 70 d.[4] According to other data, LD_{50} is 582 mg/kg BW in rats, 665 mg/kg BW in mice, and 620 mg/kg BW in rabbits.[1] Lung emphysema is reported to develop rapidly, and 2 to 5 min after HDA is administered, a yellow fluid is discharged from the mouth and nose of the animals. Death occurs in 10 min with clonic and tonic convulsions. Dissection reveals serous emphysema in the lungs and congestion in the liver, kidneys, and spleen; the gastric mucosa seems to be edematous but without necrosis and hemorrhaging.

Repeated Exposure. Mice and rats were given the dose of 100 mg/kg. Toxic effects developed after two to three administrations. A majority of the animals died after six to seven administrations, which indicates pronounced cumulative properties. With doses of 10 and 30 mg/kg BW, toxic effects developed after 2 to 3 weeks. An increase in BW gain ceased in the fourth week, and in the sixth to eighth week, there was a 10 to 35% BW loss, changes in the peripheral blood formula, an increase in the activity of cholinesterase, a change in the ratio of protein fractions in the blood serum, and a reduction in the content of SH-groups in the blood.[1] In a 2-week study in Fisher 344/N rats (0.75 to 6.7 mg/ml) and B6C3F$_1$ mice (0.2 to 3.0 mg/ml), no clinical signs of toxicity and no gross or microscopic pathological changes were found.[5]

Short-term Toxicity. Sprague-Dawley rats given daily doses of up to 500 mg/kg in their feed for 13 weeks experienced no changes in BW gain and clinical chemistry parameters examined.[3] In a 13-week inhalation study, the NOAEL for respiratory damage was 5 mg/m^3 for rats and mice.[5]

Immunotoxicity. In a 12-month study, rats were given HDA in their drinking water at concentrations of 0.1 to 10 mg/l. The inhibition of antibody production and reduction in the volume of the splenic lymphoid tissue were observed in approximately 40% of animals.[6] The doses of 0.1 and 1 mg HDA/kg, but not 10 mg HDA/kg BW given in drinking water for 1 to 2 years caused an increase in the mitotic index of the lymphoid tissues.[7]

Reproductive Toxicity. *Embryotoxicity.* No adverse effect on reproduction of rats was noted following inhalation exposure.[5] No treatment-related mortality was noted in Sprague-Dawley rats given the diet containing average daily doses of 50, 150, and 500 mg/kg BW over two generations. BW and litter size were slightly reduced in the high-dose group. There was no adverse effect on survival during lactation; doses up to 150 mg/kg BW did not affect reproduction or fertility in rats.[8] *Teratogenic* effect was not observed in mice[9] and rats.[10] *Genotoxicity.* HDA was not mutagenic in four strains of *Salmonella;* it did not increase SCE or CA in cultured Chinese hamster ovary cells. Negative results were obtained in a micronuclei test in mice *in vivo.*[5]

Chemobiokinetics. In human volunteers, orally given HDA was completely excreted with the urine within 15 h. The primary urinary metabolites found were 6-**aminohexanoic acid** and *N*-**acetyl-1,6-HDA.**[11]

Standards. EEC (1990). SML: 2.4 mg/kg. **Russia** (1988). MAC and PML: 0.01 mg/l.

Regulations. USFDA (1993) listed HMDA for use as a component in resinous and polymeric coatings that may be safely used as a food-contact surface of articles intended for use in producing, manufacturing, packing, transporting, or holding food.

References:
1. See **BUTYLENE,** 69.
2. Vernot, E. H., MacEwen, J. D., Haus, C. C., et al., Acute toxicity and skin corrosion data for some organic and inorganic compounds and aqueous solutions, *Toxicol. Appl. Pharmacol.,* 42, 417, 1977.
3. Johannsen, F. R. and Levinskas, G. J., Toxicological profile of orally administered **1,6**-hexane diamine in the rat, *J. Appl. Toxicol.,* 7, 259, 1987.
4. Ceresa, C. and De Blasiis, M., Experimental research on intoxication with hexamethylenediamine, *Med. Lavoro,* 4, 78, 1950 (in Italian).
5. NTP, *Toxicity Studies of 1,6-Hexanediamine Dihydrochloride Administered by Drinking Water and Inhalation to F344/N Rats and B6C3F$_1$ Mice,* (Draft) NIH, NTP, U.S. Deptartment of Health and Human Services, 1992, 74.
6. Shubik, V. M., Nevstruyeva, M. A., Kal'nitsky, S. A., et al., A comparative study of changes in immunological reactivity during prolonged exposure to radioactive and chemical substance in drinking water, *J. Hyg. Epidemiol. Microbiol. Immunol.,* 22, 408, 1978.
7. Ponomareva, T. V., and Merkushev, G. N., Effects of some nonradioactive and radioactive chemical compounds on the structure of the spleen, *Arch. Anat. Histol. Embryol.,* 74, 47, 1978 (in Russian).
8. Short, R. D., Johannsen, F. R., and Schardein, J. L., A two generation reproduction study in rats receiving diets containing hexamethylenediamine, *Fundam. Appl. Toxicol.,* 16, 490, 1991.
9. Manen, G. A., Hood, R. D., and Farina, J., Ornithine decarboxylase inhibition and fetal growth retardation in mice, *Teratology,* 28, 237, 1983.

10. Johannsen, F. R. and Levinskas, G. I., Toxicological profile of orally administered **1,6**-hexane diamine in the rat, *Fundam. Appl. Toxicol.*, 7, 259, 1987.
11. Brorson, T., Soapring, G., Sandstrom, J. F., et al., Biological monitoring of isocyanates and related amines. I. Determination of **1,6**-hexamethylenediamine in hydrolyzed human urine after oral administration of HDA, *Int. Arch. Occup. Environ. Health*, 62, 79, 1990.

HEXAMETHYLENEDIAMINE ADIPATE

Synonyms. Salt AG; Nylon-**6,6** salt.

Properties. Colorless, oily crystals with a faint specific odor. Solubility in water is up to 50% (room temperature). Concentrations of 5 and 1 g/l give water an unrequired odor and taste respectively.

Applications. A monomer in the production of polyamide 6,6 (Anid).

Acute Toxicity. 5 g/kg dose is reported to be lethal to both rats and rabbits. According to other data, LD_{50} is 6.7 g/kg BW for rats, and 1.14 to 3.6 g/kg BW for mice. Single administration of 0.5 or 2 g/kg BW doses to rats caused a transient decrease in the body temperature, labored breathing, and leukopenia.

Repeated Exposure failed to reveal cumulative properties in the study when a 1 g/kg BW dose was administered for 56 d (Babayev, 1981).

Short-term Toxicity revealed moderate cumulative properties. Ten administrations of 700 mg/kg BW to mice led to a pronounced toxic effect. Three mice out of twenty died. Rats were dosed with 300 and 500 mg/kg BW for 40 d. None of the animals died, but they displayed agitation, aggressiveness, retardation of BW gain, and decrease in cholinesterase activity and in the content of serum SH-groups. Gross pathological examination revealed dystrophic changes in the liver and kidneys.

Long-term Toxicity. Administration of 5 mg/kg BW to rabbits for 7 months resulted in decreased activity of blood cholinesterase and increased content of protein and SH-groups in the blood. Doses of 0.05 and 0.5 mg/kg BW appeared to be ineffective. There were no morphological changes in the visceral organs.

Standards. Russia (1986). PML: 5 mg/l.

Reference:
See **BUTYLENE**, 56.

HEXAMETHYLENEDIAMINE SEBACATE

Synonyms. Salt SG; Nylon **6,10** salt.

Properties. Light, odorless, sweetish, amorphous powder with a faint rosy tinge. Poorly soluble in water. The threshold organoleptic concentration is 1 g/l.

Applications and **Exposure.** Used as a monomer in the production of polyamide 6,8 (nylon 6,10).

Acute Toxicity. LD_{50} is found to be 1.88 g/kg BW for mice, and 11 g/kg BW for rats. Treatment-related effects included convulsions, mucosal hyperemia, and nasal bleeding in some animals in 15 min after administration. Death occurred within 3 d after exposure; survivors recovered in 7 to 9 d.

Repeated Exposure revealed no cumulative properties. Rats were dosed with a 1.3 g/kg BW dose for 45 d. The treatment resulted in reduced BW gain and changes of hematological indices and of gastric function (Babayev, 1980).

Reference:
See **ACRYLONITRILE**, #1, 230.

HEXAMETHYLENE DIISOCYANATE (CAS No 822–06–0)

Synonyms. Desmodur G; 1,6-Diisocyanate hexane; Hexane diisocyanate; Tolnate HD.

Properties. Pale-yellow, slightly volatile liquid with a pungent unpleasant odor (threshold concentration 0.001 mg/l). Reacts with water and alcohol.

Applications. A cross-linking agent or hardener in the production of polyurethane materials.

Acute Toxicity. LD_{50} is found to be 738 to 960 mg/kg in rats, 350 to 1080 mg/kg in mice, and 1100 mg/kg in cats (HSDB, 1984). Ingestion caused irritation of the mouth and GIT, headache, nausea, and vomiting.

Long-term Toxicity. HMDI inhibits acetycholinesterase. Signs of nephrotoxicity were observed in mice with chronic feeding at dose-levels of 108 mg/kg BW (NTP-86).

Chemobiokinetics. In the presence of water, diisocyanates rapidly disassociate to form a primary amine, urea, and chlorine dioxide. This led to denaturation of proteins, loss of enxzyme function and/or formation of haptens, and immunological reactiveness (Tornling et al., 1990).

Regulations. EEC (1990). MPQ in the finished material or article: 1 mg/kg (expressed as isocyanate moiety). **USFDA** (1993) listed H. for use in polyurethane resins that may be safely used as the food-contact surface of articles intended for use in contact with dry food.

ISOBUTYLENE (CAS No 115–11–7)

Synonym. 2-Methylpropene.

Properties. Colorless gas. Water solubility is 375 mg/l. Volatilizes rapidly from the open surface of water. Odor perception threshold is 0.5 mg/l. I. has no effect on the color or transparency of water.[1]

Acute Toxicity. Mice tolerate administration of 8.4 mg/kg BW dose (given as a 0.5 ml of I. solution at a concentration of 335 mg/l) without any manifestations of the toxic action.[1]

Short-term Toxicity. Mice were dosed by gavage with 3.75 mg/kg BW for 4 months. The treatment caused no changes in the behavior, BW gain, or oxygen consumption. Gross pathological examination revealed no changes in the relative weights of the visceral organs or their histological structure. Accumulation is impossible because of its rapid removal from the body.[1]

Genotoxicity. I. (gas) at 5 to 100% is found to be negative in *Salmonella* mutagenicity assay with or without metabolic activation.[2]

Carcinogenicity. I. exhibited no carcinogenic effect in mouse lymphoma cells.

Chemobiokinetics. I. is rapidly excreted unchanged through the lungs.

Standards. Russia (1988). MAC and PML: 0.5 mg/l (organolept.).

Regulations. USFDA (1993) listed I. for use (1) as a component of adhesives to be safely used in food-contact surfaces, and (2) as a component of the uncoated or coated food-contact surface of paper and paperboard that may be safely used in producing, manufacturing, packing, transporting, or holding dry food.

References:
1. See **BUTYLENE,** 28.
2. Staab, R. J. and Sarginson, N. J., Lack of genetic toxicity of isobutylene gas, *Mutat. Res.,* 130, 259, 1984.

ISOPRENE (CAS No 78–79–5)

Synonym. 2-Methyl-**1,3**-butadiene.

Properties. Colorless, mobile, volatile, combustive liquid with a pungent unpleasant odor. Natural rubber is an I. polymer that can be depolymerized by heating to yield I. monomer. Water solubility is 1200 mg/l at 20°C. Odor perception threshold is 0.005 mg/l; the concentration of 1000 mg/l does not alter the color or taste of water.[1]

Acute Toxicity. Doses of 4.5 to 5 g/kg BW appeared to be lethal for rats. LD_{50} for rats is 2.04 g/kg (females) and 1.4 g/kg BW (males). LD_{50} for mice is 1.4 g/kg (females) and 1.7 g/kg BW (males).[2] Toxic effects include NS depression, slow movements, and labored breathing.

Repeated Exposure. Ten daily administrations of 25 mg/kg BW produced no effect on the general condition, BW gain, or behavior of mice and rats. Decreased catalase activity was noted. Rats and rabbits were dosed by gavage for 2 months. The treatment caused no changes in the general condition of animals. With a 2.5 mg/kg BW dose, the glycogen-forming function of the liver was disturbed, catalase activity was reduced, and there were changes in conditioned reflex activity. Histological examination revealed no specific changes in the visceral organs.[1]

Reproductive Toxicity. I. did not produce a teratogenic effect after inhalation exposure in rats and mice.[3]

Genotoxicity. I. did not induce CA and SCE in Chinese hamster ovary cells (NTP–85).

Carcinogenicity classification: IARC: Group 2B.

Chemobiokinetics. Distribution in the brain and parenchymatous organs was similar but the level in the fatty tissues was substantially higher than that in the brain, liver, kidney, or spleen.[4] When incubated in presence of microsomes of the murine liver, the main products of biotransformation of I. were 3,4-**epoxy**-3-**methyl**-1-**butene** and 3,4-**epoxy**-2-**methyl**-1-**butene.** In presence of microsomes, the latter underwent further epoxidation with the formation of mutagenic **isoprene oxide** (Monte et al., 1985). Rats eliminate six times more I. metabolites in the urine than mice at low inhalation exposure concentrations (NTP-91).

Standards. Russia (1988). MAC and PML: 0.005 mg/l (organolept.).

Regulations. USFDA (1993) listed I. for use as a component in the uncoated or coated food-contact surface of paper and paperboard that may be safely used in producing, manufacturing, packing, transporting, or holding dry food.

References:

1. Klimkina, N. V., Hygienic norm-setting of harmful substances from the production of synthetic isoprene rubber in water bodies, *Gig. Sanit.,* 6, 8, 1959 (in Russian).
2. *Occupational and Environmental Hygiene and Health Protection in Workers of the Oil Industry,* Moscow, 1987, p. 105 (in Russian).
3. Mast, T. J., Rommereim, R. I., Weigel, R. J., et al., Inhalation developmental toxicity of isoprene in mice and rats, *Toxicologist,* 10, 42, 1990.
4. Shugaev, B. B., Concentrations of hydrocarbons in tissues as a measure of toxicity, *Arch. Ind. Health,* 18, 878, 1969.

MALEIC ANHYDRIDE (MAn) (CAS No 108–31–6)

Synonyms. *cis*-Butanedioic anhydride; **2,5**-Furandione; Maleic acid, anhydride; Toxilic anhydride.

Properties. MAn occurs as a white, crystalline powder with a pungent odor. Readily soluble in water, forming MAc (1 mg of anhydride is equivalent to 1.18 mg of acid). Odor and taste perception threshold appeared to be 1 mg/l (Lisovskaya et al., 1963).

Applications. Used as a monomer in the production of polyester plastics.

Acute Toxicity. LD_{50} in rats is reported to be 625 mg/kg BW or 1050 mg/kg (Monsanto Co.).

Long-term Toxicity. Rabbits were dosed by gavage with 2.5 mg/kg BW for 6 months. The treatment affected liver glycogen-forming function and leukocyte phagocytic activity. Histological examination revealed signs of parenchymatous dystrophy in the liver, kidneys, spleen, and GI tract mucosa (Lisovskaya et al., 1963).

Allergenic Effect has been observed in tests on guinea pigs (Kryzhanovskaya et al., 1968).

Reproductive Toxicity. Short et al. observed no treatment-related fetal effects in rats given 140 mg MAn/kg BW on days 6 to 15 of gestation. No ill effects were found in a two-generation study with the 55 mg/kg BW dose.

Standards. EEC (1990). SML: 30 mg/kg (expressed as maleic acid). **Russia** (1988). MAC and PML (maleic acid): 1 mg/l (organolept. odor and taste).

Regulations. USFDA (1993) approved the use of MAn (1) as a component of paper and paperboard for contact with aqueous and fatty food (polymer with ethyl acrylate and vinyl acetate, hydrolyzed) and dry food (a deposit control additive to the sheet-forming operation at a level not to exceed 0.075% by weight of the dry paper and paperboard), and (2) in the polyurethane resins to be safely used in the food-contact surface of articles intended for use in contact with bulk quantities of dry food. MAc is listed for use (3) in resinous and polymeric coatings to be safely used as a food-contact surface of articles intended for use in producing, manufacturing, packing, transporting, or holding food, (4) as an ingredient of cellophane for food packaging (1%), (5) as a component of adhesives for food-contact surface, and (6) in cross-linked polyester resins for repeated use as articles or components of articles coming in contact with food.

Reference:
Short, R. D., Johannsen, F. R., Levinskas, G. J., et al., Teratology and multigeneration reproduction studies with maleic anhydride in rats, *Fundam. Appl. Toxicol.*, 7, 359, 1986.

MELAMINE (CAS No 108–78–1)

Synonyms. Cyanuramide; Cyanurotriamine; Isomelamine; **2,4,6**-Triamino-**1,3,5**-triazine.

Properties. Colorless-to-white, monoclinic crystals or prisms. Water solubility is 9.5% at 20°C. Readily soluble in hot water and alcohol. In aqueous solutions, M. is transformed into **cyanuric acid.** Decomposes at 600°C to form **cyanamide.** A concentration of 3.5 g/l does not alter the taste, odor, color, or transparency of water.[1]

Applications and **Exposure.** M. is used in the manufacture of melamino-formaldehyde amino resins, a cross-linked polymeric material. It is also used in laminates, surface-coating resins, plastic molding compounds, lacquers, and adhesives as a monomer, stabilizer, etc. Melamine-based amino resins are used in housewares (dinnerware, table coverings, etc.).

Acute Toxicity. LD_{50} is 3.2 and 3.8 g/kg BW in male and female Fisher 344 rats, respectively. In $B6C3F_1$ mice given M. in corn oil by gavage, LD_{50} is found to be 3.3 (males) and 7 g/kg BW (females).[2] Toxic doses produced a diuretic effect, and crystals of dimelamino-monophosphate appear in the urine. Diuretic effect is reported in rats and dogs. According to other data, mice and rats tolerate 10 g/kg BW, and this dose as well as a threefold administration of 5 g/kg BW to rats for 6 h caused no changes in their behavior.[1]

Repeated Exposure revealed pronounced cumulative properties.[3] K_{acc} is 1.34 (by Kagan). Fisher 344 rats and $B6C3F_1$ mice were fed the diets containing 5 to 30 g/kg for 14 d. The treatment produced a hard crystalline solid in the urinary bladder in the majority of male rats receiving 10 g/kg or more and in all treated male mice. In females, this effect was found at levels of 20 g/kg or more; it was noted in rats and in 40% of mice given 30 g/kg.[2]

Short-term Toxicity. Rats and mice were given doses of 6 to 18 g/kg BW over a period of 13 weeks.[2] Stones and ulcerations were found in the urinary bladders of animals receiving 12 to 15 g/kg. Administration of 10, 100, and 1000 mg/kg BW for 3 months had no toxic effect.

Long-term Toxicity. Rats and mice were administered 2 g/kg BW doses for 6 months. The treatment resulted in reduced phagocytic index and increased cholinesterase activity and blood urea level. Histological examination revealed mild dystrophic changes in the liver, kidneys, and myocardium, and a slight hyperplastic reaction in the spleen of some animals. The NOAEL of 0.1 g/kg BW was identified in this study.[2]

Allergenic Effect. Skin applications produced a sensitizing effect.

Reproductive Toxicity. No toxic effect or gross malformations were observed in fetuses of pregnant rats treated *i/p* with 70 mg/kg BW on gestation day 5 and 6, 8 and 9, or 12 and 13.[4]

Genotoxicity. As regards its mutagenic effect, M. is believed to present no risk in production and application. It was not mutagenic to *Salmonella,* nor to *Drosophila melanogaster.* Negative results were observed also in Chinese hamster ovary cell assay, in a culture of rat hepatocytes, or in a micronuclear test.[5,6]

Carcinogenicity. No treatment-related tumors were observed after oral administration of M. to $B6C3F_1$ mice. (Animals were fed diets containing 2.25 and 4.5 g/kg BW for 103 weeks.)[2]

Male Fisher 344/N rats given the diet containing M. developed transitional cell tumors of the urinary bladder. Rats were administered doses of 2.25 and 4.5 g/kg BW (males), 4.5 and 9 g/kg BW (females) for 103 weeks. Survival was significantly reduced in the high-dose males. There was a dose-related increased incidence of bladder stones in male rats.[2] There seems to be a relationship between bladder stone formation in 7/8 rats and bladder tumor development in this study.[7] *Carcinogenicity classification.* IARC: Group 3; NTP: P—N—N—N

Chemobiokinetics. M. can be accumulated for a short time in the kidneys and bladder. 24 h after administration of [14]C- tagged M., it is not found in the tissues.[5] In rats and humans, M. is a metabolite of the antineoplastic agent, **hexamethylmelamine.** M. can be decomposed to **urea** in the body. However, M. is not metabolized in male Fisher 344 rats.[8] No metabolites were found in the urine and blood plasma. In 6 h after oral administration of 250 mg/kg BW, 50% of the dose was found in the urine of rats.[9] Of the administered dose, 93% was removed via the urine.[8] M. increased diuresis and the removal of fine crystals of low-toxicity **dimelaminomonophosphate** with the urine.[05]

Standards. EEC (1990). SML: 30 mg/kg. **Russia** (1988). PML in the food 1 mg/l; PML in drinking water n/m.

Regulations. USFDA (1993) regulates the use of M. (1) as a component of adhesives for a food-contact surface. FDA approved the use of *melamine polymers* (2) as components of articles intended for use in contact with food, (3) as ingredients of resinous or polymeric coatings for polyolefin films as the basic polymer or modified with methanol, (4) in epoxy resins as the basic polymer, (5) in paper or paperboard for contact with dry food, (6) in cellophane as the basic polymer, or (7) in the modified form as a resin to anchor coatings to the substrate. *Melamino-formaldehyde resins* may be used as surfaces in contact with food in molded articles, providing the yield of chloroform-soluble extractives does not exceed 80 $\mu g/cm^2$ of food contact surface under specific solvent and temperature parameters.

References:

1. *Environmental Factors and Their Significance for Health of Population,* Zdorov'ya, Kiev, 2, 1970, p. 115 (in Russian).
2. *Carcinogenesis Bioassay of Melamine in F344/N Rats and B6C3F$_1$ Mice (Feed Study),* NTP Technical Report No. 245, Research Triangle Park, NC, 1983.
3. Babayan, E. A. and Alexandryan, A. V., *Pharmacol. Toxicol.,* 14, 122, 1973 (in Russian).
4. Thiersch, J. B., Effect of **2,4,6**-triamino-'**S**'-triasine, **2,4,6**-'tris'(ethyleneimino)-'**S**'-triasine and **N,N',N"**- triethylene phosphoramide on rat litter in utero, *Proc. Soc. Exp. Biol. Med.,* 94, 36, 1957.
5. Mast, R. W., Jeffcoat, A. R., Sadler, B. M., et al., Metabolism, disposition, and excretion of ^{14}C-melamine in male Fisher 344 rats, *Food Chem. Toxicol.,* 21, 807, 1983.
6. Mirsalis, J., Tysen, C. K., and Butterworth, B. E., Detection of genotoxic carcinogens in the *in vivo* and *in vitro* hepatocyte DNA repair assay, *Environ. Mutat.,* 4, 553, 1982.
7. Melnick, R. L., Boorman, G. A., Haseman, J. K., et al., Urolithiasis and bladder carcinogenicity of melamine in rodents, *Toxicol. Appl. Pharmacol.,* 72, 292, 1984.
8. Mast, R. W., Jeffcoat, A. R., Sadler, B. M., et al., Metabolism and excretion of ^{14}C-melamine in Fisher 344 rats, *Toxicol. Lett.,* 18 (Suppl. 1), 68, 1983.
9. Lipschitz, W. L. and Stockey, E., Mode of action of three new diuretics: melamine, adenine and formoguanidine, *J. Pharmacol. Exp. Ther.,* 83, 235, 1945.

METHACRYLIC ACID (CAS No 79–41–4)

Synonyms. α-**Methacrylic acid;** 2-Methylpropenoic acid.

Properties. Readily mobile, colorless liquid with an acetic acid odor. Soluble in water (readily soluble in hot water) and in alcohol. Odor perception threshold of MA stabilized by hydroquinone (0.1%) is 106 mg/l. Concentrations up to 10 mg MA/l do not give water any foreign taste.[1]

Acute Toxicity. LD_{50} is reported to be 1.06 g/kg BW in rats, 1.3 g/kg BW in mice, and 1.2 g/kg BW in rabbits. A local irritating and resorptive effect is observed. Poisoning is accompanied by an increase in general debility, decreased BW gain, and reduction in diuresis.[2] Gross pathological examination indicated dystrophic changes in the parenchymatous organs and inflammation and edema along the GI tract.

Repeated Exposure failed to reveal pronounced cumulative properties. K_{acc} is 6 (by Cherkinsky). Exposure to 270 mg/kg BW dose produced toxic effect on the liver cells and decreased the antitoxic function of the liver in mice. Habituation to MA was not observed.[1]

Long-term Toxicity. The treatment caused a considerable reduction in the acidity level and variation in the chloride content of the blood in rabbits, diffuse damage to the liver, and a number of pathomorphological changes in the visceral organs, primarily in the liver and kidneys.[1]

Reproductive Toxicity. Concentrations of 1.5 μM caused malformations and growth retardation in *in vitro* rat embryocultures.[3]

Standards. Russia (1988). MAC and PML: 1 mg/l.

Regulations. USFDA (1993) listed MA as an ingredient (1) of semirigid and rigid acrylic and modified acrylic plastics that may be safely used as articles intended for use in contact with food,

(2) in polyethylene phthalate polymers that may be safely used as articles or components of plastics intended for use in contact with food, and (3) in cross-linked polyester resins for food-contact surfaces.

References:

1. Klimkina, N. V., Ekhina, R. S., Sergeev, A. N., et al., Data on hygiene substantiation of MAC for methacrylic acid in water bodies, *Gig. Sanit.*, 8, 13, 1973 (in Russian).
2. *Toxicology of High Molecular Weight Compounds,* Proc. Sci. Conf., S. L. Danishevsky, Ed., Leningrad, 1961, p. 46 (in Russian).
3. Rogers, G. J., Greenaway, J. C., Mirkes, P. E., et al., Methacrylic acid as a teratogen in rat embryo culture, *Teratology,* 33, 113, 1986.

METHACRYLONITRILE (CAS No 126–98–7)

Synonyms. 2-Cyanopropene; Isocrotononinitrile; Methacrylic acid, nitrile; 2-Methylpropenenitrile.

Properties. Colorless, transparent, volatile liquid with a specific odor. Solubility in water is 3.5 to 4.5%. Readily miscible with alcohol. Water solutions are colorless and transparent, and when left to stay for a long time do not cloud or form any residue. Solutions exhibit unpleasant specific odor that becomes more pungent with increasing temperature. Odor perception threshold, is 1.05 mg/l (practical threshold, 3.17 mg/l). Taste perception threshold is significantly higher.[1]

Acute Toxicity. The minimum lethal dose in mice appeared to be 15 mg/kg BW. The LD_{50} values vary in rats from 240 mg/kg[2,3] to 25 to 50 mg/kg BW;[05] in mice, they are reported to be 20 to 25 mg/kg BW.[05] According to other data, LD_{50} is 184 mg/kg BW in mice, 167 mg/kg BW in rats, and 216 mg/kg BW in guinea pigs.[1] Poisoning is accompanied by NS inhibition, tremor, cyanosis, and convulsions. An increase in the content of thiocyanates in the blood serum and urine of rats has been found. Death occurs within 1 to 3 d from respiratory arrest. Gross pathohistological examination failed to reveal pronounced changes. Toxicity of M. is likely to be much higher than that of orally and *i/p* administered **acrylonitrile.**

Repeated Exposure failed to reveal cumulative properties when M. was administered at doses of ⅕ and ¹⁄₁₀ LD_{50}. Habituation developed.[1] K_{acc} is identified to be 5.6 (by Lim).[4] Rats received M. in sunflower oil during pregnancy. Within 1 h following ingestion of 50 or 100 mg/kg BW (½ LD_{50}), rats displayed dose-related mild to severe conditions, including ataxia, trembling, convulsions, salivations, and irregular breathing. The rats recovered from these signs at various times depending on the dose administered.[5]

Long-term Toxicity. Manifestations of toxic action included changes in cholinesterase and serum oxidase activity, an increase in the serous index of the urine, and disorder of the hemocoagulative function of the liver in rats. Gross pathological examination revealed mild dystrophic changes in the visceral organs.[1]

Reproductive Toxicity. M. caused fetotoxicity in Sprague-Dawley rats on inhalation exposure at a concentration of 100 ppm during days 6 to 20 of gestation.[6] Rats were administered 50 mg M./kg during the first week of pregnancy (group 1); 50 mg M./kg (group 2) and 100 mg M./kg (group 3) during the second week of pregnancy. BW gain was affected in all the treated animals, pregnancy was disrupted in 1 and 3 groups (100%) and in 2 group (84%). Edema of Fallopian tubes was noted.[10]

Genotoxicity. M. was negative in *Salmonella* mutagenicity assay, in *Dr. melanogaster,* and in the mutation test in mice lymphoma cells.[7]

Chemobiokinetics. M. is a potent neurotoxin; it depletes glutathione both *in vivo* and *in vitro.*[8] The tissue distribution and characteristics of M. metabolism to **cyanide** are reported.[9] The resemblance in features of acute poisoning to that observed following treatment with inorganic cyanides is an indirect confirmation of detachment of **cyanide groups** of M. in the body. Two pathways of M. metabolism are indicated: hydrolysis and then **thiocyanate** formation; conjugation of other metabolic products with glucuronic acid and their removal in the form of glucuronides.[4] A small quantity of cyanide ions is passed with the urine within 18 h in the form of thiocyanate ions. The majority of M.-derived radioactivity is eliminated in the exhaled air as CO_2 or in the urine. Apparent saturation of metabolism was seen at 115 mg/kg, the highest dose studied (NTP–91).

Standards. EEC (1990). SML: 'not detectable' (detection limit 0.02 mg/kg, analytical tolerance included).

Regulations. USFDA (1993) listed M. as an ingredient (1) of semirigid and rigid acrylic and modified acrylic plastics to be safely used as articles intended for use in contact with food, (2) in resinous and polymeric coatings for food-contact surfaces, and (3) in polyethylene phthalate polymers to be safely used as articles or components of plastics intended for use in contact with food.

References:

1. Loskutov, N. F. and Piten'ko, N. N., Characteristics of the toxic action and safe levels of crotononitrile and isocrotononitrile in water bodies, *Gig. Sanit.*, 4, 10, 1972 (in Russian).
2. McOmie, W. A., Comparative toxicity of methacrylonitrile and acrylonitrile, *J. Ind. Hyg. Toxicol.*, 31, 113, 1949.
3. Pozzani, U. C., Kinkead, E. R., King, J. M., et al., The mammalian toxicity of methacrylonitrile, *Am. Ind. Hyg. Assoc. J.*, 29, 202, 1968.
4. Kurzaliev, S. A., Peculiarities of the toxic effect of methacrylonitrile, *Gig. Truda Prof. Zabol.*, 5, 35, 1985 (in Russian).
5. Farooqui, M. Y. N. and Villarreal, M. I., Maternal toxicity of methacrylonitrile in Sprague-Dawley rats, *Bull. Environ. Contam. Toxicol.*, 48, 696, 1992.
6. Saillenfait, A., Bonnet, P., Guenier, J. P., et al., Relative developmental toxities of inhaled aliphatic mononitriles in rats, *Fundam. Appl. Toxicol.*, 20, 365, 1993.
7. See **ACRYLAMIDE**, #21.
8. Farooqui, M. Y. N., Cavazos, R., Villarreal, M. I., et al., Toxicity and tissue distribution of methacrylonitrile in rats, *Ecotoxicol. Environ. Safety*, 20, 185, 1990.
9. Day, W. W., Cavazos, R., and Farooqui, M. Y. N., Interaction of methacrylonitrile with glutathione, *Res. Commun. Chem. Pathol. Pharmacol.*, 62, 267, 1988.
10. Farooqui, M. Y. H. and Villarreal, M. I., Maternal toxicity of methacrylonitrile in Sprague-Dawley rats, *Bull. Environ. Contam. Toxicol.*, 5, 696, 1992.

METHYL ACRYLATE (MA) (CAS No 96–33–3)

Synonyms. Acrylic acid, methyl ester; Methoxycarbonyl ethylene; Methyl propenoate; 2-Propeonic acid, methyl ester.

Properties. Colorless, transparent liquid with an acrid odor. Water solubility is 5.2% at 20°C, soluble in alcohol. Odor perception threshold is reported to be 0.01 mg/l[1] or 0.0021 mg/l.[02]

Applications and **Exposure.** Comonomer in the production of all types of vinyl monomers. Used for "internal plasticization" of rigid plastics.

Acute Toxicity. LD$_{50}$ was established at the level of 230 to 545 mg/kg BW for rats, 826 mg/kg BW for mice, and 200 mg/kg BW for rabbits. Administration of 280 mg/kg BW to rabbits resulted in the lethal effect preceded by dyspnea, cyanosis, convulsions, and drop of temperature.[2,3,03]

Long-term Toxicity. Manifestations of the toxic action included effects on the CNS, cardiovascular system, GI tract, and urinary ducts.[1] In a 6-month study, the treatment affected liver functions, acid-alkali balance of the blood and urine, and conditioned reflex activity of rats.[1]

Genotoxicity. MA did not show mutagenic activity in *Salmonella* and in *E. coli;*[4] however, it caused an increase in the percentage of CA in the culture of Chinese hamster lung cells.[5]

Carcinogenicity. No tumor was found in a 2-year inhalation study (53 to 475 mg/m^3) in rats.[6]

Carcinogenicity classification. IARC: Group 3.

Chemobiokinetics. Two h after oral administration of 34 mg/kg of ^{14}C-tagged MA to guinea pigs, it was discovered in the liver, bladder, and brain of the animals. Connecting to SH-groups seems to be important for MA detoxication metabolism in guinea pigs. MA is removed with the urine in the form of **thioether.**[7]

Standards. Russia (1988). MAC and PML: 0.02 mg/l (organolept., odor).

Regulations. USFDA (1993) listed MA, its polymers, and copolymers for use (1) in resinous and polymeric coatings in the food-contact surface of articles intended for use in producing, manufacturing, packing, transporting, or holding food, (2) as a solvent in polyester resins for food-contact surfaces, (3) as a component of adhesives to be safely used in food-contact surfaces, (4) in paper and paperboard in contact with dry, aqueous, or fatty foods, (5) in semirigid and rigid acrylic and modified acrylic plastics, (6) in cross-linked polyester resins for repeated use as articles or a component of articles coming in the contact with food, and (7) in polyethylene phthalate polymers to be safely used as articles or components of articles intended for use in contact with food.

58

References:

1. See **BUTYL METHACRYLATE,** #1.
2. Paulet, G. and Vidal, M., On the toxicity of some acrylic and methacrylic esters of acrylamides and polyacrylamides, *Arch. Mal. Prof. Med. Trav. Sec. Soc.,* 36, 58, 1975.
3. See **ETHYL ACRYLATE,** #3.
4. Waeggemaekers, T. H. J. M. and Bensink, M. P. M., Non-mutagenicity of 27 aliphatic acrylate esters in the Salmonella-microsome test, *Mutat. Res.,* 137, 682, 1984.
5. See **ETHYL ACRYLATE,** #12.
6. Klimish, H.-J. and Reininghaus, W., Carcinogenicity of acrylates: Long-term inhalation studies on methyl acrylate and n-butyl acrylate in rats (Abstract No. 211), *Toxicologist,* 4, 53, 1984.
7. Seutter, E. and Rijnties, N., Whole-body autoradiography after systemic and topical administration of methyl acrylate in the guinea-pig, *Arch. Dermatol. Res.,* 270, 273, 1981.

METHYL METHACRYLATE (MMA) (CAS No 80–62–6)

Synonyms. Methacrylic acid, methyl ester; Methyl-2-methyl-2- propenoate.

Properties. Colorless liquid. Water solubility is 1.9% at 20°C. Miscible with alcohol at any ratio. Forms **methacrylic acid** and **methyl alcohol** when hydrolyzed. Odor perception threshold is 0.45 mg/l[1] or 0.025 to 0.05 mg/l.[02,010]

Applications. Used for latex coatings and in the production of acrylic polymeric materials, including copolymers with acrylonitrile, α-methylstyrene, and butadiene.

Acute Toxicity. LD_{50} was found to be 8.7 g/kg BW for rats and rabbits, 3.6 to 5.2 g/kg BW for mice, 6.3 g/kg BW for guinea pigs, and 5 g/kg BW for dogs. The minimum lethal dose is 6.6 g/kg BW for rabbits.[1,2]

Repeated Exposure. Rats received ⅕ and ¹⁄₁₀ LD_{50} for 30 d. At the end of the experiment, LD_{50} was administered to animals. Slight signs of accumulation were manifested.[1]

Long-term Toxicity. Rats were dosed by gavage with a total dose of 8.125 g MMA/kg BW (administered twice a week for 8 months). The treatment caused a reduction in the content of glycoproteins and albumins and an increase in the activity of leucinoaminopeptidase and β-glucuronidase in the blood serum and reversible toxic effects on the liver. The kidneys were much less affected (Motoc et al., 1971). In an 8- to 9-month rat study, there was liver damage; oxidizing enzymes and erythrocytes were affected. The functional condition of the brain cortex was impaired.[1]

Reproductive Toxicity. There was no *teratogenic effect* when **polymethylmetacrylate** was used for implantation in humans. *Embryotoxicity.* Rats were exposed *i/p* to ⅓, ⅕, and ¹⁄₁₀ LD_{50} on day 5, 10, and 15 of gestation. The treatment produced increased embryolethality and developmental abnormalities. Reduced fetal weights were observed. A dose dependence was noted.[3] A *gonadotoxic* effect of MMA was found in mice and rats in response to inhalation exposure above MAC.[4] Administration of ¹⁄₂₀ LD_{50} to rats for 2 months did not affect the estral cycle.[5]

Genotoxicity. No mutagenic effect was established in people with **polymethylmethacrylate** implants. MMA appeared to be positive in a *Salmonella* mutagenicity bioassay.

Carcinogenicity. In a 2-year study, Wistar rats were given drinking water with concentrations of 6 to 2000 mg MMA/l. The treatment did not cause tumor development.[6] IARC considers that the carcinogenic properties of MMA are not proved. *Carcinogenicity classification.* IARC: Group 3; NTP: NE—NE—NE—NE.

Chemobiokinetics. Accumulation in the blood, brain, and lungs was noted following acute inhalation exposure. MMA seems to affect the capillary network of the lung tissue (Raje et al., 1985). MMA is hydrolyzed in the body to form **methacrylic acid,** which undergoes further oxidation to CO_2 and H_2O. The main part of the dose is removed with the exhaled air in the form of CO_2, irrespective of the route of MMA administration. A smaller part of MMA escapes in the form of **methylmelonate, succinate,** and possibly, β-**hydroxyisobutyrate.**[7] According to the data available, MMA metabolite was not found in the urine.[8] MMA is evidently completely oxidized in the body. Its relatively weak toxicity for rats is due to the fact that following oral administration, the monomer is rapidly converted to **pyruvic acid.**

Standards. Russia (1988). MAC and PML in drinking water: 0.01 mg/l; PML in food: 0.25 mg/l.

Regulations. USFDA (1993) regulates the use of MMA (1) as an ingredient for resinous and polymeric coatings of food-contact surfaces, (2) in acrylic and modified acrylic plastics for single and repeated use in contact with food, (3) in cross-linked polyester resins for repeated use as articles or components of articles coming in contact with food, (4) as a component of paper and paperboard used in contact with food, and (5) in polyethylene phthalate polymers to be safely used as articles or components of plastics intended for use in contact with food.

References:
1. See **BUTYL METHACRYLATE, #1.**
2. Deichmann, W., Toxicity of methyl, ethyl and *n*-butyl methacrylate, *J. Ind. Hyg. Toxicol.*, 23, 343, 1941.
3. See **ACRYLIC ACID, #7.**
4. See **ETHYLENEIMINE, #6,** 149.
5. Sheftel, V. O., Hygienic Aspects of the Use of Polymeric Materials in Water Supply, Author's abstract of thesis, VNIIGINTOX, Kiev, 1977, p. 158 (in Russian).
6. See **ETHYL ACRYLATE, #8.**
7. Bratt, H. and Hathway, D. E., Fate of methyl methacrylate in rats, *Br. J. Cancer,* 36, 114, 1977.
9. Pantucek, M., Methyl methacrylate metabolism, *Food Cosmet. Toxicol.*, 8, 105, 1970.

METHYLOLMETACRYLAMIDE (CAS No 924–42–5)

Synonym. *N*-Hydroxymethylacrylamide.

Properties. White, odorless, crystalline substance. Readily soluble in water. Concentrated solutions are colorless and transparent and do not cloud when left to stand. Taste perception threshold is 560 mg/l.

Applications. Monomer in the production of plastics, organic glass, lacquers, and enamels.

Acute Toxicity The LD_{50} values are reported to be 312 to 474 mg/kg BW in rats, 400 to 420 mg/kg BW in mice, and 328 mg/kg BW in rabbits. Poisoning results in NS affection.

Repeated Exposure revealed no cumulative properties in mice given $\frac{1}{5}$ and $\frac{1}{10}$ LD_{50} for a month.

Long-term Toxicity. In a 6-month rat study, the CNS and liver functions were found to be the target systems affected.

Chemobiokinetics. M. is hydrolyzed in the body with **formaldehyde** formation (0.26 mg formaldehyde after complete hydrolysis of 1 mg M.).

Standards. Russia (1988). MAC and PML: 0.1 mg/l.

Regulations. USFDA (1993) regulates M. for use as a monomer in the production of semirigid and rigid acrylic and modified acrylic plastics intended for use in contact with food.

Reference:
See **METHACRYLAMIDE, #1.**

α-METHYL STYRENE (CAS No 98–83–9)

Synonyms. Isopropenylbenzene; 1-Methyl-1-phenylethylene; 2-Phenylpropene.

Properties. Colorless or slightly yellow liquid with a pungent specific odor. Water solubility is 560 mg/l at 25°C, soluble in alcohol. Odor perception threshold is reported to be 0.11 mg/l[1] or 0.043 to 0.052 mg/l;[02,010] taste perception threshold is 0.08 mg/l. Practical thresholds are 0.2 and 0.14 mg/l, respectively.[1]

Acute Toxicity. LD_{50} in rats is 4.9 g/kg BW (Oglesnev, 1964). According to other data, LD_{50} is 10 g/kg BW in rats and 5 g/kg BW in mice.[1] A dose of 3 g/kg BW seems to be LD_{30} in mice and produced considerable morphological damage. Death was preceded by agitation, ataxia, tremor, and convulsions. Gross pathological examination revealed severe congestion in the visceral organs, especially in the lungs. Lethal doses caused dystrophic changes in the parenchymatous organs. Toxic encephalopathy developed (Veselova and Oglznev, 1963).

Repeated Exposure. In a 1-month study, $\frac{1}{10}$ LD$_{50}$ in mice as well as a 0.5 g/kg BW dose in rats caused retardation of BW gain. Decrease in the leukocyte count and increased level of protein and coproporphyrins in the urine were also found in rats. Histological examination revealed mild congestion and circulation disorders in the parenchymatous organs and brain, as well as bronchitis (Oglesnev).

Long-term Toxicity. No changes in the blood morphology and activity of blood cholinesterase were observed in rats given up to 0.5 mg/kg BW for 6 months. The highest dose produced reticulocytosis (Oglesnev).

Reproductive Toxicity. No teratogenicity or maternal toxicity was observed in rats after *i/p* administration of 250 mg/kg BW on days 1 to 15 of gestation. The treatment caused some fetal toxicity.[3]

Allergenic Effect. Sensitization was found to develop in guinea pigs on subcutaneous and cutaneous applications.[4]

Chemobiokinetics. Oral administration results in disorder of calcium exchange. One of the metabolism products of α-M. in the body is **atrolactic** (α-**hydroxy**-α-**phenylpropionic**) **acid** in the urine (normally absent). α-**Methyl**-α-**phenylethylene glycol** is also formed and is eliminated with the urine in the form of **glucuronide.** α-M. is oxidized with participation of microsomal enzymes of the liver to products of higher toxicity than α-M. itself. When up to 10 mg/kg BW is administered, α-M. is completely metabolized.

Standards. Russia (1988). MAC and PML: 0.1 mg/l (organolept.).

Regulations. USFDA (1993) regulates the use of α-M. (1) as a component of adhesives to be safely used in food-contact surfaces, (2) as an ingredient of paper and paperboard for contact with dry food, (3) in cross-linked polyester resins for repeated use as articles or a component of articles coming in contact with food, and (4) in resinous and polymeric coatings for articles intended for use in producing, manufacturing, packing, transporting, or holding food.

References:

1. Wolf, M. A., Rowe, V. K., McCollister, D. D., et al., Toxicological studies of certain alkylated benzenes and benzene, *Am. Med. Assoc. Arch. Ind. Health.*, 14, 387, 1956.
2. Aizvert, L. G., Data on Toxicity of α-Methyl-styrene, Author's abstract of thesis, Alma-Ata, 1979, p. 26 (in Russian).
3. Hardin, B. D., Bond, G. P., Sikov, M. R., et al., Testing of selected workplace chemicals for teratogenic potential, *Scand. J. Work Environ. Health*, 7, 66, 1981.
4. *Proc. Ryazan' Medical Institute*, 80, 56, 1963 (in Russian).

PENTAERYTHRITOL (PER) (CAS No 115–77–5)

Synonym. Pentaerythrite 2,2′-di(hydroxymethyl)-**1,3**-propane-diol.

Properties. Colorless crystals. Water solubility is 7.1% at 20°C and 19.3% at 55°C. Poorly soluble in ether and acetone. Taste perception threshold is 3 g/l.

Acute Toxicity. LD$_{50}$ is 19.5 g/kg BW for rats, and 18.5 g/kg BW for mice and rabbits. Poisoning is accompanied by a narcotic effect. Death occurs in 2 to 4 h after administration. Gross pathological examination revealed hemorrhaging zones in the gastric mucosa and intestines, dystrophic changes in the liver. However, according to other data, even 40 g/kg BW caused no changes in the rat.[05]

Repeated Exposure. Rats of both sexes were dosed with 1 g P./kg for 28 d. Morphological and biochemical investigation revealed no changes in the blood and blood serum. There were no histological findings as well. The dose of 1 g/kg was considered to be ineffective.[2]

Short-term Toxicity. PER is known to affect the motor system of the intestines and to regulate lipid exchange. Mice and rats were given $\frac{1}{5}$ to $\frac{1}{10}$ LD$_{50}$ for 20, 30, and 90 d. The exposure increased the blood histamine level, caused a reduction in animals' capacity to work and lengthened the time of barbiturate narcosis.[1]

Long-term Toxicity. *Humans.* PER is used in France as a medicine, a daily dose of 2.5 to 15 g being harmless during prolonged use. *Animals.* Rats and rabbits were exposed to doses of up to 5 mg/kg BW for 7 months. The treatment revealed a number of changes in the functional condition of the brain cortex cells, in mediator and nuclear exchange, in immune reactivity, etc.

Chemobiokinetics. PER is readily absorbed in the intestines and excreted unchanged by the kidneys, which distinguishes it from other polyols.[05] According to other data, PER is partly oxidized in the body and partly released in the urine in the form of **aldehydes** and **ketones.**

Standards. Russia (1988). MAC and PML: 0.1 mg/l.

Regulations. USFDA (1993) approved the use of **PER** (1) in resinous and polymeric coatings for food-contact surfaces, (2) in resinous and polymeric coatings for polyolefin films to be safely used as a food-contact surface, (3) in cross-linked polyester resins that may be safely used as articles or components of articles intended for repeated use in contact with food. **PER** and its **stearate ester** may be used (4) as stabilizers in rigid polyvinyl chloride and/or in rigid vinyl chloride copolymers provided that the total amount of PER and/or **PER stearate** (calculated as free PER) does not exceed 0.4% by weight of such polymers.

References:

1. Plitman, S. I., Experimental validation of the MAC for pentaerythritol and xylitol in water bodies, *Gig. Sanit.*, 2, 25, 1971 (in Russian).
2. Hayashi, S., Toyoda, K., Furuta, K., et al., Study of pentaerythritol toxicity at repeated 28-day administration to F-344 rats, *Bull. Natl. Inst. Hyg. Sci.*, 110, 32, 1992 (in Japanese).

PHENOL (CAS No 108–95–2)

Synonyms. Benzenol; Carbolic acid; Hydroxybenzene; Oxybenzene; Phenyl alcohol; Phenyl hydroxide.

Properties. White, crystalline solid that liquifies after absorption of water from air. Turns red in the air, particularly in the light. Has an acrid odor and a sharp burning taste. Water solubility is 82 g/l at 15°C; with water forms a hydrate, $C_6H_5OH \cdot H_2O$. Soluble in acetone and ethanol. Technical P. (**carbolic acid**) is a reddy-brown, sometimes black, liquid. Odor perception threshold for P. is 5.9 mg/l[010] or 7.9 mg/l;[02] for **chlorophenols** it is 0.001 mg/l. The taste and odor of P. are detected more rapidly at low water temperatures when phenol oxidation and volatility are decreased.

Applications and **Exposure.** A basic feedstock for production of phenolic resins (phenolaldehyde plastics), bisphenol A, and caprolactam. Disinfectant and antiseptic. P. derivatives may, by *in vivo* conversion, form a source of endogenous human P. exposure. P. is available in smoked meat, fish products, and in outdoor air. Hazardous contamination is unlikely to occur because of the conspicuous P. smell and taste. Chlorophenol formation in chlorinated drinking water greatly enhances smell and taste.

Acute Toxicity. LD_{50} is 340 to 650 mg/kg BW for rats and 300 to 430 mg/kg BW for mice. Species sensitivity in laboratory animals is negligible. Cats are the most sensitive; guinea pigs the most resistant. Poisoning is accompanied by respiratory function impairment. A local irritative effect of P. was noted.[1] Gross pathological examination revealed lesions in the parenchymatous organs and also in the large arteries.

Repeated Exposure failed to reveal cumulative properties of P. K_{acc} is identified to be about 20 (by Cherkinsky). There were no changes in the indices studied following oral administration of 0.5 mg P./kg BW to rats for 3 weeks (Kretov, 1974). Rats were dosed by gavage with 8 to 80 mg P./kg BW for 1.5 months. The treatment did not affect blood morphology, blood ALT, AST, cholinesterase, and peroxidase activity.[2]

Long-term Toxicity. Fisher 344 rats and $B6C3F_1$ mice received 2.5 or 5 g P./l in their drinking water for 103 weeks. A dose-related reduction in BW and water consumption were only observed.[3] Adequate studies are not available. Limit could be based on organoleptic data.

Allergenic effect. Skin sensitization was not observed in guinea pigs.

Reproductive Toxicity. Sprague-Dawley rats were exposed to 30 to 120 mg P./kg BW by oral intubation on days 6 to 15 of gestation. There was no maternal toxicity or teratogenicity, but fetal growth was retarded at the 120 mg/kg level. CD-1 mice were dosed with 70 to 280 mg/kg BW by oral intubation on days 6 to 15 of gestation. Maternal and fetal toxicity but not teratogenicity were observed. Dose dependence was noted.[4]

Genotoxicity. *Humans.* CA was observed in the peripheral lymphocytes of 50 workers exposed to formaldehyde, styrene, and phenol.[5] *Animals.* Positive results were noted in a mouse lymphoma cells test.[3] P. induced micronuclei in female mice and SCE in cultured human cells, mutations but not DNA damage in cultured animal cells (IARC 47–275).

Carcinogenicity. Fisher 344 rats and B6C3F$_1$ mice received P. at concentrations of 2.5 or 5 g/l in their drinking water for 103 weeks. There was no treatment-related increase in the incidence of tumors. A dose-related reduction in BW gain and in water consumption was only observed. In male rats, there was an increase in leukemia at the lower dose but not at the higher dose.[3] These data are considered to be inadequate for evaluation of carcinogenicity in experimental animals (IARC 47–275). *Carcinogenicity classification.* IARC: Group 3; NTP: N—N—N—N; USEPA: Group D.

Chemobiokinetics. P. metabolism is likely to involve binding with proteins, oxidation of the aromatic ring, and its breakdown to CO_2 (the amount of CO_2 is on average 10%, of **hydroquinone** 10%, and of **pyrocatechine** 1%). Unchanged **phenols** and **diphenols** bind with sulfur and glucuronic acids.[6] Elimination of P. and its oxidation products occurs rapidly via the urine, predominantly in the bound state. The main metabolites found in the urine are *p*-**cresol, phenol,** traces of **resorcinol, hydroquinone,** etc.

Standards. Russia (1988). MAC and PML in drinking water: 0.001 mg/l, PML in food: 0.05 mg/l. In the absence of water chlorination, the recommended PML: 1 mg/l.

Regulations. USFDA (1993) regulates P. for use (1) as a component of adhesives intended for use in food-contact articles, (2) in resinous and polymeric coatings in a food-contact surface of articles intended for use in producing, manufacturing, packing, transporting, or holding food, and (3) in phenolic resins for use as a food-contact surface of molded articles intended for repeated use in contact with nonacid food (pH above 5).

References:

1. Nagorny, P. A., Industrial Toxicology of Phenol Formaldehyde Resins, Phenol and Formaldehyde, and Industrial Hygiene in Areas of Exposure, Author's abstract of thesis, Kiev, 1981, p. 32 (in Russian).
2. Korolev, A. A., Aibinder, A. A., Bogdanov, M. V., et al., Hygiene and toxicity profile of phenol destruction products formed during ozone treatment of water, *Gig. Sanit.*, 8, 6, 1973 (in Russian).
3. *Bioassay of Phenol for Possible Carcinogenicity*, NCI-CG-TR-203, NTP No. 80–15, U.S. Department of Health and Human Services, Research Triangle Park, NC, 1980.
4. Price, C. J., Ledoux, T. A., Reel, J. R, et al., Teratologic evaluation of phenol in rats and mice, *Teratology*, 33, 92C, 1986.
5. Mierauskiene, J. R. and Lekevicius, R. K., Cytogenetic studies of workers occupationally exposed to phenol, styrene and formaldehyde. Abstract No. 60, *Mutat. Res.*, 147, 308, 1985.
6. *Conversion and Determination of Industrial Organic Poisons in the Body*, Meditsina, Leningrad, 1972, p. 193 (in Russian).

PROPYLENE (CAS No 115–07–1)

Synonyms. 1-Propene; Methylethylene.

Properties. A colorless gas with aromatic odor. Water solubility is 350 mg/l at 25°C. Volatilizes rapidly from the free surfaces. Odor perception threshold is 22.5 mg/l,[010] 0.5 mg/l,[1] or 0.028 mg/l.[02] P. does not alter the color or transparency of water.

Applications. Used as a starting material in the manufacture of polypropylene plastics and other basic chemicals, such as acrylonitrile, isopropyl alcohol, and propylene oxide.

Acute Toxicity. Rats tolerated administration of 0.5 ml of a solution with concentration of 340 mg P./l without behavioral changes.[1]

Repeated Exposure. Accumulation is impossible because of rapid removal from the body.[1]

Short-term Toxicity. Mice were administered a dose of 3.75 mg/kg BW for 4 months. The treatment caused no changes in the behavior of animals, BW gain, oxygen consumption, or capacity to work. Gross pathological examination failed to reveal alterations in the relative weights and histology of the visceral organs.

Long-term Toxicity. In a 6-month study, there were no changes in behavior, BW gain, phago-cytic activity of the leukocytes, and cholinesterase activity in the blood serum after oral adminis-tration of 0.05 mg/kg BW.[1]

Genotoxicity. Negative results were received in *Salmonella* mutagenicity assay (NTP–90).

Carcinogenicity. There are no data on carcinogenic or mutagenic effects of P. (IARC). Inhalation exposure to 8.6 mg/m^3 and 17.2 mg/m^3 produced no carcinogenic effect in mice and rats.[2] *Carcinogenicity classification.* IARC: Group 3; NTP: NE—NE—NE—NE.

Chemobiokinetics. A direct chemical interaction between P. and biological media of the body is unlikely. Elimination occurs rapidly in an unaltered form through the lungs.

Standards. Russia (1988). MAC and PML: 0.5 mg/l (organolept., odor).

Regulations. USFDA (1993) approved the use of P. in adhesives as a component (monomer) of articles intended for use in packaging, transporting, or holding food.

References:

1. See **BUTYLENE.**
2. Quest, J. A., Tomaszewski, J. E., Haseman, J. K., et al., Two-year inhalation toxicity study of propylene in F344 rats and B6C3F$_1$ mice, *Toxicol. Appl. Pharmacol.*, 76, 288, 1984.

PROPYLENE OXIDE (PO) (CAS No 75–56–9)

Synonyms. 1,2-Epoxypropane; Methyloxirane; Propenoxide.

Properties. Colorless, volatile liquid with a sweet, ether odor. Readily soluble in water (40.5% at 20°C) and alcohol but is likely to evaporate to a great extent. Odor perception threshold is 9.9 mg/l,[010] 11.3 mg/l[1] or 31 mg/l;[02] taste perception threshold is 0.43 mg/l.

Applications and **Exposure.** Initial product for the synthesis of polyoxypropylene, propylene glycol and its esters and ethers, polyesters, and polyurethane resins and foams, etc. Solvent for resins. Food additive. Fumigant for medical plastics and foodstuffs.

Acute Toxicity. No adverse effects have been reported due to the ingestion of PO and its reaction products with food.[2] The oral LD$_{50}$ was reported to be 380 to 1140 mg/kg BW in rats, 440 to 630 mg/kg BW in mice, 660 to 690 mg/kg BW in guinea pigs, and 1245 mg/kg BW in rabbits. Poisoned animals displayed agitation followed by depression, increased salivation, and diarrhea. Death is preceded by short-term clonic–tonic convulsions.

There were certain changes in hematological and biochemical analyses, an increase in the content of chlorides in the blood. An increased level of free histamine in the blood is considered to be the main mechanism of a toxic effect development. Gross pathological examination revealed damage to the stomach mucosa and liver in rats. The liver cells exhibited edema and signs of fatty dystrophy.[1,3,4]

Repeated Exposure revealed slight cumulative properties. Administration of 0.2 g/kg BW as a 10% PO solution in oil for 24 d produced no toxic effect in rats.[2] However, according to other data, oral doses of 25 and 100 mg/kg BW given to rats for 1.5 months affected the peripheral blood and CNS functions, the liver and kidneys.[4]

Long-term Toxicity. In a 6-month rat study, an effect of the hematological indices was noted. The treatment produced phase changes in the chloride content of the blood and protein fractions of the blood serum and other lesions.[4]

Allergic contact dermatitis was diagnosed in three cases of exposure to a PO solution.[5]

Reproductive Toxicity. *Embryotoxic* and *teratogenic effects* were not observed when pregnant rabbits were exposed to PO through inhalation. Pregnant rats given 260 mg PO/kg BW during 2 weeks of gestation exhibited higher embryotoxicity and lower offspring BW compared to the controls. Doses of 1/5 and 1/20 LD$_{50}$ given to rats for 6 months also had a toxic effect on embryos.[4] Teratogenic properties of PO were not established, although there was a reduction in size and BW, retardation of the ossification process, and distortion of the ribs in the offspring.[6] Hardin et al. did not note teratogenic effects in rats and rabbits.[7] *Gonadotoxicity.* There were no sperm abnormalities in monkeys following inhalation exposure.[8] A reduced sperm motility, damage to spermatocytes, and reduced fertility were found in rats treated with oral LD$_{50}$ (520 mg/kg BW).[4] In female rats, a single oral dose of 260 mg/kg caused disturbance of the estral cycle without affecting the generation function. A dose of 0.052 mg/kg BW given to rats for 6 months did not affect the gonads.[4]

Genotoxicity. PO was shown to produce a mutagenic effect in assays employing microorganisms and insects. According to Rapoport,[9] PO is mutagenic to *Dr. melanogaster.* An increase in micronuclei was observed following *i/p* administration of 300 mg/kg; 150 mg/kg or less did not produce such an effect.[8] PO caused CA in mammalian cells *in vitro,* in particular, chromatid gaps and breaks,[10,11] but no CA and SCE were shown in monkeys exposed via inhalation.[12] Negative results were observed in DLM assay in male mice following administration of 50 and 250 mg/kg BW for 2 weeks prior to mating.[10] Slight mutagenic activity in *in vivo* experiments was attributed to rapid detoxication of PO in the liver of mammals.[10]

Carcinogenicity. Rats were gavaged with 15 and 60 mg/kg BW doses in salad oil for 112 weeks. There was an increased incidence of squamous cell carcinoma of the forestomach; malignant tumors were found mainly at the site of entry into the body.[13] Female Wistar rats exposed by inhalation to 300 ppm for 28 months exhibited an increase in the rate of both benign and malignant mammary tumors.[14] *Carcinogenicity classification.* IARC: Group 2B.

Chemobiokinetics. Following absorption, PO is found in the blood. It is metabolized to **formaldehyde,** the content of which in the blood after administration of $^1/_5$ and $^1/_{20}$ LD$_{50}$ was 0.137 to 0.24 mg% (0.067 mg% in the controls).[4] PO metabolism is described as transformation into *S*-(2-**hydroxy**-1-**propyl**)-**glutathione** under the action of glutathione-epoxide-transferase with subsequent transformation into 1,2-**propanediol** under the action of epoxidehydrolase. Diol can subsequently be oxidized into **lactic** and **pyruvic acids.**[2]

Standards. EEC (1990). MPQ in the material or article: 1 mg/kg. **Russia** (1990). Tentative MAC: 0.1 mg/l.

Regulations. USFDA (1993) regulates the use of PO as a direct /and indirect food additive. It may be used in products that come in contact with food; the use of PO is approved for the following purposes: (1) as an etherifying agent in the production of modified food starch (at the levels of 25% max or less) and (2) as a package fumigant for certain fruit products and as a fumigant for bulk quantities of several food products, provided residues of PO or propylene glycol do not exceed specified limits.

References:

1. See **ACRYLONITRILE,** #7, 162.
2. *Propylene Oxide,* Environmental Health Criteria 56, WHO, Geneva, 1985, p. 53.
3. See **ETHYLENE OXIDE,** #2.
4. Antonova, V. I., Zommer, E. E., Kuznetzova, A. D., et al., Toxicology of propylene oxide and regulations for surface water, *Gig. Sanit.,* 7, 76, 1981.
5. Jensen, O., Contact allergy to propylene oxide and isopropyl alcohol in a skin disinfectant swab, *Contact Dermatit.,* 7, 148, 1981.
6. Meylan, W., Papa, L., De Rosa, C. T., et al., Chemical of current interest—propylene oxide: health and environmental effect profile, *Toxicol. Ind. Health,* 2, 219, 1986.
7. Hardin, B. D., Niemeier, R. W., Sikov, M. R., et al., Reproductive toxicologic assessment of the epoxides ethylene oxide, propylene oxide, butylene oxide, and styrene oxide, *Scand. J. Work Environ. Health,* 9 (2 Spec. No.), 94, 1983.
8. Hardin, B. D., Schuler, R. L., McCormic, P. M., et al., Evaluation of propylene oxide for mutagenic activity in 3 *in vivo* test systems, *Mutat. Res.,* 117, 337, 1983.
9. Rapoport, I. A. Alkylation of gene molecule, *Rep. Acad. Sci. USSR,* 59, 1183, 1948 (in Russian).
10. Bootman, J., Lodge, D. C., and Whalley, H. E. Mutagenic activity of propylene oxide in bacterial and mammalian systems, *Mutat. Res.,* 67, 101, 1979.
11. Dean, B. J. and Hodson-Walker, G., An *in vitro* chromosome assay using cultured rat liver cells, *Mutat. Res.,* 64, 329, 1979.
12. Lynch, D. W., Lewis, T. R., Moorman, W. J., et al., Sister chromatid exchanges and chromosome aberrations in lymphocytes from monkey exposed to ethylene oxide and propylene oxide by inhalation, *Toxicol. Appl. Pharmacol.,* 76, 85, 1984.
13. Dunkelberg, H., Carcinogenicity of ethylene oxide and **1,2**-propylene oxide upon intragastric administration to rats, *Br. J. Cancer,* 46, 924, 1982.
14. Kuper, C. F., Reuzel, P. G. J., Feron, V. J., et al., Chronic inhalation toxicity and carcinogenicity study of propylene oxide in Wistar rats, *Food Chem. Toxicol.,* 26, 159, 1988.

SEBACIC ACID (SA) (CAS No 11–20–6)

Synonym. Decanedioic acid.

Properties. White or cream, crystalline powder. Water solubility is 1 g/l at 17°C and 20 g/l at 100°C. Readily soluble in alcohol. Has no effect on the taste or odor of water; no frothing effect was noted. According to Andreyev,[1] organoleptic perception threshold is 250 mg/l.

Applications and **Exposure.** Used in the production of cold-resistant plasticizers, polyester and polyamide resins, and polyurethanes; a stabilizer for alkyd resins, antiscorching in the production of food-grade resins. Migration from SA-containing rubber samples into distilled water (40°C, for 24 h) has been studied by Prokof'yeva.[2] Concentrations of 0.025 to 0.03 mg SA/l are reported.

Acute Toxicity. LD_{50} is 3.4 g/kg BW in rats and 6 g SA/kg BW in mice. However, according to Kropotkin et al. (1980), administration of 11 mg/kg BW appeared to be not lethal to mice, rats, or rabbits, but in 2 d the animals experienced significant BW loss.

Short-term Toxicity. Oral administration of 110 and 1000 mg/kg BW to rats resulted in the reduction in BW gain and in STI value, a change in the enzymatic activity in the blood serum and liver homogenate. A reduction in protein content of the blood serum and pathological changes in the GI tract, liver, and kidneys were also observed. The NOEL of 11 mg/kg BW was identified in this study.[1]

Long-term Toxicity. In a 7-month study, the treatment caused similar changes in rats.[3]

Reproductive Toxicity. The NOEL for gonadotoxic effect is likely to be 110 mg/kg BW.[1]

Chemobiokinetics. Of the dose 30 to 46% is removed with the urine.

Standards. Russia (1988). MAC and PML: 2.5 mg/l.

Regulations. USFDA (1993) regulates the use of SA (1) in adhesives as components of articles intended for use in packaging, transporting, or holding food, (2) in resinous and polymeric coatings for articles and surfaces coming in contact with food, and (3) in cross-linked polyester resins that may be safely used as articles or components of articles intended for repeated use in contact with food.

References:
1. See **ADIPIC ACID**, #1.
2. *Hygiene of Use, Toxicology of Pesticides and Polymeric Materials,* Coll. Works VNIIGINTOX, A. V. Pavlov, Ed., Kiev, 14, 1984, p. 114 (in Russian).
3. See **ADIPIC ACID**, #3.

STYRENE (CAS No 100–42–5)

Synonyms. Cinnamene; Cinnamol; Ethylenylbenzene; Phenylethylene; Styrol; Styrolene; Vinylbenzene.

Properties. Colorless, viscous liquid with a characteristic penetrating unpleasant odor. On exposure to air or light undergoes polymerization and oxidation. Water solubility is 125 mg/l at 20°C and 320 mg/l at 25°C, soluble in alcohol. Odor perception threshold is 0.05 to 0.12 mg/l, or 0.011 mg/l;[02,010] taste perception threshold is 0.06 mg/l.[1]

Applications and **Exposure.** S. is used as a monomer in the production of various plastics, resins, and vulcanizates: styrenebutadiene rubber, acrylonitrile-butadiene-styrene (ABC) polymer, and styrene-acrylonitrile copolymer (SAN) resins. It is also used as a cross-linking agent and a solvent for unsaturated polyester resins. S. has been found in food packaged in polystyrene containers (in yogurt up to 0.02 mg/l),[2] and in water (0.05 to 0.1 mg/l).[3,4] It was shown to migrate from ABC-polymers into water: 0.05 mg/l for 24 h at 37°C.[4] Concentrations reported in drinking water are 0.01 to 0.05 mg/l.

Acute Toxicity. In rats, LD_{50} varies from 5 to 8 g/kg BW.[5] Poisoned animals displayed NS impairment, convulsions, loss of reflexes, cyanosis, and body temperature drop. On administration of the lethal oral doses, rats became comatose and died. Gross pathological examination revealed hepatic dystrophic changes and, incidentally, renal changes,[6] but also diffuse damage to the upper sections of the CNS. The maximum tolerated doses appear to be 0.5 mg/kg BW for mice and 2 g/kg BW for rats. These doses caused no appreciable morphological changes.[1]

Repeated Exposure. Male rats were given 1 ml/kg BW by oral intubation for 15 d. There was an increase in the serotonin and noradrenaline contents in the brain, but no changes in the amount of dophamine were marked. The monoaminooxidase activity was suppressed, but that of ACE was unaltered.[7] Doses of 100 and 200 mg/kg BW administered to rats over the same period of time caused no disruption of behavioral reactions. An increase was observed in the serotonin level in the hippocampus and hypothalamus as a result of exposure to the 200 mg/kg BW dose.[7]

Short-term Toxicity. Doses up to 500 mg/kg BW caused irritation of the esophagus and stomach and hyperkeratosis of the forestomach. No hematological changes were observed in the short-term oral studies in rats.[6] The dose of 400 mg/kg BW given orally for 100 d produced elevated levels of hepatic AST and ALT. In addition, significantly decreased activity of hepatic acid phosphatase and of other enzymes was found. Histopathological examination revealed liver focal necrosis, which was supported by the biochemical analysis described earlier.[8]

Long-term Toxicity. In a 6-month oral toxicity study (Wolf et al., 1956), the NOEL of 133 mg/kg BW was suggested.[5] In the more recently reported experiment, beagle dogs were given S. in a peanut oil suspension by gavage for up to 561 d.[9] The NOAEL of 200 mg/kg BW was identified in this study. Parameters investigated in another oral rat study were clinical signs, mortality, growth, food and water intake, hemograms, clinical chemistry, urinalysis, gross necropsy, and histopathology. The NOEL was found to be 125 mg/l of drinking water, that corresponds to 7.7 mg/kg BW for males and 12 mg/kg BW for females.[10]

Reproductive Toxicity. *Embryotoxicity.* Negative results were reported in rats following oral dosing with 90 to 150 mg/kg BW.[8] No adverse effects were observed in Sprague-Dawley rats in the three generation reproductive study: there were no treatment-related changes in rats exposed for 2 years to 125 or 250 ppm in drinking water.[12] *Teratogenicity.* The oral teratogenicity study in rats did not reveal maternal toxicity, teratogenic, or embryotoxic effect at dose-levels up to and including 300 mg/kg BW.[13] Teratogenic effect for chicken embryos, rats, and rabbits, as well as fetotoxicity in mice and hamsters following inhalation was reported in some studies.[11,13] Inhalation of S. at low concentrations affects the gonads, embryogeny, and the offspring of mammals. S. penetrates through the placenta into the milk of feeding females.[14]

Genotoxicity. S. requires metabolic activation to produce genotoxic action. Mutagenic effect is due to the formation of **styrene oxide** (q.v.) in the body. S. is transformed into styrene oxide in the blood of humans under the action of oxyhemoglobin of erythrocytes. S. is mutagenic in a variety of test systems, results being sometimes rather equivocal. It induces gene mutations in prokaryotic and eukaryotic microorganisms, in *Dr. melanogaster,* and in mammalian cells *in vitro*. Reports on chromosomal abnormalities are also contradictory. When positive results were observed, mainly high doses were used.[15] According to Simula and Priestly,[16] S. produced weak genotoxic responses in the bone marrow micronucleus, sperm morphology, and SCE assays in Swiss mice and Porton rats.

Carcinogenicity. Carcinogenicity studies in mice and rats with various routes of administration did not provide the evidence of S. being a carcinogen. In the oral carcinogenicity study with B6C3F$_1$ mice, a significantly increased incidence of lung tumors (adenoma and carcinomas) was seen in males at the highest dose-level (300 mg/kg BW administered in corn oil). However, in this study the control group was rather small.[17] In a study with Fisher 344 rats,[14] the doses of 500 to 2000 mg/kg BW given in corn oil did not cause any significantly increased tumor incidence. Sprague-Dawley rats received doses of 9 to 250 mg/kg BW in olive oil over a period of 52 weeks. The study was terminated after 140 weeks. No significantly increased tumor incidences were observed.[18]

A 2-year oral toxicity/carcinogenicity study in conjunction with a three-generation reproduction study was carried out.[10] Rats received 0.125 and 250 mg S./ℓ in their drinking water. No increase in tumor incidence was observed. BW loss was noted in a group of females exposed to 250 mg S./l. No other treatment-related effects were seen. Long-term effects of oral administration were reported. S. was given in olive oil to pregnant female O$_{20}$ mice (1350 mg/kg BW), C57BL mice (300 mg/kg BW), and BD IV rats (1350 mg/kg BW) on day 17 of gestation. Their offspring were treated weekly throughout their life-span. The incidence of tumors occurring at the sites other than the lung was higher in the untreated mice than in the S.-treated animals. Very high doses, which caused earlier mortality, were used in this study.[19]

The evidence of genotoxicity in short-term animal tests and in humans occupationally exposed to S., along with the data on the metabolite **styrene**-7,8-**oxide,** seems to be supportive of carcinogenicity of S. *Carcinogenicity classification.* IARC: Group 2B; USEPA: Group C; NTP: N—N—E—N.

Chemobiokinetics. S. is readily absorbed and distributed in the body tissues, predominantly in fats. Following inhalation exposure, its content in paranephric fatty cells is ten times higher than in any other tissue.[20] S. is shown to enter the liver through the portal vein. It is believed that S. metabolism occurs under the action of liver monooxidase with participation of cytochrome P-450. The monomer is transformed into **styrene oxide,** which is covalently combined to the macromolecules of hepatocytes and then metabolized to several metabolites, the major ones being **mandelic** and **phenylglyoxylic acids.** Both of these metabolites have been detected in the urine of exposed rodents and humans. The nature of the S. metabolites varies according to the species of mammals.[21] Unchanged S. in rats is excreted with the exhaled air.

Guidelines. WHO (1992): Guideline value for drinking water 0.02 mg/l. The levels of 0.004 to 2.6 mg/l are likely to give rise to consumer complaints of foreign odor and taste.

Standards. USEPA (1991). MCL and MCLG: 0.1 mg/l. **Russia** (1988). MAC and PML: 0.1 mg/l.

Regulations. USFDA (1993) approved the use of S. (1) as a component of paper and paperboard for contact with dry food, (2) in semirigid and rigid acrylic and modified acrylic plastics intended for use in food-contact articles, (3) as a component of adhesives to be safely used in food-contact surface, (4) in resinous and polymeric coatings that may be used safely as a food-contact surface of articles intended for use in producing, manufacturing, packing, transporting, or holding food, (5) in polyethylene phthalate polymers to be used safely as articles or components of plastics intended for use in contact with food, and (6) in cross-linked polyester resins used as articles or components of articles intended for repeated contact with food.

References:

1. See **ACETONE, #3,** 137.
2. Wharton, F. D. and Levinskas, G. I., *Chemistry and Industry,* 11, June 1976, p. 470.
3. Kataeva, S. Ye., Regulation of the application of polymeric materials in water supply systems used for economic and drinking purposes, *Gig. Sanit.,* 10, 8, 1988 (in Russian).
4. Petrova, L. I., Investigation of Possible Use of Polystyrene Plastics of Different Composition in Contact with Foodstuffs, Author's abstract of thesis, Leningrad, 1979, p. 20 (in Russian).
5. See α-**METHYL STYRENE, #1.**
6. van Apeldoorn, M. E., van der Heijden, C. A., Heijena-Markus, E., et al., Styrene Criteria Documents, Air Effects Project No. 668310, National Institute of Public Health and Environmental Protection, The Netherlands, 1985.
7. Hussain, R., Srivastava, S. P., Mushtaq, M., et al., Effect of styrene on levels of serotonin, noradrenaline, dopamine, and activity of acetyl cholinesterase and monoamine oxidase in rat brain, *Toxicol. Lett.,* 7, 47, 1980.
8. Srivastava, S. P., Das, M., Mushtaq, M., et al., Metabolism and genotoxicity of styrene, *Adv. Exp. Med. Biol.,* 136A, 1982.
9. Quast, J. F., Humiston, C. G., Kalnins, R. V., et al., *Results of Toxicity Study of Monomeric Styrene Administered to Beagle Dogs by Oral Intubation for 19 Months,* Final report, Dow Chemical Co., Midland, MI, 1979.
10. *Toxicology Study on Styrene Incorporated in Drinking-Water of Rats for Two Years in Conjunction With a Three Generation Reproduction Study,* Styrene, Revised Final Report Weeks 1–105, Vol. 1, Litton Bionetics to Chemical Manufacturers Association, 1980.
11. Murray, F. J., John, J. A., Balmer, M. F., et al., Teratologic evaluation of styrene given to rats and rabbits by inhalation or by gavage, *Toxicology,* 11, 335, 1978.
12. Beliles, R. P., Butala, J. H., Stack, C. R., et al., Chronic toxicity and three-generation reproduction study of styrene monomer in the drinking water of rats, *Fundam. Appl. Toxicol.,* 5, 855, 1985.
13. Hemminki, K., Paasivirta, J., Kurnirinne, T., et al., Alkylation products of DNA bases by simple epoxides, *Chem. Biol. Interact.,* 30, 259, 1980.
14. Ragul'e, N., Problems concerning the embryotropic effects of styrene, *Gig. Sanit.,* 11, 85, 1974 (in Russian).
15. IPCS, Environmental Health Criteria 26, *Styrene,* WHO, Geneva, 1983.
16. Simula, A. P. and Priestly, B. G., Species differences in the genotoxicity of cyclophosphamide and styrene in three *in vivo* assays, *Mutat. Res.,* 271, 49, 1992.

17. Bioassay of Styrene for Possible Carcinogenicity, National Cancer Institute Technical Report Series No. 185, NIH-Publ. No. 79–1741, 1979.
18. Maltoni, C., Study of the Biological Effects (Carcinogenicity Bioassay) of Styrene, Report of Bologna Tumour Centre Department of Experimental Oncology, Bologna 1978.
19. Ponomarkov, V. and Tomatis, L., Effects of long-term oral administration of styrene to mice and rats, *Scand. J. Work Environ. Health*, 4 (Suppl. 2), 127, 1978.
20. Withey, J. R., The toxicology of styrene monomer and its pharmacokinetics and distribution in the rat, *Scand. J. Work Environ. Health*, 4 (Suppl. 2), 31, 1978.
21. Bond, J. A., Review of the toxicology of styrene, *CRC Crit. Rev. Toxicol.*, 19, 227, 1989.

TEREPHTHALIC ACID (CAS No 100–21–0)

Synonyms. **1,4**-Benzenedicarboxylic acid; Dicarboxybenzene; **4**-Formylbenzoic acid; *p*-Phthalic acid.

Properties. White powder. Poorly soluble in water. Odor perception threshold is about 100 mg/l, taste perception threshold is 150 mg/l (Prusakov).

Applications and **Exposure.** Used in the production of polyethylene terephthalate (PET). Levels of migration of PET oligomers from PET plastics are 0.02 to 2.73 mg/kg, depending on the food and temperature attained during cooking.[1]

Acute Toxicity. In 20 min after administration of 0.5 and 5 g/kg BW to mice, there was irritation of the upper respiratory tract and motor stimulation. Lethal dose is not less than 10 g/kg BW. Death occurs as a result of a blood and lymph circulation disorder.[2] Histological examination revealed congestion in the visceral organs and parenchymatous dystrophy in the renal tubular epithelium.

Repeated Exposure. Oral exposure to 2 to 4 g/kg BW for 10 d did not affect serum cholinesterase activity. Gross pathological examination failed to reveal changes in the visceral organs of rats. Dietary intake of TA may result in bladder stone formation. According to Chin et al.,[3] for stones to form under the action of TA, a critical saturated concentration of TA and calcium in the urine is necessary, stone formation evidently being preceded by the development of hyperplasia of the bladder epithelium.

Long-term Toxicity. In a 6-month study, the doses of 0.125 and 15 mg/kg BW produced no changes in conditioned reflex activity as well as in the histological structure of the visceral organs of rats (Prusakov).

Reproductive Toxicity. Produced no teratogenic effect in rats.[4]

Chemobiokinetics. TA is rapidly absorbed and excreted from the body. Accumulation in the tissues seems to be negligible. TA is found in the kidneys of humans and rats. Up to 30% of the *i/v* dose administered to chickens is removed unchanged via the urine.[5]

Standards. EEC (1990). SML: 7.5 mg/kg. **Russia.** PML: 5 mg/l.

Regulations. USFDA (1993) regulates TA for use (1) in resinous and polymeric coatings of surfaces and articles coming into contact with food and (2) in cross-linked polyester resins for repeated use in contact with food.

References:
1. See 1,4-**BUTANEDIOL**, #2, 91.
2. See **DIMETHYL TEREPHTHALATE**, #2.
3. Chin, T. I., Tyl, R. W., Papp, J. A., et al., Chemical urolithiasis. I. Characteristics of bladder stone induction by terephthalic acid and dimethylterephthalate in weanling Fisher 344 rats, *Toxicol. Appl. Pharmacol.*, 58, 307, 1981.
4. Ryan, B. M., Hatoum, N. S., and Jernigan, J. D., A segment II inhalation teratology study of terephthalic acid in rats, *Toxicologist*, 10, 40, 1990.
5. Tremaine, L. M. and Queblemann, A. J., The renal handling of terephthalic acid, *Toxicol. Appl. Pharmacol.*, 77, 165, 1985.

TETRAFLUOROETHYLENE (TFE) (CAS No 116–14–3)

Synonym. Tetrafluoroethene.

Properties. Colorless gas. Poorly soluble in water.

Applications. Used primarily in the synthesis of polytetrafluoroethylene polymers.

Acute Toxicity. Only data on inhalation studies are available. Irritation to the respiratory tract and the lungs and kidney injury were reported.[1]

Short-term Toxicity. Only data on inhalation studies are available. No other clinical signs of toxicity were seen, but kidney damage, and to a lesser extent changes in the lungs, colon, hematopoietic system, and endocrinal glands were apparent.[2]

Genotoxicity. Based on the structure–mutagenicity relationship, it was predicted that TFE would not be genetically active (Jones and Mackrodt, 1982).

Chemobiokinetics. An increase of urinary fluoride level was determined following inhalation of 3500 ppm for 30 min. It returned to normal when exposure was discontinued.[3] TFE was metabolized to *S*-(1,1,2,2-**tetrafluoroethyl**)-*L*-**cysteine** by rat liver fractions *in vitro*. Compound may undergo conjugation with glutathione followed by degradation of the *S*-**conjugate** to produce cytotoxicity. Reactions with renal cells and reactive thiols produce cell injury, but reaction with DNA is not likely.[2]

Regulations. TFE is cleared by **USFDA** for food-related uses in perfluorocarbon resins.

References:
1. Kennedy, G. L., Toxicology of fluorine-containing monomers, *CRC Crit. Rev. Toxicol.*, 21, 149, 1990.
2. Lock, E. A., Studies on the mechanism of nephrotoxicity and prenephrocarcinogenicity of halogenated alkenes, *CRC Crit. Rev. Toxicol.*, 19, 23, 1988.
3. Ding, R. S. and Kwon, B. K.. The inhalation toxicity of perolysis products of polytetra-fluoroethylene heated below 500°C, *Am. Ind. Hyg. Assoc. J.*, 29, 19, 1968.

TETRAHYDROFURAN (CAS No 109–99–9)

Synonyms. Butylene oxide; Cyclotetramethylene oxide; **1,4-**Furanidine; Epoxybutane; Oxacyclopentane; Oxolane; Tetramethylene oxide.

Properties. Colorless, mobile liquid with a pungent ether odor. Readily soluble in water. Odor perception thresholds of 3 or 1 mg/l (Pozdnyakova, 1965) were reported; coloring occurs at a concentration of 100 mg/l.

Applications. T. is used to produce polyethers and polytetramethylene oxide and also as an industrial solvent, primarily in the dissolution of plastic resins such as PVC and VDC copolymers.

Acute Toxicity. LD_{50} is 2.8 to 3 g/kg BW in rats, 2.5 g/kg BW in mice, and 2.3 g/kg BW in guinea pigs. LD_{50} was lower in 14-d-old than in the young rats.[1] In 3 to 5 min after poisoning, animals became motionless and experienced labored breathing, hyperemia, and cyanosis of the skin. Gross pathological examination revealed severe congestion in the visceral organs[2] as well as inflammation, necrosis, and hemorrhage of the GI tract, injury to the kidney tubules, and inflammation of the liver.[3]

Repeated Exposure revealed evidence of functional accumulation when T. was inhaled. Mice seem to be more sensitive to T. than rats.

Immunotoxic effects were not observed in tests on guinea pigs due to the impossibility of T. to combine with proteins *in vivo*.[4]

Reproductive Toxicity. Doses of 600 to 500 ppm were given on days 6 to 9 of gestation to rats, and on days 6 to 17 to mice. The NOAEL for maternal toxicity appeared to be 1800 ppm for both rats and mice; for developmental toxicity, 1800 ppm for rats and 600 ppm for mice.[5]

Genotoxicity. *In vitro* tests for genotoxic potential were negative in *Dr. melanogaster*[6] and in *Salmonella*/microsome assay.[7]

Chemobiokinetics. T. is metabolized by mammalian hepatic enzymes to the oxygen followed by cleavage to a straight-chain **fatty acid.**[8]

Standards. EEC (1990). SML in food: 0.6 mg/kg. **Russia** (1988). MAC: 0.5 mg/l.

Regulations. USFDA (1993) approved T. for use (1) as a component of adhesives intended for use in food-contact articles. T. may be used safely in the fabrication of articles intended for packing, transporting, or storing foods, (2) as a solvent in the casting of film from a solution of polymeric resins of vinyl chloride, vinyl acetate, vinylidene chloride or polyvinylchloride copolymers. The residual amount of T. in the film should not exceed 1.5% by weight. T. may be used safely (3) in cellophane for packaging food (residue limit 0.1%).

70

References:

1. See **EPICHLOROHYDRIN**, #2.
2. Fujita, T. and Suzuki, Z., Enzymic studies on the metabolism of the tetrahydrofurfuryl mercaptan moiety of thiamine tetrahydro-furfuryl disulfide, *J. Biochem.*, 74, 733, 1973.
3. Sax, N. Y., *Hazardous Chemicals,* Information Annual, Vol. 1, Van Nostrand Reinhold Information Service, New York, 1986, p. 640.
4. See **FURAN**, #2.
5. Mast, T. J., Weigel, R. J., Bruce, R., et al., Evaluation of the potential developmental toxicity in rats and mice following inhalation exposure to tetrahydrofuran, *Fundam. Appl. Toxicol.*, 18, 255, 1992.
6. Valencia, R., Mason, J. M., Woodraff, R. C., et al., Chemical mutagenesis testing in Drosophila, *Environ. Mutat.*, 7, 325, 1985.
7. Maron, D., Katzenellenbogen, J., and Ames, B. N., Compatibility of organic solvents with the *Salmonella*/microsome test, *Mutat. Res.*, 88, 343, 1981.
8. *Toxicology of New Industrial Chemical Substances,* Medgiz, Moscow, 5, 1963, p. 21 (in Russian).

2,4- and 2,6-TOLUENE DIISOCYANATES (TDI) (CAS No 26471–62–5)

Properties. Transparent, colorless, or slightly yellow liquids with a sharp, pungent odor. Darken on exposure to light; are rapidly polymerized by basis. In water, 92% of TDI hydrolyzes with the formation of diaminotoluenes. Soluble in diethyl ether, acetone, and other organic solvents. Miscible with olive oil.

Applications and **Exposure.** TDI are used in the production of rigid polyurethane foam, urethane rubbers, and various polymeric materials. 2.6-TDI can be used as a component of coatings and elastomer systems. Migration from polyurethane coatings into water (exposure 3 d, 37°C) is found at the level of 0.02 mg/l.[1]

Acute Toxicity. LD_{50} is 5.5 to 7.5 g/kg BW in rats and 4.7 g/kg BW in mice (Zapp; Union Carbide Corp.). Poisoning is accompanied by a short-term agitation, with subsequent adynamia and respiratory disorder. Death within 3 to 4 d.[2]

Repeated Exposure revealed pronounced cumulative properties.[2] K_{acc} is 2.87 (by Kagan). Rats were administered the dose of 1.5 g/kg BW for 10 d. Damage to the liver and GI tract was observed.

Short-term Toxicity. Rats were exposed to oral doses of up to 240 mg TDI/kg BW by gavage for a 13-week period. Mild to moderate bronchopneumonia was reported to develop.[3] Administration of $1/20$ LD_{50} (247 mg/kg BW) to rats for 4 months caused dysbalance in protein fractions of the blood serum and deviations in the leukocyte count. Erythrocyte count and Hb level in the blood were decreased.[2]

Immunotoxicity. TDI produces sensitization in animals and humans. A systemic immune response to TDI conjugated with dog serum albumin was observed in dogs exposed to 1 mg/kg BW TDI delivered as an aerosol intratracheally once every second week for 41 weeks.[4] Antibodies to TDI were produced in guinea pigs exposed by inhalation, dermally or by *i/p* injections.[5]

Genotoxicity. TDI is reported to show equivocal results in *Salmonella* mutagenicity bioassay. It was found to be positive in the mouse lymphoma assay; it caused CA and SCE.[6,7]

Carcinogenicity. Rats received doses of 30 and 60 mg/kg BW (males), 60 and 120 mg/kg BW (females) in the diet for 2 years. The treatment produced a dose-dependent increase in the rate of subcutaneous fibromas, fibrosarcomas, and adenomas of the pancreas in males. In females there were nodular neoplasms in the liver and fibroadenomas in the mammary glands. Mice were given 60 and 120 mg/kg BW (females), 120 and 240 mg/kg BW (males) in the diet for 105 weeks. The treatment caused hemangiomas, hemangiosarcomas, and hepatic cellular adenomas to develop in the females. No growth of tumors was observed in males.[3,8] Inhalation of TDI is not reported to cause tumors. *Carcinogenicity classification.* NTP: P—P*—N—P.

Chemobiokinetics. On ingestion by rats in corn oil, 2,6-TDI formed polymers in the GI tract. Excretion occurs predominantly via the feces.[8]

Standards. EEC (1990). MPQ in finished material or article: 1 mg/kg (expressed as isocyanate moiety).

71

Regulations. USFDA (1993) has determined that 2,4- and 2,6-TDI isomers may be used (1) as components of adhesives that come in contact with food and (2) as a component of polyurethane resins that form a surface in contact with food.

References:
1. See **BISPHENOL A,** #2.
2. See **SEBACIC ACID,** #2, 122.
3. NTP, *Technical Report on the Carcinogenesis Studies of Commercial Grade 2,4(86%)- and 2,6(14%)- Toluene Diisocyanate in F344/N Rats and B6C3F₁ Mice (Gavage Studies),* NTP Technical Report No. 251, Research Triangle Park, NC, 1983.
4. Patterson, R., Zeiss, C. R., and Harris, K. E., Immunologic and respiratory responses to airway challenges of dogs with toluene diisocyanate, *J. Allergy Clin. Immunol.,* 71, 604, 1983.
5. Karol, M. H., Study of guinea pig and human antibodies to toluene diisocyanate, *Am. Respir. Dis.,* 122, 965, 1980.
6. Anderson, M. and Styles, J. A., Appendix II, The bacterial mutation test, *Br. J. Cancer,* 37, 924, 1978.
7. See **BISPHENOL A,** #6.
8. NTP, *Toxicology and Carcinogenesis Studies of Commercial Grade 2,4(80%)- and 2,6(20%)- Toluene Diisocyanate in F344/N Rats and B6C3F₁ Mice (Gavage Studies),* NTP Techincal Report No. 251, NIH Publ. No. 86–2507, Research Triangle Park, NC, 1986.

TRIETHYLENE GLYCOL, DIMETACRYLIC ESTER (CAS No 109–16–0)

Properties. Clear liquid of a yellow or green to dark-green or dark-brown color. Contains up to 3.5% toluene and 0.71% hydroquinone. Readily soluble in alcohol. Gives water an aromatic odor and a specific astringent taste. Taste perception threshold is lower than odor perception threshold. The practical taste threshold of purufied T. is reported to be 22 mmol/l, that of the technical grade product, 23.2 mmol/l. Chlorination does not enhance the odor. The lower limits of reliability of the taste perception threshold are 7.74 (purified substance) and 8.49 mmol/l (technical grade product).

Applications. T. is used for copolymerization with methylmethacrylate and other monomers to obtain unfusible and insoluble heat-resistant polymers, and also in the manufacture of unsaturated polyesters. A plasticizer for polar rubbers: acrylonitrile-butadienes and chloroprenes. Used in the production of glass-reinforced plastics.

Acute Toxicity. LD_{50} of pure or technical grade T. is 10.84 g/kg BW in rats and 10.75 BW g/kg in mice. Changes in behavior were observed. Administration produced general inhibition, apathy, and urinary and fecal incontinence. Death occured in 1 to 2 d from respiratory failure.

Repeated Exposure revealed little evidence of cumulative properties. The highest dose tested (¹⁄₁₀ LD_{50}) produced disturbances of oxidation-reduction processes and of liver function.

Long-term Toxicity. Administration of ¹⁄₂₅₀ and ¹⁄₁₂₅₀ LD_{50} caused anemia and leukopenia, as well as a decline in serum cholinesterase and peroxidase activity in rats.

Allergenic Effect was not observed on *i/p* injection of up to 8 mg/kg to guinea pigs.

Standards. Russia (1982). Recommended MAC: 0.2 mg/l.

Reference:
Manenko, A. K., Kravets-Bekker, A. K., Sakhnovskaya, N. N., et al., Hygienic substantiation of maximum allowable concentration for triethyleneglycol dimethacrylate in water bodies, *Gig. Sanit.,* 4, 17, 1982 (in Russian).

UREA (CAS No 57–13–6)

Synonyms. Carbamide; Carbonic acid, diamide; Isourea; Pseudourea.
Properties. Colorless, odorless, crystalline prisms or granules. Readily soluble in water and alcohol. Undergoes hydrolysis when heated in aqueous solutions of acids or alkalis. U. does not render water a considerable odor and does not alter its color, but does give it an astringent taste (Mazaev and Skachkova, 1966). However, according to other data, taste and odor perception threshold is 80 mg/l.[1]

Applications and **Exposure.** U. is used in the manufacture of urea-formaldehyde polymers and plastics, adhesives, synthetic fibers, dyes, etc. It is a naturally-occurring constituent of the body.

Acute Toxicity. Rats tolerate doses of up to 12 g/kg BW. LD_{50} is found to be 14.3 or 16.3 g/kg BW in rats and 11 g/kg BW in mice (Mazaev and Skachkova, 1966). In rabbits, LD_{50} was not attained.

Repeated Exposure revealed slight cumulative properties. Rats developed U. blood levels up to 40 to 45 mg% after dietary intake of a 2 g/kg BW dose for a month. A dose of 50 mg/kg BW does not alter the balance of U.[2]

Long-term Toxicity. High doses are reported to affect bioenergetic and cholinergic processes and cause changes in the CNS and visceral structure. The concentration of 10 mg/l in drinking water was found to be safe.[1]

Reproductive Toxicity. *Humans.* Effect on fertility in women is reported. *Animals.* Administration of 2 g/kg BW to rats on day 12 and to mice on day 10 of gestation caused no increase in the defect rates or resorptions.[3]

Genotoxicity. U. could produce DNA damage in the culture of human leukocytes.

Carcinogenicity. Negative results are reported in C57B1/6 mice given 674 to 6750 mg U./kg BW and in Fisher 344 rats receiving 225 to 2250 mg U./kg in their diet for a year.[4]

Chemobiokinetics. Of administered ^{15}N-labeled U., 92% was excreted in the urine of young pigs after 48 h and 1.9% during the subsequent 48 h;[5] U. is an excretory end-product of amino acid metabolism in mammals. The formation of U. occurs in the liver. Average daily urinary excretion of U. in adults is likely to be about 20.6 g.

Recommendations. JECFA (1993). Use at levels of up to 3% in chewing gum not of toxicological concern.

Regulations. USFDA (1993) approved the use of U. (1) in resinous and polymeric coatings to be safely used in food-contact surface and (2) in cellophane for packaging food.

Standards. Russia (1988). Recommended MAC: 10 mg/l.

References:

1. Mironets, N. V., Savina, R. V., Kucherov, I. S., et al., MAC of urea in reclaimed potable water and its biological action, *Space Biol. Aviacosmic Med.,* 22, 63, 1988 (in Russian).
2. Kotova, N. I., Data on substantiation of MAC of carbamate in the work-place air, *Gig. Truda Prof. Zabol.,* 3, 43, 1986 (in Russian).
3. Teramoto, S., Kaneda, M., Aoyama, H., et al., Correlation between the molecular structure of *N*-alkylureas and *N*-alkylthioureas and their teratogenic properties, *Teratology,* 23, 335, 1981.
4. Fleischman, R. W., Hayden, D. W., Smith, E. R., et al., Carcinogenesis bioassay of acetamide, hexanamide, adipamide, urea and p-toluurea in mice and rats, *J. Environ., Pathol. Toxicol.,* 1, 149, 1980.
5. Grimson, R. E., Bowland, J. P., and Milligan, L. P., Use of nitrogen-15 labeled urea utilization by pigs, *Can. J. Anim. Sci.,* 51, 103, 1971.

VINYL ACETATE (VA) (CAS No 108–05–4)

Synonyms. Acetic acid, ethenyl ester; Acetic acid, ethylene ether; Acetic acid, vinyl ester; 1-Acetooxyethylene; Vinyl ethanoate.

Properties. Colorless, transparent liquid with a pungent odor. Water solubility is 2.5% at 20°C. Soluble in ethanol, ether, acetone, chloroform and carbon tetrachloride. Odor threshold concentrations are 0.25 mg/l,[1] 0.12 mg/l,[010] or 0.088 mg/l.[02] Solutions with a concentration of up to 50 mg/l do not alter the pH.

Applications. Used for the production of polyvinyl acetate homopolymer emulsions and resins, polyvinyl alcohol, etc.

Acute Toxicity. LD_{50} is 1.6 g/kg BW in mice,[1] and 2.92 g/kg BW[2] or 2.9 g/kg BW in rats (Union Carbide Corp., 1958).

Repeated Exposure revealed slight cumulative effect when mice were administered 0.3 g/kg BW for 3 weeks.

Reproductive Toxicity. *Gonadotoxicity.* VA produced sperm abnormalities in mice.[3]

Teratogenicity. Exposure to 1000 ppm on days 6 to 15 of gestation resulted in minor skeletal defects and smaller weights of rat fetuses. No adverse fetal effects were observed in the offspring of rats receiving 5000 ppm VA in their drinking water.[4]

Genotoxicity. VA effectively induces CA in mammalian cells, as well as micronuclei in bone marrow erythrocytes and DNA–DNA cross-links.[3,5,6] It was tested for ability to induce SCE in cultured (72-h) human lymphocytes with a 48-h treatment, starting at 24 h after culture initiations. VA induced a clearly dose-dependent increase in the number of SCE. The product of transformation of VA—**acetaldehyde** (q.v.)—leads to the same cytogenic effect as VA in a culture of lymphocytes. It is likely to be a metabolite responsible for the SCE induction.[7] VA was negative in *Salmonella* mutagenicity assay.

Carcinogenicity. Fisher 344 rats received 1 and 2.5 g/kg BW with their drinking water for 2 years. The treatment resulted in increased animal mortality. Tumor types were similar in the control and test animals. Thyroid adenoma and uterine carcinoma were found with the higher dose.[8] However, IARC does not consider these results to be completely conclusive. *Carcinogenicity classification.* IARC: Group 3; NTP: N—N—XX—X.

Chemobiokinetics. After oral administration or inhalation, VA was incorporated metabolically into hepatic DNA; specific alkylation products were not detected.[9] VA rapidly breaks down in the body to form **acetic acid** and **vinyl alcohol.** It is assumed also to be metabolized (hydrolyzed) in mammalian cells via conversion to **acetaldehyde** and **acetic acid.** After VA inhalation, rats exhaled acetaldehyde.

Standards. EEC(1990). SML: 12 mg/kg. **Russia** (1988). MAC and PML: 0.2 mg/l.

Regulations. USFDA (1993) approved the use of **VA-monomer** as a component of surfaces in contact with food, including (1) adhesives used as components of articles intended for use in packaging, transporting, or holding food, (2) resinous and polymeric coatings in a food-contact surface of articles intended for use in producing, manufacturing, packing, transporting, or holding food, (3) paper and paperboard intended for use in contact with dry food, (4) cellophane to be safely used for packaging food, and (5) substances employed in the production of or added to textiles and textile fibers.

Subsequently, **polyvinylacetate** has been approved for use as a component of surfaces in contact with food, including adhesives, resinous and polymeric coatings, components of paper or paperboard in contact with dry, fatty, or aqueous foods, constituents of cellophane, and textiles and textile fibers. **VA-vinyl chloride copolymer** resins are regulated for use as components of resinous and modified acrylic plastics. Copolymers of **ethylene–VA–vinyl alcohol** are approved for use in articles in contact with food, with limitations on thickness based on polymer composition. **Ethylene–VA–copolymers** may also be used in this way. **VA–crotonic acid copolymer,** the surface of polyolefin film that is in contact with food, may be used as a coating.

References:
1. See **ACROLEIN,** #1, 64.
2. Dernehl, C. U., Clinical experiences with exposures to ethylene amines, *Ind. Med. Surg.,* 20, 541, 1951.
3. Lahdetie, J., Effect of vinyl acetate and acetaldehyde on sperm morphology and meiotic micronuclei in mice, *Mutat. Res.,* 202, 171, 1988.
4. The Society of the Plastics Industry, Inc., *Vinyl Acetate: Oral and Inhalation Teratology Studies in the Rat,* FYI-AX-1283–0278 (Sequence F), U.S. Environmental Protection Agency, Washington, D.C., 1983.
5. Norppa, H., Yursi, F., Pfaffi, P., et al., Chromosome damage induced by vinyl acetate through *in vitro* formation of acetaldehyde in human lymphocytes and Chinese hamster ovary cells, *Cancer Res.,* 45, 4816, 1985.
6. Maki-Paakkanen, J. and Norppa, A., Induction of micronuclei by vinyl acetate in mouse bone marrow cells and cultured human lymphocytes, *Mutat. Res.,* 190, 41, 1987.
7. Siri, P., Jarventaus, H., and Norppa, H., Sister chromatide exchanges induced by vinyl esters and respective carboxylic acids in cultured human lymphocytes, *Mutat. Res.,* 279, 75, 1992.
8. Lijinsky, W. and Reuber, M. D., Chronic toxicity studies of vinyl acetate in Fisher rats, *Toxicol. Appl. Pharmacol.,* 68, 43, 1983.
9. Simon, P., Fisher, J. G., and Bolt, H. M., Metabolism and pharmacokinetics of vinyl acetate, *Arch. Toxicol.,* 57, 19, 1985.

VINYL CHLORIDE (CAS No 75–01–100)

Synonyms. Chloroethene; Chloroethylene; Monochloroethylene; VC-monomer (VCM).

Properties. Colorless gas with a chloroform odor. Solubility in water is 1.1 g/l at 25°C. Odor perception threshold is reported to be 2 or 3.4 mg/l.[02]

Applications and **Exposure.** A monomer in the production of PVC and VC copolymers. Residual VCM levels in food and drinks are now estimated to be well below 0.01 mg/kg. Maximum exposures from food and drinks are estimated to be less than 0.0001 mg/d. According to FDA data,[1] a total maximum daily intake from oil, liquor, and wine bottles, food packaged in PVC or VC-VDC copolymer film, and other sources is 0.000025 mg.

Acute Toxicity. In rats, LD_{50} is reported to be 500 mg/kg BW (Dow Chemical Co.).

Repeated Exposure. Rats were orally exposed to 300 mg/kg BW dose for 4 weeks. Increased liver and kidney weights as well as histopathological changes in these organs were found (Lefaux, 1975).

Short-term Toxicity. Rats were administered 100 and 300 mg/kg BW in solution of soybean oil for 13 weeks. Hematological indices and clinical chemistry were affected. Gross pathological examination revealed elevated organ weights but failed to detect morphological liver changes.[2] The dose of 30 mg/kg BW in a 3-month study appeared to be ineffective in Wistar rats (Lefaux, 1975).

Long-term Toxicity. Wistar rats received VCM in their feed and *i/g* as a solution in soybean oil at doses of 1.7, 14.1, and 300 mg/kg BW for 144 weeks.[2] The treatment increased mortality in the test animals. Two higher doses shortened prothrombin time and increased hemopoietic function of the spleen. All doses caused morphological changes in the liver.

Immunotoxic effect was registered in workers employed in VCM production.[3]

Reproductive Toxicity. The data available to date cannot be considered conclusive, and mechanism for the possible reproduction-related risk posed by paternal exposure to VCM is not certain.[1] *Embryotoxicity.* VCM was ineffective in the offspring of mice, rats, and rabbits at maternal toxicity levels. However, a concentration of 4.8 mg/m³ was found to produce a selective embryotoxic effect manifested by the loss of permeability of the fetal vessels and functional abnormalities in the NS, liver, and peripheral blood, and changes in the weight coefficients of the internal organs of the rat offspring.[4] *Teratogenicity.* Teratology studies, after VCM inhalation, have been carried out in mice, rats, and rabbits. No significant effects on malformations or anomaly rates resulted from the exposure to 130 to 6470 mg VCM/m³. Other experiments have suggested some signs of embryotoxicity of VCM in rats and mice. No studies for VCM administered by the oral route are available.[5,6] *Gonadotoxicity.* Chronic inhalation of VCM at concentrations of 4.8 and 35 mg/m³ had a toxic effect on the gonads.[7]

Genotoxicity. VCM caused a significant degree of DNA binding in short-term studies. *Humans.* A number of cytogenic studies have demonstrated an increased frequency of CA and SCE in the peripheral lymphocytes of workers exposed to VCM at levels of 5 to 500 ppm (13 to 1300 mg VCM/m³).[7,8] Two studies reported negative results for SCE in exposed workers, while in another study, a weak positive response was found. *Animals.* VCM appeared to be mutagenic in various test systems (IARC 19–231). It induced CA, SCE and micronuclei in rodents exposed *in vivo* but did not induce mutations in the mouse spot test or DLM in rats or mice. It induced mutations in Chinese hamster cells and unscheduled DNA synthesis in rat hepatocytes *in vitro* and caused transformation of BALB/c 3T3 cells and virus-infected Syrian hamster cells. It brought about sex-linked recessive lethal mutations but not aneuploidy, heritable translocations, or DLM in *Dr. melanogaster.*[8]

Carcinogenicity. VCM has been shown to have carcinogenic effects in humans and animals. In 1974, a connection was reported between VCM and the instances of a rare tumor (angiosarcoma of the liver) in human beings and animals (Greech and Johnson, 1974). It is likely that VCM is not carcinogenic, and that it is a product of its metabolic bioactivation (**chloroethylene oxide**), that primarily exhibits carcinogenic activity. It is still not clear whether this product is the only carcinogenic metabolite of VCM. *Humans.* There is a sufficient evidence of carcinogenicity of VCM to humans. All studies refer to inhalation exposure, and positive findings come from an industrial population exposed to high concentrations of VCM. Worldwide there have now been about 70 proved cases in people exposed to very high concentrations of VCM over an average period of 20 years before it was recognized as a health hazard. The cause–effect relationship between VCM exposure and angiosarcoma of the liver is commonly accepted. Opinions vary as related to other tumors. In the Russian VCM and PVC industry, the monomer has been found to have a positive effect on the rate of malignant neoplasms (Fedotova, 1984). According to IARC, Suppl. 7–373, VCM can be also associated with hepatocellular carcinoma, brain tumors, lung tumors, and malignancies of the lymphatic and hematopoietic tissues. On the other hand, there is insufficient evidence to establish any relationship between exposure to VCM and an increased incidence of cancer of the brain, lung, and lymphatic or hematopoietic tissues.[1] According to Doll's opinion,[9] no positive evidence of a hazard of any type of cancer other than angiosarcoma of the liver has been found except possibility for a small hazard of lung cancer when exposure was heavy.

Human data on carcinogenic risk following oral exposure to VCM are not available. The carcinogenic risk of VCM administered orally should not be exaggerated.[10] *Animals.* There is a sufficient evidence for VCM-induced carcinogenicity in animals. When administered by inhalation, it induced angiosarcomas of the liver in rats, mice, and hamsters, Zymbal gland tumors in rats and hamsters, nephroblastomas in rats, and pulmonary and mammary gland tumors in mice. Sprague-Dawley rats were given doses of 3.35, 16.65, and 50 mg VCM/kg BW as a solution in olive oil four to five times a week for 52 weeks.[11] The treatment caused angiosarcomas of the liver, nephroblastomas, thymus tumors, etc. A dose dependence was noted. These studies proved VCM to be an undoubted carcinogen for animals, causing the development of tumors in different organs and various types of tumors in the same organ. In the opinion of Maltoni, not even a dose of 0.3 mg/kg BW is the threshold. In the study of Feron et al.,[2,12] development of both hepatoangiosarcomas and angiosarcomas of the lungs and abdominal cavity and mesotheliomas and adenocarcinomas of the mammary glands of rats was observed at doses of 5 mg/kg BW and more. Wistar rats were exposed to 0.014, 0.13, and 1.3 mg/kg BW in their diet. A variety of VCM-related liver lesions was found in the high-dose group. There was no evidence that feeding with VCM affected the incidence of tumors in organs other than the liver. The NOAEL for tumor induction in the rat was identified to be 0.13 mg/kg BW.

The cancer risk of oral daily intakes of 0.00002 or 0.0001 mg VCM/person/d is estimated to be neglibly small.[13] Such intakes can be considered virtually safe. Assuming that the number of carcinomas in other sites may be equal to that of hepatic angiosarcomas,[5] in the estimates for excess risk it was considered that a 10^{-5} risk occurs as a result of a lifetime exposure to 1 mg VCM/person/d. *Carcinogenicity classification.* IARC: Group 1; USEPA: Group A.

Chemobiokinetics. After ingestion of the small doses of VCM with water and food, it is rapidly excreted from the body, and therefore metabolites evidently have no time to form. Only persistent presence of VCM in the industrial environment enables it to be accumulated in the blood and to form carcinogenic metabolites.[10] The metabolism of VCM is a dose-dependent and saturable process. VCM is metabolized in the microsomal-mixed function oxidase system, forming **chloroethylene oxide,** which can be transformed spontaneously to **chloroacetaldehyde.** These two metabolites (particulary chloroethylene oxide) are highly reactive and mutagenic. A major route of metabolism of chloroacetaldehyde involves oxidation to **chloroacetic acid.**

Chloroethylene oxide, chloroacetaldehyde, and chloroacetic acid can be conjugated to glutatione and/or cysteine and excreted in the urine.[14] It is known that metabolism of chemical substances in small animals is more rapid than in large animals. Therefore, the accumulation of carcinogenic metabolites of VCM is more rapid in small animals, and there is an increased risk of angiosarcomas developing in their liver compared with that of a human being.[15] Low doses of VCM administered by gavage are metabolized and eliminated primarily in the urine. In contrast, higher doses are mainly excreted unchanged via the lung.

Guidelines. WHO (1992). Guideline value for drinking water: 0.005 mg/l.

Standards. USEPA (1991). MCL: 0.002 mg/l; MCLG: zero. **Russia** (1988). MAC: 0.05 mg/l; PML in food 0.01 mg/l.

Regulations. USFDA eliminated the use of **VCM** in drug products and proposed to alert food manufacturers to the need for monitoring packaging materials that may contain it. Later this proposal was withdrawn by FDA's Center for Food Safety and Applied Nutrition. VCM may be used (1) as a component (monomer) of adhesives, (2) in paper and paperboard coming in contact with food, (3) in polyethylene phthalate polymers to be safely used as or as components of plastics intended for use in contact with food, and (4) in PVC pipes in water supply. **EEC** (1989): presence of VCM in materials and articles prepared with VCM polymers or copolymers, which are intended to come in contact with foodstuffs, is limited to 1 mg/kg in final product. Materials and articles must not pass on to foodstuffs any VCM detectable by the method possessing detection limit of 0.01 mg/kg.

76

References:

1. *The Mutagenicity and Carcinogenicity of Vinyl Chloride: a Historical Review and Assessment,* ECETOC Technical Report No. 31, European Chemical Industry Ecology and Toxicology Centre, Brussels, 1988.
2. Feron, V. J., Speek, A. J., Willems, M. I., et al., Observation on the oral administration and toxicity of vinyl chloride in rats, *Food Cosmet. Toxicol.,* 13, 633, 1975.
3. Peneva, M. and Kis'ova, Kr., Repercussion of chronic vinyl chloride effect on some immunological parameters, *Khigiene i zdraveopazvane,* 2, 29, 1987 (in Bulgarian).
4. Sal'nikova, L. S. and Kitzovskaya, M. A., Effect of vinyl chloride on embryogenesis in rats, *Gig. Truda Prof. Zabol.,* 3, 46, 1980 (in Russian).
5. Vinyl Chloride, Air Quality Guidelines for Europe, European Series No. 23, WHO, Regional Office for Europe, WHO Regional Publications, Copenhagen, 1987, p. 158.
6. *Vinyl Chloride,* JECFA, WHO, Geneva 1986, p. 197.
7. *Hygienic Significance of Low-Intensity Factors in Industry and in Populated Areas,* Moscow, 1983, p. 60 (in Russian).
8. Muratov, M. M. and Gus'kova, S. I., Problem of vinyl chloride mutagenicity, *Gig. Sanit.,* 7, 111, 1978 (in Russian).
9. Doll, R., Effects of exposure to vinyl chloride, *Scand. J. Work Environ. Health,* 14, 61, 1988.
10. Sheftel, V. O., On the risk of vinyl chloride migration into water and foodstuffs, *Gig. Sanit.,* 2, 63, 1980 (in Russian).
11. Maltoni, C., Lefemine, C., Gilberti, A., et al., Carcinogenicity bioassays of vinylchloride monomer: a model of risk. Assessment on an experimental basis, *Environ. Health Perspect.,* 41, 3, 1981.
12. Feron, V. J., Hendriksen, C. F. M., Speek, A. J., et al., Lifespan oral toxicity study of vinyl chloride in rats, *Food Cosmet. Toxicol.,* 19, 317, 1981.
13. Til, H. P., Feron, V. J., and Immel, H. R., Lifetime (149 weeks) oral carcinogenicity study of vinyl chloride in rats, *Food Chem. Toxicol.,* 29, 713, 1991.
14. *Vinyl Chloride,* Criteria Document (Publikatiereeks Luckt, No. 34) The Hague, Ministerie van Volkshuisvesting Ruimtelijke Ordening en Milienbeheer, 1984 (in German).
15. Dietz, F. K., Ramsey, J. C., and Watanabe, P. G., Relevance of experimental studies to human risk, *Environ. Health Perspect.,* 52, 9, 1983.

4-VINYLCYCLOHEXENE (VCHE) (CAS No 100–40–3)

Synonyms. 1,3-Butadiene, dimer; 4-Ethenyl-1-cyclohexene.

Properties. Colorless, transparent liquid with a pungent aromatic odor. Water solubility is 95 mg/l at 20°C, soluble in diethyl ether, and petroleum ether. Contained in commercial polybutadiene rubber.

Applications. An intermediate in the production of plastics and rubber; a diluent for epoxy resins; a comonomer in styrene polymerization. Antioxidant.

Acute Toxicity. LD$_{50}$ was reported to be 2.63 g/kg BW in Carworth-Wistar rats.[1] According to other data, LD$_{50}$ is 7 g/kg BW in rats. The maximum tolerated dose is 4 g/kg BW.[2] A dose of 14 g/kg BW kills all the animals. Poisoning is accompanied by clonic convulsions and digestive disorder. A dose of 5 g/kg BW causes erythropenia, reticulocytosis, and thrombocytosis, and an increase in the activity of catalase and cholinesterase in the blood.[2]

Repeated Exposure B6C3F$_1$ mice were given oral doses of 200 and 400 mg/kg BW. Survival among high-dosed animals was poor.[3] In a 14-d study, doses of 300 to 5000 mg/kg BW were administered by gavage in corn oil. All rats and most mice died on ingestion of up to 1250 mg/kg BW, but there were no gross or histopathological effects.[4]

Short-term Toxicity. In a 13-week study, mice received 0 to 1.2 g VCHE/kg BW by gavage; rats were given the doses of 0 to 0.8 g/kg BW according to the same schedule. There was hyaline droplet degeneration of the proximal convoluted tubules of the kidney, the severity of which appeared to be dose-related in the rats. No compound-related gross or histopathological effects were evident in female rats or in male mice.[3,4]

Long-term Toxicity. Prolonged exposure to VCHE caused high mortality in mice and rats.[3]

Reproductive Toxicity. *Gonadotoxicity.* VCHE is shown to be ovarian toxicant in mice but not in rats. A reduction in the number of primary and mature Graafian follicles was seen in the ovaries of B6C3F$_1$ mice given the 1.2 g/kg BW dose over a period of 3 months.[3] VCHE at the doses up to 500 mg/kg reduced the gamete pool in both ovary and testis of B6C3F$_1$mice, but did not affect reproduction in either F$_0$ or F$_1$ generation.[8]

Genotoxicity. VCHE was not mutagenic to *Salmonella* in the presence or absence of exogenous metabolic system.[3]

Carcinogenicity. B6C3F$_1$ mice and Fisher 344 rats were given 200 and 400 mg VCHE/kg BW in corn oil by gastric intubation for 103 weeks. No clear evidence of carcinogenicity was noted in rats. Treatment-related increase in the incidence of granulosa-cell and mixed-cell tumors of the ovary were found in mice. High mortality was observed in male mice; increases in the incidence of lung tumors and lymphomas were statistically significant only by lifetable analysis.[3] These carcinogenicity studies were considered to be inadequate due to extensive and early mortality (in rats and male mice) as well as lack of conclusive evidence of carcinogenic effect.[5] *Carcinogenicity classification.* IARC: Group 2B; NTP: IS—IS—IS—CE.

Chemobiokinetics. Administration of a 0.5 g/kg dose to mice caused cytochrome P-450, cytochrome b$_5$, NADPH-cytochrome-*c*-reductase and aminopyrine-*N*-demethylase and epoxide hydrolase to be induced, and rapidly depleted hepatic glutathione levels, suggesting that glutathione is probably involved in metabolism of VCHE.[6] Wistar rat and Swiss mouse liver microsomal mixed-function oxidases metabolize VCHE to 4-**vinyl**-1,2- **epoxycyclohexane,** 4-**epoxyethylcyclohexene,** and traces of 4-**epoxyethyl**-1,2-**epoxy-cyclohexane**, which are further hydrolyzed to the corresponding diols.[7]

References:
1. Smith, H. F., Carpenter, C. P., and Weil, C. S., Range-finding toxicology data, List VII, *Am. Ind. Hyg. Assoc. J.,* 30, 470, 1969.
2. See **BISPHENOL A,** #2, 235.
3. NTP, *Toxicology and Carcinogenicity Studies of 4-Vinylcyclohexene in Fisher 344/N Rats and B6C3F$_1$ Mice (Gavage Studies),* NTP Technical Report Series No. 303, Research Triangle Park, NC, 1985.
4. Collins, J. J. and Manus, A. G., Toxicological evaluation of 4-vinylcyclohexen. I. Prechronic (14 day) and subchronic (13 week) gavage studies in Fisher 344 rats and B6C3F$_1$ mice, *J. Toxicol. Environ. Health,* 21, 493, 1987.
5. Collins, J. J., Montali, R. J., and Manus, A. G., Toxicology evaluation of 4-vinylcyclohexen. II. Induction of ovarian tumors in female B6C3F$_1$ mice by chronic oral administration of 4-vinylcyclohexen, *J. Toxicol. Environ. Health,* 21, 507, 1987.
6. Giannarini, C., Citty, L., Gervasi, G., et al., Effect of 4-vinylcyclohexene and its main oxirane metabolite on mouse hepatic microsomal enzymes and glutathione levels, *Toxicol. Lett.,* 8, 115, 1981.
7. Watabe, T., Hiratsuka, A., Ozawa, N., et al., A comparative study on the metabolism of *d*-limonene and 4-vinylcyclohex-1-ene by hepatic microsomes, *Xenobiotica,* 11, 333, 1981.
8. Grizzle, T. B., George J. D., Fail, P. A. et al., Reproductive effects of 4-vinylcyclohexene in Swiss mice assessed by a continuous breeding protocol, *Fund. Appl. Toxicol.,* 22, 122, 1994.

1-VINYL-2-PYRROLIDONE (CAS No 88–12–0)

Synonyms. *N*-Vinylbutyrolactam; 1-Vinyltetrahydropyrrol-2-one.

Properties. Colorless, transparent liquid. Soluble in water, alcohols, ethers, and esters.

Acute Toxicity. In rats, LD$_{50}$ is 1.37 to 1.47 g/kg BW. Poisoning is accompanied by signs of narcotic and irritating effects.

Repeated Exposure revealed moderate cumulative properties.

Reproductive Toxicity effects are not found after inhalation exposure to a concentration of 5 mg/m^3 (Kvasov, 1974).

Genotoxicity. V. is found to be positive in *Salmonella typhimurium, Dr. melanogaster,* and in tests for gene mutations in mice lymphoma cells.[1]

Carcinogenicity classification. IARC: Group 3.

Chemobiokinetics. Following ingestion, V. is mainly distributed in the liver and small intestine.

78

Partly it is excreted in the urine in an acetate form, but it is mostly (88%) combined with water-soluble acid compounds.[2]

Regulations. USFDA (1993) regulates V. for use in adhesives as a component of articles intended for use in packaging, transporting, or holding food.

References:

1. See **ACRYLAMIDE,** #21.
2. McClanahan, J. S., Lin, Y. C., and Digenis, G. A., Disposition of N-vinyl-2-pyrolidone in the rat, *Drug Chem. Toxicol.*, 7, 129, 1984.

VINYLTOLUENES (mixed isomers) (CAS No 25013–15–4)

Synonyms. Ethenylmethylbenzenes; Methylstyrenes; Methylvinyl-benzenes.

Properties. Oily liquids with a pungent odor. Technical-grade V. are mixtures of isomers. Water solubility is ~100 mg/l at 25°C. Odor perception threshold is 0.42 mg/l.[02] According to other data, organoleptic threshold is 0.005 mg/l.[01]

Acute Toxicity. In young rats, LD_{50} is 4 g/kg BW.[1] The same LD_{50} was identified for the mixture of *m*-V. (55 to 70%) and *p*-V. (30 to 45%). Poisoning is accompanied by excitation, followed by depression, side position, clonic convulsions, and hyperemia of paws and tails. High doses cause narcosis in mice.[2]

Long-term Toxicity. V. was given by gastric intubation in olive oil to Sprague-Dawley rats (10, 50, or 250 mg/kg BW, 78 weeks) and Swiss mice (50, 250, or 500 mg/kg BW, 107 weeks). There was no effect on BW gain; survival of males was reduced.[6]

Reproductive Toxicity. Embryotoxic effect is observed after inhalation exposure of guinea pigs.

Genotoxicity. V. caused no CA and SCE in Chinese hamster ovary cells (NTP-85). No mutagenic activity was found in the *Salmonella* bioassay, *Dr. melanogaster,* and in the tests for gene mutations in mouse lymphoma cells.[3]

Carcinogenicity. No increase in the incidence of tumors was observed in rats and mice.[6] *Carcinogenicity classification.* IARC: Group 3.

Chemobiokinetics. V. are the structural analogues of styrene. They are metabolized via reactive intermediates, which bind hepatic nonprotein thiols; in the rat liver V. caused a dose-dependent decrease in hepatic glutathione, with a concomitant excretion of **thioethers** in the urine.[4] However, repeated daily *i/p* administrations were not followed by any appreciable increase in metabolite excretion, i.e. no enzyme induction phenomenon was apparent.[5]

Regulations. USFDA (1993) approved the use of V. (1) in resinous and polymeric coatings for food-contact surface of articles intended for use in producing, manufacturing, packing, transporting, or holding food and (2) in cross-linked polyester resins for repeated use as articles or components of articles coming in contact with food.

References:

1. α-**METHYL STYRENE,** #1.
2. *Hazardous Chemicals,* Hydrocarbons, halogenated hydrocarbons, Handbook, V. A. Filov et al., Eds., Khimia, Leningrad, 1990, p. 207 (in Russian).
3. See **ACRYLAMIDE,** #21.
4. Vainio, H. and Heinonen, T., Metabolism and toxicity of vinyltoluene, *Toxicol. Lett.*, 18 (Suppl. 1), 151, 1983.
5. Bergemalm-Rynell, K. and Steen, G., Urinary metabolites of vinyltoluene in the rat, *Toxicol. Appl. Pharmacol.*, 62, 19, 1982.
6. Conti, B., Maltoni, C., Perino, G., and Ciliberti, A., Long-term carcinogenicity bioassays on styrene administered by inhalation, ingestion and injection and styrene oxide administered by ingestion in Sprague-Dawley rats, and para-methylstyrene administered by ingestion in Sprague-Dawley rats and Swiss mice, *Ann. N.Y. Acad. Sci.*, 534, 203, 1988.

Chapter 2
PLASTICIZERS

ACETYL TRIBUTYL CITRATE (ATBC)

Synonym. Acetylcitric acid, tributyl ester.

Properties. Liquid. Poorly soluble in water. Odor perception threshold is 50 mg/l. Taste perception threshold is 25 mg/l.

Applications and **Exposure.** ATBC is used as a plasticizer for films, in particular, vinyl chloride–vinylidene chloride copolymer films. Daily intake is reported to be 1.5 mg.[1]

Migration in the cheese wrapped in VDC copolymer films (exposure 5 d, temperature 5°C) was found at the level of 6.1 ppm, and into wrapped cake at the level of 3.2 ppm.[2] Migration from plasticized vinylidene chloride–vinyl chloride copolymer film in fatty or aqua-type foods was determined at the levels from 0.4 mg/kg after minimal contact during microwave cooking of a soup to 79.8 mg/kg for use of the film during the microwave cooking of peanut-containing cookies.[3]

Acute Toxicity. Rats and mice tolerate doses of up to 20 g/kg BW ATBC without signs of intoxication. Some animals died when given 22 to 24 g/kg BW.[4]

Repeated Exposure. Rats were dosed by gavage with 0.125 to 2.5 g ATBC/kg BW. The treatment caused retardation of BW gain, a reduction in STI, and an increase in the relative weights of the liver.[4]

Short-term Toxicity. Rats and mice were exposed to 0.4 and 1 g/kg BW for 4 months. The administration caused no changes in general condition or in BW gain of animals. No effect of ATBC was found on blood coagulation time, content of blood serum calcium, prothrombin time, or detoxication- and protein-forming functions of the liver. There were no changes in the relative weights of the visceral organs (Bidnenko, 1973).

Long-term Toxicity. Rats and mice received 50 and 250 mg ATBC per kilogram BW as a milk solution with their feed. The 250 mg/kg dose caused an increase in BW gain and a reduction in the STI value in mice. A reduction in blood peroxidase activity of rats, followed by an increase, was observed. There were no deviations in the indices studied at the end of the experiment. A 50 mg/kg BW dose was identified to be the NOAEL.[4]

Reproductive Toxicity. *Gonadotoxicity.* Exposure to ACTB produced no significant effect on the male gonads. Only some exfoliation of the spermatogenic epithelium was reported.[4] *Embryotoxicity.* An increase in the fetal weight and size and also in the weight of the placenta was noted in a long-term study with administration of a 250 mg/kg BW dose. Growth and development of the progeny were unchanged.[4]

Genotoxicity. ATBC was shown to be negative in *Salmonella* mutagenicity assay.

Standards. Russia (1988). MAC and PML: **n/m.**

Regulations. USFDA (1993) listed ATBC for use as a component (1) of adhesives for articles intended for use in packaging, transporting, or holding food, (2) of resinous and polymeric coatings to be safely used as a food-contact surface of articles, and (3) of resinous and polymeric coatings for polyolefin films for use as a food-contact surface. **British Food Agency** required labeling of food cling wraps.

References:

1. Anon., Plasticizers migration in food, *Food Chem. Toxicol.*, 29, 139, 1991.
2. *Plasticizers: Continuing Surveillance,* The 30th Report Steering Group on Food Surveillance, The Working Party on Chemical Contaminants from Food Contact Materials: Sub-Group on Plasticizers, Food Surveillance, Paper No. 30, HMSO, London, 1990, p. 55.
3. Castle, L., Jickells, S. M., Sharman, M., et al., Migration of the plasticiser acetyltributyl citrate from plastic film into foods during microwave cooking and domestic use, *J. Food Prot.*, 51, 916, 1988.
4. Larionov, L. N. and Cherkasova, T. E., Hygiene evaluation of acetyltributylcitrate, *Gig. Sanit.*, 4, 102, 1977 (in Russian).

BENZYL OCTYL ADIPATE

Synonym. Adipic acid, benzyloctyl ester.

Properties. Viscous liquid. Poorly soluble in water, soluble in oil.

Applications. Used as a plasticizer in the production of PVC, nitro-, ethyl-, and benzylcellulose, chlorinated rubber, etc. Renders the materials resistance to low temperatures.

Acute Toxicity. In rats, rabbits, and dogs, the LD_{50} is 20 ml/kg BW (24-h observation) and 10 ml/kg BW (7-d observation).

Long-term Toxicity. In a 1-year study, rats were given 0.5 and 1 ml/kg BW twice a week. The treatment produced no effect on growth, hematology, or reproductive function of animals.[05]

Standards. Russia. Recommended PML: **n/m.**

BUTYL BENZYL ADIPATE (BBA)

Synonyms. Adipic acid, butylbenzyl ester.

Properties. Clear yellowish liquid with a characteristic odor. Poorly soluble in water, readily soluble in alcohol.

Applications. Used as a plasticizer in the production of PVC and other plastics.

Acute Toxicity. In rats, the LD_{50} is 19.4 g/kg BW. Poisoning is accompanied by excitation with subsequent inhibition, lassitude, and impairment of motor coordination. Four to five h before death, the animals assume side position and do not react to external stimuli. In 5 to 7 d after a single exposure, signs of intoxication have completely disappeared in surviving animals. Mild anemia was found.

Reference:

Hygiene Aspects of the Use of Polymeric Materials and Articles Made of Them, Coll. Works, VNIIGINTOX, Medved', L. I., Ed., Kiev, 1969, p. 36 (in Russian).

BUTYL BENZYL PHTHALATE (BBP) (CAS No 85–68–7)

Synonyms. 1,2-Benzenedicarboxylic acid, butyl phenylmethyl ester; Phthalic acid, benzylbutyl ester.

Properties. Colorless, oily liquid. Solubility in water is 2.9 mg/l.

Applications. Used as a plasticizer in the production of polymeric materials. Compatible with synthetic polymers: PVC, polyacrylates, PVA, and nitrocellulose. Plasticizer in the production of vinyl floor tiles and adhesives, in food containers and wrapping materials. Increases the water- and ageing- resistance of materials.

Acute Toxicity. LD_{50} in rats seems to be more than 10 g/kg BW when BBP is given in the feed and 2.33 g/kg BW when it is given *i/v*.[1] Other latest data are also reported: in Fisher 344 rats given BBP in corn oil, LD_{50} is found to be 2.33 g/kg BW; in B6C3F$_1$ mice, it is 4.17 g/kg (females) and 6.16 g/kg BW (males).[2]

Repeated Exposure. Administration of high doses caused exhaustion, adynamia or aggression, tonic convulsions with subsequent pareses, and paralyses of the extremities. Gross pathological examination revealed irritation of the lower respiratory pathways, dystrophic changes in the liver and brain, and spinal cord lesions with demyelinization of the antero-lateral columns and peripheral nerves.[3] Fisher 344 rats were exposed to BBP for 14 d (0.625 to 5% in the diet). BW gain and thymus weights were reduced in the 2.5 and 5% dose groups. Enlargement of the liver and kidneys, thymic atrophy, and morphological changes in these organs were observed. Prolonged exposure to BBP can affect the hematopoietic system.[2,4]

Short-term Toxicity. In a subchronic study, rats were dosed with the doses of 300 to 600 mg/kg BW. The treatment resulted in reduced BW gain, pareses of the rear extremities, and death, all to a greater extent in males. Different breeds of animals of a single species exhibited different toxicity. In dogs ingesting diets with 1 to 5% BBP for 90 d, there were no alterations in urinary or hematological parameters. No gross or histopathological effects were found.[5] In a 90-d study, depressed BW gain and testicular degeneration were noted, but no compound-related mortality occurred in rats given the doses of 25 g/kg BW.[2]

Reproductive Toxicity. *Gonadotoxicity.* BBP produced a direct toxic action on the testes, with secondary effects on other reproductive organs. The weights of testes, epididymis, prostate, and seminal vesicles were reduced in rats exposed to 2.5 and 5% BBP in their feed.[5] Histological examination showed atrophy of the testes, prostate, and seminal vesicles, atrophy of the thymus and epididymis in 2.5 and 5% dose groups. Plasma testosterone concentration was decreased at a higher dose-level. Ability to cause testicular atrophy depends on the route of administration and the species. Mice seem to be less sensitive than rats and guinea pigs for the effect on the gonads. In dogs, testicular lesions were not observed, even with the high doses.[1] *Embryotoxicity* and *teratogenicity.* Pregnant Wistar rats were exposed to BBP (2% in the diet) during pregnancy. Decrease in food consumption and BW in pregnant rats was noted. Administration during pregnancy (on days 0 to 20) caused complete resorption of all the implanted embryos; administration on days 0 to 7 and 7 to 16 produced increased postimplantation loss. Exposure on days 7 to 16 of pregnancy revealed striking teratogenicity: cleft palate in 95% of fetuses and fusion of sternebrae were predominantly observed.[7] Administration during the first and second halves of pregnancy produced embryolethality and teratogenicity effects, respectively.[7]

Genotoxicity. BBP was not shown to be mutagenic in *Salmonella* and mouse lymphoma assays; it did not cause CA and SCE in mammalian cultured cells.[8]

Carcinogenicity. No differences in tumor incidence were reported in the control animals and in B6C3F₁ mice given doses of 6 or 12 g/kg BW for 103 weeks. Fisher 344 rats given the same doses over the same period showed increased incidence of myelomonocytic leukemia in high-dosed females.[2,9] *Carcinogenicity classification.* USEPA: Group C; NTP: IS—P—N—N.

Chemobiokinetics. Fisher 344 rats were given doses of 20 to 2000 mg ^{14}C-BBP/kg. Plasticizer was found to be rapidly metabolized, with subsequent excretion as glucuronides in the urine and feces. The main route of excretion of metabolites is biliary. These metabolites are reabsorbed and ultimately eliminated in the urine.[10]

Standards. USEPA (1991). Proposed MCL: 0.1 mg/l, MCLG: 0.

Regulations. USFDA (1993) regulates BBP as a component of the following materials used in contact with food products, provided that the BBP contains not more than 1% by weight of dibenzylphthalate and provided further that the finished food-contact article, when extracted with the solvent or solvents characterizing the type of food and under the conditions of time and temperature characterizing the condition of its intended use will yield net chloroform-soluble extractives not to exceed 0.5 mg/in.[2]: (1) adhesives in articles for packaging, transporting or holding food, (2) paper and paperboard for contact with dry food, (3) cross-linked polyester resins used in articles intended for repeated use, (4) polymeric substances used in the manufacture of articles, and (5) resinous and polymeric coatings for polyolefin films to be used safely as a food-contact surface of articles intended for use in producing, manufacturing, packing, transporting, or holding food.

References:

1. Hammond, B. G., Levinskas, G. J., Robinson, E., et al., A review of subchronic toxicity of butylbenzyl phthalate, *Toxicol. Ind. Health,* 3, 79, 1987.
2. *Bioassay of Butylbenzyl Phthalate for Possible Carcinogenicity,* NTP No. 81–25, DHHS Publ. No NIN 80–1769, U.S. Department of Health and Human Services, Washington, D.C., 1981.
3. Aldyreva, M. V. and Gafurov, S. A., *Industrial Safety in the Production of Synthetic Leathers,* Medgiz, Moscow, 1980, p. 138 (in Russian).
4. Agarwal, D. K., Maronpot, R. R., Lamb, J. C., et al., Adverse effects of butylbenzyl phthalate on the reproductive and hematopoietic systems of male rats, *Toxicology,* 35, 189, 1985.
5. Hammond, B. G., Toxicology of butylbenzyl phthalate, *Toxicologist,* 1 (Abstr. No. 414), 14, 1981.
6. Ema, M., Itami, T., and Kawasaki, H., Effect of period of exposure on the developmental toxicity of butyl benzylphthalate in rats, *J. Appl. Toxicol.,* 12, 57, 1992.
7. Ema, M., Itami, T., and Kawasaki, H., Embryolethality and teratogenicity of butyl benzylphthalate in rats, *J. Appl. Toxicol.,* 12, 179, 1992.

8. See **BISPHENOL A,** #6.
9. Tarone, R. E., Chu, K. C., and Ward, J. M., Variability in the rates of some common naturally occuring tumors in Fisher 344 rats and (C57BL/6N × C3H/HeN)F$_1$ (B6C3F$_1$) mice, *J. Natl. Cancer Inst.,* 66, 1175, 1981.
10. Eigenberg, D. A., Bozigian, H. P., Carter, D. E., et al., Distribution, excretion and metabolism of butylbenzyl phthalate in the rat, *J. Toxicol. Environ. Health,* 17, 445, 1986.

BUTYLENE GLYCOL ADIPATE

Synonym. Adipic acid, butyleneglycol ester.
Applications. Used as a plasticizer in the production of PVC and other plastics.
Long-term Toxicity. Rats and dogs were dosed with 1.5 and 10 mg/kg BW for 2 years. The treatment produced no effect on BW gain, food consumption, or mortality. Gross pathological examination failed to reveal macro- and microchanges in the structure of the visceral organs and tissues.
Reproductive Toxicity. In a three-generation study, rats were administered 1 to 10 mg B./kg BW. Reproductive function or development of the offspring were not affected.
Regulations. USFDA (1993) approved the use of 1,3-**butyleneglycoladipate** terminated with a 16 wt% mixture of myristic, palmic, and stearic acids, at levels not exceeding 33% by weight of PVC homopolymers used in contact with food (except foods that contain more than 8% of alcohol), at temperatures not to exceed room temperature.
Reference:
Fancher, O. E., Kennedy, G. L., Plank, J. B., et al., Toxicity of a butylene glycol adipic acid polyester, *Toxicol. Appl. Pharmacol.,* 26, 58, 1973.

BUTYLPHTHALYL BUTYL GLYCOLATE (BPBG) (CAS No 85–70–1)

Synonym. Phthalic acid, butyl(butoxycarbonylmethyl) ester.
Properties. Colorless liquid without taste and odor. Solubility in water is 0.0012% (30°C).
Applications. Used as a plasticizer in the production of PVC, PVA, etc. to render the materials heat- and light-resistant.
Acute Toxicity. In rats, LD$_{50}$ is found to be 14.6 g/kg BW.[1] Administration of 2.1 g/kg BW dose to rabbits and 4.7 g/kg BW dose to rats produced no toxic effects.[04]
Long-term Toxicity. Rats were exposed to 0.02, 0.2, and 2% of BPBG in their feed. Only the highest dose-level caused retardation of BW gain. No pathohistological changes in the viscera were observed.[04,05]
Standards. Russia. Recommended PML: **n/m.**
Regulations. USFDA (1993) regulates the use of BPBG (1) in adhesives as a component of articles intended for use in packaging, transporting, or holding food, (2) in resinous and polymeric coatings in a food-contact surface, and (3) in resinous and polymeric coatings for polyolefin films to be safely used as a food-contact surface of articles intended for use in producing, manufacturing, packing, transporting, or holding food. Permitted in some EC countries (Italy and Netherlands) for use in materials coming into contact with food products.
Reference:
Fishbein, L. and Albro, P. W., Chromatographic and biological aspects of the phthalate esters, *J. Chromatogr.,* 70, 365, 1972.

BUTYL STEARATE (BS) (CAS No 123–95–5)

Synonym. Stearic acid, butyl ester.
Properties. Colorless or pale-yellow liquid, almost odorless at the temperatures above 20°C. Soluble in alcohol, miscible with vegetable oils.
Applications. Used as a plasticizer in the production of polystyrene and other polymeric materials. Migration from ABC-polymers into sunflower oil at 20°C for 3 months is reported to be 0.04 mg/l.[1]

Acute Toxicity. Young rats (BW of 60 to 75 g) tolerate doses of 32 g/kg.[2]

Repeated Exposure failed to reveal cumulative properties. Rats and mice were exposed to 1 to 4 g/kg BW for 1.5 months. No toxic effect was noted to develop.[3]

Long-term Toxicity. Rats received 0.25 to 6.25 g/kg BW in their feed. No changes in growth, hematology, or histology of the visceral organs have been reported.[2] Administration of 20 and 400 mg/kg BW produced no toxic effect.[3]

Reproductive Toxicity. Addition to the feed did not affect the reproductive function in rats.[2]

Standards. Russia. PML in drinking water: **n/m.**

Regulations. USFDA (1993) listed BS for use (1) in adhesives as a component of articles intended for use in packaging, transporting, or holding food, (2) in resinous and polymeric coatings in a food-contact surface, (3) in resinous and polymeric coatings for polyolefin films to be safely used as a food-contact surface, (4) as a plasticizer for rubber articles intended for repeated use in contact with food (content up to 30% by weight of the rubber product), (5) as a defoaming agent that may be safely used as a component of food-contact articles, (6) in cross-linked polyester resins to be safely used as articles or components of articles intended for repeated use in contact with food, and (7) as a component of the uncoated or coated food-contact surface of paper and paperboard intended for use in producing, manufacturing, packing, transporting, or holding dry, aqueous and fatty food.

References:

1. *Toxicology and Sanitary Chemistry of Polymerization Plastics,* Coll. Sci. Proc., B. Yu. Kalinin, Ed., Leningrad, 1984, p. 64 (in Russian).
2. Smith, C. C., Toxicity of butylstearate, dibutyl-sebacate, dibutylphthalate, and methoxyethyloleate, *Arch. Ind. Hyg. Occup. Med.,* 4, 310, 1953.
3. Komarova, E. N., Toxic properties of some additives for plastics, *Gig. Sanit.,* 12, 30, 1976 (in Russian).

CHLOROPARAFFINS (CP) (CAS No 63449–39–8)

Synonyms. Cerechlor; Chlorinated paraffin waxes and hydrocarbon waxes; Chlorowax; Paraffin waxes.

Properties. Viscosity and density of CP increase with the Cl content. CP differ in the amount of Cl contained in the molecule (from 28 to 70%). There are many grades of CP. CP of short and medium chain length are oily liquid materials at ambient temperature. The long-chain highly chlorinated paraffins are solid, waxy materials. At 65 to 70% of Cl, they are whitish solids. They have an odor when the Cl content exceeds or is equal to 55%. Insoluble in water and alcohol, soluble in plasticizers, vegetable oil, and fat.

Applications. Used as plasticizers in the production of rubber and PVC. Flame retardants. *Carcinogenicity classification.* IARC: Group 2B.

Regulations. USFDA (1993) approved the use of CP as cross-linking agent in polysulfide polymer–polyepoxy resins used as the surface contacting dry food.

CHLORINATED PARAFFINS, 48.5% chlorination

Acute Toxicity. In mice, LD_{50} of CP containing 48.5% chlorine and 0.0024% iron is 26 g/kg BW.

Repeated Exposure failed to reveal cumulative effect in mice given 2.6 g/kg BW for a month. Consumption of 1 g CP with the diet over 6 weeks caused no signs of intoxication in rats.[05]

Reference:

Proc. Azerbaijan Research Institute Occup. Hygiene, Baku, 5, 1970, p. 180 (in Russian).

CHLORINATED PARAFFINS C $_{10-13}$, 58% chlorination

Acute Toxicity. LD_{50} exceeds 4 g/kg BW.[1] LD_{50} for CP C_{12}, 59% chlorination, exceeds 21.5 ml/kg BW in rats.[2]

Repeated Exposure. Fisher-344 rats were given the doses of 900 to 27,300 ppm in the diet or 30 to 3000 mg/kg BW by gavage over a period of 14 d. The liver was found to be the target organ (its weight was increased at dose-levels of 100 mg/kg and above). Histological examination revealed hepatocellular hypertrophy. The NOAEL was considered to be 30 mg/kg.[3,4] In a 16-d study, Fisher 344 rats were dosed with 7.5 g/kg BW (CP C_{12}, 60% chlorination). The treatment caused animal mortality. B6C3F$_1$ mice died when they received 3.75 g/kg BW and more. The livers of animals were found to be enlarged.[5]

Short-term Toxicity. In a 90-d study, Fisher 344 rats were exposed to the CP doses of 10 to 625 mg/kg BW (in the diet or by gavage). The treatment showed the kidney and parathyroid glands to be the target organs, in addition to the liver. No overt signs of toxicity were noted. There was an increase in the liver and kidney weights at doses 100 mg/kg and above. Thyroid-parathyroid weights were increased at 625 mg/kg BW. Microscopically, hepatocellular hypertrophy was shown to develop. No treatment-related microscopic changes were found in the tissues. The NOEL by oral route appeared to be 10 mg/kg BW.[3,4] In a 90-d study, rats tolerated the doses up to 5 g/kg; mice tolerated up to 2 g/kg BW. An increase in the liver weight was noted. Histological examination revealed liver hypertrophy.[5]

Reproductive Toxicity. Fisher 344 rats received 100 to 200 mg/kg BW, rabbits 10 to 100 mg/kg BW. The doses of 30 and 100 mg/kg caused maternal toxicity in rats; a 100 mg/kg dose increased postimplantation loss, early and late resorptions, and decreased the number of viable fetuses per dam. No effects on dams or fetuses were noted at the lowest dose. Rabbits were more sensitive, but no teratogenic effects were observed in any dose group. The doses of 28 to 1000 ppm in the diet were administered to young ducks. The NOEL for reproduction was considered to be 166 ppm in the diet.[3,4] However, the significance of these studies is questioned by the IARC Workgroup.

Genotoxicity. Fisher 344 rats were given 250 to 2500 mg/kg BW orally (by gavage) for 5 consecutive days. The exposure did not increase the frequency of chromosomal or chromatide aberrations in the bone marrow cells. CP was shown to be negative in DLM assay; being administered at dose-levels of 250 to 2000 mg/kg BW, it did not induce any increase in early fetal deaths or a decrease in viable embryos during the ten matings of the study.[3] CP C_{10-13}, 50% chlorination,[1] and CP C_{12}, 60% chlorination,[3] are not found to be mutagenic in several strains of *Salmonella,* with and without metabolic activation.

Carcinogenicity. The doses of 312 and 625 mg/kg BW (CP C_{12}, 60% chlorination, length of exposure 104 weeks) produced clear evidence of carcinogenicity, namely, increased incidence of hepatocellular neoplasms, adenomas, or adenocarcinomas of the kidney tubular cells in male Fisher 344 rats, follicular cell adenomas, or carcinomas of the thyroid gland in female rats.[3] In B6C3F$_1$ mice, an increase in the incidence of hepatocellular tumors and of alveolar/bronchiolar carcinomas in males, and of follicular-cell tumors of the thyroid gland in females are reported.[5]

As usual, in the NTP studies, extremely large doses to provide adequate information were used. These doses could produce a marked liver and kidney toxicity.[3] Anyway, carcinogenicity of CP may occur through a nongenotoxic mechanism, according to mutagenicity data. ***Carcinogenicity classification.*** C_{12}, 60% chlorination—NTP: CE*—CE*—CE*—CE*.

Chemobiokinetics. A part of the dose was absorbed, but the main part was excreted with the feces. Radiolabel assay has shown distribution mainly in the liver, fatty tissue, and ovary.[3]

References:
1. Birtley, R. D. N., Conning, D. M., Daniel, J. W., et al., The toxicological effects of chlorinated paraffins in mammals, *Toxicol. Appl. Pharmacol.,* 54, 514, 1980.
2. Howard, P. H., Santodonato, J., and Saxena, J., Investigation of Selected Potential Environmental Contaminants: Chlorinated Paraffins, EPA-560/2-75-007; PB 248634, U.S. Environmental Protection Agency, Washington, D.C., 1975.
3. Serrone, D. M., Birtley, R. D. N., Weigand, W., et al., Toxicology of chlorinated paraffins, *Food Chem. Toxicol.,* 25, 553, 1987.
4. *Toxicology and Carcinogenesis Studies of Chlorinated Paraffins (C$_{12}$, 60% chlorine average content),* NTP Technical Report 308, U.S. Department of Health and Human Services, Publ. Health Service, NIH, 1986.
5. Bucher, J. R., Alison, R. H., Montgomery, C. A., et al., Comparative toxicity and carcinogenicity of two chlorinated paraffins in Fisher 344/N rats and B6C3F$_1$ mice, *Fundam. Appl. Toxicol.,* 9, 454, 1987.

CHLORINATED PARAFFINS C$_{14-17}$, 52% CHLORINATION

Acute Toxicity. LD$_{50}$ exceeds 4 g/kg BW.[1]

Repeated Exposure. In a 14-d study, dietary administration of 150 to 15,000 ppm increased liver weights in Fisher 344 rats. Histological examination revealed diffuse hepatocellular hypertrophy at 5000 and 15,000 ppm in the diet. Dietary level of 500 ppm was considered to be the NOAEL.[2,3]

Short-term Toxicity. In a 90-d study, increased liver and kidney weights at doses of 100 mg/kg and higher were observed. At 625 mg/kg BW dose-level, thyroid and parathyroid weights were increased in male rats, and adrenal weights were increased both in males and females. Histological examination revealed hepatocellular hypertrophy in the liver in the high-dose group. The NOEL was identified to be 10 mg/kg BW.[2,3]

Reproductive Toxicity. Despite a high level in the ovary, no morphological changes were observed in this organ. Rats and rabbits were given 500 to 5000 and 10 to 100 mg/kg BW, respectively. There were signs of maternal toxicity in the rats at the high dose-level. Rabbits seem to be more sensitive. No teratogenicity effect was noted.[2] Mice were fed doses of 100 to 6250 ppm for 28 d before and during mating, and, in the case of females, continuously up to postnatal day 21. Pups were given the same diet as their parents from weaning until the pups were 70 d of age. No impairment of reproductive function was noted.[2] However, the significance of these studies is questioned by the IARC Workgroup.

Genotoxicity. CP was administered orally to Fisher 344 rats by gavage for 5 consecutive days at doses of 500 to 5000 mg/kg BW. The exposure did not increase the frequency of chromosomal or chromatide aberrations in bone marrow cells. CP C$_{14-17}$, 52% chlorination, was not found to be mutagenic in several strains of *Salmonella* with and without metabolic activation.[1]

Carcinogenicity. No evidence of carcinogenicity was noted in male rats. In females, there was an increased incidence of adrenal gland medullary neoplasms.[3] The liver seems to be the target organ in this study, but no carcinogenicity in the liver was reported in the NTP Report. An increased incidence of malignant lymphoma was noted in B6C3F$_1$ mice. However, the induction of lymphomas in mice as an index of carcinogenic activity might be questioned. In any case, this provides no clear evidence of carcinogenic potential.[2]

Chemobiokinetics. Absorption from the GI tract is limited; a part of the dose was absorbed, but mostly it was excreted with the feces. Radiolabeled assay has shown distribution mainly in the liver, adipose tissue, and ovary. Fecal excretion includes both unabsorbed material and radiolabeled material excreted in the bile. Tissue concentrations were the highest initially in the liver and kidney and later in the adipose tissue and ovary.[2,3]

References:

1. See **CHLORINATED PARAFFINS C$_{10-13}$, 58% chlorination, #1.**
2. See **CHLORINATED PARAFFINS C$_{10-13}$, 58% chlorination, #3.**
3. *Toxicology and Carcinogenesis Studies of Chlorinated Paraffins (C$_{23}$, 43% chlorine, average content),* NTP Technical Report 305, U.S. Department of Health and Human Services, Publ. Health Service, NIH, 1986.

CHLORINATED PARAFFINS C$_{20-30}$, 43% CHLORINATION

Acute Toxicity. LD$_{50}$ exceeds 4 g/kg BW.[1] Rats tolerated 10 ml/kg BW (CP C$_{24}$, 40% chlorination).[2]

Repeated Exposure. Fisher 344 rats received by gavage 30 to 3000 mg/kg BW for 14 d. No treatment-related effects were found in organ weights or in tissue histology.[3]

Short-term Toxicity. In a 90-d study, Fisher 344 rats and B6C3F$_1$ mice were given CP C$_{23}$, 43% chlorination, by gavage at dose levels of 3750 and 7500 mg/kg BW, respectively, in corn oil. No manifestation of the toxic action was noted in the treated animals.[4] In another 90-d study, 100 to 3750 mg CP/kg BW administered by gavage produced no adverse effect on BW gain, water or food consumption, or clinical biochemical indices. There were the treatment-related effects (inflammatory changes and necrosis on histological examination) on the liver in female rats at all doses, but no damage was observed in the livers of males.[3]

Long-term Toxicity. Similar hepatic lesions were found in female Fisher 344 rats in a 90-d study, and in male rats receiving much larger doses of Chlorowax 40 for 6 to 12 months.[5] In a 2-year oral rat study, only lymphocytic infiltration and granulomatous inflammation in the liver mesenteric and pancreatic lymphoid nodes were found. Mice given 5000 mg/kg BW dose displayed no non-neoplastic lesions.[4]

Reproductive Toxicity. *Teratogenicity.* Rats and rabbits were given doses of 500 to 5000 mg/kg BW. There were the signs of maternal toxicity in rats at high dose-level without any fetal malformations. Rabbits exhibited similar sensitivity. No teratogenic response was shown.[3] However, the significance of these studies is questioned by the IARC work group.

Genotoxicity. CP was administered orally by gavage to Fisher 344 rats for 5 consecutive days at doses of 500 to 5000 mg/kg BW. The exposure did not increase the frequency of chromosomal or chromatide aberrations in bone marrow cells.[3] CP C_{20-30}, 42% chlorination, is not found to be mutagenic in several strains of *Salmonella,* with and without metabolic activation.[1]

Carcinogenicity. B6C3F$_1$ mice were dosed with 2500 and 5000 mg CP C_{23}, 43% chlorination per kilogram BW, for 103 weeks. There was an increase in the incidence of malignant lymphomas in males. Fisher 344 rats received 1875 and 3750 mg/kg (males) and 100, 300, and 900 mg/kg BW (females) for 103 weeks. The treatment caused phaeochromocytomas of the adrenal medulla in females.[4] *Carcinogenicity classification.* C_{23}, 43% chlorination—NTP: NE—EE—CE—EE.

Chemobiokinetics. A higher level of radioactivity was detected in the ovary than in the blood or adipose tissue during the first 7 d after administration of the labeled material.[3]

References:

1. See **CHLORINATED PARAFFINS** C_{10-13}, **58% chlorination,** #1.
2. See **CHLORINATED PARAFFINS** C_{10-13}, **58% chlorination,** #2.
3. See **CHLORINATED PARAFFINS** C_{10-13}, **58% chlorination,** #3.
4. See **CHLORINATED PARAFFINS** C_{10-13}, **58% chlorination,** #5.
5. Bucher, J. R., Montgomery, C. A., Thompson, R., et al., Hepatic lesion associated with administration of chlorowax 40 to F344/N rats, *Toxicologist,* 5, 158, 1984.

CHLORINATED PARAFFINS C_{22-26}, 70% CHLORINATION

Acute Toxicity. LD$_{50}$ exceeds 4 g/kg BW.[1] Rats tolerate 50 g/kg BW (CP C_{24}, 70% chlorination).[2]

Repeated Exposure. Fisher 344 rats received CP for 14 d by dietary administration of 150 to 15,000 ppm. No treatment-related effects were found in organ weights or in tissue histology.[3]

Short-term Toxicity. Dietary administration of 100 to 3750 mg/kg BW for 90 d caused a slight decrease in BW gain at the highest dose-level. Increased ALT and AST activity and liver weights in the high-dose group were noted. Hepatocellular hypertrophy and cytoplasmic fat vacuolation were found to develop.[3]

Reproductive Toxicity. Rats received doses of 500 to 5000 mg/kg BW, and rabbits were administered 100 to 1000 mg/kg BW. Neither rats nor rabbits exhibited maternal toxicity or developmental abnormalities.[3] However, the significance of these studies is questioned by the IARC Workgroup.

Genotoxicity. CP was administered orally by gavage to Fisher 344 rats for 5 consecutive days at doses of 500 to 5000 mg/kg BW. The exposure did not increase the frequency of chromosomal or chromatide aberrations in bone marrow cells. CP C_{23}, 43% chlorination,[4] and CP C_{10-20}, 70% chlorination,[5] are not found to be mutagenic in several strains of *Salmonella,* with and without metabolic activation.

Chemobiokinetics. A small part of radiolabeled material was absorbed after oral administration. The highest content of radioactivity was noted in the liver. Radioactivity did not appear to concentrate in the ovary, unlike other chlorinated paraffins. The NOEL of 900 mg/kg was identified in this study.[3]

References:

1. See **CHLORINATED PARAFFINS** C_{10-13}, **58% chlorination,** #1.
2. See **CHLORINATED PARAFFINS** C_{10-13}, **58% chlorination,** #2.
3. See **CHLORINATED PARAFFINS** C_{10-13}, **58% chlorination,** #3.
4. See **CHLORINATED PARAFFINS** C_{10-13}, **58% chlorination,** #4.

5. Meijer, J., Rundgren M., Astrom, A., et al., Effects of chloroparaffins on some drug-metabolizing enzymes in the rat liver and in the Ames test, *Adv. Exp. Med. Biol.* 136, 821, 1981.

CRESYL DIPHENYL PHOSPHATE (CAS No 26444–49–5)

Synonym. Phosphoric acid, cresyl ester.
Properties. Odorless liquid.
Applications. Used as a plasticizer in the production of PVC and cellulose acetate and acetobutyrate.
Acute Toxicity. Administration of 4 g/kg BW caused no mortality in rats. In 2 to 3 d after administration, the animals became lethargic with profuse diarrhea. Death was preceded by signs of paralysis. Gross pathological examination revealed capillary atonia with edema and brain hemorrhages. Toxicity depends on the presence of *o*-cresol. According to other data, LD_{50} is 6.4 to 12.8 g/kg BW in rats and mice. These doses do not cause paralysis.[05]
Reference:

Mallette, F. S., Studies on toxicity and skin effects of compounds used in rubber and plastics industries: accelerators, activators, and antioxidants, *AMA Arch. Ind. Hyg. Occup. Med.*, 5, 311, 1952.

DIALKYL ADIPATE (DAA-C_{789})

Synonym. Adipic acid, dialkyl esters, a mixture.
Composition. A mixture of dialkyl esters of adipic acid.
Properties. Oily liquid with a yellowish color.
Applications. DAA is used for plasticizing PVC and its copolymers, rendering them greater elasticity and lower temperature resistance than DEHP. One of the principal plasticizers of PVC.
Acute Toxicity. In rats, LD_{50} (DAA-C_{8-10}) is ~40 g/kg BW; in mice, it is 8 to 12 g/kg BW. Poisoned animals are depressed, with unkempt fur.[1,2]
Repeated Exposure revealed moderate cumulative properties. K_{acc} is 8.33 (by Lim).
Short-term Toxicity. Rats were exposed to 0.125 to 1% DAA in their feed for 98 d. Treatment at higher dietary levels produced a reduction in food consumption and blood Hb level.[2]
Standards. Russia. Recommended PML: **n/m.**
Regulations. USFDA (1993) approved the use of **DAA-C_{789}**, in which the C_{789}-alkyl groups are derived from linear α-olefins by the oxoprocess. It is listed in CFR for exclusive use as a component of resinous and polymeric coatings, (1) at levels not to exceed 24% by weight of the permitted vinyl chloride homo- and/or copolymers used in contact with nonfatty foods (thickness <0.005 in.), (2) at levels not exceeding 24% by weight of the permitted PVC homo- and/or copolymers used in contact with fatty foods having a fat and oil content not exceeding a total of 40% by weight (thickness of polymers <0.005 in.), (3) at levels not exceeding 35% by weight of the permitted PVC homo- and/or copolymers used in contact with nonfatty foods (thickness of polymers <0.002 in.), (4) at levels not exceeding 35% by weight of the permitted PVC homo- and/or copolymers used in contact with fatty foods having a fat and oil content not exceeding a total of 40% by weight (thickness of polymers <0.002 in.); FDA also approved the use of DAA as (5) an ingredient in paper and paperboard intended for use in contact with dry food and (6) components of the uncoated or coated food-contact surface of paper and paperboard intended for use in producing, manufacturing, packing, transporting, or holding aqueous and fatty food.

DAA-C_{8-10}, made from C_{6-8-10} or C_{8-10} synthetic fatty alcohols, are listed in CFR to be used (7) at levels not exceeding 24% by weight of the permitted PVC homo- and/or copolymers used in contact with nonfatty foods (thickness <0.005 in.), (8) at levels not exceeding 24% by weight of the permitted PVC homo- and/or copolymers used in contact with fatty foods having a fat and oil content not exceeding a total of 40% by weight (thickness of polymers <0.005 in.), (9) at levels not exceeding 35% by weight of the permitted PVC homo- and/or copolymers used in contact with nonfatty foods (thickness of polymers <0.002 in.), and (10) at levels not exceeding 35% by weight of the permitted PVC homo- and/or copolymers used in contact with fatty foods having a fat and oil content not exceeding a total of 40% by weight (thickness of polymers <0.002 in.).

References:

1. Mel'nikova, N. N., Toxicologic characteristics of plasticizer dialkyladipate-810, *Gig. Truda Prof. Zabol.*, 12, 57, 1984 (in Russian).
2. Gaunt, I. F., Grasso, P., Landsdown, A. B., et al., Acute (rat and mouse) and short-term (rat) toxicity studies on dialkyl (C_{789}) adipate, *Food Cosmet. Toxicol.*, 7, 35, 1969.

DIALKYL(C_{789}) PHTHALATE (DAP-C_{789}) (CAS No 83968–18–7)

Synonym. Phthalic acid and alcohols, dialkyl esters.

Properties. Clear oily liquid with a specific weak odor. Poorly soluble in water. Soluble in organic solvents. Odor perception threshold is 0.46 g/l; taste perception threshold is 0.3 g/l. Foam-forming ability threshold is 20 mg/l.[1]

DAP-C_{8-10} does not color water but imparts it a slight aromatic odor and a sweetish, astringent taste.

Applications. DAP-C_{789} is used for plasticizing PVC and its copolymers, being similar to DEHP in plasticizing properties. It is widely used for making film materials and artificial leather.

Acute Toxicity. In rats and mice, LD_{50} appeared to be more than 20 g/kg BW. In 24 h after poisoning, the animals experienced depression, labored breathing, and adynamia, motor coordination disorder, paresis of the extremities, and exhaustion. However, according to other data, a single dose of 20 g/kg BW produced no death or clear signs of poisoning in rats and mice.[2,3] LD_{50} of **DAP-C_{8-10}** is reported to be higher than 30 g/kg BW in male and more than 18 g/kg BW in female rats, 20.5 g/kg BW in mice, and 17 g/kg BW in guinea pigs.[1]

Repeated Exposure revealed moderate cumulative properties. K_{acc} is 5 (by Lim).[3] Rats tolerate 3 g/kg BW doses for 10 d,[2] but 2.5 g/kg BW administered to Wistar rats for 7 to 21 d[4] increased the liver size and changed its enzyme activity. DAP-C_{789} is found to enhance xenobiotic metabolism in females but to depress it in males. Degenerative changes are noted in the liver cells of males but not of females. Male rats were given *i/g* oily solutions of **DAP-C_{8-10}** for 30 d. There was a polymorphism of the toxic action with predilection to affect the liver function (hydrocarbon, protein, and lipid metabolism), myocardium, CNS, and kidneys (at the dose of 1000 mg/kg, but not 100 mg/kg BW). Decreased activity of several enzymes is reported.[1]

Short-term Toxicity. Rats tolerated administration of 10 to 500 mg/kg BW doses for 4 months.[2] Addition of 1% DAP-C_{789} to the diet for 90 d caused retardation of BW gain and reduction in blood Hb level, erythrocyte count, and hematocrit.[5] Addition of 0.5 and 1% DAP-C_{789} to the diet of rats resulted in increased absolute and relative weights of the liver and kidneys. The relative weights of the brain, spleen, heart, and other organs were unchanged. Histological examination revealed the amount of hemosiderin in the spleen to be increased at the maximum dose.[3]

Allergenic Effect is not observed on skin application (Braun et al., 1970).

Reproductive Toxicity. *Gonadotoxicity.* Wistar rats were exposed *s/c* to 4 mg DAP-C_{789}/kg BW. The exposure caused sperm damage and reduced sperm cell motility. The relative weights of the testes were unchanged.[6] According to other data, following oral administration of DAP-C_{789}, the testes were reduced in size and weight, and there was atrophy of the seminiferous tubules. ***Embryotoxic effect*** was not found when 1000 mg/kg dose of **DAP-C_{8-10}** was given to rats.[1]

Standards. Russia (1988). MAC: 2.0 mg/l. For **DAP-C_{8-10}** recommended value: 0.3 mg/l (1990). Recommended PML: 12.0 mg/l.[5]

Regulations. USFDA (1993) regulates DAP-C_{789} as a component of adhesives to be safely used in a food-contact surface. It may be used only in polymeric substances at levels not to exceed 24 to 35% by weight of the permitted vinyl chloride homo- and/or copolymers used in contact with non-fatty foods depending on average thickness of such polymers.

References:

1. Zaitsev, N. A., Korolev, A. A., Baranov, Yu. B., et al., Hygienic regulation of diethyl phthalate, di-*n*-hexylphthalate and dialkylphthalate-810 in water medium, *Gig. Sanit.*, 9, 26, 1990 (in Russian).
2. *Pharmacol. and Toxicol.*, Republ. Issue, Kiev, 10, 151, 1980 (in Russian).
3. Timofiyevskaya, L. A., Toxicity of dialkylphthalate in mixture of alcohols C_7–C_8–C_9 (DAP-789), *Gig. Sanit.*, 10, 89, 1982 (in Russian).

4. Mangham, B. A., Foster, J. R., and Lake, B. G., Comparison of the hepatic testicular effects of orally administered di(2- ethylhexyl) phthalate and dialkyl C_{7-9} phthalate in the rat, *Toxicol. Appl. Pharmacol.*, 61, 205, 1981.
5. Gaunt, I. F., Colley, J., Grasso, P., et al., Acute (rat and mouse) and short-term (rat) toxicity studies on dialkyl C_{7-9} phthalate (a mixture of phthalate esters of alcohols having 7–9 carbon atoms), *Food Cosmet. Toxicol.*, 6, 609, 1968.
6. Bainova, M. et al., Toxicology effects of dialkyl phthalates, *Rep. Bulg. Acad. Sci.*, 35, 121, 1982 (in Bulgarian).

DIALKYL(C_{10}-C_{13}) PHTHALATE

Synonym. *o*-Phthalic acid ester mixed with C_{10}-C_{13} alcohols.
Properties. Clear, light-yellow liquid.
Acute Toxicity. Rats and mice tolerate the maximum dose of 50 g/kg BW (divided administrations). Poisoned animals experienced depression, adynamia, and after a week, decreased BW and skin turgor.
Repeated Exposure failed to reveal cumulative properties: no rats died on administration of 5 g/kg by Lim method. Signs of toxicity included CNS depression. Animals recovered on day 4 or 5, despite continuing administration.[1]
Reference:
Timofiyevskaya, L. A. and Kuz'mina, A. N., Mechanism of neuroparalytic effect of phthalic ethers, *Gig. Truda Prof. Zabol.*, 1, 50, 1983 (in Russian).

DI(2-BUTOXYETHYL) PHTHALATE (CAS No 117–83–9)

Synonyms. Butylglycol phthalate; Butylcellosolve phthalate.
Properties. Liquid with a faint, characterictic odor. Solubility in water is 0.03% (25°C). Resistant to UV rays and hydrolysis.
Applications. Used as a plasticizer for vinyl resins and in the production of plastisols, chlorinated rubber, ethyl cellulose, polystyrene, and polymethylmethacrylate.
Acute Toxicity. LD_{50} appeared to be 6 ml/kg BW for guinea pigs and 8.38 g/kg for rats. Acute poisoning of guinea pigs (3.2 ml/kg BW) is accompanied by ataxia, lassitude, and loss of reflex activity.[05]
Repeated Exposure revealed marked cumulative properties: 50% of animals died from a total dose of 4.5 ml/kg BW.[05]
Reproductive Toxicity. *Teratogenicity* effect is observed following administration of 0.1 to 0.025 ml into the yolk sac of the chick embryo. CNS lesions appeared in the postnatal period.[1,2]
Regulations. USFDA (1993) regulates DBEP as a component of adhesives intended for use in articles coming into contact with food.
References:
1. Haberman, S., Guess, W. L., Rowan, D. F., et al., Effects of plastics and their additives on human serum proteins, antibodies and developing chick embryos, *Soc. Plastic Eng. J.*, 24, 62, 1968.
2. Bower, R. K., Haberman, S., and Mintin, P.D., Teratogenic effects in chick embryo caused by esters of phthalic acid, *J. Pharmacol. Ther.*, 171, 314, 1970.

DIBUTYL ADIPATE (DBA) (CAS No 105–99–7)

Synonyms. Butyl adipate; Adipic acid, dibutyl ester; Hexanedioic acid, dibutyl ester.
Properties. Clear, yellowish liquid with a characteristic odor. Solubility in water 0.025% (20°C).
Applications and **Exposure.** Used as a plasticizer in the production of PVC, ethyl and nitro-cellulose, polystyrene, and synthetic rubber. Estimated daily intake is 1.5 mg.[1]
Acute Toxicity. LD_{50} for rats is found to be 12.9 g/kg BW.[05] Symptoms of narcotic action are present. Animals became immobile, irresponsive to food and water, with rapid, shallow respiration,

and diarrhea. Survivors recovered on days 5 to 7 after a single exposure. Gross pathological examination revealed mild anemization of the visceral organs.[2]

Repeated Exposure. Rats were given $^1/_{10}$ LD$_{50}$ for 2 months. The treatment caused labored respiration, neutrophilia (with the leukocyte formula shifted to the left), and monocytosis, eosinophilia, and dystrophic changes in the liver.[1]

Short-term Toxicity revealed mild toxic effect in dogs given a DBA isomer, **diisobutyladipate** (DIBA), for 3 months. The NOAEL was considered to be 2.5% DIBA in the diet.[3]

Long-term Toxicity. In a 2-year rat study, the NOEL was identified to be about 0.5% DIBA by the weight in the diet.[3]

Reproductive Toxicity. DBA produced teratogenic effect in rats.[4]

Genotoxicity. *Diisononyladipate* is shown to be negative in *Salmonella* mutagenicity assay and in mouse lymphoma cells. It does not cause morphological transformations in the culture of Syrian hamster embryo cells.[5]

Standards. Russia. PML in food and water: **n/m.**

References:

1. *Industrial Hygiene and Protection of Health of Workers in the Petroleum and Petrochemical Industries,* Moscow, 1982, p. 181 (in Russian).
2. See **BUTYL BENZYL ADIPATE,** 36.
3. Weil, C. S. and McCollister, D. D., Relationship between short- and long-term feeding studies in designing an effective toxicity test, *J. Agric. Food Chem.,* 11, 486, 1963.
4. Singh, A. R., Lawrence, W. H., and Autian, J., Embryonic-fetal toxicity and teratogenic effects of adipic acid esters in rats, *J. Pharmacol. Sci.,* 62, 1596, 1973.
5. McKee, R. H., Lington, A. W. Traul, K. A., et al., An evaluation of the genotoxic potential of diisononyladipate, *Environ. Mutat.,* 6, 461, 1984.

DIBUTYLCARBITOL ADIPATE

Synonym. Diethylene glycol, dibutyl ether, adipate.

Properties. Colorless, oily liquid.

Acute Toxicity. LD$_{50}$ is 7 g/kg BW in rats and 6 g/kg BW in mice. Poisoning is accompanied by distinct manifestations of narcotic effect.

Repeated Exposure failed to reveal cumulative properties; development of habituation was noted. Poisoning was accompanied by excitation. CNS inhibition occurs in 1 to 2 h after repeated administration.

Reference:

See **DIALKYL (C_{10}_C_{13}) PHTHALATE.**

DIBUTYLCARBITOL FORMAL (CAS No 143–29–3)

Synonym. Bis[2-(2-butoxyethoxy)ethoxy]methane.

Properties. Yellowish liquid.

Acute Toxicity. In rats, LD$_{50}$ is 1.75 g/kg BW; in mice, it is 2.7 g/kg BW. Interspecies susceptibility is not evident.

Repeated Exposure failed to reveal cumulative properties. K$_{acc}$ >10 (by Lim). Habituation develops.

Reference:

Timofiyevskaya, L. A., Toxicity and hazard of some plasticizers, *Gig. Sanit.,* 5, 87, 1981 (in Russian).

DIBUTYL DIPHENATE

Synonym. 2,2′-Biphenyldicarboxylic acid, dibutyl ester.

Properties. Oily liquid. Poorly soluble in water; mixes with oil at all ratios.

Acute Toxicity. LD_{50} is 17 g/kg BW in rats and 38 g/kg BW in mice. Toxicity of DBDP increases when it is given in oil solutions: in this case, LD_{50} is 5.4 g/kg BW for rats and 17.5 g/kg BW for mice. Gross pathological examination revealed pulmonary congestion, hemorrhages, and edema, as well as congestion in other visceral organs.[1,2]

Repeated Exposure. Rats and mice were dosed by gavage with $1/10$ LD_{50} of the pure chemical and its oil solution every other day for 1 month. The treatment caused retardation of BW gain and increased liver relative weights.

Short-term Toxicity. Similar changes were noted in rats and mice exposed to $1/20$ LD_{50} for 135 d. Histological examination failed to find any lesion in the visceral organs.[2]

Standards. Russia. PML recommended: **n/m.**

References:
1. *Current Problems of Environmental Hygiene,* Erisman Research Hygiene Institute, Moscow, 1976, p. 53 (in Russian).
2. *Toxicology and Hygiene of High Molecular Weight Compounds and of the Chemical Raw Material Used in Their Synthesis,* Proc. 3rd All-Union Conf., S. L. Danishevsky, Ed., Khimiya, Moscow-Leningrad, 1966, p. 89 (in Russian).

DIBUTYL MALEATE (DBM) (CAS No 105–76–0)

Synonym. Maleic acid, dibutyl ester.

Properties. Viscous liquid. Poorly soluble in water, soluble in alcohol.

Acute Toxicity. LD_{50} is 2.7 g/kg BW for rats and 4.4 g/kg BW for mice.[1] According to other data, these values are 4.4 and 9.3 g/kg BW, respectively.[2] Gross pathological examination showed severe distension of the stomach and intestine, with congestion in their serous membrane and mesentery. Histological examination revealed congestion in all the visceral organs, with pulmonary hemorrhages in some animals.

Repeated Exposure revealed no marked cumulative properties. Rats and mice were dosed by gavage with $1/5$ and $1/10$ LD_{50}. The treatment caused exhaustion, increase in the relative weights of the kidneys, as well as death of some animals. Gross pathological examination failed to reveal changes in the viscera.[1]

Short-term Toxicity. Rats and mice were orally exposed to $1/100$ and $1/50$ LD_{50} (in an oil solution). Administration caused no changes in enzyme and conditioned reflex activity, in BW gain, hematological indices, and urinalysis. Histological examination failed to reveal changes in the viscera.[1]

Standards. Russia. Recommended PML: **n/m.**

Regulations. USFDA (1993) approved the use of DBM (1) in adhesives as a component of articles intended for use in packaging, transporting, or holding food and (2) as a monomer in the manufacture of the uncoated or coated food-contact surface of paper and paperboard that may be safely used in producing, manufacturing, packing, transporting, or holding dry food.

References:
1. See **BISPHENOL A,** #2, 226.
2. See **DIBUTYL DIPHENATE,** #1.

DIBUTYL PHTHALATE (DBP) (CAS No 84–74–2)

Synonyms. Phthalic acid, dibutyl ester; *n*-Butyl phthalate; **1,2**-Benzenedicarboxylic acid, dibutyl ester.

Properties. Colorless, odorless, oily liquid. Solubility in water is 0.1% (20°C). Aromatic odor and bitter taste perception threshold is 5 mg/l; practical threshold is 10 mg/l.

Applications and Exposure. Used as a plasticizer in the production of PVC, polymethylmetacrylate, polyvinylacetate, and cellulose esters. It makes the materials light-stable. A solvent for chlorinated rubbers. Migration from PVC (DBP content 0.04%, exposure 8 d) into potato snacks was reported to be 3.5 mg/kg, into coated candy (0.5% DBP, 5 d) 1.23 mg/kg, and in covered chocolate (0.21% DBP, 5 d) 0.9 mg/kg.[1]

Acute Toxicity. LD_{50} is ~10 g/kg BW in rats, 5 to 20 g/kg BW in mice, and 8 to 10 g/kg BW in chickens.[2,03] Guinea pigs and rabbits are found to be less sensitive. The acute effect threshold appeared to be 26 mg/kg BW for the effect on behavior.[2]

Repeated Exposure revealed moderate cumulative properties. K_{acc} is 7 (by Lim). Following oral exposure to the dose of 3 g/kg BW, animals died in 3 to 14 d. Histological examination revealed stomach paresis and intestinal and mesenteric lesions.[3] Oral administration of ^{14}C-labeled DBP failed to demonstrate accumulation of radioactivity in the gonads.[4]

Short-term Toxicity. Wistar rats were gavaged with 120 mg/kg BW for 3 months. This treatment caused lesions in the GI tract mucosa, development of pneumonia and endometritis, and increase in the liver weight. A dose of 1200 mg/kg BW caused death of 5% of animals.[5]

Long-term Toxicity. No signs of intoxication are observed when 0.05% DBP is added to the diet of rats for a year. Addition of 0.125% DBP to the feed caused 15% mortality in rats.[4] Retardation of BW gain, leukocytosis, and NS affection were observed in rats given 20 mg/kg DBP.[3]

Immunotoxicity effects are revealed at the dose-level of 200 mg/kg BW. Tissue antibodies determine the cytotoxic effect on the fetus and progeny.[6]

Reproductive Toxicity. *Embryotoxicity.* DBP is capable of passing through the placenta; it accumulates in the brain and subcutaneous tissues. It is distributed in the fetal blood and in the amniotic fluid.[7] Exposure of Wistar rats to 120 and 600 mg/kg BW in the diet for 3 months had no effect on the number of litters. Administration of these doses during pregnancy resulted in decreased litter size and increased neonatal mortality.[5,8] Administration of 250 mg/kg BW dose throughout the gestation period produced no selective fetotoxicity.[9] However, Zinchenko[6] reported an embryotoxic effect to occur when rats were given 20 and 200 mg/kg BW throughout the pregnancy. CD-1 mice of both sexes were dosed with 0.03 to 1% DBP in their diet for 7 d prior to and during a 98-d cohabitation period. Reductions in the number of litters per pair and of live pups per litter and in the proportion of pups born alive were found at the 1% dose but not at a lower dose-level. A crossover mating trial demonstrated that female mice but not males were affected by DBP.[10] Though high doses of DBP are reported to be embryotoxic, the NOEL in mice is more than 2000 times the estimated level of human intake via the food chain.[10] *Teratogenicity.* Shiota et al. have found the development of neural cord abnormalities in the fetuses of mice given 400 mg/kg BW and above during pregnancy.[10] Neural tube defects, namely exencephaly and spina bifida, were found in this study. The NOEL for the embryotoxic effect appeared to be 70 mg/kg or 1% DBP in the feed; the NOEL for the teratogenic effect appeared to be 0.2% in the feed. *Gonadotoxicity.* DPB is known to be a testicular toxicant.[11] Oral administration of 0.5 and 1 g/kg BW significantly reduced relative testes weights within 6 and 4 d, respectively. **Monobutyl phthalate,** the major urinary metabolite, causes even more marked reduction in the testes weights. The testicular lesion produced by DBP in immature rats is characterized by early sloughing of spermatids and spermatocytes and severe vacuolation of Sertoli cell cytoplasm.[12] Urinary zinc levels following DBP treatment are increased.[13,14]

Genotoxicity. No mutagenic effect is noted in *Dr. melanogaster*.[13] DBP is shown to cause mutations in mouse lymphoma cells (NTP-88).

Carcinogenicity. In the long-term studies,[04] oral doses of 100, 300, and 500 mg DBP/kg BW administered for 15 to 21 months revealed no signs of tumor growth over six generations in rats. It was noted that the doses tested correspond to 60, 180, and 300 mg/kg BW in man. *Carcinogenicity classification.* USEPA: Group D.

Chemobiokinetics. DBP metabolism involves hydrolysis to the monoester, with subsequent oxidation of the remaining alkyl chain. Main metabolites: **phthalic acid, monobutyl phthalate, mono(3-hydroxybutyl)phthalate, mono(4-hydroxybutyl)phthalate.** Phthalic acid was found to be a minor metabolite (less than 5% to the total urinary metabolites). No accumulation of DBP or its metabolites was found in rats given feed containing 1 mg DBP/kg BW for 3 months.[14] According to Jaeger and Rubin, DBP is unlikely to be fully hydrolyzed in the body. Possibly, it may accumulate, mainly in the blood and lungs.[15] Excretion of ^{14}C-labeled DBP occurs mainly through the urine.

Standards. Russia (1988). PML in food: 0.25 mg/l, PML in drinking water: 0.1 mg/l.

Regulations. USFDA (1993) approved the use of DBP (1) in adhesives used as components of articles intended for use in packaging, transporting, or holding food, (2) in cross-linked polyester resins to be safely used as articles or components of articles intended for repeated use in contact with food, (3) as a plasticizer for rubber articles intended for repeated use in contact with food (content up to 30% by weight of the rubber product), (4) as an ingredient of cellophane for food packaging alone or in combination with other phthalates where total phthalates do not exceed 5%, (5) as a component of the uncoated or coated food-contact surface of paper and paperboard intended for use in producing, manufacturing, packing, transporting, or holding aqueous and fatty food, (6) in resinous and polymeric coatings in a food-contact surface of articles intended for use in producing, manufacturing, packing, transporting, or holding food only for containers having a capacity of 1000 gal or more when such containers are intended for repeated use in contact with alcoholic beverages containing up to 8% alcohol, and (7) in slimicide in the manufacture of paper and paperboard that may be safely used in contact with food.

References:

1. See **ACETYL TRIBUTYL CITRATE,** #2.
2. Balynina, E. S. and Timofiyevskaya, L. A., On the problem of application of behavioural reactions in the toxicity studies, *Gig. Sanit.,* 7, 54, 1978 (in Russian).
3. See **BUTYL STEARATE,** #3.
4. Gangolli, S. D., Testicular effects of phthalate esters, *Environ. Health Perspect.,* 45, 77, 1982.
5. Nikanorov, M., Mazur, H., and Piekacz, H., Effect of orally administered plasticizers and polyvinyl chloride stabilizers in the rat, *Toxicol. Appl. Pharmacol.,* 25, 253, 1973.
6. Zinchenko, T. M., Studies of autoallergenic effect of dibutylphthalate and di(2-ethylhexyl)-phthalate, *Gig. Sanit.,* 2, 80, 1986 (in Russian).
7. *Effect of Occupational Factors on the Specific Functions of Female Body,* Sverdlovsk, 1978, p. 101 (in Russian).
8. Peters, J. W. and Cook, R. M., Effect of phthalate esters on reproduction in rats, *Environ. Health Perspect.,* 3, 91, 1973.
9. See **ETHYLENEIMINE,** #6, 39.
10. Shiota, K., Chou, M. I., and Nishimura, H., Embryotoxic effects of di(2-ethylhexyl)phthalate (DEHP) and di-*n*-butylphthalate (DBP) in mice, *Environ. Res.,* 54, 342, 1980.
11. Lamb, J. C., Chapin, R. E., Teague, J., et al., Reproductive effects of four phthalic acid esters in the mouse, *Toxicol. Appl. Pharmacol.,* 88, 255, 1987.
12. Cater, B. R., Cook, M. W., Gangolli, S. D., et al., Studies on dibutyl phthalate-induced testicular atrophy in the rat: effect on zinc metabolism, *Toxicol. Appl. Pharmacol.,* 41, 609, 1977.
13. Sheftel, V. O., Zinchenko, T. M., and Katayeva, S. Ye., Sanitary toxicology of phthalates as water pollutants, *Gig. Sanit.,* 8, 64, 1981 (in Russian).
14. See **ETHYLENEIMINE,** #6, 43.
15. Jaeger, R. J. and Rubin, R. J., Plasticizers from plastic devices: extraction, metabolism, and accumulation by biological systems, *Science,* 170, 460, 1970.

DIBUTYL SEBACATE (DBS) (CAS No 109–43–3)

Synonym. Sebacic acid, dibutyl ester.

Properties. Pale-yellow, oily liquid with the odor of ether. Solubility in water is 0.005% at 25°C, soluble in alcohol.

Applications. Used as a plasticizer in the production of PVC and its copolymers, acetobutyrate cellulose, polyvinyl butyral, chlorinated rubber, polyvinylidene chloride. Compatible with nitro- and ethylcellulose, polystyrene, and synthetic rubber. Renders the materials high flexibility and low temperature resistance.

Acute Toxicity. LD_{50} is reported to be 16 to 32 g/kg BW in rats and 18 to 25.5 g/kg BW in mice.[1,2,04] General condition and BW gain in the survivors did not differ from that in the controls. Histological examination revealed pulmonary hemorrhage, edema, and congestion in the myocardium, brain, intestinal wall, and stomach.

Repeated Exposure. There were no signs of toxicity in mice dosed with 0.9 to 3.6 g/kg BW and rats dosed with 0.9 to 34 g/kg BW for 1.5 months.[2]

Short-term Toxicity. Decreased BW gain, reticulocytosis, decline in the erythrocyte count, and blood Hb level were observed in rats given 600 mg/kg BW for 13 weeks.[3]

Long-term Toxicity. Exposure to $1/50$ LD_{50} for 9 months caused no changes in rats[2] while 0.01, 1.25, and 6.25% DBS in the diet caused no signs of intoxication to develop in 2-year studies in rats.[04]

Reproductive Toxicity. Decreased BW gain in the second generation of rats was found with the maximum dose in the diet.[4]

Genotoxicity. Cytogenic effect is not reported in the somatic (brain) cells of mice given doses of 6 to 24 g/kg BW.[3]

Chemobiokinetics. DBS is rapidly hydrolyzed by lipase *in vitro*. It is suggested to be hydrolyzed in the body in the same way as fats.[05]

Standards. Russia. PML in food: 4 mg/l, PML in drinking water: **n/m.**

Regulations. USFDA (1993) regulates the use of DBS (1) as a component of adhesives for articles intended for contact with food, (2) as a plasticizer for rubber articles intended for repeated use in contact with food (content up to 30% by weight of the rubber product), (3) in resinous and polymeric coatings in a food-contact surface of articles intended for use in producing, manufacturing, packing, transporting, or holding food, (4) as a component of resinous and polymeric coatings for polyolefin films for food-contact surface, and (5) as a component of the uncoated or coated food-contact surface of paper and paperboard intended for use in producing, manufacturing, packing, transporting, or holding aqueous and fatty food.

References:
1. See **BUTYL STEARATE, #2.**
2. See **DIBUTYL PHTHALATE, #3.**
3. *Scientific and Technical Progress at Medical Prophylaxis,* Part 1, Moscow, 1971, p. 60 (in Russian).

DICYCLOHEXYL PHTHALATE (DCHP) (CAS No 84–61–7)

Synonym. Phthalic acid, dicyclohexyl ester.

Properties. White, crystalline powder with a characteristic faint odor. Poorly soluble in water, soluble in alcohol and fats.

Applications. Used as a plasticizer in the production of nitro- and benzylcellulose; combines with PVC, polyvinylbutyral, polystyrene, and acrylic plastics. Migration from PVC (DCHP content 0.33%, exposure 5 d) into potato snacks was reported to be 6.2 mg/kg, into coated candy (0.50% DBP, exposure 5 d) 0.1 mg/kg, and in covered chocolate (0.21% DBP, exposure 5 d) 0.38 mg/kg.[1]

Acute Toxicity. LD_{50} is not attained with a single administration. After seven administrations, LD_{50} is likely to be about 30 ml/kg BW.[2] LD_{33} is less than 60 g/kg BW. Poisoning is accompanied by adynamia, general inhibition, and lack of response to external stimuli.[3]

Repeated Exposure failed to reveal cumulative properties. K_{acc} is >10 (by Lim). Habituation develops.[3] Rats given 0.5 to 2.5 g/kg BW for 7 d exhibited liver enlargement, increase in 7-ethoxy-cumarin-*o*-diethylase activity and in the content of P-450 cytochrome in the hepatic microsomes in young animals.[3]

Long-term Toxicity. Rats were exposed to oral doses of 0.5 and 1 mg DCHP/kg BW in a 20%-solution in vegetable oil for 1 year. There were no changes in biochemical and hematological analyses. Gross pathological examination failed to reveal changes in the viscera or the progeny.

Reproductive Toxicity. In a long-term study, the doses of 5, 10, and 100 mg DCHP/kg BW given in the feed for 18 months produced no effect on reproduction or any carcinogenic activity in rats.[04,05]

Chemobiokinetics. DCHP seems to be a weak, medicinal type inducer of metabolism in the rat liver.[3]

Regulations. USFDA (1993) approved the use of DCHP (1) as an ingredient of cellophane for food packaging, alone or in combination with other phthalates where total phthalates do not exceed 5%, (2) as a component of the uncoated or coated food-contact surface of paper and paperboard, (3) as a component of adhesives for articles intended for use in contact with food, for use only alone or in combination with other phthalates, (4) as a plasticizer in polymeric substances used in the manufacture of food-contact articles, and (5) in plastic film or sheet prepared from polyvinylacetate, polyvinyl chloride, and/or vinyl chloride copolymers at temperatures not to exceed room temperature, containing no more than 10% by weight of the total phthalates, calculated as phthalic acid.

Standards. Russia. Recommended PML in drinking water: **n/m.**

References:
1. See **ACETYL TRIBUTYL CITRATE**, #2.
2. See **DIALKYL(*C₇₈₉*) PHTHALATE**, #3.
3. Lake, B. G., Foster, J. R., Collins, M. A., et al., Studies on the effects of orally administered dicyclohexyl phthalate in the rat, *Acta Pharmacol. Toxicol.*, 51, 217, 1982.

DIDECYL ADIPATE (mixed isomers)

Synonym. Adipic acid, didecyl ester.

Properties. Colorless or light-amber colored liquid with a weak characteristic odor. Solubility in water is ~0.01% at 20°C. Slightly leached by water and detergents.

Applications. Used as a plasticizer in the production of PVC, nitro- and ethylcellulose, polystyrene, and chlorinated rubber.

Acute Toxicity. LD_{50} appeared to be 20.5 ml/kg BW in mice, and 21 g/kg BW in rats.[04]

Repeated Exposure. No changes were found in rats dosed with 0.5% DDA in the diet for 30 d.

Regulations. USFDA (1993) approved the use of DDA as a plasticizer for rubber articles intended for repeated use in contact with food (content up to 30% by weight of the rubber product).

DIDECYL GLUTARATE (CAS No 3634–94–4)

Synonyms. Glutaric acid, didecyl ester; Pentanedioic acid, didecyl ester.

Properties. Light-yellow solid.

Applications. Used as a plasticizer in the production of film materials and synthetic leathers.

Acute Toxicity. LD_{50} was not attained in rats and mice. LD_{16} appeared to be 60 g/kg BW being identified by administration of divided doses to mice. Poisoning produced decreased BW gain, adynamia and unkemptness.

Repeated Exposure revealed moderate cumulative properties. K_{acc} is 4.2 (by Lim). The treatment caused depression, adynamia, unkemptness and reduced skin turgor in rats given the dose of $^{1}/_{10}$ LD_{16}.

Reference:
See **DICYCLOHEXYL PHTHALATE**, #2.

DIDECYL PHTHALATE (CAS No 84–77–5)

Synonym. Phthalic acid, didecyl ester.

Properties. Colorless, odorless, viscous liquid. Poorly soluble in water.

Applications. Used as a plasticizer in the production of PVC; used in plastisol manufacture.

Acute Toxicity. A dose of 63 ml/kg BW caused no mortality in rats.[04]

Regulations. USFDA (1993) regulates the use of DDP as (1) a plasticizer in rubber articles intended for repeated use in producing, manufacturing, packing, processing, treating, packaging, transporting, or holding food (total not to exceed 30% by weight of the rubber products), and (2) an adjuvant in the preparation of slimicides in the manufacture of paper and paperboard that may be safely used in contact with food.

DIDODECYL PHTHALATE (CAS No 2432–90–8)

Synonyms. Dilauryl phthalate; Phthalic acid, didodecyl ester.

Properties. Colorless, odorless, oily liquid. Poorly soluble in water, soluble in ethanol.

Applications. Used mainly as a PVC plasticizer, rendering materials highly elastic. Exhibits low volatility.

Acute Toxicity. LD_{50} was not attained even after administration of 15 g/kg BW to mice and rats. The calculated LD_{50} is 29.8 g/kg BW for rats and above 7.9 g/kg BW for mice.

Repeated Exposure failed to reveal cumulative properties when 2 and 5 g/kg BW were administered to rats for 2 months.

Standards. Russia. PML in food: 2 mg/l.

Regulations. USFDA (1993) regulates the use of DDP as an adjuvant in the preparation of slimicides in the manufacture of paper and paperboard that may be safely used in contact with food.

Reference:

Antonyuk, O. K., On the toxicity of didodecyl phthalate, *Gig. Truda Prof. Zabol.*, 11, 51, 1973 (in Russian).

DIETHYLENE GLYCOL (DEG) (CAS No 111–46–6)

Synonyms. Carbintol; DiEG; Diglycol; Digol; **2,2**-Dihydroxyethyl ester; **2,2**-Hydroxydiethanol.

Properties. Thick, colorless liquid with a faint odor. Mixes with water and alcohol at all ratios. Odor perception threshold is 3.28 g/l; taste perception threshold is 2.05 g/l.[1] According to other data, organoleptic threshold is 240 mg/l.[01]

Applications and **Exposure.** DEG is used as a plasticizer in the production of materials based on regenerated cellulose; it is a solvent or a component of synthetic coatings, lacquers, and cosmetics. In 1985, DEG was reported to be found in certain German and Austrian wines and juices at concentrations of 0.01 to 48 g/l, but no cases of intoxication in man were reported.[8]

Acute Toxicity. DEG is likely to be less toxic than **ethylene glycol** (see above). *Humans.* Man is more sensitive to DEG than rodents. Poisonings are known as a result of the occasional consumption of products containing DEG. Lethal dose is reported to be about 1 ml/kg BW. A content of 0.5 to 1 mg/kg is considered to be permissible in wine (Altman, 1986). *Animals.* LD_{50} values are reported to be 13.3 to 23.7 g/kg BW in mice, 12.5 to 32 g/kg BW in rats, 2.69 to 4.4 g/kg BW in rabbits, 7.8 to 14 g/kg BW in guinea pigs, 3.3 g/kg BW in cats, and 9 g/kg BW in dogs.[1,06] Poisoning is accompanied by a short period of excitation and aggression with subsequent inhibition, disturbance of motor coordination, refusal of food, and vomiting. An important signs of DEG acute intoxication is renal damage (increased thirst and urination in the early stage, followed by severely reduced urine production with heavy protein excretion). Shortness of breath and coma were observed prior to death in 1 to 5 d. Gross pathological examination revealed hydropic degeneration, particularly in the renal tubular epithelium and in the centrolobular portion of the liver, point hemorrhages in the stomach and intestinal walls.

Repeated Exposure. When a 3.1 g/kg BW dose is given to rats for 20 d, no cumulative effect is found.[1] Levels found in contaminated wine, usually less than 3 g/l, would seem unlikely to represent a hazard to those consuming part or even all of a bottle on a on-off basis.[2]

Short-term Toxicity. Kidneys are likely to be a target organ in the picture of DEG intoxication. DEG produced hydropic swelling and degeneration of the epithelial cells and the development of necrosis. **Calcium oxalate** crystals are formed in the bladder, leading to the onset of hematuria. A 0.5 mg/kg BW dose given to rats for 4 months had no harmful effect on the experimental animals.[3]

Long-term Toxicity. DEG is shown to produce renal damage, formation of bladder calcium oxalate stones, and liver damage in a number of species, including man. In many studies, toxic effect could be influenced by MEG (**monoethylene glycol**) contamination. Bladder stones are formed at dietary levels above 20 g/kg food.

In an unpublished BIBRA study,[2] rats were given 0.4 to 4% DEG in the diet for 99 d, or 2% in the diet for 225 d. DEG used contained less than 0.01% MEG (the lowest reported MEG contamination). At 4% concentration (3 g/kg BW in males and 3.7 g/kg BW in females), DEG caused the death of 6 out of 15 males, with the signs of kidney damage. Dietary levels of 0.4% DEG (about 300 mg/kg BW) produced **oxalate crystals** in the urine, particularly in females, and mild defect in kidney function in the males, but no histological damage. At a 100 mg/kg BW dose, the only treatment-related finding was the presence of a small amount of **oxalic acid** in the urine. The NOAEL of 100 mg/kg BW therefore seems justified (presence of an urinary metabolite was not considered by authors as a toxic finding). Bearing in mind that man might be very much more susceptibile than the rat to DEG short-term toxicity, a choice of conservative uncertainty factor (200) would lead to a TDI of 0.5 mg/kg BW.

Allergenic Effect is moderate (skin tests).[4]

Reproductive Toxicity. *Embryotoxicity* and *Teratogenicity.* Administration of 11.2 g/kg BW on days 7 to 14 of gestation to CD-1 mice caused 4% maternal mortality and reduced pup BW gain on postpartum days 1 to 3.[5] CD-1 mice were dosed orally via drinking water with a total dose of approximately 8 ml/kg (9 g/kg BW). Decreased fetal weights and some malformations were observed.[6] *Gonadotoxicity.* DEG exhibits a very pronounced gonadotoxic effect, similar to that of other glycols, with resulting reduction of sperm motility time, increased number of immobile forms, and lowered resistance to NaCl solution. DEG considerably decreases the spermatogenesis index, the number of cells with generative changes increases, and cytochrome-*c*-oxidase activity in the testicular tissues and alkaline phosphatase activity in the tissues of the epididymis are also increased (0.5 mg/kg BW).[7]

Carcinogenicity. Incidences of malignant bladder tumors in rats and mice due to DEG exposure are reported. Sanina described the development of malignant mammary gland tumors in rats on chronic exposure to DEG8 (Sanina, 1968).

Chemobiokinetics. Oral doses of 1 and 5 ml ^{14}C-DEG/kg BW given to rats were rapidly and almost completely absorbed and distributed from the blood into kidneys, brain, spleen, liver, and muscle fat. The main metabolite is 2-hydroxyethoxyacetate. DEG produces metabolic acidosis, hydropic degeneration of the tubuli, oliguria, anuria, accumulation of urea-N, and death in uraemic coma. Of ^{14}C-DEG, 73 to 96% is excreted with the urine, 0.7 to 2.2% with the feces.[8]

Standards. Russia (1988). MAC and PML: 1 mg/l. **EEC** (1990). SML: 30 mg/kg alone or with **ethylene glycol.**

Regulations. USFDA approved the use of DEG (1) as a component of adhesives intended for use in articles coming into the contact with food, (2) in resinous and polymeric coatings for a food-contact surface, (3) in cross-linked polyester resins forarticles or components of articles intended for repeated use in contact with food, and (4) as a substance employed in the production of or added to textiles and textile fibers intended for use in contact with foods.

References:
1. Plugin, V. P., Ethylene glycol and diethylene glycol as a goal of hygiene standard-setting in the sanitary protection of water bodies, *Gig. Sanit.,* 3, 16, 1968 (in Russian).
2. Hesser, L., Diethylene glycol toxicity, *Food Chem. Toxicol.,* 24, 261, 1986.
3. Sheftel, V. O., Bardik, Yu. V., and Petrusha, V. G., On effect of harmful substances administered to animals in decreasing regimen, *Gig. Sanit.,* 12, 73, 1982 (in Russian).
4. *Hygienic Aspects of the Use of Polymeric Materials,* Proc. 2nd All-Union Meeting on Health and Safety Monitoring of the Use of Polymeric Materials in Construction, Kiev, 1976, p. 102 (in Russian).
5. Schuler, R. L., Hardin, B. D., Niemeier, R. W., et al., Results of testing 15 glycol ethers in a short-term in vivo reproductive toxicity assay, *Environ. Health Perspect.,* 57, 141, 1984.
6. Williams, J., Reel, J. R., George, J. D., et al., Reproductive effects of diethyleneglycol and diethyleneglycol monoethyl ether in Swiss CD-1 mice assessed by continuous breeding protocol, *Fundam. Appl. Toxicol.,* 14, 622, 1990.
7. Byshovets, T. F., Barilyak, I. R., Korkach, V. I., et al., Gonadotoxic effect of glycols, *Gig. Sanit.,* 9, 84, 1987 (in Russian).
8. Heilmair, R., Lenk, W., and Lohr, D., Toxicokinetics of diethyleneglycol (DEG) in the rat, *Arch. Toxicol.,* 67, 655–666, 1993.

DI- and TRIETHYLENEGLYCOL, MONOETHYL ESTERS, mixture

Synonym. Plasticizer VTG.

Acute Toxicity. In mice, LD_{50} is 4.2 g/kg BW. Administration of the lethal doses caused immediate CNS inhibition with subsequent adynamia in 1 to 2 h. The furry coat became moist, and labored breathing was noted. Mice given high doses refused feed and died in 1 to 2 d. Main manifestations of poisoning included adynamia, convulsions, and respiratory arrest. Gross pathological examination revealed congestion, circulatory disturbances, and parenchymatous and fat dystrophy in the visceral organs.

Repeated Exposure revealed slight cumulative properties. K_{acc} is 4.5 (by Lim).

Reference:
See **BUTYL STEARATE,** 48.

DI(2-ETHYLHEXYL) ADIPATE (DEHA) (CAS No 103–23–1)

Synonym. Dioctyladipate; DOA; Hexanedioic acid, bis(2-ethylhexyl) ester.

Properties. Colorless or light-amber liquid with a characteristic odor. Solubility in water is 0.025% at 20°C, soluble in alcohol.

Applications and **Exposure.** DEHA combines well with PVC and its copolymers, as well as with the majority of polar thermoplastics. Gives the materials a good low-temperature resistance (down to −60°C). Food is the most important source of human exposure. The following DEHA migration levels from PVC films were reported:[1] 1 to 20 mg/kg in fresh chicken, 8 to 48 mg/kg in cooked chicken, 5 to 9 mg/kg in fresh beef, and 9 to 94 mg/kg in sandwiches. DEHA has been found occasionally in drinking water at the concentrations of 0.001 to 0.005 mg/l.

Acute Toxicity. In rats, LD_{50} is reported to be 9 to 45 g/kg BW.[2] Poisoning is accompanied by excitation with subsequent general inhibition, apathy, and motor coordination disorder. By 4 to 5 h before death, the animals assume a side position and are comatose. In the survivors, signs of intoxication have completely disappeared in 5 to 7 d. No death occurred when rats were given a 2.5 g/kg BW dose. In mice, the LD_{50} values ranged from 15 g/kg in males to 24.6 g/kg BW in females.[3] It was reported to be 8.4 g/kg BW in rabbits and 12.2 g/kg BW in guinea pigs; the latter tolerated a 4.5 g/kg BW dose, but all died at 18 g/kg BW. Death occurs after progressive paralysis.[05] Doses higher than 6 g/kg BW induced **peroxisome proliferation** (PP) in the liver of rodents.

Repeated Exposure. Repeated administration of high doses to rodents produced hepatomegaly, hepatic PP, induction of liver catalase, and enzymes involved in the oxidation of fatty acids as well as hypolipidemia.[4] Exposure to 1% DEHA in the feed for 7 weeks resulted in reversible decrease of triglyceride and cholesterol concentration in the blood plasma. Exposure of mice to 5% DEHA in the diet for a month caused retardation of BW gain. Doses of 1 and 1.8 g/kg BW increased the content of SH-groups in the blood serum.[5]

Short-term Toxicity. In a 13-week study, Fisher 344 rats and B6C3F$_1$ mice were exposed to dietary concentrations up to 25 g/kg. The treatment with a high dose caused retardation of BW gain in all the animals, while 6.30 mg DEHA per kilogram produced this effect only in male rats and female mice. No compound-related increased mortality, histopathological changes, or reduction in feed consumption was noted in this study.[3]

Long-term Toxicity. Introduction of 2% DEHA in the diet increased liver size and induced hypolipidemia and hepatic *PP* in male Fisher 344 rats.[6] Fisher 344 rats and B6C3F$_1$ mice received DEHA at dietary levels of 12 and 25 g/kg. The treatment caused no effect on longevity. Retardation of BW gain was observed in mice and in rats in the high-dose group. Histological examination revealed no changes in the viscera of rats and mice, with the exception of mice livers, where tumors were found.[3] Rats were exposed orally to DEHA for 6 months. The treatment with the highest dose caused retardation of BW gain. Administration of 100 and 200 mg/kg BW led to a rise in the content of free SH-group in the blood serum. The NOAEL was identified to be 10 mg/kg BW.[5]

Reproductive Toxicity. *Embryotoxicity.* Wistar rats were given DEHA in the diet from 10 weeks premating up to 36 d postpartum. The treatment with 12 g DEHA per kilogram diet resulted in reduction of total litter weights, BW gain of pups, and mean litter size. Histological examination failed to reveal changes in the pups.[7] ICR mice were exposed to a single *i/p* dose of 0.5 to 10 ml/kg BW that produced an antifertility and mutagenic effect, as indicated by reduced percentage of pregnancies and increased number of early fetal deaths. A reduction in the number of implantations was also noted. However, these findings were questioned,[8] since the study lacked data on the number of pregnancies per treated male and the number of corpora lutea per female. A *teratogenicity* study in the rat has demonstrated a dose-dependent increase in minor skeletal defects (slightly poorer ossification) in addition to an increased incidence of ureter abnormalities. The NOEL was identified to be 28 mg/kg BW. ADI (TDI) of 0.3 mg/kg BW was calculated for this NOEL.[8]

Genotoxicity. DEHA was tested for genotoxic activity in a large number of tests. With one exception these studies have shown that DEHA is not a genotoxic agent. Clastogenic activity was not found in the *in vitro* assay in human leukocytes. Orally administered DEHA does not bind covalently to mouse liver DNA;[9] it was negative in the mouse lymphoma assay and in the *Salmonella* test.[10] However, positive results were shown in a DLM study.[11]

Carcinogenicity. A study in mice where high doses of DEHA (up to 25,000 ppm in the diet) were fed for the lifetime, showed an increased rate of hepatic tumors, though a similar study in rats,[3] which were exposed to lower doses than mice, did not show any increase in the rate of hepatic tumors. NOEL was not defined because of very high doses applied (1.8 to 3.8 g/kg BW). DEHA is a hepatic peroxysomal proliferator in rodents, and carcinogenicity of DEHA may be explained by its activity as a peroxysomal proliferater. In contrast to reported findings 10 years ago of a carcinogenicity bioassay,[3] Keith et al. revealed recently a higher sensitivity of Fisher 344 rats than B6C3F$_1$ mice to hepatic *PP* caused by DEHA.[12] In spite of assuming that *PP* is associated with cancer development, the dose-dependency of *PP* would lead to the conclusion that the threshold exists for DEHA carcinogenicity. Such a threshold can be established at 0.01 mg/kg BW as a NOEL based on the peroxisomal parameter in the rat.[4,13,14] *Carcinogenicity classification.* IARC: Group 3; USEPA: Group C; NTP: N—N—P—P*.

Chemobiokinetics. A rapid and almost complete absorption from the GI tract is reported. The highest proportion of a single *i/g* dose was recovered in the stomach and intestine.[15] Studies indicated little if any prolonged retention of DEHA or its metabolites in the blood and tissues after oral administration.[8] DEHA is rapidly hydrolyzed. **Adipic acid** and **2-ethylhexanol** have been observed as DEHA metabolites. DEHA inhibits the synthesis of cholesterol from acetone. It has a hypocholesterolic action and affects the profile of the phospholipids synthesized in the liver. Bell[16] reported that DEHA and DEHP had a similar effect. Guest et al.[17] found that B6C3F$_1$ mice dosed orally with 50 or 500 mg/kg BW excreted 91, 7, and 1 to 2% of the radioactive dose in the urine, feces, and expired air. According to other data, DEHA is excreted in the bile and in the urine.[18]

Healthy men received 45 mg DEHA, tagged with deiterium. Main metabolites in the urine were found to be 2-ethylhexanoic acid and 2-ethyl-5-hydroxyhexanoic acid, which was detected also in the blood plasma. Metabolites were found in the urine during first day at quantity of about 13%.[19]

Guidelines. WHO (1992). Guideline value for drinking water: 0.09 mg/l.

Standards. USEPA (1991). MCL and MCLG: 0.5 mg/l. **England** (1990). Maximum intake: 8.2 mg/d.[1] **Russia** (1990). PML: **n/m.**

Regulations. USFDA (1993) regulates DEHA as a plasticizer in polymeric substances used in the manufacture of articles that are used in contact with food, and up to 50% may occur as a component of the following products when used in contact with food: (1) adhesives, (2) cellophane, (3) closures with sealing gaskets (up to 2%) for food containers, (4) water-insoluble hydroethyl cellulose film, and (5) rubber articles (up to 30%) intended for repeated use. DEHA may be used in films for contacting nonfatty foods at all temperatures and for fatty foods stored in a refrigerator. The level of DEHA allowed depends on the film thickness. In **England,** DEHA may be used at levels of up to 35% in flexible PVC for all applications. In **Germany,** DEHA may be used at levels of up to 35% in general purpose film. This film may not be used in contact with powdered and fine-grain foods, milk and milk derivatives (including cheese), fatty foods and foods containing alcohol or essential oils. DEHA may be used at levels of up to 22% in films for wrapping fresh meat, provided migration into the meat does not exceed 60 ppm. In **Belgium, The Netherlands, France,** and **Spain** DEHA may be used in all food-contact applications, provided that under normal conditions of use its concentration in the foodstuff does not exceed 60 ppm. In **Italy,** DEHA may be used in film for wrapping fruit and vegetables, either fresh or dried. DEHA may be used in sealing gaskets and closures for virtually all foodstuff except meat and milk.

References:

1. See **ACETYL TRIBUTYL CITRATE,** #2, 55.
2. Smyth, H. F., Carpenter, C. P., and Weil, C. S., Range-finding toxicity data: list IV, *AMA Arch. Ind. Hyg.,* 4, 119, 1951.
3. *Carcinogenesis Bioassay of Di(2-ethylhexyl) Adipate in F344 Rats and B6C3F$_1$ Mice (Feed Study),* NTP Technical Report Series No. 212, Research Triangle Park, NC, 1982.
4. *A 21 Day Feeding Study of Di(2-ethylhexyl) Adipate to Rats: Effect on the Liver and Liver Lipids,* BIBRA Report No. 0542/1/85, 1986.
5. *Safe Application and Toxicology of Pesticides, and Clinical Poisoning,* L. I. Medved', Ed., Kiev, 9, 1971. p. 373 (in Russian).
6. Reddy, J. K., Reddy, M. K., Usman, M. I., et al., Comparison of hepatic peroxisome proliferative effect and its implication for hepatocarcinogenicity of phthalate esters, di(2-ethylhexyl) phthalate, and di(2-ethylhexyl) adipate with a hypolipidimic drug, *Environ. Health Perspect.,* 65, 317, 1986.

7. Imperial Chemical Industries, *DEHA: Fertility Study in Rats,* ICI Report No. CTL/P. 2229, 1989.
8. Imperial Chemical Industries, *DEHA: Teratogenicity Study in the Rat,* ICI Report No. CTL/p. 2119, 1988.
9. von Daniken, A., Lutz, W. K., Jackh, R., et al., Investigation on the potential for binding of di(2-ethylhexylphthalate (DEHP) and di(2-ethylhexyl)adipate (DEHA) to liver DNA *in vivo, Toxicol. Appl. Pharmacol.,* 73, 373, 1984.
10. See **BISPHENOL A,** #6.
11. Singh, A. R., Lawrence, W. H., Autian, J., Dominant lethal mutation and antifertility effects of di(2-ethylhexyl)adipate and diethyl adipate in mice, *Toxicol. Appl. Pharmacol.,* 32, 566, 1975.
12. Keith, Y., Cornu, M. C., Canning, P. M., et al., Peroxysomal proliferation due to di(2-ethylhexyl)adipate, 2-ethylhexanol and 2-ethylhexanoic acid, *Arch. Toxicol.,* 66, 321, 1992.
13. Final report on the safety assessment of dioctyl adipate and diisopropyl adipate, *J. Am. Coll. Toxicol.,* 3, 101, 1984.
14. MRI (Midwest Research Institute), *Toxicological Effects of Diethylhexyladipate,* Final Report, MRI Project No. 7343-B, CMA Contract No. PE-14.0-BIO-MRI, 1982.
15. Takahashi, T., Tanaka, A., and Yamaha, T., Elimination, distribution and metabolism of di(2-ethylxehyl) adipate (DEHA) in rats, *Toxicology,* 22, 223, 1981.
16. Bell, F. P., Di(2-ethylhexyl)adipate (DEHA): effect on plasma lipids and hepatic cholesterol genesis in the rat, *Bull. Environ. Contam. Toxicol.,* 32, 20, 1984.
17. Guest, D., Pallas, F., Northup, S., et al., Metabolic studies with di(2-ethylxehyl)adipate in the mouse, *Toxicologist,* 5, 237, 1985.
18. Bergman, K. and Ü Albanus, L., Di(2-ethylxehyl) adipate: gabsorption, autoradiographic distribution and elimination in mice and rats, *Food Chem. Toxicol.,* 25, 309, 1987.
19. Steel, G. T., Woollen, B. H., Loftus, N. I., et al., Biological monitoring of exposure to plasticisers, Abstr. Brit. Toxicol. Soc. Meet., Edinburgh 18th–20th Sept., 1991, *Hum. Exp. Toxicol.,* 11, 387, 1992.

DI(2-ETHYLHEXYL) AZELATE (CAS No 103–24–2)

Synonym. Azelaic acid, bis(2-ethylhexyl)ester; Dioctyl azelate; Statlex DOX; Truflex DOX.
Properties. Colorless, odorless liquid. Poorly soluble in water, soluble in alcohol.
Applications. Used as a plasticizer in the production of PVC, polystyrene and other plastics. Compatible with nitro-, ethyl-, and acetobutyrate cellulose, styrene-butadiene co-polymers, polychloroprene and butadiene-nitrile-acrylic rubber.
Acute Toxicity. LD_{50} is reported to be 8.7 to 9.1 g/kg BW in rats and 11.5 g/kg BW in mice.[04] Poisoning is accompanied by depression, which is followed with adynamia, labored breathing, and spasms of the separate muscle groups. Death is preceded by a lethargic condition.
Repeated Exposure failed to reveal cumulative properties. Habituation develops. K_{acc} is 6 (by Lim). The treatment caused signs of intoxication to develop, including excitation and irritation of the upper respiratory tract mucosa and eyes. Recovery occurs within days.[1]
Short-term Toxicity. There were no symptoms of intoxication in rats dosed with 0.5 to 2.5 g/kg BW over 3 months.[05]
Standards. Russia. Recommended PML in drinking water: **n/m.**
Regulations. USFDA (1993) approved the use of D. as a plasticizer in polymeric substances and as a component of adhesives intended for use in contact with food: (1) at levels not exceeding 24% by weight of the permitted PVC homo- and/or copolymers used in contact with nonfatty, nonalcoholic food (thickness of polymers <0.003 in.), (2) at levels not exceeding 24% by weight of the permitted PVC homo- and/or copolymers used in contact with fatty, nonalcoholic food having a fat and oil content not exceeding a total 30% by weight (thickness of polymers <0.003 in.).
Reference:
Timofiyevskaya, L. A., Toxicity of di(2-ethylhexyl) azelainate, *Gig. Truda Prof. Zabol.,* 1, 52, 1983 (in Russian).

DI(2-ETHYLHEXYL) PHENYLPHOSPHATE (DEHPP) (CAS No 16368–97–1)

Properties. Clear, viscous liquid of low volatility. Poorly soluble in water.

Acute Toxicity. In mice given DEHPP in oil solution, LD_{50} is 6.5 g/kg BW; pure substance LD_{50} appeared to be 9.1 g/kg BW. Rats and guinea pigs tolerate the administration of 15 g/kg BW. Poisoning is accompanied by depression, followed by excitation, aggressiveness, diarrhea, and BW loss. Some animals experienced convulsions. In the survivors, there is reversible alopecia in the area of the back and haunches.

Repeated Exposure showed weak cumulative properties. Exposure to DEHPP by Lim method revealed a mild cholinesterase action, mainly affecting the NS and parenchymatous organs. The treatment caused a reduction in BW, increased CNS excitability, a dysbalance in the blood serum protein fractions, an increased content of residual nitrogen in the blood serum, and the appearance of erythrocytes in the urine. Histological examination of the visceral organs revealed changes in the brain, liver, and kidneys. Increased relative weights of the kidneys and liver were noted.

Regulations. USFDA (1993) approved the use of DEHPP (1) as a component of adhesives intended for use in contact with food, and (2) as an ingredient for resinous and polymeric coatings for polyolefin film to be safely used as a food-contact surface of articles.

Reference:

Kalinina, N. I., Toxicity of organophosphorous plasticizers tributylphosphate and di(2-ethylhexyl) phenyl phosphate, *Gig. Truda Prof. Zabol.*, 8, 30, 1971 (in Russian).

DI(2ETHYLHEXYL) PHTHALATE (CAS No 117–81–7)

Synonyms. 1,2-Benzenedicarboxylic acid, bis(2-ethylhexyl) ester; Bis(2-ethylhexyl)phthalate; DEHP.

DIOCTYL PHTHALATE (CAS No 117–84–0)

Synonyms. DOP; Phthalic acid, dioctyl ester.

Properties. Clear, colorless, oily liquid with a faint odor. DEHP forms colloidal solutions. Because of this phenomenon, the "true solubility in water" is believed to be 0.025 to 0.05 mg/l. Detection threshold for odor and bitter-salt taste is 2.5 mg/l.

Applications and **Exposure.** Used as a principal plasticizer in the production of PVC and its copolymers. It gives the materials high elasticity and low-temperature resistance ($-45°C$), which are combined with resistance to ultraviolet radiation. It is also used as a plasticizer for cellulose nitrate, polystyrene, and other polymers. Because of its solubility in fats, DEHP is used in packaging materials for vegetables, fruit, juices, and milk. The likely human intake in Europe seems to be approximately 20 to 32 mg/person/annum. Migration levels in food and water are 0.05 to 2.5 mg/l. In man, after transfusion of blood kept in DEHP-plasticized PVC container, the plasticizer content in the spleen, liver, lungs, and omentum was 2.5 to 27 mg/l.[1] USATSDR (1988), quoted in a USFDA survey of various foods, showed that DEHP levels in most foods were less than 1 mg/kg (margarine, cheese, meat, cereal, eggs, milk, white bread, etc.).

Acute Toxicity. *Humans.* Two volunteers were dosed with 10 or 5 g DEHP, respectively. Ten grams caused mild gastric disturbances and moderate catharsis, while five grams did not.[2] *Animals.* LD_{50} is reported to be 30 to 33 g/kg BW in rats, 6.4 to 17.7 BW g/kg in mice, and 34 g/kg BW in rabbits. However, no mortality in mice was reported, even with the dose of 60 g/kg BW.[07] The acute action threshold for behavioral effect is 47 mg/kg BW, and the 'no effect' dose is reported to be 19 mg/kg BW.[3] Single oral administration of 2 g/kg BW to dogs apparently caused no toxicity.[4]

Repeated Exposure revealed cumulative properties. K_{acc} is 2.3 to 2.7 (by Lim). Liver and testes appeared to be the main target organs. DEHP can cause functional hepatic damage, as reflected by morphological changes, alterations in the activity of energy-linked enzymes, as well as changes in metabolism of lipids and carbohydrates. The most striking effect is proliferation of hepatic peroxisomes (*PP*).[5] *Humans.* A month after *i/v* administration of approximately 150 mg DEHP per week, dialysis patients were reported to have no liver changes that might have resulted from their treatment. Peroxisomes were reported to be "significantly higher".[6] *Animals*. Administration of the dose of 2 g/kg BW for 21 d increases hepatic alcohol dehydrogenase, microsomal protein, and cytochrome P-450 concentrations in rats.[7] The NOEL of 25 mg/kg BW was identified for *PP*, based on changes in peroxisomal-related enzyme activities and/or ultramicroscopic changes in a 14-d gavage study in Sprague-Dawley rats (effect level 100 mg/kg BW).[8,9] The NOEL of 2.5 mg/kg BW was observed in a similar 7-d study;[8] the NOEL of 10 mg/kg BW was defined in a 3-week study.[10] In addition to the effects on the liver, associated changes in the kidney and thyroid of Wistar rats were reported.[11]

Short-term Toxicity. Doses equal or more then 50 mg/kg BW caused a significant, dose-related increase in liver weights, a decrease in serum triglyceride and cholesterol levels, and dose-related microscopic changes in the liver, i.e. periportal accumulation of fat and mild centrilobular loss of glycogen. A significant increase in hepatic peroxisomal enzyme activities and in the number of peroxisomes in the liver was found.[8–10] Rats fed 20 to 60 mg DEHP/kg BW for 104 d exhibited growth retardation and increased liver and kidney weights. No histopathological findings were noted in the visceral organs.[4]

Long-term Toxicity. *Humans* are less sensitive for chemically induced *PP* compared to rodents. Provisional ADI appeared to be 0.025 mg/kg BW. *Animals. PP* in the liver seems to be the most sensitive effect, and the rat appears to be the most sensitive species. There appeared to be a threshold for *PP* and peroxisome-associated enzyme induction. The NOEL for the induction of *PP* in the rat is 2.5 mg/kg BW.[5] In a 2-year feeding study, Fisher 344 rats and B6C3F$_1$ mice were fed the diets containing 6 to 12 g/kg BW (rats) and 3 to 6 g/kg BW (mice) for 103 consecutive weeks. Hepatic and testicular effects were observed. Scientific interpretation of these chronic studies seems to be problematic because of use of MTD (maximum tolerated dose).[12,13] In a 1-year feeding study in guinea pigs, there were no signs of hepatic changes. In a 2-year experiment, rats received the diet containing 0.4 and 0.13% DEHP. The liver morphology was altered, as evidenced by excessive pigmentation, congestion, and some fatty degeneration.[14] No hepatic histological alterations were reported with 0.02% DEHP administered in the diet.[15] The hepatic effect of DEHP remains to be resolved.

Reproductive Toxicity. *Gonadotoxicity.* Testicular effects, consisting of atrophy, tubular degeneration, inhibition or cessation of spermatogenesis, were seen in mice, rats, guinea pigs, and ferrets. The testicular injury was accompanied by decreased zinc content in the gonads and increased urinary excretion of zinc.[16,17] At the dose-level of 5 g/kg BW, testosterone level in blood was decreased. The dose of 20 g/kg BW reduced content of triglycerides and cholesterol in the sperm. Atrophy and decreased zinc contents in testes were caused by 0.2 and 2 g/kg BW.[14,18–20] Administration of 0.15 to 2 g/kg BW to rats for 2 weeks caused reduction of spermatozoan count in epididymis. Decreased weights of the testes were observed at the maximum dose level. Increased activity of γ-glutamyltrans-peptidase, LDH (at all doses), β-glucuronidase (at maximum dose), decreased activity of acid phosphatase, and sorbitoldehydrogenase (1.2 g/kg BW) were found in the testicular tissue.[18] Complete suppression of fertility in mice was seen at 430 mg/kg BW; it was significantly reduced at 140 mg/kg. The NOEL of 15 mg/kg BW was identified in this study.[21] *Embryotoxicity.* Oral administration produced a significant reduction in placental weight. Reproduction effect is dose-dependent and is influenced by the duration of phthalate administration.[15] At oral doses equal to and higher than 200 mg/kg BW, decreased fetal weights and an increased number of resorptions were observed in rats. Maternal toxicity was observed with the doses up to 340 and 500 mg/kg BW.[15,22,23] A single administration on day 7 of gestation caused a reduction in BW of living mouse fetuses. However, no significant changes in the number of live fetuses and no gross or skeletal abnormalities are reported.[24] *Teratogenicity.* Rats were less susceptible to DEHP-related adverse effects on fetal development than mice. In a carefully designed experimental protocol, DEHP lacks birth defect actions. Teratogenicity occurs principally in mice at doses that are exceedingly high. Fetal mortality, fetal resorption, decreased fetal weight, neural tube defects, and skeletal disorders, viz. exencephaly, spina bifida, open eye-lid, exophthalmia, major vessel malformations, club-foot, and delayed ossification, were seen in several DEHP teratogenicity studies. The NOAEL for these effects in mice was found to be 0.025% DEHP in the diet (equal to 35 mg/kg BW).[22] According to other data, the LOEL in mice appeared to be 0.05 ml/kg BW.[25] However, according to other data, teratogenic effects were not observed in Fisher 344 rats even at dose-levels of 0.5 to 2% DEHP in the diet (equal to 250 to 1000 mg/kg BW).[25]

Immunotoxicity. Male rats were *i/p* exposed to 0.002 to 0.012 LD_{50} DOP. The treatment affected *B*-lymphocyte population by suppressing the synthesis of cellular DNA, thus modulating the immune system.[26]

Genotoxicity. In the majority of mutagenicity studies *in vitro* and *in vivo,* DEHP showed negative results, viz. no induction of gene mutations in the bacterial system, eukaryotic systems, or mammalian systems *in vitro,* no induction of CA in somatic or germ cells *in vitro,* and no induction of CA in somatic or germ cells *in vivo.* No evidence for a covalent interaction of DEHP with DNA, no induction of single-strand breaks in DNA or unscheduled DNA repair by DEHP was reported. However DEHP induced aneuploidy in eukariotic cells *in vitro* and cell transformations in mammalian cells *in vitro.*[6,27–29.07]

Carcinogenicity. An oral 2-year study in mice revealed increased incidence of hepatocellular carcinoma in males and females at both dose-levels (3 and 6 g/kg in the diet). Rats, administered 6 or 12 g/kg in the diet for 2 years, showed also an increased incidence of hepatocellular carcinomas and neoplastic nodules in the liver.[30] An increased incidence of liver tumors in mice and rats in chronic bioassays is hypothesized to be caused by prolonged hepatocellular *PP* and enhanced production of the peroxisomal metabolic by-product, hydrogen peroxide. Nonhuman primates and humans seem to be far less sensitive to *PP* than mice and rats.[31] The differences in the metabolic fate of DEHP in rodents compared to primates are reported. No *PP* was observed in marmosets after feeding with 2 g DEHP/kg BW for 14 d, whereas in a parallel study in rats, marked *PP* was observed. *In vivo* rat studies did not show tumor initiating or promoting activity or sequential carcinogenic activity of DEHP in the liver.[21] *Carcinogenicity classification.* IARC: Group 2B; USEPA: Group B2; NTP: P—P*—P*—P*.

Chemobiokinetics. In rats, DEHP is well absorbed from the GI tract after oral administration. Metabolism involves hydrolysis to the monoester, with subsequent oxidation of the remaining alkyl chain. Phthalic acid was found to be a minor metabolite (less than 5%) to the total urinary metabolites. Hydrolysis to **mono(2-ethylhexyl)phthalate** (MEHP) with release of 2-**ethylhexanol** largely occurs prior to intestinal absorption.[32] The highest levels of the dose were found in the liver and adipose tissue. No or little accumulation in rats was observed. Estimated half-life for DEHP and its metabolites in rats is 3 to 5 d for fat and 2 d for other tissues.[6] Of a single oral dose, 56% is excreted in rats unchanged in feces within 4 d, whereas the excretion rate for continuous feeding is not known.[32] In mice and rats, urinary metabolites consist primarily of terminal oxidation products.[33]

Guidelines. WHO (1992). Guideline value for drinking water: 0.008 mg/l.

Standards. USEPA (1992). MCL: 0.006 mg/l, MCLG: zero. **Russia** (1988). PML in drinking water: 0.1 mg/l.

Regulations. USFDA (1993) approved the use of DEHP (1) as a component of adhesives intended for use in contact with food, (2) as a plasticizer in polymeric substances, (3) in rubber articles intended for repeated use in producing, manufacturing, packing, processing, treating, packaging, transporting, or holding food (total not to exceed 30% by weight of the rubber products), (4) in resinous and polymeric coatings in a food-contact surface of articles intended for use in producing, manufacturing, packing, transporting, or holding food, (5) in acrylic and modified acrylic plastics for single and repeated use in contact with foodstuffs, (6) in manufacture of paper and paperboard, (7) as a cellophane ingredient for packaging food, alone or in combination with other phthalates, where total phthalates do not exceed 5%, and (8) as a defoaming agent that may be used safely in the manufacture of paper and paperboard intended for use in producing, manufacturing, packing, transporting, or holding food. In **Belgium, The Netherlands,** and **Spain,** DEHP may be used in all applications, provided the level migrating into the foodstuff is less than 40 ppm. In **France** it may not be used to contact alcoholic beverages and fatty foods. In **Italy** the possible applications are limited in exactly the same way as for DEHA. In **Germany,** DEHP may be used at up to 35% in general purpose film, with the same foodstuff restrictions as for DEHA. It may also be used in conveyor belts intended to be in contact with food. In this application, the level may be up to 12% for fatty foods, including milk products and up to 40% for other foodstuffs. In the **U.K.,** it may be used at up to 40% in items for contact with nonfatty foods.

References:

1. Di(2-ethylhexyl) Phthalate, IPCS, Environmental Health Criteria, WHO, Geneva, 1992, p. 142.
2. *Some Industrial Chemicals and Dyestuffs,* IARC monographs, Lion, 29, 1982, p. 269.
3. See **DIBUTYL PHTHALATE, #2.**

4. Schaffer, C. B., Carpenter, C. P., and Smyth, H. F., Acute and subacute toxicity of di(2-ethyl-hexyl) phthalate with note upon its metabolism, *J. Ind. Hyg. Toxicol.*, 27, 130, 1945.
5. Seth, P. K., Hepatic effects of phthalate esters. *Environ. Health Perspect.*, 45, 27, 1982.
6. Thomas, R. D., Ed., *Drinking Water and Health, Board on Toxicology and Environmental Health Hazards,* Vol. 6, Commission on Life Sciences, National Research Council, National Academic Press, Washington, D.C., 1986, p. 338.
7. Thomas, J. A. and Thomas, J. M., Biological effects of di-(2-ethylhexyl)phthalate and other phthalic acid esters, *CRC Crit. Rev. Toxicol.*, 13, 283, 1984.
8. Morton, S. J., The Hepatic Effects of Dietary Di(2-ethylhexyl) Phthalate, Ph.D. thesis, The Johns Hopkins University, Ann Arbor, MI, 1979, p. 135.
9. Lake, B. G., Pels Rijcken, W. R., Gray, T. J. B., et al., Comparative studies on the hepatic effects of di- and mono-*n*-octyl phthalates, di(2-ethylhexyl) phthalate and clofibrate in the rat, *Acta Pharmacol. Toxicol.*, 54, 167, 1986.
10. Barber, E. D., Astill, B. D., Moran, E. J., et al.,Peroxisome induction studies on seven phthalate esters, *J. Toxicol. Environ. Health,* 3, 7, 1987.
11. Hinton, R. H., Mitchell, F. E., Mann, A., et al., Effects of phthalic acid esters on the liver and thyroid, *Environ. Health Perspect.*, 70, 195, 1986.
12. *Carcinogenesis Bioassay of Di(2-ethylhexyl) phthalate in F344 Rats and B6C3F₁ Mice (Feed Study),* NTP, Research Triangle Park, NC, 1982.
13. Phthalic Acid Esters: Toxicology Evaluation and Suggestions for Additional Safety Testing, NTP, Research Triangle Park, NC, 1981.
14. Carpenter, D., Weil, C. S., and Smyth, H. F., Chronic oral toxicity of di-(2-ethylhexyl) phthalate for rats, guinea pigs and dogs, *AMA Arch. Ind. Hyg. Occup. Med.*, 8, 219, 1953.
15. See **DIBUTYL PHTHALATE, #5.**
16. See **DIBUTYL PHTHALATE, #4.**
17. Gray, T. J. and Gangolli, S. D., Aspects of testicular toxicity of phthalic esters, *Environ. Health Perspect.*, 65, 229, 1986.
18. Parmar, D., P., Srivastava, S. P., Seth, P. K., et al., Effect of di(2-ethylhexyl)phthalate on spermatogenesis in adult rats, *Toxicology, 42, 47, 1986.*
19. Agarwal, D. K., Eustis, S., Lamb, J. C., et al., Influence of dietary zinc on di(2-ethylhexyl) phthalate testicular atrophy and zinc depletion in adult rats, *Environ. Health Perspect.*, 84, 12, 1986.
20. Saxena, D. K., Srivastava, S. P., Chandra, S. V., et al., Testicular effects of di(2-ethyl-hexyl)phthalate: histochemical and histopathological alterations, *Ind. Health,* 23, 191, 1985.
21. *Toxicological Evaluation of Certain Food Additives and Contaminants,* Food Additives Series No. 24, 33rd Meeting of JECFA, WHO, Geneva, 1989, 222.
22. Tyl, R. W., Price, C. J., Marr, M. C., et al., Developmental toxicity evaluation on dietary di(2-ethylhexyl) phthalate in Fisher 344 rats and CD-1 mice, *Fundam. Appl. Toxicol.*, 10, 395, 1988.
23. Onda, H. et al., Effect of phthalate ester on reproductive performance in rats, *Jpn. J. Hyg.*, 31, 507, 1976.
24. Tomita, I., Nakamura, Y., Yagi, Y., et al., Teratogenicity/fetotoxicity in mice, *Environ. Health Perspect.*, 45, 71, 1982.
25. Nakamura, Y., Yagi, Y., Tomita, I., et al., Teratogenicity of di(2-ethylhexyl) phthalate in mice, *Toxicol. Lett.*, 4, 113, 1979.
26. Dogra, R. K. S., Khanna, S., Nagale, S. L., et al., Effect of dioctylphthalate on immune system of rat, Indian, *J. Exp. Biol.*, 23, 315, 1985.
27. Turnbull, D. and Rodricks, J. V., Assessment of possible carcinogenic risk to humans resulting from exposure to di(2-ethylhexyl) phthalate (DEHP), *J. Am. Coll. Toxicol.*, 4, 11, 1985.
28. Putman, D. L., Moore, W. A., Schechtman, L. M., et al., Cytogenic evaluation of di(2-ethyl-hexyl) phthalate and its major metabolites in Fisher 344 rats, *Environ. Mutat.*, 5, 227, 1983.
29. Tomita, I., Nakamura, Y., Aoki, N., et al., Mutagenic/carcinogenic potential of DEHP and MEHP, *Environ. Health Perspect.*, 45, 119, 1982.
30. Zeiger, E., Howarth, S., Speck, W., et al., Phthalate esters testing in the NTP's environmental mutagenesis test development program, *Environ. Health Perspect.*, 45, 99, 1982.
31. Stott, W. T., Chemically induced proliferation of peroxisomes: implications for risk assessment, *Regul. Toxicol. Pharmacol.*, 8, 125, 1988.

32. Kluwe, W. M., Overview of phthalate ester pharmacokinetics in mammalian species, *Environ. Health Perspect.*, 45, 3, 1982.
33. Albro, P. W., Chapin, R. E., Corbett, J. T., et al., Mono(2-ethylhexyl) phthalate, a metabolite of di(2-ethylhexyl phthalate), causally linked to testicular athrophy in rats, *Toxicol. Appl. Pharmacol.*, 100, 193, 1989.

DI(2-ETHYLHEXYL) SEBACATE (CAS No 122-62-3)

Synonyms. Dioctylsebacate; Sebacic acid, dioctyl ester.
Properties. Colorless, practically odorless, oily liquid. Solubility in water is 0.02% (20°C). Soluble in alcohol.
Applications. D. seems to be one of the best low-temperature-resistant plasticizers to produce PVC and PVA and other elastomers; it is compatible with nitro- and ethyl cellulose and with polyvinylidene chloride.
Acute Toxicity. LD_{50} is 26.2 g/kg BW in rats, and 19.6 g/kg BW in mice. Poisoning is accompanied by adynamia and lethargy; in 1 h after administration, the animals assume a side position. The next day there are coordination disorder, dyspnea, and diarrhea. Death occurs in 2 to 4 d. Gross pathological examination reveals dystrophic and necrotic changes in the liver, myocardium, spleen, and brain.[1]
Repeated Exposure failed to reveal accumulation in rats. Administration of up to 4 g/kg BW caused no mortality in animals, even when the LD_{100} was exceeded.[2]
Long-term Toxicity. Retardation of growth or signs of intoxication were not found in rats given 500 mg/kg BW for 6 months and 200 mg/kg BW for 16 months. Gross pathological and histological examination revealed no abnormalities in the viscera.[05]
Standards. Russia. PML in food: 4 mg/l. Recommended PML in drinking water: **n/m.**
Regulations. USFDA (1993) regulates the use of D. (1) as a component in resinous and polymeric coatings of a food-contact surface, (2) as a plasticizer for rubber articles intended for repeated use in contact with food (up to 30% by weight of the rubber product), (3) in adhesives used as components of articles intended for use in packaging, transporting, or holding food, and (4) as an ingredient for closures with sealing gaskets on containers intended for use in producing, manufacturing, packing, processing, preparing, treating, packaging, transporting, or holding food (up to 2%).
References:
1. *Toxicity of Synthetic Lubricants,* Erisman Research Hygiene Institute, Moscow, 1977, p. 60 (in Russian).
2. *Proc. Research Hygiene Institute,* Sofia, 10, 1966, p. 105 (in Bulgarian).

DI(2-ETHYLHEXYL) TEREPHTHALATE (CAS No 6422-86-2)

Synonyms. Benzenedicarboxylic acid, bis(2-ethylhexyl) ester; Terephthalic acid, dioctyl ester.
Properties. Colorless liquid.
Acute Toxicity. LD_{50} is not attained in mice, but a part of animals died after administration of 20 g/kg BW. Poisoning was accompanied by excitation, followed by CNS inhibition.
Reference:
Timofiyevskaya, L. A., Toxicity of di(2-ethylhexyl) terephthalate, *Gig. Sanit.*, 8, 91, 1982 (in Russian).

DIETHYL PHTHALATE (DEP) (CAS No 84-66-2)

Synonyms. 1,2-Benzenedicarboxylic acid, diethyl ester; Ethyl phthalate; Phthalic acid, diethyl ester.
Properties. Colorless, oily liquid, odorless or with a faint odor. Solubility in water is 0.15% at 20°C, soluble in alcohol and fats. Does not color water but imparts a slight aromatic odor and a

sweetish, astringent taste to it. Odor perception threshold is 1.62 g/l, taste perception threshold is 7 g/l, foam-forming ability threshold is 80 mg/l.[1]

Applications. DEP is used in the production of cellulose esters (cellulose acetate films). Blends well (up to 70%) with cellulose nitrate as well as with polyacrylates and polymethyl metacrylates. It is used extensively as a denaturant for cosmetic alcohol and in hair spray preparations.

Acute Toxicity. The LD$_{50}$ is reported to vary from 9.5 g/kg BW or 10.3 g/kg BW (males) to 10.8 g/kg BW (females) in rats, 6.2 to 9.5 g/kg in mice; it was 3 g/kg BW in guinea pigs and 1 g/kg BW in rabbits.[1,2,05] The poisoning is accompanied by depression, apathy, adynamia, side position, respiratory distress, and convulsions. Death within 2 to 3 d. Gross pathological examination revealed congestion in the abdominal cavity organs and cerebral meninges.

Repeated Exposure revealed marked cumulative properties: rats died when given 280 to 1400 mg/kg BW. Male rats were exposed *i/g* to oily solutions of 100 and 1000 mg/kg BW over a period of 30 d. The treatment showed polymorphism of toxic action with predilection to affect the liver function (hydrocarbon, protein, and lipid metabolism), myocardium, CNS, and kidneys (a 1000 mg/kg BW dose). There was also a decreased activity of several enzymes. A pronounced dose-dependence in the toxic action of DEP was found.[1]

Short-term Toxicity. Rats were given the diets containing 0.2, 1, and 5% DEP for 16 weeks. Food consumption was decreased at the highest dose-level. Hematological analyses, water intake, serum enzyme levels, urinary cell-excretion rate, and histological examination showed no significant changes.[1]

Reproductive Toxicity. *Embryotoxicity.* High embryolethality is reported as a result of exposure to high doses given one or three times during pregnancy.[3] However, embryotoxicity was not observed following oral administration of 100 mg/kg BW.[1] CD-1 mice of both sexes were dosed for 7 d prior to and during a 98-d cohabitation period. These animals were given the diets with 0.25 to 2.5% DEP. There was no apparent effect on reproductive function, despite significant effects on BW gain and liver weights.[4] *Teratogenic* effect is reported to develop in the chick embryo.[5] *Gonadotoxicity.* Rats given 7.2 mmol/kg BW dose for 4 d exhibit gonadotoxic effect. The treatment did not alter the zinc content in the testis and urine.[6]

Genotoxicity. DEP appeared to be positive in the *Salmonella* mutagenicity assay, it caused SCE but not CA in Chinese hamster ovary cells (NTP–91). It has been shown to be toxic to cultured cells.[5] *Carcinogenicity classification.* USEPA: Group D.

Chemobiokinetics. DEP metabolism involves hydrolysis to the monoester with subsequent oxidation of the remaining alkyl chain. **Phthalic acid** was found to be a minor metabolite (less than 5%) among the total urinary metabolites. The principal metabolite is **monoethylphthalate**.[2] A small part of ^{14}C-labeled DEP was passed with the urine in 24 h after application. After 72 h, activity was detected in the lungs, heart, gonads, spleen, and brain but not in the fatty tissue.[7]

Standards. Russia (1990). Recommended MAC: 0.1 mg/l.

Regulations. USFDA (1993) regulates the use of DEP (1) in adhesives as a component of articles intended for use in packaging, transporting, or holding food, (2) as an ingredient of acrylic and modified acrylic plastics in articles intended for use in contact with food, and (3) in resinous and polymeric coatings for polyolefin films to be safely used as a food-contact surface of articles intended for use in producing, manufacturing, packing, transporting, or holding food.

References:
1. See **DIALKYL(C_{789}) PHTHALATE,** #1.
2. *Hygiene and Toxicology,* Proc. Sci. Conf., Kiev, 1967, 90 (in Russian).
3. See **DIBUTYL PHTHALATE,** #8.
4. See **DIBUTYL PHTHALATE,** #11.
5. Brown, D., Butterworth, K. R., Gaunt, I. F., et al., Short-term oral toxicity study of diethyl-phthalate in the rats, *Food Chem. Toxicol.,* 16, 415, 1978.
6. Foster, P. M. D., Lake, B. G., Cook, M. W., et al., Structure–activity requirements for the induction of testicular atrophy by butyl phthalates in immature rats: effects on testicular zinc content, *Adv. Exp. Med. Biol.,* 136A, 445, 1982.
7. Autian, J., Toxicity and health threats of phthalate esters: review of the literature, *Environ. Health Perspect.,* 4, 3 1973.

DIHEPTYL PHTHALATE (DHP) (CAS No 84–75–3)

Synonym. Phthalic acid, diheptyl ester.

Properties. Colorless, almost odorless liquid. Solubility in water is 0.1 mg/l.

Applications. Used as a plasticizer in the production of vinyl polymers.

Acute Toxicity. In rats, LD_{50} is 29.6 g/kg BW.[1]

Repeated Exposure. Fisher 344 rat received 1 and 5 g DHP/kg BW for 28 d. No mortality was noted. Retardation of BW gain and an increase in the kidney (females) and liver (both sexes) weights was evident only in high-dosed animals (5 g/kg BW). Ineffective dose appeared to be <0.2 g/kg BW.[1]

Reproductive Toxicity. *Gonadotoxicity.* Rats were exposed orally to 7.2 mmol/kg BW dose for 4 d. The treatment caused testicular atrophy, an increase in the excretion of zinc in the urine, and a reduction in the zinc content of the testes.[2] In the above described study,[1] a decrease in testes weight, edema, and necrotization of hepatocits were reported in animals given 1 and 5 g DHP/kg BW. Atropy of seminal ducts and cessation of spermatogenesis were reported. *Teratogenic* effect in mice was reported by Nakashima et al.[3] However, Plasterer et al. (1985) observed no congenital defects and no increase in lethality in the offspring of mice given 3.5 g DMP/kg BW on days 7 to 14 of gestation.

References:
1. Matsushima, Y., Onodere, H., Kunitoshi M., et al. Study of diheptylphthalate toxicity after repeated administration for 28 days, *Bull. Natl. Inst. Hyg. Sci.*, 110, 26, 1992 (in Japanese).
2. See **DIETHYL PHTHALATE, #6.**
3. Nakashima, K., Kishi, K., Nishikiori, M., et al., Teratogenicity of di-*n*-heptyl phthalate, *Teratology*, 16, 117, 1977.

DIHEXYL ADIPATE

Synonym. Adipic acid, dihexyl ester.

Properties. Yellowish, oily liquid.

Applications. Used as a plasticizer in the production of PVC and other polymeric materials.

Acute Toxicity. In rats, LD_{50} was not attained; in mice it appeared to be about 20 g/kg BW. Poisoning gives way to inhibition, with subsequent excitation. Complete adynamia occurs on the other day.

Repeated Exposure revealed no signs of accumulation. $K_{acc} > 10$ (by Lim).

Reference:
Mel'nikova, N. N., Toxicity of plasticizer dihexyl adipinate, *Gig. Sanit.*, 12, 57, 1984 (in Russian).

DIHEXYL AZELATE (DHA) (CAS No 103–24–2)

Synonym. Azelaic acid, dihexyl ester.

Properties. Liquid.

Applications. Used as a plasticizer to produce vinyl copolymers and Saran film, giving them low-temperature resistance.

Acute Toxicity. LD_{50} is found to vary from 15 g/kg to 24 g/kg BW in rats, from 15 g/kg to 45 g/kg BW in mice, and from 6 g/kg to 10 g/kg BW in guinea pigs.[04]

Long-term Toxicity. Addition of 15% DHA to the diet caused retardation of BW gain and an increase in kidney weights. Dogs were dosed with 0.1 to 3 g DHA/kg BW. The treatment produced no changes in the general condition of animals or in the microscopic structure of the visceral organs.

Chemobiokinetics. DHA metabolism comprises its hydrolysis to form **azelaic acid** and **hexyl alcohol.** Azelaic acid is known to be rapidly excreted with the urine in the dog, rabbit and man. Another pathway of DHA decomposition in the body occurs via formation of **adipic acid,** from which **succinoic acid** is obtained through β-oxidation. Both acids have no effect on the kidneys.

Standards. Russia. Recommended PML in drinking water: **n/m.**

Regulations. USFDA (1993) regulates DHA for use only (1) in polymeric substances used in contact with nonfatty food, (2) in polymeric substances used in contact with fatty food and limited to use at levels not exceeding 15% by weight of such polymeric substances, and (3) at levels of 15 to 24% by weight of the permitted vinyl chloride homo- and/or copolymers used in contact with fatty food, having a fat and oil content not exceeding 30% by weight (the average thickness of such polymers should not exceed 0.003 in.).

Reference:

Hodge, H. C., Maynard, E. A., Downs, E. L., et al., Chronic oral toxicity studies of di-*n*-hexylazelate in rats and dogs, *Toxicol. Appl. Pharmacol.*, 4, 247, 1962.

DIHEXYL PHTHALATE (DHP) (CAS No 84–75–3)

Synonym. Phthalic acid, dihexyl ester.

Properties. Does not color water but imparts a slight aromatic odor and a sweetish, astringent taste to it. Odor and taste perception thresholds are 2.8 and 1 mg/l, respectively. Foam-forming ability threshold is 20 mg/l.[1]

Applications. Used as a plasticizer in the production of vinyl polymeric materials.

Acute Toxicity. In rats, LD_{50} is 20 g/kg BW or even more than 30 g/kg BW. In mice and guinea pigs, it is 22.5 and 12 g/kg BW, respectively.[1] Oral administration to rats revealed accumulation in the tissue of large fat droplets of DHP. There was necrosis and an increased activity of glucose-6-phosphatase in the central lobular part of the liver.[2]

Repeated Exposure. Administration of 2% DHP in the diet of rats for 21 d caused hepatomegaly, decreased liver glycogen content, and the foci of necrosis and fatty accumulations. Endoplasmic reticulum proliferation, increased content of trijodine tironine, and decreased content of tiroxine in the blood serum are reported in this study.[3] Male rats were *i/g* exposed to oily solutions (100 and 1000 mg/kg BW doses) for 30 d. Polymorphism of toxic action with predilection to affect the liver function (hydrocarbon, protein, and lipid metabolism), myocardium, CNS, and kidneys was observed with the 1000 mg/kg BW dose. Decreased activity of several enzymes was found.[1]

Reproductive Toxicity. *Embryotoxicity.* CD-1 mice of both sexes were given the diets with 0.3 to 1.2% DHP for 7 d prior to and during 98 d of cohabitation. A dose-related adverse effect was found on the number of litters per pair, and of live pups per litter, and on the proportion of pups born alive. A crossover mating study demonstrated that both sexes were affected.[4] According to other data, at a dose-level of 1000 mg/kg, an embryotoxic effect was not found.[1] *Gonadotoxicity.* DHP was found to produce testicular injury. The testicular lesion produced by DHP in immature rats is characterized by early sloughing of spermatids and spermatocytes and severe vacuolation of Sertoli cell cytoplasm.[5]

Regulations. USFDA (1993) listed DHP for use as a plasticizer in polymeric substances used in the manufacture of food-contact articles.

Standards. Russia (1990). Recommended MAC: 0.5 mg/l (organolept.).

References:

1. See **DIALKYL(C_{789}) PHTHALATE**, #1.
2. Mann, A. H., Price, S. C., Mitchell, E. F., et al., Comparison of short-term effects of di(2-ethylhexyl)phthalate, di(*n*-hexyl)phthalate, and di(*n*-octyl)phthalate in rats, *Toxicol. Appl. Pharmacol.*, 77, 116, 1985.
3. See **DI(2-ETHYLHEXYL) PHTHALATE**, #11.
4. See **DIETHYL PHTHALATE**, #4.
5. See **DIBUTYL PHTHALATE**, #12.

DIISOBUTYL PHTHALATE (DIBP) (CAS No 84–69–5)

Synonym. Phthalic acid, diisobutyl ester.

Properties. Colorless, almost odorless liquid.

Applications. Used as a plasticizer in the production of PVC and other polymers.

Acute Toxicity. LD_{50} is 15 to 25 g/kg BW in rats, 12.8 g/kg BW in mice, and 10 g/kg BW in guinea pigs.[03]

Short-term Toxicity. Rats received 5% DIBP in their diet for several months. The treatment caused no animal mortality.

Long-term Toxicity. Dogs received 2 g DIBP/kg BW in their diet for several months. The treatment caused no mortality and produced no toxic effect.[03]

Reproductive Toxicity. DIBP produced teratogenic effect in rats.[1]

Regulations. USFDA (1993) regulates the use of DIBP (1) in adhesives used as components of articles intended for use in packaging, transporting, or holding food, (2) in cellophane to be safely used for packaging food, alone or in combination with other phthalates, where total phthalates do not exceed 5%.

Reference:

Singh, A. R., Lawrence, W. H., and Autian, L., Teratogenicity of phthalate esters in rats, *J. Pharmacol. Sci.*, 61, 51, 1972.

DIISOOCTYL ADIPATE

Synonym. Adipic acid, diisooctyl ester.

Properties. Colorless or pale-amber liquid with a weak characteristic odor.

Applications. Used as a plasticizer in the production of PVC and other polymers.

Toxicity is found to be negligible.

Regulations. USFDA (1993) approved the use of D. (1) as a plasticizer for rubber articles intended for repeated use in contact with food (up to 30% by weight of the rubber product) and (2) as an ingredient for closures with sealing gaskets on containers intended for use in producing, manufacturing, packing, processing, preparing, treating, packaging, transporting, or holding food (up to 2%).

DIISOOCTYL PHTHALATE (DIOP) (CAS No 27554–26–3)

Synonym. Phthalic acid, diisooctyl ester.

Properties. Oily, colorless liquid with a characteristic weak odor. Solubility in water is 0.1% (25°C).

Applications. Used as a plasticizer in the production of PVC and other polymeric materials.

Acute Toxicity. LD_{50} is reported to be 15 g/kg[1] or 17.3 ml/kg BW (Union Carbide Corp.) for rats and 2.8 g/kg BW for mice.[1]

Reproductive Toxicity and **carcinogenicity** effects were not found during 15 to 21-month oral exposure of Wistar rats. The doses of DIOP tested in this study are 100 to 500 mg/kg in the diet.[05]

Chemobiokinetics. DIOP was added to the diet of Sprague-Dawley rats, beagle dogs, and miniature pigs at a dose of 50 mg/kg BW for 21 to 28 d. Before oral administration, a single dose was given, labeled with [14]C. Half a dose was excreted in feces, half in the urine. In the dogs, excretion occurred predominantly in the feces, in pigs, mainly via the urine.[2]

Note. The isomer *Di(1-methylheptyl) phthalate*, known incorrectly by the name *Dicapryl phthalate*, is also of low toxicity: daily exposure to 2.5 g/kg BW causes no signs of intoxication in rats.

Regulations. USFDA (1993) approved the use of DIOP (1) in adhesives as a component of articles intended for use in packaging, transporting, or holding food and (2) as an ingredient of resinous and polymeric coatings of the food-contact surface of articles intended for use in producing, manufacturing, packing, transporting, or holding food.

References:

1. Antonyuk, O. K., Toxicity of phthalic esters. Review of the literature, *Gig. Truda Prof. Zabol.*, 1, 32, 1975 (in Russian).
2. Ikeda, G. J., Sapienza, P. P., Convillion, J. L., et al., Distribution and excretion of two phthalate esters in rats, dogs, and miniature pigs, *Food Cosmet. Toxicol.*, 16, 409, 1978.

DI(2-METHOXYETHYL) PHTHALATE (DMEP) (CAS No 117–82–8)

Synonyms. Methylcellosolve phthalate; Phthalic acid, methylglycol ester.

Properties. Colorless, oily liquid with a faint odor. Water solubility is 0.8% (20°C), insoluble in organic oils.

Applications. Used as a plasticizer in the production of nitro- and acetyl cellulose, PVA, PVC, and PVDC, giving polymeric materials good light-resistance.

Acute Toxicity. LD_{50} is reported to be 2.75 g/kg (mixed with a small amount of ethyl alcohol) or 4.4 g/kg BW in rats, 3.2 to 6.4 g/kg BW in mice, and 1.6 to 3.2 g/kg BW in guinea pigs.[03] Administration caused no visible signs of poisoning. Animals that did not die in the first 36 h recovered. Histological examination failed to reveal changes in the viscera (IARC 19-392).

Long-term Toxicity. Rats were given 0.3 to 0.9 g/kg BW in their diet for 21 months. Treatment with the highest dose caused decrease of BW gain. There were no visible signs of poisoning and no deviations in the relative weights of the visceral organs.

Reproductive Toxicity. DMEP is known to be the most *teratogenic* and *embryotoxic* of all the phthalates. The rat embryos do not hydrolyze DMEP to monoester. However, after a single *i/v* administration to pregnant Wistar-Porton rats on day 14 of gestation, ester and diester are found in the fetal tissues. DMEP and 2-**methoxyethanol,** derived from DMEP, seem to be teratogens.[1] DMEP was administered to Wistar rats on day 12 of gestation. The treatment caused hydronephrosis, heart defects, and short limb and tail. This suggests DMEP to be a proximate teratogen.[2] Embryotoxic and teratogenic effects of DMEP are also reported.[3] DMEP passes rapidly through the placenta. It causes decreased zinc content in the fetal tissues.

References:

1. Campbell, J., Holt, D., and Webb, M., Dimethoxyethyl phthalate: teratogenicity of the diester and its metabolites in the pregnant rats, *J. Appl. Toxicol.,* 4, 35, 1984.
2. Ritter, E. J., Scott, W. J., Randall, J. L., et al., Teratogenicity of dimethoxyethyl phthalate and its metabolites methoxyethane and methoxyacetic acid in the rat, *Teratology,* 32, 25, 1985.
3. See **DIISOBUTYL PHTHALATE.**

DIMETHYL MALEAT (DMM) (CAS No 624–48–6)

Synonyms. 2-Butenedioic acid, dimethyl ester; Maleic acid, dimethyl ester.
Properties. A clear, colorless liquid.
Applications. Used as an intermediate in the production of copolymers and films.
Acute Toxicity. LD_{50} is 1410 mg/kg BW in rats and 1340 mg/kg BW in mice.[1]
Genotoxicity. Negative results were observed in *Salmonella* mutagenicity assay at concentrations up to 5 mg DMM per plate. Mice given 1 g DMM/kg BW by gavage exhibited no clastogenic effect.[2]
References:

1. Smyth, H. F., Carpenter, C. P., and Weil, C. S., Range-finding toxicity data, List VI, *Am. Ind. Hyg. Assoc. J.,* 23, 95, 1962.
2. Miltenburger, H. G., Maleinsauredimethyl ester-Prufung der akuten oralen Toxizitat bei der Maus Laboratorium fur Mutagenitatsprufungen, Technische Hochschule Darmstadt, Report No. 23 on Behalf of the BG Chemie, 1982 (in German).

DIMETHYL PHTHALATE (DMP) (CAS No 131–11–3)

Synonym. Phthalic acid, dimethyl ester.
Properties. Clear, colorless, oily liquid, without or with a faint odor. Solubility in water is 0.45% (20°C). Soluble in alcohol and fats. Taste threshold concentration is 3.5 mg/l.[1]
Applications. DMP is used as a plasticizer for cellulose esters, particularly for cellulose acetate and, less frequently, for cellulose nitrate. It is used in mixtures with other plasticizers, for example, with triphenylphosphate, triacetine, diethyl- and dibutyl phthalates. DMP is used as a plasticizer for PVC, polyvinyl acetate, rubber, polyacrylates, and polymethyl-methacrylates.
Acute Toxicity. LD_{50} is 5.5 to 7.2 g/kg BW in mice, 6.9 to 8.7 g/kg BW in rats, and 2.4 g/kg BW in guinea pigs.[1,03] The toxicity manifestations included depression, apathy, and adynamia. After an exposure of 3 to 4 d, the animals developed labored respiration and convulsions and died within 2 to 3 h. Gross pathological examination revealed congestion in the visceral organs and cerebral meninges.

Repeated Exposure revealed marked cumulative properties. Administration of $1/5$ and $1/20$ LD_{50} has shown K_{acc} to be 2.9 (by Lim).[2] The treatment with $1/10$ and $1/50$ LD_{50} resulted in decreased BW gain, functional disturbances in the liver, kidneys and NS, changes in hematological analysis and in the morphology and relative weights of the visceral organs.[3] In a 30-d study, increased ascorbic acid content in the suprarenals was marked on day 5 but not on day 30 of the exposure. Animals displayed reduction in liver and brain cholinesterase acitivity, behavioral changes, and erythrocytosis. Doses below $1/50$ LD_{50} did not produce histological changes. The NOAEL was identified to be 0.17 mg/kg BW.[1]

Long-term Toxicity. Young rats received 8% DMP in their feed. The treatment caused chronic renal impairment. A dose of 4% caused retardation of BW gain, while a 2% dose appeared to be harmless.[05]

Reproductive Toxicity. *Gonadotoxicity.* DMP is likely to produce no testicular injury.[6] A 4-d administration of 7.2 mmol/kg BW to rats had no gonadotoxic effect. *I/g* administration at the LD_{50} level did not alter spermatozoan motility time.[4] *Embryotoxicity.* The implantation process was affected after DMP administration on day 3, 6, and 9 of gestation. The treatment caused a reduction in fetal weights and sizes; bleeding during parturition led to maternal death in several cases.[5,6] Significant embryolethality and death of neonates were observed on dermal application of 1250 mg DMP/kg BW in pregnant rats (Gleyberman et al., 1975). *Teratogenicity.* DMP is shown to produce an increased incidence of skeletal abnormalities in neonatal rats.[7] No teratogenic effect is reported in mice.[8]

Allergenic Effect not observed.[1]

Genotoxicity. DMP is found to be positive in *Salmonella TA 100* mutagenicity assay.[9] It caused no CA but produced SCE in Chinese hamster ovary cells (NTP-88). Skin application of 20 to 40% DMP solution or oral exposure to pure DMP (200 to 2000 mg/kg BW) for 1 to 1.5 months produced neither cytogenic effect nor increased DLM and changes in fertility of animals.[4] There was no increase in the frequency of DLM in mice given orally 1250 mg DMP/kg BW for 5 weeks, even 10 weeks after experiment onset.[10]

Carcinogenicity. Kozumbo et al. consider DMP to represent a great mutagenic and carcinogenic hazard.[9] *Carcinogenicity classification.* USEPA: Group D.

Chemobiokinetics. Metabolism involves hydrolysis to the monoester with subsequent oxidation of the remaining alkyl chain. **Monomethyl phthalate, phthalic acid, benzoic acid** and **formaldehyde** are found in the blood and urine of rats after skin application of a dose of 1.5 mg/kg.[11] Phthalic acid was found to be a minor metabolite (14%) among the total urinary metabolites.

Standards. Russia (1988). MAC: 0.3 mg/l; PML in drinking water: 0.5 mg/l.

Regulations. USFDA (1993) regulates the use of DMP (1) in adhesives as a component of articles intended for use in packaging, transporting, or holding food, (2) as a solvent in polyester resins intended for use in articles coming in contact with food, (3) as an ingredient of acrylic and modified acrylic plastics for single and repeated use in contact with food, and (4) as a component in cross-linked polyester resins.

References:

1. Shatinskaya, I. G., Comparative Description of Methods of Studying Accumulation in Resolving Problems of the Safe Regulation of Harmful Chemicals, Author's abstract of thesis, Kiev, 1986, p. 24 (in Russian).
2. See **ETHYLENEIMINE,** #6, 39.
3. Current Problems of Applications of Polyvinylchloride Materials, All-Union Institute Med. Inform., 1978, p. 4 (in Russian).
4. See **ETHYLENEIMINE,** #6, 43.
5. See **DIBUTYL PHTHALATE,** #8.
6. See **DI(2-BUTOXYETHYL) PHTHALATE,** #2.
7. See **DIISOBUTYL PHTHALATE.**
8. Plasterer, M. R., Bradshaw, W. S., Booth, G. M., et al., Developmental toxicity of nine selected compounds following prenatal exposure in the mouse. Naphthalene, *p*-nitrophenol, sodium selenite, dimethyl phthalate, ethylene thiourea, and four glycol ether derivatives, *J. Toxicol. Environ. Health,* 15, 25, 1985.
9. Kozumbo, W. J., Kroll, R., and Rubin, R. J., Assessment of the mutagenicity of phthalate esters, *Environ. Health Perspect.,* 45, 103, 1982.

10. Yurchenko, V. V. and Gleyberman, S. E., *Med. Parasitol. Parasitol. Dis.*, 1, 58, 1980 (in Russian).
11. Surina, T. Ya., Gleyberman, S. E., and Nikolayev, G. M., Metabolism of dimethylphthalate on skin applications, *Med. Parasitol. Parasitol. Dis.*, 4, 67, 1984 (in Russian).

DINONYL PHTHALATE (DNP) (CAS No 84–76–4)

Synonym. Phthalic acid, dinonyl ester.

Properties. Colorless, almost odorless, oily liquid. Poorly soluble in water, soluble in alcohol.

Applications. Used as a plasticizer for PVC and other plastics, and in the manufacture of plastisols and pastes.

Acute Toxicity. LD_{50} is found to be 2 g/kg BW in rats and 21.5 g/kg BW in mice.[1]

Repeated Exposure failed to reveal cumulative properties. K_{acc} is 10 (by Lim).[2] There were no toxic manifestations in chickens on five administrations of a 1 mg/kg BW dose.[1]

Short-term Toxicity. Cats were given 2 mg/kg BW oral doses for 3 months. The treatment did not cause retardation of BW gain or changes in the hematological analysis or in liver and kidney functions.[3]

Regulations. USFDA (1993) approved *Diisononylphthalate* for use as a plasticizer in polymeric substances in the manufacture of food-contact articles only at levels not exceeding 43% by weight of the permitted PVC homo- and/or copolymers used in contact with food only of the specified type (**21 CFR**) at temperatures not exceeding room temperature (thickness of polymers <0.005 in.).

References:
1. See **BUTYLPHTHALYL BUTYL GLYCOLATE.**
2. See **ETHYLENEIMINE,** #6, 39.
3. Hoffmann, H. T., *Z. Arbeitsmed. Arbeitsschutz.*, 11, 240, 1961 (in German).

DIPHENYL(2-ETHYLHEXYL) PHOSPHATE (DPEHP) (CAS No 1241–94–7)

Synonym. Phosphoric acid, diphenyl-2-ethylhexyl ester.

Properties. A liquid with a faint odor.

Applications. Used as a plasticizer in the production of PVC and vinyl polymers. Gives elasticity at low temperatures, light stability, and resistance to combustion. Mixes well with acrylonitrile-butadiene synthetic rubber, nitro- and ethylcellulose, cellulose acetobutyrate, polymethylmethacrylate, polystyrene, and polyvinylbutyral.[05]

Acute Toxicity. LD_{50} was not attained. Rabbits tolerate a single administration of 24 g/kg BW. Higher doses caused diarrhea. The same dose caused only emaciation in rats. The animals recovered. Histological examination failed to reveal changes in the viscera.

Long-term Toxicity. No mortality or decrease in BW gain was noted in rats exposed to 0.125% DPEHP in the diet for 2 years. No signs of intoxication were observed in dogs fed the diet with 1.5% DPEHP for the same period of time. Gross pathological examination failed to reveal changes in the viscera.

Chemobiokinetics. DPEHP is shown to be poorly absorbed in the intestine. Its metabolism in the body occurs through hydrolysis to form **phenol metabolites.** Intake of 5 ml/kg BW by volunteers enabled **phenol** (a product of DPEHP hydrolysis) in all forms to be detected in the urine and feces. 96% DPEHP is excreted as **phenol derivatives:** 83% in the feces and 13% with the urine.

Standards. Russia. Recommended PML: **n/m.**

Reference:
Treon, J. F., Cappel, J., and Sigmon, H., Toxicity of 2-ethylhexyl diphenylphosphate, metabolic fate in man and animals, *Arch. Ind. Hyg. Occup. Med.*, 8, 268, 1953.

DIPHENYL PHTHALATE (DPP) (CAS No 84–62–8)

Synonym. Phenyl phthalate.

Properties. White powder with a yellowish tint. Poorly soluble in water, alcohol, and fats.

Applications. Used as a plasticizer in the manufacture of nitro- and ethylcellulose, polystyrene, phenol and vinyl resins. Forms strong films with nitro- and ethylcellulose.

Short-term Toxicity. Dogs were given 1.2% and 5% DPP in their diet for 90 d. The higher dose-level caused decreased BW gain due to loss of appetite in animals (Erikson, 1965).

Chemobiokinetics. Intestinal absorption is insignificant. Of DPP, 90% is passed with the feces; 4 to 5% DPP is found in the urine as **phthalic acid** and the remainder in the form of **free** or **bound phenols.**

Regulations. USFDA (1993) approved the use of DPP (1) in adhesives as a component of articles intended for use in packaging, transporting, or holding food, (2) in cellophane for food packaging, (3) for use alone or in combination with other phthalates, in plastic film or sheet prepared from polyvinyl acetate, polyvinyl chloride, and/or vinyl chloride copolymers. Such film or sheet shall be used in contact with food at temperatures not to exceed room temperature and shall contain no more than 10% by weight of the total phthalates, calculated as phthalic acid and (4) as a plasticizer in polymeric substances used in the manufacture of food-contact articles.

DIVINYL ADIPATE (DVA) (CAS No 4074–90–2)

Synonym. Adipic acid, divinyl ester.

Properties. Clear, colorless liquid. Poorly soluble in water, readily soluble in alcohol. Odor and taste perception thresholds are 1 and 0.5 mg/l, respectively; practical organoleptic threshold is 2 mg/l. Concentrations of up to 20 mg/l affect neither color nor transparency of water.

Applications. Used as a plasticizer to give particular low-temperature resistance to polymeric materials.

Acute Toxicity. LD_{50} is found to be 6.4 g/kg BW in rats and mice and 4.3 g/kg BW in guinea pigs and rabbits. Poisoning is accompanied by excitation with subsequent depression, the animals become irresponsive to tactile or pain stimuli.

Repeated Exposure revealed marked cumulative properties. K_{acc} is 4.4 (by Cherkinsky). The treatment with $^1/_{20}$ and $^1/_{10}$ LD_{50} produced reticulocytosis, an increase with subsequent decline in liver cholinesterase activity, a decline in SH-group content, a rise in liver glycogen content, in the relative weights of the suprarenals and a reduction of their ascorbic acid level. Gross pathological examination revealed inflammatory and dystrophic changes in the viscera.

Long-term Toxicity. In a 7-month rat study, the toxic manifestations consisted of reticulocytosis, reduction in phagocytic leukocyte activity, increase with subsequent decline in the blood cholinesterase activity. Dystrophic and inflammatory changes were found on autopsy.

Standards. Russia. Recommended PML: 0.2 mg/l.

Reference:

Mironets, N. V., Comparative toxicologic characteristics of the plasticizers vinylmethyladipate and divinyladipate and their hygienic standard in water bodies, *Gig. Sanit.,* 10, 88, 1970 (in Russian).

(2-ETHYLHEXYL) DIPHENYL PHOSPHATE (CAS No 1241–94–7)

Synonyms. Phosphoric acid, 2-Ethylhexyldiphenyl ester; 2-Ethyl-1-hexanol ester with diphenyl phosphate.

Properties. Viscous liquid with a faint odor.

Applications. Used as a plasticizer in the production of PVC and vinyl polymers. Renders materials elastic at low temperatures, light stability and resistant to combustion. Mixes well with acrylonitrile-butadiene synthetic rubber, nitro- and ethylcellulose, cellulose acetobutyrate, polymethylmethacrylate, polystyrene, and polyvinylbutyral.

Acute Toxicity. LD_{50} is not attained. 24 g/kg appears to be the tolerable dose for rats and rabbits. Gross pathological examination did not reveal changes in the viscera.[05]

Long-term Toxicity. 1 and 2% in the rat feed caused increased mortality. 0.125% in the diet of rats and 1.5% in the diet of dogs seemed to be harmless for animals (Freon et al., 1953).

Reproductive Toxicity. Charles River rats were dosed with 300 to 3000 mg/kg BW by gavage on days 6 through 15 of gestation. Maternal BW was reduced on administration of higher doses. No signs of teratogenicity were noted.[1]

Chemobiokinetics. E. metabolism occurs through its hydrolysis in the body. Intake of 5 ml E./kg BW by volunteers caused all forms of **phenol** to appear in their urine.

Regulations. USFDA (1993) approved the use of EHDPP (1) in resinous and polymeric coatings in a food-contact surface and (2) in resinous and polymeric coatings for polyolefin films to be safely used as a food-contact surface of articles intended for use in producing, manufacturing, packing, transporting, or holding food.

Reference:

Robinson, E. C., Hammond, B. G., Johnnsen, F. R., et al., Teratogenicity studies of alkylary phosphate ester plasticizers in rats, *Fundam. Appl. Toxicol.*, 7, 138, 1986.

ETHYLPHTHALYL ETHYL GLYCOLATE (CAS No 84-72-0)

Synonym. Ethyl(ethoxycarbonylmethyl) phthalate.

Properties. Almost colorless, oily liquid with a very faint odor. Water solubility is 0.17% (30°C).

Applications. Used as a plasticizer for PVC, cellulose acetate, nitrocellulose, ethylcellulose, PVA, polyvinylidene chloride, acrylonitrile rubbers. Renders materials heat- and light-resistant.

Acute Toxicity. LD_{50} is not attained.

Repeated Exposure. Administration of 10% E. in the diet caused death in rats within 7 to 15 d.

Long-term Toxicity. Rats received 0.05, 0.5, and 5% E. in their feed for 2 years. Only the highest dose caused mortality. The lower doses exerted no visible harmful effect. Dogs were given up to 250 mg/kg BW for 12 months. No retardation of BW gain or signs of poisoning were reported.[05]

Chemobiokinetics. Calcium oxalate crystals were found in the kidneys.

Regulations. USFDA (1993) approved the use of E. (1) in adhesives as a component of articles intended for use in packaging, transporting, or holding food and (2) in resinous and polymeric coatings in a food-contact surface of articles intended for use in producing, manufacturing, packing, transporting, or holding food.

Reference:

See **BUTYLPHTHALYL BUTYL GLYCOLATE.**

GLYCEROL TRIHEPTANOATE (GT) (CAS No 620-67-7)

Synonyms. Glycerin, triheptyl ether; Triheptylin.

Properties. Odorless, light, almost colorless liquid.

Applications. Used as a plasticizer in PVC production. Gives good low-temperature resistance to plastic materials. Used as a component of plastisols for reducing the viscosity of pastes.

Long-term Toxicity. No retardation of BW gain was observed in Wistar rats that consumed the diet with 0.05, 0.1, and 1 g GT/kg BW (Le Breton).

Reproductive Toxicity was not observed.

HEPTYL NONYL PHTHALATE

Properties. Clear, colorless liquid.

Acute Toxicity. Administration of 20 g/kg BW dose of Alfanole (H. trademark) caused diarrhea in rats and mice.

Short-term Toxicity. Rats exposed to 60 mg/kg BW for 3 months developed no signs of toxicity.

Long-term Toxicity, Reproductive Toxicity, and **Carcinogenicity.** Wistar rats were gavaged with the doses of 0.5, 1, and 3 mg H. per day for 15 to 21 months. Gross pathological examination failed to reveal treatment-related lesions. In a five-generation study, reproductive function of animals was unaffected. No carcinogenic effect was reported.[05]

Reference:

Gaunt, J., Colley, J., Grasso, P., et al., Acute (rat and mouse) and short-term (rat) toxicity studies on dialkyl-79-phthalate (a mixture of phthalate esters of alcohols having 7 to 9 carbon atoms), *Food Cosmet. Toxicol.*, 6, 609, 1968.

HEXACHLOROBENZENE (HCB) (CAS No 118–74–1)

Synonyms. Anticarie; Bant-cure; Bant-no-more; Pentachlorophenyl chloride; Perchlorobenzene.

Properties. Needles. Solubility in water is 0.007 mg/l at 20°C, sparingly soluble in cold alcohol, soluble in hot alcohol. Odor perception threshold is 0.06 mg/l. Heating does not increase odor intensity.

Applications. Used as a plasticizer, as a solvent, and as a fungicide.

Acute Toxicity. LD_{50} is reported to be 1.7 to 4 g/kg BW in rats and mice; in guinea pigs and rabbits, it is 1 g/kg BW.

Repeated Exposure. *Humans.* A misuse of HCB in Turkey in 1955 through 1959 caused about 3000 cases of porphyria with a mortality rate of 10%.[1] *Animals.* HCB is highly lipophilic and thus has a propensity to bioaccumulate.[2] K_{acc} is reported to be 1 (by Kagan). Degree of accumulation depends on the dose administered and on the length of the period during which HCB is given.[3] Wistar rats were given 100 mg/kg BW every other day for 6 weeks. The treatment caused porphyria to develop as a result of disturbance of polyporphyrin biosynthesis in the liver.[4] After five administrations of the same dose of HCB in corn oil to Sprague-Dawley rats, hepatic porphyria in female rats was found to develop after a delay period of 6 weeks, whereas toxicity was minimal in male rats.[5] A total dose of 1500 mg HCB/kg BW in corn oil was administered to Sprague-Dawley rats as 50 mg/kg BW for 6 weeks or 100 mg/kg BW for 3 weeks. In males, HCB caused porphyria, measured as urinary uroporphyrin and hepatic porphyrin levels. This total dose given to females for 3 weeks was not porphyrinogenic. A total dose of 600 mg/kg BW induced porphyria after 6 weeks. The minimally effective cumulative dose inducing porphyria was determined to be 400 mg/kg BW.[6]

Short-term Toxicity. HCB administration results in hepatomegaly, porphyria, and interference with hematopoiesis and CNS function. Manifestations of toxic action are highly dependent on species, dose, and time of exposure. Pigs were given 50 mg/kg BW in their feed for 90 d. The treatment caused an increase in coproporphyrin excretion, the induction of microsomal liver enzymes, histological changes, increased relative weights of the liver, and death. A dose of 0.05 mg/kg BW is considered to be ineffective.[7] Gross pathological examination of rats fed >2 mg/kg in the diet for 15 weeks revealed changes in the liver and spleen. A dose of 0.5 mg/kg BW was ineffective.[8] Third-litter sows were given HCB at concentrations of 1 and 20 ppm in their food throughout gestation and nursing. Accumulation in the fatty tissue was noted at the higher dose. Animals displayed neutrophilia, gastric irritation, fatty replacement of Brunner's gland, and pancreatic periductal fibrosis.[9]

Long-term Toxicity. Rats were dosed with 1 mg/d for a year. The treatment caused exhaustion, digestive and metabolic disorders, and death. In dogs, the same dose produced nodular hyperplasia of the lymphoid tissue of the GI tract.[3]

Immunotoxicity. HCB is found to cause immunodepression.

Reproductive Toxicity. *Teratogenic* effect is not found as a result of administration of 120 mg/kg BW to pregnant Wistar rats at the time of organogenesis.[9,10] However, Andrews and Courtney observed hydronephrosis in the newborns of CD rats and CD-1 mice treated with 10 mg HCB/kg BW on days 15 to 20 and 6 to 16 of gestation, respectively.[11] *Embryotoxicity.* Rats received 60 to 140 mg/kg BW with their feed. The treatment produced a significant dose-related increase in the mortality of the first and second generation of the offspring.[12] HCB transport across the placenta has been described in rats and mice.[8] A dose-related accumulation of HCB in the tissues, placenta, and fetal gall bladder was revealed in the treated pregnant hamsters and guinea pigs.

Genotoxicity. HCB does not cause DLM in rats or CA in Chinese hamster cell culture nor in bacteria (IARC, Suppl. 6–331). It is not shown to be genotoxic in *Salmonella*, *E. coli*, and human peripheral blood lymphocytes *in vitro*.[13]

Carcinogenicity. In a two-generation rat study, significant linear trends in the rate of parathyroid adenomas, neoplastic liver nodules, and other changes in the liver and kidneys were observed.[14] Later, HCB was reported to induce liver cancer in female rats.[5] Hamsters received 4, 8, and 16 mg/kg BW in the diet throughout their lifetime. A significant dose-dependent increase in the incidence of hepatomas and thyroidal hemoangioepitheliomas and adenomas was found in the treated animals. Mice received doses of 6.5, 13, and 26 mg/kg BW for their life-span. An increased incidence of hepatomas was observed only at two higher dose-levels.[14] *Carcinogenicity classification.* IARC: Group 2B.

Chemobiokinetics. Absorbed HCB is distributed mainly to the liver, kidneys, and brain. Fat contains 500 times more HCB than blood. HCB is biotransformed to **tetra-** and **pentachlorobenzenes** and **pentachlorophenol,** as well as to sulfur-containing metabolites. Biliary excretion of **pentachlorothiophenol,** a metabolite originating from glutathione conjugation of HCB, was higher in male than female rats.[5] Nine weeks after the start of administration of 0.5 g HCB/kg BW, an equilibrium was established between the HCB intake into the body and its excretion.[4] There was a high accumulation tendency in the rhesus monkeys given a daily oral dose of 0.11 mg ^{14}C-HCB for 15 months. The excretion storage pattern showed a very slow approach to a saturation level.[16] However, HCB is poorly absorbed and excreted unchanged, predominantly with the feces. Small amounts of HCB are metabolized by the intestinal microflora. HCB excretion was observed for many months after administration had ceased.[4] It is readily transferred in the milk of lactating dams to their suckling neonates.

Guidelines. WHO (1992). Guideline value for drinking water: 0.001 mg/l.

Standards. USEPA (1991): Proposed MCL: 0.001 mg/l, MCLG: zero. **Russia** (1988): MAC and PML: 0.05 mg/l. PML in wheat grain: 0.01 mg/kg.

Regulations. EEC(1992). Banned to certain uses owing to its effects on health and the environment.

References:

1. Peters, H. A., Hexachlorobenzene poisoning in Turkey, *Fed. Proc.*, 35, 2400, 1976.
2. Foster, W. G., McMahon A., Jarrel, J. F., et al., Hexachlorobenzene suppresses circulating progesterone concentrations during the lutea phase in the cynomologus monkey, *J. Appl. Toxicol.*, 12, 13, 1992.
3. Gralla, E. J., Fleischman, R. W., Luthra, Y. K., et al., Toxic effects of hexachlorobenzene after daily administration to beagle dogs for one year, *Toxicol. Appl. Pharmacol.*, 20, 227, 1977.
4. Koss, G., Seubert, S., Seubert, A., et al., Studies on the toxicology of hexachlorobenzene. V. Different phases of porphyria during and after treatment, *Arch. Toxicol.*, 52, 13, 1983.
5. D'Amour, M. and Charbounean, M., Sex-linked differences in hepatic glutathione conjugation of hexachlorobenzene in the rat, *Toxicol. Appl. Pharmacol.*, 112, 229, 1992.
6. Krishnan, K., Brodeur, J., Charbounean, M., et al., Development of an experimental model for the study of hexachlorobenzene-induced hepatic porphyria in the rat, *Fundam. Appl. Toxicol.*, 17, 433, 1990.
7. den Tonkelaar, E. M., Verschuuren, H. G., Bankovska, J., et al., Hexachlorobenzene toxicity in pigs, *Toxicol. Appl. Pharmacol.*, 43, 137, 1978.
8. Courtney, K. D., Andrews, J. E., and Graddy, M. A., Placental transfer and fetal deposition of hexachlorobenzene in the hamster and guinea pig, *Environ. Res.*, 37, 239, 1985.
9. Hansen, L. G., Simon, J., Dorn, S. B., et al., Hexachlorobenzene distribution in tissues of swine, *Toxicol. Appl. Pharmacol.*, 51, 1, 1979.
10. Khera, K. S., Teratogenicity and dominant lethal studies of hexachlorobenzene in rats, *Food Cosmet. Toxicol.*, 12, 471, 1986.
11. Andrews, J. E. and Courtney, K. D., Hexachlorobenzene-induced renal maldevelopment in CD-1 mice and CD rats, *Hexachlorobenzene,* Proc. Intern. Symposium, C. R. Morris and J. R. P. Cabral, Eds., IARC Sci. Publ., 77, 381, 1986.
12. Kitchin, K. T., Linder, R. E., Scotti, T. M., et al., Offspring mortality and maternal lung pathological in female rats fed hexachlorobenzene, *Toxicology,* 23, 33, 1982.
13. Siekel, P., Chalupa, J., Beno, J., et al., A geno- toxicological study of hexaclorobenzene and pentachloranizole, *Teratogen. Carcinogen. Mutagen.*, 11, 55, 1991.
14. Arnold, D. L., Moodie, C. A., Charbonneau, S. M., et al., Long-term toxity of hexachlorobenzene in the rat and the effect of dietary vitamin A, *Food Chem. Toxicol.*, 23, 779, 1985.
15. Cabral, J. R. P., Shubik, P., Mollner, T., et al., Carcinogenic activity of hexachlorobenzene in hamsters, *Nature,* 269, 510, 1977.
16. Rozman, K., Mueller, W., Coulston, F., et al., Long-term feeding study of hexachlorobenzene in rhesus monkeys, in The 16th Annual Meeting Soc. Toxicol., Abstracts, Toronto, March 27–30, 1977, p. 173.

ISOPHTHALIC ACID (CAS No 121–91–5)

Synonym. *m*-Phenylenedicarboxylic acid.

Properties. White, odorless, finely crystalline powder. Water solubility is 0.013 g per 100 at 25°C; solubility in hot water is 2.2 g/l; soluble in alcohol.

Applications. Used in the production of lacquers and coatings. An ingredient in the manufacture of amide resins and PVC plasticizers.

Acute Toxicity. LD_{50} is 10.8 g/kg BW in rats and 9.58 g/kg BW in mice. Primary target organ and tissue are the liver and CNS.[1]

Repeated Exposure revealed pronounced cumulative properties in rats.

Allergenic effect was found in *in vitro* experiments and skin tests.[2]

Standards. Russia (1988). MAC and PML: 0.1 mg/l.

Regulations. USFDA (1993) regulates the use of IA (1) in adhesives as a component of articles intended for use in packaging, transporting, or holding food, (2) in resinous and polymeric coatings in a food-contact surface of articles intended for use in producing, manufacturing, packing, transporting, or holding food, and (3) in cross-linked polyester resins to be safely used as articles or components of articles intended for repeated use in contact with food.

References:
1. See **ADIPIC ACID,** #2, 217.
2. *Toxicology of New Chemicals and Occupational Hygiene in Their Production and Application,* Rostov-na-Donu, 1974, p. 113 (in Russian).

METHOXYETHYL OLEATE (CAS No 11–10–4)

Synonyms. Methylcellosolvoleate; Oleic acid, 2-methoxyethyl ester.

Properties. Oily liquid with a characteristic odor. Insoluble in water (25°C). Soluble in alcohol.

Applications. Used as a plasticizer in the production of PVC, polyvinylbutyral, ethyl cellulose, chlorinated, natural and synthetic rubber, etc. Gives materials transparency and elasticity.

Acute Toxicity. LD_{50} in young rats (BW of 70 to 75 g) is identified to be 16 g/kg BW.

Long-term Toxicity. Rats received 0.01 to 1.25% M. in their diet for a year. More than half of animals died at the dose-level of 1.25% M. in the diet by the end of the ninth month of the experiment; lower levels caused retardation of growth. Gross pathological examination revealed kidney stones. Addition of 0.01% M. to the diet appeared to be harmless.

Chemobiokinetics. M. is shown to be hydrolyzed by lipase *in vitro.* It is likely to be hydrolyzed in the body similar to fats.

Reference:
See **BUTYL STEARATE,** #2.

METHYLCYCLOHEXYL PHTHALATE

Synonym. Phthalic acid, methylcyclohexyl ester.

Properties. Viscous, light-yellow liquid with a faint odor. Poorly soluble in water, soluble in alcohol.

Applications. Used as a plasticizer in the production of nitrocellulose, PVC, PVA, polystyrene, polymethylmethacrylate, etc. Used to obtain cellophane.

Toxicity is likely to be insignificant.

MINERAL PETROLEUM OIL (MPO) (CAS No 8012–95–1)

Synonyms. Petroleum liquid; White petroleum mineral oil.

Composition. Mineral (petroleum) oils are a mixture of methanenaphthenic, aromatic, and naphtheno-aromatic compounds. They contain 0.2 to 2% of sulfur (elemental or in the form of H_2S). Purified oil contains it in organic forms. White mineral oil is refined to meet specified requirements

(US Pharmacopeia XX, 1980, etc.). Some species of MPO are characterized by a high degree of removal of aromatic hydrocarbons.

Properties. Clear, colorless, odorless, viscous liquids. Naphthene compressor oil-40 (NCO-40) occurs as a colorless liquid with a specific faint odor. High-purity mineral oil (HPMO) is a light-straw, oily liquid with an indefinite odor of petroleum products.

Applications. Used in the production of polyethylene and as a plasticizer for high-impact polystyrene, PVC, ABC plastics, polyamides, polyurethanes, ethylcellulose, etc. Polystyrene plasticized with compressor oil does not alter the characteristics of water or simulant media in contact with it.

Acute Toxicity. Administration to mice and rats produced no functional or morphological damage. No manifestations of the toxic effect were found on administration of 25 g NCO-40 or HPMO/kg BW to rats and mice. No acute effect threshold was established.[1,2] However, according to other data, *i/g* administration of petroleum oil to mice caused a toxic effect.[3] LD_{50} is reported to be 22 g/kg BW.[06]

Repeated Exposure failed to reveal cumulative properties. Mice and rats received 2 g/kg BW for 2 months. The treatment produced reversible changes in motor activity.

Short-term Toxicity. JECFA reviewed two 90-d dietary studies with MPO in rats.[4] Hematological changes and MPO deposition in the liver and spleen were reported.

Long-term Toxicity. Rats and mice were dosed by gavage with 50 and 200 mg MPO/kg BW for 10 months. No harmful effect was found.[2] In a 6-month study, guinea pigs were administered IS-45 oil at a dose of 0.5 g/kg BW. The treatment caused an increase in the phagocytic index and in the γ-globulin blood serum content.[5]

Reproductive Toxicity. No gonadotoxic action is reported; the reproductive function was not affected.

Genotoxicity. An increased frequency of CA was observed in the peripheral blood lymphocytes of glass workers exposed to mineral oil mists (Sram, Hola, Kotesovec, 1985). Two insulation oils from highly-refined mineral-base oils induced transformation of Syrian hamster embryo cells and enhanced transformation of mouse C3H 10TI/2 cells (IARC 33–87). Unused new, re-refined, and used crankcase oils induced transformation in Syrian hamster embryo cells (IARC 33–87).

Carcinogenicity. The carcinogenicity of MPO depends on the source and formulation of the petroleum. Analysis of MPO used for medicinal and cosmetic purposes reveals the presence of several carcinogenic polycyclic aromatic hydrocarbons. *Humans.* Considerable mortality or morbidity from stomach cancer was seen in workers exposed to MPO (IARC, Suppl. 7–252). *Animals.* The macroscopical and histopathological examination of Swiss mice treated by skin application for up to 18 months showed a definite tumorogenic skin effect of the ''aromatic extract'' and ''the distillate''. Gradiski et al.[3] pointed out that the observed effect was associated with the PAH concentration in the samples of white petroleum oil. *Carcinogenicity classification.* IARC: Group 1.

Standards. Russia (1988). Proposed PML: **n/m.**

Regulations. USFDA (1993) regulates MPO as a direct and indirect food additive. MPO may be used (1) in adhesives used as components of articles intended for use in packaging, transporting, or holding food, (2) as a component of resinous and polymeric coatings of a food-contact surface, (3) in cellophane for food packaging, (4) as a defoaming agent that may be used safely as a component of articles and in the manufacture of paper and paperboard intended for use in contact with food, (5) as an ingredient of paper and paperboard for contact with dry food, (6) in acrylate ester copolymer coating to be used safely as a food-contact surface of articles intended for packaging and holding food, including heating of prepared food, (7) as a substance employed in the production of or added to textiles and textile fibers intended for use in contact with food, and (8) as a plasticizer for rubber articles intended for repeated use in contact with food up to 30% by weight of the rubber product, alone or in combination with waxes, petroleum, total not exceeding 45% by weight of the rubber articles that contain at least 20% by weight of the ethylene- propylene copolymer elastomer. The FDA also regulates MPO as additives in animal feed. The FDA recommends warning labels for drugs containing MPO that are taken internally and classifies over-the-counter drug products containing MPO as GRAS.

Recommendations JECFA (1991). ADI for mineral oil (food grade): not specified.

References:

1. *Toxicology and Sanitary Chemistry of Plastics,* Abstracts, NIITEKHIM, Moscow, 1, 1979, p. 28 (in Russian).

2. See **ADIPIC ACID, #2, 263.**
3. Gradiski, D., Vinit, J., Zissu, D., et al., The carcinogenic effect of a series of petroleum-derived oils on the skin of mice, *Environ. Res.,* 32, 258, 1983.
4. The 33rd Report of JECFA, Technical Report Series 776, 24, 1989.
5. Krasovsky, G. N. and Friedland, S. A., Toxicology characteristics of mineral oils as flotation reagent for iron ores, *Gig. Sanit.,* 7, 17, 1969 (in Russian).

PHTHALIC ANHYDRIDE (PAn) (CAS No 85–44–9)

Synonyms. 1,2-Benzene-*o*-dicarboxylic anhydride; **1,3**-Isobenzofurandione; Phthalic acid anhydride; Phthalide.

PHTHALIC ACID (PAc) (CAS No 88–99–3)

Synonym. 1,2-Benzene-*o*-dicarboxylic acid.
Properties. *Phthalic anhydride.* Flaky crystals of white to light-brown color with an odor of naphthalene; readily volatizes. Poorly soluble in cold water, readily soluble in hot water with phthalic acid formation (1 mg PAn is equivalent to 1.12 mg of PAc). Soluble in alcohol. *Phthalic acid.* White crystals. Solubility in water is 0.57 g/100 ml at 20°C and 18 g/100 ml at 99°C. Odor and taste perception threshold is 56 to 57 mg/l.[1]
Applications. Widely used in the production of plasticizers and pigments, also as an anti-scorching agent and as a light- and heat-stabilizer of polyolefins.
Acute Toxicity. LD_{50} of PAn was identified to be 1.1 g/kg BW in rats and 1.5 g/kg BW in mice. LD_{100} seems to be 2 g/kg BW. According to other data, LD_{50} in mice is 2.2 g/kg BW.[2] The doses of 2.5 to 5 g/kg BW cause death in rats from necrosis of the renal tubules. Exposure to lethal doses results in adynamia, dyspnea, and damp, dishevelled fur. Gross pathological examination revealed distension of the stomach and intestine with ulcerations of their walls and pulmonary hemorrhages.[3]
Repeated Exposure failed to reveal cumulative properties. A dose of 680 mg/kg BW caused the majority of rats to die within a month (Pludro et al., 1969). Guinea pigs received a 500 mg/kg BW dose every 2 to 3 d for 42 d. The treatment caused no mortality or retardation of BW gain. The blood parameters were unchanged. Histological examination revealed signs of surface necrosis in the GI tract.[3]
Short-term Toxicity. Cats were given 68 mg PAn/kg BW for 90 d. Administration caused meteorism (tympanites) and diarrhea, and some excitation of the NS without other signs of intoxication (Pludro et al., 1969). Rabbits were given the dose of 20 mg/kg BW for 3 months. PAn administration produced leukocytosis and increased aldolase activity. On gross pathological examination, there was moderate parenchymatous dystrophy of the liver cells, with slight perivascular lymphoid infiltration.[2]
Long-term Toxicity. Rats and rabbits were given PAn for 6 months. The treatment resulted in increased bilirubin content (rabbits) and a decreased number of thrombocytes. Gross pathological examination showed dystrophic and reactive changes in the liver, kidneys, stomach, and intestine.[1]
Reproductive Toxicity. Gonadotoxicity. A gonadotoxic effect[4] on chronic inhalation exposure to PAn was observed at concentrations of 0.2 to 1 mg/m3. *Embryotoxic* effect (congenital defects) was shown in chickens after administration of 3 to 20 mg PAn/kg into the yolk sac.[5]
Carcinogenicity classification. NTP: N—N—N—N.
Standards. EEC (1990). SML: 7.5 mg/kg. **Russia** (1983). PML in drinking water: 0.5 mg/l.
Regulations. USFDA (1993) approved PAn for use (1) as an accelerator for rubber articles intended for repeated use in contact with food at quantities up to 1.5% by weight of the rubber product, (2) as a component of adhesives intended for contact with food, and (3) in polyurethane resins used as the food-contact surface of articles intended for use in contact with dry food.
References:
1. Meleshchenko, K. F., *Prevention of the Pollution of Water Bodies by Liquid Effluents of Chemical Plants,* Zdorov'ya, Kiev, 1971, p. 70 (in Russian).
2. *Toxicology of New Chemical Substances Used in Rubber and Tire Industry,* Meditsina, Moscow, 1968, p. 157 (in Russian).

3. See **EPICHLOROHYDRIN,** #3, 63.
4. Protsenko, E. I., Gonadotropic action of phthalic anhydride, *Gig. Sanit.,* 1, 105, 1970 (in Russian).
5. Verrett, M. J., Mutcher, M. K., Scott, W. F., et al., Teratogenic effect of capstan and related compounds in the developing chicken embryo, *Ann. N.Y. Acad. Sci.,* 160, 334, 1969.

POLYETHYLENE GLYCOLS (PEG) (CAS No 25322–68–3)

Synonyms. Carbowaxes; Poly(ethylene oxide); Polyoxes.
Properties. Liquids (MW = 200 to 600) or waxiform (MW = 1000 to 6000) products. Solubility in water is inversely proportional to molecular weight. Liquid PEG are colorless, almost odorless, and miscible with water. Waxiform PEG (carbowaxes) are soluble in water (50 to 73%). At a concentration of 1 g/l, they do not alter the color, odor, or taste of water.
Applications. Used as wetting or softening agents, antistatics in the production of urethane rubber, components of detergents, etc.
Acute Toxicity. After *i/v* administration, PEG are excreted unchanged. Waxiform PEG with MW = 4000 and 6000 are nontoxic when administered to rats; PEG with MW = 1000 and 1540 exhibited slight toxicity. The mean lethal doses of PEG are presented in the following table.

Mean Lethal Doses of Polyethylene Glycols (g/kg BW)

Polyethylene Glycols	Mice	Rats	Guinea Pigs	Rabbits
200 (insoluble)	33.9–38.3	28.9	16.9	14.1–19.9
300 (insoluble)	31.0	27.5–31.1	19.6–21.1	17.3–21.1
400 (insoluble)	28.9–35.6	12.9–30.2	15.7–21.3	22.3
600 (insoluble)	35.6–47.0	38.1	28.3	18.9
1000 (50% aq. sol.)	>50	42.0	22.5–41.0	>50
4000 (50% aq. sol.)	>50	>50	46.4–50.9	>50
6000 (50% aq. sol.)	>50	>50	>50	>50
9000 (50% aq. sol.)	>50	>50	>50	>50

Short-term Toxicity. There were no signs of toxicity in rats given 16% PEG (MW = 6000) or 4% PEG (MW = 200 to 4000) in their diet for 90 d.[1] Administration of PEG with MW = 200 to monkeys (2 to 4 ml/kg BW) and rats (2.5 to 5 ml/kg BW) for 13 weeks led to the deposition of a small quantity of oxalates in the lumen of the proximal tubules of the renal cortex (in monkeys only). No other morphological, biochemical, and hematological changes were found.[2]
Long-term Toxicity. There were no signs of toxicity in the hematological analysis or gross pathological examination in dogs given 10 to 90 mg PEG/kg BW (MW = 4000) for 43 to 178 d. Unchanged PEG are excreted in the urine. Addition of 4% PEG with M = 1500 to 4000 for 2 years and of 2% PEG with MW = 400, 1540, and 4000 to the feed of dogs for a year appeared to be harmless.[1,3]
Reproductive Toxicity. See **Diethylene glycol.**
Standards. Russia. Proposed PML: **n/m.**
Regulations. USFDA (1993) regulates PEG for use (1) in resinous and polymeric coatings in a food-contact surface, (2) as defoaming agents that may be safely used as components of articles intended for use in contact with food, (3) in cross-linked polyester resins for repeated use in articles or components of articles coming in the contact with food (PEG-6000), (4) as a component of the uncoated or coated food-contact surface of paper and paperboard intended for use in producing, manufacturing, packing, transporting, or holding aqueous and fatty food, (5) in adhesives used as components of articles intended for use in packaging, transporting, or holding food (PEG 200 to 6000), and (6) as a substance employed in the production of or added to textiles and textile fibers intended for use in contact with food (PEG 400 to 6000). PEG may be safely used (7) if the additive is an addition polymer of ethylene oxide and water with a mean molecular weight of 200 to 9500, (8) if PEG contains no more than 0.2% by weight of the ethylene and diethylene glycols and if its molecular weights are 350 or higher and no more than 0.5% by weight of the total of ethylene and diethylene glycols and if its mean molecular weight is below 350. **PEG monolaurate** (PEG-400)

containing not more than 0.1% by weight of the ethylene and/or ethylene glycol may be used at a level not to exceed 0.3% by weight of the twine, as a finish on twine to be used for tying meat, provided the twine fibers are produced from nylon resins (21 CFR #178.3760). **PEG alginate** is listed for use as a component of the uncoated or coated food-contact surface of paper and paperboard intended for use in producing, manufacturing, packing, transporting, or holding aqueous and fatty food.

References:
1. See **ETHYLENE OXIDE**, #2.
2. Prentice, D. E. and Majeed, S. K., Oral toxicity of polyethylene glycol (PEG 200) in monkeys and rats, *Toxicol. Lett.*, 2, 119, 1978.
3. See **ACRYLIC ACID**, #3.

1,2-PROPYLENE GLYCOL (PG) (CAS No 57–55–6)

Synonyms. **1,2**-Dihydroxypropane; 2-Hydroxypropanol; Isopropylene glycol; Methylethylene glycol; **1,2**-Propanediol.

Properties. Colorless, odorless, viscous liquid with a sweet taste. Mixes with water and alcohol at all ratios. Organoleptic perception threshold is 340 mg/l.[01]

Acute Toxicity. LD_{50} is 26.4 g/kg BW in rats and 20.3 g/kg BW in mice.[1]

Repeated Exposure. Rats were given 5 and 10% aqueous solutions of PG for 5 weeks. Administration resulted in increased liver weight and blood glucose level. The treatment reduced blood urea concentration and the erythrocyte count but did not affect the weights of other visceral organs.[2] Male Sprague-Dawley rats received undiluted PG at a dose of 4 mg/kg BW by gastric intubation for 30 d. The treatment produced no significant differences in plasma concentration of total phospholipids, cholesterol, triglycerides, and free fatty acids, and in the liver concentration of phospholipids, triglycerides, and gangliosides. Liver total cholesterol was moderately but significantly increased.[3]

Short-term Toxicity. Rats received 3.28 g PG/kg BW for 3 months. The treatment caused no impairment of their general condition and BW gain or alteration in the hematological analysis and renal function. Histological examination failed to reveal any changes in the visceral organs.[1]

Long-term Toxicity. Dogs were dosed with 5 mg/kg BW for 2 years. The treatment increased the rate of erythrocyte hemolysis and lowered Hb content (reversible changes). A dose of 2 mg/kg appeared to be ineffective.[4] No toxic manifestations were found to develop in rats given 200 to 2100 mg/kg BW for 2 years. The 25 mg/kg BW dose is suggested to be acceptable in man.[5]

Reproductive Toxicity. Rats received 0.2 ml of 10%-solution during the first 10 d of pregnancy. The treatment produced no adverse fetal effects.[6]

Allergenic Effect is not observed.

Genotoxicity. Produced a negative response in the SCE test system.[7]

Carcinogenicity. Negative results are reported in rats.[09]

Chemobiokinetics. PG metabolism is likely to occur through partial oxidation in the body to **lactic acid** with subsequent formation of **glucuronic acid.** PG is partially removed unchanged.

Standards. JECFA (1964) considered the dose of 20 mg/kg as safe to man. **Russia.** PML in drinking water: **n/m.**

Regulations. USFDA (1993) regulates the use of PG (1) as a component of adhesives used in a food-contact surface, (2) as a defoaming agent that may be used safely as a component of food-contact articles, (3) in resinous and polymeric coatings for polyolefin films to be used safely as a food-contact surface, (4) in polyurethane resins to be used safely in articles intended for use in contact with bulk quantities of dry food, (5) in cross-linked polyester resins for repeated use in articles or as components of articles coming in contact with food, (6) as a plasticizer for rubber articles intended for repeated use (up to 30% by weight of the rubber product), and (7) as a substance employed in the production of or added to textiles and textile fibers intended for use in contact with food. PG (MW above 1000) and PG alginate are listed (8) as components of paper and paperboard for use in contact with aqueous and fatty foods. PG is used in foods at levels not to exceed current GMP that result in maximum levels of 5% for alcoholic beverages, 2.5% for frozen dairy products, and 2% for all other food categories.

References:
1. *Cryobiology and Cryomedicine,* Vol. 8 Kiev, 1981, p. 46 (in Russian).
2. Vaille, Ch., Debray, C., Koze, C., et al., Hyperglycemic action of propylene glycol, *Ann. Pharmacol. Franc.,* 29, 577, 1971 (in French).
3. Hoenig, V. and Werner, F., Is propylene glycol an inert substance? *Toxicol. Lett.,* 5, 389, 1980.
4. Weil, C. S., Woodside, M. D., Smyth, H. F., et al., Results of feeding propylene glycol in the diet to dogs for two years, *Food Cosmet. Toxicol.,* 9, 479, 1971.
5. Gaunt, I. F., Carpanini, F. M., Grasso, P., et al., Long-term toxicity of propylene glycol in rats, *Food Cosmet. Toxicol.,* 10, 151, 1972.
6. El-Shabrawy, O. A. and Arbid, M., Evaluation of some drug solvents for teratological investigations in rats, *Egypt. J. Vet. Sci.,* 24, 143, 1988.
7. See **FURFURAL,** #5.

TRIACETIN (CAS No 102–76–1)

Synonyms. Glycerin triacetate; **1,2,3**-Propanetiol.
Properties. Colorless, odorless, oily liquid. Solubility in water is 7%; readily soluble in alcohol.
Applications. Used as a plasticizer in the production of cellulose derivatives and as a solvent for dye stuffs.
Acute Toxicity. LD_{50} is 6.4 to 12.8 g/kg BW in rats and 3.2 to 6.4 g/kg BW in mice. Poisoning is accompanied by weakness and ataxia.[03]
Regulations. USFDA (1993) approved the use of T. as (1) a component of resinous and polymeric coatings and (2) as a component of adhesives to be safely used in a food-contact surface.

TRI(2-BUTOXYETHYL) PHOSPHATE (CAS No 78–51–3)

Synonyms. 2-Butoxyethanol phosphate; Phosphoric acid, butylcellosolve ester.
Properties. Liquid with a sweetish odor. Solubility in water is 1.1 g/l; soluble in fats.
Applications. Used as a fire-resistant and light-stable plasticizer in the production of vinyl resins, rubber, nitrocellulose, and cellulose acetate and synthetic rubber.
Acute Toxicity. LD_{50} is 2.4 to 3 g/kg[4] or 5 g/kg BW in rats and 3.94 g/kg BW in mice.[1] Poisoning with lethal doses is characterized by an anticholinesterase action. Signs of intoxication include adynamia, ataxia, clonic-tonic spasms, disturbance of the rhythm, and rate of respiration, salivation, and tremor. According to Laham et al.,[2] a single administration of 1 to 3.3 g/kg BW to Sprague-Dawley female rats and 1 to 9 g/kg BW to males affects the peripheral NS. There was a significant reduction in caudal nerve conduction velocity. Light and electron microscopic examination revealed degenerative changes in myelinated and unmyelinated fibers.
Repeated Exposure failed to reveal cumulative properties. K_{acc} is 7 (by Lim).
Short-term Toxicity. Sprague-Dawley rats were dosed with 0.25 or 0.5 ml/kg BW for 18 weeks. The treatment caused a decline in cholinesterase activity in male but not in female rats. Histological examination revealed necrotic changes in the myocardium, with inflammatory cell infiltration.[2]
Allergenic effect is not observed.
Regulations. USFDA (1993) approved the use of T. in adhesives as a component of articles intended for use in packaging, transporting, or holding food.

References:
1. *Current Problems of Industrial Hygiene and Occupational Pathology,* Voronezh, 1975, p. 46 (in Russian).
2. Laham, S., Szabo, J., Long, G., et al., Dose-response toxicity of tributoxyethyl phosphate orally administered to Sprague-Dawley rats, *Am. Ind. Hyg. Assoc. J.,* 46, 442, 1985.

TRIBUTYL CITRATE (TBC)

Synonyms. *n*-Butyl citrate; Citric acid, tributyl ester; **2**-Hydroxy-**1,2,3**-propanetricarboxylic acid, tributyl ester.

Properties. Colorless or pale-yellow, odorless liquid. Poorly soluble in water. Miscible with most organic solvents.

Applications. Used as a plasticizer in PVC production. Solvent for nitrocellulose, lacquers.

Acute Toxicity. Single administration of 10 to 30 ml/kg BW caused no signs of intoxication.

Repeated Exposure failed to show signs of accumulation. TBC is likely to be of low toxicity, due to its insolubility in the body fluids. Rats received 5% TBC in their diet for 8 weeks. The treatment did not affect growth of animals. However, 10%-dose-level of TBC reduced BW gain and caused diarhhea. Histological examination revealed no changes in the viscera.

Compare to *Acetyl tributyl citrate.*

Standards. Russia. Recommended PML: **n/m.**

Regulations. USFDA (1993) approved the use of TBC in adhesives as a component of articles intended for use in packaging, transporting, or holding food.

Reference:

Finkelstein, M. and Gold, H., Toxicology of the citric acid esters: tributylcitrate, acetyltributyl-citrate, triethylcitrate, and acetyltriethylcitrate, *Toxicol. Appl. Phamacol.,* 1, 283, 1959.

TRIBUTYL PHOSPHATE (TBP) (CAS No 126–73–8)

Synonym. Phosphoric acid, tributyl ester.

Properties. Colorless, oily liquid with a sharp odor. Solubility in water is 397 mg/l at 19°C, completely soluble in mineral oil. Miscible with organic solvents. Odor perception threshold is 0.014 mg/l; taste perception threshold is 0.019 mg/l (Zyabbarova and Teplyakova, 1968).

Applications. TBP is used as a primary plasticizer in the manufacture of nitrocellulose, plastics, and vinyl resins; as a solvent for cellulose esters, lacquers, and natural gums.

Acute Toxicity. LD_{50} is 1.4 to 3.35 g/kg BW in rats, 0.9 to 1.24 g/kg BW in mice, and 1.8 g/kg BW in chickens.[1] Administration of 1 mg TBP/kg BW is accompanied by CNS excitation without convulsions, as well as its transient impairment.[4] Approximately the same dose causes impairment of the renal function in rabbits, with the appearance of protein in the urine.

LD_{50} of *triisobutyl phosphate,* which is of similar toxicity, appears to be in the range of 3.2 to 6 g/kg BW.[05]

Repeated Exposure revealed moderate cumulative properties. $K_{acc} > 3$. TBP exerts a neurotoxic effect on the peripheral nerves; it produces a weak anticholinesterase action, affecting mainly the NS and parenchymatous organs. The kidney seems also to be a target organ (Kalinina and Peresadov, 1970). Wistar male rats were fed 0.5 and 1% TBP in the diet for 10 weeks.[2] BW gain and food consumption in the treated animals were significantly lower than those in the controls. There were no differences in cholinesterase activity in the serum, but that in the brain was significantly increased. The blood coagulation time was found to be prolonged. Delayed neuropathy was observed following oral administration of 0.42 ml/kg BW dose for 14 d. In this study, no axonal degeneration or other overt signs of toxicity were reported.[3]

Short-term Toxicity. In a 18-week study, doses of 200 and 300 mg/kg BW produced diffuse hyperplasia of the urinary bladder epithelium in Sprague-Dawley rats.[4]

Long-term Toxicity. The treatment of rats and rabbits caused dystrophic changes in the liver, necrosis of separate cells or of groups of cells, and fatty dystrophy. Functional effects are not reported (Zyabbarova and Teplyakova).

Reproductive Toxicity. TBP is reported to be slightly teratogenic in chickens at high dose-levels. A dose of 0.42 mg/kg BW (14 d) causes degenerative changes in the seminiferous tubules to develop.[4] According to Schroeder et al., TBP produced no teratogenic effect in rats and rabbits.[5]

Genotoxicity. TBP is shown to be negative in bacterial tests and in *Dr. melanogaster,* but it was positive in *Salmonella* mutagenicity assay.[6]

Chemobiokinetics. Following ingestion, TBP is distributed mainly to the blood, GI tract, and liver.[7] More than 50% of an orally administered dose was absorbed within 24 h.[8] TBP undergoes oxidation of the butyl moieties. Oxidized methyl groups are removed as glutathione conjugates and subsequently excreted as *N*-**acetyl cysteine derivatives.** Excretion occurs predominantly via the urine.

Standards. Russia. MAC: 0.01 mg/l (organolept.).

Regulations. USFDA (1993) regulates the use of TBP (1) in adhesives as a component of articles intended for use in packaging, transporting, or holding food and (2) as a defoaming agent that may be safely used in the manufacture of paper and paperboard intended for use in producing, manufacturing, packing, transporting, or holding food.

References:

1. IPCS, *Tri-n-butyl Phosphate,* Environmental Health Criteria 112, WHO, Geneva, 1991, p. 80.
2. Oishi, H., Oishi, S., and Hiraga, K., Toxicity of tri-*n*-butyl phosphate, with special reference to organ weights, serum components and cholinesterase activity, *Toxicol. Lett.,* 6, 81, 1980; Toxicity of several phosphoric acid esters in rats, *Toxicol. Lett.,* 13, 29, 1982.
3. Laham, S. and Long, G., Subacute oral toxicity of tri-*n*-butyl phosphate in the Sprague-Dawley rat, *J. Appl. Toxicol.,* 4, 150, 1984.
4. Laham, S., Long, G., and Broxup, B., Induction of urinary bladder hyperplasia in Sprague-Dawley rats orally administered tri-*n*-butyl phosphate, *Arch. Environ. Health,* 40, 301, 1985.
5. Schroeder, R. E., Gehart, J. ., and Kneiss, J., Developmental toxicity studies of tributyl phosphate (TBP) in the rat and rabbit, *Teratology,* 43, 455, 1991.
6. Gafieva, Z. A. and Chudin, V. A., Evaluation of the mutagenic activity of tributylphosphate on Salmonella typhimurium, *Gig. Sanit.,* 9, 81, 1986 (in Russian).
7. Khalturin, G. V. and Andryushkeyeva, N. I., Toxicokinetics of tributyl phosphate following single and chronic intragastric intake by rats, *Gig. Sanit.,* 2, 87, 1986 (in Russian).
8. Suzuki, T., Sasaki, K., Takeda, M., et al., Metabolism of tributylphosphate in male rats, *J. Agric. Food Chem.,* 32, 603, 1984.

TRI(2-CHLOROETHYL) PHOSPHATE (CAS No 115–96–8)

Synonyms. Phosphoric acid, tri(2-chloroethyl) ether; Tris(2-chloroethyl)phosphate.

Properties. Clear, colorless liquid of low volatility with a faint odor. Solubility in water is 0.7%, soluble in numerous organic solvents.

Applications. Used as a plasticizer in the production of cellulose derivatives. A plasticizer for fire-resistant lacquers (coatings) and plastics based on ethylcellulose, polyester resins, polyacrylates, and polyurethanes.

Acute Toxicity. LD$_{50}$ is reported to be 0.2 to 0.4 g/kg BW,[05] 0.63 to 0.74 g/kg BW,[1] or 6.8 g/kg BW (Dvorkin) in rats, 0.74 g/kg BW in mice, and 1 g/kg BW in guinea pigs.[1] According to other data, LD$_{50}$ in rats is 1250 mg/kg BW (sex unspecified), or 500 mg/kg BW in males and 430 to 800 mg/kg BW in females.[2] T. produces a weak anticholinesterase action. Poisoning is accompanied by tonic and tetanus-like convulsions but not paralysis. The skin of animals becomes cyanotic, and they assume a side position. Death within 1 to 2 d. Gross pathological examination revealed celiectasia, visceral congestion, and tuberosity of the spleen.

Repeated Exposure revealed slight cumulative properties. T. exhibits a polymorphous toxic action, with the CNS as the primary target, followed by the liver, kidneys, and myocardium. Rats were given an oily solution of T. (dose-level of 6.3 and 63 mg/kg BW) for 1.5 months. No deaths occurred. The greater dose resulted in decreased activity of some enzymes (AST, ALT, LDH, creatinekinase, X-hydroxybutyrate dehydrogenase) and creatinine content in the blood serum and other changes. The dose of 6.3 mg/kg BW caused less pronounced reversible changes.[1]

Short-term Toxicity. Administration of 350 mg/kg BW by oral gavage over a period of 16 weeks resulted in necrosis of pyramidal neurons in the CA I region of the hippocampus of Fisher 344 rats but not in B6H3F$_1$ mice.[3]

Long-term Toxicity. Fisher 344 rats received 44 and 88 mg T./kg BW by gavage for up to 103 weeks and B6H3F$_1$ mice received 175 and 350 mg T./kg BW for 66 weeks. The principal toxic effect occurred in the brain and kidney.

Reproductive Toxicity. Embryo- and gonadotoxic effects were not demonstrated with the 6.3 mg/kg BW dose.[1] Wistar rats were gavaged with 50 to 200 mg/kg BW suspended in olive oil on days 7 to 15 of gestation. There were no changes in maternal body weight, food consumption, or general appearance up to 100 mg/kg BW level. A dose of 200 mg/kg BW caused weakness and decreased maternal food consumption; 7 out of 30 dams died. No malformations were registered at any dose; there was normal development of the offspring of all groups.[4]

Genotoxicity. T. is not found to react with DNA *in vivo*,[5] but it caused an increased number of SCE in Chinese hamster V79 cell line and DLM in rats after inhalation exposure.[6,7]

Carcinogenicity. In the above described study,[3] renal tubular hyperplasia and adenoma were observed. Renal neoplasms ware found in 10% of low-dose and in 50% of high-dose male rats. Mice were less sensitive.

Chemobiokinetics. *In vitro* metabolism by rat and human liver slices led to formation of bis(2-chloroethyl)hydrogen phosphate and 2-chloroethanol and three unidentified minor metabolites. T. metabolism was approximately twice as rapid in rat liver slices as human liver slices (NTP-92).

Standards. Russia. Proposed MAC: 1 mg/l.

Regulations. USFDA (1993) listed T. for use in adhesives as a component of articles intended for use in packaging, transporting, or holding food.

References:

1. Zaitsev, N. A. and Skachkova, I. N., Hygienic regulation of tri(chloroethyl)phosphate in water, *Gig. Sanit.,* 9, 77, 1989 (in Russian).
2. Ulsamer, A. G., Osterberg, R. E., McLaughlin, J., et al., Flame retardant chemicals in textiles, *Clin. Toxicol.,* 17, 101, 1980.
3. Matthews, H. B., Eustis, S. L., and Haseman, J., Toxicity and carcinogenicity of chronic exposure to tris(2-chloroethyl)phosphate, *Fund. Appl. Toxicol.,* 20, 477, 1993.
4. Kawashima, K., Tanaka, S., Nakaura, S., et al., Effect of oral administration of tris(2-chlorethyl)phosphate to pregnant rats on prenatal and postnatal developmental, *Bull. Natl. Inst. Hyg. Sci.,* 101, 55, 1983.
5. Lown, J. W., Joshua, A. V., and Melaughlin, L. W., Novel antitumor nitrosoureas and related compounds and their reactions with DNA, *J. Med. Chem.,* 23, 798, 1980.
6. Sala, M., Gu, Z. G., Meons, G., et al., *In vivo* and *in vitro* biological effects of the flame retardant tris(2,3-dibromopropyl)phosphate and tris(2-chlorethyl)phosphate, *Eur. J. Cancer,* 18, 1337, 1982.
7. Shepelskaya, N. R. and Dyshinevich, N. Ye., Experimental study of the gonadotoxic effect of tris(2-chloroethyl)phosphate, *Gig. Sanit.,* 6, 20, 1881 (in Russian).

TRIETHYL CITRATE (TEC) (CAS No 77–93–0)

Synonyms. Citric acid, triethyl ester; Ethyl citrate.

Properties. Colorless, odorless, oily liquid with a bitter taste. Readily soluble in water (6.9%). Miscible with alcohol and ethers.

Applications. Used as a plasticizer in the production of PVA and cellulose citrate.

Acute Toxicity. LD_{50} is found to be 5.9 to 8 g/kg BW in rats and 4 g/kg BW in cats. Poisoning is accompanied by convulsions and respiratory disorders. Changes in behavior and in the GI tract were observed. The toxic action of TEC is likely to occur due to the binding of **calcium** in the body fluids. Similar data were obtained in relation to **Acetyl triethyl citrate.** However, its LD_{50} in rats is half that of TEC.

Repeated Exposure. Small amounts of TEC in the feed are harmless for rats and cats. Cats were given 280 mg/kg BW for 8 weeks. The treatment did not affect BW gain, hematological analysis, or sugar and nitrogen level in the blood. Nevertheless, after four to five doses, weakness, ataxia, and depression occurred.[1]

Long-term Toxicity. According to results of a 2-year feeding study, rats can tolerate TEC at a dose of up to 2 g/kg BW. Dogs tolerated up to 0.25 ml/kg BW for 6 months without any adverse effect.[2]

Reproductive Toxicity. The doses of 0.5 to 10 mg/kg BW are not teratogenic for chick embryos (Verett, 1980).

Genotoxicity. TEC is not mutagenic in bacterial tests.[1]

Chemobiokinetics. TEC is likely to be hydrolyzed *in vivo* to yield **citrate** and **ethyl alcohol** by usual biochemical routes.

Standards. Russia. Recommended PML: **n/m.**

Regulations. USFDA (1993) listed TEC for use (1) as a component of resinous and polymeric coatings in a food-contact surface and (2) as a plasticizer in resinous and polymeric coatings for polyolefin films to be safely used as a food-contact surface of articles intended for use in producing, manufacturing, packing, transporting, or holding food.

Recommendations. JECFA (1989). ADI for man: 0 to 10 mg/kg.

References:

1. See **TRIBUTYL CITRATE.**
2. See 1,3-**BUTANEDIOL,** #1, 18.

TRIETHYLENE GLYCOL (TEG) (CAS No 112–27–6)

Synonyms. Triglycol; Ethylene glycol, di-2-hydroxyethyl ester.

Properties. Glycerin-like, colorless, odorless liquid with a burning taste. Miscible with water and alcohol, very hygroscopic. A concentration up to 1 g/l does not affect color, odor, and taste of water. At levels up to 500 mg/l, foam formation does not occur. According to other data, organoleptic perception threshold is 700 mg/l.[01]

Applications. TEG is used in the manufacture of solvents and plasticizers. An antistatic.

Acute Toxicity. *Humans.* LD_{50} is reported to be 5 g/kg BW. *Animals.* LD_{50} is in the range of 15.5 to 17.5 g/kg BW in rats, 18.5 to 20.8 g/kg BW in mice, 7.9 to 14.7 g/kg BW in guinea pigs, and 8.4 to 9.5 g/kg BW in rabbits (Stenger et al., 1968). Toxic manifestations comprise convulsions, exophthalmia, dyspnea, ataxia, and hematuria. Gross pathological examination revealed pronounced swelling of the renal tubular epithelium, vacuolar dystrophy, and focal hepatocyte necrobiosis, a marked congestion in the viscera.[1]

Repeated Exposure failed to reveal cumulative properties. Female rats were dosed by gavage with 1 g/kg BW for 20 d. The treatment produced changes in the liver, kidney, and CNS function. Main signs of intoxication appeared to be an increase in blood Hb level and eosinophil count and in serum ALT activity, a decline in the activity of cytochrome-*c*-oxidase, and a rise in the STI value. Gross pathological examination revealed congestion of the viscera and parenchymatous dystrophy in the liver and kidney.[1]

Long-term Toxicity. Rats received up to 5 mg TEG/kg BW. The treatment lowered the activity of transaminase in the liver, led to disruption of the amino acid conversion process and to subsequent restriction of their involvement in the synthesis of proteins, glycogen, and fatty acids.[2] Administration of 5 mg/kg BW to rats affected kidney and liver functions (a decline in some enzyme activity, reduction of glycogen content).[1]

Reproductive Toxicity. *Gonadotoxicity.* TEG is found to disturb energy exchange in the testicular tissue and produce dystrophic changes in the spermatogenic epithelium (see ***Diethylene glycol***). TEG caused a reduction in sperm motility; it increased the number of immobile forms and lowered their resistance. It induced DLM and damaged the sex cells of male rats, but produced little effect on the female gonads.[2] *Teratogenic* effect is not observed in mice, rats, and rabbits administered 1 to 4 ml TEG/kg BW throughout pregnancy.

Genotoxicity. TEG was shown to be positive in the *Salmonella* mutagenicity assay (NTP–91).

Carcinogenicity. Negative results are reported in male rats.[09]

Chemobiokinetics. Intestinal absorption is very low. TEG does not cause oxalate stone formation in the bladder and kidneys. Almost entirely excreted unchanged with the urine and feces (90 to 97%).[3]

Standards. Russia (1988). MAC: 1 mg/l; PML in drinking water: **n/m.**

Regulations. USFDA (1993) regulates the use of TEG (1) as a component in resinous and polymeric coatings for a food-contact surface, (2) as a plasticizer in polymeric substances used in the manufacture of food-contact articles, (3) in cellophane for food packaging, (4) as a component of polyester resins for coatings not exceeding a coating weight of mg/in.² and those intended for contact under specified conditions, and (5) as a component of adhesives (diethylene glycol content in TEG not to exceed 0.1%).

References:

1. Tolstopyatova, G. V., Korcach, V. I., Barilyak, I. R., et al., Hygienic regulations of tri-, tetra-, and pentaethylene glycol in waterbodies, *Gig. Sanit.,* 12, 77, 1987 (in Russian).
2. Korkach, V. I. and Spitkovskaya, L. D., The effect of triethylene glycol on biochemical processes in animals, *Gig. Sanit.,* 5, 91, 1986 (in Russian).

3. McKennis, H., Turner, R. A., Turnbull, L. B., et al., The excretion and metabolism of triethylene glycol, *Toxicol. Appl. Pharmacol.*, 4, 441, 1962.

TRIS(2-ETHYLHEXYL) PHOSPHATE (TEHP) (CAS No 78–42–2)

Synonyms. 1-Hexanol, 2-ethyl, phosphate; Phosphoric acid, tris(2-ethylhexyl) ester.

Properties. Thick, clear liquid. Water solubility is >0.1 g/l at 20°C.

Applications. Used as a plasticizer in the production of PVC, giving it low-temperature resistance. Dispersing agent of plastisols.

Acute Toxicity. TEHP appears to be a substance of low acute toxicity. LD_{50} in rats is 37.8 g/kg BW. Some animals die only after administration of 46 g/kg dose.[05] However, Akhlustina[1] reported the LD_{50} to be 7.2 g/kg BW in mice; a dose of 10 g/kg BW in rats had no lethal effect. After administration of the lethal dose, mice exhibited disturbances in respiratory and GI tract function. Death occurred within 18 to 24 h. Gross pathological examination revealed visceral congestion, marked perivascular and pericellular brain edema, and parenchymatous dystrophy of the liver and kidneys. TEHP causes no neurotoxic (demyelinization) effect.

Repeated Exposure failed to reveal cumulative properties. Administration of $^1/_{10}$ LD_{50} caused no mortality in rats and mice.[1] The fur falls out around the anal orifice, and there is skin erosion. Changes are noted in the STI, blood cholinesterase activity, leukocyte count, urinary hippuric acid content, and the glycemic curve.

Short-term Toxicity. Exposure to 430 mg/kg BW (rats) for 30 d revealed no signs of intoxication. A 1.55 g/kg dose caused retardation of BW gain.[05]

Genotoxicity. TEHP is found to be negative in *Salmonella* mutagenicity bioassay (NTP–92).

Carcinogenicity. Formation of liver carcinomata is reported on administration of 500 and 1000 mg/kg BW to mice.[2] *Carcinogenicity classification.* NTP: E—CE—EE—SE.

Chemobiokinetics. Following inhalation exposure to a labeled aerosol, TEHP is detected in the brain, liver, and stomach contents. After 2 d, the content of TEHP excreted in the feces is significantly higher than in the urine.

Regulations. USFDA (1993) approved the use of TEHP (1) in adhesives used as components of articles intended for use in packaging, transporting, or holding food and (2) as a defoaming agent that may be used safely in the manufacture of paper and paperboard intended for use in producing, manufacturing, packing, transporting, or holding food.

References:

1. *Hygiene and Toxicology of Polymeric Construction Materials*, A. N. Bokov, Ed., Medical Institute, Rostov-na-Donu, 1973, p. 353 (in Russian).
2. Haseman, J. K., Crowford, D. D., Huff, J. E., et al., Results from 86 two-year carcinogenicity studies conducted by the National Toxicology Program, *J. Toxicol. Environ. Health*, 14, 661, 1984.

TRIETHYL PHOSPHATE (CAS No 78–40–0)

Synonyms. Ethyl phosphate; Phosphoric acid, triethyl ester.

Properties. Colorless liquid, soluble in water and alcohol.

Applications. Used as a plasticizer in the production of plastics. Blends well with PVC and its copolymers. Gives products high elasticity and low-temperature resistance. T. is used to make artificial leather and film materials.

Acute Toxicity. In rats and mice, LD_{50} is 1.4 g/kg BW. Administration of the lethal doses led to excitation, with subsequent CNS inhibition, motor coordination disorder, paresis of the hind legs, and respiratory disturbances. Death occurred in 24 h. Gross pathological examination revealed visceral congestion, particularly in the liver, and hyperemia of the gastric mucosa.

Repeated Exposure revealed no cumulative properties.

Standards. Russia. PML in drinking water: 0.5 mg/l.

Regulations. USFDA (1993) approved the use of T. in adhesives as a component of articles intended for use in packaging, transporting, or holding food.

128

Reference:

Proc. Research Institute Epidemiol. Hygiene, Kuibyshev, 5, 117, 1968 (in Russian).

TRIPHENYL PHOSPHATE (TPP) (CAS No 115–86–6)

Synonym. Phosphoric acid, triphenyl ester.

Properties. Crystalline, slightly aromatic solid. Water solubility is about 2.1 mg/l, moderately soluble in ethanol. A marked change is noted in the taste and smell of solutions stored in containers made of etrol that has been plasticized with TPP.

Applications and **Exposure.** Used as a plasticizer in the production of cellulose acetate articles and in the manufacture of lacquers. Because of its limited compatibility and low effectiveness, it is used in blends with other plasticizers. A flame retardant and a solvent. Average daily intake in food was found to be several ng/kg BW (USFDA).

Acute Toxicity. LD_{50} is 3.5 to 10.8 g/kg BW in rats, 1.3 to 5 g/kg BW in mice, more than 4 g/kg BW in guinea pigs, and about 8 g/kg BW in rabbits.[1] After administration, animals are depressed. The urine and expired air smell of TPP. Death occurs within 4 to 5 d. Gross pathological examination revealed distention of and inflammatory changes in the stomach and intestine, a clay-like consistency of the liver and kidneys, and brain hemorrhages.[2,3] Sutton et al. reported 3 to 4 g/kg BW dose to be harmless in rats and mice.[4]

Repeated Exposure failed to reveal cumulative effect.[2,3] The early studies[5] on neurotoxicity of TPP have been questioned. In a 35-d feeding study in male Holtzman rats (doses applied 1 and 5 g/kg BW), retardation of BW gain and increase in liver weights were found only with a 5 g/kg BW dose, but no hematological changes were noted. A pronounced reduction in the BW of rabbits on repeated administration of 2.4 g/kg BW was observed.[4] Although ineffective in the CNS of rats and young chickens, TPP caused limb paralysis in cats.[05]

Short-term Toxicity. Oral administration of 380 to 1900 mg/kg BW for 3 months to rats caused no deaths; there was no evidence of abnormal growths; cholinesterase activity was unaltered.

Long-term Toxicity. Administration in the diet of rats at the level of 5, 10, and 100 mg TPP/kg BW for 6 months caused no retardation of BW gain.[05]

Immunotoxicity. Sprague-Dawley rats were orally exposed to TPP at doses of 2.5 to 10 g/kg BW for 120 d. No significant effect on the humoral response was found. The only effect noted was increased level of globulin at the dose of 10 g/kg BW.[6]

Reproductive Toxicity. There was no overt maternal toxicity or embryotoxicity in the Sprague-Dawley rats after dietary exposure to 166 to 690 mg/kg BW over a period of 91 d, including the mating and gestation periods.[7]

Genotoxicity. There were negative results in several *in vitro* studies.[1]

Chemobiokinetics. TPP is likely to be hydrolyzed to form **phenol metabolites,** which are excreted in the urine. TTP is quite possibly excreted unchanged by the same route.

Standards. Russia (1988). PML in food: 0.5 mg/kg. PML in drinking water: 1 mg/l.

Regulations. USFDA (1993) approved the use of TPP (1) in adhesives as a component of articles intended for use in packaging, transporting, or holding food and (2) in cross-linked polyester resins to be used safely in articles or as a component of articles coming in contact with food.

References:

1. IPCS, *Triphenyl Phosphate,* Environmental Health Criteria No. 111, WHO, Geneva, 1991, 80.
2. Antonyuk, O. K., Hygienic evaluation of the plasticizer triphenyl phosphate added to polymer composition, *Gig. Sanit.,* 8, 98, 1974 (in Russian).
3. *Problems of Industrial Hygiene, Occupational Pathological and Toxicology in the Production and Use of Organophosphorus Plasticizers,* Meditsina, Moscow, 1983, p. 96 (in Russian).
4. Sutton, W. L., Terhaar, C. J., Miller, F. A., et al., Studies on the industrial hygiene and toxicology of triphenyl phosphate, *Arch. Environ. Health,* 1, 45, 1960.
5. Smith, M. I., Evolve, E., and Frazier, W. H., Pharmacological action of certain phenol esters with special reference to the etiology of the so-called ginger paralysis., *Public Health Rep.,* 45, 2509, 1930.
6. Hinton, D. M., Jessop, J. J., Arnold, A., et al., Evaluation of immunotoxicity in subchronic feeding study of triphenyl phosphate, *Toxicol. Ind. Health,* 3, 71, 1987.

7. Welsh, J. J., Collins, T. F. X., Whitby, K. E., et al., Teratogenic potential of triphenyl phosphate in Sprague-Dawley (Spartan) rats, *Toxicol. Ind. Health,* 3, 357, 1987.

TRI(2-PROPOXYETHYL) PHOSPHATE

Synonyms. Propyl glycol phosphate; Propylcellosolve phosphate.

Properties. Colorless liquid with a faint odor.

Applications. Used as a plasticizer in the production of cellulose acetate.

Acute Toxicity. In Wistar rats, LD_{50} of T. mixed with pure ethyl alcohol (small amounts to reduce viscosity) appeared to be 4 to 6 g/kg BW. Rats tolerated doses of 2 to 2.5 g/kg BW.

Long-term Toxicity. No signs of intoxication or histological changes were noted in rats dosed with up to 0.9 g T./kg diet for 21 months. The treatment caused no growth retardation.[04,05]

Reproductive Toxicity. At the dose-level of 50 to 368 mg T./kg, it did not affect development of the progeny over three to five generations.

Standards. Russia. PML: **n/m.**

TRIXYLYL PHOSPHATE (CAS No 25155–23–1)

Synonyms. Phosphoric acid, trixylylester; Phosphate xylenol.

Composition. A mixture of isomers differing in the position of two CH_3 groups in the xylyl radicals.

Properties. Dark, oily liquid. Poorly soluble in water; readily soluble in oil. Perception threshold for the effect on the organoleptic properties of water is 0.05 mg/l.

Acute Toxicity. LD_{50} values of *m*-T. and *p*-T. are 24 and 25 g/kg BW, respectively.[1] According to other data, the maximum tolerated dose in mice is 5 g/kg BW, and LD_{50} is 12 g/kg BW; rats tolerate 20 g/kg BW. Administration of lethal doses led to weakness and adynamia. The fur of animals becomes disheveled. Histological examination revealed diffuse congestion in the renal tissue.[2] Administration in vegetable oil increases toxicity. LD_{50} of *di-3,5-xylylphenylphosphate* for mice is 7.4 g/kg BW. Administration did not cause paralysis.[3]

Repeated Exposure. T. has marked cumulative properties and is retained almost entirely in the body. Rats received $^1/_{30}$ LD_{50} for a month. The treatment led to exhaustion and decline in cholinesterase activity. Gross pathological examination revealed areas of congestion in the brain, gliosal nodules, invagination of the vessel walls, nerve cell changes, and fatty dystrophy of the renal tubular epithelium.[2] No paralyses were observed in chickens in the course of 40 daily administrations of 1 g/kg into the crop of rooster.[3]

Standards. Russia (1983). MAC: 0.05 mg/l (organolept.).

References:
1. *Problems of General and Industrial Toxicology,* Leningrad, 1965, p. 135 (in Russian).
2. See **TRIETHYL PHOSPHATE, 94.**
3. *Problems of Occupational Hygiene, Occupational Pathological and Toxicology in Phosphorganic Plasticizers Production and Use,* Moscow, 1973, p. 80.

VASELINE OIL (VO)

Composition. A mixture of hydrocarbons. Purified fractions obtained after distillation of kerosene. Polycyclic aromatic hydrocarbons, in particular benzo[a]pyrene, are found in technical grade VO. These compounds are not present in VO that had undergone double purification.

Properties. A colorless, oily liquid without odor or taste. Poorly soluble in water.

Acute Toxicity. 15 g/kg BW of technical grade VO caused no mortality in rats and mice.[1]

Repeated Exposure revealed slight cumulative properties.

Long-term Toxicity. No signs of intoxication or carcinogenicity effect were noted in rats given 0.1 g/kg BW over a period of 10 months.[2]

Carcinogenicity. No carcinogenic action was observed on subdermal administration to rats in the groin region.

Regulations. Russia: Medical grade VO is provisionally permitted by Ministry of Health for use in the production of rubbers intended for contact with food products. PML: **n/m.**

References:

1. Vysheslavova, M. Ya. Toxicity of vaseline oil, *Probl. Nutr.,* 3, 73, 1976 (in Russian).
2. *Toxicology of the Components of Rubber Mixes and of Rubber and Latex Articles,* Central Research Institute of Petroleum Chemistry, Moscow, 1974, p. 20 (in Russian).

VINYLMETHYL ADIPATE (CAS No 2969–87–1)

Synonyms. Adipic acid, vinylmethyl ester.

Properties. A clear, colorless liquid. Poorly soluble in water, readily soluble in alcohol. Aromatic odor and taste are detectable at a concentration of 2 mg/l. Practical organoleptic threshold is in the range of 5 to 5.5 mg/l. A concentration of up to 20 mg V./l does not affect the color or transparency of water.[1]

Applications. A plasticizer giving plastics a special low-temperature resistance.

Acute Toxicity. In rats and rabbits, LD_{50} is 6.2 g/kg BW.[1] Lethal doses produce excitation with subsequent CNS inhibition and death. Guinea pigs are insensitive to V.: a single administration caused none of them to die.

Repeated Exposure revealed moderate cumulative properties. K_{acc} is 6.4 (by Cherkinsky). A decrease in leukocyte phagocytic activity and in ascorbic acid contents in the suprarenals was found in rats that received $1/20$ LD_{50}. Histological examination showed inflammatory and dystrophic changes in the viscera.[1]

Genotoxicity. V. was negative in mutagenicity assay in *Dr. melanogaster.*[2]

Standards. Russia. PML in drinking water: 3 mg/l.

References:

1. Mironets, N. V., Comparative toxicologic characteristics of the plasticizers vinylmethyl adipate and divinyl adipate and their hygienic standards in water bodies, *Gig. Sanit.,* 10, 88, 1970 (in Russian).
2. Sheftel, V. O., Shkvar, L. A., and Naumenko, G. M., The use of some genetic methods in hygienic studies, *Vrachebnoye Delo,* 7, 120, 1969 (in Russian).

XYLITOL (CAS No 87–99–0)

Synonyms. Klinit; Pentane pentol; Xylit.

Properties. Colorless, hygroscopic crystals. Water solubility is 30% (20°C). Soluble in alcohol. Has a cooling, sweet taste. Renders water a foreign taste at a concentration of 1 g/l.[1]

Applications and **Exposure.** Used as a plasticizer in the production of cellophane and other plastics. X. is a sweet, five-carbon sugar alcohol, which has been recommended as a sugar substitute for special dietary uses.

Acute Toxicity. LD_{50} is 17.3 or 22 g/kg BW in rats, 12.5 g/kg BW in mice, and 25 g/kg BW in rabbits.[2] Lethal doses produce narcotic action. Death within 6 h. Gross pathological examination revealed hemorrhagic foci in the mucosa of the stomach and intestine.[1]

Repeated Exposure. Long Evans male rats were fed a X.-containing diet for 4 weeks. The amount of polyol in the diet was increased from 5% to the final 20% level within 3 weeks. The treatment produced retardation of BW gain and a fourfold increase in the titratable acid excretion. Urinary pH was lowered (from 6.6 to 5.6). Increase in daily urine volumes by 49% was noted.[3]

Short-term Toxicity. Mice and rats were dosed by gavage with $1/10$ and $1/5$ LD_{50} for 20, 30, and 90 d. The treatment caused a decline in capacity to work, a rise in the blood histamine level, and a shortening of the period of barbiturate narcosis.[1] A reversible enlargement of the caecum was reported.[2]

Long-term Toxicity. Rats and rabbits received X. by gavage for 7 months. The treatment caused disturbances in the functional state of the cerebral cortex cells, in mediator and nuclear metabolism.[1] Young Wistar and Sprague-Dawley rats, CD-1 and NMRI mice were found capable of adapting 20% dietary X.[2]

Immunotoxicity. In a long-term study, administration of X. did not affect immunoreactivity of rats.

Reproductive Toxicity. In the three-generation reproduction study, 20% X. in the diet caused no impairment in NMRI mice.[2]

Genotoxicity. Produced a doubling of the SCE frequency, and a three-point monotonic increase with at least the highest dose at the p <0.001 significance level.[4]

Chemobiokinetics. X. metabolism occurs through intensive oxidation within a few hours after administration. Neither X. nor its metabolites are found in the urine. Increased amounts of **methylmalonic acid** and 2-**oxoglutaric acid** are noted in the urine of polyol-fed rats. The urinary excretion of **citric acid** and **maleic acid** was also increased significantly. The increased levels of urinary organic acids may be explained in terms of impaired mitochondrial oxidation of these acids and of impaired conversion of methymalonic acid to **succinic acid.**[3]

Standards. Russia. Recommended MAC and PML: 1 mg/l.

References:

1. Plitman, S. I., Experimental validation of the Maximum Allowable Concentration of pentaerythritol and xylitol in water bodies, *Gig. Sanit.,* 2, 25, 1971 (in Russian).
2. Third Int. Cong. on Toxicology, San Diego, Abstracts, *Toxicol. Lett.,* 18 (Suppl. 1), 37, 1983.
3. Hamalainen, M. M., Organic aciduria in rats fed high amounts of xylitol or sorbitol, *Toxicol. Appl. Pharmacol.,* 90, 217, 1987.
4. See 1,2-**PROPYLENE GLYCOL,** #7.

Chapter 3
STABILIZERS

AKRIN *MD*

Composition and **Properties.** A mixture of products of the high-temperature condensation of diphenylamine with acetone. Dark-brown, viscous liquid. Insoluble in water, partially soluble in alcohol.

Applications. Used as a stabilizer in the manufacturing of synthetic rubber and lattices, a thermostabilizer for polypropylene, PVC, and polyamides.

Acute Toxicity. In mice, LD_{50} is 23 g/kg BW.[08] High doses affect the osmotic stability of erythrocytes and cause methemoglobinemia. Histological examination revealed circulatory disturbances in the visceral organs, necrotic changes in the small intestine mucosa, and parenchymatous dystrophy of the renal tubular epithelium.

Short-term Toxicity. Rats were dosed with 250 mg/kg BW for 3.5 months. Manifestations of toxic action included decreased BW gain, mild methemoglobinemia, and stimulation of conditioned reflex activity. Gross pathological examination revealed surface necrosis of the villi and desquamation of the epithelium of the gastric and intestinal mucosa.[08]

4-ALK(C_7–C_9)OXY-2-HYDROXYBENZOPHENONE

Synonym. Benzone OA.

Properties. Viscous, light-orange liquid. Almost insoluble in water.

Applications. Used as a light-stabilizer of polyolefins, polystyrene, pentaplast, PVC, cellulose acetobutyrate, and other polymers.

Acute Toxicity. LD_{50} is reported to be 10 g/kg BW in rats and 8.7 g/kg BW in mice.

Repeated Exposure failed to reveal cumulative effect. Mice tolerate 0.75 to 3 g/kg BW and rats 1 g/kg BW doses without signs of intoxication.

Long-term Toxicity. Treatment with 150 and 200 mg/kg BW (mice) or 200 mg/kg BW (rats) doses revealed no changes in BW gain, hematological analyses, conditioned reflex activity, and structure of the visceral organs.

Allergic Effect not found.

Standards. Russia (1988). MAC and PML: **n/m.**

Reference:

Mikhailets, I. B. Toxicity of some stabilizers for plastics, *Plast. massy,* 12, 41, 1976 (in Russian).

N-ALKYL(C_7–C_9)-*N*-PHENYL-*p*-PHENYLENEDIAMINE

Composition and **Properties.** A mixture of products where *R* is C_7H_{15} to C_9H_{19}. Thick, dark-brown liquid. Insoluble in water, readily soluble in alcohol.

Applications. Used as a thermostabilizer of butadiene-styrene and polyisoprene rubber.

Acute Toxicity. LD_{50} is 4 g/kg BW in rats and 3.16 g/kg BW in mice. Poisoning is accompanied by CNS inhibition. Histological examination revealed fatty dystrophy and perivascular infiltration by lymphohistocytic elements in the liver, moderate fatty dystrophy of the kidneys, and diffuse infiltration by the interstitial cells of the intermuscular spaces of the myocardium.

Reference:

See **PHTHALIC ANHYDRIDE,** #2, 35.

2-AMINO-4-*p*-HYDROXYPHENYLAMINO-1,3,5-TRIAZINE

Properties. White, odorless powder. Poorly soluble in water and alcohol.

Applications. Used as a thermo- and light-stabilizer of polyamides and polyurethanes.

Acute Toxicity. LD_{50} was reported to be 520 mg/kg BW in mice and 1666 mg/kg BW in rats. Manifestations of the toxic effect in rats develop in 2 d and include edema and cyanosis of the tail and ears and ulcers on the paws. Death occurs within 3 to 12 d. The lethal dose causes weakness, adynamia, and hind leg paresis in mice in 10 to 12 h after administration. Death occured in 2 to 4 d. Gross pathological examination revealed distention of the stomach and intestine and vascular and liver congestion (Vlasyuk, 1967).

Repeated Exposure revealed evident cumulative properties. K_{acc} is 1.4 (by Kagan). Rats were exposed to A. for 2 months. In this study, 10 out of 11 rats given 166 mg/kg BW died by the end of the administration period, while only 7 out of 18 animals died when treated with 83 mg/kg BW. Rats tolerated administration of 33 mg/kg BW without external signs of intoxication. The first and the second groups developed exhaustion, changed hematological indices, and increased protein content in the urine. The animals in the third group exhibited these changes only to the end of the treatment.

Reference:
See **BUTYL BENZYL ADIPATE**, 314.

2-AMINO-4-α-NAPHTHYLAMINO-1,3,5-TRIAZINE

Properties. White, odorless powder. Poorly soluble in water and alcohol.

Applications. Used as a thermo- and light-stabilizer of polyamides and polyurethanes.

Acute Toxicity. LD_{50} was reported to be 1.35 g/kg BW in mice and 5.34 g/kg BW in rats.[1] General weakness and reduced pain response developed in animals in 1 to 5 d after poisoning. Death within 1 to 10 d. Gross pathological examination revealed congestion in the visceral organs and brain.[2]

Reference:
See **BUTYL BENZYL ADIPATE**, 314.

4-(*o*-AMINOPHENYLIMINOMETHYL)-2,6-DI-*tert*-BUTYLPHENOL

Synonym. *N*-(3,5-Di-*tert*-butyl-4-hydroxybenzylidene)-*o*-phenylenediamine.

Applications. Used as a stabilizer in plastic manufacturing.

Acute Toxicity. In mice, the LD_{50} is 15 g/kg BW.[08] The treatment caused slight alteration of osmotic stability of the erythrocytes and mild methemoglobinemia. Histological examination revealed circulatory disturbances in the visceral organs, necrotic changes in the intestinal mucosa and parenchymatous dystrophy of the renal tubular epithelium.

Short-term Toxicity. Rats received 100 mg/kg BW for 3.5 months. The treatment caused mild methemoglobinemia and stimulation of conditioned reflex activity. Retardation of BW gain was noted. Gross pathological examination revealed superficial necrosis of the villi and desquamation of the intestinal mucosal epithelium[08.]

BARIUM COMPOUNDS

Properties. *Ba* **chloride.** Colorless crystals (usually + H_2O). Solubility in water is 36.2 per 10 g at 20°C, insoluble in alcohol. Threshold for effects on the organoleptic properties of water is 4 mg/l.[1] *Ba-Cd* **laurate** contains 13.8 to 14.5% *Cd* and 8 to 9.2% *Ba*. *Ba* **stearate.** White powder. Poorly soluble in water. *Ba-Cd* **stearate.** Finely dispersed, white or yellow powder. Difficult to wet and poorly soluble in water. *Ba* **sulfate.** White, odorless powder. Solubility in water is 0.22 mg per 100 g at 18°C and 0.41 mg per 100 g at 100°C. *Ba* **acetate** and *Ba* **nitrate** are soluble in water, but *Ba* **chromate, fluoride, oxalate, phosphate,** and **sulfate** are quite insoluble.

Applications and **Exposure.** Ba^{2+} compounds are used in plastics and rubber, in ceramic glazes and enamels, in glass-making, and in cosmetics. Some *Ba* salts are used as stabilizers of PVC and other plastics. *Ba* **sulfate** is used as a filler for rubber. Food and water seem to be the main sources of exposure to Ba^{2+}. The long-term mean dietary intake is about 1 mg *Ba*/d. Intake from the air is negligible.

Acute Toxicity. *Humans.* Soluble *Ba* salts possess high acute toxicity. Doses of 0.2 to 0.5 g/kg BW ingested by man caused acute poisoning, and 3 to 4 g/kg BW or even 0.8 to 0.9 g/kg BW doses are fatal. The symptoms of intoxication include general weakness, dyspnea, and impaired cardiac activity. Toxicity of *Ba* salts is likely to be a function of their aqueous solubility; depending on the dose and solubility of the **barium** compounds, death may occur in a few hours or a few days. Ba^{2+} causes strong vasoconstriction by its direct stimulation of the smooth muscles, and convulsions and paralysis following stimulation of the CNS.[2] *Animals.* Main manifestations of the acute toxic action comprise cardiovascular, gastrointestinal, NS, hemapoietical, and skeletal muscle changes.

Ba chloride. In rats, LD_{50} is reported to be 118 to 150 mg/kg BW[3,06] or 400 mg/kg BW;[4] the last value was recently confirmed: 419 mg/kg BW in male rats and 408 mg/kg BW in females. Signs of poisoning develop in 20 to 30 min after administration and include apathy and adynamia with subsequent smooth muscle spasm, which manifests itself as vomiting, diarrhea, and ejaculation. Paresis of the hind limbs and convulsions are observed in some rats. Death is preceded with complete adynamia and side position of animals. A single administration of 300 mg/kg decreased kidney weights.[5]

Ba caprylate. LD_{50} is reported to be 1 g/kg BW in rats, 1.1 g/kg BW in mice, and 1.25 to 1.9 g/kg BW in guinea pigs.[6] The hemopoietic system appeared to be the target tissue.

Ba stearate. LD_{50} is 1.5 to 4 g/kg BW in rats, 2.3 to 5.5 g/kg BW in mice, and 1.9 to 3.6 g/kg BW in guinea pigs.[6–8] Poisoning is accompanied by immobility, apathy, and refusal of food. Death occurs in 2 to 8 d. Gross pathological examination revealed parenchymatous and fatty dystrophy of the liver and kidneys.

Ba-Cd stearate. LD_{50} is 1.98 g/kg BW in mice; a dose of 0.6 g/kg BW causes no mortality. In rats, LD_{50} is 3.17 g/kg BW. Gross pathological examination showed lesions in the GI tract and testes and dystrophic changes in the heart, liver and kidneys.

Ba sulfate. Pure $BaSO_4$ is nontoxic, since it is hardly absorbed at all, but the technical product often contains poisonous admixtures of $BaCO_3$ and $BaCl_2$.

Repeated Exposure. *Humans.* The toxicity of Ba^{2+} is determined primarily by its reaction with *Ca* and *K* salts, as well as by the neurotropic character of the action of Ba^{2+} and its compounds. Repeated exposure to *Ba chloride* seems to have caused recurrent outbreaks of ''Pa-Ping'' disease (a transient paralysis resembling familial periodic paralysis) in China.[9] On a short-term study in a small number of volunteers, there was no consistent indication of adverse cardiovascular effects following exposure to up to 10 mg Ba^{2+}/l in water.[10] *Animals.* Pronounced cumulative properties of the inorganic *Ba* salts are reported.[4]

Ba chloride has K_{acc} equal to 1.8. In a 10-d study, survival in rats given 300 mg/kg dose was substantially lower.[5] Rats were given 100 ppm *Ba chloride* in their feed for a month.[11] The treatment caused no cardiomyopathy or increase in the blood pressure.

Ba-Cd stearate has marked cumulative properties. On administration of $1/5$, $1/10$, and $1/20$ LD_{50}, K_{acc} appeared to be 1.89, 1.54, and 1.1 respectively. There was, however, little evidence of cumulative properties of *Ba stearate.*

Short-term Toxicity. No clinical signs of toxicity (except for a decrease in the relative weights of the adrenals at the highest dose) or microscopic alterations are seen in rats given tap water containing up to 250 mg $BaCl_2$/l for 13 weeks.[12] No adverse histological and hematological changes or effects on serum enzymes were found in rats given the doses of 1.7 to 45.7 mg *Ba chloride*/kg BW via drinking water over the same period of time.[12] No effects on the blood pressure were reported in the rat studies after a 20-week exposure to a dose of 15 mg *Ba chloride*/kg BW.[13] There was, however, an increase in the systolic blood pressure of rats exposed to relatively low concentrations of $BaCl_2$ in drinking water. The NOAEL of 2000 ppm was identified in a 3-month study[14] in Fisher 344/N rats and B6C3F$_1$ mice given $BaCl_2$ in their drinking water.

Long-term Toxicity. Available epidemiological studies have not shown cardiovascular effect in humans; the available animal studies have demonstrated that hypertension is associated with exposure to Ba^{2+}. *Humans.* In the most sensitive epidemiological study conducted to date, there were no significant differences in blood pressure or the prevalence of cardiovascular disease between a population exposed to drinking water containing 7.3 mg Ba^{2+}/l compared to one ingesting water containing 0.1 mg Ba^{2+}/l. Thus, the NOAEL in this study appeared to be 7.3 mg Ba^{2+}/kg BW.[15] This value is within the range of that derived from the toxicological studies in animals. *Animals.* Significant increases in the mean systolic blood pressure were observed in rats exposed to drinking water containing **Ba chloride** over a long period of time: 5.1 mg/kg BW for 16 months and 0.5 mg/kg BW for 8 months.[16,17] The NOAEL was identified to be 0.51 mg/kg BW,[16] and ADI/TDI hence appeared to be 0.051 mg/kg (uncertainty factor was considered to be 10, since the results of a well-conducted epidemiological study indicated that humans are not more sensitive than rats to Ba^{2+} in drinking water). More recent studies of the same authors[11] failed to reveal changes in BW gain, appearance, selected weights or morphology of rats given 1 to 100 ppm $BaCl_2$ in drinking water for 16 months. Drinking water containing 10 ppm **Ba chloride** given for 8 months caused no cardiomyopathy or increase in the blood pressure.

Reproductive Toxicity. Some **Ba** salts may cross the placenta barrier and produce adverse reproductive and *teratogenic* effects. *Gonadotoxic* effect was reported to occur at a dose-level of 0.5 mg/kg, and the NOAEL of 0.005 mg/kg BW was identified for total reproductive effects.[18]

Genotoxicity. A dose of 0.5 mg **Ba chloride**/kg BW was found to be genotoxic (increase in the number of CA without affecting the mitotic index).[18] Ba^{2+} did not increase the frequency of mutations in repair-deficient strains of *Bac. subtilis*[19] and did not induce errors in viral DNA transcription *in vitro*.[20]

Carcinogenicity. A carcinogenic effect was not observed in a extremely limited life-time bioassay in rats and mice exposed to 5 mg Ba^{2+}/l in drinking water, based on gross examination only of tumors at autopsy.[21] *Carcinogenicity classification.* USEPA: Group D.

Chemobiokinetics. Not only soluble **Ba** salts but insoluble **barium** compounds may also be absorbed to a significant extent.[22] The degree of absorption of Ba^{2+} from the GI tract depends on the solubility of the compound, the species, the contents of the GI tract, diet, and age. Moreover, insoluble **barium** salts may be partially solubilized in the acid medium of the stomach. Less than 10% of an ingested quantity seems to be absorbed in adults. In spite of the poor solubility of **Ba sulfate**, it is not insoluble, and no data indicate that dissolved **Ba sulfate** is not absorbed from the GI tract. Ba^{2+} is rapidly distributed in the blood plasma, principally to the bones,[23] but it may accumulate in the kidney, liver, and myocardium as well. Its metabolism is similar to that of **calcium;** however, unlike it, **barium** has no known biological function. Excretion of **Ba** compounds occurs mainly with the feces and to a lesser extent with the urine, elimination varying according to the route of administration and the solubility of the compound.[24]

Guidelines. WHO (1991). Guideline value for drinking water: 0.7 mg/l.

Standards. EEC (1982). MAC in drinking water: 0.1 mg/l. **USEPA** (1991). MCL and MCLG: 2 mg/l; MPC for colors that may be used in food, drugs, and/or cosmetics (1986): 0.5 g/kg. **Russia** (1988). MAC and PML in food and drinking water : 0.1 mg/l. **Canada** (1987). MAC in drinking water: 1 mg/l.

Regulations. USFDA (1993) approved the use of **Ba compounds** (1) in resinous and polymeric coatings for a food-contact surface (**Ba sulfate**), (2) as a filler for rubber articles intended for repeated use in contact with food, (3) as components of adhesives for a food-contact surface (**Ba acetate** and **Ba sulfate**), (4) as colorants in the manufacturing of articles intended for use in contact with food (**Ba sulfate**), and (5) as a catalyst in phenolic resins (**Ba hydroxide**).

References:

1. See **ACRYLONITRILE,** #1, 54.
2. The Metals, in *Patty's Ind. Hyg. Toxicol.,* Clayton, G. D. and Clayton, F. E., Eds., John Wiley & Sons, New York, 1981, p. 1531.
3. See **ACRYLONITRILE,** #7, 34.
4. Akinfieva, T. A. and Gerasimova, I. L., Comparative toxicity of some barium compounds, *Gig. Truda Prof. Zabol.,* 6, 45, 1984 (in Russian).
5. Borzelleca, J. F., Condie, L. W., Egle, J. L., et al., *J. Am. Coll. Toxicol.,* 7, 675, 1988.
6. Mitin, L. S., Comparative toxicity investigation of some barium compounds, *Gig. Sanit.,* 11, 91, 1974 (in Russian).
7. Antonovich, L. A. and Bake, M. Ya., Toxicity of barium stearate, *Pharmacol. Toxicol.,* 49, 117, 1986 (in Russian).
8. Antonovich, L. A. and Sprudzhas, D. P., Synthesis and study of physiologically active substances, *Reports Republ. Sci. Conf.,* December 14, 1984, Vil'nus, 1984, p. 12 (in Russian).
9. Shankle, R. and Keane, J. R., Acute paralysis from barium carbonate, *Arch. Neurol.,* 45, 579, 1988.

10. Wones, R. G., Stadler, B. L., and Frohman, L. A., Lack of effect of drinking water barium on cardiovascular risk factors, University of Cincinnati College of Medicine (manuscript), 1987.
11. Perry, H. M., Kopp, S. J., Perry, E. F., et al., Hypertension and associated cardiovascular abnormalities induced by chronic barium feeding, *J. Toxicol. Environ. Health,* 28, 373, 1989.
12. Tardiff, R. G., Robinson, M., and Ulmer, N. S., Subchronic oral toxicity of $BaCl_2$ in rats, *J. Environ. Pathol. Toxicol.,* 4, 267, 1980.
13. McCauley, P. T., Douglas, B. H., Laurie, R. D., et al., Investigations into the effects of drinking water barium on rats, in *Advances in Modern Environmental Toxicology,* Princeton Publishing Co., New York, 9, 1985, p. 197.
14. Dietz, D. D., Elwell, M. R., Davis, W. E., et al., Subchronic toxicity of barium chloride dianhydride administered to rats and mice in the drinking water, *Fundam. Appl. Toxicol.,* 19, 527, 1992.
15. *Advances in Modern Environmental Toxicology,* Princeton Publishing Co., New York, 9, 1985, p. 231.
16. *Trace Substances Environ. Health.,* Hemphil, D. D., Ed., Columbia, University of Missouri, 16, 155, 1983.
17. *Advances in Modern Environmental Toxicology,* Calabrese, E., Ed., Princeton Publishing Co., New York, 1985, chap. 20, p. 221.
18. Krasovsky, G. N. and Sokolovsky, N. G., Genetic effects of heavy metals, *Gig. Sanit.,* 9, 56, 1979 (in Russian).
19. Nishioka, H., Mutagenic activities of metal compounds in bacteria, *Mutat. Res.,* 31, 185, 1975.
20. Loeb, L., Sirover, M., and Agarval, S., Infidelity of DNA synthesis as related to mutagenesis and carcinogenesis, *Adv. Exp. Med. Biol.,* 91, 103, 1978.
21. Schroeder, H. A. and Mitchener, M., Life-term effects on mercury, methyl mercury and nine other trace elements on mice, *J. Nutr.,* 105, 452, 1975.
22. McCauley, P. T. and Washington, I. S., Barium bioavailability as the chloride, sulphate, or carbonate salt in the rat, *Drug Chem. Toxicol.,* 6, 209, 1983.
23. National Academy of Sciences, *Drinking Water and Health,* Vol. 1, National Research Council, Washington, D. C., 1977.
24. IPCS, *Barium,* Health and Safety Guide, WHO, Geneva, 1991, 28.

p-BENZYLIDENAMINOPHENOL

Synonym. *N*-Benzylidene-*p*-hydroxyaniline.
Properties. Colorless, crystalline substance. Poorly soluble in water, readily soluble in alcohol.
Acute Toxicity. In mice, LD_{50} is 3.5 g/kg BW.[08] Administration of lethal doses slightly affected osmotic stability of the erythrocytes but caused methemoglobinemia. Gross pathological examination revealed circulatory disturbances in the visceral organs, necrotic changes in the small intestine mucosa, and parenchymatous dystrophy of the renal tubular epithelium.
Short-term Toxicity. Rats received 100 mg/kg BW for 3.5 months. Manifestations of the toxic effect included retardation of BW gain, slight methemoglobinemia, and stimulation of conditioned reflex activity. Gross pathological examination revealed necrosis of the villi and sloughing of the epithelium of the small intestinal mucosa.[08]

BIS-(β-CARBOBUTOXYETHYL)TIN DICHLORIDE (CBETC)

Applications. Used as a stabilizer for PVC plastics.
Repeated Exposure. In a 14-d study, rats were given CBETC in their feed. A dietary level of 450 ppm did not affect BW, but at the 1350-ppm level, growth retardation and decrease in the relative liver weights were evident. CBETC also reduced the relative weights of the thymus and spleen. A diminished amount of liver glycogen was the only treatment-related histopathological change observed.
Immunotoxicity. CBETC shows a particularly high degree of lymphotoxicity *in vitro,* but in contrast to dialkyltins, the estertins do not induce lymphocytotoxicity when administered *in vivo* and are less toxic than DOTC or DBTC.

Chemobiokinetics. In lymphocyte metabolism studies, CBETC induced a dose-dependent stimulation of glucose consumption.
Reference:
Penninks, A. H. and Seinen, W., Comparative toxicity of alkyltin and estertin stabilizers, *Food Chem. Toxicol.,* 20, 909, 1982.

BIS-(β-CARBOMETHOXYETHYL)TIN DICHLORIDE (CMETC)

Applications. Used as a stabilizer for PVC plastics.
Repeated Exposure. In a 14-d feeding study in rats, a dietary level of 450 ppm did not affect BW gain, but at the 1350-ppm level, growth retardation and a decrease in the relative weight of the liver were noted. A diminished amount of liver glycogen was the only treatment-related histopathological change observed.
Immunotoxicity. In contrast to **dialkyltins,** the estertins do not induce lymphocytotoxicity when administered *in vivo* and are less toxic than DOTC or DBTC.
Chemobiokinetics. In lymphocyte metabolism studies, CMETC induced a dose-dependent stimulation of glucose consumption.
Reference:
See **BIS-(β-CARBOBUTOXYETHYL)TIN DICHLORIDE.**

BIS[(3,5-DI-*tert*-BUTYL-4-HYDROXYPHENYL) ETHOXYCARBONYLETHYL]SULFIDE

Synonyms. Thiodiethyleneglycol bis-3-(3,5-di-*tert*-butyl-4- hydroxyphenyl)propionate; fenosan 30.
Properties. White crystalline powder. Poorly soluble in water, soluble in alcohol.
Applications. Used as a thermostabilizer in the production of polyolefins and other plastics.
Acute Toxicity. In male and female rats and in female mice, LD_{50} is reported to be in the range of 6.3 to 23.8 g/kg BW. Male mice tolerate Fenozan administration even at higher dose-levels. Acute action threshold (for STI) is 1.2 g/kg in female mice, 2.5 g/kg in male mice, and 0.3 g/kg BW in female rats. STI appeared to be unchanged in male rats.
Repeated Exposure revealed no cumulative properties. Administration of low doses (even 10 LD_{50} *in toto*) did not cause animal mortality. BW gain was unaltered. Gross pathological examination revealed thyroid changes in rats.
Short-term Toxicity. Rats tolerate administration of an overall dose equivalent to 10 LD_{50}. Changes in the thyroid structure were noted. Mice exhibit no retardation of BW gain and have no pathomorphological lesions.
Long-term Toxicity. Mice received 50 and 200 mg/kg BW for 10.5 months without toxic manifestations. An increase in the relative weights of the thyroid is reported in rats.
Reproductive Toxicity. Reproductive function of the gonads is not found to be affected following skin application of B.
Allergenic Effect was not observed in acute experiments on guinea pigs.
Standards. Russia. PML in drinking water: 3 mg/l; PML in food: **n/m.**
Reference:
Toxicology and Sanitary Chemistry of Plastics, Abstracts, NIITEKHIM, Moscow, 4, 1979, p. 26 (in Russian).

1,1'-BIS[(2-METHYL-4'-HYDROXY-5-*tert*-BUTYL)PHENYL]-PENTANE

Applications. Used as an antioxidant in the production of polyolefins and synthetic rubber.
Acute Toxicity. LD_{50} is approximately 17 g/kg BW in rats. Administration does not affect the GI tract (Haskel).
Repeated Exposure. Rats received 0.05% B. in their feed for 90 d. The treatment resulted in fatty infiltration of the liver. A dose of 0.005% in the diet had no toxic effect.[05]

2,2'-BIS(*p*-PHENYLAMINOPHENOXY)DIETHYL ETHER

Properties. Finely dispersed, odorless, grey powder. Insoluble in water and alcohol.

Applications. Used as a stabilizer in the production of polyamides and polyurethanes.

Acute Toxicity. Rats tolerate a single 10 g/kg BW dose without signs of intoxication.

Repeated Exposure caused 50% mortality in rats given 1 g/kg BW by day 36 of the exposure. Symptoms of intoxication were evident. 0.5 g/kg BW dose produced no visible signs of intoxication.

Short-term Toxicity. No animals given 0.5 g B./kg BW died in a 3-month experiment.

Reference:

See **ACRYLONITRILE,** #1, 138.

BIS(TRIBUTYLTIN) OXIDE (TBTO) (CAS No 56–35–9)

Synonyms. Butinox; Hexabutyldistannoxane; Hexabutylditin; Oxybis(tributyltin); Tributyltin oxide.

Applications. TBTO in particular is an effective biocidal preservative. It has been used as a PVC stabilizer and an antifouling agent in marine paints, representing up to 20% of total ingredients.

Acute Toxicity. In rats, LD_{50} is 55 mg/kg BW. The range reported for acute oral LD_{50} in rodents is 10 to 234 mg/kg.[1]

Repeated Exposure revealed material and functional accumulation in mice given TBTO for 20 d. K_{acc} is 3.5 to 5.4 ($1/5$ LD_{50}) and 1.5 to 1.6 ($1/25$ LD_{50}). Wistar rats fed dietary levels of 5 to 320 mg/kg (0.25 to 16 mg/kg BW) exhibited decreased food and water consumption and clinical signs of toxicity. Doses of 4 and 16 mg/kg BW caused thymus weight reduction. At a dose-level of 16 mg/kg, histopathological changes in the thymus, mesenteric lymph nodes, and liver were noted. Hematological effects were also found.[2] Administration of 0.25 mg/kg BW for 30 d caused anemia and leukocytosis in rats.[3]

Short-term Toxicity. In a 13-week feeding study in rats, the NOEL was established to be 0.2 mg/kg BW.[4] Doses of 0.5, 5, and 50 mg/kg BW administered to Wistar rats for 106 weeks[5] produced increased mortality, reduction of BW, and hematological changes, including anemia, lymphocytopenia, and thrombocytosis. Decreased kidney function and increased plasma enzyme activity were also noted. The NOAEL appeared to be 0.5 mg/kg BW.

Long-term Toxicity. In a 6-month study, rabbits received 0.04 mg TBTO/kg BW. The treatment caused no statistically reliable changes in hematological analyses, in serum proteins and protein fraction contents, or in the CNS condition of animals.[3]

Reproductive Toxicity. In NMRI mice given 35 mg TBTO/kg BW on days 6 to 15 of gestation, an increase in embryolethality was revealed. Developmental abnormalities were increased in a dose-dependent fashion at a 17.7 mg/kg dose-level and above. However, there was also the evidence of maternal toxicity at this and higher dose-levels.[6] An oral dose of 16 mg/kg BW administered to pregnant Long-Evans rats on days 6 to 20 of gestation[7] resulted in retarded fetal growth. A reduction in the numbers of live births and an effect on postnatal growth and behavior were observed at doses of 10 mg/kg BW and above. Maternal toxicity was also evident at these dose-levels. No evidence of teratogenicity was found in a rabbit study at doses of up to 2.5 mg/kg BW.[8]

Immunotoxicity. When rats were exposed to 20 or 80 ppm in their feed for 6 weeks, TBTO was found to cause changes in the blasto-transformation processes of the thymus and spleen cells. There was a decline in the content of T-lymphocytes and a rise in the content of B-lymphocytes in the spleen. The number of viable cells in the thymus and spleen was reduced.[9]

Genotoxicity. Administration of a single LD_{50} caused a mutagenic effect in mouse bone marrow cells. Mice given single oral doses of 31 to 125 mg/kg failed to show an increased incidence of micronuclei in bone marrow polychromatic erythrocytes.[4] TBTO is recognized as a genotoxic chemical by the rec-assay.[10]

Carcinogenicity. In a 106-week feeding study in rats, the incidence of benign tumors of the pituitary was significantly elevated at 0.5 mg and 50 mg TBTO/kg of the diet (2.5 mg/kg BW). At a dietary level of 50 mg TBTO/kg, an increase in the incidence of adrenal and parathyroid adenomas was noted. This increase was not dose related.[5]

Chemobiokinetics. When administered with drinking water, TBTO is distributed primarily to the kidneys with only low levels in the blood.[11] Given by *i/p* injection to mice, the major part of the dose was rapidly eliminated in the feces.

Regulations. USFDA (1993) regulates the use of TBTO in adhesives as a component of articles intended for use in packaging, transporting, or holding food for use as a preservative only. The use of triorganotin compounds in antifouling paints has been banned in the **U.K.** and **France,** and an **EEC** ban has been proposed.

References:

1. Oakley, S. D. and Fawell, J. K., Toxicity of Selected Organotin Compounds to Mammalian Species, Water Research Centre, Report ER 1307-M, 1986.
2. Krajnc, E. I., Wester, P. W., Loeber, J. G., et al., Toxicity of bis(tri-*n*-butyltin)oxide in the rat. I. Short-term effects on general parameters and on the endocrine and lymphoid systems, *Toxicol. Appl. Pharmacol.*, 75, 363, 1984.
3. Belyaeva, N. N., Bystrova, T. A., Revazova, Iu. A., et al., Comparative evaluation of toxicity and mutagenicity of organotin compounds, *Gig. Sanit.*, 5, 10, 1976 (in Russian).
4. Schweinfurth, H. A., Toxicology of tributyltin compounds, in Tin and its uses, *Q. J. Int. Tin Res. Inst.*, 143, 9, 1985.
5. Wester, P. W., Krajnc, E. I., van Leeuwen, F. X. R., et al., Chronic toxicity and carcinogenicity of bis(tri-*n*-butyltin)oxide (TBTO) in the rat, *Food Chem. Toxicol.*, 28, 179, 1990.
6. Davis, A., Barale, R., Brun, G., et al., Evaluation of the genetic and embryotoxic effects of bis(tri-*n*-butyltin)oxide, a broad spectrum pesticide in multiple *in vivo* and *in vitro* short-term tests, *Mutat. Res.*, 188, 65, 1987.
7. Crofton, K. M., Dean, K. F., Boncec, V. M., et al., Prenatal or postnatal exposure to bis(tri-*n*-butyltin)oxide in the rat: postnatal evaluation of teratology and behaviour, *Toxicol. Appl. Pharmacol.*, 97, 113, 1989.
8. Schweinfurth, H. A. and Gunzel, P., The tributyltins: mammalian toxicity and risk evaluation for humans, in *Ocean 87, The Ocean "An International Workplace"*, 1987, p. 1421.
9. Vos, J. G., de Klerk, A., Krajnc, E. I., et al., Toxicity of bis(tri- *n*-butyltin)oxide in rat. II. Suppression of thymus-dependent immune responses and of parameters of nonspecific resistance after short-term exposure, *Toxicol. Appl. Pharmacol.*, 75, 387, 1984.
10. Hamasaki, T., Sato, T., Nagase, H., et al., The genotoxicity of organotin compounds in SOS chromotest and res- assay, *Mutat. Res.*, 280, 195, 1990.
11. Evans, W. N., Cardarelli, N. F., and Smith, D. J., Accumulation and excretion of [^{14}C]bis(tri-*n*-butyltin)oxide in mice, *J. Toxicol. Environ. Health*, 5, 871, 1979.

BUTYLATED HYDROXYANISOLE (BHA) (CAS No 25013–16–5)

Synonyms. *tert*-Butyl hydroxianisole; **2,6**-Di-*tert*-butyl-*p*-cresol; *tert*-Butyl-4-methoxyphenol; **(1,1**-Dimethylethyl)-4-methoxyphenol; Ionol *CP*; Topanol *OC*.

Properties. White or slightly yellow, waxy solid. Has a faint odor. Produced in the form of small white tablets. A mixture of two isomers (*o*- and *m*-isomers at the ratio of 85:15). Insoluble in water even at 50°C. Soluble in ethanol, other alcohols, fats, and oils. Commercial BHA is known to exist as an isomeric mixture of 3-*tert*-butyl-4-hydroanisole and 2-*tert*-butyl-4-hydroanisole.

Applications and **Exposure.** The primary use for BHA is as an antioxidant and preservative in cosmetics, in food packaging and rubber. Antioxidant for polyolefins. Fifty countries permit BHA as a food additive. A synthetic phenolic antioxidant. Estimated average daily dietary intake (in the Netherlands) does not exceed the ADI except in extreme cases in 1- to 6-year-olds.

Acute Toxicity. In rats, LD_{50} is 2.2 g/kg BW; in mice, it is 2 g/kg BW.[1] Rodents given a BHA diet are less vulnerable to acute toxicity.[2]

Repeated Exposure. Consumption of diets containing 2% BHA (Wistar Han/BGA rats) or 0.5 to 2% BHA (Fisher 344 rats) for a month caused superficial necrosis, ulceration, and hyperplasia of the squamous epithelium of the forestomach.[3]

Long-term Toxicity. *Humans.* At the low concentrations permitted in food, BHA has been a part of the human diet for many years without evidence of adverse effect (WHO, 1987). Oral administration of the ADI (0.5 mg/kg BW) to man and a 400 times higher dose to the rat (200 mg/kg) resulted in plasma BHA levels within one order of magnitude.[4] It indicates that the current ADI might not be sufficient to protect man from possible adverse effect. *Animals.* Fisher 344 rats were fed 0.5 and 2% BHA in the diet over a period of 104 weeks. Increased incidence of hyperplasia of the forestomach was observed in a dose-dependent manner.[5] There were similar findings in hamsters.[6] The NOEL of 62.5 mg/kg BW was identified for the induction of proliferative changes in the rat forestomach.[7] Monkeys received 125 or 500 mg/kg BW in corn oil by gavage for 17 weeks without demonstrating adverse effects. BHA given orally to monkeys at about maximum tolerated dose failed to induce the massive changes noted with rats given 2% dietary BHA.[8]

Reproductive Toxicity. BHA exhibits slight *embryotoxicity* but no *teratogenicity* (rabbits, pigs, rhesus monkeys). The dose of 750 mg/kg BW given on days 1 to 20 of gestation or for 70 d before conception and through gestation caused no abnormalities in rats.[9] *ICI-SPF* mice were given the dose of 500 mg/kg, 7 d before conception and until day 18 of gestation. Maternal mortality reached up to 25% but no signs of embryotoxicity or teratogenicity were found.[9] Pregnant Danish Landrace swine were fed diets containing 50 to 400 mg/kg from insemination to day 110 of gestation: BW was lower in dams fed 400 mg/kg BW, but no other significant signs of maternal toxicity, embryotoxicity, or teratogenicity were reported.[10] Pregnant New Zealand *SPF* rabbits were administered doses of 50 to 400 mg/kg BW by gavage on days 7 to 18 of gestation: this produced no treatment-related effects in dams or offspring.[10] Pregnant Danish Landrace swine were fed diets containig up to 3.7% BHA for 16 weeks: no treatment-related lesions of the forestomach or glandular stomach were reported.[11]

Genotoxicity. No evidence of genotoxicity in the hepatocyte primary culture/DNA repair test, the *Salmonella*/microsome test, the adult rat liver epithelial cell/hypoxanthine–guanine phosphoribosyl transferase test, and also in the Chinese hamster ovary cells and SCE test were reported for BHA.[12–14]

Carcinogenicity. There is some evidence of carcinogenic potential of BHA in the forestomach of rodents. Fisher 344 rats were fed 100 to 6000 ppm in the diet for 76 weeks. No increase in neoplasms at any site was noted. In the diet, 12,000 ppm BHA resulted in a small increase in squamous cell papilloma of the nonglandular squamous portion of the stomach.[15] Administration of up to 2% BHA in the diet for 104 weeks induced benign and malignant tumors of the forestomach in Fisher 344 rats.[5,6]

The mechanism of BHA carcinogenic activity is not understood. Its inhibitory activity in relation to carcinogenicity has been demonstrated.[16–18] BHA was also shown to be a potent enhancer of mutagenicity and carcinogenicity induced by chemicals (IARC 40–123). *Carcinogenicity classification.* IARC: Group 2B.

Chemobiokinetics. BHA is rapidly absorbed and metabolized. Accumulates in the fatty tissue but has not been found elsewhere. Main metabolites were 4-*O*-conjugates: the *O*-**sulfate** and the *O*-**glucuronide** (astil). BHA can be oxidized by tissue peroxydases to a derivative that can bind to DNA. Excretion occurs in the urine. Of the administered dose, 95% is excreted in days in the form of glycuronides.[19,20]

Recommendations. JECFA (1989). Temporary ADI: 0 to 0.5 mg/kg BW. **EEC** (1987). ADI: 0 to 0.5 mg/kg.

Regulations. USFDA (1993) regulates BHA as a direct and indirect food ingredient. According to the U.S. concept, it is GRAS as a result of its relatively low toxicity. FDA classified BHA as GRAS when total content is not more than 0.2% the total fat or oil content of the food. BHA is listed for use (1) as an antioxidant for rubber articles intended for repeated use in contact with food (up to 1.5 wt% rubber product), (2) in adhesives used as components of articles intended for use in packaging, transporting, or holding food, (3) as a defoaming agent (not to exceed 0.1% defoamer), (4) in resinous and polymeric coatings of articles coming in contact with food. BHA may be used (5) in the manufacture of closures with sealing gaskets for food containers. **British Standard** (1992): BHA is listed as an antioxidant for polyethylene and polypropylene compositions used in contact with foodstuffs or water intended for human consumption (max. permitted level for the final compound, 0.2%).

References:

1. Lehman, A. J., Fitzhugh, O. G., Nelson, A. A., et al., The pharmacological evaluation of antioxidants, *Adv. Food Res.,* 3, 197, 1951.

2. Miranda, C. L., Reed, R. L., Cheek, P. R., et al., Protected effect of hydroanisole against the acute toxicity of monocrotaline in mice, *Toxicol. Appl. Pharmacol.*, 59, 424, 1981.
3. Altmann, H. J., Wester, P. W., Metthiaschk, G., et al., Induction of early lesion in the forestomach of rats by 3-*tert*-butyl-4-hydroxyanisole, *Food Cosmet. Toxicol.*, 23, 723, 1985.
4. Verhagen, H., Thijissen, H. H. W., den Hoor F, et al., Disposition of single oral doses of butylated hydroxyanisole in man and rat, *Food Chem. Toxicol.*, 27, 151, 1989.
5. Ito, N., Fukushima, S., Hagiwara, A., et al., Carcinogenicity of butylated hydroxyanisole and butylated hydroxytoluene in Fisher 344 rats, *J. Natl. Cancer Inst.*, 70, 343, 1983.
6. Ito, N., Fukushima, S., and Tsuda, H. Carcinogenicity and modification of the carcinogenic response by butylated hydroxyanisole, butylated hydroxytoluene, and other antioxidants, *Crit. Rev. Toxicol.*, 15, 109, 1985.
7. JECFA, Butylated Hydroxyanizole, Food Additives Series 21, WHO, Geneva, 1986, p. 3.
8. Iverson, F., Truelove, Y., Nera, E., et al., An 85-day study of butylated hydroxyanisole in the cynomologus monkey, *Cancer Lett.*, 26, 43, 1985.
9. Clegg, D. J., Absence of teratogenic effect of butylated hydroxyanisole and butylated hydroxytoluene in rats and mice, *Food Cosmet. Toxicol.*, 3, 387, 1965.
10. Hansen, E. and Meyer, O., A study of teratogenicity of butylated hydroxyanisole in rabbits, *Toxicology*, 10, 195, 1978.
11. Olsen, P., The carcinogenic effect of butylated hydroxyanisole on the stratified epithelium of the stomach in rat versus pig, *Cancer Lett.*, 21, 115, 1983.
12. Williams, G. M., McQueen, C. A., and Tong, C., Toxicity studies of BHA and BHT. I. Genetic and cellular effects, *Food Chem. Toxicol.*, 28, 793, 1990.
13. Rogers, C. G., Nayak, B. N., and Herouks-Metcalf, C., Lack of induction of sister chromatid exchanges and mutations to 6-thioguanine resistance in V79 cells by butylated hydroxyanisole with and without activation by rat or hamster hepatocytes, *Cancer Lett.*, 27, 61, 1985.
14. Abe, S. and Sasaki, M., Chromosomal aberrations and sister chromatid exchanges in Chinese hamster cells exposed to various chemicals, *J. Natl. Cancer Inst.*, 58, 1635, 1977.
15. Williams, G. M., Wang, C. K., and Iatropoulos, M. J., Toxicity studies of BHA and BHT. II. Chronic feeding studies, *Food Chem. Toxicol.*, 28, 799, 1990.
16. Hocman, G., Chemoprevention of cancer: phenolic antioxidants (BHT, BHA), *Int. J. Biochem.*, 7, 639, 1988.
17. Wattenberg, L. W., Chemoprevention of cancer, *Cancer Res.*, 45, 1, 1985.
18. Ito, N. and Hirose, M., The role of antioxidants in chemical carcinogenesis, *Gann*, 78, 1011, 1987.
19. Rahimtula, A., *In vitro* metabolism of 3-*tert*-butyl-4-hydroxyanisole and its irreversible binding to proteins, *Chem. Biol. Interact.*, 45, 125, 1983.
20. Truhaut, R., *L'Alimentation et la Vie*, 50, 57, 1962 (in French).

BUTYLATED HYDROXYTOLUENE (BHT) (CAS No 128–37–0)

Synonyms. Advastab; Agidol 401; Alkofen *BP;* 2,6-Bis(1,1-dimethylethyl)-4-methylphenol; Butylhydroxytoluene; CAO-3; Ionol.

Properties. White, crystalline solid. Solubility in water is 0.2 mg/l at 20°C; soluble in alcohols, food oils, and fats.

Applications and **Exposure.** Synthetic phenolic antioxidant used widely to prevent oils, fats. Antioxidant in production of plastics (mainly polyolefins), cosmetics, rubber, elastomers. A thermostabilizer for polyethylene, polypropylene, polyesters, polystyrene, polyurethanes, PVC, and copolymers of vinyl and vinylidene chloride. It cannot be excluded that ADI for BHT is exceeded in all ages and sex groups.

Acute Toxicity. The LD_{50} values are reported to be 1.7 to 1.97 g/kg BW in albino Wistar rats, 0.94 to 2.1 g/kg BW in cats, 2.1 to 3.2 g/kg BW in New Zealand rabbits, and 10.7 g/kg BW in guinea pigs.[1] According to other data, LD_{50} is 2.45 g/kg BW in rats and 2 g/kg BW in mice. With a single administration, there is depression, loss of mobility, and hypothermia.[08]

Repeated Exposure. Administration of 1/5 LD_{50} to rats (in the diet) caused growth retardation, a reduction in blood peroxidase, catalase and cholinesterase activity, and an increase in the relative weights of the liver and in its fat content.[2] The NOEL for (transient) hemorrhagic effects following oral administration is about 5 mg/kg BW in a short-term experiment with rats.[3]

Short-term Toxicity. Ninety administrations of 50 mg/kg BW to rats caused functional and morphological changes.[4]

Long-term Toxicity. *Humans.* At low concentrations, BHT is permitted in food; it has been a part of the human diet for many years without evidence of adverse effect (WHO, 1987). *Animals.* On prolonged exposure, there is an increase in diuresis, liver weights, and the activity of cytochrome P-450-dependent macrosomal liver enzymes. Pulmonary hypertrophy and hyperplasia are noted.[5] Daily ingestion of fat containing 4% BHT (calculated at 4 mg/kg) by dogs for a year produced no adverse effects.[2] According to other data, addition of 1% BHT to the diet resulted in enchanced formation of oxidizing enzymes (BHT-oxidase) and of other enzymes (amino-purinemethylase), which are involved in nitrogen metabolism.[6] The life-span of animals kept under unfavorable living conditions is in this case extended.

Reproductive Toxicity. *Gonadotoxicity.* At high dose-levels, BHT is shown to induce sperm abnormalities in mice. *Embryotoxicity.* Mice were given 0.5 to 20% BHT in the diet from weaning for the life-span. Animals were allowed to mate and produce offspring for ~18 months. The number of live-born progeny was reduced by about 10% in the high-dose group, but no other signs of toxicity in the parents or offspring were noted.[7] Mice were exposed to 750 mg BHT/kg BW in arachis oil administered by gavage throughout gestation (five dams), or before and throughout gestation (seven dams). Normal litter was produced.[8] Wistar SPF rats received 500 mg BHT/kg BW in the diet for 13 weeks before conception and throughout gestation and lactation. No adverse effects in the offspring were observed.[9] The NOAEL for reproductive effects was identified to be 25 mg/kg BW.[10] *Teratogenic* effects were not observed in rats and mice.[8]

Genotoxicity. No evidence of BHT genotoxicity was reported. BHT is not shown to cause point mutations, it lacks clastogenic potential and does not interact with or damage DNA.[11-13] BHT does not represent a relevant genotoxic risk to man.

Carcinogenicity. BHT is likely to be a potent enhancer of mutagenicity and carcinogenicity induced by chemicals (IARC 40–123). There is some evidence of carcinogenicity in the liver of rodents.[14,15] However, no increase in neoplasm incidences at any site was noted in Fisher 344 rats fed up to 12 g BHT per kilogram of the diet for 76 weeks.[16] The mechanism of BHT carcinogenic activity is not well understood. Inhibitory activity in relation to carcinogenicity was shown.[17,18]

Carcinogenicity classification. IARC: Group 3; NTP: N—N—N—N.

Chemobiokinetics. *Humans.* Only traces of BHT are found in the blood, and there are no metabolites 24 h after ingestion. In the urine, there are derivatives of BHT: the acid and glucuronide.[5] *Animals.* After ingestion, BHT is widely distributed to various tissues (primarily to the small intestine, stomach, liver, kidney). The lung showed no greater accumulation than most of the other tissues.[19] The predominant metabolic pathway involves oxidation of the 4-methyl group.[20] In rabbits the urinary metabolites consisted of **glucuronides, sulfates,** and **free phenols** excreted in the urine and feces. Unchanged BHT was isolated from the feces but not from the urine.[21]

Recommendations. JECFA (1991). Temporary ADI: 0 to 0.125 mg/kg. **EEC** (1986). ADI: 0 to 0.05 mg/kg.[22]

Standards. Russia (1978). PML in food: 2 mg/l.

Regulations. USFDA (1993) approved BHT for use (1) as a component of food-contact adhesives. According to the U.S. concept, BHT is GRAS as a result of its relatively low toxicity. BHT may be also used (2) as a defoaming agent (not to exceed 0.1% defoamer), (3) in resinous and polymeric coatings of articles coming in the contact with food, and (4) in the manufacture of closures with sealing gaskets for food containers.

References:

1. Deichmann, W. B., Clemmer, J. J., Rakoczy, R., et al., Toxicity of ditertiary butylmethylphenol, *Arch. Ind. Health,* 11, 93, 1955.
2. Karplyuk, K. A., Toxicity of BHT, *Prob. Nutr.,* 26, 378, 1971 (in Russian).
3. Report Sci. Comm. Food Commission of EC on Antioxidants, CS/ANT/20-Final, 1987.
4. Gilbert, D. and Goldberg, L., BHT oxidase. A liver- microsomal enzyme induced by the treatment of rats with BHT, *Food Cosmet. Toxicol.,* 5, 481, 1967.
5. Jori, A., Toxico-kinetic aspects of butylated hydroxytoluene: a review, *Ann. 1st Super. Sanit.,* 19, 271, 1983.

6. Hirose, M., Shibata, M., Hagiwara, A., et al., Chronic toxicity of butylated hydroxytoluene in Wistar rats, *Food Cosmet. Toxicol.*, 19, 147, 1981.

7. Johnson, A. R., A re-examination of the possible teratogenic effects of butylated hydroxytoluene and its effect in the reproductive capacity of the mouse, *Food Cosmet. Toxicol.*, 3, 371, 1965.

8. See **BUTYLATED HYDROXYANISOLE, #9.**

9. Meyer, O. and Hansen, E., Behavioural and developmental effects of butylated hydroxytoluene dosed to rats in utero and in the lactation period, *Toxicology*, 16, 247, 1980.

10. Evaluation of Certain Food Additives and Contaminants, The 33rd Report of the JECFA, WHO, Geneva, 1991, 7.

11. Williams, G. M., McQueen, C. A., and Tong, C., Toxicity studies of BHA and BHT. I. Genetic and cellular effects, *Food Chem. Toxicol.*, 28, 793, 1990.

12. Bomhard, E. M., Bremmer, J. N., and Herbold, B. A., Review of the mutagenicity/genotoxicity of butylated hydroxytoluene, *Mutat. Res.*, 277, 187, 1992.

13. Ishidate, M., Yoshikawa, K., and Sofuni, T., A primary screening for mutagenicity of food additives in Japan, *Mut. Toxicol.*, 3, 82, 1980.

14. Clayson, D. B., Iverso, F., Nera, E., et al., Histopathological and autoradiopathological studies on the forestomach of Fisher-344 rats treated with BHA and related chemicals, *Food Chem. Toxicol.*, 24, 1171, 1986.

15. Chapp, N. K., Tyndall, R. L., Satterfield, L. C., et al., Selective sex-related modification of diethylnitrosamine-induced carcinogenesis in BALB/c mice by concomitant administration of butylated hydroxytoluene, *J. Natl. Cancer Inst.*, 61, 177, 1978.

16. Williams, G. M., Wang, C. K., and Iatropoulos, M. J., Toxicity studies of BHA and BHT. II. Chronic feeding studies, *Food Chem. Toxicol.*, 28, 799, 1990.

17. Hocman, G., Chemoprevention of cancer: phenolic antioxidants (BHT, BHA), *Int. J. Biochem.*, 7, 639, 1988.

18. Wattenberg, L. W., Chemoprevention of cancer, *Cancer. Res.*, 45, 1, 1985.

19. Daugherty, J. P., Beach, L., Franks, H., et al., Tissue distribution and excretion of radioactivity in male and female mice after a single administration of [^{14}C]butylated hydroxytoluene, *Res. Commun. Subst. Abuse*, 1, 99, 1980.

20. Daniel, J. M., Gage, J. C., Jones, D. I., et al., Excretion of butylated hydroxyanizole and butylated hydroxytoluene by man, *Food Cosmet. Toxicol.*, 5, 475, 1967.

21. Akagi, M. and Aoki, I., Studies on food additives. VI. Metabolism of **2,6**-di-*tert*- butyl-**p**-cresol in a rabbit. I. Determination and paper chromatography of a metabolite, *Chem. Pharmacol. Bull.*, 10, 101, 1962.

22. Anon. *Food Chem. Toxicol.*, 29, 73, 1991.

1,4-BUTYLENE GLYCOL, β-AMINOCROTONATE

Properties. Powder. Poorly soluble in water.

Applications. Used as a stabilizer in the production of PVC.

Acute Toxicity. In rats, LD$_{50}$ significantly exceeds the 4 g/kg BW dose.[05]

Repeated Exposure. Rats received 4% A. in their diet. The treatment resulted in decreased BW gain due to deterioration of the organoleptic properties of the feed.

Reproductive Toxicity effects are not observed.

Standards. Russia: Recommended PML in drinking water **n/m**.

BUTYL GALLATE

Synonym. Gallic acid, butyl ester.

Properties. Greyish, odorless powder, rather bitter. Poorly soluble in water, soluble in alcohol.

Applications. Stabilizer of PVC and linear polyesters.

Acute Toxicity. In mice, LD$_{50}$ is 860 mg/kg BW. Gross pathological examination revealed congestion, lesions of the GI tract mucosa, and parenchymatous dystrophy of the renal tubular epithelium.

Repeated Exposure failed to show cumulative properties. No mice died during administration of 45, 90, and 170 mg/kg BW for 2 months. There was retardation of BW gain and some changes were found on histological examination of the visceral organs.

Short-term Toxicity. No manifestations of toxic action or morphological changes were observed in rats given 200 mg/kg BW for 4.5 months.

Reproductive Toxicity. Affects the reproductive processes in fish.[08]

Reference:

Karplyuk, I. A., On the problem of the possible use of dodecylgallate as an antioxidant of dietary fats, *Gig. Sanit.,* 12, 34, 1962 (in Russian).

2-(3'-*tert*-BUTYL-2'-HYDROXY-5'-METHYLPHENYL)-5'-CHLORBENZO-TRIAZOLE

Synonym. Tinuvin 326.

Properties. Pale-yellow, crystalline powder. Poorly soluble in water and alcohol.

Applications. Used as a light-stabilizer in the production of polypropylene, polyethylene, polybutylene, etc; stabilizer for unsaturated polyesters and vinyl plastics.

Acute Toxicity. Mice tolerate 15 g/kg BW; rats do not die after administration of 10 g/kg BW.

Repeated Exposure. Rats and mice tolerate exposure to 1 to 3 g/kg BW without any deviation in their general condition or structure of the viscera.

Long-term Toxicity. Treatment with 150 and 300 mg/kg BW (mice) and 200 mg/kg BW (rats) produced no effect on BW gain, hematological analyses, blood enzyme composition, and conditioned reflex activity. Gross pathological examination revealed no changes in the visceral organs. Unlike 2-(3,5-di-*tert*-butyl-2-hydroxyphenyl)-5-chlorobenzo-triazole or *Tinuvin 327* (q.v.), B. in which a **methyl** replaces one *tert*-**butyl** group, is virtually nontoxic.

Allergenic effect was not observed.

Standards. Russia. PML in drinking water: **n/m.**

Reference:

See 4-**ALK**(C_7–C_9)**OXY**-2-**HYDROXYBENZOPHENONE.**

N-BUTYLIDENEANILINE

Synonym. Antox.

Properties. A thick, dark-brown liquid, poorly soluble in water.

Applications. Used as a thermostabilizer for vulcanizates based on butyl rubber and neoprene.

Acute Toxicity. LD_{50} is 2.2 g/kg BW in mice and 2.6 g/kg BW in rats. Manifestations of toxic action include depression and immobility. Bloody diarrhea was noted in 2 d. Histological examination revealed tiny hemorrhagic foci in the lungs, small areas of destruction and inflammatory type microfoci in the liver, and signs of engorgement in the kidney medulla.

Repeated Exposure. The treatment with $\frac{1}{10}$ LD_{50} produced pronounced functional changes in the liver and a decline in catalase activity and gas exchange indices. Gross pathological examination revealed hepatocyte granularity, tiny hemorrhagic foci in the lungs, bronchitis, and engorgement and congestion in the brain and spleen.

Reference:

Kel'man, G. Ya., *Toxic Properties of Chemical Additives for Polymeric Materials,* Meditsina, Moscow, 1974, p. 53 (in Russian).

2-*tert*-BUTYL-6-METHYL-4–α-METHYLBENZYLPHENOL

Acute Toxicity. In mice, LD_{50} is 1.9 g/kg BW.[08] Gross pathological examination revealed congestion, edema, and sloughing of the esophageal and gastric mucosa, necrosis of the small intestine walls, and parenchymatous dystrophy of the renal tubular epithelium.

Short-term Toxicity. Rats given 75 mg B./kg BW for 4 months exhibited stimulation of the cortical areas of the CNS. Histological examination revealed congestion, cellular infiltration, and edema of the gastric mucosa, congestion and necrosis of the small intestinal mucosa, and parenchymatous dystrophy of the renal tubular epithelium.[08]

2-*tert*-BUTYL-4-METHYLPHENOL (CAS No 2409–55–4)

Synonym. 2-*tert*-Butyl-*p*-cresol.

Properties. Colorless, crystalline powder. Poorly soluble in water, soluble in alcohol.

Acute Toxicity. LD_{50} is reported to be 0.8 g/kg BW in mice, 2.5 g/kg BW in rats, and 1.2 g/kg BW in guinea pigs.

Regulations. USFDA (1993) listed B. as an antioxidant in semirigid and rigid acrylic and modified acrylic plastics to be used safely as articles intended for use in contact with food.

Reference:

See *N,N'*-**BIS**(1,4-**DIMETHYLPENTYL**)-*p*-**PHENYLENEDIAMINE, 124.**

4-*tert*-BUTYLPHENYL SALICYLATE

Synonyms. 2-Hydroxybenzoic acid, 4-(**1,1**-dimethylethyl)phenyl ester; Salicylic acid, 4-*tert*-butylphenyl ester; Salol B; TBS inhibitor.

Properties. Light crystalline powder with a slight odor resembling that of salol. Solubility in water up to 0.1% at 20°C.

Applications. Used as a light-stabilizer and plasticizer in the production of PVC, polyesters, polystyrene, polyolefins, cellulose esters, and polyurethanes. Used to make films for packaging food products.

Acute Toxicity. In rats, LD_{50} exceeds 1.2 g/kg BW; in mice, it is 2.9 g/kg BW. Histological examination revealed congestion, edema and sloughing of the GI tract mucosa, and parenchymal dystrophy of the renal tubular epithelium.[05,08]

Long-term Toxicity. Approximately 1 g/kg in the diet caused some retardation of BW gain in rats. Dogs were not affected with 0.5 g/kg administered in their feed.[05] Rats received 145 mg/kg BW over a period of 5 months. Functional manifestations of the toxic effect consisted of stimulation of the CNS cortical areas. Gross pathological examination revealed congestion in the visceral organs, cellular infiltration and edema of the gastric mucosa, necrosis of the small intestinal mucosa, parenchymal dystrophy of the renal tubular epithelium, and hyperplasia of the spleen pulp.[08]

BUTYLTIN TRILAURATE (BTTL) (CAS No 25151–00–2)

Applications. Used as a stabilizer in the production of PVC plastics.

Acute Toxicity. In rats, LD_{50} is 325 mg/kg BW.

Repeated Exposure. Cumulative properties were evident in mice given 65 and 13 mg/kg BW for 20 d, and cumulative effect increased as the dose was reduced. Exposure to 5 mg/kg BW for 30 d led to development of signs of anemia and leukocytosis.

Long-term Toxicity. Rabbits received 2 mg/kg BW for 6 months. The treatment did not affect hematological analyses and serum protein fractions ratio in animals.

Genotoxicity. Cytogenetic investigation revealed mutagenic effect in the bone marrow cells of mice following a single administration of LD_{50}.

Reference:

See **BIS(TRIBUTYLTIN) OXIDE,** #3.

CALCIUM BENZOATE

Synonym. Benzoic acid, calcium salt trihydrate.

Properties. Orthorhombic crystals or powder. Solubility in water is 2.67 g per 100 g at 0°C. Highly soluble in boiling water.

Applications. Used as a stabilizer in PVC production. A food preservative (fats, fruit juices, etc.).

Acute Toxicity. LD$_{50}$ is found to be 4 g/kg BW in rats, and 2.3 g/kg BW in mice. Manifestations of toxic effect are absent. Death within few days. Survivors do not differ from controls.

Repeated Exposure failed to reveal evident cumulative properties. The treatment with a dose of 0.4 g/kg BW for 2 months led to mortality of less than 50% of animals. Gross pathological examination revealed parenchymatous dystrophy of the cardiac and hepatic cells, with the loss of separate cells.

Short-term Toxicity. In a 4-month study, an 80 mg/kg BW dose caused no retardation of BW gain or changes in hematological analyses, in the liver and kidney functions and in the relative weights of the visceral organs.

Standards. Russia. PML in drinking water: **n/m.**

Regulations. USFDA (1993) approved the use of *Ca benzoate* as a stabilizer in polymers used in the manufacture of articles or components of articles intended for use in contact with food.

Reference:
See **BISPHENOL A,** #2, 198.

CALCIUM STEARATE (CAS No 1592–23–0)

Properties. White powder. Poorly soluble in water. Unstable in the presence of lipolythic microorganisms, for which it is a nutrient medium.

Applications. Used as a thermostabilizer of PVC; can be used as a lubricant in plastic manufacturing.

Acute Toxicity. Rats and mice tolerate administration of the 5 g/kg BW dose without signs of intoxication.

Repeated Exposure. A 250 mg/kg BW dose increases BW gain in rats. *Ca* compounds are likely to suppress oxidation processes in the body.

Long-term Toxicity. Mice were dosed by gavage with 20 and 100 mg *Ca* stearate/kg BW for 9 months. The treatment resulted in CNS inhibition due to the known physiological features of **calcium.** Gross pathological examination failed to find morphological changes in the visceral organs.

Chemobiokinetics. *Ca* makes up about 1.6% of human total BW. A relatively small amount of *Ca* lost from the body is excreted in the urine. The major part of *Ca* eliminated from the body is excreted in the feces.

Standards. Russia. PML in drinking water: **n/m.**

Regulations. USFDA (1993) approved the use of *Ca* stearate (1) as a lubricant in phenolic resins used as the food-contact surface of molded articles intended for repeated use in contact with non-acid food (pH above 5), (2) as a stabilizer in polymers used in the manufacturing of articles or components of articles intended for use in contact with food, and (3) as a plasticizer in rubber articles intended for repeated use in producing, manufacturing, packing, processing, treating, packaging, transporting, or holding food (total not to exceed 30% by weight of the rubber products).

Reference:
See **BUTYL STEARATE,** #3.

DIALKYL-3,3′-THIODIPROPIONATE

Synonym. Thiodipropionic acid, dialkyl ester.

Properties. White, crystalline powder or clear, colorless, oily liquid. Poorly soluble in water, soluble in alcohol.

Applications. Antioxidant with plasticizing properties. A synergistic additive for various thermo- and light-stabilizers of polypropylene and PVC. Stabilizer of packaging film for food products.

Acute Toxicity. Rats and mice tolerate the dose of 20 g/kg BW.

Repeated Exposure failed to reveal cumulative properties. Rats were dosed with 200 mg/kg BW and mice with 150 to 300 mg/kg BW. The treatment produced no changes in the CNS, hematological analyses, or blood enzyme activity. Histological examination failed to reveal changes in the visceral organs.

Allergenic Effect. Does not cause sensitizitaion when applied to the skin of guinea pigs.

Standards. Russia. PML in drinking water: **n/m.**

Reference:

See 4-**ALK(C$_7$-C$_9$)OXY**-2-**HYDROXYBENZOPHENONE.**

DIALKYL-3,3'-THIODIVALERIATE

Synonym. Thiodivaleric acid, dialkyl ester.

Toxicity and **Allergenic Effect:** See **Dialkylthiodipropionate.**

Standards. Russia. PML in drinking water: **n/m.**

Regulations. British Standard (1992): Didodecyl 3,3'-thiodipropionate, ditetradecyl 3,3'-thiod-ipropionate, dioctadecyl 3,3'-thiodipropionate are listed as antioxidants for polyethylene and poly-propylene compositions used in contact with foodstuffs or water intended for human consumption (maximum permitted level for the final compound 1%).

Reference:

See 4-**ALK(C$_7$-C$_9$)OXY**-2-**HYDROXYBENZOPHENONE.**

DIBENZYLTIN-S,S'-BIS(ISOOCTYLMERCAPTOACETATE) (DBTIG) (CAS No 28675–83–4)

Synonyms. *D*-Benzyl-*TG;* Dibenzyltin-*S,S'*-diisooctylthio-glycolate.

Applications. Used as a stabilizer in the production of PVC plastics.

Acute Toxicity. In rats, LD$_{50}$ is reported to be 2 g/kg[1] or 1.25 g/kg BW.

Repeated Exposure failed to reveal cumulative properties. An increased relative weight of the liver is found at the dose level of 200 mg/kg BW; a rise in the activity of ALT and AST was observed when the dose of 100 mg/kg BW was administered.[2]

Short-term Toxicity. In a 3-month study, Wistar rats received 180 mg DBTIG per kilogram in the diet. Poisoning resulted in 15% animal mortality. The relative liver weights were increased.[3]

Long-term Toxicity. In a 1.5-year study, the liver, enzymatic activity, and serum proteins content were predominantly affected.[2]

Reproductive Toxicity. An increase in embryolethality and a reduction in the number of embryos were found in rats given DBTIG before mating and during pregnancy.[3] Doses of 18 and 90 mg/kg BW given to rats throughout the pregnancy led to a reduction in litter size and to an increase in neonatal mortality.[4]

References:

1. Klimmer, O. R. und Nebel, I. U., Experimentelle Untersuchungen zur Frage der Toxicitat einiger Stabilizatoren in Kunstoffen aus Polyvinylchlorid, *Arzneimittel. Forsch.,* 10, 44, 1960 (in German).
2. Mazur, H., Effect of oral administration of dioctyltin bis-iso-octylthioglycolate and dibenzyltin bis-iso-octylthio-glycolate on rat organism, *Roczn. Panstv. Zakl. Hig.,* 22, 39, 1971 (in Polish).
3. Mazur, H., Effect of oral administration of dioctyltin bis-iso-octylthioglycolate and dibenzyltin bis-iso-octylthioglycolate on rat organism. II. Influence on fertility and fetal development, *Roczn. Panstv. Zakl. Hig.,* 5, 509, 1971 (in Polish).
4. Nikonorow, M., Mazur, H., and Piekacz, H., Effect of orally administered stabilizers in the rat, *Toxicol. Appl. Pharmacol.,* 26, 253, 1973.

2-(3,5'-DI-*tert*-BUTYL-2'-HYDROXYPHENYL)-5-CHLOROBENZO-TRIAZOLE

Synonyms. Benazole *BH;* Tinuvin 327.

Properties. Light-yellow, crystalline powder. Poorly soluble in water and alcohol.

Applications. Used as a light-stabilizer in the production of polyolefins and also as a stabilizer for polyacetate and cellulose esters.

Acute Toxicity. Mice given 7.5 and 10 g/kg BW and rats administered 5 g/kg BW exhibited only decrease in BW gain.

Repeated Exposure. Up to 50% of mice died in response to 40 administrations of 0.5 and 2 g D./kg BW. The treatment caused a reduction in BW gain, increased excitability, an increase in the relative weights of the liver and kidneys, hypertrophy and parenchymatous and fatty dystrophy of the liver, discomplexing of liver proteins.

Long-term Toxicity. Mice were fed 100 and 200 mg D./kg BW given in suspension in sunflower oil. Retardation of BW gain and an increase in excitability were noted. Mortality reached 45% in animals fed 200 mg/kg BW and 64% in animals fed 100 mg/kg BW. Gross pathological examination revealed pronounced liver hypertrophy.

Reproductive Toxicity and **Carcinogenicity.** Reproductive function of rats given 5 mg/kg for 2 years had not been affected over three generations. The treatment did not cause tumor growth.[05]
Reference:
See **BISPHENOL A,** #2, 193.

2,5-DI-*tert*-BUTYLHYDROQUINONE (CAS No 88–58–4)

Synonyms. Dibug; Santovar O.
Properties. Light-grey crystalline powder. Poorly soluble in water, soluble in hot alcohol.
Applications. Used as a stabilizer of polyolefins, polyformaldehyde, and synthetic rubber.
Acute Toxicity. In mice, the LD_{50} is reported to be 15 g/kg BW; however, according to other data, it is 4.3 g/kg BW. In rats, it is 0.8 g/kg BW (Kodak Co., Rep.). No manifestations of the toxic effect were found in rats given 5 g/kg BW. Drowsiness and adynamia developed after administration of larger doses. Gross pathological examination revealed focal bronchopneumonia, pulmonary vascular stasis, and congestion in the liver.

Repeated Exposure failed to reveal cumulative properties or signs of intoxication in rats. BW gain was decreased and restored by the end of the second month of treatment. The treatment caused inhibition of antitoxic and oxidizing functions of the liver, the balance of protein fractions in the blood serum was changed, and erythrocyte catalase activity was reduced. Gross pathological examination revealed congestion in the brain vessels, fatty dystrophy of the liver, parenchymatous dystrophy of the kidneys, and bronchopneumonia. Mice exhibited the same changes with the 2 g/kg BW dose administered for 2 months.
Standards. Russia. PML in drinking water: **n/m.**
Reference:
See *N*-**BUTYLIDENEANILINE,** 16.

2,6-DI-*tert*-BUTYL-4-METHOXYMETHYLPHENOL (CAS No 87–97–8)

Synonym. 3,5-Di-*tert*-butyl-4-hydroxybenzyl alcohol, methyl ester.
Properties. White, crystalline powder, turning yellow when exposed to light. Poorly soluble in water, soluble in alcohol.
Applications. Stabilizer for synthetic rubber; a thermostabilizer for polyolefins.
Acute Toxicity. In rats, the LD_{50} is 10.6 g/kg BW. The stabilizer exhibited a narcotic effect. Death within 3 d. Gross pathological examination revealed congestion in the visceral organs and brain.
Repeated Exposure. Investigation of cumulative properties showed four out of ten rats to die after four to six administrations in 16 d.
Reference:
See 2,6-**DI-***tert*-**BUTYL-4-METHOXYMETHYLPHENOL,** 154

DIBUTYLTIN-*S,S'*-BIS(ISOOCTYLMERCAPTOACETATE) (CAS No 25168–24–5)

Synonyms. Dibutyltin-S,S'bis(isooctylthioglycolate); BTS-70.
Properties. Oily liquid with a strong, unpleasant odor. Poorly soluble in water. Odor perception threshold is 0.05 mg/l; taste perception threshold is 0.23 mg/l.[1]
Applications. Used as a stabilizer in the production of PVC plastics.

Acute Toxicity. LD_{50} is 570 mg/kg in mice and 500 mg/kg BW in rats.[2] According to other data, LD_{50} in rabbits and male and female rats is found to be 500, 340, and 220 mg/kg BW, respectively.[1] Administration of high doses caused refusal of food and diarrhea in the animals. Rats become photophobic, adynamic, and highly aggressive. Death occurred on days 1 to 11 after poisoning. Gross pathological examination revealed distension of the intestine, hyperemia with necrotic areas in the intestinal walls. There were also gall bladder engorgement and inflammation of the bile duct walls.

Repeated Exposure revealed evident cumulative properties. K_{acc} is 0.3 (by Cherkinsky). Male rats received 8.5, 17, and 34 mg/kg BW. The treatment affected the content of SH-groups in the blood. The highest dose caused changes in STI and hematological analyses. No changes were found in enzyme activity.[1]

Short-term Toxicity. Histological examination revealed parenchymatous dystrophy of the renal tubular epithelium, cellular breakdown in the lung tissue, the loss of parenchyma cells in the liver, and lymphocyte and plasma cell infiltration in rats given 0.5 and 5 mg/kg BW for 3 months.[2]

Long-term Toxicity. In a 7-month study, changes were observed only in conditioned and unconditioned reflex activity.

Reproductive Toxicity. *Gonadotoxic* effect in rats is not observed.[1] Wistar rats were given 120 mg/kg BW on day 9 and 13 of pregnancy.[2] The dose of 0.01 mg/kg BW appeared to be ineffective for *embryotoxic* and *teratogenic* effects.

Standards. Russia. (1988). MAC: 0.01 mg/l; PML in drinking water: 0.001 mg/l.

References:
1. Igumnov, A. S., Data on the Maximum Allowable Concentration for dibutyltin dithioglycolate in water bodies, *Gig. Sanit.*, 2, 13, 1975 (in Russian).
2. Hygienic Aspects of the Use of Polymeric Materials, Proc. 2nd All-Union Meeting on Health and Safety Monitoring of the Use of Polymer Materials in Construction, Kiev, 1976, 36 (in Russian).

DIBUTYLTIN BIS(TRIFLUOROACETATE) (CAS No 52112–09–1)

Applications. Used as a stabilizer in the production of PVC.

Acute Toxicity. In rats, LD_{50} is reported to be 53.6 mg/kg BW.

Repeated Exposure revealed cumulative properties. On administration of $1/5$ or $1/25$ LD_{50} to mice for 20 d, K_{acc} (by Cherkinsky) appeared to be 3.6 to 5.4 and 1.5 to 1.6, respectively. Anemia and leukocytosis were observed in rats that received 0.25 mg/kg BW for 30 d.

Long-term Toxicity. In a 6-month study, significant increase in the cholinesterase activity was noted in rabbits. The catalase activity has fallen sharply by the end of the experiment.

Genotoxicity. A mutagenic effect in bone marrow cells is reported in mice on a single administration of LD_{50}.

Chemobiokinetics. When $1/5$ and $1/25$ LD_{50} was given to mice for 20 d, material accumulation was found. Excretion from the body was slow.

Reference:
See **BIS(TRIBUTYLTIN) OXIDE, #3.**

DIBUTYLTIN DICHLORIDE (DBTCl) (CAS No 683–18–1)

Properties. Light-brown, crystalline substance. Solubility in water is 80 mg/l. Odor perception threshold is 3 mg/l.

Applications. Used in the production of PVC stabilizers. A catalyst in polyurethane foam products. A vulcanizing agent for silicone rubber.

Acute Toxicity. LD_{50} is found to be 35 mg/kg BW in mice, 112 to 182 mg/kg BW in rats, 125 mg/kg BW in rabbits, and 190 mg/kg BW in guinea pigs.[1] Lethal doses caused motor disturbances in the poisoned animals. Pathological changes were detected in the bile duct of rats after acute oral dose of 20 to 160 mg/kg BW.[2] DBTCl affects cytochrome P-450 activity in the small intestine, liver, and kidney of rats in 24 h after a single oral dose administration.[3] A single administration of 5 to

35 mg/kg BW to rats resulted in reduced thymus weights but produced no effect on spleen weights. Recovery was completed in 9 d post-treatment.[4]

Repeated Exposure revealed pronounced cumulative properties of DBTCl. A 20 mg/kg BW dose caused 100% mortality in rats by day 13. LD_{50} of ¹⁄₁₀ resulted in apathy, depression of BW gain, and reduced skeletal muscle tone from days 2 to 5 of the treatment.[1] Administration of 50 to 150 mg/kg of the diet (2.5 to 7.5 mg/kg BW) for 2 weeks induced marked changes in the lymphoid tissues, including dose-related reduction in thymus weights and lymphocyte depletion in the thymus.[5]

Short-term Toxicity. Rats were fed dietary levels of 10 to 80 ppm (0.5 to 4 mg/kg BW) for 90 d. The effect included depression of BW gain at the highest dose. Thymus weights were not monitored, but no pathological changes were revealed on examination of a section of this organ. Anemia was caused in rats by the dose of 80 mg/kg diet but not by the 40 mg/kg dose. The NOEL appeared to be 2 mg/kg BW.[6]

Immunotoxicity. The immunotoxic potential of dialkyltins in species other than the rat is less clear. Decreased weights of thymus and burza Fabricii were found in rats and chickens exposed over a period of 10 to 14 d.[7] However, Seinen[5] did not find any evidence of lymphoid toxicity in mice, guinea pigs, or Japanese quail fed DBTCl.

Reproductive Toxicity. In a 3-month study, the doses of 20 and 40 mg/kg BW caused an embryolethal effect. No teratogenic effect was reported.[8] However, gastric intubation with 5 to 10 mg/kg BW resulted in a significant dose-dependent increase in the incidence of fetuses with malformations (anatomy of tail, anal atresia, club foot, deformity of the vertebral column, defects of the ribs and anophthalmia), even at a dose that did not induce apparent toxicity in maternal rats.[9,10]

Genotoxicity. Positive results were found in *in vitro* mammalian cell mutation assays. DBTCl did not require the presence of an exogenous metabolizing system to express genotoxic activity.[11,12] DBTCl showed high SOS-inducing potency in the SOS-chromotest with *Escherichia coli*. It is recognized as a genotoxic chemical by the rec-assay.[13]

Chemobiokinetics. At low concentrations DBTCl blocks α-ketoacid dehydrogenase and, at high concentrations, causes disruption of oxidative phosphorylation (Penninks et al., 1983). It is excreted unchanged in the bile (60%) and as **monoethyltin** (25%) and **diethyltin** (5%) in the urine.[14]

Standards. Russia. (1986). MAC and PML: 0.002 mg/l.

Regulations. USFDA (1993) approved the use of DBTCl as a catalyst in polyurethane resins for articles intended for use in contact with dry food alone or in combination, at levels not to exceed a total of 3% by weight in resin used.

References:

1. See **ACRYLIC ACID,** #1, 15.
2. Barness, J. M. and Stoner, H. B., Toxic properties of some dialkyl and trialkyl tin salts, *Br. J. Ind. Med.,* 15, 15, 1958.
3. Rosenberg, D. W. and Kappas, A., Actions of orally administered organotin compounds on metabolism and cytochrome P-450 content and function in the intestinal epithelium, *Biochem. Pharmacol.,* 38, 1155, 1989.
4. Snoeij, N. J., Penninks, A. H., and Seinen, W., Dibutyltin and tributyltin compounds induce thymus atrophy in rats due to a selective action on thymic lymphoblasts, *Int. J. Immunopharmacol.,* 10, 891, 1988.
5. Seinen, W., Voss, J. G., van Spanje, I., et al., Toxicity of organotin compounds, *Toxicol. Appl. Pharmacol.,* 42, 197, 1977.
6. Gaunt, I. F., Colley, J., Grasso, P., et al., Acute and short-term toxicity studies on di-*n*-butyltin dichloride in rats, *Food Cosmet. Toxicol.,* 6, 599, 1968.
7. Renhof, M., Kretzer, V., Schurmeyer, T., et al., Toxicity of organotin compounds in chicken and rats, *Arch. Toxicol.,* Suppl. 4, 148, 1980.
8. See **DIBENZYLTIN-***S,S′***- BIS(ISOOCTYLMERCAPTOACETATE,** #4.
9. Ema, M., Itamy, T., Kawasali, H., et al., Toxicology susceptible period for teratogenicity of di-*n*-butyltin dichloride in rats, *J. Appl. Toxicol.,* 73, 81, 1992.
10. Ema, M., Itamy, T., Kawasali, H., et al., Teratogenicity of di-*n*-butyltin dichloride in rats, *Toxicol. Lett.,* 58, 347, 1991.
11. Li, A. P., Dahl, A. R., and Hill, J. O., *In vitro* cytotoxicity and genotoxicity of dibutyltin dichloride and dibutylgermanium dichloride, *Toxicol. Appl. Pharmacol.,* 64, 482, 1982.

12. Westendorf, J. and Marquardt, H., DNA interaction and mutagenicity of the plastic stabilizer di-*n*-octyl dichloride, *Arzneim. Forsch.*, 36, 1263, 1986 (in German).

13. Hamasaki, T., Sato, T., Nagase, H., et al., The genotoxicity of organotin compounds in SOS chromotest and res-assay, *Mutat. Res.*, 280, 195, 1992.

14. Bridges, J. W., Davies, D. S., and Williams, R. T., The fate of ethyltin and diethyltin derivates in the rat, *Biochem. J.*, 105, 1261, 1967.

DIBUTYLTIN DILAURATE (DBTL) (CAS No 77–58–7)

Synonyms. Butynorate; Dibutylbis(lauroyloxy)tin; Dibutylbis [(1-oxododecyl)-oxy]stannate; Laustan *B;* Tinostat.

Properties. Soft crystals or yellow liquid. A clear, oily liquid at higher temperatures. Poorly soluble in water but gradually hydrolyzes.

Applications. A stabilizer in the production of PVC and chlorinated rubber; a light-stabilizer for polystyrene; a light- and thermostabilizer of polyamides and foam plastics; a catalyst in the cold vulcanization of organosilicon rubber.

Acute Toxicity. The oral LD_{50} was reported to be 1.7 g/kg BW in mice and 13 g/kg BW in rats. A single administration of 5 g/kg BW caused liver and stomach damage. Evidence for neurotoxicity was reported: rats were given orally the doses of 20 to 80 mg/kg BW in 0.2 ml of groundnut oil for 3 consecutive days.

Administration caused a decrease in spontaneous locomotor activity and learning, and a decreased level of the neurotransmitters, noradrenaline, dopamine, and serotonin, particularly in the hypo-thalamus and frontal cortex.[1]

Repeated Exposure revealed marked cumulative properties in mice given 85 to 340 mg/kg BW and in rats, receiving 65 to 130 mg/kg BW. The treatment caused a depression of BW gain and a change in the weight coefficients of the visceral organs, CNS excitation, and lesions in the stomach, intestines, and liver.[2] Rabbits were fed 15 to 40 mg DBTL/kg BW over a period of 6 weeks. Indication of liver damage through elevated serum SGOT, SGPT, and lactate dehydrogenase was found.[3] Rats received oral doses of 17.5 mg/kg BW for 15 days. Of the animals, 20% died during the experiment, and the survivors showed reduced BW gain. Some hepatic enzymes had been inhibited.[4]

Reproductive Toxicity. Pregnant rats were treated orally on day 8 of gestation with 0.08 mmol DBTL/kg. DBTL has been shown to cause malformations such as cleft mandible, ankyloglossia, fused ribs, etc. in rat fetuses.[5]

Standards. Russia. (1988). MAC and PML in drinking water: 0.01 mg/kg.

Regulations. USFDA (1993) approved the use of DBTL (1) as a curing (cross-linking) catalyst for silicone adhesives used as components of articles intended for use in packaging, transporting, or holding food (the maximum amount of tin catalyst used shall be that required to effect optimum cure but shall not exceed 1% siloxane resins solids), (2) as a catalyst in polyurethane resins for the food-contact surface of articles intended for use in contact with dry food alone or in combination, at levels not to exceed a total of 3% by weight in the resin used, and (3) in resinous and polymeric coatings in a food-contact surface of articles intended for use in producing, manufacturing, packing, transporting, or holding food.

References:

1. Alam, M. S., Hussain, R., Srivastava, S. P., et al., Influence of dibutyltin dilaurate on brain neuro-transmitter systems and behaviour in rats, *Arch. Toxicol.*, 61, 373, 1988.

2. Maksimova, N. S., Krynskaya, I. L., and Yevsyukov, V. I., Toxicity of organotin and cadmium stabilizers, *Plast. massy*, 12, 33, 1976 (in Russian).

3. Tanaka, S., Experimental studies on the effects of dibutyltin dilaurate, dibutyltin dichloride and dimethyltin dichloride to serum enzymes and lipids of rabbits, *Tokyo Ika Daigaku Zasshi*, 38, 607, 1980 (in Japanese.)

4. Mushtaq, M., Mukhtar, H., Datta, K. K., et al., Toxicological studies of a leachable stabilizer di-*n*-butyltin dilaurate (DBTL): effects on hepatic drug metabolizing enzyme activities, *Drug Chem. Toxicol.*, 4, 75, 1981.

5. Noda, T., Morita, S., and Baba, A., Teratogenic effects of various di-*n*-butyltins with different anions and butyl(3-hydroxybutyl)tin dilaurate in rats, *Toxicology*, 85, 149, 1993.

DI-[(3,5-DI-*tert*-BUTYL-4-HYDROXYPHENYL) PROPYL]THIODIACETATE

Synonym. Tioalkofen P.

Properties. Highly dispersed white powder. On heating, soluble in alcohols, poorly soluble in water.

Acute Toxicity. LD_{50} is 1.98 g/kg BW in rats and 1.38 g/kg BW in mice. Lethal doses lead to CNS excitation, including restlessness, convulsions, and subsequent adynamia. Death within 2 to 4 d.

Reference:

Kirichek, L. T. and Derkach, Z. M., Acute toxicity of new light-stabilizers, *Gig. Sanit.*, 6, 87, 1985 (in Russian).

DIETHYLENE GLYCOL BIS-3-(3,5-DI-*tert*-BUTYL-4-HYDROXYPHENYL)- PROPIONATE

Synonym. Fenozan 28.

Properties. White, crystalline powder. Poorly soluble in water.

Applications. Used as a stabilizer in the production of polyolefins and other polymeric materials.

Acute Toxicity. Rats tolerate administration of the 10 g/kg BW dose without visible manifestations of toxic effect.

Repeated Exposure. Young rats tolerated daily administration of D. (overall dose 90 g/kg BW). The treatment did not affect BW gain or other indices of body condition.

Chemobiokinetics. A significant amount of D. is unlikely to be absorbed from the GI tract into the blood. It is excreted unchanged from the body.

Reference:

See **ADIPIC ACID**, #2, 269.

DIETHYLTIN DICHLORIDE (DETDCl) (CAS No 866–55–7)

Repeated Exposure. DETDCl decreased cytochrome P-450 activity but induced hemoxygenase activity in the liver and kidney of rats.[1] Doses of 2.5 to 7.5 mg/kg BW administered for 2 weeks induced pronounced changes in the lymphoid tissues, including dose-related reduction in thymus weights and lymphocyte depletion in the thymus.[2]

Reproductive Toxicity. Administration of 10 mg DETDCl/kg BW to adult rats and to neonates induced weight loss, but there were no indications of neurotoxicity.[3]

Chemobiokinetics. DETDCl is poorly absorbed in the GI tract in rats;[4] it is distributed mainly to the liver, and excreted primarily with the bile.[5]

References:

1. Rosenberg, D. W., Drummond, G. S., and Kappas, A., The influence of organometals on heme metabolism, *In vivo* and *in vitro* studies with organotins, *Mol. Pharmacol.*, 21, 150, 1981.
2. See **DIBUTYLTIN DICHLORIDE**, #2.
3. Bouldin, T. W., Goines, N. D., Bagnell, C. R., et al., Pathogenesis of trimethyltin neuronal toxicity. Ultrastructural and cytochemical observations, *Am. J. Pathol.*, 104, 237, 1981.
4. Mazayev, V. T., Tikhonova, Z. I., and Shlepnina, T. G., Distribution and elimination of tin taken up into the body in the form of organotin compounds, *J. Hyg. Epidemiol. Microbiol. Immunol.*, 20, 392, 1976 (in Russian).
5. See **DIBUTYLTIN DICHLORIDE**, #15.

2,6-DIISOBORNYL-4-METHYLPHENOL

Synonym. Alkofen *DIP*.

Applications. Used as a stabilizer in the production of polyethylene and polypropylene.

Acute Toxicity. Mice tolerate the dose of 7.7 g/kg BW. Gross pathological examination revealed no morphological changes in the visceral organs.

Long-term Toxicity. Rats were dosed by gavage with 1.5 g/kg BW for 6.5 months. The treatment caused an impairment of conditioned reflex activity. Gross pathological examination revealed congestion in the visceral organs, edema, and desquamation of the gastric and esophageal mucosa, and parenchymatous dystrophy of the renal tubular epithelium.[08]

Reproductive Toxicity. Affects the reproduction of fish but does not affect the development of chick embryos.

Standards. Russia. Recommended PML in drinking water: **n/m.**

3,5-DIISOPROPYL SALICYLAMIDE

Properties. Grey, finely dispersed, odorless powder. Poorly soluble in water.

Applications. Used as a stabilizer of polyamides and polyurethanes.

Acute Toxicity. Rats and mice tolerate administration of a 10 g/kg BW dose without any manifestations of toxic action.

Repeated Exposure. 0.5 and 1 g/kg BW doses produced consistent mortality and marked toxicity effects in rats. The animals tolerated administration of 0.2 g/kg BW for 2 months without visible signs of toxicity.

Reference:
See **BUTYL BENZYL ADIPATE,** 314.

2,6-DIISOPROPYL-4-*p*-TOLUENESULFAMIDOPHENOL

Properties. Finely dispersed, odorless, grey powder. Poorly soluble in water.

Applications. A stabilizer. Used in the production of polyamides and polyurethanes.

Acute Toxicity. Rats and mice tolerate administration of a 10 g/kg BW dose without any manifestations of toxic action.

Repeated Exposure. The doses of 0.5 and 1 g/kg BW caused 25% animal mortality during 5 weeks of treatment and produced visible symptoms of intoxication.

Short-term Toxicity. Rats tolerate administration of a 0.5 g/kg BW dose during 3 months of the experiment without any manifestations of toxic action.

Reference:
See **BUTYL BENZYL ADIPATE,** 314.

DILAURYL-3,3'-THIODIPROPIONATE (DLTDP) (CAS No 123–28–4)

Synonym. 3,3'-Thiodipropionic acid, didodecyl ester.

Properties. White, crystalline powder. Poorly soluble in water.

Applications. Used as a synergistic additive to thermo- and light-stabilizer in the production of polyolefins; an antioxidant with plasticizing properties.

Acute Toxicity. Maximum tolerated dose in mice and rats is reported to be 2 g/kg[05] or even 15 g/kg BW.[08] Gross pathological examination failed to reveal changes in the visceral organs.

Repeated Exposure. In a 100-d study, dogs were given a 3% mixture of thiodipropionic acid and DLTDP (1:9) in their diet; the mixture was first heated to 190°C. Gross pathological examination failed to reveal changes in the visceral organs. Mice received 20 administrations of 625 mg/kg BW. The treatment caused no mortality or morphological changes in the visceral organs.[08]

Long-term Toxicity. In a 2-year study, rats received 1 and 3% DLTDP in their diet. This treatment caused retardation of growth. Addition of 0.5 to 1.3% DLTDP to the diet for 2 years had no toxic effect on growth, survival rate, and structure of the viscera in rats.[05] In a 6-month study, rats were dosed by gavage with 430 and 1420 mg DLTDP/kg BW. The treatment caused excitation of the CNS cortical areas only. There were no gross pathological changes in the visceral organs.[08]

Reproductive Toxicity. Affects fish reproduction processes and the development of chick embryos.[08]

Chemobiokinetics. D. is removed from the body in the urine in the form of 3,3'-**thiodipropionic acid.**

Regulations. USFDA (1993) approved the use of DLTDP in resinous and polymeric coatings for the food-contact surface of articles intended for use in producing, manufacturing, packing, transporting, or holding food.

4,4'-DIMETHOXY-4-DIPHENYLAMINE (CAS No 101–70–2)

Properties. White, crystalline disks. Poorly soluble in water, readily soluble in hot alcohol.

Applications. D. is used in combination with other antioxidants to prevent rubber ageing and fatigue.

Acute Toxicity. In rats and mice, LD_{50} is 2.5 g/kg BW. Administration of lethal doses to mice was accompanied by agitation and body tremor. Some animals exhibited hard, labored breathing with subsequent convulsions. Death occurred within days. Gross pathological examination revealed changes in the liver, kidneys, and spleen.

Repeated Exposure. No cumulative properties were evident in rats dosed by gavage with 410 mg/kg BW for 2 months. The treatment affected CNS and liver functions; gas exchange was decreased.

Reference:

See *N*-**BUTYLIDENEANILINE,** 51.

4–α,α-DIMETHYL BENZYL PYROCATECHOL

Properties. Crystalline powder. Solubility in water is 0.001%.

Acute Toxicity. In mice, LD_{50} is 15 g/kg BW.[08]

Short-term Toxicity. Mice were dosed by gavage with 8.75 mg/kg BW for 4 months. The treatment caused some growth retardation. Gross pathological examination revealed necrosis in the stomach and intestinal walls, parenchymatous dystrophy of the renal tubular epithelium, and hyperplasia of the spleen pulp.

N-(1,3-DIMETHYLBUTYL)-N'-PHENYL-*p*-PHENYLENEDIAMINE

Synonyms. Santoflex 13; Diafen FDMB; Antioxidant 4020.

Properties. Crystalline powder. Poorly soluble in water, soluble in alcohol.

Applications. Used as a stabilizer in the production of synthetic rubber and vulcanizates.

Acute Toxicity. LD_{50} is found to be 2.5 g/kg BW in rats and 3.2 g/kg BW in mice. High doses produced a narcotic effect. An acute action threshold appeared to be 1 g/kg BW.

Repeated Exposure failed to reveal cumulative properties. No animal mortality was noted, but there were retardation of BW gain and impairment of the liver and kidney functions.

Reference:

See *N*-(1,3-**DIMETHYLBUTYL)**-*N'*-**PHENYL**-*p*-**ENYLENEDIAMINE,** 32.

N,N'-DI-α-NAPHTHYL-*p*-PHENYLENEDIAMINE

Properties. Light-grey powder. Poorly soluble in water.

Applications. Used as a stabilizer in synthetic rubber manufacturing. D. is a thermostabilizer in the production of vulcanizates based on natural and synthetic rubber, polypropylene, pentaplast, polyamides, etc.

Acute Toxicity. Rats tolerate 5 g/kg BW, and mice do not die from 7.5 and 10 g/kg BW. A single administration of high doses caused only retardation of BW gain. According to other data, LD_{50} is 7.5 g/kg BW for mice. A single oral administration is reported to cause mild changes in the osmotic resistance of erythrocytes and a slight methemoglobinemia.[08-9] Gross pathological examination revealed circulatory disturbances in the visceral organs, necrotic changes in the intestinal mucosa, and parenchymatous dystrophy in the renal tubular epithelium.

Repeated Exposure. Rats were exposed by gavage to 40 administrations of 0.5 g/kg BW and mice to 0.5 and 2 g/kg BW. The treatment increased excitability in the animals. In some mice, mild liver hypertrophy was noted. Gross pathological examination revealed no morphological changes in the viscera.

Short-term Toxicity. Mice were given 100 and 200 mg D./kg BW doses as an oil suspension. The treatment resulted in retardation of BW gain and an increase in the relative weights of the liver and kidneys. In some mice, mild wasting and slight liver hypertrophy were noted. Histological examination revealed no abnormalities in the viscera. Rats were dosed by gavage with 75 mg/kg BW for 4.5 months. Manifestations of toxic action included changes in BW gain, slight methemoglobinemia, and increased conditioned reflex activity. Gross pathological examination revealed superficial villous necrosis and desquamation of the intestinal mucosa epithelium.

Reference:
See **BISPHENOL A**, #2, 193.

N,N'-DI-β-NAPHTHYL-*p*-PHENYLENEDIAMINE (CAS No 93–46–9)

Synonyms. Antioxidant 123; Diafen *NN*; Santowhite *Cl*.

Properties. A light-yellow powder. Poorly soluble in water.

Applications. Used as a stabilizer in the synthesis of rubber and vulcanizates and in the production of polypropylene, pentaplast, propylene oxide, etc.

Acute Toxicity. LD_{50} for mice is 5 g/kg[1] or 20 g/kg BW.[08] Administration of lethal doses caused slight changes in erythrocyte osmotic resistance and slight methemoglobinemia. Histological examination revealed circulatory disturbances in the visceral organs, necrotic changes in the small intestinal mucosa, and parenchymatous dystrophy of the renal tubular epithelium.

Repeated Exposure caused retardation of BW gain, anemia, leukocytosis, and an increase in the methemoglobin content. CNS inhibition and liver function impairment were noted.

Short-term Toxicity. Rats were dosed by gavage with 120 mg/kg BW three times a week for 4 months. Signs of CNS stimulation were observed. Gross pathological examination revealed profound necrosis of the intestinal mucosa. A dose of 100 mg/kg BW administered for 3.5 months led to decreased BW gain, slight methemoglobinemia, and changes in conditioned reflex activity.[08]

Reproductive Toxicity. D. affects fish reproduction but has no effect on the development of chick embryos.[08]

Carcinogenicity. D. is likely to be a weak carcinogen (see *N*-**Isopropyl**-*N'*-**phenyl**-*p*-**phenylenediamine**).

Regulations. USFDA (1993) approved the use of D. (1) in adhesives, as a component of articles intended for use in packaging, transporting, or holding food, (2) in closures with sealing gaskets on containers (not exceeding 1% by weight of the closure-sealing gasket composition), and (3) as an antioxidant in rubber articles intended for repeated use in producing, manufacturing, packing, processing, treating, packaging, transporting, or holding food (total not to exceed 5% by weight of the rubber product).

Reference:
See *N*-**BUTYLIDENEANILINE**, 126.

4,4'-DI-*tert*-OCTYLDIPHENYLAMINE (CAS No 101–67–7)

Synonym. Octamine.

Properties. White, crystalline powder.

Applications. Used as a stabilizer in the production of butadiene-styrene, polychloroprene, and acrylonitrile-butadiene rubber. An antioxidant for polyolefins.

Acute Toxicity. Rats and mice tolerated the dose of 20 mg/kg BW. Following administration of this dose, there were no signs of intoxication in animals.

Repeated Exposure to 4 g D./kg BW caused a reduction in blood catalase and acetylcholine activity in rats. Gross pathological examination revealed no damage to the visceral organs. D. does not affect eyes and skin.

Reference:
Methodology Aspects of Investigation on the Biological Action of Chemicals, Erisman Research Hygiene Institute, Moscow, 1982, p. 40 (in Russian).

DIOCTYLTHIOTIN (DOTT) (CAS No 15535–79–2)

Synonyms. Dioctyltin thioglycolate; Stanklear 176.

Composition and **Properties.** A sulfur-containing compound of dioctyltin. Yellowish, oily liquid with an unpleasant odor (perception threshold 0.5 mg/l at 60°C).[1]

Applications. Used as a thermostabilizer to produce clear PVC materials.

Acute Toxicity. LD_{50} is 1.75 g/kg BW for rats and 2.25 g/kg BW for mice.[1]

Repeated Exposure revealed evident cumulative properties. K_{acc} is 1.45 (by Cherkinsky).[2] On administration of $1/20$ and $1/5$ LD_{50}, all rats died in 3 weeks.

Standards. Russia. (1976). PML in drinking water: 0.05 mg/l.

References:
1. See **DIBUTYLTIN-*S,S'*-BIS(ISOOCTYLMERCAPTO- ACETATE)**, #2, 36.
2. Sheftel, V. O., Sahm, Z. S., Martson, L. V., et al., On permissible migration levels of PVC stabilizers, *Gig. Sanit.*, 3, 109, 1977 (in Russian).

DI-*n*-OCTYLTIN-*S,S'*-BIS(ISOOCTYLMERCAPTOACETATE) (DOTBIOMA) (CAS No 26401–97–8)

Synonym. Di-*n*-octyltin bis(isooctylthioglycolate).

Composition and **Properties.** A light-yellow, viscous liquid containing 11 to 12% tin and 25% epoxydized soy oil. Poorly soluble in water, readily soluble in plasticizers.

Applications and **Exposure.** Used as a stabilizer in the production of PVC plastics. Migration from PVC film to food and food-simulating liquids is found to be continuous, with no indication of cessation.[1]

Acute Toxicity. LD_{50} was reported to be 1.1 g/kg BW in mice and 1.5 g/kg[2] or 2.1 g/kg BW[3] in rats. Poisoning is characterized by the symptoms of CNS impairment with adynamia and respiratory difficulty. Administration of 4 g/kg BW into the stomach of mice resulted in dilatation, ishemia of the stomach walls, and intestinal wall hyperemia.[4]

Repeated Exposure revealed reasonably evident cumulative properties in rats on daily administration of $1/5$ LD_{50} for 20 d.[5] K_{acc} is 6.9 (by Cherkinsky). Wistar rats were dosed by gavage with 100 and 200 mg/kg BW. All the animals died in 10 d from acute inflammation of the GI tract. The first administration impaired mobility, caused drowsiness, diarrhea, and a reduced BW gain.

Short-term Toxicity. Rats were given a 20 mg/kg BW dose for 3 months. The treatment caused 15% animal mortality.[6]

Long-term Toxicity. Rats were given 20 and 200 mg DOTBIOMA/kg BW as a suspension in olive oil for 90 d and also with the feed (0.02% in the diet) for 7 to 18 months. Treatment-related effects included retardation of BW gain, increased liver and kidney weights, enhanced ALT and AST activity, and changes in the ratio of the serum protein fractions.[7]

Allergenic Effect was found when $1/3000$ LD_{50} had been applied to the skin of guinea pigs.[3]

Reproductive Toxicity. A reduction in the number of embryos and increased fetal mortality were reported in rats administered DOTBIOMA before mating and during pregnancy.[7] The dose of 21 mg/kg caused an *embryotoxic* effect.[5]

Genotoxicity. A mixture of organotins (though principally comprised of DOTBIOMA) exhibited mutagenicity in one of five strains of *S. typhimurium* when tested in the absence of metabolic activation but was negative in its presence.[8]

Chemobiokinetics. DOTBIOMA seems to be slowly absorbed in the GI tract. As all thioalkyltin compounds, it is likely to undergo diamination and hydroxylation in the body and is partly excreted in the feces.

Standards. Russia. Recommended PML in food: 0.03 mg/l, PML in drinking water: 0.5 mg/l.

Regulations. USFDA (1993) permits the use of DOTBIOMA alone or in combination, at levels not to exceed a total of 3% by weight in resin used to stabilize plastic bottles and films that are in contact with food.

References:

1. Schwope, A. D., Till, D. E., Ehutholt, D. J., et al., Migration of an organotin stabilizer from PVC film to food and food-simulating liquids, *Deutsch. Leben-Rundsch.*, 82, 277, 1986 (in German).
2. See **DIBENZYLTIN-*S,S'*-BIS(ISOOCTYLMERCAPTO ACETATE**), #1.
3. See **DIBUTYLTIN-*S,S'*-BIS(ISOOCTYLMERCAPTO ACETATE**), #2, 159.
4. Pelikan, Z., Cerny, E., and Polster, M., Toxic effects of some di-*n*-octyltin compounds in white mice, *Food Cosmet. Toxicol.*, 8, 655, 1970.
5. 14 Sci. Session on Chemistry and Technology of Organic Compounds in Sulphate Mineral Oils, Abstracts of Reports, Riga, 1976, p. 96 (in Russian).
6. See **DIBUTYL PHTHALATE**, #5.
7. See **DIBENZYLTIN-*S,S'*-BIS(ISOOCTYLMERCAPTO- ACETATE**, #2.
8. See **BISPHENOL A,** #6, 1.

DIOCTYLTIN CARBOXYLATE (DOTC)

Synonym. Stanklear 80.

Composition and **Properties.** A dioctyltin compound that does not contain sulfur. A pale-yellow, oily liquid with a faint odor (perception threshold is 3 mg/l at 60°C).[1]

Applications. Used as a thermo- and light-stabilizer for PVC and for food packaging films.

Acute Toxicity. LD_{50} is reported to be 5.4 g/kg BW in rats and 3.6 g/kg BW in mice.[1]

Repeated Exposure revealed cumulative properties. Dosing with ¹/₅ or ¹/₂₀ LD_{50} caused all rats to die by day 18.[1] K_{acc} is 1.34 (by Cherkinsky). Rats received water extracts of PVC stabilized with 0.5 to 2 ppm DOTC over a period of 5 weeks. The extract from the sample containing the lowest level of the stabilizer caused no pathological changes. The remainder affected hematological analyses, as well as protein, enzyme, and detoxifying function of the liver, produced reduction in the STI and increases in urinary protein and blood urea.[2]

Standards. Russia. (1976). PML in drinking water: 0.1 mg/l.

References:

1. See **DIOCTYLTHIOTIN**, #2.
2. *The Condition and Prospects for the Development of Medical Technique*, Part 2, Moscow, 1975, p. 154 (in Russian).

DIOCTYLTIN DICHLORIDE (DOTDCl) (CAS No 3542–36–7)

Applications. A starting substance for stabilizer production.

Acute Toxicity. Administration of single doses of 240 to 1200 mg/kg BW resulted in a dose-dependent reduction of thymus weights and of thymocyte number in mice.[1,2]

Repeated Exposure. Rats were given doses of 50 and 150 ppm in the diet for 1.5 months. The treatment caused some reduction in BW gain and in the relative weights of the thymus, spleen, and liver, in urine density, in blood Hb and serum ALT and AST activity.[3] Evidence of lymphoid toxicity in mice, guinea pigs, or Japanese quail was not found.[4] A weekly oral dose of 500 mg/kg BW (but not 100 mg/kg) given to mice for 8 weeks led to the reduction of thymus weights and induction of immune deficiency.[1] A dose of 75 mg/kg BW fed to rats for 8 to 12 weeks caused thymus atrophy and lymphocytopenia.[5]

Short-term Toxicity. Administration of 200 ppm in the diet over a period of 4 months caused no signs of toxicity in the rats.[6] However, according to other data, 50 to 150 ppm in the diet (2.5 to 7.5 mg/kg BW) administered for 2 weeks induced marked changes in the lymphoid tissues, including a dose-related reduction in thymus weight and lymphocyte depletion in the thymus.[4] Rats were fed the diets containing 50 and 150 ppm DOTDCl (2.5 and 7.5 mg/kg BW) for 6 weeks. Depressed growth was reported at the highest dose; elevated serum alkaline phosphatase activity was apparent at both 2.5 and 7.5 mg/kg BW. The principal effect of treatment was decreased thymus weight at both dose levels.[3]

Immunotoxicity. DOTDCl produces a specific immunodeficiency in rats. The immunotoxic potential of dialkyltins in species other than the rat is less clear. T-lymphocytes are the main target, and hence cell-mediated immunity and T-cell-dependent humoral immunity tend to be suppressed.[7,8] Reduction of thymus weights and of immunocompetence was observed in rats fed a diet containing 75 ppm (3.75 mg/kg BW) for 8 or 12 weeks.[7] After dosing for 10 to 14 d, decreased weights of thymus and burza Fabricii were found in rats and chickens, respectively.[9]

Genotoxicity. Negative results were noted in the Ames test and in induction of unscheduled DNA synthesis in the primary culture of rat hepatocytes.[10] There were positive results in *in vitro* mammalian cell mutation assays. DOTDCl was shown to interact with the DNA of cultured V79 Chinese hamster cells. Neither compound required the presence of an exogenous metabolizing system in order to express activity.[10,11]

Chemobiokinetics. In rats, DOTDCl is distributed primarily to the liver and kidneys, with lower levels in the adrenals, pituitary, and thyroid.[12] It is excreted principally in the feces.

References:

1. Miller, K. and Scott, M. P., Immunological consequences of dioctyltin dichloride-induced thymic injury, *Toxicol. Appl. Pharmacol.*, 78, 395, 1985.
2. Henninghousen, G. and Lange, P., Toxic effects of di-*n*-octyltin dichloride on the thymus in mice, *Arch. Toxicol.*, Suppl. 2, 315, 1979.
3. Seinen, W. and Willems, M. I., Toxicity of organotin compounds. I. Atrophy of the thymus and thymus-dependent lymphoid tissue in rats fed di-*n*-octyltin dichloride, *Toxicol. Appl. Pharmacol.*, 35, 63, 1976.
4. See **DIBUTYLTIN DICHLORIDE, #5.**
5. Miller, K., Scott, M. P., and Foster, J. R., Thymic involution in rats given diets containing dioctyltin dichloride, *Clin. Immunol. Immunopathol.*, 30, 62, 1984.
6. See **DIBUTYLTIN DICHLORIDE, #2.**
7. Miller, K., Maisey, J., and Nicklin, S., Effect of orally administered dioctyltin dichloride on murine immunocompetence, *Environ. Res.*, 39, 434, 1986.
8. Seinen, W., Vos, J. G., van Krieken, R. V., et al., Toxicity of organotin compounds. III. Suppression of thymus-dependent immunity in rats by di-*n*-butyltin dichloride and di-*n*-octyltin dichloride, *Toxicol. Appl. Pharmacol.*, 42, 213, 1977.
9. Renhof, M., Kretzer, V., Schurmeyer, T., et al., Toxicity of organotin compounds in chicken and rats, *Arch. Toxicol.*, Suppl. 4, 148, 1980.
10. Westendorf, J., Marquardt, H., et al., DNA interaction and mutagenicity of the plastic stabilizer di-*n*-octyltin dichloride, *Arzneim.-Forsch. Drug Res.*, 36, 1263, 1986 (in German).
11. Li, A. P., Dahl, A. R., and Hill, J. O., *In vitro* cytotoxicity and genotoxicity of dibutyltin dichloride and dibutylgermanium dichloride, *Toxicol. Appl. Pharmacol.*, 64, 482, 1982.
12. Penninks, A. H., Hilgers, L., and Seinen, W., The absorption, tissue distribution and excretion of di-*n*-octyltin dichloride in rats, *Toxicology*, 44, 107, 1987.

DIOCTYLTIN DI(MONOISOBUTYLMALEATE)(DOTDIBM) (CAS No 15571–59–2)

Properties. Oily, yellowish liquid with an unpleasant odor. Poorly soluble in water. Odor perception theshold is 0.1 mg/l; taste perception threshold is 0.4 mg/l.

Applications. Used as a stabilizer in PVC plastics production.

Acute Toxicity. LD_{50} is found to be 2.7 g/kg BW in rats and 2.25 g/kg BW in rabbits and guinea pigs.

Repeated Exposure revealed evident cumulative properties. After eight to ten administrations, of $^1/_5$ and $^1/_{20}$ LD_{50}, clinical signs of intoxication became evident in rats. The treatment resulted in convulsions, immediately followed by death. An effect on the blood content of SH-groups was produced by $^1/_{40}$ to $^1/_{10}$ LD_{50}, as was a decline in the STI value. There were no changes in the enzyme systems of animals.

Long-term Toxicity. In a 6-month study in rats, manifestations of toxic action included deviations in the hematological analyses and a reduction in the SH-groups level in the blood, as well as pronounced physiological changes: adynamia, impairment of conditioned reflex activity, a decrease in muscle tonus and in reactions to painful and tactile stimuli.

Standards. Russia. (1986). MAC and PML: 0.02 mg/l.

Reference:

Golovanov, O. V., Substantiation of maximum allowable concentration for PVC octyl stabilizers in water bodies, *Gig. Sanit.*, 5, 36, 1975 (in Russian).

DIOCTYLTIN-*S,S*'-ETHYLENEBISMERCAPTOACETATE (DOTEBMA)

Synonym. Dioctyltin-*S,S*'-ethylenebis thioglycolate.

Properties. A viscous, yellowish liquid, poorly soluble in water. Taste perception threshold is 0.1 mg/l; odor perception threshold is 0.4 mg/l.

Applications. Used as a stabilizer in the production of PVC plastics.

Acute Toxicity. The LD_{50} is found to be 640 mg/kg BW in male and 840 mg/kg BW in female rats, 340 mg/kg BW in rabbits, and 450 mg/kg BW in guinea pigs.

Repeated Exposure revealed pronounced cumulative properties. After five to seven administrations of $1/10$ and $1/20$ LD_{50}, clinical signs of intoxication became evident in rats. Administration of $1/10$ LD_{50} caused 80% animal mortality. The treatment led to a reduction in the erythrocyte count and STI.

Long-term Toxicity. Rats were dosed by gavage with 0.1 mg/kg BW for 6 months. The treatment caused adynamia, decrease in muscle tonus and in a response to painful and tactile stimuli, as well as changes in hematological analyses and a decline in the SH-group content and in the STI at the end of the experiment.

Standards. Russia. (1986). MAC and PML: 0.002 mg/l.

Reference:

See **DIOCTYLTIN DI(MONOISOBUTYLMALEATE)**.

DIOCTYLTIN MALEATE (DOTM)

Properties. Odorless, light yellow, oily liquid. Soluble in organic solvents. Tin content up to 25%.

Applications. Used as a thermostabilizer for production of PVC and polymer packaging films used in food industry.

Acute Toxicity. LD_{50} is reported to be 2.95 g/kg BW for rats, 1.55 g/kg BW for mice, and 2.6 g/kg BW for guinea pigs. The lethal doses led to a brief motor excitation with subsequent CNS depression, salivation and lacrimation, adynamia, acute respiratory distress, and asphyxia. A change in hepatocyte ultrastructure and in the activity of the isoenzyme spectrum of the lactate-dehydrogenase of the liver and, to a lesser extent, of the blood serum in rats was caused by administration of $1/5$ LD_{50}. Urokaninase and sorbitoldehydrogenase appear in the blood serum. Depression in amidopyrine demethylation activity was found.

Repeated Exposure revealed evident accumulation. K_{acc} is 1.34 (by Kagan). Administration of $1/20$ LD_{50} caused 70% rat mortality during a month of treatment.

Allergenic Effect is noted at the threshold dose level.

Regulations. USFDA (1993) permitted the use of DOTM at a level not to exceed 3% by weight in resin used to stabilize plastic bottles and films that are in contact with food.

Standards. Russia. PML in food: 0.1 mg/l.

Reference:

Levitskaya, V. N. and Badayeva, L. N., Effect of some organotin stabilizers on ultrastructure of hepatocytes in rat study, *Gig. Sanit.*, 6, 74, 1984 (in Russian).

DIPHENYLAMINE (CAS No 122–39–4)

Synonyms. Fenam; *N*-Phenylbenzeneamine.

Properties. White and yellowish crystals with a floral odor.

Solubility in water is 30 mg/l. Soluble in alcohol. Odor perception threshold is 0.05 mg/l.

Applications. D. is used in the production of plastics, paints, dyes, and lacquers. A stabilizer for formaldehyde co-polymers, polyolefins, epoxy resins, PVC, etc. A stabilizer for polyoxyethylene.

Acute Toxicity. LD_{50} is reported to be 11.5 g/kg BW in rats and 2.9 g/kg BW in mice.[1] However, according to Korolev et al.,[2] LD_{50} is 2 g/kg BW in rats and 1.75 g/kg BW in mice. Manifestations of toxic action included symptoms of CNS impairment and cyanosis development. Up to 13% metHb, up to 3% sulfHb, Heinz bodies, and lower Hb level were found in the blood[1] in 1.5 to 55 h following administration of 1/2 LD_{50}. LD_{33} is reported to be 0.6 g/kg BW. Gross pathological examination revealed dehydration and acute nephrosis. No signs of intoxication in survivors were noted.

Repeated Exposure revealed low to moderate cumulative properties. No animal morbidity was caused by adminstration of 1/5, 1/25, and 1/125 LD_{50} given to rats for 25 d. The NOEL for change in the erythrocyte count and peroxidase activity appeared to be 1/25 LD_{50}. Male mice were given 10 weekly doses of 1.4 g/kg BW by gavage. This treatment resulted in 50% survival for longer than 4 months. Severe kidney damage (all normal structure in the renal cortex had been lost) and changes in liver morphology have been reported.[3]

Long-term Toxicity. Rats were dosed by gavage with 5 mg/kg for 6 months. The treatment did not affect conditioned reflex activity, liver excretory function, peroxidase and ceruloplasmin activity, or content of SH-groups in the blood.[2]

Allergenic Effect was not found at the maximum dose tested, namely 0.5 g/kg.

Reproductive Toxicity. *Gonadotoxicity* effect is not observed at a dose of 0.5 mg/kg BW. The functional state of the spermatozoa and the morphology of seminiferous tubular epithelium were not affected.[2] Rats received 2.5% D. in their diet during the last 6 d of pregnancy. The treatment produced cystic dilation of the renal collecting ducts and degeneration of the proximal tubules in the fetuses.[4]

Genotoxicity. No mutagenic effect was found.

Carcinogenicity Classification. NTP: N—N—E—P.

Standards. Russia. (1988). MAC: 0.05 mg/l (organolept., odor).

Regulations. USFDA (1993) approved the use of D. (1) as an adjuvant to control pulp content in the manufacturing of paper and paperboard prior to the sheet-forming operation, (2) as an antioxidant for fatty-based coatings and adjuvants, provided it is used at a level not to exceed 0.005% by weight of the coating solids, and (3) in resinous and polymeric coatings for the food-contact surface of articles intended for use in producing, manufacturing, packing, transporting, or holding food.

References:

1. *Hazardous Substances in Industry,* Handbook for Chemists, Engineers and Physicians, 7th ed., Lazarev, N. V. and Levina, A. N., Eds., Khimiya, Leningrad, 1976, Vol. 2, p. 289 (in Russian).
2. Korolev, A. A., Arsen'yev, M. V., and Vitvitskaya, B. R., et al., Experimental data on hygiene substantiation of MAC for diphenylamine and diphenyldiethylurea in water bodies, *Gig. Sanit.,* 5, 21, 1976 (in Russian).
3. Kronevy, T. and Holmberg, B., Acute and subchronic kidney injuries in mice induced by Diphenylamine, *Exp. Pathol.,* 17, 77, 1979.
4. Crocker, J. F. S. and Vernier, R. L., Chemically induced polycystic disease in the newborn, *Pediatr. Res.,* 4 (Abstr.), 448, 1970.

N,N'-DIPHENYL-*p*-PHENYLENEDIAMINE (CAS No 74–31–7)

Synonyms. Antioxidant *DIP*; **1,4**-Dianilinobenzene; Diafen *FF*; *N,N*'-Diphenyl-**1,4**-benzenediamine; Nonox *DPPD*.

Properties. Dark-grey, crystalline powder. Poorly soluble in water, slightly soluble in alcohol.

Applications. Used as a stabilizer in the production of synthetic rubber, polyamides and polyethylene; a polymerization inhibitor; a stabilizer for foodstuffs.

Acute Toxicity. LD_{50} is 2.37 g/kg BW in rats and 18.5 g/kg BW in mice.[08]

Short-term Toxicity. Mice were dosed by gavage with 120 mg/kg BW, three times a week for 4 months. Gross pathological examination revealed hemodynamic disorders and parenchymatous dystrophy in the kidneys. Sloughing of the mucosal epithelium was noted in the small intestine.[08]

Long-term Toxicity. Fisher 344 rats were fed the diet containing 0.5 and 2% D. for 104 weeks. There was a dose-dependent reduction of BW gain but no lower survival rate. Hematological indices and urinalysis showed no remarkable treatment-related changes.

Reproductive Toxicity. Embryotoxicity and effect on fertility are reported.

Genotoxicity. D. appeared to be positive in *Salmonella* mutagenicity assay and in the test of gene mutations in mammals.

Carcinogenicity. In a 2-year feeding study, there was no increase in tumor incidence.

Chemobiokinetics. Calcium deposition in the kidney of males was the only significant histological change relating to treatment.

Regulations. USFDA (1993) approved the use of D. in adhesives used as components of articles intended for use in packaging, transporting, or holding food.

Reference:

Hasegawa, R., Fukushima, S., Hagiwara, A., et al., Long-term feeding s tudy of *N,N'*-diphenyl-*p*-phenylenediamine in Fisher 344 rats, *Toxicology,* 54, 69, 1989.

N,N'-DIPHENYLTHIOUREA (CAS No 102–08–9)

Synonyms. *sym*-Diphenylthiourea; Stabilizer C; Sulfocarb-anilide; Thiocarbanilide.

Properties. Crystalline leaflets. On decomposition may form phenyl mustard oil. Poorly soluble in water, freely soluble in alcohol and ether.

Applications. Used primarily as a thermostabilizer in the production of PVC and as a vulcanization accelerator for rubber.

Acute Toxicity. LD_{50} is 1500 mg/kg BW in rats and 720 mg/kg BW in cats.

Long-term Toxicity. Wistar rats received 0.1% D. in the diet over a period of 1 year.[05] The treatment caused no toxic effect. Histological examination revealed no changes in the visceral organs.

Carcinogenic properties are not revealed.

DODECYL GALLATE (CAS No 1166–52–5)

Synonyms. Gallic acid, dodecyl ester; Gallic acid, lauryl ester; Lauryl gallate; **3,4,5**-Trihydroxybenzoic acid, dodecyl ester.

Acute Toxicity. In rats, LD_{50} is 6.6 g/kg BW.[1]

Repeated Exposure. Feeding a diet containing 2 g/kg BW for 70 d produced no effect on BW gain in rats.[12] However, there was no survival in rats after feeding 25 or 50 g/kg BW for 10 d.[2,3]

Short-term Toxicity. In a 13-week study, there was no adverse effect in pigs consuming a dietary level of 2 g/kg BW.[3,4]

Long-term Toxicity. No signs of toxicity were found in rats fed 350 to 2000 mg/kg BW in the feed.[4] Male Wistar rats were exposed by gavage to 10, 50, and 250 mg/kg BW for 150 d. Both higher doses produced changes in serum lipids and enzymes and reduction in the spleen weight. Gross pathological examination revealed changes in the liver, kidney, and spleen. The NOAEL appeared to be 10 mg/kg BW.[6]

Reproductive Toxicity. Rats were dosed via their feed with 350 to 5000 mg/kg over three generations. Growth retardation and loss of litters were found, probably due to underfeeding at 5 g/kg BW dose-level; slight hypochromic anemia was observed at 2000 mg/kg.[4]

Carcinogenicity. In the above-described study,[4] no increase in tumor incidence was reported.

Recommendations. JECFA (1993). ADI: 0 to 0.05 mg/kg BW.

References:

1. Tollenaar, F. D., Prevention of rancidity in edible oils and fats with special reference to the use of antioxidants, *Proc. Pacific. Sci. Congr.,* 5, 92, 1957.
2. *Joint FAO/WHO Expert Committee on Food Additives,* WHO Food Additives Series 10, WHO, Geneva, 1979, p. 45.
3. van der Heijden, C. A., Janssen, P. J. C. M., and Strik, J. J. T. W., Toxicology of gallates: a review and evaluation, *Food Chem. Toxicol.,* 24, 1067, 1986.

4. van Esch, G. J., The toxicity of the antioxidants propyl, octyl and dodecyl gallate, *Voeding,* 16, 683, 1955.
5. Sluis, K. J. H., The higher alkyl gallates as antioxidants, *Food Manuf.,* 26, 99, 1951.
6. Mikhailova, Z. N., et al., Toxicological studies of the long term effects of the antioxidant dodecyl gallate on albino rats, *Voprocy Pitaniya,* 2, 49, 1992 (in Russian).

EPOXIDIZED SOYBEAN OIL (ESBO) (CAS No 8013–07–8)

Synonym. Epoxom.

Properties. Clear, viscous, light-yellow liquid. Polysiccative oil with a high fatty acid content. Insoluble in water, soluble in fats.

Applications and **Exposure.** Used as a light- and thermostabilizer with plasticizing properties for PVC and chlorinated rubber (in combination with other substances). The stabilizing value of ESBO is proportional to the epoxy index. The more highly epoxidized vegetable oils produce a better stabilizing effect but are correspondingly more toxic. It is used in conjunction with organic acid salts. Daily intake is estimated to be 0.4 mg/d.[1]

Acute Toxicity. In rats, LD_{50} is found to be 28.7 or even 40 g/kg BW.

Long-term Toxicity. Rats consumed 1% ESBO and more in their feed over a period of 1 year. The treatment lowered the nutritional protein utilization coefficient and increased basal metabolism and water consumption. Accumulation of epoxy acids was noted in fat deposits. Retardation of BW gain became evident. Dogs received 5% ESBO in the diet. Manifestations of toxic action included an increase in the relative liver and kidney weights. There were no special features in the results of hematological analyses or in visceral organ histology.[04] No signs of intoxication were observed in rats given 1.4 g/kg BW twice a week for 16 months and in dogs treated for 12 months.[04] No evidence of systemic toxicity was found after administration of 100 mg/kg BW.[2]

Genotoxicity. ESBO is not found to be mutagenic in the *Salmonella* test.

Carcinogenicity. No evidence of carcinogenicity was reported following administration of a 100 mg/kg BW dose.[2]

Regulations. USFDA (1993) approved the use of ESBO (1) as a component of adhesives intended for use in a food-contact surface, (2) in the manufacturing of closures with sealing gaskets for food containers, (3) as a plasticizer in polymeric substances used in the manufacturing of food-contact articles, (4) as a plasticizer in resinous and polymeric coatings (iodine number maximum 14; oxirane oxygen content 6% minimum), as the basic polymer, and (5) in polysulfide polymer-polyepoxy resins in articles safely used as the dry food-contact surface in accordance with the specified conditions.

References:
1. Kieckebush, W., u. a., *Fette Seifen Austrichmittel,* Bd. 65, 11, 919, 1963 (in German).
2. Anon. Plasticizers migration in food, *Food Chem. Toxicol.* ,29, 139, 1991.

(2-ETHYLHEXYL) EPOXYSTEARATE

Synonyms. Drapex 2,3; Octylepoxystearate.

Properties. Liquid of low viscosity, insoluble in water. Odor and taste perception threshold is 1 mg/l.

Applications. Used as a light- and thermostabilizer with plasticizing properties in the production of plastics. Gives high low-temperature resistance to PVC.

Acute Toxicity. LD_{50} was not attained when the dose of 14 g/kg was administered to rats. In mice, LD_{50} is found to be 11.8 g/kg BW.

Short-term Toxicity. Rats were dosed by gavage with 600 mg/kg BW dose for 3 months. Manifestations of toxic action were not evident. In 2 months after onset of administration,'there was leukocytosis and an increase in relative liver weights.

Standards. Russia. PML in drinking water: 1 mg/l (organolept., odor and taste).

Reference:
See **ADIPIC ACID,** #2, 297.

164

4-(2-HYDROXYETHOXY)DIPHENYLAMINE

Synonym. 2-(*p*-Phenylaminophenoxy) ethanol.

Properties. Finely dispersed, odorless, grey powder. Insoluble in water.

Applications. Used as a stabilizer in the production of polyamides and polyurethanes.

Acute Toxicity. LD_{50} is reported to be 3 g/kg BW in rats and 1.75 g/kg BW in mice. Administration of high doses led to a sharp decline in muscle tonus, to paresis, ataxia of the hind legs, and lethargy. Death was preceded by adynamia and shallow respiration.

Repeated Exposure revealed evident cumulative properties. K_{acc} is 2.4 (by Kagan). The treatment with 0.4 g/kg BW resulted in almost total mortality. The lower doses caused anemia and leukocytosis and changes in the hippuric acid synthesis. Other manifestations of toxic effect included changes in the duration of hexenal sleep, prothrombin time, residual blood nitrogen, diuresis, urinary protein, and the relative weights of the visceral organs.

Reference:

See **ACRYLONITRILE**, #1, 147.

2-HYDROXY-4-METHOXYBENZOPHENONE (CAS No 131–57–7)

Synonyms. Benzophenone-3; 2-Hydroxy-4-methoxybenzophenone (HMB); (2-Hydroxy-4-methoxyphenyl)phenylmethanone; 4-Methyl-2- hydroxybenzophenone; Oxybenzone.

Properties. Light-yellow or white, crystalline powder. Almost insoluble in water, soluble in alcohol.

Applications. Used in the production of plastic surface coatings, polymers, and cosmetics. Light-stabilizer for polystyrene, pentaplast, PVC, cellulose acetobutyrate, polyamide, and polyolefin fibers, paints, and varnishes based on perchlorovinyl resins.

Acute Toxicity. LD_{50} is reported to be 7.4 g/kg BW or even more than 12.8 g/kg in rats.[1]

Repeated Exposure. In a 14-d study, Fisher 344 rats were given feed containing 3125 to 50,000 ppm HMB. The treatment caused an increase in the liver and kidney weights. Renal lesions (consisting of dilated tubules and regeneration of tubular epithelial cells) were noted in high-dosed rats. Marked hepatocyte cytoplasmic vacuolization was observed in liver. The NOAEL of 6250 ppm HMB in the diet is identified in rats for microscopic kidney lesions. Histological examination revealed kidney damage in mice given 50,000 ppm HMB in their diet for 14 d.[2]

Short-term Toxicity. Rats were fed 0.5 to 1% HMB in the diet for 90 d. Animals displayed retardation of BW gain, nephrotoxicity, reduction in the relative weights of the hypophysis, thymus, heart, and adrenals.[3] In a 13-week study, Fisher 344 rats and B6C3F$_1$ mice received 3125 to 50,000 ppm in their feed. Rats displayed retardation of BW gain, increased liver and kidney weights, and kidney lesions (papillary degeneration or necrosis and inflammation).

Reproductive Toxicity. In a NTP–92 study, a markedly lower epididymal sperm density and an increase in the length of the estrus cycle at the end of the experiment were observed in rats given 50,000 ppm HMB in the diet. The NOAEL was not reached, but it must be more than 23 mg HMB/kg BW.[2]

Genotoxicity. HMB was weakly mutagenic in *Salmonella* (with metabolic activation). In a 13-week mice study, SCE and CA were noted in Chinese hamster ovary cells in presence of a metabolic activation system. No increase in the frequency of micronucleated erythrocytes in the blood of mice was reported.[2]

Chemobiokinetics. Following administration of ^{14}C-HMB, two major components found in the bile are glucuronides and **demethylated HMB**; the third metabolite was probably a **sulfate ester** of **hydroxylated HMB**. Excretion occurs primarily through the kidney. After oral administration of ^{14}C-HMB, nine radioactive components are noted in the urine.[4]

Standards. Russia. PML in food: 2 mg/l.

References:

1. Lewerenz, H. J., Lewerenz. G., and Plass, R., Acute and subchronic toxicity of UV-absorber MOB in rats, *Food Cosmet. Toxicol.*, 10, 41, 1972.

2. NTP, Toxicity Studies of **2-Hydroxy-4-methoxybenzophenone** Administered Topically and in Dosed Feed to F344/N Rats and B6C3F$_1$ Mice, Technical Report, U.S. Department of Health and Human Services, Washington, D.C., October 1992, p. 52.

3. Cosmetic Ingredient Review Expert Panel, Final report on the safety assessment of the benzo-phenones-1, -3, -4, -5, -9, and -11, *Am. Coll. Toxicol.*, 2, 35, 1983.
4. El Dareer, S. M., Kalin, J. R., Tillery, K. F., et al., Disposition of 2-hydroxy-4-methoxyben-zophenone in rats, dosed orally, intravenously, or topically, *J. Toxicol. Environ. Health,* 19, 491, 1986.

2-(2'-HYDROXY-5'-METHYLPHENYL)BENZOTRIAZOLE (CAS No 2440–22–4)

Synonyms. Benazole P.; 2-(*2H*-Benzotriazol-2-yl)-4-methylphenol; 2-(*2H*-Benzotriazol-2-yl)-*p*-cresol; Tinuvin P.

Properties. Pale-yellow, crystalline powder. Poorly soluble in water, soluble in alcohol.

Applications. Used as a thermostabilizer in the production of polyamides, a light stabilizer for polystyrene, PVC, polymethacrylate, polypropylene, and cellulose acetate.

Acute Toxicity. In mice, LD_{50} is 6.5 g/kg BW. Lethal doses resulted in adynamia with subsequent death on day 2. Gross pathological examination revealed pulmonary hemorrhages and fatty liver dystrophy. According to other data, mice and rats tolerated administration of 5 and 10 g/kg BW in vegetable oil suspension.[1,2]

Repeated Exposure revealed slight cumulative properties. The treatment with $^1/_5$ LD_{50} resulted in decreased BW gain, lowered threshold of neuromuscular excitability, and impairment of the detoxifying liver function. Gross pathological examination revealed initial signs of fatty dystrophy in the liver, signs of congestion in the kidneys, and local hemorrhages in the spleen pulp.[3]

Long-term Toxicity. In a 6-month study, rats given 120 mg/kg BW exhibited slight erythrocyte hemolysis. Gross pathological examination revealed desquamation of the epithelium of the small intestine.[1] Mice and rats were given 10 and 50 mg/kg BW for 12 months. Gross pathological ex-amination failed to reveal changes in the visceral organs. A 50 mg/kg BW dose appeared to be the NOAEL.[2]

Reproductive Toxicity effects are not observed in rats exposed to a dose of 5 mg/kg BW for 2 years over three generations.[05]

Carcinogenicity. Tumor growth was noted in a 2-year oral study in rats.[05]

Standards. Russia. PML in food: 2 mg/l.

Regulations. USFDA regulates H. for use (1) as a component of nonfood articles at levels not to exceed 0.25% by weight of rigid PVC and/or rigid vinyl chloride copolymers, (2) in polystyrene that is limited to use in contact with dry food in compliance with CFR requirements at levels not to exceed 0.25% by weight of polystyrene and/or rubber-modified polystyrene polymers intended to contact nonalcoholic food, provided that the finished basic rubber-modified polystyrene polymers in contact with fatty foods shall contain not less than 90% by weight of the total polymer units derived from styrene monomers, (3) at levels not to exceed 0.5% by weight of the polycarbonate resins, provided that the finished polycarbonate resins contact food only of specified types and under the specified conditions of use, and (4) at levels not to exceed 0.5% by weight of the ethylene-1,4-cyclohexylene dimethylene terephthalate copolymers and of ethylene phthalate polymers that contact food only under conditions specified in CFR.

References:
1. Putilina, L. V., Data on toxicity of some aromatic amines and azomethines, *Gig. Truda Prof. Zabol.*, 3, 49, 1966 (in Russian).
2. See **MINERAL PETROLEUM OIL**, #1, 25.
3. See *N*-**BUTYLIDENEANILINE**, 78.

2-HYDROXY-4-PROPOXYPHENYLTHENOYL KETONE

Acute Toxicity. LD_{50} is found to be 10 g/kg BW in rats and 10.2 g/kg BW in mice.

Repeated Exposure revealed no cumulative properties. Administration of 0.5 and 2 g/kg to mice was not accompanied by any signs of intoxication.

Long-term Toxicity. Mice and rats were dosed by gavage with 100 and 200 mg/kg BW. The treatment produced no changes in BW gain or hematological analyses. Gross pathological exami-nation failed to reveal changes in the viscera.

Allergenic Effect was not found.
Reference:
See 4-**ALK**(C_7-C_9)**OXY**-2-**HYDROXYBENZOPHENONE**.

2-ISOBORNYL-4-METHYLPHENOL

Composition and **Properties.** Contains admixtures of 2-**isobornyl**-5-**methyl-phenols** and 4-**isobornyl**-3-**methylphenols**. Light-yellow, viscous liquid. Almost insoluble in water, readily soluble in vegetable oils.

Applications. Used as a stabilizer in the production of polyethylene, polypropylene, and PVC and ABS plastics.

Acute Toxicity. In mice, the LD_{50} is 4 to 4.8 g/kg BW. Poisoning is accompanied by CNS inhibition, weakness, and subsequent death in 1 to 2 d. Gross pathological examination revealed congestion in the visceral organs, edema and sloughing of the esophageal and gastric mucosa, necrosis of the intestinal walls, and parenchymatous dystrophy of the renal tubular epithelium.[08]

Long-term Toxicity. Rats were exposed orally to a 175 mg/kg BW dose, 44 times during 4 months. Manifestations of the toxic action included lymphocytosis and impairment of conditioned reflex activity. Gross pathological examination revealed an increase in the relative weights of the liver and kidneys and circulatory disturbances.[08]

Reproductive Toxicity. Does not affect the development of chick embryos.

N-ISOPROPYL-N'-PHENYL-p-PHENYLENEDIAMINE

Synonyms. Antioxidant *4010 NA*; Diafen *FP;* Santoflex *IP*.

Properties. White, crystalline powder that turns pink on storage. Soluble in ethanol, almost insoluble in water.

Applications. Used as a stabilizer in the production of synthetic rubber; antioxidant and antiozone agent for vulcanizates; thermostabilizer for polyethylene, polystyrene, and polyamides.

Acute Toxicity. LD_{50} is reported to be in the range of 0.8 to 1.1 g/kg BW for rats and of 3 to 3.9 g/kg BW for mice.[1,2,08] Administration of high doses led to lethargy. The treatment caused methemoglobinemia and slight changes in the osmotic resistance of erythrocytes. Histological examination revealed circulatory disturbances in the visceral organs, necrotic changes in the small intestinal mucosa and in the renal tubular epithelium.

Repeated Exposure failed to reveal cumulative properties. Rats received 180 mg/kg (as a suspension in 1 ml of 3% starch solution) for 35 d. The treatment caused no mortality in animals or pronounced changes in their general condition. BW gain seemed to be slightly increased.[1] In rats, manifestations of the toxic effect were characterized by leukocytosis and CNS inhibition, an increase in urinary protein amount, a reduction in the percentage of bromosulfalein excretion. The relative weights of the liver, kidneys, spleen, and testes were increased.

Short-term Toxicity. Rats were given 200 mg/kg BW for 4 months. The treatment caused retardation of BW gain, slight methemoglobinemia, and an increased conditioned reflex activity. Gross pathological findings included surface necrosis of the villi and desquamation of the intestinal mucosal epithelium.[08] Rabbits were dosed with 20 mg/kg BW over a period of 4 months. Poisoning was accompanied by liver and kidney lesions.[2]

Carcinogenicity. I. is related to the aromatic aminocompounds, which have a potential to form carcinogenic substances in the body. While excreted through the kidneys, they might give rise to bladder malignant tumors.

Weak carcinogenic activity is reported in rats that received I. in the diet for 2 years (Pliss, 1966).

References:
1. See **ANILINE**, #1, 126.
2. See **PHTHALIC ANHYDRIDE**, #1, 175.

ISOPROPYL PYROCATECHOL PHOSPHITE

Synonym. Pyrocatechol phosphorous acid, isopropyl ester.
Applications. Used as a stabilizer in the production of plastics.

Acute Toxicity. In mice, LD_{50} is 2.2 g/kg BW.[08] Gross pathological examination revealed circulatory disturbances in the visceral organs, particularly in the lungs, necrotic changes in the gastric and small intestinal mucosa, and parenchymatous dystrophy of the renal tubular epithelium.

Short-term Toxicity. Mice were given 200 mg I./kg BW for 3.5 months. The treatment caused mild changes in BW gain and CNS activity. Gross pathological examination revealed necrotic changes in the esophageal and intestinal mucosa, parenchymatous dystrophy of the renal tubular epithelium, and fatty dystrophy of the liver cells.[08]

LAST *A*

Composition and **Properties.** A complex stabilizer based on **barium amylphenolate, alkylphenol**, and **triphenylphosphate**. Colorless or cream-colored, highly mobile liquid. Poorly soluble in water, soluble in ethanol.

Applications. Used as a secondary thermostabilizer of PVC together with *Ca-Cd-*, *Ba-Cd-Zn-* and epoxy-stabilizers.

Acute Toxicity. LD_{50} is reported to be 5.9 g/kg BW in rats and 8.7 g/kg BW in mice.[1] According to other data, LD_{50} is 11 g/kg BW in rats and 5.7 g/kg BW in mice. High dose administration led to convulsions and, subsequently, to side body position. Gross pathological examination revealed lesions of the GI tract.[1,2]

Repeated Exposure revealed evident cumulative properties according to the data published by Rumyantsev et al. (1981) but in contrast to other data.[2]

Short-term Toxicity. No reduction in BW gain or change in STI value was observed in rats exposed to the oral dose of 550 mg/kg BW for 3 months. After 2 months of treatment, leukocytosis and an increase in the relative renal weights were noted. No mortality was observed.[1]

Allergenic Effect was reported.

References:
1. Krynskaya, I. L., Yevsyukov, V. I., and Maksimova, N. S., Toxicity of some stabilizers for PVC-compositions, *Plast. massy,* 12, 45, 1976 (in Russian).
2. See **ADIPIC ACID**, #2, 297.

LAST B-94

Composition. A complex stabilizer containing coprecipitated **calcium** and **zinc stearates, epoxidized** soy oil (see above), **glycerine, topanol,** and **trihexylcyanurate.**

Properties. White, viscous paste.

Applications. Used as a thermostabilizer in the production of rigid and plasticized PVC.

Acute Toxicity. Rats and mice tolerate administration of 10 g/kg BW.

Repeated Exposure failed to reveal cumulative properties. Rats were gavaged with the dose of 2 g/kg BW. An increase in relative liver weights was found.

Long-term Toxicity. Mice were given 200 mg/kg BW for 10 months. The treatment produced a temporary increase in motor activity and in the relative lung weights in mice.

Allergenic Effect was not observed.

Reference:
See **LAST A**, #1.

LAST *DP-4*

Composition. Mechanical mixture of coprecipitated *Ba* and *Cd* **laurates** (1:2), *Zn* **laurate, glycerine, epoxidized soy oil** (see above), **bisphenol** A (see above), and an **ethylene glycol ester** based on **phthalic** and **methacrylic acids.**

Properties. A white, hygroscopic powder with the odor of vegetable oil.

Applications. Used as a thermo- and light-stabilizer in the production of vinyl plastics and of plasticized PVC and its copolymers.

Acute Toxicity. LD_{50} is found to be 3.5 g/kg BW in rats and 2.1 g/kg in mice. A mixture of **Last DP-4** and **Last A** (1:5) has LD_{50} of 2 g/kg BW in mice and of 2.5 g/kg BW in rats. Gross pathological examination revealed damages to the GI tract.

Repeated Exposure revealed marked cumulative properties. K_{acc} is 1.75. Some male mice died when given a mixture of **Last DP-4** and **Last A** (1:5) at a dose of 400 mg/kg BW. The treatment caused an impairment of liver and kidney functions. Doses of 100 and 200 mg/kg produced an increase in the relative liver weights only.

Long-term Toxicity. Some of the male rats died when exposed to 10 and 50 mg **Last DP-4**/kg BW. Toxicity of the stabilizer may be enhanced by the coprecipitated **Ba-Cd laurate** contained in the stabilizer.

Allergenic Effect was not noted.

Reference:

See **LAST A**, #1.

MANNITE (CAS No 69–65–8)

Synonyms. Cordycepic acid; Diosmol; **1,2,3,4,5,6**-Hexanehexol; Manicol; Manna sugar; Mannidex; *D*-Mannitol; Osmitrol; Osmosal.

Properties. White, odorless, crystalline powder with a sweet taste. A stereo-isomer of sorbitol.

Applications and **Exposure.** Used as a stabilizer in the production of PVC plastics, artificial resins, and plasticizers.

M. occurs naturally in algae, fungi, bacteria, and variety of higher plants.[1]

Acute Toxicity. *Humans.* A laxative effect could be initiated with the single oral dose greater than 20 g.[2] *Animals.* LD_{50} is reported to be 17.3 g/kg BW in rats and 22 g/kg BW in mice.[1] High doses administered to mice are followed by CNS inhibition, degeneration of the GI tract mucosa, and diarrhea.

Repeated Exposure. M. is used as an osmotic diuretic. The daily dose in man is 50 mg/kg BW (unrestricted).

Short-term Toxicity. Rhesus monkeys were given 3 g M. in their diet for 3 months (IARC 19–237). No signs of intoxication were noted.

Long-term Toxicity. Fisher 344 rats and $B6C3F_1$ mice of each sex were fed the diet containing 2.2 and 5% M. for 103 weeks. No significant differences between treated animals were found. The incidence of dilatation of the gastric fundal gland was higher (46%) in treated female rats than in the controls (12%). There was a mild nephrosis that was considered to be related to the treatment.[3]

Genotoxicity. M. showed negative results in *Salmonella* mutagenicity assay, rat bone marrow studies, and cultured human cells.[4]

Carcinogenicity. No increase in tumor incidence under the conditions of the above-described study[3] was found. *Carcinogenicity classification.* NTP: N—N—N—N.

Chemobiokinetics. Following oral administration, M. is rapidly metabolized. In humans, an estimated one-third of the dose to have been absorbed was excreted intact in the urine, and the rest of the absorbed material was oxidized to **carbon dioxide** and **water.**[5]

Standards. Russia. PML: n/m.

Regulations. USFDA (1993) listed M. as a food additive and regulates it for use (1) in resinous and polymeric coatings for the food-contact surface of articles intended for use in producing, manufacturing, packing, transporting, or holding food and (2) in cross-linked polyester resins used as articles or components of articles coming in contact with food.

References:

1. Joint FAO/WHO Expert Committee on Food Additives, *Toxicological Evaluation of Some Antimicrobials, Antioxidants, Emulsifiers, Stabilizers, Flour-treatment Agents, Acids and Bases,* FAO Nutr. Mutag. Report, Series No 40A, BC, WHO, Geneva, 1967, p. 160.

2. Ellis, F. W. and Krantz, J. C., Sugar alcohols. XXII. Metabolism and toxicity studies with mannitol and sorbitol in man and animals, *J. Biol. Chem.*, 141, 147, 1941.

3. Abdo, K. M., Haseman, J. K., and Boorman, G., Absence of carcinogenic response in F344 rats and $B6C3F_1$ mice given *D*-mannitol in the diet for two years, *Food Chem. Toxicol.*, 21, 259, 1983.

4. NTP, *Carcinogenesis Bioassay of D-Mannitol in F344 Rats and B6C3F1 Mice,* Technical Report 81–52, NIH Publ. No 82–1792, Department Health and Human Services, Washington, D.C., 1981.

5. Nasrallah, S. M. and Iber, F. L., Mannitol absorption and metabolism in man, *Am. J. Med. Sci.*, 258, 80, 1969.

α-METHYLBENZYLPHENOLS, mixture

Synonym. Alkofen *MB*.
Composition. A product of alkylation of phenol by styrene.
Properties. Viscous liquid with a color ranging from light-yellow to dark-brown and with unpleasant odor. Poorly soluble in water, soluble in alcohol and in aqueous solutions of acids and alkalis. Forms stable emulsions with water.
Applications. Used as a stabilizer in the production of synthetic rubber and polyolefins; an antioxidant for vulcanizates.
Acute Toxicity. In rats, LD_{50} is 3.5 g/kg BW. Administration of high doses is characterized by a narcotic effect. Gross pathological examination revealed inflammatory changes in the GI tract, fatty dystrophy of the renal tubular epithelium, and a significant increase of polynuclear cell number in the spleen.[1]
Repeated Exposure failed to reveal cumulative properties.
Long-term Toxicity. In a 6-month study, mild retardation of BW gain and CNS stimulation were found in rats given the 175 mg/kg dose.
Reproductive Toxicity. Skin applications to rats caused growth retardation and a decline in birth rate in succeeding generations not exposed to the product.[1] The treatment produced an embryotoxic effect in guinea pigs (blind, unviable offspring).
Allergenic Effect was not observed in experiments on guinea pigs.[2]
References:
1. *Toxicology and Hygiene of High Molecular Weight Compounds and of the Chemical Raw Material Used for Their Synthesis*, Proc. 2nd All-Union Conf., Letavet, A. A. and Danishevsky, S. L., Eds., Leningrad, 1964, p. 72 (in Russian).
2. See *N,N'*-**BIS**(1,4-**DIMETHYLPENTYL**)-*p*-**PHENYLENEDIAMINE**, 60.

4–α-METHYL BENZYL PYROCATECHOL

Properties. Crystalline powder. Solubility in water is 0.001%.
Acute Toxicity. Histological examination of animals that died after a single administration of the lethal dose showed congestion, lesions in the mucosa of the GI tract walls, and parenchymatous dystrophy of the renal tubular epithelium. No morphological changes were found in the survivors.
Long-term Toxicity. Rats were dosed by gavage with 62.5 mg/kg BW for 4 months. The treatment caused changes in conditioned reflex activity. Gross pathological examination showed lesions in the GI tract mucosa, as well as parenchymatous dystrophy of the renal tubular epithelium.[08]

4-METHYL-2,6-DI-(α-METHYLBENZYL)PHENOL

Synonym. Alkofen *MBP*.
Properties. Viscous, oil-like liquid. Poorly soluble in water.
Applications. Used as a thermo- and light-stabilizer in the synthesis of vulcanizates based on natural, butadiene, and styrene-butadiene rubber; a thermostabilizer of polyolefins, PVC, and cellulose esters.
Acute Toxicity. In mice, LD_{50} is 4.3 g/kg BW.[08] Manifestations of toxic action include excitation with subsequent CNS inhibition. M. exerts an irritating and necrotizing effect on the tissues.
Long-term Toxicity. Rats were exposed to 215 mg/kg BW for about 5 months. The treatment resulted in CNS excitation and impairment of kidney function. Histological examination revealed necrosis of the gastric and intestinal mucosa and dystrophy of the renal tubular epithelium.[08]
Reproductive Toxicity. Affects fish reproductive processes but does not affect the development of chick embryos.

4-METHYL-DI(1-METHYLHEPTADECYL)PHENOL

Applications. Used as a stabilizer in the manufacture of polyolefins.

Acute Toxicity. LD_{50} is found to be 6.4 ml/kg BW in rats, and 3.2 ml/kg BW in mice.

Repeated Exposure. 1% solution of the product added to the feed of puppies appeared to be harmless (Terhaer, 1967).

2,2′-METHYLENE-BIS(6-*tert*-BUTYL-4-METHYLPHENOL)

Synonyms. Agidol 2; Antioxidant 2246; Bisalkofen *BP;* Bis(3-*tert*-butyl-2-hydroxy-5-methyl-phenyl)methane; *CAO-5*.

Properties. White, crystalline powder. Almost insoluble in water, readily soluble in alcohol.

Applications. Used as a thermo- and light-stabilizer in the production of natural and synthetic rubber, polyolefins, pentaplast, and polystyrene.

Acute Toxicity. Adult rats tolerate administration of 25 g/kg, young rats of 36 g/kg, and young and adult mice of 30 g/kg BW without manifestations of toxic effect. An increased thirst was noted.[1] According to Stasenkova et al.,[2] however, the LD_{50} seems to be 5 g/kg BW in rats and 11 g/kg BW in mice. Marked CNS depression and refusal of food were accompanied by severe weight loss.

Repeated Exposure failed to reveal marked cumulative properties. When technical and pure M. were administered over a period of 1 month, the animals died from the total doses of 15.5 and 28 g/kg BW, respectively. K_{acc} (by Lim) appeared to be 3 and 6.2 for technical grade product and pure chemical, respectively.[2]

Short-term Toxicity. Mice received 250 mg/kg BW for 17 weeks. This treatment caused slight stimulation of conditioned reflex activity. Gross pathological examination revealed mild changes in the gastric and intestinal mucosa.[08]

Long-term Toxicity. Rats were dosed by gavage with 1 g/kg BW for 5 months. The treatment did not produce clear symptoms of intoxication. Slightly decreased motor activity and BW gain and transient leukopenia were noted. Gross pathological examination revealed dystrophic changes in the epithelial cells of the liver, myocardium, renal tubular epithelium, and bronchi. There was parenchymatous dystrophy in the lungs. Capillary congestion was present in all the organs.[1] In a 10-month experiment,[08] rats received a 50 mg/kg BW dose of pure substance in the diet. Manifestations of toxic action included reversible functional changes in the NS (STI), liver, and oxygen consumption. A dose of 10 mg/kg BW caused changes in the detoxicant function of the liver. Gross pathological examination failed to reveal significant changes in the visceral organs.[2]

Reproductive Toxicity. A decrease in fertility of rats was observed after prolonged administration of doses exceeding 50 mg/kg BW.

Carcinogenicity. B. has been shown to cause an inhibiting effect on the induction of tumors by benzo[a]pyrene.[3]

Allergenic Effect is not observed.

Standards. Russia. PML in food: 4 mg/l; recommended PML in drinking water: 1 mg/l.

Regulations. USFDA (1993) regulates M. for use (1) at levels not to exceed 0.1% by weight of the olefin polymers used in articles that contact food, (2) at levels not to exceed 1% by weight of the polyoxymethylene copolymers, and (3) at levels not to exceed 0.5% by weight of the polyoxymethylene homopolymers in compliance with existing regulation (CFR).

References:
1. See **BISPHENOL A,** #2, 57.
2. Stasenkova, K. P., Shumskaya, N. I., Sheveleva, T. A., et al., Toxicity of bisalkofen *BP,* a stabilizer for polymeric materials, *Kauchuk i Rezina,* 1, 24, 1977 (in Russian).
3. Braun, D. D., Effect of polyolefin antioxidants on induction of neoplasms by benzo(a)pyrene, *Gig. Sanit.,* 6, 18, 1975 (in Russian).

2,2′-METHYLENE-BIS(6-*tert*-BUTYL-4-METHYLPHENYL)-α-NAPHTHYL-PHOSPHITE

Synonym. Stafor-10.

Properties. White powder. Soluble in alcohol.

Applications. Used as a Templene stabilizer.

Acute Toxicity. Rats and mice tolerated administration of 5 g/kg without signs of toxic action.

Repeated Exposure of mice to 250 to 1000 mg/kg BW and rats to 500 mg/kg BW did not produce any functional or morphological changes.

Long-term Toxicity effects were not observed in mice and rats fed 25 and 100 mg/kg BW for 5.5 months.

Regulations. Russia. PML: **n/m** when the content of M. in the polymer material is not more than 0.3%.

Reference:

See **MINERAL PETROLEUM OIL,** #1, 18.

4,4′-METHYLENE-BIS(2,6-DI-*tert*-BUTYLPHENOL)

Synonyms. Antioxidant 702; Bis(**3,5**-di-*tert*-butyl-**4**- hydroxyphenylmethane); Ionox 220, *MB-1*.

Properties. Light-yellow, crystalline powder. Poorly soluble in water, of limited solubility in alcohol.

Applications. MB-1 is used as a stabilizer in the production of polyolefins and as an antioxidant for vulcanizates; it is used also in the production of epoxy-, phenol-, and ion-exchange resins.

Acute Toxicity. Rats and mice tolerate administration of 5 g/kg BW.[1]

Repeated Exposure. Rats and mice tolerate administration of 1 and 5 g/kg BW for 10 d. Histological examination revealed moderate fatty dystrophy of the liver cells, renal tubular epithelium, and signs of irritation of the spleen pulp.[1] Administration to male Sprague-Dawley rats at the level of 1.135 mmol% for 1 week induced fatty liver and increased triglyceride, diglyceride, and cholesteryl-ester concentration 7.1-, 5.8-, and 6.1-fold, respectively.[2]

Regulations. USFDA (1993) approved the use of M. in adhesives used as components of articles intended for use in packaging, transporting, or holding food at levels not to exceed (1) 0.25% by weight of the petroleum hydrocarbon resins, (2) 0.25% by weight of the terpene resins, (3) 0.5% by weight of the polyethylene, provided that the polyethylene and product contact foods only of the types identified in CFR, (4) 0.5% by weight of polybutadiene used in rubber articles, provided that the rubber end-product contacts food only of the types identified in CFR, (5) 0.5% by weight of the polybutadiene used in rubber articles in compliance with correspondent regulations (CFR).

References:

1. See **BISPHENOL A,** #2, 180.
2. Takahashi, O. and Hiraga, K., Effect of 4 bisphenolic antioxidants on lipid contents of rat liver, *Toxicol. Lett.,* 8, 77, 1981.

2,2′-METHYLENE-BIS-[4-METHYL-6-(1-METHYLCYCLOHEXYL)PHENOL]

Synonyms. Bisalkofen *MCP;* **6,6′**-Bis(1-methylcyclohexyl)-**2,2′**-methylenedi-*p*-cresol; Bis-**2**-hydroxy-**5**-methyl-**3**-(1-methyl- cyclo-hexyl)phenylmethane; Nonox *WSP*; Permanax *WSP*.

Properties. White, crystalline powder. Soluble in alcohol, poorly soluble in water.

Applications. Used as a thermostabilizer in the synthesis of polyethylene, polyesters, and impact-resistant polystyrene.

Acute Toxicity. Rats and mice tolerate administration of 10 g/kg BW. Administration of a dose of 3.2 g/kg BW had no toxic effect. The treatment produced no morphological changes in the viscera.[1,05,08]

Repeated Exposure failed to reveal cumulative properties. Rats tolerated administration of 0.5 to 2 g/kg BW over a period of 2 months. No morphological changes were found.

Long-term Toxicity. Rats received 5% M. in their diet for 2 years. The treatment resulted in liver damage, hepatocyte degeneration, and proliferation of the bile duct cells of rats. A 0.5% level of M. in the diet had no toxic effect.[05] No signs of intoxication were observed in dogs fed 0.01 g/kg BW for 2 years, but a 5 g/kg BW dose caused severe liver damage (Hodge et al.). Excitation of the cortical areas of the CNS and increased BW gain were reported in rats given 2 g/kg BW for 5 months. No morphological changes were found in the visceral organs.[08]

Standards. Russia. Recommended PML: **n/m.**

Regulations. USFDA (1993) approved the use of M. (1) in closures with sealing gaskets that may be used safely on containers intended for use in producing, manufacturing, packing, processing, treating, packaging, transporting, or holding food (1%) and (2) as an antioxidant and/or stabilizer in polymers used in the manufacturing of food-contact articles (for use at levels not to exceed 0.2% by weight of the polyethylene). **British Standard** (1992): M. is listed as an antioxidant for polyethylene and polypropylene compositions used in contact with foodstuffs or water intended for human consumption (maximum permitted level for the final compound = 0.2%).

Reference:

See *N*-**BUTYLIDENEANILINE**, 29.

4-METHYL-2-α-METHYLBENZYLPHENOL

Synonym. 2-(α-methylbenzyl)-*p*-cresol.

Properties. Yellowish liquid. Soluble in alcohol, forms stable emulsions with water.

Acute Toxicity. In mice, LD_{50} is 2 g/kg BW. Histological examination revealed congestion, edema, and sloughing of the esophageal and gastric mucosa, necrosis of the small intestine walls, and parenchymatous dystrophy of the renal tubular epithelium.[08]

N-β-NAPHTHYL-*p*-AMINOPHENOL (CAS No 93–45–8)

Synonyms. *N*-(*p*-Hydroxyphenyl)-β-naphthylamine; *p*-Hydroxyneozone.

Properties. Light-grey, fine, crystalline powder. Poorly soluble in water, soluble in alcohol.

Acute Toxicity. LD_{50} is found to be 5.2 g/kg BW in rats and 2.2 g/kg BW in mice.[1] Manifestations of toxic action included marked degenerative changes and congestion in the liver, lungs, and spleen.

Repeated Exposure. Rats received 1/5 LD_{50} for 2 months. This treatment caused mild functional disturbances, retardation of BW gain, an increase of arterial pressure, and neuromuscular excitability. Histological examination revealed changes in the liver and kidneys.[1,2]

References:

1. See *N*-**BUTYLIDENEANILINE**, 42.
2. See *N*-(1,3-**DIMETHYLBUTYL**)-*N'*-**PHENYL**-*p*-**PHENYLENEDIAMINE**, 26.

α-NAPHTHYLPYROCATECHOL PHOSPHITE

Synonym. Pyrocatechol phosphorous acid, α-naphthyl ester.

Properties. A solid.

Applications. Used as a stabilizer in the production of polyamides, polyolefins, and polyethylene terephthalate.

Acute Toxicity. In mice, LD_{50} is 1.4 g/kg BW.[08] Histological examination revealed circulatory disturbances in the visceral organs, necrotic changes in the gastric and intestinal mucosa, and parenchymatous dystrophy in the renal tubular epithelium.

Long-term Toxicity. Mice received 12 mg/kg BW for 3.5 months. The treatment caused threshold changes in BW gain and conditioned reflex activity. Gross pathological examination revealed necrotic lesions in the esophageal and intestinal mucosa, parenchymatous dystrophy of the renal tubular epithelium, and fatty dystrophy of the liver cells.[08]

β-NAPHTHYLPYROCATECHOL PHOSPHITE

Synonym. Pyrocatechol phosphoric acid, β-naphthyl ester.

Applications. Used as a stabilizer of plastics.

Acute Toxicity. LD_{50} is 1.3 g/kg BW in rats and 0.7 g/kg BW in mice.[08] Gross pathological examination revealed edema and cell infiltration of the submucosal and muscle layers of the stomach wall and necrosis of the intestinal mucosa.

Repeated Exposure. No cumulative effect was observed in rats given doses of 170 and 520 mg/kg BW for a month. Rats were dosed by gavage with 130 to 260 mg/kg BW for 2 months. Of mice 50 to 80% died. Animals tolerated a dose of 65 mg/kg BW.

Long-term Toxicity. Mice were given 13 and 65 mg/kg BW for 6 months. The LOAEL was identified to be 13 mg/kg BW for BW gain and effect on the NS.[08]

Reproductive Toxicity. Does not affect the development of chick embryos.

OCTADECYL-3-(3,5-DI-*tert*-BUTYL-4-HYDROXYPHENYL)PROPIONATE

Synonyms. Alkofen *BP 18;* 3-(3,5-Di-*tert*-butyl-4-hydroxy-phenyl)propionic acid, octadecyl ester; Irganox 1076.

Properties. Yellowish, crystalline powder. Poorly soluble in water and in vegetable oil.

Applications. Used as an antioxidant and thermostabilizer in the production of polypropylene, as a stabilizer for polyethylene, polyamides, polystyrene, PVC, and polyesters.

Acute Toxicity. Administration of 5 g/kg BW to rats and 7.5 and 10 g/kg BW to mice caused retardation of BW gain. The LD_{50} in rats exceeded 15 g/kg BW. Gross pathological examination failed to reveal changes in the viscera.[1,05,08]

Repeated Exposure. Forty administrations of 0.5 and 2 g/kg BW doses to mice and 0.5 g/kg dose to rats led to increased excitability in the treated animals. There were no gross pathological findings.[08]

Short-term Toxicity. Mice received doses of 100 and 200 mg/kg BW in suspension in sunflower oil. The treatment resulted in retardation of BW gain. Histological examination failed to reveal any structural changes.[08]

Reproductive Toxicity and **Carcinogenicity.** Rats given orally 5 mg/kg BW for 2 years did not exhibit any reproductive abnormalities or tumor growth.[05]

Standards. Russia. Recommended PML: **n/m.**

Regulations. British Standard (1992): O. is listed as an antioxidant for polyethylene and polypropylene compositions used in contact with foodstuffs or water intended for human consumption (maximum permitted level for the final compound = 0.5%).

Reference:
See **BISPHENOL A,** #2, 193.

OCTYL GALLATE (OG) (CAS No 1034–01–1)

Synonym. 3,4,5-Trihydroxybenzoic acid, *n*-octyl ester.

Acute Toxicity. In rats, LD_{50} is 4.7 g/kg BW.[1] According to other data, it is 2.7 g/kg in male and 2 to 2.3 g/kg BW in female rats.[6] In humans, after drinking beer containing 20 mg/l OG, severity of erythema and edema in the oral mucosa was greater than after drinking untreated beer.[2]

Short-term Toxicity. In a 13-week study in rats, using dietary levels of 1, 2.5, and 5 g OG/kg BW feed, the only effect observed was a slight elevation of serum glutamic-oxalacetic transaminase. No histological changes were found. A similar result in dogs was observed at 5 g/kg feed but not at 1 or 2.5 mg/kg feed.[2] In another 13-week study in pigs, no effect on growth, hematological analyses, organ weights, or histology was noted after using a dose of 2 g/kg feed.[3]

Long-term Toxicity. No increase in tumor incidence was noted in rats given 350 to 5000 mg/kg feed over a period of 2 years.[3]

Reproductive Toxicity. In three-generation studies in rats,[3,4] no adverse effect was observed as a consequence of consumption of 350 to 5000 mg OG/kg. In the two-generation test,[2] rats were dosed with 1 and 5 g/kg feed. Dose-dependent reduction in implantation sites, as well as a reduction in the number of corpora lutea in P_2 parent animals, was observed. In the third-litter pups of the second generation, the incidence of gross kidney alterations was increased at the 5 g/kg dose-level. No histological changes were found.

Chemobiokinetics. Octyl and propyl gallates are metabolized similarly. Esch detected the unchanged compound as a major urinary component in rats.[3,5]

Standards. JECFA (1993). ADI: 0 to 0.1 mg/kg BW (temporary).

174

References:
1. Joint FAO/WHO Expert Committee on Food Additives, WHO Food Additives Series 10, WHO, Geneva, 1986, p. 45.
2. van der Heijden, C. A., Janssen, P. J. C. M., and Strik, J. J. T. W., Toxicology of gallates: a review and evaluation, *Food Chem. Toxicol.*, 24, 1067, 1986.
3. **DODECYL GALLATE**, #4.
4. Sluis, K. J. H., The higher alkyl gallates as antioxidants, *Food Manuf.*, 26, 99, 1951.
5. Koss, G. and Koransky, W., Enteral absorption and biotransformation of the food additive octyl gallate in the rat, *Food Chem. Toxicol.*, 20, 591, 1982.
6. Brun, R., Contact eczema due to an antioxidant of margarine (gallate) and change of occupation, *Dermatologica*, 140, 390, 1970.

PENTAERYTHRITE TETRA-3-(3,5-DI-*tret*-BUTYL-4-HYDROXYPHENYL)-PROPIONATE

Synonyms. Fenosan 23; Irganox 1010; Tetralkofen *PE*.

Properties. White, crystalline powder. Poorly soluble in water.

Applications. Used as an antioxidant and thermostabilizer for polypropylene, polyethylene, impact resistent polystyrene, poly-4-methylpentene. A stabilizer for natural and synthetic rubber, PVC. A copolymer of acrylonitrile with butadiene and styrene, polyacetals, alkyde resins, polyamides, and polyesters.

Acute Toxicity. LD_{50} was not attained. Rats tolerate administration of 5 g/kg BW; mice did not die following a single intake of 7.5 and 10 g/kg BW. Neither retardation of BW gain was reported[08] nor were there gross pathological findings.[1]

Repeated Exposure failed to reveal cumulative properties.[2] Rats were exposed to the overall dose of 53 g/kg BW during a month. In the group of ten rats, one animal died. No changes in BW gain, blood morphology, or weight coefficients of the visceral organs were noted. K_{acc} exceeds 10. Fifty administrations of 2 and 5 g/kg BW to mice and rats over a period of 2 months caused no mortality or reduction in BW gain. Histological examination revealed no changes in the visceral organs.[08] Forty administrations of 0.5 and 2 g/kg BW to mice and of 0.5 g/kg BW to rats showed no gross pathology to develop.[1]

Long-term Toxicity. Doses of 100 and 200 mg/kg BW given to mice as a suspension in sunflower oil caused decrease of BW gain. Histological examination revealed no lesions in the organs and tissues.[1,08]

Reproductive Toxicity and **Carcinogenicity.** The dose of 100 mg/kg BW given for 2 years did not affect reproductive function of rats and has not been shown to provide tumor growth.[05]

Standards. Russia. PML in drinking water: **n/m.**

Regulations. British Standard (1992). P. is listed as an antioxidant for polyethylene and polypropylene compositions used in contact with foodstuffs or water intended for human consumption (maximum permitted level for the final compound = 0.5%).

References:
1. See **BISPHENOL A**, #2, 193.
2. Mel'nikova, V. V., Toxicity of Fenozane 23, *Gig. Sanit.*, 12, 92, 1987 (in Russian).

N-PHENYL-*N'*-CYCLOHEXYL-*p*-PHENYLENEDIAMINE

Synonym. Diafen *FC*.

Properties. Light-grey powder with a pleasant odor. Solubility in water is 0.0015% (12°C), soluble in alcohol.

Applications. Used as a stabilizer in the production of synthetic rubber, polypropylene, and polyacetaldehyde.

Acute Toxicity. In mice, LD_{50} is 3.9 g/kg BW.[08] Manifestations of the toxic effect following a single administration included methemoglobinemia and slight changes in the osmotic resistance of erythrocytes. Histological examination revealed circulatory disturbances in the visceral organs, necrotic changes in the intestinal mucosa, and parenchymatous dystrophy of the renal tubular epithelium. In some animals, nephritis, culminating in nephrosclerosis, was observed.

Short-term Toxicity. Rats were dosed by gavage with 60 mg/kg BW for 4.5 months. The treatment resulted in changes in BW gain, methemoglobinemia, stimulation of conditioned reflex activity. Gross pathological examination revealed superficial necrosis of the villi and desquamation of the epithelium of the small intestinal mucosa.[08]

Reproductive Toxicity. Affects reproduction in fish.

N-PHENYL-α-NAPHTHYLAMINE (N-P-1) (CAS No 90–30–2)

Synonyms. Naftam-1; Neozone *A;* *N*-Phenylnaphthylamine-1.

Properties. Yellowish, crystalline powder. Poorly soluble in water, soluble in alcohol.

Applications. Used as a stabilizer for various synthetic rubbers. A thermostabilizer for vulcanizates and polyethylene.

Acute Toxicity. In mice, LD$_{50}$ is 1.8 g/kg BW. A single high dose of N-P-1 caused methemoglobinemia and mild changes in the osmotic stability of the erythrocytes. Histological examination revealed circulatory disturbances, necrotic changes in the intestinal mucosa, and parenchymatous dystrophy of the renal tubular epithelium.[08]

Short-term Toxicity. Rats were dosed by gavage with 50 mg/kg BW N-P-1 for 3.5 months. The treatment led to retardation of BW gain, mild methemoglobinemia, and stimulation of conditioned reflex activity. Superficial necrosis and desquamation of the epithelium of the small intestinal mucosa were found on autopsy.[08]

Reproductive Toxicity. Affects the reproduction of fish but not the development of chick embryos.[08]

Carcinogenicity. Admixture of β-**naphthylamine** may render N-P-1 carcinogenicity. ICR mice were dosed with repeated *i/v* injections of both technical and pure N-P-1 in dimethylsulfoxide. The treatment resulted in a high percentage of malignant tumors (as well as in the case of N-P-2). Similar results are found in the studies with TA-1 mice.[1] N-P-1 is reported to cause tumors in the lungs, kidney, ureter, and bladder. Is considered to be carcinogenic by RTECS criteria.

Chemobiokinetics. The similar carcinogenic potential of N-P-1 and N-P-2 suggests other routes of metabolic activation besides dephenylation for both chemicals.[1]

Regulations. USFDA (1993) approved the use of N-P-1 as an antioxidant in rubber articles intended for repeated use in producing, manufacturing, packing, processing, treating, packaging, transporting, or holding food (total not to exceed 5% by weight of the rubber product).

Reference:

Wang, H., Wang, D., and Dzeng, R., Carcinogenicity of *N*-phenyl-1-naphthylamine and *N*-phenyl-2-naphthylamine in mice, *Cancer Res.*, 44, 3098, 1984.

N-PHENYL-β-NAPHTHYLAMINE (CAS No 135–88–6)

Synonyms. 2-Anilinonaphthalene; 2-Naphthylphenylamine; Neozone *D*; Nonox *D*; *N*-Phenylnaphthylamine-**2**.

Properties. Light-grey powder. Poorly soluble in water and in cold alcohol.

Applications. Used as a stabilizer in the production of the synthetic rubber, as a thermostabilizer for polyethylene, ethylene-vinyl acetate copolymer, and polyisobutylene. An anti-aging agent for vulcanizates.

Acute Toxicity. LD$_{50}$ is 3.73 g/kg BW in rats and 1.45 g/kg BW in mice.[1] Manifestations of toxic action in mice included reduced muscle tonus, drowsiness, apathy, and CNS depression; in rats, there was general apathy and a slight reduction in motor activity. Guinea pigs tolerate 1 g P./kg BW.[2] The same dose caused all the rabbits to die.

Repeated Exposure. Guinea pigs tolerated six administrations (at weekly intervals) of 1 g/kg BW without retardation of BW gain. Poisoning was accompanied by a slight rise of temperature. Gross pathological examination revealed moderate congestion in the visceral organs, parenchymatous dystrophy of the renal tubular epithelium, and pneumonia.[2] All rabbits died given 1 g/kg BW for 5 to 15 d.[3] This dose killed 50% of treated rats. Functional and morphological changes were noted. According to other data, there were only the signs of intoxications in rats given 1.75 g/kg BW for a month. None of the animals died.

Short-term Toxicity. Rats were dosed by gavage with 100 mg/kg and 1000 mg/kg BW for 4 months. The treatment caused anemia and an increase in urinary albumen content.[4]

Long-term Toxicity. A transient liver and kidney dysfunction was found in rats given 100 mg/kg BW over a period of 12 months. A 20 mg/kg dose appeared to be ineffective.[3]

Reproductive Toxicity. SNK mice were dosed with 9 mg P. per animal during all periods of gestation and postnatally. The treatment caused malignant tumors in the offspring. Embryotoxic action was not found.[5]

Allergenic Effect was noted.

Genotoxicity. The naphthylamines have been shown to be genetically active compounds that exhibit postmetabolic activity. They appeared to be positive in *Dr. melanogaster* and produced sperm abnormalities in mice. Some other positive findings have been revealed to justify their ability to induce genotoxic effects in mammals.

However, P. did not cause CA in cultured cells and has been reported to be negative in *Salmonella* mutagenicity assay.[6]

Carcinogenicity. Rats received 20 mg/kg in the antenatal and then in postnatal periods of ontogenesis. An increase in the incidence of malignant and benign neoplasms was observed.[7] At present there is no sufficient proof of carcinogenicity of P. and no grounds for banning it. Admixture of β-naphthylamine may render P. carcinogenic, but a product with the minimum admixture should be used. According to Russian recommendations, materials may be regarded as safe from the point of view of carcinogenicity when the technical product contains 0.005% or less of β-naphthylamine.[8]

Carcinogenicity classification. NTP: NE—NE—NE—EE. See *N*-**Phenyl**-1-**naphthylamine.**

Chemobiokinetics. In man, P. metabolism may result in formation of β-**naphtylamine.** See *N*-**Phenyl**-1-**naphthylamine.**

Standards. Russia. (1986). PML in food and water: 0.2 mg/l.

Regulations. USFDA (1993) approved the use of N-P-2 (1) in resinous and polymeric coatings for polyolefin films to be safely used as a food-contact surface, (2) as a component of the uncoated or coated food-contact surface of paper and paperboard intended for use in contact with aqueous and fatty food (for use only as an antioxidant in dry rosin size and limited to use at a level not to exceed 0.4% by weight of the dry rosin size), and (3) as an antioxidant in rubber articles intended for repeated use in producing, manufacturing, packing, processing, treating, packaging, transporting, or holding food (total not to exceed 5% by weight of the rubber product).

References:

1. See *N*-**BUTYLIDENEANILINE,** 40.
2. Verkhovsky, G. Ya., Toxicity of Some Chemicals Used in the Production of Vulcanizates, Author's abstract of thesis, Moscow, 1965, p. 14 (in Russian).
3. See **BISPHENOL A,** #2, 195.
4. *Current Problems of Industrial Hygiene and Occupational Pathology,* Voronezh, 1975, p. 74 (in Russian).
5. Sal'nikova, L. C., Vorontsov, R. S., Pavlenko, G. I., et al., Mutagenic, embryotropic and blastomogenic effect of neozon D (phenyl-β-naphthylamine), *Gig. Truda Prof. Zabol.,* 9, 57, 1979 (in Russian).
6. See **BISPHENOL A,** #6.
7. See **ADIPIC ACID,** #2, 285.
8. Pliss, G. B. and Zabezhinsky, M. A., On carcinogenic hazard of Neozone D (N-phenyl-β-naphthylamine), *Prob. Oncol.,* 29, 91, 1983 (in Russian).

PHENYL SALICYLATE (PS) (CAS No 118–55–8)

Synonyms. 2-Hydroxybenzoic acid, phenyl ester; Salicylic acid, phenyl ester; Salol.

Properties. White, small crystals or crystalline powder with a pleasant aromatic odor. Solubility in water is 0.0015% at 12°C, soluble in alcohol.

Applications. Used as a light-stabilizer of polyesters, cellulose ethers, PVC, polyvinylidenechloride, polystyrenes, polyolefins, and polyurethanes. An antiseptic for the GI and urinary tracts.

Acute Toxicity. In rats and rabbits, LD_{50} is reported to be 3 g/kg BW.

Repeated Exposure. The maximum curing daily dose in man is 6 g/kg BW.

Chemobiokinetics. In the intestine, PS is broken down by lipase or simply by the alkaline environment into **phenol** and **salicylic acid.**

Standards. EEC (1990). SML: 2.4 mg/kg.

PROPYL GALLATE (PG) (CAS No 121–79–9)

Synonyms. Gallic acid, propyl ester; Progallin *P;* Tenox *PG;* **3,4,5**-Trihydroxybenzoic acid, *n*-propyl ester.

Properties. Colorless, odorless, white to pale brownish-yellow, crystalline powder having a slightly bitter taste. Solubility is 3.5 g/l in water at 25°C, 1030 g/l in alcohol, and 830 g/l in ether.

Applications. PG is used as an antioxidant in fats and oils, in cosmetics and packaging materials, and as a food additive.

Acute Toxicity. In mice, rats, hamsters, and rabbits, LD_{50} varies from 2 to 3.8 g/kg BW.[1] In cats, it is 0.4 g/kg BW.

Repeated Exposure In a 4-week feeding study, increased liver enzyme activities were found at a dose-level of 5 g/kg feed; no effect was noted at 1 g/kg feed (Strik, 1986).[3]

Short-term Toxicity. Exposure to 117 to 5000 mg/kg BW feed caused no adverse effects.[1] No changes were observed in mice after 90-d consumption of the diet containing 5 or 10 g PG/kg BW.[2] In a 13-week study, rats received 490, 1910, or 7455 mg PG/kg feed. At a high dose-level, changes in hematological analysis and increased activity of hepatic ethoxy-resorufin-O-deethylase were noted. Pathological changes were observed in the spleen. The NOAEL of 1910 mg PG/kg feed (135 mg/kg BW) was determined in this study (Speijers et al., 1993).

Long-term Toxicity. In rats given 10 g PG/kg diet for 2 years, no adverse effects were reported, but a 50 g PG/kg dose caused growth retardation and patchy hyperplasia of the forestomach.[4] In mice that received 5 and 10 g/kg feed, only a reduced spleen weight at a 10 g/kg dose was found.[2]

Allergenic Effect. In guinea pigs, sensitizing properties were observed, especially after intradermal applications. Sensitization did not occur when there was oral pre-exposure.[5]

Reproductive Toxicity. A three-generation reproduction study revealed no treatment-related effects when 350 to 5000 mg/kg was added to the diet of rats.[6] No teratogenic effect occurred in rats with up to 25 g/kg in the diet.[7]

Genotoxicity. PG is found to be negative in *Salmonella* mutagenicity assay.[8] It failed to demonstrate genotoxic effect in the DLM test in rats or in *in vitro* cytogenic tests in human embryonic lung cells and in *in vivo* bone marrow metaphase analysis in Sprague-Dawley rats.[9,10]

Carcinogenicity. Fisher 344 rats and B6C3F$_1$ mice were exposed to the dietary levels of 5 to 20 g/kg feed. No dose-related increase in tumor incidences was observed.[11] Oral administration of PG was found to decrease the incidence of tumors induced in rats by **dimethylbenz[a]anthracene.**[12] *Carcinogenicity classification.* NTP: E—N—E—N.

Chemobiokinetics. Following oral administration, more than 70% PG is absorbed in the intestines. Biokinetics of PG apparently differs from that of **octyl** and **dodecyl gallates,** those being absorbed and hydrolyzed to a lesser degree than the PG. In rats and rabbits, PG is hydrolyzed to **gallic acid** and undergoes further methylation, yielding 4-O-**methylgallic acid** as a biotransformation product. Excretion occurs in the urine, either as the free compound or as the glucuronide conjugate. In rats, PG may be eliminated via the urine as unchanged ester.[6,13]

Recommendations. JECFA (1993). ADI: 0 to 1.4 mg/kg BW.

Regulations. USFDA (1993) regulates the use of PG in resinous and polymeric coatings in a food-contact surface of articles intended for use in producing, manufacturing, packing, transporting, or holding food. In the U.S. permitted use level of PG (as total antioxidants) is 0.02% for all foods.

Standards. The Netherlands. Maximum permitted level: 0.01% in fats and oils (total antioxidants).

References:
1. See **DODECYL GALLATE,** #2.
2. Dacre, J. C., Long-term toxicity study of *n*-propyl gallate in mice, *Food Cosmet. Toxicol.,* 12, 125, 1974.
3. See **DODECYL GALLATE,** #3.
4. Lehman, A. G., Fitzhugh, O. G., Nelson, A. A., et al., The pharmacological evaluation of antioxidants, *Food Res.,* 3, 197, 1951.

5. *Propyl Gallate COLIPA Antioxidant No. 10,* COLIPA Monograph CSC/386/83, 1983.
6. See **DODECYL GALLATE,** #4.
7. Tanaka, S., Kawashima, K., Nakaura, S., et al., Effect of dietary administration of propyl gallate during pregnancy on prenatal and postnatal development of rats, *Tanaka J. Food Hyg. Soc. Jap.,* 20, 378, 1979 (in Japanese).
8. Rosian, M. and Stich, H., Enhancing and inhibiting effect of propyl gallate on cocarcinogen-induced mutagenesis, *J. Environ. Pathol. Toxicol.,* 4, 159, 1980.
9. Litton Bionetics, Inc., Mutagenic Evaluation of Compound FDA 71–39, Propyl Gallate, NTIS Report No. 245–441, 1974.
10. NTP, Carcinogenesis Bioassay of Propyl Gallate in F344/N Rats and B6C3F$_1$ Mice (Feeding Study), Technical Report Series No. 240, 1982.
11. Abdo, K. M., Huff, J. U. E., Haseman, J. K., et al., Carcinogenesis bioassay of propyl gallate in F344 rats and B6C3F$_1$ mice, *J. Am. Coll. Toxicol.,* 2, 425, 1983.
12. King, M. M. and McCay, P. B., Modulation of tumor incidence and possible mechanisms of inhibition of mammary carcinogenesis by dietary antioxidants, *Cancer Res.,* Suppl. 43, 2485S, 1983.
13. Dacre, J. C., Metabolic pathways of the phenolic antioxidants, *Jl. N. Z. Inst. Chem.,* 24, 161, 1960.

SORBITOL (CAS No 50–70–4)

Synonyms. Diakarmon; *D*-Glucitol; *L*-Gulitol; Sionon; Sorbilande; Sorbit; Sorbol.

Properties. Odorless, white granules, powder or crystalline powder having a sweet taste (~60% as sweet as sugar, w/w), with a cooling sensation in the mouth. Hygroscopic. Readily soluble in water, poorly soluble in cold ethyl alcohol.

Applications and **Exposure.** Used in the manufacturing of propylene glycol, synthetic plasticizers, and resins. A stabilizer in PVC production. S. occurs naturally in many plant materials.

Acute Toxicity. In rats, mice, and rabbits, LD$_{50}$ values are in excess of 15 mg/kg BW (Tanaka and Gomi).[1] The only treatment-related effects from high doses are soft stools and a laxative action. If adequate time is allowed for adaptive changes in the microflora of the lower bowel, then animals have been shown to tolerate diets containing up to 20% S. without exhibiting overt laxation.

Repeated Exposure. Increase in daily urine volumes was caused by 63% S. in the diet.[2]

Short-term Toxicity. S. is added to food (for example, that of diabetics) and causes no side effects, since it does not raise the blood glucose level or raises it to a considerably lesser extent than glucose.

Long-term Toxicity. In a 2-year feeding study of xylitol in rats, a group exposed to 20% S. was included as the control.[1] Induced changes consisted of caecal enlargement, reduced BW gain up to week 78, decreased efficiency of food utilization, increased water intake and urinary output, etc. Adrenal medullary hyperplasia and decreased absolute thyroid weights were also reported. JECFA (1982) concluded that a diet containing as much as 20% sorbitol ''produced gross dietary imbalance, which may produce metabolic imbalance.''

Reproductive Toxicity. S. is shown to be negative in unpublished *teratogenicity* studies in rats, rabbits, mice, and hamsters.[1] Charles River rats were fed diets with 2.5, 5, and 10% S. Treatment was associated with no consistent adverse effect on any index of reproductive performance. No abnormal pups were observed in any generations. MacKenzie and Hauck[1] concluded that S. administered in the diet to three succesive generations of rats at levels up to 10% had no adverse effect on growth or reproductive performance.

Genotoxicity. S. appeared to be negative in a majority of tests (*Salmonella* mutagenicity assay, the rat bone marrow test, and DLM assay in mice).[1]

Chemobiokinetics. The absorption of S. occurs by passive diffusion and is slower than that of glucose or fructose. It is metabolized basically in the liver, and only very small quantities are found in the blood or urine.[3] Two metabolic pathways are described: oxidation after conversion to **glucose** or direct oxidation of **fructose.** Herz considered S. converted first to fructose and then to glucose.[4] Increased amounts of **methylmalonic acid** and 2-**oxoglutaric acid** were found in the urine of polyol-fed rats. The urinary excretion of **citric acid** and **maleic acid** was also increased significantly. The increased levels of urinary organic acids in the polyol-fed rats may be explained in terms of impaired mitochondrial oxidation of these acids and of impaired conversion of methymalonic acid to succinic acid.[2]

Regulations. USFDA (1993) regulates S. for use (1) in resinous and polymeric coatings in a food-contact surface, (2) in resinous and polymeric coatings for polyolefin films to be used safely as a food-contact surface of articles intended for use in producing, manufacturing, packing, transporting, or holding food, and (3) in cross-linked polyester resins to be used safely as articles or components of articles intended for repeated use in contact with food.

References:

1. MacKenzie, K. M. and Hauck, W. N., Three-generation reproduction study of rats ingesting up to 10% sorbitol in the diet — and a brief review of the toxicological status of sorbitol, *Food Chem. Toxicol.*, 24, 191, 1986.
2. Hamalainen, M. M., Organic aciduria in rats fed high amounts of xylitol or sorbitol, *Toxicol. Appl. Pharmacol.*, 90, 217, 1987.
3. Adcock, L. H. and Gray, C. H., The metabolism of sorbitol in the human subject, *Biochem. J.*, 65, 554, 1957.
4. Hers, H. G., The conversion of fructose-1-C^{14} and sorbitol-1-C^{14} to liver and muscle glycogen in the rat, *J. Biol. Chem.*, 214, 373, 1955.

TETRAETHYLTIN (TET) (CAS No 597–64–8)

Synonym. Tetraethylstannate.

Properties. Colorless, oily liquid with a specific, sharp odor. Solubility in water is 10 to 12 mg/l at 20°C. Odor perception threshold is 0.2 mg/l; practical threshold is 0.5 mg/l. The metallic taste perception threshold is 1.5 mg/l.[1]

Applications. Used as a stabilizer in PVC material manufacturing.

Acute Toxicity. LD_{50} is reported to be 9 to 15 mg/kg BW in rats, 40 mg/kg BW in mice and guinea pigs, and 7 mg/kg BW in rabbits.[1,2] Administration of lethal doses caused BW loss, adynamia, convulsions, and muscle weakness, with subsequent death in 5 to 7 d in a comatose condition.

Repeated Exposure revealed marked cumulative properties. Rats given $1/10$ LD_{50} died in 2 to 3 weeks. Manifestations of toxic action included adynamia, muscle weakness, refusal of food, and exhaustion.[1,2]

Long-term Toxicity. Anemia and reticulocytosis were observed in rats. Other toxic effects included a change in the activity of some enzymes and in SH-group content in the blood serum.

Chemobiokinetics. TET metabolism in isolated hepatocytes includes formation of **ethane** and **ethylene,** the latter being the principal metabolite (95%). No toxic effects on hepatocytes have been noted.[3]

Standards. Russia. (1988). MAC and PML: 0.0002 mg/l.

References:

1. Skachkova, I. N., Hygienic substantiation of Maximum Allowable Concentration for tetraethyltin in water bodies, *Gig. Sanit.*, 4, 11, 1967 (in Russian).
2. Mazayev, V. G., Korolev, A. A., and Skachkova, I. N., Experimental study of toxic effect of tetraethyltin and dichlorodibutyltin in the animals, *J. Hyg. Epidemiol. Microbiol. Immunol. (Prague)*, 15, 104, 1971 (in Russian).
3. Wiebkin, P., Prough, R. A., and Bridges, J. W., The metabolism and toxicity of some organotin compounds in isolated rat hepatocytes, *Toxicol. Appl. Pharmacol.*, 62, 409, 1982.

2,2′-THIOBIS(6-*tert*-BUTYL-4-METHYLPHENOL)

Synonyms. Advastab 406; Bis(3-*tert*-butyl-2-hydroxy-methyl-phenyl)sulfide; *CAO-6;* Tioalkofen *BP.*

Properties. Colorless, crystalline powder. Poorly soluble in water, soluble in alcohol.

Applications. Used as a thermostabilizer in the synthesis of vulcanizates based on natural and butyl rubbers, polyolefins, pentaplast, and PVC.

Acute Toxicity. In rats, LD_{50} has been found to be up to 25 g/kg BW. Gross pathological examination failed to reveal changes in the visceral organs.[08]

Short-term Toxicity. Rats were dosed by gavage with 3.25 g/kg BW for 4 months. Administration of T. caused no morphological changes. The only treatment-related effect observed was the change in conditioned reflex activity.[08]

Reproductive Toxicity. T. affects the reproductive processes in fish but not the development of chick embryos.[08]

Carcinogenicity. T. is likely to produce an inhibiting effect upon tumor induction by benzo[a]pyrene.[1]

Standards. Russia. PML in food: 4 mg/l.

Reference:

Braun, D. D., Effect of polyolefin antioxidants on induction of neoplasms by benzo[a]pyrene, *Gig. Sanit.*, 6, 8, 1975 (in Russian).

4,4′-THIOBIS(2-*tert*-BUTYL-5-METHYLPHENOL) (CAS No 96–69–5)

Synonyms. Bis(5-*tert*-butyl-4-hydroxymethylphenyl)sulfide; Santonox; Santonox *R;* Tioalkofen *BM;* 4,4′-Thiobis(2-*tert*- butyl-*m*-cresol); 4,4′-Thiobis(6-*tert*-butyl-3-methylphenol).

Properties. Santonox (technical product) is a light-grey, crystalline powder. Purified product is fine, white crystals. Solubility in water is 0.08%, soluble in alcohol.

Applications. Used as a light-stabilizer of polystyrene, PVC, polypropylene, and cellulose acetate; a thermostabilizer of polyamides, etc.

Acute Toxicity. In mice, LD_{50} is 4.88 g/kg BW. Lethal doses caused adynamia. Death occurred in 2 d. Rats appeared to be less sensitive: administration of up to 5 g/kg BW caused no evident signs of intoxication. Gross pathological examination revealed coarse, fatty dystrophy of the liver.[1]

Repeated Exposure. Rats were dosed by gavage with 1/30 LD_{50} for 2 months. The treatment caused an increase in alkaline phosphatase activity in the blood and inhibition of phagocytic activity. Rats were given 1/5 LD_{50}; this killed a half of animals by day 41 of the treatment. Manifestations of toxic action included retardation of BW gain, lowering of the threshold of neuromuscular excitability, and impairment of the detoxifying function of the liver.[1] Histological examination revealed early signs of fatty dystrophy in the liver, signs of congestion in the lungs, and hemorrhagic foci in the spleen pulp.

Short-term Toxicity. In a 3-month rat study, administration of 0.05 and 0.005% T. in the diet caused a decrease in BW gain (with the higher dose). Gross pathological examination failed to reveal changes in the viscera.[05]

Chemobiokinetics. ^{14}C-labeled 4,4′-thiobis(6-*tert*-butyl- *m*-cresol) has been shown to be incompletely absorbed in male rats. Rate of absorption is proportional to the administered dose. Distribution occurs mainly in the liver, blood, muscles, and adipose tissues. The major metabolites: glucuronide conjugates of the parent compound. It is rapidly cleared from all tissues except the adipose one.[2]

Regulations. British Standard (1992). T. is listed as an antioxidant for polyethylene and polypropylene compositions used in contact with foodstuffs or water intended for human consumption (maximum permitted level for the final compound = 0.25%).

References:

1. See *N*-BUTYLIDENEANILINE, 78.
2. Birnbaum, L. S., Eastin, W. C., Johnson, L., et al., Disposition of 4,4′-thiobis(6-*tert*-butyl-*m*-cresol) in rats, *Drug Metab. Disposit.*, 11, 537, 1983.

2,2′-THIOBIS(6-ISOBORNYL-4-METHYLPHENOL)

Properties. White, crystalline powder. Poorly soluble in water, soluble in alcohol.

Acute Toxicity. Rats and mice tolerate 10 and 15 g/kg BW, respectively.

Repeated Exposure failed to reveal cumulative properties. Manifestations of toxic action were not found on administration of 1 to 3 g/kg BW.

Long-term Toxicity. In a 7-month study, rats were given 200 mg T./kg BW, and mice 150 and 300 mg T./kg BW. The treatment did not affect BW gain, hematological analyses, or conditioned reflex activity. Histological examination failed to reveal differences from the controls.

Allergenic Effect has not been observed.
Standards. Russia. Recommended PML: **n/m.**
Reference:
See 4-**ALK(C_7-C_9)OXY**-2-**HYDROXYBENZOPHENONE.**

4,4′-THIOBIS(2-METHYL-6-CYCLOHEXYL)PHENOL

Acute Toxicity. In mice, LD_{50} is 4.8 g/kg BW. Gross pathological examination revealed congestion in the viscera.[08]

2,2′-THIOBIS(4-METHYL-6-α-METHYLBENZYLPHENOL)

Synonym. Thioalkofen *MBP.*
Properties. White, crystalline powder. Poorly soluble in water.
Applications. Used as a thermo- and light-stabilizer in the production of PVC and polyolefins.
Acute Toxicity. Mice tolerate a 15 g/kg BW dose. Gross pathological examination showed no lesions.
Repeated Exposure failed to reveal cumulative properties.
Long-term Toxicity. In a 6-month study, rats were dosed with 0.43 and 1.42 g/kg BW. The treatment caused only stimulation of the CNS cortical areas. There were no pathological changes in the viscera.[08]
Reproductive Toxicity. Does not affect the reproduction in fish.

4,4′-THIODIMETHYLENEBIS(2,6-DI-*tert*-BUTYLPHENOL)

Synonym. *TB-3.*
Properties. White, crystalline powder. Poorly soluble in water and alcohol.
Acute Toxicity. Rats and mice tolerate administration of 10 g/kg BW.[08] Gross pathological examination revealed no changes in the viscera.
Repeated Exposure failed to reveal marked cumulative properties. One-month exposure to 1.5 g/kg BW caused 30% mortality in mice. Administration of a 10 g/kg dose for a month appeared to be nonlethal for rats. Gross pathological and histological examination revealed no lesions in the visceral organs.
Short-term Toxicity. Mice were dosed by gavage with 75 and 150 mg/kg BW over a period of 4 months. The treatment was shown to cause only excitation of the CNS cortical areas in animals. Gross pathological examination revealed no changes in the visceral organs.[08]
Standards. Russia. Recommended PML: **n/m.**

4,4′-THIODIPHENOL (CAS No 2664–63–3)

Synonym. 4,4′-Dihydroxydiphenylsulfide.
Properties. Light, fine crystalline powder. Poorly soluble in water, soluble in alcohol.
Acute Toxicity. LD_{50} is reported to be 6.4 g/kg BW in rats and 5.5 g/kg BW in mice. Poisoning is accompanied by lethargic condition. Death occurred in 3 d. Gross pathological examination revealed congestion in the visceral organs.
Repeated Exposure. All the animals died within 16 d during treatment by Lim method.
Reference:
See 2,6-**DI**-*tert*-**BUTYL-4-METHOXYMETHYLPHENOL,** 154.

THIOHYDROXYETHYLENEDIOCTYLTIN (TOEDO)

Properties. Odorless, white, crystalline powder. Poorly soluble in water.
Applications. Used as a stabilizer in the production of PVC materials.

Acute Toxicity. LD$_{50}$ is found to be 2.23 g/kg BW in rats, 1.8 g/kg BW in mice, and 2.02 g/kg BW in guinea pigs. Manifestations of toxic action: see DOMO. Lesions are observed following a single administration of ¹/₅ LD$_{50}$.

Repeated Exposure revealed evident cumulative properties. K$_{acc}$ is 0.7 (by Kagan). Animals given ¹/₂₀ LD$_{50}$ and some of those given ¹/₁₀₀ had died within a month.

Reproductive Toxicity. *Embryotoxic* effect is not observed.

Standards. Russia. PML in food: 0.05 mg/l.

Reference:
See **DIOCTYLTIN MALEATE.**

1,1,3-TRI-(5-*tert*-BUTYL-4-HYDROXY-2-METHYLPHENOL)BUTANE

Synonyms. Topanol *CA;* **6,6′,6″**-tri-*Tert*-butyl-**4,4′,4″**-(**1**-methylpropan-1-yl-3-ylidene)tri-*m*-cresol; Trisalkofen *BMB.*

Properties. White, crystalline powder. Solubility in water is 0.03% at 25°C, soluble in alcohol.

Applications. Used as an antioxidant in the production of polypropylene, as a stabilizer for polyethylene and PVC, and for synthetic rubber.

Acute Toxicity. LD$_{50}$ is 14 g/kg BW in rats and 16.1 g/kg BW in mice. Poisoning was accompanied by CNS inhibition. Gross pathological examination revealed gastric paresis.[08]

Repeated Exposure revealed marked cumulative properties. On administration of 3 g/kg BW to mice, LD$_{50}$ was attained at a total dose of 11.2 g/kg. Of animals, 90% died after three administrations of 6 g/kg BW.[08] However, female rats are reported to tolerate a dose of 2 g/kg BW for 2 weeks without any visible changes in their general condition.[05]

Long-term Toxicity. General condition, hematological analyses, and conditioned reflex activity were not affected in mice given the doses of 230 or 690 mg/kg BW three times a week for 6 months. Capacity to work was lowered, and relative weights of the heart and spleen were increased. Gross pathological examination failed to reveal changes in the viscera.[08]

Reproductive Toxicity. Affects reproduction in fish.[08]

Chemobiokinetics. After ingestion, T. is poorly absorbed in the GI tract of experimental animals. It is removed predominantly in the feces (98% of the dose).

Standards. Russia. Recommended PML: **n/m. British Standard** (1992). T. is listed as an antioxidant for polyethylene and polypropylene compositions used in contact with foodstuffs or water intended for human consumption (maximum permitted level for the final compound = 0.25%).

2,4,6-TRI-*tert*-BUTYLPHENOL

Synonyms. Alkofen *B;* Antioxidant *P-21.*

Properties. White, crystalline powder with a yellowish tint. Solubility in water is 0.00005%. Soluble in ethanol.

Applications. Used as a stabilizer in the production of polyolefins, impact resistant polystyrene, and synthetic rubber, and also of acetyl cellulose and acetobutyrate cellulose etrols.

Acute Toxicity. In mice, the LD$_{50}$ is 1.6 g/kg BW.[1] Histological examination revealed congestion, edema, and desquamation of the esophageal and gastric mucosa, necrosis of the small intestinal wall, and parenchymatous dystrophy of the renal tubular epithelium.

Short-term Toxicity. In a 4-month rat study, a dose of 88 mg/kg BW caused slight CNS excitation. Histological examination revealed parenchymatous dystrophy of the renal tubular epithelium.[1]

Reproductive Toxicity. *Embryotoxic* effect was not observed in the experiment with chick embryos. Uncertain evidence of *teratogenicity* has been reported.

Genotoxicity. T. is structurally close to butylated hydroxytoluene, which exhibits a mutagenic effect in the form of pathological changes in the sperm. It exerts a modulating effect as related to mutagens and teratogens and antimutagen action as related to benzo[a]pyrene.[2]

Chemobiokinetics. T. is shown to be absorbed rapidly in the GI tract of Sprague-Dawley fasting rats. It appears immediately in the blood, where its maximum concentration is attained in 15 to 60 min. T. or its metabolites is not found in the urine of rats; one metabolite is found in the feces and bile.[3]

Standards. Russia. PML in food: 2 mg/l.

References:

1. Broitman, A. Ya., Comparative toxicity of some stabilizers applied to raise thermo- and photoresistance of polymeric materials, *Gig. Truda. Prof. Zabol.*, 2, 20, 1962 (in Russian).
2. Malkinson, A. M., Review: putative mutagens and carcinogens in foods. III. Butylated hydroxytoluene (BHT), *Environ. Mutat.*, 5, 353, 1983.
3. Takahasi, O. and Hiraga, K., Metabolic studies in the rat with **2,4,6**-tri-*tert*-butylphenol: a haemorrhagic antioxidant structurally related to BHT, *Xenobiotika*, 13, 319, 1983.

2,4,6-TRI-*tert*-BUTYLPHENYLPYROCATECHOL PHOSPHITE

Synonyms. Alkofen B phosphite; Pyrocatecholphosphorous acid, **2,4,6**-tri-*tert*-butylphenolic ester.

Acute Toxicity. LD_{50} is reported to be 3 g/kg BW in rats and 1.5 g/kg BW in mice.[08] Gross pathological examination revealed edema and cellular infiltration of the submucosal and muscle layers of the stomach wall. Necrotic areas were found in the small intestinal mucosa.

Repeated Exposure. Partial dose-dependent mortality was observed in rats given a 100 mg/kg BW dose over a period of 2 months. Mice tolerate administration of 75 to 300 mg/kg BW. Lethal doses caused copious hemorrhages in the limbs, tips of the ears, tail, scrotum, and mesentery but did not alter blood cholinesterase activity.[08]

Reproductive Toxicity. Affects the reproductive processes in fish.

TRIBUTYLTIN METACRYLATE (TBTMA) (CAS No 2135–70–6)

Synonyms. Tributyl(metacryloyloxy)stannate; Tin tributylmetacrylate.

Properties. Colorless or light-yellow, oily liquid with a distinct, unpleasant odor. Taste perception threshold is 0.43 mg/l, odor perception threshold is 0.064 mg/l, and practical threshold is 1.154 mg/l.

Applications: Used as a stabilizer in the production of PVC materials.

Acute Toxicity. LD_{50} is reported to be 210 mg/kg BW in rats, 150 mg/kg BW in guinea pigs, and 150 mg/kg BW in rabbits. Manifestations of toxic action included lethargy, loss of appetite, photophobia, adynamia, and, in many animals, paresis of the hind limbs.

Repeated Exposure revealed cumulative properties. Rats were given 5, 10, and 20 mg/kg BW for three weeks. Lethality was observed from the second day of the treatment. Muscular weakness, paresis of the hind limbs, abdominal swelling, lacrimation, and salivation were observed. All the doses tested affected hematological analyses; a 20 mg/kg dose affected unconditioned reflex activity (STI).

Long-term Toxicity. In a 6-month study, rats received T. in an oil solution. The treatment caused a decrease in the excretory function of the liver and an increase in the vitamin C content of the suprarenals. Higher doses caused depression of BW gain.

Standards. Russia. (1988). MAC and PML: 0.0002 mg/l.

Reference:

Tsai, V. N., Maximum allowable concentration of tributyltin methacrylate in water bodies, *Gig. Sanit.*, 4, 42, 1975 (in Russian).

2,4,6-TRI-(3′,5′-DI-*tert*-BUTYL-4′-HYDROXYBENZYL)MESITYLENE

Synonyms. Agidol-40; *AO-40;* Ionox 330; **1,3.5**-Trimethyl-**2,4,6**-tris(3′,5′-*tert*-butylhydroxybenzyl) benzene.

Properties. White, crystalline powder. Poorly soluble in water. Solubility in oil is 4%.

Applications. Used as an antioxidant in the production of polyolefins and polyamides.

Acute Toxicity. Administration of 10 g/kg BW does not cause death in rats. Poisoning is characterized by a weak narcotic effect and adynamia 1 to 3 h after administration. Acute effect threshold appeared to be 1 g/kg BW for STI and body temperature.[1]

Repeated Exposure. Administration of 1 g/kg BW by the Lim method did not cause functional changes or death. There was BW gain decrease and reduction in diuresis and hippuric acid content in the urine after exposure to sodium benzoate.[1]

Short-term Toxicity. Addition of 1% T. to the diet for 90 d produced no hematological and histological changes in rats.[2]

Long-term Toxicity. A dose of 100 mg/kg BW caused functional changes in the CNS, liver, and kidneys of rats. Pathological changes were noted in the stomach. A dose of 20 mg/kg BW is considered to be the NOAEL.[1]

Reproductive Toxicity. A dose of 5 mg/kg was subthreshold in terms of effect upon the body of pregnant animals, fetus, and progeny; in terms of gonadotoxic effect, a 100 mg/kg BW dose was ineffective.[1]

Allergenic Effect was not found.[1]

Chemobiokinetics. T. is not absorbed in the body. After administration, ^{14}C-labeled T. is detected unchanged in the feces. Urine, expired air, bones, and other tissues show absence of radioactivity (Wright et al., 1964).

Standards. Recommended PML in milk: 1.25 mg/l.[1]

Regulations. USFDA (1993) approved T. for use (1) at levels not to exceed 0.5% by weight of the polymers except nylon resins identified in CFR and (2) at levels not to exceed 1% by weight of nylon resins identified in CFR.

References:
1. Stasenkova, K. P. and Samoilova, L. M., Data on substantiation of permissible migration level for stabilizer *AO-40* (agidol) from rubbers and plastics, *Gig. Sanit.*, 9, 31, 1978 (in Russian).
2. Stevenson, D. E., Chambers, P. L., and Hunter, C. G., Toxicity studies with **2,4,6**-tri-(**3′,5′**-di-*tert*-butyl-**4′**-hydroxybenzyl) mesitylene in the rat, *Food Cosmet. Toxicol.*, 3, 281, 1965.

TRIETHYLTIN CHLORIDE (TETCl) (CAS No 994-31-0)

Synonyms. Chlorotriethyl stannate; Chlorotriethyl tin.

Properties. Colorless liquid.

Applications. Used as a stabilizer in the production of PVC materials.

Acute Toxicity. *Humans.* Consumption of a drug containing triethyltin as an impurity resulted in 209 patients who developed neurological symptoms: increased intracranial pressure as a consequence of cerebral edema; 110 patients died.[1] *Animals.* LD_{50} values have been reported to be 3.45 mg/kg BW in male and 2.88 mg/kg BW in female rats, 20 mg/kg BW in mice, 15 mg/kg BW in guinea pigs, and 7 mg/kg BW in rabbits. Lethal doses affected the NS and caused animals to die on day 2 to 15.[07] Changes in behavior and in conditioned reflex activity were observed in rats following *i/a* administration of a 3 mg/kg BW dose.[2]

Repeated Exposure revealed cumulative properties. Rats received TETCl via drinking water with a concentration of 20 mg/l for 28 d. The treatment affected muscle tonus and caused hind limb paralysis. Lower doses led to a reduction in motor activity. This phenomenon appeared after 2 weeks and disappeared after 1 month.[2,3]

Rats fed diets containing 15 ppm triethyltin (0.75 mg/kg BW) for 2 weeks exhibited retardation of BW gain, decreased relative weights of the thymus and spleen, but increased relative adrenal weights (at 50 ppm level). Of the animals, 30% died, while survivors exhibited brain edema.[4] TETCl has a neurotoxic effect.

Chemobiokinetics. Following absorption in the GI tract, TETCl is distributed predominantly to the liver, kidneys, and brain. Edema is noted in the white matter of the brain. The mechanism by which tri-substituted *Sn* compounds exert a toxic effect consists in inhibition of oxidative phosphorylation in the mitochondria as a result of the penetration therein of *Sn* and *Cl* ions and also in the suppression of mitochondrial ATP-ase activity.

References:
1. Alajouanine, T., Derobert, B. L., and Thieffry, S., Etude clinique d'ensemble de 210 cas d'intoxication par les sels organiques de l'etain, *Rev. Neurol.*, 98, 85, 1958 (in Italian).

2. Wenger, G. R. and McMillan, D. E., Effect of triethyltin-SO_4 in mice responding under a multiple schedule of food presentation, *Toxicologist,* 5, 24, 1985.
3. Reiter, L., Kidd, K., Heavner, G., et al., Behavioural toxicity of acute and subacute exposure to triethyltin in the rat, *Neurotoxicology,* 2, 97, 1981.
4. Snoeij, N. J., van Iersel, A. A. J., Penninks, A. H., et al., Toxicity of triorganotin compounds: comparative *in vivo* studies with a series of trialkyltin compounds and triphenyltin chloride in male rats, *Toxicol. Appl. Pharmacol.*, 81, 274, 1985.

TRI(α-METHYLBENZYL) PHOSPHITES, mixture

Synonym. Phosphite *P-24*.

Properties. Light-yellow, viscous liquid. Poorly soluble in water, soluble in alcohol.

Applications. Used as a thermostabilizer in the production of polyethylene, polypropylene, and ethylene-propylene copolymers.

Acute Toxicity. In mice, LD_{50} is 20 g/kg BW (laboratory sample) and 5.6 g/kg BW (industrial sample).[08] Manifestations of toxic action included excitation and clonic convulsions. Recovery in 2 to 3 h after administration. Gross pathological examination revealed a stretched stomach and flabby walls of the small intestine.

Repeated Exposure. Mice were given 1/5 and 1/20 LD_{50} (industrial sample) for 3 weeks. The treatment caused no animal mortality.

Short-term Toxicity. Mice were treated with 400 mg/kg BW (laboratory sample) for 4 months. BW gain and conditioned reflex activity were affected. Gross pathological examination failed to reveal changes in the viscera.[08]

Long-term Toxicity. In a 6-month study, a 55 mg/kg BW dose (industrial sample) appeared to be ineffective in mice. Gross pathological examination failed to reveal changes in the visceral organs.[08]

Reproductive Toxicity. Affects the reproductive processes in fish.

TRI-*p*-NONYLPHENYL PHOSPHITE

Synonyms. Phosphite *NF;* Phosphoric acid, tri-*p*-nonylphenylester; Polygard.

Properties. Light-yellow, viscous liquid. Hydrolyzes in water, soluble in alcohol.

Applications. Used as a stabilizer in the production of synthetic rubber, impact-resistant polystyrene, polyurethane elastomers, and polyvinylchloride plastics.

Acute Toxicity. LD_{50} values are found to be 23 g/kg BW with the English product or 5 g/kg BW (German), 4.2 g/kg BW (Japanese), or 9.3 g/kg BW (Russian). Manifestations of the toxic effect included a brief period of stimulation, with subsequent CNS inhibition, narcosis, and death.[1] Gross pathological examination revealed congestion, edema, and detachment of the esophageal and gastric mucosa, as well as necrosis in the small intestinal walls.

Short-term Toxicity. Administration of 0.5 g/kg BW for 4 months produced retardation of BW gain and stimulation of the CNS cortical areas. Other functional indices tested were unchanged. Gross pathological examination revealed necrosis of the small intestinal mucosa.[08] The lower dose (100 mg/kg BW) caused dystrophic changes in the liver and some deviations in the relative weights of the thyroid and spleen. The NOEL of 20 mg/kg BW was reported.[2]

Standards. Russia. PML in food: 4 mg/l, PML in drinking water: 3 mg/l.

References:

1. See **MINERAL PETROLEUM OIL**, #1, 23.
2. See **ADIPIC ACID**, #2, 266.

TRIPHENYL PHOSPHITE (CAS No 101–02–0)

Synonyms. Phosphoric acid, triphenyl ester; Phosphite *F*.

Properties. Colorless crystals or clear, oily liquid, with a light odor of phenol. Poorly soluble in water, soluble in organic solvents.

Applications. Used as an antioxidant and stabilizer for synthetic rubber; a secondary stabilizer for polyolefins, polyorganosiloxanes, and epoxy resins; a thermo- and light-stabilizer of cellulose ethers.

Acute Toxicity. LD_{50} is reported to be 1.5 g/kg BW in rats and 1.3 g/kg BW in mice.[1,2] When P. is given in an oil suspension, LD_{50} is 1.92 and 1.08 g/kg BW, respectively (Talakin et al., 1987). Lethal doses led to an immediate excitation, with subsequent CNS inhibition, body tremor, equilibrium disturbances, and convulsions. Death within 2 to 3 d. Histological changes included circulatory disturbances and peripheral nerve degeneration (granular breakdown of myelin).

Repeated Exposure failed to reveal cumulative properties. TPP was shown to be a potent neurotoxicant; it produced a characteristic delayed neurotoxicity.[3] Rats exposed to repeated *s/c* injections of TPP displayed dysfunctional changes. Two *s/c* administrations of 1.2 g/kg BW caused tail rigidity and hind limb paralysis. Gross pathological examination showed lesions in the spinal cord.[4] The same total dose administered *s/c* during 2 weeks caused inhibition of neurotoxic esterase in rats.[05]

Allergenic Effect was reported on skin application.

Chemobiokinetics. TPP administration affects mitochondrial metabolism in skeletal muscle, especially red muscle of chickens.[2]

Standards: EEC (1990). SML: 2.4 mg/kg.

References:
1. Bol'shakov, A. M. and Baranov, V. I., On hygiene norm-setting of the MAC for triphenylphosphite at workplace, *Gig. Tryda Prof. Zabol.*, 9, 55, 1977 (in Russian).
2. Golubev, A. A., Andreeva, N. B., Dvorkin, E. A., et al., Toxicity characteristics of trioctyl phosphite and triphenyl phosphite, *Gig. Truda Prof. Zabol.*, 10, 38, 1973 (in Russian).
3. Kouno, N., Katou, K., Yamauchi, T., et al., Delayed neurotoxocity of triphenylphosphite in hens: pharmacokinetic and biochemical studies, *Toxicol. Appl. Pharmacol.*, 100, 440, 1989.
4. Veronesi, B., Padilla, S. S., and Newland, D., Biochemical and neurological assessment of triphenylphosphite in rats, *Toxicol. Appl. Pharmacol.*, 83, 203, 1986.
5. Padilla, S. S., Grizzle, S. B., and Lyerby, D., Triphenyl phosphite: *in vivo* and *in vitro* inhibition of rat neurotoxic esterase, *Toxicol. Appl. Pharmacol.*, 87, 249, 1987.

ZINC compounds

Properties and **Applications. Zinc** salts, such as **Zn chloride, Zn nitrate,** and **Zn sulfate,** are soluble in water; **Zn carbonate** and **Zn oxide** are poorly soluble. Levels in drinking water above 3 mg/l give an undesirable astringent taste and may result in discoloration. The threshold perception concentration for the effect on the organoleptic properties of water is 5 mg/l. Cohen et al.[1] showed that 5% of the population distinguish water not containing Zn^{2+} from water containing it (as **Zn sulfate**) at a concentration of 4.3 mg/l. At a concentration of more than 5 mg/l, water becomes opalescent, and an oily film may form on boiling.

Zinc compounds are widely used as stabilizers. **Zn acetate dihydrate:** Crystals with faint acetic odor and astringent taste. Readily soluble in water and alcohol. **Zn caprylate:** Crystalline substance. Sparingly soluble in boiling water. Used as a PVC stabilizer. **Zn carbonate:** Solid of rhombohedral structure. At 140°C is dissociated with the formation of **Zn oxide.** Solubility in water is 10 mg/l at 15°C. Used in rubber manufacturing; a pigment. **Zn oxide:** White or cream, odorless, amorphous powder. Hexagonal crystals. Poorly soluble in water and ethanol. **Zn stearate:** White amorphous powder with a slight characteristic odor. Repels water (poorly soluble). Soluble in ethanol. Used in the rubber and plastic industry. **Zn sulfate monohydrate:** Powder or granules. Soluble in water, poorly soluble in alcohol. Used for bleaching paper and in the manufacture of other **Zinc** salts.

Exposure. The diet is normally the principal source of Zn^{2+} exposure. Its content in a typical mixed diet of North American adults has been reported to vary between 10 mg/d and 15 mg/d.

Acute Toxicity. *Humans.* A concentration in water of 0.7 to 2.3 g/l can cause vomiting in humans. Poisoning results in pulmonary distress, fever, chills, and gastroenteritis. Generally, a protective mechanism occurs after consumption of more than 500 mg **Zn,** corresponding to 2 g **Zn** in the form of **sulfate.**[2] *Animals.* **Zn chloride.** LD_{50} is identified to be 350 to 480 mg/kg BW in rats and 250 mg/kg BW in mice. The acute action threshold (for anemia, adynamia, and labored breathing) appeared to be 0.1 g/kg BW. **Zn sulfate:** LD_{50} is found to be 2.95 g/kg BW in rats. **Zn acetate hydrate:** LD_{50} is 2.5 g/kg BW in rats. **Zn caprylate:** LD_{50} is 2.37 g/kg BW in mice; in rats it was not attained. **Zn carbonate:** Rats tolerate administration of 10 g/kg BW; mice do not die after ingestion of 7 g/kg BW dose. **Zn nitrate:** LD_{50} is 0.34 g/kg BW in mice, 1.56 g/kg BW in male rats, and 1.4 g/kg BW in female rats. **Zn oxide:** In mice, LD_{50} is 7.95 g/kg BW. Acute effect threshold is 1 g/kg BW. Toxic action manifestations included increased blood Hb concentration, changed motor activity, and reduced ceruloplasmin activity in the blood plasma. **Zn stearate:** Rats tolerate administration of 6 g/kg BW as 50% emulsion in sunflower oil. This dose caused only a reduction in the catalase index. Rats and mice tolerate a 5 g/kg BW dose without any sign of intoxication.[3,5–7]

Repeated Exposure. Accumulation of the *Zn* salts is unlikely to occur due to delayed absorption and rapid excretion from the body. Adult male rabbits were exposed to **lead** (0.2%), **zinc** (0.5%), or to both *Pb* and *Zn* (0.2 and 0.5%), which were given as **acetates** in their drinking water for 2 or 4 weeks. Zn^{2+} did not prevent peripheral neuropathy or metabolic alterations caused by **lead.**[8] Female rats received 250 mg *Zn* **stearate** per kilogram for 7 weeks. The total dose (10.5 g/kg BW) seems to be more than 12 times greater than that consumed over the same period with the diet. No manifestations of toxic action were observed.[9] Retardation of BW gain was noted at the dose levels of 250 and 1000 mg Zn^{2+}/kg BW.

Short-term Toxicity. Mice and rats were dosed by gavage with *Zn* **sulfate** at dose-levels of 0.3, 3, and 30 g/kg BW in their feed for 13 weeks. The maximum dose caused retardation of BW gain, reduction in food and water consumption (mice), erythrocyte count, enzyme activity, cholesterol, and glucose content. Gross pathological examination revealed morphological changes in the GI tract, spleen, and kidneys, and decreased relative weights of the visceral organs. The NOAEL appeared to be 458 mg/kg BW in male mice, 479 mg/kg BW in female mice, and 240 BW mg/kg in rats.[10] In a 4-month study in rats, the LOAEL is reported to be 3.5 mg *ZnO*/kg BW.[11] Female rats received 160 to 640 mg *Zn* **acetate**/kg BW in their drinking water for 3 months. The highest dose caused death of some animals. Some other manifestations of toxicity were also evident. Zn^{2+} content has been found to be the highest in the blood, bones, liver, and kidneys, showing a dose-related effect.[12]

Long-term Toxicity. *Humans.* The major consequence of chronic ingestion of Zn^{2+} for medical purposes is copper deficiency and anemia as a result of the intestinal interaction of **zinc** and **copper.**[13] Long-term administration of 150 mg daily produced the same result,[14] but some opposite results were reported after an equal dose given over a period of 6 weeks to healthy volunteers.[15] *Animals.* Zn^{2+} toxicosis has been observed in various mammalian species, including ferrets, sheep, cattle, horses, and dogs.[16] Signs of toxicosis among 95 calves began to appear when calves were fed approximately 1.2 to 2 g Zn^{2+} per day and exposed to a cumulative Zn^{2+} intake of 42 to 70 g per calf.[17] In a 6-month study, male rats were dosed by gavage with 5 mg $ZnCl_2$/kg BW. In this study carried out in Russia, the treatment is reported to cause changes in the activity of liposomal enzymes in the liver, gonads, and, to a lesser extent, kidneys.[18] Rats were given 25 mg $ZnCl_2$/kg BW for 9 months. Manifestations of toxic action included retardation of BW gain at the beginning of the experiment, a persistent lowering of the STI and of motor activity, and an increase in the blood erythrocyte count. This dose also caused a significant and persistent rise in arterial pressure and changes in ECG. Histological examination revealed foci of vacuolar, large fatty dystrophy in the liver. The NOAEL of 1 mg/kg BW was established in this study.[4] **Zn oxide:** Rats were dosed with 5 mg *ZnO*/kg BW for 6 months. Histological examination revealed hyaline cylinders in the kidneys and moderate hyperplasia in the white pulp of the spleen. A dose of 5 mg/kg BW was considered to be the LOAEL.[4] **Zn stearate:** Rats and mice tolerate administration of 4 and 100 mg/kg without manifestations of toxic effect. Gross pathological examination failed to reveal lesions in the viscera.[07]

Immunotoxicity. An intake of 150 mg Zn^{2+} by healthy males twice a day in food and drink for 1.5 months caused a reduction in lymphocyte and bacterial phagocytic activity.[19] A high Zn^{2+} content in water and food depressed the immune system, whereas *i/v* administration of 1 mg/kg stimulated the immune response.[20] Zn^{2+} slows down histamine release and alleviates allergic effect.

Reproductive Toxicity. *Gonadotoxicity.* Rats received 5 mg $ZnCl_2$/kg BW in their drinking water for 6 months. The treatment caused a 40% reduction in the activity of the lactate dehydrogenase in the seminal fluid and a 43% reduction of pyruvic acid activity. The gonadotoxic

188

effect of Zn^{2+} could be explained by its selective accumulation in the sperm.[21] In a 4- month study, functional and morphological changes in spermatogenesis were observed in rats given 2.5 mg Zn^{2+} or ZnO/kg BW.[22] *Embryotoxicity.* Addition of 0.4% Zn^{2+} to the diet of pregnant animals caused a decrease in the fetal weights and reduced the activity of cytochrome oxidase in the liver. There were no deviations in the levels of *Zn*, *Cu*, *Fe*, *Ca*, and *Mg* in the fetal tissues. Implanting *Zn* before mating and during gestation led to a decrease in the number of implantation sites in rabbits.[23] Embryolethal effect was noted in rats given drinking water containing 10 mg $ZnCl_2$/l during pregnancy.[24]

Genotoxicity. *Humans.* Subtoxic doses of *Zn* salts have been reported to cause severe CA in human lymphocyte culture and may be the cause of the mutagenic effect found among workers in the **zinc** industry experiencing *Ca* deficiency in their food intake.[25] Concentrations of 0.07 to 0.21 mg *Zn* acetate/ml is shown to increase the number of CA in a culture of human leukocytes.[26] However, there are no direct indications that Zn^{2+} is a mutagen in man or in mammals. *Animals.* Mutagenic effect was noted in rats given 5 mg Zn^{2+}/kg BW (6 months) and 100 mg Zn^{2+}/kg BW (1 month).[17] $ZnCl_2$ does not increase the percentage of DLM.

Carcinogenicity. *Humans.* No increased mortality was shown in any type of cancer among **zinc**-refinery workers. *Animals.* **Zinc** compounds are unlikely to be carcinogenic, no matter what route of administration, except by intratesticular injection, was applied.[27] However, according to other data, pulmonary reticulosarcomata, seminomata, and testicular tumors were observed 18 to 24 months after administration of *Zn* metal and *Zn* chloride to rats *i/p*, *i/a*, *s/c*, and into the testis.[28] Tumor localization was not related to the route of administration. Forelimb fibromata have been reported to occur in three out of five rats given $ZnCl_2$ with water at a dose of 5 mg/kg for 16 to 18 months.[29]

Chemobiokinetics. Man absorbs about 20 to 30% of the dietary Zn^{2+}. Absorption occurs throughout the small intestine; the jejunum has the highest rate of absorption.[30,31] According to other data (Engstrom et al.), Zn^{2+} is readily absorbed in the GI tract of newborn and sexually mature rats. In the GI tract, **zinc** compounds have been shown to form carbonates, which are poorly soluble and slowly absorbed. Increased *Zn* intake with food does not considerably raise Zn^{2+} content in the blood because of its relatively rapid excretion in the feces.[32] Zn^{2+} is distributed mainly in the liver, kidneys, bones, retina, prostate, and muscle.[30] It has been shown to be accumulated in the mitochondria of the liver and to affect mitochondrial function. In rats given a daily dose of 1.4 mg $ZnCl_2$, content of Zn^{2+} in the liver increased in the first 17 d but more prolonged administration resulted in normalization of its concentration. A mechanism by which Zn^{2+} absorption is reduced is associated with the exhaustion of metallothionein synthesis in the intestinal mucosa.[33] Erythrocytes contain most of the **zinc** in blood; 10 to 20% is found in plasma.[34] The excretion takes place via GI tract (1 to 2 mg/d) and via urine (about 0.5 mg/d) and by sweating (about 0.5 mg/d).[31] The bulk of Zn^{2+} taken orally is excreted in the feces (90%) and the remainder in the urine (10%). On average, 3 to 4 mg of Zn^{2+} are excreted in 24 h.

Guidelines. WHO (1992). Guideline value in drinking water = 3 mg/l (aesthetic criterion). Taking into account recent studies on humans, this concentration would not be likely to present a hazard to health but could give rise to consumer complaints of taste and appearance of the water. JECFA proposed a provisional maximum TDI of 0.3 to 1 mg/kg (1982).

Standards. USEPA (1991). SMCL: 5 mg/l. **Russia** (1988). MAC (at tap) and PML in drinking water: 5 mg/l.

Regulations. USFDA (1993) approved the use of **Zinc** compounds in (1) resinous and polymeric coatings and (2) as the components of adhesives to be used safely in a food-contact surface. *Zn* oxide may be used (3) as a component of waterproof coatings at a level not to exceed 1% by weight of the dry product. *Zn* stearate may be used (4) as a component of paper and paperboard for contact with dry food, (5) as antioxidants and/or stabilizers in polymers used in the manufacturing of food-contact articles, and (6) in phenolic resins for the food-contact surface of molded articles intended for repeated use in contact with nonacid food (pH above 5). *Zn* carbonate and *Zn* sulfide may be used (7) as fillers for rubber articles intended for repeated use in contact with food and (8) *Zn* hydrosulfide is regulated as a substance employed in the production of or added to textiles and textile fibers intended for use in contact with food. *Zn* salicylate is listed for use (9) in rigid PVC and/or in rigid vinylchloride copolymers, provided that total salicylates (calculated as the acid) do not exceed 0.3% by weight of such polymers and (10) as antioxidants and/or stabilizers in polymers used in the manufacturing of food-contact articles. **Zn carbonate** is listed as (11) a colorant in the manufacturing of melamine-formaldehyde resins in molded articles, in xylene-formaldehyde resins condensed with 4,4′-isopropylidene diphenol-epichlorohydrin epoxy resins, ethylene-vinyl acetate copolymers, and in urea-formaldehyde resins in molded articles. **Zn chromate** and **Zn oxide** are listed as colorants in the manufacturing of rubber articles intended for repeated use; total use is not to exceed 10% by weight of the rubber article.

References:

1. Cohen, J. M., et al., Taste threshold concentrations of metals in drinking water, *J. Am. Water Work Assoc.*, 52, 660, 1960.
2. *Trace Elements in Health and Disease, Vol. 1, Zinc and Copper,* Prasad, A. S., Ed., Academic Press, New York, 1976.
3. Sax, N. J. and Lewis, R. J., Sr., *Dangerous Properties of Industrial Materials,* 7th ed., Van Nostrand Reinhold, New York, 1989.
4. Shumskaya, N. I., Mel'nikova, V. V., Zhilenko, V. N., et al., Hygienic assessment of zinc ions in rubber extracts contacting with food products, *Gig. Sanit.,* 4, 89, 1986 (in Russian).
5. Vorob'yova, R. S., Spiridonova, V. S., and Shabalina, L. P., Hygienic assessment of PVC stabilizers containing heavy metals, *Gig. Sanit.,* 4, 18, 1981 (in Russian).
6. Spiridonova, V. S. and Shabalina, L. P., Experimental study on zinc nitrate and zinc carbonate toxicity, *Gig. Truda Prof. Zabol.,* 3, 50, 1986 (in Russian).
7. See **BUTYL STEARATE,** #3.
8. Hietanen, E., Aitio, A., Koivusaari, U., et al., Tissue concentrations and interaction of zinc with lead toxicity in rabbits, *Toxicology,* 25, 113, 1982.
9. See **BISPHENOL A,** #2, 196.
10. Keizo, M., Masahiro, H., Kunitoshi, M., et al., *J. Pest. Sci.,* 8, 327, 1981.
11. *Proc. Kharkov Medical Institute,* 114, 156, 1974 (in Russian).
12. Llobet, J. M., Domingo, J. L., Colomina, M. F., et al., Subchronic oral toxicity of zinc in rats, *Bull. Environ. Contam. Toxicol.,* 41, 36, 1988.
13. *Modern Nutrition in Health and Disease,* Shils, M. E. and V. R. Young, Eds., Lea and Febinger, Philadelphia, 1988.
14. Festa, M. D., Anderson, H.-L., Dowdy, R. P., et al., Effect of zinc intake on copper excretion and retention in man, *Am. J. Clin. Nutr.,* 41, 285, 1985.
15. Samman, S. and Roberts, D. C. K., The effect of zinc supplements on plasma zinc and copper levels and the reported healthy symptoms in healthy volunteers, *Med. J. Aust.,* 146, 246, 1987.
16. Torrance, A. G. and Fulton, R.., Jr., Zinc-induced hemolytic anemia in a dog, *J. Am. Vet. Med. Assoc.,* 191, 443, 1987.
17. Graham, T. W., Goodger, W. J., Christiansen, V., et al., Economic losses from an episode of zinc toxicosis on a California veal calf operation using a zinc sulphate-supplemented milk replacer, *J. Am. Vet. Med. Assoc.,* 190, 668, 1987.
18. Merkur'yeva, R. V., Koras, G. N., Koganova, Z. I., et al., Biochemical studies on enzyme activity of lysosomes in tissues and biological fluids in the course of oral zinc treatment, *Gig. Sanit.,* 10, 17, 1979 (in Russian).
19. Chandra, R. K., Excessive intake of zinc impairs immune response, *JAMA,* 252, 1443, 1984.
20. *Coll. Problems of Immunology,* Vol. 4, Zdorov'ya, Kiev, 1969, p. 66 (in Russian).
21. Leonov, V. A. and Dubina, T. L., *Zinc in the Body of Man and Animals,* Nauka i Tekhnika, Minsk, 1971, p. 120 (in Russian).
22. Reports All-Union Toxicol. Conf., Abstracts, November 25–27, 1980, Moscow, 1980, p. 119 (in Russian).
23. Zipper, J., Medel, M., and Prager, R., Suppression of fertility by intrauterine copper and zinc in rabbits. A new approach to intrauterine contraception, *Am. J. Obstet. Gynecol.,* 105, 529, 1969.
24. Dinerman, A. A., *Role of Environmental Contaminants in Impaired Embryo Development,* Meditsina, Moscow, 1980, p. 192 (in Russian).
25. De Knudt, G. and Deminatti, M., Chromosome studies in human lymphocytes after *in vitro* exposure to metal salts, *Toxicology,* 10, 67, 1978.

26. Voroshilin, S. I., Plotko, E. G., Fink, T. V., et al., Cytogenic effects of inorganic compounds of tungsten, zinc, cadmium and cobalt on human and animal somatic cells, *Cytol. Genet.*, 12, 241, 1978 (in Russian).

27. *Environmental Carcinogenesis,* Elsevier/North-Holland, Amsterdam, 1979, p. 165.

28. Dvizhkov, P. P., Blastomogenic properties of industrial metals and its compounds, *Arch. Pathol.*, 29, 3, 1967 (in Russian).

29. *Metals. Hygienic Aspects of the Assessment and Improvement of the Environment,* Collected works, Moscow, 1983, p. 66 (in Russian).

30. Prasad, A. S., Clinical and biochemical manifestations of zinc deficiency in human subjects, *J. Am. Coll. Nutr.*, 4, 65, 1985.

31. Lee, H. H., Prasad, A. S., Brewer, C. J., et al., Zinc absorption in the human small intestine, *Am. J. Physiol.*, 256, G87, 1989.

32. See **ACRYLONITRILE,** #1, 277.

33. *The Biological Role of Trace Elements and Their Use in Agriculture and Medicine,* Vol. 1, Collected works, Ivano-Frankovsk, 1978, p. 112 (in Russian).

34. Sted, L. and Cousins, R. J., Kinetics of zinc absorption by luminally and vascularly perfused rat intestine, *Am. J. Physiol.*, 248, 646, 1985.

ZINC 2-BENZIMIDAZOLTHIOLATE (CAS No 3030–80–6)

Properties. Whitish, crystalline powder. Poorly soluble in alcohol.

Applications. Used as a stabilizer in the synthesis of vulcanizates based on natural and synthetic rubber.

Acute Toxicity. LD_{50} is reported to be 540 mg/kg BW in rats and 740 to 860 mg/kg BW in mice.[1] Lethal doses produce severe circulatory impairment in the lungs, liver, and brain.

Repeated Exposure failed to reveal cumulative properties. Rats were administered 1/5 LD_{50} for 1 month. The treatment caused retardation of BW gain, CNS depression, anemia, and a rise in the blood metHb concentration.[1] Histological examination revealed parenchymatous dystrophy of the liver and hyperplasia of the lymphoid follicles in the spleen.

Short-term Toxicity. Rats received 20 mg/kg BW for 4 months. Manifestations of toxic action included impairment of liver function, expressed as a reduction in the content of amine nitrogen in the urine, a rise in serum lipids content, and a reduction in cholinesterase activity.[2]

References:

1. See *N*-**BUTYLIDENEANILINE.**

2. See **PHTHALIC ANHYDRIDE,** #2, 184.

ZINC DIBUTYLDITHIOCARBAMATE (ZD) (CAS No 136–23–2)

Synonym. Butyl ziram.

Properties. White or slightly yellowish powder. Poorly soluble in alcohol.

Applications. Used as a vulcanization accelerator for natural rubber and latex, as a stabilizer for rubber-based adhesive systems, isobutylene-isoprene copolymers and polypropylene, as an antioxidant in the production of butyl rubber. It is used in a number of rubber and food handling.[1]

Acute Toxicity. Administration of 10 g/kg BW dose caused no mortality in mice.

Short-term Toxicity. In a 3-month study, rabbits were given 50 mg/kg BW, rats received 20 mg/kg BW. The treatment did not cause changes (Flinn, 1938). Rat were fed diets containing 100, 500, or 2500 ppm ZD for 17 weeks.[1] Hematological indices, clinical chemistry, and urinalysis were not affected. At 2500 mg/kg dose-level, an increase in the relative weights of the liver and kidneys was observed, and food intake was reduced. There were no histological changes in the tissues examined. The NOAEL of 500 ppm in the diet was identified in this study, providing the intake is between 41 and 47 mg/kg BW.

Long-term Toxicity. Gross pathological examination failed to reveal lesions in the visceral organs of Wistar rats that received 5 mg/kg BW for 2 years.[05] In an 18-month study, mice were initially exposed by oral intubation to a dose-level of 1000 mg/kg BW for 3 weeks and thereafter in the diet at a level of 2600 ppm. No evidence of carcinogenicity was observed.[2]

Carcinogenicity. Negative results have been reported in mice.[09]

Regulations. USFDA (1993) regulates ZD for use (1) in can-end cement, (2) in adhesives used as components of articles intended for use in packaging, transporting, or holding food, (3) as a stabilizer in resinous and polymeric coatings, (4) in coatings for polyolefin films for use only at levels not to exceed 0.2% by weight of isobutyleneisoprene copolymers, provided that the finished copolymers contact food only of the types identified in CFR and at levels not to exceed 0.02% by weight ofthe polypropylene polymers, and (5) in closures with sealing gaskets on food containers (not exceeding 0.8% by weight of the closure-sealing gasket composition and for use only in vulcanized natural or synthetic rubber gasket compositions), and (6) as an accelerator of rubber articles intended for repeated use in contact with food (total not to exceed 1.5% by weight of the rubber product).

References:

1. Gray, T. J. B., Butterworth, K. R., Gaunt, I. F., et al., Short-term toxicity study of zinc dibutyl dithiocarbamate in rats, *Food Chem. Toxicol.,* 16, 237, 1978.
2. See *N*-**CYCLOHEXYL**-2-**BENZOTHIAZOLE SULPHENAMIDE, #7.**

ZINC DIETHYLDITHIOCARBAMATE (CAS No 14324–55–1)

Synonyms. Carbamate *EZ;* Ethyl zimate; Ethyl ziram.

Properties. White powder. Poorly soluble in water and alcohol.

Applications and **Exposure.** Used as a vulcanization accelerator in the production of rubber and latices, as a stabilizer for butyl rubber, and for butadiene and urethane rubbers. Migration of dithiocarbamates from smoked sheet NR vulcanizates into water after 3-d exposure at 20°C was determined to be 0.6 mg/l.[1]

Acute Toxicity. In mice, LD_{50} is 0.7 g/kg BW.

Short-term Toxicity. Rabbits received 450 mg/kg BW for 3.5 months (daily during the first month, three times a week thereafter). The treatment affected metabolic processes and liver function, balance of protein fractions in the blood serum, and hematological analyses.[2]

Reproductive Toxicity. No teratogenic effect or postnatal behavioral changes were observed in rats gavaged with up to 250 mg/kg BW on days 7 to 15 of gestation.[3]

Carcinogenicity. Negative results have been reported in mice.[09]

Standards. Russia. PML in food: 0.03 mg/l.

Regulations. USFDA (1993) approved the use of *Zn* D. (1) as a component of adhesives for articles coming into contact with food and (2) as an accelerator in rubber articles intended for repeated use in producing, manufacturing, packing, processing, treating, packaging, transporting, or holding food (total not to exceed 1.5% by weight of the rubber product).

References:

1. *Handbook on the Hygiene of Polymer Use,* K. I. Stankevich, Ed., Zdorov'ya, Kiev, 1984, p. 106, (in Russian).
2. See α-**METHYLBENZYLPHENOLS,** mixture, #1, 78.
3. Nakaura, S., Tanaka, S., Kawashima, K., et al., Effects of zinc diethyldithiocarbamate on the prenatal and postnatal development of the rat, *Bull. Natl. Inst. Hyg. Sci.,* 102, 55, 1984.

ZINC DIISOPROPYLDITHIOCARBAMATE

Synonym. Zinc *N,N'*-propylene-*1,2*-bisdithiocarbamate.

Properties. White powder. Poorly soluble in water and alcohol.

Applications. Used as a vulcanization accelerator in rubber production.

Acute Toxicity. In mice, LD_{50} is 4.25 g/kg BW. Manifestations of toxic action include apathy, adynamia, and unkempt fur.

Short-term Toxicity. There were no deviations in hydrocarbon and fat metabolism or liver function of the rabbits given 20 mg/kg BW for a month. Histological examination failed to reveal dystrophic changes in the liver and kidneys.

Reference:

See **PHTHALIC ANHYDRIDE,** #2, 127.

ZINC PHENYLETHYLDITHIOCARBAMATE (CAS No 14634–93–6)

Synonyms. Accelerator *P* extra *N*; Bis(*N*-ethyldithiocarbamato) zinc; Carbamate *EFZ*.

Properties. White, odorless powder. Poorly soluble in water and alcohol.

Applications and **Exposure.** Used as a vulcanizing agent (ultra-accelerator) in the production of vulcanizates. Migration from butadiene-nitrile vulcanizates into water after 3-d exposure at 20°C was determined at the level of 0.08 mg/l.[1]

Acute Toxicity. In mice, LD_{50} is reported to be 17 g/kg BW. Three administrations of 2 g/kg BW altered activity of some enzymes. Sensitivity to alcohol was increased.[2]

Short-term Toxicity. Rats were dosed by gavage with 500 mg/kg BW for 1.5 months. Manifestations of toxic action comprised severe impairment of hematological indices and of a number of biochemical parameters in rats. Gross pathological examination revealed congestion in the visceral organs, acute enlargement of the spleen, and distension of the stomach and intestine. In a 31-month study, 10 and 100 mg/kg BW doses were administered to rats. The treatment resulted in retardation of BW gain, anemia, and an increase in serum ceruloplasmin activity. The high dose resulted in increased kidney and spleen weights.[2]

Genotoxicity. Was shown to be mutagenic in the *Salmonella* bioassay.[3]

Standards. Russia. PML in food: 1.0 mg/l.

References:

1. *Hygiene of Use and Toxicology of Pesticides and Polymeric Materials,* Vol. 17, Coll. Sci. Proc. VNIIGINTOX, Kiev, 1987, p. 153 (in Russian).
2. Yalkut, S. I., On the toxicology of vulkacit *P* extra *N,* a rubber vulcanization accelerator, *Hyg. Sanit.,* 6, 31, 1971 (in Russian).
3. International Symposium on Prevention of Occupational Cancer, Finnish Institute of Occupational Health, Helsinki, 1982.

Chapter 4

CATALYSTS, INITIATORS, HARDENERS, CURING, AND CROSS-LINKING AGENTS

ALUMINUM COMPOUNDS

Synonyms. *Al* hydrate; *Al* trihydrate; Alkagel; Alternagelis; Antacid; Antidiar; Creamalin; Fluagel; Hydrolum; Vanogel.

Properties. *Al* **trialkyl:** Colorless liquids at room temperature. Alkylaluminum halides are colorless, volatile liquids. Bond cleaved by water and alcohol. *Al* **oxide** is in the form of colorless crystals. The solubility in water is 0.098 mg% (18 to 37°C). *Al* **hydroxide:** A white powder, poorly soluble in water. *Al* **chloride.** White when a pure substance. Ordinarily gray or yellow to greenish. Combines with water with explosive violence and liberation of much heat. Freely soluble in many organic solvents (Merk). Gives water a bitter, astringent taste, with perception threshold at a concentration of 0.1 mg/l; odor perception threshold is 0.5 mg/l (calculating on the Al^{3+} ion). Odor perception threshold for *Al* **nitrate** appears to be the concentration of 0.1 mg/l. The turbidity index threshold is 0.5 mg/l.

Applications. *Al* **chloride** and **nitrate, trialkyl aluminum,** etc., are used as catalysts in the production of polymeric materials (Ziegler-Natta polymerization catalysts, etc.). *Al* **hydroxide** is used as an emulsifier, ion-exchanger, detergent, in waterproofing fabrics, paper, and lubricating compositions. It is a fire-retardant for PE, PVC, polyester, and latex compositions; an active filler. *Al* **oxide** is used as a filler and catalyst. *Al* **chloride** is used as an acid catalyst in the manufacturing of rubber and as a lubricant.

Exposure. *Al* is found in drinking water at concentrations of up to 2.5 mg/l. Concentrations and types of *Al* complexes may vary with the pH of water. Exposure to *Al* compounds comprises their use as food containers and packaging constituents, pharmaceutical preparations, food additives, and coagulants in drinking water. *Al* is present in treated drinking water in the form of low molecular weight reactive species.

Acute Toxicity. Sensitivity to different *Al* compounds in acute experiments in rats, guinea pigs, and rabbits have been shown to be almost equal. Korolev and Krasovsky reported LD_{50} values for certain *Al* salts.

LD_{50} Values (g/kg BW) for Al Salts

	Al chloride		Al nitrate		Al sulfate	
	Salt	Al	Salt	Al	Salt	Al
Rats	0.32	0.18	0.28	0.21	0.41	0.32
Mice	0.39	0.20	0.37	0.19	0.82	0.41
Guinea pigs					0.49	0.40

From Korolev, A. A. and Krasovsky, G. N., *Gig. Sanit.*, 4, 12, 1978 (in Russian), with permission.

There are other data reported:[2] LD_{50} of $Al(NO_3)_3 \cdot 9H_2O$ is found to be 3.6 g/kg BW in rats and 3.98 g/kg BW in mice; of $AlCl_3 \cdot 6H_2O$ is 3.3 g/kg and 1.99 g/kg BW, respectively; LD_{50} of $Al(SO_4)_3 \cdot 18H_2O$ is identified to be more than 9 g/kg BW in rats and mice. Death occurred in 1 to 4 d. Lethal doses of *Al* salts exhibit a broad spectrum of toxic action, affecting the CNS, blood, liver, kidneys, and gonads. *Al* compounds are shown to cause porphyria in rats (but not in man). Alkaline phosphatase levels in rats decreased within 3 h following oral administration of 17 mg $AlCl_3$/kg.[3]

Repeated Exposure. *Al* salts have not been found to have evident cumulative properties. Exposure to the doses of 200 to 600 mg/kg BW for 28 d shortens the time rats can remain on a revolving column and increases their activity in open space.[4] Rats fed a diet containing 2665 mg *Al* **sulfate** per kilogram (133 mg/kg BW) exhibited a negative phosphorus balance.[5] Male Sprague-Dawley rats were fed diets containing **basic sodium *Al* phosphates** and *Al* **hydroxide** (mean doses of 67 to 302 mg/kg BW) for 28 d. The treatment caused no changes in BW gain, hematology, and clinical chemistry. Accumulation of *Al* in the bone was not found.[6] The U.S. National Academy of Sciences[7] developed a suggested NOAEL for 1- and 7-d exposure to 35 and 5 mg/l, respectively. These values were based on an 18-d study in rats receiving an oral dose of 1 g/kg BW. After 18 d, decreased liver glycogen and coenzyme A levels were reported.[8] Histological changes in the liver and spleen and elevated *Al* levels were noted in the heart and spleen of rats that received drinking water containing *Al* **nitrate** (calculated doses of 54 and 108 mg/kg BW). The NOAEL of 27 mg/kg BW was identified in this study.[9] In a 1-month study, rats were dosed with Al^{3+} via drinking water. The treatment produced changes in serum alkaline phosphatase activity, content of adenine nucleotides in the blood, and functional state of the spermatozoa. The LOAEL was established to be 17 mg/kg BW for rats and mice and 9 mg/kg for rabbits.[10]

Long-term Toxicity. *Humans.* Average dietary intake of *Al* compounds is 9 to 36 mg/d. Individuals consuming *Al*-containing antacid may intake up to 3 g/d. For many years, *Al* was not considered to be toxic in humans. However, recent reports suggest that exposure to *Al* may pose a health hazard. *Al* has been associated with dialysis osteomalacia and dialysis encephalopathy in persons with chronic renal failure who have undergone long-term hemodialysis. Its role in other neurological disorders remains controversial (Alzheimer's disease, amyotropic lateral sclerosis, and Parkinsonism dementia in Guam). It is safe to use *Al* ware for the preparation of food. *Al* intake is thought to reach 80 mg/d. *Animals.* Available data are controversial. In a 7-month rat study, signs of general toxicity were reported only at a dose-level of 5 mg/kg BW, but not at lower doses. An impairment of conditioned reflex activity and changes in behavior and in serum alkaline phosphatase activity were observed in rats given the dose of 2.5 mg/kg BW.[3,10] Rats exposed to *Al* salt at a concentration of 5 mg/l in their drinking water had no deviations in clinical and biochemical indices.[11] The treatment did not affect their life-span.

Allergenic Effects. Addition of Al^{3+} to the feed led to the development of allergic asthma in laboratory animals. *Al* **stearate** has not been reported to exhibit such effect.[5]

Reproductive Toxicity. *Humans.* The health of workers consuming daily (in food and beer) 300 to 400 mg *Al* (calculated as elementary *Al*) was monitored for 3 years. No signs of harmful effect in their health or (in women) in relation to sexual function were noted.[12] *Animals.* **Embryotoxicity.** *Al* exposure has been associated with neurotoxicity in mouse dams and their offspring. Golub et al.[13] reported weight loss and neurotoxicity on days 12 to 15 postpartum in dams fed 0.5 g or 1 g *Al* **lactate**/kg (75 mg/kg or 150 mg/kg BW). Offspring of rats given oral doses of 155 mg or 192 mg Al/kg BW from day 8 of gestation through parturition experienced a delay in weight and neuromotor development.[14] Pregnant Swiss mice were given by gavage 30 mg *Al* **hydrate**/kg on gestation days 6 to 15. No embryotoxic or fetotoxic effects, no malformations or variations related to the treatment were observed.[15] **Gonadotoxicity.** In a 6-month study, rats were exposed to 2.5 mg $AlCl_3$/kg BW dose. A gonadotoxic effect was found in these animals. Meanwhile, consumption of Al^{3+} in drinking water at concentrations of 0.5 to 2 g/l for 90 d did not affect the reproductive function of male rats.[16,17] **Teratogenic** effects in rats were caused by *Al* **fluoride** and *Al* **nitrate** but not by *Al* **lactate** and *Al* **trichloride**.[18]

Genotoxicity. *Al* compounds have been shown to produce CA and alter cell division in animals and humans. The dose of 0.025 mg/kg BW was reported to be ineffective in the DLM test.[15]

Carcinogenicity. A concentration of 5 mg/l in drinking water presents no risk of tumor development in rats.[11]

Chemobiokinetics. Metabolism of *Al* in humans is not well understood, but it seems to be poorly absorbed (less than 1%). Ingestion by volunteers of up to 125 mg/d did not reveal its accumulation in the body. Absorption of *Al* occurs predominantly in the stomach and duodenum, in an acidic environment. The normal distribution of *Al* is the greatest in the lungs, followed by bone, and, to a lesser extent, soft tissues and brain.[19] It was also found in the liver, muscles, and testes.[20] Probably, *Al* salts could exhibit their toxic effect due to some diminution in phosphorus absorption and a speeding-up of phosphorylation reactions. *Al* is specifically absorbed by hepatocyte nuclei. The DNA of the nuclei have a specific affinity for *Al*. Most of the absorbed *Al* appears to be rapidly excreted with the urea (<0.02 mg/d).

Guidelines. WHO (1992): Guideline value for drinking water = 0.2 mg/l. (This level is likely to give rise to consumer complaints of depositions, discoloration.) **JECFA** (1988) established PTWI of 7 mg/kg based on studies of aluminum phosphate (acidic). The chemical form of **aluminum** in drinking water is different.

Standards. USEPA (1991). Secondary MCL: 0.05 to 0.2 mg/l. **Russia** (1988). MAC and PML in food and water: 0.5 mg/l.

Regulations. USFDA (1993) approved the use of *Al* salts (1) in resinous and polymeric coatings for the food-contact surface of articles intended for use in producing, manufacturing, packing, transporting, or holding food (*Al* **butyrate**), (2) as components of adhesives applied in contact with foodstuffs, (3) in cellophane to be used safely for packaging food (*Al* **hydroxide**), (4) as a defoaming agent in acrylate ester copolymer coating to be used safely as a food-contact surface of articles intended for packaging and holding food, including heating of prepared food (*Al* **hydroxide** and **stearate**), (5) as colorants in the manufacture of articles intended for use in contact with food (*Al*, *Al* **hydrate**, *Al* **mono-, di-,** and **tristearate**, *Al* **silicate** (**mica** and **China clay**), and (6) as a component of the uncoated or coated food-contact surface of paper and paperboard intended for use in producing, manufacturing, packing, transporting, or holding aqueous and fatty food (*Al* **acetate**).

References:

1. Korolev, A. A. and Krasovsky, G. N., Hygienic substantiation of MAC for a new reagent aluminum oxychloride in drinking water, *Gig. Sanit.,* 4, 12, 1978 (in Russian).
2. Llobet, J. M., Domingo, J. L., Gomez, M., et al., Acute toxicity studies of aluminum compounds: antidotal efficacy of several chelating agents, *Toxicol. Appl. Pharmacol.,* 60, 280, 1987.
3. Krasovsky, G. N., Vasyukovich, L. Y., Charyev, O., Experimental study of the biological effects of lead and aluminum following oral administration, *Environ. Health Perspect.,* 30, 47, 1979.
4. Synthesis and Study of Pharmacologically Active Substances, Republ. Sci. Conf., Abstracts, Vil'nus, December 14, 1984, 12 (in Russian).
5. Ondreicka, R., Ginter, E., and Kortus, J., Chronic toxicity of aluminum in rats and mice and its effects on phosphorus metabolism, *Br. J. Ind. Med.,* 23, 305, 1966.
6. Hicks, J. S., Haskett, D. S., and Sprague, G. L., Toxicity and aluminum concentration in bone following dietary administration of two sodium aluminum phosphate formulations in rats, *Food Chem. Toxicol.,* 25, 533, 1987.
7. U.S. National Academy of Science, *Drinking Water and Health,* Vol. 4, National Academy Press, Washington D.C., 1982, p. 155.
8. Kortus, J., The carbohydrate metabolism accompanying intoxication by aluminum salts in rat, *Experientia,* 23, 912, 1967.
9. Gomez, M. et al., Short-term oral toxicity study of aluminum in rats, *Arch. Farmacol. Toxicol.,* 12, 144, 1986 (in Spanish).
10. Materials of the 2nd Joint Soviet-American Symp. Environ. Health Protection, Moscow, 1977, p. 34 (in Russian).
11. Elliott, H. L., Dryborgh, F., Fell, G. S., et al., Aluminum toxicity during regular haemodialysis, *Br. Med. J.,* 1, 1101, 1978.
12. Evenstein, Z. M., Allowable content of aluminum in the food intake of adults, *Gig. Sanit.,* 2, 109, 1971 (in Russian).
13. Golub, M., Gershwin, M., Donald, J., et al., Maternal and developmental toxicity of chronic aluminum exposure in mice, *Fundam. Appl. Toxicol.,* 8, 346, 1987.
14. Bernuzzi, V., Desor, D., and Lehr, P. R., Effects of prenatal aluminum exposure on neuromotor maturation in the rat, *Neurobehavior. Toxicol. Teratol.,* 8, 115, 1986.
15. Colomina, M. T., Gomez, M., Domingo, J. L., and Corbella, J., Lack of maternal and developmental toxicity in mice given high doses of aluminum hydroxide and ascorbic acid during gestation, *Pharm. Toxicol.,* 1994, 74, 236.
16. Materials of the 2nd Soviet-American Symp. on Environ. Health Protection, Moscow, 1977, p. 41 (in Russian).
17. Dickson, R. L., Sherins, R. L., and Lee, I. P., Assessment of environmental factors affecting male fertility, *Environ. Health Perspect.,* 30, 53, 1979.
18. See **BISPHENOL A,** #5, 843.
19. Ganrot, P. O., Metabolism and possible health effects of aluminum, *Environ. Health Perspect.,* 65, 363, 1986.
20. Talakina, N. V., Novikov, Yu, V., Plitman, S. I., et al., Setting norms of the aluminum in drinking water of various hardness, *Gig. Sanit.,* 12, 75, 1988 (in Russian).

ANTIMONY COMPOUNDS

Properties. Sb_2O_3 is in the form of colorless crystals. Solubility of Sb_2O_3 in water is 1.6 mg per 100 g (15°C). Taste perception threshold for Sb^{3+} and Sb^{5+} **compounds** is 0.6 mg/l.

Applications and **Exposure**. *Sb* compounds are used as flame retardants for plastics and elastomers, as regeneration activators for vulcanizates, and as pigments. *Sb* **trioxide** is used as a polycondensation catalyst in **Lavsan** (Polyethylene terephthalate) production and as a filler for increasing the fire-resistance of vulcanizates. *Sb* **sulfate** could be applied as a vulcanization accelerator and a dye for vulcanizates. Estimated dietary intake for adults is about 0.018 mg/d.[1] Concentrations found in drinking water were less than 0.005 mg/l.[2]

Acute Toxicity. *Humans*. Doses of 0.1 to 1 g of *Sb* compounds are lethal to man.[3] Acute intoxication is characterized by abdominal and muscular pains, diarrhea, and hemoglobinuria. A poisoning has been described of 250 children who drank lemonade contaminated with *Sb* and suffered from vomiting and diarrhea. Sb^{3+} is more toxic than Sb^{5+}. *Animals*. Exposure to *Sb* compounds affects the CNS with nausea and vomiting development. Gross pathological examination revealed mucosal irritation and inflammation, particularly in the distal region of the large intestine. Doses of 115 to 120 mg/kg BW of **tartar emetic** (calculated as Sb^{3+}) are lethal to rabbits. Doses of 450 to 600 mg BW cause vomiting in cats. In mice, LD_{50} for *Sb* **trioxide** (calculated as Sb^{3+}) is found to be 209 mg/kg BW; that of Sb_2O_5 is 978 mg/kg BW.[4] In dogs, 33 mg *Sb* **tartrate**/kg BW causes emesis. Cats appear to be more sensitive to the emetic effect.[5]

Repeated Exposure. BW gain is decreased in rats on daily administration of 20 mg **tartar emetic**/kg BW—the main potassium antimonate salt of tartaric acid being $Sb[KOOC(CHOH)_2COO]$ $SbO \cdot H_2O$. In a 1-month study, the toxic effect was evident in rabbits, given Sb_2O_3 at a dose of 150 mg/kg BW.[6] Four rabbits were dosed with *Sb* **tartrate** at a dose-level of 5.6 mg *Sb*/kg BW for 7 to 22 d. The treatment slightly increased nonprotein nitrogen in the blood and urine. Gross and histological examination revealed hemorrhagic lesions in the kidney cortex, stomach, and small intestine, liver atrophy with fat accumulation and congestion. This dose is suggested as a LOAEL.[7]

Short-term Toxicity. In a 91-d study, rats were exposed to two *Sb*-containing pigments with oral daily doses of 36 and 22 mg/kg BW, respectively. Adverse effects were not observed.[8] No manifestations of toxic action were noted in rats given the doses of 0.1 to 4 mg *Sb* compounds/kg BW with food over a period of 3.5 months.[5]

Long-term Toxicity. In a life-time study in which rats were exposed to *Sb* **potassium tartrates** (the most toxic of the common antimony compounds) via drinking water at a dose of 0.43 mg/kg/d, effects observed were decreased longevity and altered blood levels of glucose and cholesterol (LOAEL). ADI/TDI appeared to be 0.00043 mg/kg BW.[9] However, according to Elinder and Frieberg, even consumption of 100 mg Sb^{3+} or Sb^{5+} oxides/kg BW by dogs and cats for several months produced no toxic effect in the animals.[10]

Reproductive Toxicity. *Embryotoxicity*. Inhalation exposure to 0.27 mg/m³ caused a pronounced embryotoxic effect in rats. Concentration of 0.082 mg/m³ produced maternal toxicity.[11] The treatment with **tartar emetic** reduced litter viability and the overall life-span in rats by 15% compared to the controls.[9] *Teratogenicity*. No developmental effects were observed in ewes' offspring whose mothers were given *Sb* **potassium tartrate** at a dose of 2 mg/kg BW for 45 d or throughout gestation.[12] Five *i/m* injections of Sb^{5+} to Wistar rats during gestation caused no abnormalities in fetuses.[13] Belyaeva found no teratogenic effect of *Sb* **trioxide** in rats.[14] *Gonadotoxicity*. *Sb* compounds are shown to impair the generative function and disturb the sexual cycle of females, although general condition remains satisfactory. Ability to conceive and fertility are lowered. In males, signs of impaired spermatogenesis are noted and fertilizing ability is reduced.[4] Inhalation exposure to *Sb* **trioxide** over 2 months resulted in sterility and fewer offspring in rats.[14]

Genotoxicity. **Potassium** and **sodium antimony tartrates** induce CA in cultured human leukocytes.[15]

Carcinogenicity. An increase in the incidence of benign and malignant tumors was not observed after life-time exposure to an average daily dose of 0.43 mg *Sb* **potassium tartrate**/kg BW via drinking water in Charles River CD mice.[9] *Carcinogenicity classification*. IARC: Group 2B (inhalation exposure to *Sb* **trioxide**) and Group 3 for *Sb* **trisulfide**.

Chemobiokinetics. Following ingestion, *Sb* compounds are absorbed slowly from the intestine of experimental animals. The highest tissue levels (approximately 50%) were detected in the intestine and bones.[16] However, according to Edel et al., *Sb* compounds are found to be accumulated in the thyroid, liver, kidneys, spleen, and in the membrane structure of erythrocytes.[17] *Sb* compounds do not bind to blood proteins. Their transfer may occur from maternal to fetal blood.[18] *Sb* compounds are excreted in the urine with only low levels in the feces.

Guidelines. **WHO** (1992). Provisional guideline value for drinking water: 0.005 mg/l.

Standards. **USEPA** (1992). MCL and MCLG: 0.006 mg/l. **Czechoslovakia.** Permissible levels: 0.3 mg/kg in foods, 0.05 mg/l in drinks. **Russia** (1988). MAC: 0.05 mg/l in water and milk, 0.3 mg/kg in meat and bread products, 0.5 mg/kg in fish.

Regulations. **USFDA** (1993) approved the use of *Sb* oxide in adhesives as a component of articles intended for use in packaging, transporting, or holding food.

References:
1. Toxicological Profile for Antimony and Its Compounds, Draft, U.S. Dept. Health and Human Service, ATSDR, Atlanta, GA, 1990.
2. Revision of WHO Guidelines for Drinking Water Quality, Report of the First Review Group Meeting on Inorganics, Bilthoven, March 18–22, 1991, p. 2; Medmenham, U.K., January 27–29, 1992.
3. See **ACRYLONITRILE,** #7, 130.
4. Gudzovsky, G. A., *Antimony and Its Compounds,* Science Review, International Project Centre GKNT, Moscow, 1984, p. 28 (in Russian).
5. Flury, F., Zur Toxicologie des Antimons, *Arch. Exp. Pathol. Pharmacol.,* 126, 87, 1927 (in German).
6. Results Scientific Studies, Sysin Institute Environ. Health, Acad. Med. Sci. USSR, Moscow, 1974, p. 20 (in Russian).
7. Pribyl, E., On the nitrogen metabolism in experimental subacute arsenic and antimony poisoning, *J. Biol. Chem.,* 74, 775, 1927.
8. Bomhard, E., Loser, E., Dornemann. A., et al., Subchronic oral toxicity and analytical studies on nickel rutile yellow and chrome rutile yellow with rats, *Toxicol. Lett.,* 14, 189, 1984.
9. Schroeder, H. A., Mitchener, M., and Nason, A. P., Zirconium, niobium, antimony, vanadium, and lead in rats: life-term studies, *J. Nutr.,* 100, 59, 1970.
10. *Handbook on Toxicology of Metals,* Amsterdam e. a., 1980, p. 283.
11. Grin', N. V., Govorunova, N. N., Bessmertny, A. N., et al., Experimental study of embryotoxic effect of antimony oxide, *Gig. Sanit.,* 10, 85, 1987 (in Russian).
12. James, L. F., Lazar, V. A., and Binns, W., Effect of sublethal doses of certain minerals on pregnant ewes and fetal development, *Am. J. Vet. Res.,* 27, 132, 1966.
13. Casals, J. B., Pharmacokinetic and toxicological studies of antimony dextran glycoside (RL-712), *Br. J. Pharmacol.,* 46, 281, 1972.
14. Belyayeva, A. P., The effect of antimony on reproductive function, *Gig. Truda Prof. Zabol.,* 11, 32, 1967.
15. Paton, G. R. and Allison, A. C., Chromosome damage in human cell cultures induced by metal salts, *Mutat. Res.,* 16, 332, 1972.
16. Gerber, G. B., Maes, J., and Eykens, B., Transfer of antimony and arsenic to the developing organism, *Arch. Toxicol.,* 49, 159, 1982.
17. Heavy Metals in the Environment, Int. Conf., Heidelberg, September 1983, p. 180.
18. Leffler, P. and Nordstroem, S., Metals in Maternal and Fetal Blood, Governmental reports, announcements, and indexes, Issue 9 (in Swedish).

CHROMIUM COMPOUNDS

Properties. In general, Cr^{6+} salts are more soluble than those of Cr^{3+}, making Cr^{6+} relatively mobile. Cr^{3+} is shown to oxidize to Cr^{6+} in drinking water systems in presence of chlorine at concentrations similar to those used to disinfect drinking water. *Cr* **chloride, nitrate,** and **sulfate** are readily soluble in water; the **chromates** and **dichromates** are also soluble. The threshold concentration of Cr^{3+} for tint is 1 mg/l, and for taste, 4 mg/l. *Cr* **hydroxide** at a concentration of 0.5 mg/l does not noticeably increase the turbidity of water; 1 mg/l reduces transparency to 15 cm in terms of Snellin scale.

Applications and **Exposure.** *Cr* compounds are used in the manufacturing of catalysts and stabilizers. Concentrations in drinking water are usually less than 0.005 mg/l of total *Cr,* although concentrations of 0.06 to 0.12 mg/l also were reported. In general, food appears to be the major source of intake, but in some cases drinking water may provide about 70% of the total *Cr* intake.

Acute Toxicity. *Humans.* Ingestion of 1 to 5 g of "chromate" (not further specified) results in severe acute effects such as GI tract impairment, hemorrhagic diathesis, and convulsions. Death may occur with a clinical picture of cardiovascular shock.[1] *Animals.* In rats, LD_{50} varies from 50 to 250 mg/kg BW for Cr^{6+}, and from 185 to 615 mg/kg BW for Cr^{3+}, based on tests with **(di)chromates** and **chromic** compounds, respectively.[1] In mice, LD_{50} of **aluminum-chromium-potassium** catalyst (type A-30) has been reported to be about 450 mg/kg BW (calculated as Cr_2O_3); in rats, it is 792 mg/kg BW. In the case of a **ferrochrome** catalyst, the LD_{50} is 695 mg/kg; in the case of a **zinc-chromium** catalyst (36% **chromium oxide,** 63% **zinc oxide**), LD_{50} is 2060 mg/kg in rats and 1110 mg/kg BW in mice (Shustova and Samoilovich, 1971).

Repeated Exposure. According to Turebayev et al.,[2] *Cr* compounds may accumulate in the reticuloendothelial system and affect hemopoiesis. A decline in diuresis with water consumption remaining the same is noted in Wistar rats given drinking water with Cr^{6+} concentration of 700 mg/l for 4 weeks. The treatment did not alter concentration of glucose, ketones, nitrites, and bilirubin in the urine. Proteinuria developed, and motor activity was lowered. No changes in diuresis and motor activity were observed with a concentration of 70 mg/l.[3] Mice were given $K_2Cr_2O_3$ in their drinking water at the concentration of 10 mg/l (calculated as *Cr*) for 30 d. The exposure resulted in altered phagocyte activity and reduced bactericidal capacity of blood; a 1 mg/l concentration exerted no harmful effect.[4]

Short-term Toxicity. Rats were given nonhydrated **chromic oxide pigment** in their feed over a period of 90 d. The only effect observed was a dose-related decrease (no statistics applied) in the liver and spleen weights at a dose equivalent to 480 and 1210 mg Cr^{3+}/kg BW.[5]

Long-term Toxicity. Rabbits were dosed by gavage with 50 mg Cr^{3+}/kg BW. The treatment did not affect general condition, BW gain, and hematological indices.[6] A brief rise in blood sugar level was noted during the third month. Gross pathological examination failed to reveal considerable lesions to the visceral organs. Doses of 1.9 to 5.5 mg/kg BW of **mono-** and **dichromates** given for 29 to 685 d caused no signs of intoxication in dogs, cats, and rabbits (IARC 23–300). A cardiotoxic and gerontological effect of *Cr* compounds is reported.[7] The latter is characterized by the change in the metabolic reactions of glucose conjugates and by morphological and functional changes in older animals. In a 1-year study, Sprague-Dawley rats were given Cr^{6+} and Cr^{3+} compounds at the concentration of 25 mg/l (2.5 mg/kg BW) in their drinking water. The only effect observed was some accumulation of *Cr* in different tissues.[8] From this study, the NOEL of 2.5 mg/kg BW and TDI of 0.005 mg/kg BW were identified (considering an uncertainty factor of 500, because of the limitation of the study). But the results of a life-time study[5] suggest that the NOEL for Cr^{3+} may be considerably higher.

Allergenic Effect. Such reactions as asthma and contact dermatitis are reported to be induced by certain *Cr* compounds.

Reproductive Toxicity. *Embryotoxicity.* A single administration of Cr^{6+} to female rats at critical periods of embryonic development had a pronounced embryotoxic (but not teratogenic) effect.[9] However, survival and growth of NMRI mice was not affected in a three-generation study where animals were given drinking water containing 135 mg Cr^{6+}/l.[11] Dinerman[12] believes that Cr^{6+} at the MAC level has no effect on embryonic development. A dose of 0.008 mg/kg BW leads to reduction in the number of live neonates and to an increase in the overall pre- and postimplantation mortality. The LOEL is reported to be 0.25 mg Cr^{3+}/kg and 0.025 mg Cr^{6+}/kg BW for embryotoxic effect and 0.25 mg Cr^{6+} and Cr^{3+}/kg BW for gonadotoxicity.[12] *Teratogenicity.* Administration of *Cr* compounds to pregnant rats on day 3 of gestation causes spontaneous changes in the neuroglia and neural tube defects in all embryos. A teratogenic effect with Cr^{3+} is obtained only at high doses, it being unclear whether this is the result of an effect on the fetus or on the body of the mother. Embryotoxic and teratogenic effect was observed in purebred and mongrel hamsters exposed to 8 mg/kg dose on day 8 of gestation.[13] Addition of 5% insoluble, nonhydrated **chromic oxide pigment** to their feed did not result in embryo- and fetotoxicity or teratogenicity.[5] Teratogenic effects caused by *Cr* **chloride,** *Cr* **trioxide,** and *Cr* **potassium dichromate** were observed in rats.[13,14] *Gonadotoxicity.* Conception rate in rats appeared to be reduced following inhalation or food-intake of Cr^{6+} or Cr^{3+} compounds for 2 months.[15]

Genotoxicity. Cr^{6+} compounds are positive in a wide range of *in vitro* and *in vivo* genotoxicity tests, whereas Cr^{3+} compounds are not. The explanation is that Cr^{6+} readily penetrates cell-membranes in contrast to Cr^{3+}. The threshold for DLM effect is identified at the 0.25 mg/kg dose-level for Cr^{6+} and Cr^{3+}.[10]

Cr^{6+}. **Humans**. People occupationally exposed to Cr^{6+} compounds had elevated incidences of CA and SCE in their peripheral blood lymphocytes; reports on SCE induction were conflicting (IARC 23–205). In human cells *in vitro*, these compounds caused CA, SCE, and DNA damage. **Animals**. Cr^{6+} induced DLM, CA, and micronuclei in rodents treated *in vivo*. In cultured rodent cells, it induced transformation, CA, SCE, mutation, and DNA damage (IARC 2–100, IARC 23–205).

Cr^{3+}. **Humans**. No data are available on the genetic and related effects in humans. Conflicting results were obtained for the induction of CA in human lymphocytes *in vitro*, and neither SCE nor unscheduled DNA synthesis were induced in human cells *in vitro*. **Animals**. Mutagenic potential of Cr^{3+} is 100 times lower than that of Cr^{6+}. Cr^{3+} did not induce micronuclei in bone marrow cells of mice treated *in vivo*. Conflicting results were obtained concerning the induction of CA, mutation, and SCE in cultured rodent cells. Insoluble crystalline **Cr oxide** induced SCE and mutation in cultured Chinese hamster cells, which were shown to contain particles of the test material (IARC 2–100; IARC 23–205).

Carcinogenicity. Cr^{6+} was reported to be carcinogenic via inhalation. The limited data available do not provide evidence for carcinogenicity via the oral route. **Humans**. In epidemiological studies, an association was demonstrated between occupational exposure to Cr^{6+} compounds and mortality due to lung cancer, especially in the chromate-producing industry. Data on lung cancer risk in other chromium-associated occupational settings and for cancer at other sites than the lungs seem to be insufficient.[1] There is also some evidence of an increased hazard of GI tract cancer. **Animals**. In a rat study with the oral route of exposure to 2 to 5% insoluble, nonhydrated **Cr oxide** in the feed, no increase in tumor incidence was observed.[5] There is limited evidence for the carcinogenicity of **Co-Cr alloy** in rats [IARC, 1980]. **Carcinogenicity classification**. IARC: Group 1 for Cr^{6+}, Group 3 for Cr^{3+}; USEPA: Group D for **Cr total**.

Chemobiokinetics. The absorption after oral exposure is relatively low and depends on the speciation. Cr^{6+} is more readily absorbed from the GI tract than Cr^{3+} and is capable of penetrating the cellular membranes. After entering the blood system, Cr^{6+} is likely to be concentrated selectively in the liver (the principal sites of **Cr** accumulation) and erythrocytes, where it undergoes metabolic inactivation. Other sites of **Cr** accumulation seem to be the kidneys and bones. Cr^{3+} is a more stable oxidation state, and under physiological conditions, it may form complexes with ligands such as nucleic acids, proteins, and organic acids. Biological membranes are thought to be impermeable to Cr^{3+}, although phagocytosis of particulate Cr^{3+} can occur. Cr^{6+} usually forms strongly oxidizing **chromate** and **dichromate ions,** which are easily reduced under physiological conditions to Cr^{3+}. The saliva and gastric juice in the upper alimentary tract of mammals, including humans, have a varied capability to reduce Cr^{6+}, with the gastric juice having a notably high capacity. Thus, the normal body physiology provides detoxification for Cr^{6+}. **Cr** is mainly excreted in the urine and in the feces.

Guidelines. WHO (1991) sets guideline value for **Cr total** (for analytical problems). Provisional guideline value of 0.05 mg/l for **Cr total** was recommended, as well as re-evaluation of **Cr,** when additional information is available.

Standards. EEC (1989). MAC for **Cr total**: 0.05 mg/l. **USEPA** (1991). MCL and MCLG for **Cr total**: 0.1 mg/l. The National Research Council in the National Academy of Sciences (NAS, 1989) established safe and adequate daily dietary intake for Cr^{3+} to range from 0.05 to 0.2 mg/d. The lower limit is based on the absence of deficiency symptoms in individuals consuming an average of 0.05 mg **Cr**/d. **Russia** (1988). MAC: 0.5 mg Cr^{3+}/l and 0.05 mg Cr^{6+}/l. **Canada** (1987). MAC for **Cr total**: 0.05 mg/l.

Regulations. USFDA regulates the use of **Cr** as an indirect food additive and the use of **Cr oxide** in drugs and cosmetics. It may be used as a colorant in olefin polymers in the manufacturing of articles intended for use in contact with food. **Cromium (Cr^{3+})** complex of *N*-ethyl-*N*-heptadecyl-fluorooctane sulfonyl glycine may be used (1) as a component of paper for packaging dry food in accordance with CFR prescribed conditions. **Cr caseinate** and **Cr nitrate** may be used (2) as components of an adhesive for a food-contact surface.

References:
1. Borneff, I., Engelhardt, K., Griem, W., et al., Carcinogenic substances in water and soil, *Arch. Hyg.*, 152, 45, 1968 (in German).
2. *Industrial Hygiene, Occupational Pathology and Toxicology in the Chemical Industry and Non-ferrous Metallurgy in Kazakhstan,* Alma-Ata, 1984, p. 215 (in Russian).

3. Diaz-Mayans, J., Laborda, R., Nunez, A., et al., Hexavalent chromium effects on motor activity and some metabolic aspects of Wistar albino rats, *Comp. Biochem. Physiol.*, 83, 191, 1986.

4. Karimova, R. I., Talayeva, U. G., and Vasyukovich L. Y., Laboratory experiments on the joint effects of metals and enterobacteria, *Gig. Sanit.*, 12, 58, 1982 (in Russian).

5. Janus, J. A. and Krajnc, E. I., Integrated Criteria Document, Chromium: Effects, Appendix to Report No. 758701001, National Institute Public Health and Environmental Protection, Bilthoven, The Netherlands, 1989.

6. *Protection of Water Reservoirs Against Pollution by Industrial Liquid Effluents*, Vol. 2, Medgiz, Moscow, 1954, p. 101, (in Russian).

7. *Current Biochemical Methods in Environmental Hygiene*, Moscow, 1982, p. 68 (in Russian).

8. MacKenzie, R. D., Byerrum, R. U., Decker, C. F., et al., Chronic toxicity studies. II. Hexavalent and trivalent chromium administered in drinking water to rats, *Am. Med. Assoc. Arch. Ind. Health*, 18, 232, 1958.

9. *Hygiene Aspect of Environment Protection*, Moscow, 6, 1978, p. 154 (in Russian).

10. Ivancovic, S. and Preussmann, R., Absence of toxic and carcinogenic effects after administration of high doses of chromic oxide pigment in subacute and long-term feeding experiments in rats, *Food Cosmet. Toxicol.*, 13, 347, 1975.

11. Dinerman, A. A., *The Role of Environmental Pollutants in Disturbance of Embryonal Development*, Meditsina, Moscow, 1980, p. 192 (in Russian).

12. Smirnov, M. E., Late effects of water-containing hexa- and trivalent chromium on the body, *Gig. Sanit.*, 7, 33, 1985 (in Russian).

13. See **BISPHENOL A,** #5, 843.

14. Gale, T. F., The embryotoxic response to maternal chromium trioxide exposure in different strains of hamsters, *Environ. Res.*, 29, 196, 1982.

15. *Industrial Hygiene and Human Health Protection*, Minsk, 1976, p. 120 (in Russian).

COBALT COMPOUNDS

Applications and **Exposure.** *Co* compounds are used primarily as catalysts and pigments. *Co* **naphthenate** (CAS No 61789–51–3) is used to enhance the adhesion of sulfur-vulcanized rubber to steel and other metals and as a catalyst. *Co* compounds occur in vegetables via uptake from soil, and vegetables account for the major part of human dietary intake of *Co.*

Acute Toxicity. Following administration to rats by gastric intubation, LD_{50} (mg/kg BW) for different anhydrous *Co* compounds appeared to be: *Co* oxide—202, *Co* phosphate—387, *Co* chloride—418, *Co* sulfate—424, *Co* nitrate—434, *Co* acetate—503.[1] Acute effects included sedation, diarrhea, and decreased body temperature. Gross pathological examination revealed hemorrhages and dystrophic changes in the liver, kidney, and heart. According to Christensen and Luginbyh, LD_{50} of *Co* compounds for rats is 180 mg/kg.[2]

Repeated Exposure. Oral administration of 2.5 to 10 mg Co^{2+}/kg BW caused polycythemia in rats.[3,4]

Rabbits were dosed by gavage with 3 mg $Co(NO_3)_2$/kg BW for 2 months. The blood system was reported to be predominantly affected. The treatment caused an increase in erythrocyte count and Hb level, reticulocytosis and hyperplasia of the red bone marrow (Yastrebov).[5]

Long-term Toxicity. *Humans.* Actual cases of acute or chronic poisoning are not common. Excessive ingestion is characterized by congestive heart failure and polycythemia and anemia.[5] Syndrome of cardiomyopathy was described in subjects consuming large volumes of beer containing added *Co.*[6,7] *Animals.* Rats received 5 mg $CoCl_2$/kg BW (calculated as Co^{2+}) for 6 months. An impairment of hemopoiesis (true polycythemia) with subsequent thymal hyperplasia was found to develop.[8]

Allergenic Effect. Co^{2+} may provoke allergic dermatitis.[9]

Reproductive Toxicity. *Gonadotoxicity.* The dose of 265 mg $CoCl_2$/kg BW given in the diet for 98 d, providing an initial dose of 20 mg/kg BW, induced degenerative and necrotic changes in the seminiferous tubules of rats.[10] $CoCl_2$ reduced fertility in male mice. *Embryotoxicity.* Rats received 12 to 48 mg $CoCl_2$/kg BW by gavage from day 14 of gestation through day 21 of lactation. The treatment caused a decrease in the number of litters, and in survival and growth of the offspring.[11]

Significant increase in the early embryonic losses was found in mice.[12] Concentrations of 0.05 to 5 mg **CoCl₂**/l in drinking water given to rats before or during pregnancy caused embryotoxicity and death.[13] **Teratogenicity.** No developmental toxicity was observed in the offspring of rats given daily doses of 25 to 100 mg **CoCl₂**/kg BW by gavage on days 6 to 15 of gestation.[14] **Co acetate** and **Co chloride** are not reported to be teratogenic in hamsters and rats, respectively.[15]

Genotoxicity. Genetic toxicology of **Co** and **Co** compounds has been reviewed.[15,16] In prokaryotic assays, **Co** salts were generally nonmutagenic. They had a weak or no genetic effect in bacteria. In mammalian cells, *in vitro*, **Co** compounds caused DNA strandbreaks, SCE, and aneuploidy, but not CA. **Co acetate** enhanced viral transformation, and **Co sulfide** induced morphological transformation in Syrian hamster embryo cells. **Co chloride** induced CA in laboratory bred mice *in vivo* after single oral administrations of $^1/_{10}$ to $^1/_{40}$ of the lethal toxic dose. The effect appeared to be dose-dependent. A clastogenic effect of **CoCl₂** has been reported.[17]

Carcinogenicity. Workers occupationally exposed to **Co** compounds exhibited significantly enhanced risk for lung cancer. Interpretation of the available evidence for the carcinogenicity of **Co** compounds in animals is difficult (IARC, 1991). When administered to rats via *i/p* injections, **Co oxide** (76.7% **Co**) has a weak carcinogenic effect. **Co chloride** appeared to induce a carcinogenic effect in mice after oral administration of subtoxic doses.[18]

Chemobiokinetics. Absorption from the GI tract in rats was found to vary between 11 and 34%, depending on administered dose value.[19] The highest concentration was found in the liver; a low concentration was noted in the kidney, pancreas, and spleen.[20,21] **Co** specifically reacts with SH-groups of **dihydrolipoic acid,** thus preventing oxidative decarboxylation of pyruvate to acetylcoenzyme A.[22] After a low dose of 2.8 mg/kg, the tissue levels of **Co** compounds were not different from those in the control animals. The half-life of **Co** compounds in plasma was approximately 24 h. Approximately 20% **Co naphthenate** administered orally were eliminated in the urine.[7]

Standards. Russia (1988). MAC and PML: 0.1 mg Co^{2+}/l.

Regulations. USFDA (1993) listed **Co** (1) as an ingredient in resinous and polymeric coatings for a food-contact surface; **Co naphthenate** may be used (2) as an accelerator in polyester resins for food-contact surfaces; **Co acetate** may be used as (3) a component of adhesives for a food-contact surface and (4) as an ingredient of paper and paperboard intended for use in contact with dry food. **Co aluminate** may be used safely as a colorant in the manufacturing of articles intended for use in contact with food only (5) in resinous and polymeric coatings and (6) melamine-formaldehyde resins, (7) in xylene-formaldehyde resins condensed with **4,4'**-isopropylidenediphenol-epichlorohydrin epoxy resins, (8) in ethylene-vinylacetate copolymers, and (9) in urea-formaldehyde resins in molded articles.

References:

1. Speijers, G. J. A., Krajnc, E. I., Berkvens, J. M., et al., Acute oral toxicity of inorganic cobalt compounds in rats, *Food Chem. Toxicol.*, 20, 311, 1982.
2. Christensen, H. E. and Luginbyh, T. T., Toxic Substances, List 218, Public Health Service, NIOSH, Rockville, MA, 1974.
3. Orten, J. M. and Bucciero, M. C., The effect of cysteine, histidine, and methionine on the production of polycythemia by cobalt, *J. Biol. Chem.*, 176, 961, 1948.
4. Oskarsson, A., Reid, M. C., and Sunderman, F. W., Effects of cobalt chloride, nickel chloride, and nickel subsulfide upon erythropoiesis in rats, *Ann. Clin. Lab. Sci.*, 11, 165, 1981.
5. Calabrese, E. J., Canada, A. T., and Sacco, C., Trace elements and public health, *Annu. Rev. Public Health*, 6, 131, 1985.
6. Taylor, A. and Marks, V., Cobalt: A review, *J. Hum. Nutr.*, 32, 165, 1976.
7. NTP Fiscal Year Annual Plan, 1991, p. 100.
8. See **ACETONE**, #3, 182.
9. Camarasa, J. M. G., Cobalt contact dermatitis, *Acta Dermat. Venerol.*, 47, 287, 1967.
10. Corrier, D. E., Mollenhauer, H. H., Clark, D. E., et al., Testicular degeneration and necrosis induced by dietary cobalt, *Vet. Pathol.*, 22, 610, 1985.
11. Domingo, J. L., Patternain, J. L., Llobet, J. M., et al., Effects of cobalt on postnatal development and late gestation in rats upon oral administration, *Rev. Esp. Fisiol.*, 41, 293, 1988 (in Spanish).
12. Pedigo, N. G., George, W. J., and Andersen, M. B., The effect of acute and chronic exposure to cobalt on male reproduction in mice, *Reprod. Toxicol.*, 2, 45, 1988.

202

13. Nadeenko, V. G., Lenchenko, V. G., Soichenko, S. P., et al., Embryotoxic effect of cobalt administration *per os, Gig. Sanit.,* 2, 6, 1980 (in Russian).
14. Paternain, J. L., Domingo, J. L., and Corbella, J., Developmental toxicity of cobalt in the rat, *J. Toxicol. Environ. Health,* 24, 193, 1988.
15. Leonard, A. and Lauwerys, R., Mutagenicity, carcinogenicity and teratogenicity of cobalt metal and cobalt compounds, *Mutat. Res.,* 239, 17, 1990.
16. Beyersmann, D. and Hartwig, A., The genetic toxicology of cobalt, *Toxicol. Appl. Pharmacol.,* 115, 137, 1992.
17. Palit, S., Sharma, A., and Taluker, G., Chromosomal aberrations induced by cobaltous chloride in mice *in vivo, Biol. Trace. Elem. Res.,* 29, 139, 1991.
18. Steinhoff, D. and Mohr, U., On the question of carcinogenic action of *Co*-containing compounds, *Exp. Pathol.,* 41, 169, 1991.
19. Taylor, D. M., The absorption of cobalt from the gastrointestinal tract of the rat, *Physiol. Med. Biol.,* 6, 445, 1962.
20. Stenberg, T., The distribution in mice of radioactive cobalt administered by two different methods, *Acta Odontol. Scand.,* 41, 143, 1983.
21. Clyne, N., Lins, L.-E., Pehrsson, S. K., et al., Distribution of cobalt in myocardium, skeletal muscle and serum in exposed and unexposed rats, *Trace Elem. Med.,* 5, 52, 1988.
22. Alexander, C. S., Cobalt-beer cardiomyopathy, *Am. J. Med.,* 53, 395, 1972.

CYANOGUANIDINE (CAS No 461–58–5)

Synonym. Dicyandiamide.
Properties. Colorless, crystalline substance. Solubility in water is 2.26% (13°C). Gives no color to water and does not affect its transparency. Taste perception threshold is 10 mg/l.
Applications. Used as a curing agent for epoxy resins and in the production of plastics and melamine.
Acute Toxicity. Mice and rats tolerate administration of 15 g/kg. Lethal doses cause respiratory center stimulation and vasodilatation in the skin. Hematological analysis is unchanged. Cases of fatal poisonings are not reported.[1] However, according to Hold (IARC 19–206), LD_{50} is 4 g/kg BW in mice and 3 g/kg BW in rabbits.
Repeated Exposure revealed no cumulative properties.
Long-term Toxicity. Rats were dosed with 1 and 2 g/kg BW for 6 months. The treatment produced some changes in cholinesterase activity, reduction in phagocytic activity, and a rise in the blood urea level in rats. Gross pathological examination revealed mild dystrophy of the liver and kidney. The LOAEL of 50 mg/kg BW was identified in this study.[1] However, according to Frawley (IARC 19–206), administration of 2.5 g/kg in the diet for 2 years did not cause intoxication as well.
Allergenic Effect. C. seems to be a contact sensitizer.
Chemobiokinetics. C. does not exhibit properties of cyanide compounds and does not form CN^--ion. It is broken down in the body, forming **cyanamide** and **urea.**
Standards. Russia (1988). MAC and MPL: 10 mg/l (organolept., taste).
Regulations. USFDA (1993) approved the use of C. (1) in resinous and polymeric coatings for a food-contact surface, (2) as a modifier for amino resins and as a fluidizing agent in starch and protein coatings for paper and paperboard intended for use in contact with dry food, (3) in adhesives used as components of articles intended for use in packaging, transporting, or holding food, (4) in rubber articles intended for repeated use in contact with food, (5) as an antioxidant in the manufacturing of articles or components of articles intended for use in contact with food, subject to the provision to use C. only at levels not to exceed 1% by weight of polyoxymethylene copolymer, and (6) as a retarder of rubber articles intended for repeated use in food-contact articles (total not to exceed 10% by weight of the rubber product).
Reference:
See **ACRYLIC ACID,** #1, 96.

CYANURIC ACID (CAc) (CAS No 108–80–5)

Synonyms. Isocyanic acid; Pseudocyanuric acid; *sym*-Triazinetriol; Tricyanic acid; Trihyodroxy-cyanidine; **2,4,6**-Trihydroxy-*S*-triasine.

MONOSODIUM CYANURATE (MC)

Properties. CAc occurs as a colorless, crystalline powder. Solubility in water is 2.7 g/l, poorly soluble in alcohol. A concentration of 6 mg/l does not change organoleptic properties of water.

Applications. CAc derivatives are used as cross-linking agents in polymerization.

Acute Toxicity. LD_{50} of **CAc** is reported to be 10 g/kg[1] or 3.37 g/kg in rats, and 7.7 g/kg BW in mice.[2] However, according to other data, rats tolerate administration of a 10 g/kg dose without toxic manifestations.[3] High doses of **CAc** are found to cause adynamia, motor coordination disorder, and hematuria in the treated animals. Death occurs in 1 to 3 d.

Repeated Exposure revealed cumulative properties. K_{acc} is 1.85 (by Kagan). No evidence of bioaccumulation was noted in rats and dogs following repeated oral administration of 5 mg **MC** per kilogram BW for 15 d.

Short-term Toxicity. Bladder calculi with accompanying hyperplasia were observed in a few animals from the group exposed orally to **MC** at a dose of 5375 mg/l for 90 d. Hematological indices, urinalysis, or histopathology were not altered in animals exposed to 896 and 1792 mg/kg.[1]

Reproductive Toxicity. No increase in skeletal or external defects was observed in the offspring of rats given 500 mg CA/kg BW with or without calcium hypochlorite on days 6 to 15 of gestation.[4] No teratogenic effects were noted in the offspring of rats (doses up to 5 g/kg BW) or rabbits (doses up to 0.5 g/kg) exposed by gavage to **MC** on days 6 to 15 (rats) or on days 6 to 18 (rabbits) of gestation.[2] No adverse treatment-related effects on reproductive performance were observed following **MC** administration to male and female rats in their drinking water at concentrations of 400 to 5375 mg/l throughout three consecutive generations. An increased incidence of calculi in the urinary bladder of high-dosed F_2 parent males was noted. The histological examination of the urinary bladder revealed microscopic changes attributed to chronic irritation from the calculi.[1]

Genotoxicity. No mutagenic and gonadoxic activity of **MC** was observed in the *Salmonella* test and in *in vitro* mouse lymphoma cell point mutation assay. It neither induced SCE in Chinese hamster ovary cells *in vitro* nor demonstrated induction of rat bone marrow cell clastogenesis *in vivo*.[1]

Carcinogenicity. In a 2-year study, **MC** was administered to mice and rats in their drinking water. No considerable signs of toxicity (except for an increased incidence of calculi in the bladders of high-dosed male rats) or treatment-related oncogenic effects were observed. The LOAELs were considered to be the concentrations of 2400 mg/l in rats and 5375 mg/l in mice.[1]

Chemobiokinetics. In rats and dogs, **MC** was completely absorbed after oral administration. It is rapidly excreted unchanged with the urine following a single oral dose of 5 mg/kg. **CAc** was shown to be rapidly and quantitatively eliminated unchanged in excreta following oral ingestion by human volunteers.[1] The radiolabeled **CAc** was rapidly eliminated unchanged in the urine of rats.

Standards. Russia (1988): MAC: 6 mg/l (organolept.).

References:
1. Monsanto Material Safety Data Sheet, 1992.
2. Babayan, E. A. and Alexandryan, A. V., Toxicity of cyanuric acid, *Pharmacol. Toxicol.*, 49, 122, 1986 (in Russian).
3. See **BUTYL BENZYL ADIPATE,** 299.
4. Canelli, E., Chemical, bacteriological and toxicological properties of cyanuric acid and chlorinated isocyanurates as applied to swimming pool disinfection, *Am. J. Public Health*, 64, 155, 1974.

DIALLYL PHTHALATE (DAP) (CAS No 131–17–9)

Synonyms. Allyl phthalate; **1,2**-Benzenedicarboxylic acid, di-2-propenyl ester; Phthalic acid, diallyl ester.

Applications. Used as a cross-linking agent in the production of unsaturated resins.

Acute Toxicity. In rats and mice, LD_{50} is reported to be 0.7 to 1.7 g/kg BW.[1]

Repeated Exposure. DAP is the most toxic phthalate ester manufactured. It is more hepatotoxic to rats than to mice.[2] Hepatic necrosis was found to develop in Fisher 344 rats but not in B6C3F₁ mice dosed orally with up to 400 mg DAP/kg BW.[3]

Genotoxicity. DAP is reported to be negative in *Salmonella* mutagenicity assay. It caused CA and SCE in cultured cells and was shown to be positive in the mouse lymphoma assay.[4]

Carcinogenicity. In NTP-81 studies, oral exposure (by gavage) showed equivocal evidence of DAP carcinogenicity in mice and no or equivocal evidence in rats. *Carcinogenicity classification.* NTP: NE—EE—E—E.

Chemobiokinetics. Toxicity of DAP depends on **allyl alcohol** cleaved from it. When single doses of 1 to 100 mg/kg BW have been given orally to Fisher 344 rats and B6C3F₁ mice, 6 to 12% DAP was excreted as CO_2, 80 to 90% via the urine in 24 h.[2]

References:
1. See **DIETHYL PHTHALATE, #7.**
2. Eigenberg, D. A., Carter, D. E., Schram, K. H., et al., Examination of the differential hepato- toxicity of diallyl phthalate in rats and mice, *Toxicol. Appl. Pharmacol.,* 1, 12, 1986.
3. Gorden, E. B., Nwagbarocha, S., Douglas, E., et al., Diallyl Phthalate: C50657 Subchronic Rat and Mouse Study, Litton Bionetics, Inc., LB1 Project No. 10608, 1978.
4. See **BISPHENOL A, #6.**

4,4′-DIAMINODIPHENYL SULFONE (CAS No 80–08–0)

Synonyms. Bis(4-aminephenyl)sulfone; Dapsone; **4,4′**-Sulfonyl-dianiline.

Properties. A powdery substance. Water solubility is 15 mg/l at 20°C, soluble in alcohol and acetone. Odor and taste threshold concentration exceeds 15 mg/l.[1]

Applications. Used as a curing agent for epoxy resins and as an antioxidant.

Acute Toxicity. LD₅₀ is 1000 to 1400 mg/kg (LD₀ being 18 to 32 mg/kg BW) in rats, 375 mg/kg in mice, and 357 mg/kg in cats. Poisoning is accompanied by narcosis.[2] Death occurs within 5 d.

Repeated Exposure revealed pronounced cumulative properties. K_{acc} is 1.4 (by Kagan).

Long-term Toxicity. Rats were dosed by gavage with 5 mg/kg for 6 months. The treatment produced a steady change in the functional condition of the liver and an increase in lactate dehy- drogenase activity. Methemoglobinemia and changes in conditioned reflex activity were observed.[1]

Chemobiokinetics. See *4,4′-Diaminodiphenylmethane.*

Standards. Russia (1988). MAC and PML: 1 mg/l.

References:
1. Tsyganovskaya, L. Kh. et al., Substantiation of MAC for **4,4′**-diaminodiphenylsulphone in water bodies, *Gig. Sanit.,* 9, 23, 1978 (in Russian).
2. See 4,4′-**DIAMINODIPHENYL ETHER, #2, 118.**

DIBENZOYL PEROXIDE (DP) (CAS No 94–36–0)

Synonyms. Acetoxyl; Benzoyl peroxide; Perbenyl.

Properties. White or yellowish, fine, hygroscopic powder. Very poorly soluble in water; solu- bility in alcohol is 12 g/l. On hydrolysis forms **benzoic acid** and **benzoyl hydroxyperoxide.**

Applications. Used as a vulcanizing agent in the production of vulcanizate mixes based on natural and synthetic rubber and polyester resins. Also used in the production of polyester, glass-reinforced plastics, and polystyrene foam. Polymerization catalyst for vinyl, acrylic, and styrene plastics and for unsaturated polyesters. It is also used as a decolorant of flour and milk in cheese manufacturing.

Acute Toxicity. LD₅₀ is 6.4 g/kg in rats and 2.1 g/kg BW in mice. Lethal doses caused immediate motor coordination disorder, with subsequent CNS inhibition and respiratory distress. Death occurs in 24 h.[1,2] According to IARC 36–272, LD₅₀ is more than 950 mg/kg BW for rats.

Repeated Exposure failed to reveal signs of accumulation.[2]

Short-term Toxicity. Rats and mice were given ¹/₅₀ and ¹/₁₀ LD₅₀ for 4 months. The treatment with the higher dose of DP caused adynamia and significant retardation of BW gain.[2] In a 3-month study in dogs, there were no manifestations of toxicity when 160 mg DP/kg was administered in the diet.[3]

Long-term Toxicity. Rats and mice were fed for 120 and 80 weeks, respectively, up to 1000 times greater quantities of DP than that consumed by humans in the diet. No signs of intoxication were found to develop in the animals. Exposure to 280 and 2800 mg/kg feed led to some retardation of BW gain in rats. It is thought that a dose of up to 40 mg DP/kg diet is harmless to man.[3]

Allergenic Effect is likely to develop as skin reactions in man after contact with DP.[4]

Reproductive Toxicity. *Gonadotoxicity.* Prolonged administration of 2800 mg DP/kg feed to rats caused testicular atrophy, which may be associated with the destruction of tocopherol in the feed. There were no such changes in mice.[3] *Teratogenic* effect as well as late death is observed in chick embryos.[5]

Genotoxicity. DP is neither mutagenic in *Salmonella* nor does it induce CA in cultured Chinese hamster lung cells[6] and DLM in Swiss mice (*i/p* injection of 54 or 62 mg/kg BW).[7] It does not seem to be directly genotoxic.

Carcinogenicity. Addition of milk containing 28 to 2800 mg DP/kg to the feed of rats for 120 weeks and to the feed of mice for 80 weeks produced no carcinogenic effect.[3] However, DP was shown to promote carcinogenic activity in the mouse skin. *Carcinogenicity classification.* IARC: Group 3.

Regulations. USFDA (1993) approved DP for use as a direct and indirect food additive: (1) in adhesives used as components of articles intended for use in packaging, transporting, or holding food, (2) as a preservative of the uncoated or coated food-contact surface of paper and paperboard intended for use in producing, manufacturing, packing, transporting, or holding aqueous and fatty food, (3) as a catalyst in the formulation of polyester resins, and (4) as an accelerator in the production of rubber articles intended for repeated use. DP may be used also (5) as a bleaching agent for flour and for milk used in the preparation of certain cheeses and lecithin, (6) as an ingredient in the preparation of hydroxylated lecithin. In 1982, it was proposed that DP be affirmed as GRAS as a direct human food ingredient when used as a bleaching agent following conditions of use of current GMP. In **the U.K.** DP may be added to all types of flour or bread (except wholemeal) at up to 50 mg/kg flour.

Standards. Russia. PML in drinking water: **n/m.**

References:

1. NIOSH, Criteria for a Recommended Standard, *Occupational Exposure to Benzoyl Peroxide,* Publ. No. 77–166, Washington, D.C., 1977.
2. See **BUTYL BENZYL ADIPATE,** 311.
3. Sharratt, M., Frazer, A. C., and Forbes, O. C., Study of the biological effects of benzoyl peroxide, *Food Cosmet. Toxicol.,* 2, 527, 1964.
4. Poole, R. L., Griffith, J. F., and MacMillan, F. S. K., Experimental contact sensitization with benzoyl peroxide, *Arch. Dermatol.,* 102, 635, 1970.
5. Korhonen, A., Hemminki, K., and Vainio, H., Embryotoxic effects of eight organic peroxides and hydrogen peroxide on three-day chicken embryos, *Environ. Res.,* 33, 54, 1984.
6. Ishidate, M., Sofuni, T., and Yoshikawa, K., A primary screening for mutagenicity of food additives in Japan, *Mutagen. Toxicol.,* 3, 80, 1980 (in Japanese).
7. Epstein, S. S., Arnold, S., Andrea, J., et al., Detection of chemical mutagens by the dominant lethal assay in the mouse, *Toxicol. Appl. Pharmacol.,* 23, 288, 1972.

DI-*tert*-BUTYLPEROXYSUCCINATE

Synonym. Diperoxysuccinic acid, di-*tert*-butyl ester.

Properties. White powder. Poorly soluble in water, readily soluble in alcohol.

Applications. Used as a polymerization initiator, in the form of 50% solution in dibutyl phthalate.

Acute Toxicity. No rats died when a single oral dose of 4 g/kg BW was administered. In mice, LD_{50} is 5.3 g/kg for females and 7.7 g/kg BW for males, but it appeared to be much higher if D. is given to mice as a 50% solution in dibutyl phthalate. In such a case, LD_{50} is 16.5 g/kg BW in females, while males tolerate 10 g/kg BW. There were no manifestations of the toxic action or retardation of BW gain in the survivors compared with the controls during a 3-week follow-up. Gross pathological examination revealed no changes in the viscera.[1]

Genotoxic effect is found on exposure to **disuccinyl peroxide** (succinic acid peroxide).[2]

References:

1. *Environmental Protection in Plastic Industry and Safety of the Use of Plastics,* Plastpolymer, Leningrad, 1978, p. 133 (in Russian).

2. Luzzarti, D. and et Chevallier, M. R., Comparison of the lethal and mutagenic action of an organic peroxide and radiation on *E. coli, Ann. Inst. Pasteur,* 93, 366, 1957 (in French).

DIBUTYLTIN DIACETATE (DBTA) (CAS No 1067–33–0)

Synonyms. Bis(acetyloxy)dibutylstannate; Diacetoxydibutyltin.

Applications. Used as a catalyst in urethane manufacturing and esterification.

Acute Toxicity. LD_{50} is reported to be 46 mg/kg in mice and 32 mg/kg BW in rats (Litton Bionetic, Inc.).

Reproductive Toxicity. DBTA caused maternal toxicity only at the highest dose-level when administered orally to rats throughout pregnancy.[1] Following administration of 1.7 to 15 mg/kg BW to Wistar rats on days 7 to 17 of gestation, dose-dependent thymic atrophy in the pregnant animals was reported.[2] There was an increase in the incidence of dead or resorbed fetuses and of total fetal resorption, as well as an increase in external malformations.

Genotoxicity. No mutations were noted in the *Salmonella* assay[3] or in *Dr. melanogaster.*[4] Positive results were shown in mouse lymphoma cells and in Chinese hamster ovary cells (SCE) (NTP–85).

Carcinogenicity. Fisher 344 rats were exposed to 66.5 or 133 ppm in their diet, while $B6C3F_1$ mice received 76 ppm or 152 ppm in the feed for 78 weeks. Increased incidence of hepatocellular adenomas in female mice and both hepatocellular adenomas and carcinomas in male mice were found, but these were not statistically significant.[5] *Carcinogenicity classification.* NTP: N—IS—N—N.

Chemobiokinetics. In mice, the major part of the dose is excreted in the feces unchanged; small quantities of the dose are eliminated in the expired air as **carbon dioxide** and **butene.**[1]

Regulations. USFDA (1993) approved the use of DBTA in polyurethane resins for the food-contact surface of articles intended for use in contact with dry food alone or in combination, at levels not to exceed a total of 3% by weight in the resin used.

References:

1. Noda, T., Morita, S., Shimizu, M., et al., Safety evaluation of chemicals for use in household products. VII. Teratology studies of di-*n*-butyltin diacetate in rats, *Annu. Rep. Osaka Inst. Publ. Health Environ. Sci.,* 50, 66, 1988.
2. Noda, T., Yamano, T., Summizu, M., et al., Comparative teratogenicity of di-*n*-butyltin diacetate and *n*-butyltin trichloride in rats, *Arch. Environ. Contam. Toxicol.,* 23, 216, 1992.
3. Tennant, R. W., Stasiewicz, S., Spalding, J. W., Comparison of multiple parameters of rodent carcinogenicity and *in vitro* genetic toxicity, *Environ. Mut.,* 8, 205, 1986.
4. Woodruff, R. C., Mason, J. M., Valencia, R., et al., Chemical mutagenesis testing in *Drosophila, Environ. Mut.,* 7, 677, 1985.
5. National Cancer Institute, *Bioassay of Dibutyltin Diacetate for Possible Carcinogenicity,* TR-183 DHEW Publ. 79–1739, Washington, D.C., 1979.

DIETHYLENETRIAMINE (DETA) (CAS No 111–40–0)

Synonyms. Aminoethylethandiamine; 2,2′-Diaminodiethylamine.

Properties. Colorless, oily liquid with a pungent, specific odor. Readily miscible with water. Odor perception threshold is 0.7 mg/l at 20°C and 0.2 mg/l at 60°C. Taste perception threshold is 1.3 mg/l.[1]

Applications. Used in the synthesis of epoxy and ion-exchange resins.

Acute Toxicity. For rats, mice and rabbits LD_{50} is in the range of 970 to 1200 mg/kg BW. However, according to other data,[07] DETA is less toxic on a single administration, and LD_{50} is 2.33 g/kg BW. Guinea pigs seem to be the most sensitive animals: LD_{50} is 600 mg/kg BW. Poisoned animals displayed behavioral changes and convulsions.

Repeated Exposure failed to reveal cumulative properties of DETA. Guinea pigs tolerated administration of 120 mg/kg BW for a month.

Long-term Toxicity. In a 6-month study, guinea pigs were dosed by gavage with 0.6 mg DETA/kg BW. The treatment did not cause retardation of BW gain; it did not affect the content of metHb in the blood and ascorbic acid in the liver or the activity of liver enzyme systems. Administration of 10 mg/kg BW to rabbits over a 6-month period caused a reduction in prothrombin blood level and an increase in the activity of glutaminate-oxalate and glutaminate-pyruvate transaminase. The dose of 1 mg/kg BW was identified to be the NOAEL in this study.[1]

Genotoxicity. DETA itself exhibits very mild effect in *Salmonella* mutagenicity assay.[2] It could produce a mutagenic action due to contamination with **ethyleneimine** and other highly active mutagens.[3] See also *Polyethylenepolyamines.*

Standards. Russia (1988). MAC and PML: 0.2 mg/l (organolept., odor).

Regulations. USFDA (1993) approved the use of DETA (1) as a modifier for aminoresins for use in paper and paperboard articles coming in the contact with aqueous, fatty, and dry food, (2) in resinous and polymeric coatings intended for use in the food-contact surface of articles, (3) as a defoaming agent that may be used safely in the manufacturing of paper and paperboard intended for use in producing, manufacturing, packing, transporting, or holding food.

References:
1. Trubko, E. I. and Teplyakova, E. V., Studies on diethylenetriamine hygienic norm-setting in water bodies, *Gig. Sanit.,* 7, 103, 1972 (in Russian).
2. Hulla, J. E., Rogers, S. J., and Warren, J. R., Mutagenicity of a series of polyamines, *Environ. Mut.,* 3, 332, 1981.
3. Hedenstedt, A., Rannug, U., Ramel, C., et al., Mutagenicity and metabolism studies on 12 thiuram and dithiocarbamate compounds used as accelerators in the Swedish rubber industry, *Mutat. Res.,* 68, 313, 1979.

N,N'-DIETHYLETHANOLAMINE (DEEA) (CAS No 100–37–8)

Synonyms. 2-Diethylaminoethanol; Diethyl-β-hydroxyethylamine; 2-Hydroxy-triethylamine.

Properties. Viscous, hygroscopic liquid. Water soluble; miscible with alcohol. Takes up CO_2 from air.

Applications. DEEA is used as a resin-curing agent and a chemical intermediate.

Acute Toxicity. In rats, LD_{50} of nonneutralized DEEA is 1.3 g/kg BW; that of neutralized DEEA is 5.6 g/kg BW.[1]

Long-term Toxicity. In a 6-month dietary study, neutralized DEEA was given to rats in their drinking water. The treatment produced a decrease of BW gain and increased kidney relative weights. Neither renal lesions, nor other treatment-related effects were found.[2]

Carcinogenicity. Oncogenic properties of aliphatic amines are still questionned. They do not appear to cause carcinogenic action without nitrosation.

Chemobiokinetics. Easily absorbed from the GI tract.

References:
1. *Dangerous Properties of Industrial Materials,* 7th ed., van Nostrand Reinhold, New York, 1989, 2, 1230.
2. Kornish, H. H., Oral and inhalation toxicity of 2-diethylaminoethanol, *Am. Ind. Hyg. Assoc. J.,* 26, 479, 1965.

DILAUROYL PEROXIDE (DP) (CAS No 105–74–8)

Synonym. Lauroyl peroxide.

Properties. Fatty, white, coarse powder with a faint odor. Poorly soluble in water, readily soluble in alcohol and vegetable oils.

Applications. Used as a polymerization catalyst in the production of polystyrenes, in vulcanization, and in plasticization of natural and synthetic rubber, for curing synthetic resins, in the production of polyesters, and as an initiator in the suspension polymerization of vinyl chloride, ethylene, and acrylates.

Acute Toxicity. LD_{50} was not attained in rats and mice. A 10 g/kg BW dose caused 30% animal mortality. High doses produced CNS inhibition, drowsiness and dyspnea. Death occurred with increasing asphyxia. Gross pathological examination revealed visceral congestion and liver enlargement. Parenchymatous dystrophy of the liver, kidneys, and myocardium and disturbed hemopoiesis were also noted.

Repeated Exposure. Cumulative properties are not pronounced. A dose of 3 g/kg BW administered to rats for 10 d caused 50% animal mortality. Leukocytosis with subsequent leukopenia was found to develop. Death occurs in 2.5 to 3 months after the treatment with signs of anemia. Gross pathological examination showed the visceral organs being pallied and the liver, kidneys, and spleen being enlarged. In a 1-month study, the NOAEL appeared to be 1 g/kg BW (Orlov, 1968).

Genotoxicity. DP is not shown to be mutagenic to *Salmonella typhimurium TA98* or *TA100* in presence of an exogenous metabolic system.

Carcinogenicity. In the mouse and rat studies that were considered inadequate for evaluation of complete carcinogenicity (IARC 36–319), DP did not produce the evidence of carcinogenic effect following subcutaneous administration or skin application.

Regulations. USFDA (1993) regulates DP for use (1) in adhesives as a component of articles intended for use in packaging, transporting, or holding food, (2) as a polymerization catalyst in components of paper and paperboard intended for use in contact with aqueous and fatty foods, and (3) as a catalyst, at a level of 1.5% by weight max, in cross-linked polyester resins to be used safely for repeated contact with foods.

Reference:

Yamaguchi, T. and Yamashita, Y., Mutagenicity of hydroperoxides of fatty acids and some hydrocarbons, *Agric. Biol. Chem.,* 44, 1675, 1980.

DIMETHYLAMINE (DMA) (CAS No 124–40–3)

Synonym. *N*-Methylmethanamine.

Properties. A gas, easily converted into a colorless liquid on cooling or under pressure, with the odor of ammonia or rotting fish. DMA is soluble in water (550 g/l at 25°C) and in ethanol. The aqueous solutions exhibit an alkaline reaction (unstable methylammonium hydroxide is formed). With acids, DMA forms salts, which are nonvolatile, odorless, water-soluble solids. The odor perception threshold is 0.67 mg/l; the practical threshold is 1 mg/l[1] or, according to other data, 0.29 mg/l.[02] The taste perception threshold is much higher.

Applications and **Exposure.** DMA is predominantly used as a curing agent or emulsifier. It is also found to be a degradation product in rubber materials that are made using dialkyldithiocarbamine acid derivatives as vulcanization accelerators. Reported migration levels of 0.1 to 0.2 mg/l usually depend on rubber and contact media compositions.

Acute Toxicity. LD_{50} is 698 mg/kg in rats, 316 mg/kg in mice, and 240 mg/kg BW in guinea pigs and rabbits. When DMA was neutralized by *HCl*, LD_{50} for rats, guinea pigs, and rabbits appeared to be 1, 1.07, and 1.6 g/kg BW, respectively. The poisoning was accompanied by excitation, with subsequent adynamia and motor coordination disorder. Gross pathological examination revealed extensive hemorrhages in the stomach and intestinal wall.[1] Amines are known to affect the activity of the mixed oxidase, the function and metabolism of biogenic amines, this culminating in impaired neurohumoral control.

Repeated Exposure failed to reveal cumulative properties. Guinea pigs and rabbits were dosed by gavage with $1/10$ LD_{50} for a month. The treatment affected blood Hb level, cholinesterase activity, serum urea, and ascorbic acid content in the visceral organs. There was an increase in the urinary coproporphyrin and in the relative liver weights.[1]

Long-term Toxicity. A temporary rise of urea level in the blood serum, an increase in the leukocyte count, and a decline in the amount of ascorbic acid in the suprarenals were noted in guinea pigs.[1]

Reproductive Toxicity. No maternal or fetal effects were found in mice given *i/p* 2.5 or 5 m*M* DMA/kg BW on days 1 through 17 of gestation.[2]

Genotoxicity effect is observed in mice.[3]

Carcinogenicity. Following inhalation exposure to DMA and nitrogen peroxide, their reaction products are reported to exhibit carcinogenic potential.[4]

Chemobiokinetics. A small amount of DMA undergoes oxidative transformation. In mammals, the secondary amines, to which DMA refers, are more resistant to deamination, compared with the primary. They may be deaminated or excreted unchanged. DMA is predominantly eliminated via the urine, and to a lesser extent, it is excreted with the bile or secreted into the intestine, where it may be reabsorbed.[5]

Standards. Russia. MAC and PML: 0.1 mg/l.

References:

1. Dzhanashvili, G. D., Hygienic substantiation of the maximum content of dimethylamine in water bodies, *Gig. Sanit.,* 6, 12, 1967 (in Russian).
2. Guest, I. and Varma, D. R., Developmental toxicity of methylamines in mice, *J. Toxicol. Environ. Health,* 32, 319, 1991.
3. Friedman, M. A., Miller, G., Sengupta, M., et al., Inhibition of mouse liver protein and nuclear RNA synthesis following combined oral treatment with sodium nitrile and dimethylamine or methylbenzylamine, *Experentia,* 28, 21, 1972.
4. Benemansky, V. V., Prusakov, V. M., and Dushutin, K. K., Biological activity of reaction products of dimethylamine and hydrogen dioxide during short-term exposures in animals, *Gig. Sanit.,* 12, 15, 1979 (in Russian).
5. Ishiwata, H., Iwata, R., and Tanimura, A., Intestinal distribution, absorption and secretion of dimethylamine and its biliary and urinary excretion in rats, *Food Chem. Toxicol.,* 22, 649, 1984.

N,N'-DIMETHYLCYCLOHEXYLAMINE (CAS No 98–94–2)

Properties. Colorless, highly volatile liquid with an unpleasant odor. Poorly soluble in water, readily soluble in oil and alcohol.

Applications. Used as a catalyst in the production of polyurethane foams.

Acute Toxicity. LD_{50} is reported to be 450 mg/kg in rats, 320 mg/kg in mice,[1] 720 mg/kg in guinea pigs, and 620 mg/kg BW in rabbits (Schmidt et al., 1983). Poisoning is accompanied by clonic–tonic convulsions with subsequent death in the first 2 h after administration. D. exhibits no aftereffect and does not produce methemoglobinemia in contrast to the structurally similar **cyclohexylamine.**

Reproductive Toxicity. No gonadotoxic effect was observed following inhalation exposure to D.

Genotoxicity. An increase in the number of CA in bone marrow cells was observed in rats as a result of D. inhalation at a concentration of 92 mg/m^3.

Reference:

Smirnova, E. S., Kasatkin, A. N., Obryadina, G. I., et al., Toxic properties of *N,N'*-dimethylcyclohexylamine, *Gig. Truda Prof. Zabol.,* 5, 54, 1984 (in Russian).

α,α'-DIMETHYL-4-CYCLOHEXYLBENZYL HYDROPEROXIDE

Synonym. *p*-Isopropylcyclohexylbenzene peroxide.

Properties. Crystals. The technical product is a yellowish-brown liquid with an unpleasant odor. Solubility in water is 20 mg/l, readily soluble in alcohols and ethers.

Applications. Used as a polymerization initiator.

Acute Toxicity. In mice, LD_{50} is 2.6 g/kg BW. The dose of 3.5 g/kg causes 100% mortality in animals. Poisoning is accompanied by adynamia.

Reference:

See **BISPHENOL A,** #2, 169.

DI(1-METHYLPROPYLIDENE) DIPEROXIDE

Properties. Colorless, oily liquid with a pleasant odor.

Applications. Used as a polymerization initiator, particularly in the production of polyesters.

Acute Toxicity. *Humans.* Intake of 50 ml of Norox catalyst (60% D., **cyclohexanone** and **dimethyl phthalate**—see above) caused fatal poisoning in man. *Animals.* LD_{50} is 470 mg/kg for rats and 260 mg/kg BW for mice.

Reference:

See **ZINC DIETHYLDITHIOCARBAMATE,** #2, 74.

ETHYLENEDIAMINE (EDA) (CAS No 107–15–3)

Synonyms. β-Aminoethylamine; **1,2**-Diaminoethane; **1,2**-Ethanediamine.

Properties. Colorless, oily liquid with an odor of ammonia. Mixes with water and alcohol at all ratios. Threshold perception concentration for odor change is 0.21 mg/l; practical odor perception threshold is 0.85 to 1 mg/l.[1,010] Taste perception threshold is 0.8 mg/l (practical threshold is 2.5 mg/l). According to other data,[01] organoleptic perception threshold is 12 mg/l.

Applications. Used in the synthesis of polyamides and as a curing agent for epoxy resins.

Acute Toxicity. LD_{50} was reported to be 0.7 to 1.4 g/kg BW in rats and 0.45 g/kg BW in mice and guinea pigs.[05,2] According to other data,[1] the median lethal doses for rats and mice are 1.2 and 1 g/kg BW, respectively. LD_{50} of **EDA dihydrochloride** was found to be 3.25 g/kg in rats and 1.62 to 1.77 g/kg BW in guinea pigs.[2] Lethal doses led to excitation followed by CNS inhibition and, often, convulsions. Asthmatic syndrome was shown to develop. Manifestations of toxic action included a reduction in the activity of mono- and diaminooxidase in the liver and of catalase and peroxidase activity. Alterations in serotonin metabolism have been noted.

Repeated Exposure revealed moderate cumulative properties. K_{acc} is 5.4 (by Lim). Rats were gavaged with $^1/_{10}$ LD_{50}. The treatment resulted in disorder of liver and kidney functions, and in their increased relative weights. The content of γ-globulin in the blood was increased.[1]

Short-term Toxicity. In a 3-month study, rats were given 0.05 to 1 g **EDA dihydrochloride**/kg BW. The highest dose caused retardation of BW gain in Fisher 344 rats. No increase in mortality was observed. Gross and histopathological examination revealed a dose-related increase in hepatocellular pleomorphism and mild hepatocellular degeneration.[2]

Reproductive Toxicity. No evidence of teratogenicity was found in rats treated with up to 1 g **EDA dihydrochloride**/kg BW on days 6 through 15 of gestation.[3]

Genotoxicity. EDA has been shown to be positive in *Salmonella* mutagenicity assay (direct mutagenic activity). Fisher 344 rats were given 0.05 to 0.5 g/kg BW for 23 d. A genotoxic effect was not observed in DLM, SCE, and unscheduled DNA synthesis tests.[4]

Chemobiokinetics. After ingestion and absorption in the GI tract of animals, **EDA** is metabolized by *N*-acetylation, which may also occur in man. The principal metabolites in the rat urine are **N-acetyl** and **N,N'-diacetyl** derivatives. A small amount of **hippuric acid** is also found. Tagged EDA is passed via the urine (47%) and expired air (18%).[5] Wistar rats were treated with ^{14}C-**EDA·2HCl** at dose-levels of 5, 50 mg/kg, and 500 mg/kg. **EDA** was mostly found in the thyroid, bone marrow, liver, and kidneys. Urinary excretion was shown to be the main route of elimination (42%). Of radioactivity, 5 to 32% was excreted via feces, 9% via expired air in the form of CO_2. The basic metabolite **N-acetylethylenediamine** provided for 50% urinary radioactivity.[6]

Standards. EEC (1990). SML: 12 mg/kg. **Russia** (1988). MAC and PML: 0.2 mg/l (organolept., odor).

Regulations. USFDA (1993) regulates the use of **EDA** (1) in resinous and polymeric coatings in a food-contact surface of articles intended for use in producing, manufacturing, packing, transporting, or holding food, (2) as a component of the uncoated or coated food-contact surface of paper and paperboard that may be used safely in producing, manufacturing, packing, transporting, or holding dry food, (3) in adhesives used as components of articles intended for use in packaging, transporting, or holding food, and (4) as an adjuvant in the preparation of slimicides in the manufacture of paper and paperboard that may be used safely in contact with food.

References:
1. *Proc. Erisman Research Hygiene Institute,* Vol. 10, Moscow, 1979, p. 141 (in Russian).
2. Raymond, S. H., Yang, R. H., Garman, R. R., et al., Acute and subchronic toxicity of ethylenediamine in laboratory animals, *Fundam. Appl. Toxicol.,* 3, 512, 1983.
3. DePass, L. R., Yang, R. S. H., and Woodside, M. D., Evaluation of the teratogenicity of ethylenediamine dihydrochloride in Fisher 344 rats by conventional and pair-feeding studies, *Fundam. Appl. Toxicol,* 9, 687, 1987.
4. See **ZINC PHENYLETHYLDITHIOCARBAMATE,** #3.
5. Coldwell, J. and Cotgreave, I. A., *Drug Determinant Therapeutic and Forensic Contexts,* New York, 1984. 47.
6. Raymond, S. H., Yang, R. H., and Tallant, M. J., Metabolism and pharmacokinetics of ethylenediamine in the rat following oral, endotracheal, or intravenous administration, *Fundam. Appl. Toxicol.,* 2, 252, 1982.

ETHYLENEDIAMINE TETRAACETIC ACID (EDTA) (CAS No 60–00–4)

Synonyms. Edetic acid; Ethylenedinitrilotetraacetic acid; Havidote.

ETHYLENEDIAMINE TETRAACETIC ACID, DISODIUM SALT (DSEDTA)

Synonym. Trilone B.

Properties. DSEDTA is a white, crystalline powder with a rather salty taste. Solubility in water is 10% at 20°C. It does not decompose in water and forms complexes with metals.[05] Water solubility is 0.5 g EDTA/l (25°C).

Applications and **Exposure.** EDTA and its salts are used as components in the production of food-contact paper and paperboard. EDTA is found in the environment as metal complexes. The levels recorded in natural water are usually less than 0.1 mg/l. DSEDTA is used in synthetic rubber manufacturing.

Acute Toxicity. LD_{50} of DSEDTA is reported to be 2 g/kg BW,[05] but according to other data, it was not attained. After a single administration of 2.5 g/kg BW, rabbits became apathetic, refused feed, and suffered from severe diarrhea. Poisoning is accompanied by acceleration of the erythrocyte sedimentation rate and reduction in serum calcium and serum cholinesterase activity. Death within 5 d.[05]

Repeated Exposure revealed no accumulation of EDTA. Clinical experience with EDTA for treating metal poisonings has provided the evidence of its safety in man. Rats were given a saturated solution of DSEDTA (a total dose of 5.8 g/kg BW). The treatment caused 50% animal mortality in the first 2 or 3 d. Administration of 9 g/kg total dose resulted in 100% mortality in rats.[05]

Short-term Toxicity. Zinc deficiency may develop as a consequence of zinc complexed by EDTA.[1] This was reaffirmed in Dutch workers.[2]

Long-term Toxicity. In a 7-month study, rabbits were given the oral dose of 2.1 mg DSEDTA/kg BW. The treatment did not alter hematological analysis, cholinesterase activity, or serum calcium level in animals.[05]

Reproductive Toxicity. Equivocal data are reported in regard to developmental effects in animals orally exposed to EDTA or its salts. A teratogenic effect was found in rats given 2 to 3% EDTA in the diet after day 6 of gestation.[1] The treatment produced cleft palate, brain and eye defects, and skeletal abnormalities in fetuses. However, Schardein et al.[3] did not observe adverse fetal changes in rats gavaged with up to 1 g EDTA or its sodium and calcium salts per kilogram BW on days 7 to 14 of gestation.

Carcinogenicity. EDTA showed no carcinogenic potential when administered, even at high doses. *Carcinogenicity classification.* NTP: N—N—N—N.

Chemobiokinetics. EDTA is poorly absorbed from the gut; it seems to be metabolically inert. Feeding studies in rats and dogs gave no substantial evidence of EDTA interference with mineral metabolism in either species. Nevertheless, EDTA is known to be capable of chelating metals, both in liquid media and in animal body. It has been suggested that EDTA may enter kidney cells and, by interfering with zinc metabolism, exacerbate the toxicity of cadmium.[4]

Guidelines. WHO (1992). Provisional guideline value of EDTA for drinking water: 0.2 mg/l (in view of the posibility of complexation with zinc, the proposed guideline value was derived assuming that a child weighing 10 kg drinks 1 liter of water). JECFA (1974) proposed ADI of 0 to 2.5 mg $CaNa_2EDTA$/kg (0 to 1.9 mg/kg as the free acid) as a food additive, but not in food of children under 2 years of age.

Regulations. USFDA (1993) listed EDTA and its sodium and/or calcium salt (1) as a component of paper and paperboard intended for use in producing, manufacturing, packing, transporting, or holding aqueous and fatty food, (2) as a substance employed in the production of or added to textiles and textile fibers intended for use in contact with food, and (3) in cellophane to be safely used for packaging food.

Standards. Russia. Recommended PML for DSEDTA: 3.5 mg/l.

References:
1. Swenerton, H. and Hurley, L. S., Teratogenic effects of chelating agent and their prevention by zinc, *Science*, 173, 62, 1971.
2. Janssen, P. J. C. M., Knaap, A. G. A. C., and Taalman, K. D. F. M., *EDTA: Litertuuronderzoek naar de orale toxiciteit voor de mens,* Report No. 718629008, Rijksinstitut voor Volksgezondheid en Milienhygiene, Bilthoven, 1990 (in Dutch).

3. Schardein, J. L., Sakowski, J., Petrere, J., et al., Teratogenesis studies with EDTA and its salts in rats, *Toxicol. Appl. Pharmacol.*, 61, 423, 1981.
4. Dieter, H. H., Implication of heavy metal toxicity related to EDTA exposure, *Toxicol. Environ. Chem.*, 27, 91, 1990.

FATTY ACIDS (SYNTHETIC), DIACYL(C_6-C_9) PEROXIDES, MIXTURE

Properties. Colorless liquid with a specific odor.

Applications. Used as an efficient initiator in the production of low-density polyethylene.

Acute Toxicity. LD_{50} is 28 g/kg in male and 15.2 g/kg BW in female rats; in mice, it is found to be 10.1 and 9 g/kg BW, respectively. Histological examination revealed vascular engorgement and slight cerebral edema. The acute action threshold was not attained.

Repeated Exposure. Rats and mice tolerate administration of 0.5 and 2.5 g/kg BW for 2 months without visible changes in their general condition or retardation of BW gain. Gross pathological examination failed to reveal changes in the viscera.

Reference:
See **MINERAL PETROLEUM OIL**, #1, 20.

ISOPROPYLBENZENE HYDROPEROXIDE (CAS No 80–15–9)

Synonyms. Bis-α,α-(dimethylbenzyl) peroxide; Cumyl hydroperoxide; Dicumyl peroxide; Di-isopropylbenzene peroxide; α,α-Dimethylbenzyl hydroperoxide; Percum.

Properties. A colorless, oily liquid with an odor reminiscent of ozone or light-yellow crystalline substance. Poorly soluble in water, readily soluble in alcohol and vegetable oils. Odor perception threshold is 3 mg/l.[1]

Applications. Used in the production of vinyl chloride and acrylates as polycondensation and polymerization initiator; an initiator in the synthesis of unsaturated polyester resins. Polymerization catalyst in the production of polystyrenes. Used in vulcanization and plasticization of natural and synthetic rubber and for curing epoxy resins. A cross-linked agent for silicone rubber, polyethylene foam, and polystyrene foam. Acetophenone is found to be a degradation product of B. (Novitskaya, 1984).

Acute Toxicity. LD_{50} is reported to be 800 to 1270 mg/kg BW in rats and 342 to 350 mg/kg BW in mice.[2] Single lethal doses affect the CNS, liver, and kidney. Poisoned animals displayed general inhibition, drowsiness, and dyspnea; death occurred from respiratory failure. Gross pathological examination revealed parenchymatous dystrophy of the liver, kidney, and myocardium, as well as hyperemia of the gastric and intestinal mucosa, congestion in the visceral organs, and liver enlargement.[3]

Repeated Exposure revealed little evidence of cumulative properties.[1]

Short-term Toxicity. Rats given a 100 mg/kg dose for 3 months displayed no changes in the blood formula or in catalase and cholinesterase activity.[1] In the course of treatment, animals became lethargic and drowsy.

Long-term Toxicity. In a 7-month study, rabbits were given doses of up to 25 mg/kg BW. The treatment affected different liver functions. Vitamin C blood level was decreased, while carotine level increased. Cholinesterase activity tended to be reduced.[1]

Reproductive Toxicity. *Embryotoxic* and *teratogenic* effects are observed in chick embryos.[4] *Genotoxic* effect has been noted in *Dr. melanogaster*.[5] No DLM in the male sex cells have been noted in rats given 0.25 mg/kg for 2 months.[6]

Carcinogenicity. Incomplete data have been reported concerning carcinogenic effect following *s/c* injections (administered amount unspecified).[7]

Standards. Russia. MAC and PML: 0.5 mg/l.

Regulations. USFDA (1993) approved the use of B. (1) in adhesives as a component of articles intended for use in packaging, transporting, or holding food, (2) in cross-linked polyester resins to be used safely as articles or components of articles intended for repeated use in contact with food, and (3) as an accelerator of rubber articles intended for repeated use in producing, manufacturing, packing, processing, treating, packaging, transporting, or holding food (total not to exceed 1.5% by weight of the rubber product).

References:
1. Smirnova, R. D. and Kosmina, L. F., Experimental data on substantiation of MAC for hydroperoxide isopropylbenzene in water bodies, *Gig. Sanit.,* 12, 17, 1971 (in Russian).
2. See **ZINC DIETHYLDITHIOCARBAMATE,** #2, 74.
3. *Proc. Research Hygiene Epidemiol. Institute,* Kuibyshev, 5, 107, 1968 (in Russian).
4. See **DIBENZOYL PEROXIDE,** #5.
5. Sheftel, V. O., Shquar, L. A., and Naumenko, G. M., Application of some genetic methods in hygiene studies, *Vrachebnoye Delo,* 7, 128, 1969 (in Russian).
6. *Hygienic aspects of the use of polymeric materials, New Methods of Hygienic Monitoring of the Use of Polymers,* Proc. 3rd All-Union Meeting, Stankevich, K. I., Ed., Kiev, December 2–4, 1981, p. 328 (in Russian).
7. Anon., Peroxides, genes and cancer, *Food Chem. Toxicol.,* 23, 957, 1985.

4,4'-METHYLENE BIS(2-CHLOROANILINE) (MOCA) (CAS No 101–14–4)

Synonyms. 4,4'-Diamino-3,3'-dichlorodiphenylmethane; 3,3'-Dichloro-4,4'-methylenedianiline.

Properties. Light-grey or cream-colored, crystalline powder. Solubility in water 13.6 mg/l (IARC 57–271), soluble in alcohol.

Applications. Used as a curing agent in the manufacturing of plastics and for vulcanization of isocyanate-containing polyurethane rubber, polyurethane foams, and glass-reinforced plastics.

Acute Toxicity. The LD_{50} values are reported to be 2.1 g/kg BW in rats and 0.88 g/kg BW in mice. In mice, poisoning is accompanied by adynamia, apathy, loss of appetite. Death occurs in 3 d. Manifestations of toxic action in rats are not very pronounced, and death occurs in 2 d.[1] The LOEL for methemoglobinemia formation appears to be 83 mg/kg. Gross pathological examination revealed distention of the stomach and intestine, traces of blood in the urinary bladder, and pleural effusion in the thorax. Histological examination detected fine-drop adiposis of the liver, tiny foci of inflammatory infiltration, and circulatory disturbances in the visceral organs.

Repeated Exposure failed to reveal evident cumulative properties. Administration of $^1/_5$ LD_{50} caused 50% mortality only on day 52. Manifestations of toxic action comprised decreased BW gain, impairment of liver functions and of gas exchange intensity. Gross pathological examination revealed morphological changes in the liver, myocardium, lungs, and spleen.

Genotoxicity. MOCA is comprehensively genotoxic in *in vitro* tests and in bacteria.[2,3]

Carcinogenicity. MOCA produces a variety of tumors in the liver, mammary glands, and urinary bladder.[3,4] According to Russfield et al.,[5] it induced lung and liver tumors in rats and mice. *Carcinogenicity classification.* IARC: Group 2A.

Chemobiokinetics. The doses of 28.1 mM ^{14}C-MOCA/kg BW were administered for 28 consecutive days. Distribution appears to be the highest in the liver, kidney, lung, spleen, testes, and urinary bladder.[6] According to Tobes et al.,[7] administration of 0.49 mg ^{14}C-MOCA/kg BW led to accumulation of radioactivity in the small intestine, liver, adipose tissue, etc. The major route of excretion seems to be via the feces.

References:
1. See *N*-**BUTYLIDENEANILINE,** 38.
2. Mori, H., Yoshimi, N., Sugie, S., et al., Genotoxicity of epoxy resin hardeners in the hepatocyte primary culture/DNA repair test, *Mutat. Res.,* 204, 683, 1988.
3. McQueen, C. A. and Williams, G. M., Review of genotoxicity and carcinogenicity of **4,4'**-methylenedianiline and **4,4'**-methylene-bis-2-chloroaniline, *Mutat. Res.,* 239, 133, 1990.
4. Stula, E. F., Barnes, J. R., Sherman, H., et al., Urinary bladder tumors in dogs from **4,4'**-methylenebis(2-chloroaniline) (MOCA), *J. Environ. Pathol. Toxicol.,* 1, 31, 1977.
5. Russfield, A. B., Homburger, F., Boger, E., et al., The carcinogenicity effect of **4,4'**-methylenebis-2-chloraniline in mice and rats, *Toxicol. Appl. Pharmacol.,* 31, 47, 1975.
6. Chever, K. L., DeBord, D. G., and Swearengin, T. E., **4,4'**-methylenebis (2-chloroaniline) (MOCA): the effect of multiple oral administration, route, and Phenobarbital induction on macromolecular adduct formation in rat, *Fundam. Appl. Toxicol.,* 16, 71, 1991.
7. Tobes, M. C., Brown, L. E., Chin, B., et al., Kinetics of tissue distribution and elimination of **4,4'**-methylene bis(2-chloroaniline) in rats, *Toxicol. Lett.,* 17, 69, 1983.

214

4,4'-METHYLENE BIS(CYCLOHEXYLAMINE)

Synonym. 4,4'-Diaminodicyclohexylmethane.
Properties. Brown, wax-like substance. Poorly soluble in water.
Applications. Used as a curing agent in the production of epoxy resins. Used in the synthesis of polyamide resins and in rubber production.
Acute Toxicity. LD_{50} is 270 mg/kg BW in rats and 197 mg/kg BW in mice. Administration of lethal doses is accompanied by NS depression, motor coordination disorder, tremor of the head and extremities, and clonic–tonic convulsions.
Short-term Toxicity. Rats tolerate a 3-month administration of $^1/_{10}$ and $^1/_5$ LD_{50} without any sign of intoxication.
Reference:
See ZINC DIETHYLDITHIOCARBAMATE, #2, 60.

4,4'-METHYLENEDIANILINE (CAS No 101–77–9)

Synonyms. Bis(aminophenyl)methane; Dianilinemethane; **4,4'**-Diaminodiphenylmethane; **4,4'**-Diphenylmethanediamine; **4,4'**-Methylene-bisbenzenamine.
Properties. Crystalline solid or pearly leaflets; colorless to pale-yellow thin flakes or lumps with a faint amine-like odor (IARC 39–348). Water solubility is 1 g/l at 25°C, highly soluble in ethanol, acetone, and diethyl ether.
Applications and **Exposure.** Used in the production of synthetic rubber, dyes, and polyamide resins. A curing agent in the synthesis of epoxy resins. A vulcanization accelerator for chloroprene rubber, a vulcanization activator. An antioxidant for vulcanizates based on natural and synthetic rubber. A monomer in the production of polydiisocyanates.
Acute Toxicity. *Humans.* Collective poisoning with bread made of flour contaminated with M. has been reported. The liver seems to be the target organ (Shoental). *Animals.* LD_{50} is 665 mg/kg BW in rats and 264 mg/kg BW in mice.[1] Lethal doses produced CNS inhibition, motor coordination disorder, tremor of the head and extremities, and clonic–tonic convulsions.
Repeated Exposure revealed moderate cumulative properties. Administration of $^1/_{10}$ and $^1/_5$ LD_{50} produced some rat mortality. Short-term excitation and aggressiveness were followed by adynamia.[2] Hepatotoxic effects were described in rats gavaged with 8 to 600 mg M./kg BW for 10 d (necrotizing cholangitis, periportal necrosis, and glycogen loss at the 200 mg/kg dose level and higher).[3] A dose-related reduction in BW gain was found in male Fisher 344 rats exposed to 1.6 and 3.2 g **M. dihydrochloride**/l in their drinking water for 14 d.[4] There were increased weights of the adrenal gland, uterus, and thyroid gland in ovariectomized female Sprague-Dawley rats given 14 daily doses of 150 mg **M. dihydrochloride**/kg BW by gavage.[5]
Short-term Toxicity. Atrophy of the liver parenchyma and increased relative spleen weights associated with hyperplasia of the lymphatic system were observed in Wistar rats given 83 mg/kg BW for 12 weeks.[6] Cirrhosis occurred in all the 21 male rats exposed to doses of 38 mg/kg by gavage for 17 weeks.[7] Bile-duct hyperplasia was observed in rats following oral administration (in the diet or in drinking water).[4,8,9]
Long-term Toxicity. In a 40-week study, bile duct proliferation was found in male Wistar rats fed a diet containing 1 g M./kg. The hepatic parenchyma was replaced by proliferating bile ducts, and eventually portal cirrhosis developed. These alterations reversed when treatment was discontinued.[10] Mineralization of the kidney was revealed in rats and mice exposed to M. in their drinking water for 2 years.[4]
Genotoxicity. M. appeared to be mutagenic to *Salmonella*. It induced DNA damage in Chinese hamster cells and in the liver of rats, and SCE in the bone marrow of mice treated *in vivo*.[11]
Carcinogenicity. B6C3F$_1$ mice were given 150 or 300 mg **M. dihydrochloride**/l in the drinking water for 103 weeks. Treatment-related increase in the incidence of thyroid follicular-cell adenomas and hepatocellular neoplasms was observed. Similar effect was found in Fisher 344 rats under the same conditions. The incidence of thyroid follicular-cell carcinomas in the high-dose group of animals was significantly increased as compared to the controls.[1,12] *Carcinogenicity classification.* NTP: P*—P—P*—P*.

Chemobiokinetics. M. metabolism occurs to form *N*-**acetyl**- and *N,N'*-**diacetyl derivatives.** Metabolites are removed in the urine.

Regulations. USFDA (1993) has approved the use of **4,4'**-M. (1) as a catalyst and a cross-linking agent for epoxy resins, (2) as a catalyst for polyurethane resins, (3) in resinous and polymeric coatings in a food-contact surface for use only in coatings for containers with a capacity of 1000 gal (37.85 1) or more, when the containers are intended for repeated use in contact with alcoholic beverages containing up to 8% alcohol by volume, and (4) as an antioxidant for rubber articles intended for repeated use in contact with food (up to 5% by weight of the rubber product).

References:
1. See **ZINC DIETHYLDITHIOCARBAMATE,** #2, 60.
2. See **ACRYLONITRILE,** #1, 60.
3. Gohike, R. and Schmidt, P., **4,4'**-Diaminodiphenylmethane: histological, enzyme histological and autoradiographic investigations in acute and subacute experiments in rats with and without additional heat stress, *Int. Arch. Arbeitsmed.,* 32, 217, 1974 (in German).
4. NTP, *Carcinogenesis Studies of* **4,4'***-Methylenedi Aniline Dihydro-chloride (CAS No 13552–44–8) in Fisher-344/N Rats and B6C3F₁ Mice (Drinking Water Studies),* Technical Report No. 248, Research Triangle Park, NC, 1983.
5. Tullner, W. W., Endocrine effects of methylenedianiline in the rat, rabbit and dog, *Endocrinology,* 66, 470, 1960.
6. Pludro, G., Karlowski, K., Mankowska, M., et al., Toxicological and chemical studies of some epoxy resins and hardeners. I. Determination of acute and subacute toxicity of phthalic acid anhydride, **4,4'**-diaminodiphenylmethane and of the epoxy resin: Epilox EG-34, *Acta Pol. Pharmacol.,* 26, 352, 1969.
7. *Bladder Cancer, A Symposium,* Deichmann, W., and Lampe, K., Eds., Aesculapius, Birmingham, AL, 1967, 187.
8. Miyamoto, J., Okuno, Y., Kadota, T., et al., Experimental hepatic lesions and drug metabolizing enzymes in rats, *J. Pest. Sci.,* 2, 257, 1977.
9. Gohike, R., **4,4'**-Diaminodiphenylmethane in a chronic experiment, *Zeitschr. Gesumte Hyg.,* 24, 159, 1978 (in German).
10. Fukushima, S., Shibata, M., Hibino, T., et al., Intrahepatic bile duct proliferation induced by **4,4'**-diaminodiphenylmethane in rats, *Toxicol. Appl. Pharmacol.,* 48, 145, 1978.
11. McQueen, C. A. and Williams, G. M., Review of genotoxicity and carcinogenicity of **4,4'**-methylenedianiline and **4,4'**-methylene-bis-2-chloroaniline, *Mutat. Res.,* 239, 133, 1990.
12. Weisburger, E. K., Murthy, A. S. K., Lilja, H. S., et al., Neoplastic response of Fisher-344 rats and B6C3F₁ mice to the polymer and dyestuff intermediates, **4,4'**-methylene-bis(*N,N'*-dimethyl) benzenamine, **4,4'**-oxydianiline, and **4,4'**-methylene-dianiline, *J. Natl. Cancer Inst.,* 72, 1457, 1984.

METHYL ETHYL KETONE PEROXIDE (MEKP) (CAS No 1338–23–4)

Synonyms. 2-Butanone peroxide; Di(1-methylethylketone) peroxide.

Properties. Colorless, oily liquid with a pleasant odor.

Applications and **Exposure.** Used in the manufacturing of acrylic resins, as a hardening agent for fiberglass-reinforced plastics, as a polymerization initiator and curing agent for unsaturated polyester resins. It is commercially available as a 40 to 60% solution in **dimethyl phthalate** (DMP).

Acute Toxicity. *Humans.* Ingestion by man of 50 ml **Norox catalyst** (60% MEKP, cyclohexanone, and DMP) caused fatal poisoning.[1] *Animals.* LD$_{50}$ is 470 to 484 mg/kg BW in rats and 260 mg/kg BW in mice.[1,2] Manifestations of toxic action included GI tract bleeding, abdominal burns, necrosis, perforation of the stomach, stricture of the esophagus, severe metabolic acidosis, rapid hepatic failure, and respiratory insufficiency.[3,4]

Repeated Exposure. Skin lesions have been noted following dermal exposure. Topical administration of MEKP + DMP resulted in necrotic inflammatory and regenerative skin lesions limited to the application site. MEKP is highly irritating and corrosive to the skin and mucous membranes.[5]

Reproductive Toxicity. MEKP administered into the air chamber was toxic to 3-d chicken embryos: lethality and malformations were noted in embryos.[6]

Genotoxicity. MEKP was not shown to be positive in *Salmonella* mutagenicity assay (MEKP and DMP 45:55 w/w). It induced SCE and CA in Chinese hamster ovary cells. No increase in the frequency of micronucleated erythrocytes was observed in the peripheral blood samples obtained from mice at the termination of the 13-week study.[6]

Carcinogenicity. MEKP induced malignant lymphomas in C57Bl mice (treatment route and regimen of administration were not specified).[7] The dose of 10 mg MEKP per mouse showed a weak tumor-promoting activity when applied topically to the skin of mice that had been UV-irradiated.[8]

Regulations. USFDA (1993) has regulated MEKP as an indirect food additive. It is permitted (1) as a catalyst in the production of resins to be used at levels not exceeding 2% of the finished resin and (2) in cross-linked polyester resins to be used safely as articles or components of articles intended for repeated use in contact with food.

References:

1. See **ZINC DIETHYLDITHIOCARBAMATE,** #2, 74.
2. Floyd, E. P. and Stockinger, H. E., Toxicity studies of certain organic peroxides and hydroperoxides, *Am. Ind. Hyg. Assoc. J.,* 19, 205, 1958.
3. Mittleman, R. E., Romig, L. A., and Gressman, E., Suicide by ingestion of methyl ethyl ketone peroxide, *J. Forensic Sci.,* 31, 312, 1986.
4. Korhunen, P. J., Ojanpera, I., Lalu, K., et al., Peripheral zonal hepatic necrosis caused by accidental ingestion of methyl ethyl ketone peroxide, *Hum. Exp. Toxicol.,* 9, 197, 1985.
5. NTP, *Toxicity Studies of Methyl Ethyl Ketone Peroxide in Dimethyl Phthalate (45:55) Administered Topically to Fisher-344/N Rats and B6C3F₁ Mice,* Technical Report, U.S. Department of Health and Human Services, Publ. Health Service, NIH, Washington, D.C., 1992, 58.
6. Korhonen. A., Hemminki, K., and Vainio, H., Embryotoxic effects of eight organic peroxides and hydrogen peroxide on three-day chicken embryos, *Environ. Res.,* 33, 54, 1984.
7. Kotin, P. and Falk, H. L., Organic peroxides, hydrogen peroxide, epoxides and neoplasma, *Radiat. Res.,* Suppl. 3, 193, 1963.
8. Logani, M. K., Sambuco, C. P., Forbes, P. D., et al., Skin-tumor promoting activity of a potent lipid peroxidizing agent, *Food Chem. Toxicol.,* 22, 879, 1984.

1,3-PHENYLENEDIAMINE (CAS No 108–45–2)

Synonyms. *m*-Aminoaniline; *m*-Benzenediamine; *m*-Diaminobenzene; *m*-Phenylenediamine.

Properties. Colorless or faintly tinted rhombic crystals, which acquire a violet color in the light and have an unpleasant odor. Readily soluble in water and alcohol.

Applications and **Exposure.** Used as a curing agent for epoxy compositions. Antioxidant in a variety of applications. Migration from the epoxy coatings (*ED*-16 resin) into water (24-h exposure at 20°C) was determined at the level of 0.013 mg/l; migration from the epoxy coatings (*ED*-20 resin) into 96% ethanol (10-d exposure at 50°C) was found to be 0.15 mg/l.[1]

Acute Toxicity. LD_{50} is in the range of 280 to 350 mg/kg BW in rats, 65 to 72 mg/kg BW in mice, 437 mg/kg BW in rabbits, and 450 mg/kg BW in guinea pigs. There are certain differences in species sensitivity. Lethal doses caused immediate respiratory disturbances and altered NS function. Death occurred in 3 to 48 h after administration. Methemoglobin formation was not observed. Gross pathological examination revealed lung and brain edema and granularity of the liver tissue.[2] A threshold dose for acute effect on coproporphyrin concentration in the urine is 10 mg/kg BW.

Repeated Exposure revealed slight cumulative properties. The possibility of functional accumulation is reported, despite there being no fatal consequences.[2] K_{acc} (by Kagan) appeared to be 3.3 (administered dose 0.5 mg/kg BW) or 1 (administered dose 0.1 mg/kg BW).

Long-term Toxicity. Rats were dosed by gavage with 10 mg/kg for 7 months. Methemoglobinemia was not noted in animals. The NOAEL in rats appeared to be 0.02 mg/kg BW for the effect on hematology parameters and for coproporphyrin content in the urine.[3] B6C3F₁ mice were exposed to 0.02 or 0.04% *m*-PDA in their drinking water for 78 weeks. There was no increase in mortality; BW was found to be significantly lower in the groups of high-dosed females and males.[3]

Allergenic Effect was observed in a chronic toxicity study in rats and guinea pigs.[2]

Genotoxicity. *m*-PDA appeared to be mutagenic in *Salmonella* assay.[4]

Carcinogenicity. A single case of sarcoma was reported in the group of five rats given *s/c* injections of ¹/₁₀ LD_{50} every other day for 11 months.[5] Male Charles River CD rats and ICR mice were fed *m*-**PDA hydrochloride** at doses of 1 to 4 g/kg of their diets. No increase in tumor incidence was noted in this study.[6]

The maximum tolerated dose tested in the study[4] severely inhibited BW gain but did not produce carcinogenic effect. *m*-PDA in this study showed no carcinogenic potential in B6C3F$_1$ mice when administered in drinking water.

Standards. EEC (1990). MPQ in finished material or article: 1 mg/kg. **Russia** (1988). MAC and PML in drinking water: 0.1 mg/l; PML in food: 0.005 mg/l.

References:

1. *Handbook on the Hygiene of Polymer Use,* Stankevich, K. I., Ed., Zdorov'ya, Kiev, 1984, 106 (in Russian).
2. Myannik, L. V., Toxicological and Hygienic Characteristics of **m**-Phenylenediamine and of Epoxy Coatings Cured with it Used in the Food Industry, Author's abstract of thesis, Kiev Medical Institute, 1981, 17 (in Russian).
3. Amo, H., Matsuama, M., Amano, H., et al., Carcinogenicity and toxicity study of **m**-phenylenediamine administered in the drinking water to (C57BL/6×C3H/He)F$_1$ mice, *Food Chem. Toxicol.,* 26, 893, 1988.
4. Ames, B. N., Kammen, H. O., and Yamasaki, E., Hair dyes are mutagenic: identification of a variety of mutagenic ingredients, *Proc. Natl. Acad. Sci. U.S.A.,* 72, 2423, 1975.
5. Saruta, N., Yamaguchi, S., and Matsuoka, T., Sarcoma produced by subdermal administration of **m**-phenylene-diamine and **m**-phenylenediamine hydrochloride, *Kyushu. J. Med. Sci.,* 13, 175, 1962 (in Japanese).
6. Weisburger, E. K., Russfield, A. B., Homburger, F., et al., Testing of twenty-one environmental aromatic amines or derivatives for long-term toxicity or carcinogenicity, *J. Environ. Pathol. Toxicol.,* 2, 325, 1978.

1,4-PHENYLENEDIAMINE (CAS No 106–50–3)

Synonyms. *p*-Aminoaniline; *p*-Benzenediamine; *p*-Diaminobenzene; *p*-Phenylenediamine; Santoflex *LC;* Ursol D.

Properties. Almost colorless, thin leaves rapidly darkening in air. Solubility in water is 3.24 g per 100 ml, soluble in alcohol. Water color-change threshold is 0.1 mg/l. A concentration of 1 g/l is odorless, and that of 0.1 g/l is colorless. The color of *p*-PDA solutions increases in the course of storage (Zhakov).

Applications. *p*-PDA is used predominantly as a curing agent in the production of epoxy materials and coatings; a vulcanizing agent and accelerator for rubber; an antioxidant.

Acute Toxicity. LD$_{50}$ was reported to be 133 to 180 mg/kg BW in rats and 145 mg/kg BW in guinea pigs. Lethal doses led to liver damage and metHb formation.[1] However, according to Lloyd et al., LD$_{50}$ was found to be 98 mg/kg in rats.[2]

Short-term Toxicity. Mice were fed dietary levels of 0.4% *p*-PDA for 12 weeks. The animals exhibited decreased BW gain and increased relative weights of the liver and kidneys. Some of the animals died. The doses of 0.2% and less produced no remarkable toxicity.[3]

Long-term Toxicity. The effect upon glycogen-forming function of the liver has been reported in rabbits.[1] In rats exposed to 0.1% *p*-PDA in the feed, increased splenic and body relative weights were noted (Katsumi et al.).

Reproductive Toxicity. No malformations were observed in a reproductive toxicity study. An *i/p* dose of 20 mg/kg caused no DLM in rats.[4]

Genotoxicity. Inadequate results were obtained in the *in vivo* micronuclei assay in bone marrow cells of rats exposed orally to *p*-PDA. Treatment of *p*-PDA solution with hydrogen peroxide revealed a strong mutagenic effect in *Salmonella* in conditions of metabolic activation.[5]

Carcinogenicity. According to NCI Report-79, *p*-PDA caused no carcinogenic response in Fisher 344 rats or B6C3F$_1$ mice given 625 and 1250 ppm in their feed. In an 80-week feeding study in Fisher 344 rats, concentration of 0.1% *p*-PDA was found to be a NOEL for carcinogenic effect.[3]

Carcinogenicity classification. NTP: N—N—N—N.

Allergenic Effect. *p*-PDA has a pronounced sensitizing potential. Sensitization seems to be caused not by *p*-PDA itself but by its partial oxidation product *p*-**quinonediamine.** Some researchers reported a sensitizing effect on dermal application (Kurlyandsky et al., 1987).

Chemobiokinetics. Following ingestion, *p*-PDA is nearly completely absorbed in Fisher 344 rats and B6C3F$_1$ mice. The highest radioactivity was present in the muscles, skin, and liver.[6] Metabolism occurs through oxidation to form **quinonediamine.** Beside it, *p*-PDA is partially acetylated to a **diacetyl derivative.** *p*-PDA or its metabolites do not bind covalently to hepatic DNA. Rats and mice rapidly cleared radioactivity from their tissues. In 24 h only 10 to 15% radioactivity was still present in the animal body. *p*-PDA is excreted predominantly through the urine (70 to 90%) and also in the feces (10 to 20%). Of radioactivity in the urine 95% is *p*-PDA metabolites.

Standards. Russia (1988). MAC and PML: 0.1 mg/l.

References:

1. See **ACRYLONITRILE,** #1, 230.
2. Lloyd, G. K., Ligett, M. P., Kynoch, S. R., et al., Assessment of the acute toxicity and potential irritancy of hair dye constituents, *Food Cosmet. Toxicol.,* 15, 607, 1977.
3. Imaida, K., Ishihara, Y., Nishio, O., et al., Carcinogenicity and toxicity tests on *p*-phenylene-diamine in Fisher-344 rats, *Toxicol. Lett.,* 16, 259, 1983.
4. Barnett, C., Loehr, R., and Corbett, J., Dominant lethal mutagenicity study on hair dyes, *J. Toxicol. Environ. Health,* 2, 657, 1977.
5. Nishi, K. and Nishioka, H., Light induces mutagenicity of hair-dye *p*-phenylenediamine, *Mutat. Res.,* 104, 347, 1982.
6. Ioannou, Y. M. and Matthews, H. B., *p*-Phenylenediamine: comparative disposition in male and female rats and mice, *J. Toxicol. Environ. Health,* 16, 299, 1985.

PIPERIDINE (CAS No 110–89–4)

Synonyms. Hexahydroperidine; Pentamethylenimine.

Properties. Colorless liquid with a penetrating ammonia-like odor. Readily miscible with water and alcohols.

Applications. Used as a curing agent in the synthesis of epoxy resins. P. derivatives are used as vulcanization accelerators for rubber.

Acute Toxicity. *Humans.* General weakness, muscular paralysis, labored breathing, and asphyxia was noted in individuals who ingested 30 to 60 mg P./kg BW.[05] *Animals.* In rats, LD$_{50}$ is 50 mg/kg BW.[1] Water solutions of P. are rather less toxic. LD$_{50}$ of 8% aqueous solution is 371 mg/kg BW.[03] A dose of about 250 mg/kg BW decreased the neuromuscular excitability threshold as well as body temperature and increased the arterial pressure. Gross pathological examination revealed acute vascular disturbances, necrosis of the gastric mucosa, and pneumonia, as well as signs of parenchymatous dystrophy of the liver and renal tubular epithelium.

Repeated Exposure revealed pronounced cumulative properties in the course of treatment with $^1/_{10}$ LD$_{50}$. K$_{acc}$ appeared to be 0.6 (by Kagan).

Reproductive Toxicity. *Embryotoxic* effect is not specific.[03]

Carcinogenicity. Negative results have been reported in rats.[09]

Chemobiokinetics. P. is readily absorbed in the GI tract.

Standards. Russia. PML: 0.07 mg/l.

Reference:

Toxicology of New Industrial Chemical Substances, Issue 9, Meditsina, Moscow, 1967, p. 91 (in Russian).

POLYETHYLENEPOLYAMINES (PEPA)

Composition and **Properties.** Mixtures of variable composition containing ethylenediamine (n = 1), triethylenetetramine (n = 3) up to 90%, diethylenetriamine (n = 2), and other more complex amines. Oily liquids of yellow to dark-brown color with a penetrating odor. Readily soluble in water (see **Ethylene diamine**). Odor perception threshold is 5.6 mg/l; taste perception threshold is 10 mg/l.[1]

Applications. Used as curing agents in the synthesis of epoxy and ion-exchange resins and rubber; a vulcanization accelerator for rubber.

Acute Toxicity. LD$_{50}$ is reported to be 2 to 3 g/kg BW in rats, 1.8 to 2 g/kg BW in mice, and 0.82 mg/kg BW in guinea pigs.[1] Gross pathological examination revealed necrotic lesions in the stomach as well as liver affection, circulatory disturbances, and changes in the biogenic amine level. Administration of a 4 g/kg BW dose to mice caused increased excitability with subsequent adynamia, labored breathing, disappearance of tendon reflexes, and convulsions. Gross pathological examination revealed visceral congestion and necrotic foci in the GI tract mucosa.[2]

Repeated Exposure failed to reveal cumulative properties. K_{acc} is 11.8 (by Cherkinsky).[2] Administration of $^1/_{10}$ to $^1/_5$ LD_{50} to rats and mice for 45 d caused increased oxygen consumption, reticulocytosis, a decline in cholinesterase activity, and other changes.[1]

Short-term Toxicity. Rats tolerate administration of $^1/_{100}$ to $^1/_5$ LD_{50} for 3 months. BW gain, hematological analyses, and STI are unchanged.[2]

Long-term Toxicity. No manifestations of toxic action were found in rats and rabbits exposed to 0.5 mg/kg BW via drinking water for 6 months.[2] Higher doses caused changes visible on gross pathological examination: increased liver and kidney weights, congestion, hemorrhages, and granular dystrophy in the parenchymatous organs.[1]

Reproductive Toxicity. *Gonadotoxicity.* In a 6-month study, sloughing of spermatogenic epithelium and testicular dystrophic changes were revealed in rats exposed to the 0.5 mg/kg BW dose.[1] The treatment disturbed the estrus cycle and caused decreased sperm motility. *Embryotoxicity.* Reduced fetal weights and increased mortality were observed.[1] However, according to other data,[3] administration of $^1/_{100}$ and $^1/_5$ LD_{50} to pregnant female rats had no embryotoxic effect.

Standards. Russia (1988). MAC and PML: 0.005 mg/l.

References:
1. Antonova, V. I., Salmina, Z. A., and Vinokurova, T. V., Toxicity of polyethylenepolyamine and substantiation of its maximum allowable concentration in water bodies, *Gig. Sanit.*, 2, 32, 1977 (in Russian).
2. See **TETRAHYDROFURAN,** # 10, 35.
3. Sheftel, V. O., Hygienic Aspects of the Use of Polymeric Materials in Water Supply, Author's abstract of thesis, VNIIGINTOX, Kiev, 1977, p. 156 (in Russian).

PYROMELLITIC DIANHYDRIDE (PD) (CAS No 89–32–7)

Properties. Triclinic plates (P. acid). Solubility in water is 15 g/l, freely soluble in alcohol.

Applications. Used as a curing agent in the production of epoxy and polyester resins. Used in the manufacturing of aromatic polyamides (synthetic fibers manufacturing) and plasticizers.

Acute Toxicity. LD_{50} is reported to be 2.25 to 3.5 g/kg BW in rats, 2.3 to 2.65 g/kg BW in mice, and 1.6 g/kg BW in guinea pigs.[1,2] Poisoning is accompanied by CNS depression, including behavioral changes (ataxia, somnolence, adynamia, muscle weakness, etc.). Death within the first 3 d.

Repeated Exposure revealed little evidence of cumulative properties.[2,3] However, Kondratyuk et al.[1] believe PD has pronounced cumulative potential: LD_{50}/LOEL = 256.

Rats were given $^1/_7$ LD_{50} for 1.5 months. Manifestations of toxic action included CNS inhibition, erythropenia, and changes in kidney relative weights. Disturbances in hydrocarbon-phosphate and nitrite-protein metabolism as well as alteration of the renal function were reported in rats gavaged with 8.8 to 220 mg/kg BW for 1 month.

Long-term Toxicity. Female rats exposed to 4.4 and 8.8 mg/kg BW for 6 months developed changes in the activities of a number of enzymes (cholinesterase, AST and ALT, cytochromoxidase, etc.). Gross pathological and histological changes were evident.[1]

Allergenic Effect. PD exhibits mild sensitizing properties on skin application of 0.2 mg/kg BW to guinea pigs.

Reproductive Toxicity. *Embryotoxicity.* No such effect was observed in rats given oral doses of 44 and 220 mg/kg BW.

Gonadotoxicity. In rats exposed to the oral dose of 44 mg/kg, a weak effect on gonad morphology was observed. The LOAEL for this effect appeared to be the 8.8 mg/kg BW dose.

Standards. Russia. Recommended PML in drinking water: 0.4 mg/l.

References:
1. Kondratyuk, V. A., Gun'ko, L. M., Pereima, V. Ya., et al., Hygienic substantiation of the maximum allowable concentration for pyromellitic dianhydride in water bodies, *Gig. Sanit.*, 11, 79, 1986 (in Russian).
2. See **BISPHENOL A,** #2, 230.

220

STYRENE OXIDE (StO) (CAS No 96–09–3)

Synonyms. (Epoxyethyl)benzene; Epoxystyrene; Phenylethylene; Phenyloxirane; Styrene-**7,8**-oxide.

Properties. Colorless to straw-colored liquid with a pleasant odor. Water solubility is 0.28%, miscible with alcohol. Odor threshold concentration is 0.063.[010]

Applications. Used in the production of epoxy plastics as a catalyst, a cross-linking agent, and a reactive diluent for epoxy resins, and also for manufacturing of coatings.

Acute Toxicity. The LD_{50} values are reported to be 2 to 4.3 g/kg BW in rats, 2.8 g/kg BW in rabbits, and 2 g/kg BW in guinea pigs.[1,2]

Reproductive Toxicity. Inhalation of StO by rats and hamsters had a certain toxic effect on embryos but caused no teratogenic effect (IARC 36–250).

Genotoxicity. In mammalian cell cultures, including Chinese hamster cell lines, mouse cell lines, Wistar rat hepatocytes, and human lymphocytes,[3-5] StO was genotoxic, producing mutations, CA, micronuclei, and anaphase bridges. In human lymphocytes (whole blood cultures), StO was clustogenic and produced CA, micronuclei, and SCE. In rodents exposed to StO, CA were seen in mice and rats but not in Chinese hamsters.[3-5] StO initiated CA in the bone marrow cells of CD-1 mice given in doses of 50 to 1000 mg/kg BW.[6] StO (~5 mg per plate) induced mutations (spot test) in *Salmonella.*[7,8]

Carcinogenicity. Male and female Sprague-Dawley rats were administered 50 and 250 mg/kg BW by gavage for 52 weeks.[9,10] There was a high incidence of forestomach epithelial tumors, including papillomas, *in situ* carcinomas, and invasive carcinomas. A dose-response was demonstrated for these tumors. No tumors were reported in the control animals. *Carcinogenicity classification.* IARC: Group 2A.

Chemobiokinetics. StO is the main metabolite of **styrene** (q.v.), formed under the action of cytochrome P-450-dependent monoxygen of the liver. StO metabolism may occur via the formation of conjugates with glutathione. Thioether metabolites of StO, which are precursors of the mercapturic acids, failed to show mutagenic activity.[11] StO is also found to be transformed into **phenylethyleneglycol** by microsomal epoxyhydrases of the liver, kidneys, and other organs.[12] Metabolites are removed by the kidneys.

Regulations. USFDA (1993) approved the use of StO as (1) a cross-linked agent for epoxy resins in coatings for containers with a volume of 1000 gal (3785 l) or more intended for repeated use in contact with alcoholic beverages containing up to 8% ethanol by volume, (2) a component of paper and paperboard for contact with dry food, (3) in cross-linked polyester resins for repeated use as articles or components of articles coming in contact with food, (4) in resinous and polymeric coatings to be used safely in a food-contact surface.

References:
1. Smyth, H. F., Carpenter, C. P., Weil, C. S., et al., Range-finding toxicity data: list VII, *Am. Ind. Hyg. Assoc. J.,* 30, 470, 1969.
2. See **BISPHENOL A, DIGLYCIDYL ETHER,** #1.
3. Norppa, H. and Vainio, H., Genetic toxicity of styrene and some of its derivatives, *Scand. J. Work Environ. Health,* 9, 108, 1983.
4. *Biological Reactive Intermediates—LL: Chemical Mechanisms and Biological Effects,* Part A Snyder, R., Parke, D. V., Kocisis, J., et al., Eds., Plenum Press, New York, 1982, 207.
5. Vainio, H., Norppa, H., and Belevedere, G., Metabolism and mutagenicity of styrene oxide, *Prog. Clin. Biol. Res.,* 141, 215, 1984.
6. Loprieno, N., et al., *Scand. J. Work Environ. Health,* 4 (Suppl. 2), 169, 1978.
7. Milvy, P. and Garro, A., Mutagenic activity of styrene oxide, a presumed styrene metabolite, *Mutat. Res.,* 40, 15, 1976.
8. de Meester, C., Poncelet, F., Roberfroid, M., et al., Mutagenic activity of styrene and styrene oxide. A preliminary study, *Arch. Int. Physiol. Biochim.,* 85, 398, 1977.
9. Maltoni, C., Failla, G., and Kassapidis, G., First experimental demonstration of the carcinogenic effects of styrene oxide, *Med. Lavoro,* 5, 358, 1979 (in Italian).
10. Maltoni, C., Early results of the experimental assessment of the carcinogenic effects of one epoxy solvent: styrene oxide, in *Occupational Health Hazards of Solvents,* Vol. 2., Englund, A., Ringen, K., and Mehlman, M. A., Eds., Princeton Science Publishers, Princeton, NJ, 1982, 97.

11. Pogano, D. A., Yagen, B., Hernandez O., et al., Mutagenicity of (R) and (S) styrene-**7,8**-oxide and the intermediary mercapturic acid metabolites formed from styrene-**7,8**-oxide, *Environ. Mut.*, 5, 575, 1982.
12. Oesch, F., Jerina, D. M., Daly, J. W., et al., Induction, activation, and inhibition of epoxide hydrolase, *Chem. Biol. Interact.*, 6, 189, 1973.

TIN INORGANIC COMPOUNDS

Properties. The most important inorganic compounds of *Sn* are **oxides, chlorides, fluorides,** and **halogenated sodium stannates and stannites.** Some *Sn* salts are water-soluble.

Applications and **Exposure.** *Sn* is a component of a number of catalysts: organotin compounds are applied as PVC stabilizers. *Sn* is used principally in the production of coatings in the food industry. Food, and particularly canned food, therefore represents the major route of human exposure to *Sn* compounds. Higher concentrations are found in canned food as a result of dissolution of the *Sn* coating or *Sn* plate. Acidity of the food, the presence of oxidants, time and temperature of storage, and presence of air in the can headspace influence this process. **Tin** concentrations in foodstuffs in unlacquered cans frequently exceed 0.1 mg/g (WHO, 1989). Although *Sn* is a natural component of food products, its physiological significance is unknown. An intake of 1 to 30 mg *Sn* per d consumed in the food, is 1000 times greater than the amount consumable in water.

Acute Toxicity. *Humans.* The main adverse effects with excessive levels of *Sn* compounds in food, such as canned fruit, has been acute gastric irritation. *Sn* compounds act as an irritant for the GI tract mucosa. Vomiting, diarrhea, fatigue, and headache were seen in a lot of cases following the consumption of canned products (*Sn* compound concentrations as low as 150 mg/kg in one incident involving canned beverages and 250 mg/kg in other canned foods).[1] *Animals.* In the mouse and rat, LD_{50} values are 250 and 700 mg/kg BW for *Sn* **chloride,** 2.14 and 2.2 g/kg BW for *Sn* **sulfate,** and 2.7 and 4.35 g/kg BW for **sodium stannate hydrate,** respectively. Both rats and mice tolerate administration of 10 g *Sn* **oxide**/kg BW (II, IV). The rabbit is less sensitive: LD_{50} is 10 g/kg BW.[2] High doses seem to affect the CNS, producing ataxia. The species-related differences illustrated by the LD_{50} values are apparent even at low levels of exposure. The cat appeared to be more sensitive to oral administration of **tin** compounds than either the dog or the rat, only in cats vomiting and diarrhea were observed after oral administration of **tin**-containing fruit beverages.[1] Poisoning by inorganic salts of *Sn* compounds is characterized by a brief period of excitation, which gives way to general inhibition. Intoxication is manifested through transient digestive upset and by apathy. Poisoning symptoms disappear in 2 to 3 d, depending on the dose. At autopsy, sacrificed animals show distention of the stomach.[3]

Repeated Exposure. *Humans.* In nine male volunteers consuming packaged military rations (*Sn* compounds contents ranging from 13 to 204 mg/kg) for successive 24-d periods and in other cases with human volunteers who ate canned food with *Sn* compounds contents varying from 250 to 700 mg/kg over a period of 6 to 30 d, no toxic effects were noted.[2] Toxic signs following the consumption of tin-containing food could be seen only after administration of tin concentrations of about 1400 ppm and higher.[4] *Animals.* Hb concentration in the blood of rats decreased significantly after feeding a diet containing 150 mg **tin** compounds per kilogram.[6]

Short-term Toxicity. In the 4- and/or 13-week feeding studies, rats were given various *Sn* salts or **tin oxide** at dose-levels of 50 to 10000 mg/kg BW. Doses of 3 g/kg and more caused anemia, changes in tissue enzyme activities, and extensive damage to the liver and kidney. These findings were especially observed with the more soluble *Sn* salts, like chloride, o-phosphate, sulfate, etc.[1,5] Biochemical effects attributable to **tin** intoxication have been observed, even after oral administration of 1 and 3 mg/kg BW.[6] These doses reflect 10 and 30 ppm **tin** in the diet. The relative weight of the femur, calcium concentration, and lactic dehydrogenase and alkaline phosphatase activity in the serum appeared to be significantly decreased in rats given the highest dose of **tin chloride.** The NOAEL was considered to be less than 0.6 mg/kg BW in this study.

Long-term Toxicity. *Humans.* There is no evidence of adverse effects in man associated with chronic exposure to **tin** compounds. *Animals.* Toxic effects can be caused by the ingestion of rather high doses of **tin.** Rats were exposed to 200 to 800 mg *Sn* **chloride**/kg food for 115 weeks. No

222

histopathological changes were observed. The NOAEL of 20 mg/kg BW was identified in this study.[1] No effects on survival or retardation of BW gain were observed in mice that received 1 and 5 g *Sn* **chlorostannate**/l or 5 g *Sn* **oleate**/l over a period of 1 year.[1]

Immunotoxicity. *I/p* administration of 167 mol $SnCl_2$/kg BW for 3 d caused a decline in the immune response to sheep erythrocytes. Liver weight was increased, and production of antibody-forming cells slowed down.[7]

Reproductive Toxicity. *Sn* has not been shown to be teratogenic or fetotoxic in mice, rats, and hamsters. Testicular degeneration was shown in rats administered 10 mg *Sn* **chloride**/kg BW in the feed for 13 weeks.[5] This *Sn* salt at doses of 200 to 800 mg/kg feed did not affect reproductive performance of rats, but transient anemia was observed in the offspring prior to weaning.[1] Low transplacental transfer of *Sn* compounds was observed after feeding of different **tin** salts (concentration of 500 mg/kg in the diet) to pregnant rats, although no effects were seen in the fetuses.[1]

Genotoxicity. Tin chloride produced a negative response in the SCE test system.[11]

Carcinogenicity. Carcinogenicity bioassays did not show an increase in tumor incidence in mice and rats,[1,8] except for one case.

Chemobiokinetics. *Sn* inorganic compounds are poorly absorbed from the GI tract. Only a small percentage of **sodium stannate hydrate,** for example, is absorbed. In presence of the citric acid, often found in fruit juice, the absorption of **tin** is increased.[9] About 50% of the dose was shown to be absorbed by man, if 0.11 mg of **tin** is ingested with the diet.[10] *Sn* compounds do not accumulate in the tissues. Administration in a 1.2 mg *Sn*/l solution for 36 d was not accompanied by accumulation in the body. The highest tissue concentration was found in the bones (principal site of distribution). Metabolism of *Sn* compounds is preceded by alkylation and dealkylation processes. Methylation of Sn^{4+} is initiated by sedimentation microorganisms, and it occurs along with abiotic methylation. *Sn* compounds are rapidly excreted from the body, primarily in the urine and feces (up to 99% of administered dose).

Recommendations. WHO (1982). ADI: 2 mg/kg. **WHO** (1989). PWTI: 14 mg/kg. **WHO** (1991) concluded that due to the low toxicity of **inorganic tin,** a tentative guideline value could be derived three orders of magnitude higher than the normal **tin** concentration in drinking water. Therefore, the presence of *Sn* in drinking water does not present a hazard to human health. For that reason, the establishment of a numerical guideline value was not deemed necessary (1991).

Regulations. Russia. The *Sn* content in canned food is regulated as 100 to 200 mg/kg for condensed milk, meat, fish, and vegetable products. **Czechoslovakia.** 10 mg/kg is permitted in milk.

References:
1. *Toxicological Evaluation of Certain Food Additives and Contaminants,* Food Additives Series 17, JECFA, WHO, Geneva, 1982, 297.
2. IPSC, *Tin and Organotin Compounds—a Preliminary Review,* Environmental Health Criteria No. 15, WHO, Geneva, 1980.
3. Bessmertny, A. N. and Grin', N. V., Acute toxicity studies on inorganic tin compounds for the purpose of hygienic norm-setting, *Gig. Sanit.,* 6, 82, 1986 (in Russian).
4. Benoy, C. J., Hooper, P. A, and Schneider, R., The toxicity of tin in canned fruit juice and solid foods, *Food Cosmet. Toxicol.,* 9, 645, 1971.
5. de Groot, A. P., Feron, V. J., and Til, H. P., Short-term toxicity studies on some salts and oxides of tin in rats, *Food Cosmet. Toxicol.,* 11, 19, 1973.
6. Yamaguchi, M., Saito, R., and Okada, S., Dose-effect of inorganic tin on biochemical indices in rats, *Toxicology,* 16, 267, 1980.
7. Hayashi, O., Chiba, M., Kikuchi, M., et al., The effect of stannous chloride on the humoral immune response in mice, *Toxicol. Lett.,* 21, 279, 1984.
8. NTP, *Carcinogenesis Bioassay of Stannous Chloride,* Technical Report, DHHS Publ., NIH No 81–1787, Bethesda, MD, 1982.
9. Kojima, S., Saito, K., and Kiyozumi, M., Studies on poisonous metals: IV. Absorption of stannic chloride from rat alimentary tract and effect of various food components on its absorption, *Yakugaku Zasshi,* 98, 495, 1978.
10. Johnson, M. A. and Greger, J. L., Effects of dietary tin on calcium metabolism of adult males, *Am. J. Clin. Nutr.,* 35, 655, 1982.
11. See **FURFURAL,** #5.

TITANIUM TETRACHLORIDE (CAS No 7550–45–0)

Properties. Colorless, mobile liquid with a penetrating acid odor, which fumes in the air. Gives no odor to water, but a faintly acidic, slightly astringent taste (rating 2) is determined at 4.5 mg/l. It hydrolyzes, whereupon Ti^{4+} **hydroxide** precipitates out, turning water milky. Color change threshold is 12.5 mg/l.[1]

Applications. A component of Ziegler-Natta polymerization catalysts.

Acute Toxicity. The LD_{50} values are 150 mg/kg BW for mice and 472 mg/kg BW for rats. Rabbits and guinea pigs are more sensitive: LD_{50} is 100 mg/kg BW.[1]

Repeated Exposure. Mice tolerate administration of 15 mg/kg BW for a month. The STI is found to be lowered in animals given 50 mg/kg BW for 10 d. Rats received 5 to 20 mg/kg BW over a period of 2 months. The treatment caused no changes in catalase activity, hematological analysis, or oxygen consumption. The higher dose caused an increase in blood cholinesterase activity.[1]

Long-term Toxicity. In a 6-month study, guinea pigs were given 20 and 5 mg/kg, rabbits received 5 and 1 mg/kg BW. The treatment with the higher dose caused retardation of BW gain and decreased blood cholinesterase activity in guinea pigs. Gross pathological examination revealed lymphohistocyte infiltrations and an increased number of Kupfer cells in the liver parenchyma. The foci of necrosis were observed in the intestinal mucosa. A dose of 1 mg/kg BW in rabbits did not cause evident changes, either in the functional state or in the histological structure of the visceral organs.[1] Mice were given 5 mg/kg BW of soluble Ti^{4+} salts in drinking water throughout their life-span. The treatment did not affect the latter and, furthermore, BW of animals exposed to Ti^{4+} was found to be greater than that of the controls (Schroeder et al., 1964).

Reproductive Toxicity. Ti^{4+} has been shown to pass through placenta into the fetal body.

Embryotoxicity. Rats and mice were given 5 mg soluble Ti^{4+} salts per liter in drinking water. The treatment caused a reduction in the number of newborns in the third generation. The male/female ratio in the progeny was reduced.[2] There are no data on the *teratogenic* effect. Ti (pure metals) showed no cytotoxicity (Takeda et al., 1989).

Genotoxicity. Ti^{4+} did not show mutagenic activity.[3]

Chemobiokinetics. Following ingestion, only about 3% administered dose is absorbed in the GI tract. Distribution occurs predominantly in the lungs, and then in the kidneys and liver. Bones are the principal site of storage. Following inhalation exposure, Ti^{4+} is immediately detected in large amounts in the blood of rats.[4] It is excreted with the urine and feces. Ti^{4+} is found in the brain of healthy people and in human embryos.

Standards. Russia. PML (Ti^{4+}): 4 mg/l (organolept., taste).

References:

1. See **CYCLOHEXANONE,** #3, 233.
2. Berlin, M. and Nordman, O., Titanium, in *Handbook on Toxicology of Metals,* Amsterdam, 1980, 627.
3. Hise, A. W., et al., *Trace Metals in Health and Disease,* Raven Press, New York, 1979, 55.
4. See **ACRYLONITRILE,** #1, 128.

TOLUENE-2,4-DIAMINE (CAS No 95–80–7)
TOLUENE-2,6-DIAMINE (CAS No 823–40–5)

Synonyms. 2,4- and 2,6-Diaminotoluenes; 2,4- and 2,6-TDA.

Properties. Colorless crystals. Freely soluble in hot water, alcohol, and ether.

Applications. Used as curing agents in epoxy resin production. Rubber antioxidants, intermediates in the manufacturing of **toluene diisocyanates.**

Acute Toxicity. In rats, LD_{50} of **2,4-** and **2,6-**TDA (mix.) is 270 to 300 mg/kg BW.[1] Poisoning is accompanied by CNS inhibition and methemoglobinemia.

Repeated Exposure. Male Fisher 344 rats were dosed by gavage with 70 mg **2,4-**TDA/kg BW for 5 d. Activities of microsomal P-450-dependent emzymes were depressed, while there was a pronounced increase in that of epoxide hydrolase.[2]

Long-term Toxicity. Renal toxicity was found to develop following oral administration of **2,4**-TDA for 2 years. Oral ingestion of 50 and 100 mg/kg BW accelerated the development of chronic renal disease in Fisher 344 rats. The treatment also resulted in decreased survival.[3]

Reproductive Toxicity. *Gonadotoxicity.* **2,4**-TDA appeared to be a potent reproductive toxicant in the male rats when given in the diet at the dose of 15 mg/kg BW for 10 weeks. Its toxic action includes an effect on spermatogenesis (66% reduction), decreased weight of seminal vesicles and epididymis, as well as a diminished level of circulating testosterone and elevation of serum-luteinizing hormone. The effect persisted after an additional 11 weeks on a normal diet, and in addition, profound testicular atrophy was noted.[4] *Embryotoxicity* and *teratogenicity.* **2,6**-TDA is found to be embryotoxic in rats and rabbits; it caused malformations in rats. The NOAEL for these effects is identified to be 10 mg **2,6**-TDA/kg BW in rats and 30 mg **2,6**-TDA/kg BW in rabbits.[5]

Genotoxicity. **2,4**- and **2,6**-TDA are shown to be positive in *Dr. melanogaster* and negative in mammalian assays.[6]

Carcinogenicity. **2,4**-TDA produced tumors in rodents: administration in the diet at a dose more than 79 mg/kg BW caused subcutaneous and mammary gland tumors in rats and hepatocellular and vascular tumors in mice.[7] **2,6**-TDA was not carcinogenic in rats and mice of both sexes.[8] *Carcinogenicity classification (2,4-TDA).* NTP: P—P*—N—P*.

Chemobiokinetics. Following ingestion, TDA are predominantly distributed in the liver, kidneys, and adrenal glands. Major metabolic pathways include acetylation of amino groups, oxidation of methyl groups, and ring hydroxylation. **Phenolic metabolites** of **2,4**-TDA and small amounts of unchanged TDA are excreted in the urine.[6,9]

Regulations. USFDA (1993) approved the use of TDA as an antioxidant in rubber articles intended for repeated use in producing, manufacturing, packing, processing, treating, packaging, transporting, or holding food (total not to exceed 5% by weight of the rubber product).

References:

1. Weisbrod, D. and Stephan, U., Studies on the toxic, methemoglobin-producing and erythrocyte-damaging effects of diaminotoluene after a single administration, *Zeitschr. Gesamte Hyg.*, 29, 395, 1983 (in German).
2. Dent, J. G. and Graichen, M. E., Effect of hepatocarcinogens on epoxide hydrolase and other xenobiotic metabolizing enzymes, *Carcinogenesis*, 3, 733, 1982.
3. Cardy, R. H., Carcinogenicity and chronic toxicity of **2,4**-toluenediamine in F344 rats, *J. Natl. Cancer Inst.*, 62, 1107, 1979.
4. Thysen, B., Varma, S., and Bloch, E., Reproductive toxicity of **2,4**-toluenediamine in the rat. I. Effect on male fertility, *J. Toxicol. Environ. Health*, 16, 753, 1985; II. Spermatogenic and hormonal effects, *J. Toxicol. Environ. Health*, 16, 763, 1985.
5. Knickerbocker, M., Re, T. A., Parent, R. A., et al., Teratogenic evaluation of *o*-toluenediamine in Sprague-Dawley rats and Dutch Belted rabbits, *Toxicologist*, 19, 89, 1980.
6. *Diaminotoluenes*, Environmental Health Criteria 74, WHO, Geneva, 1987, 67.
7. National Cancer Institute, *Bioassay of 2,4-Diaminotoluene for Possible Carcinogenicity*, Technical Report Series No. 162, Bethesda, MD, 1979.
8. National Cancer Institute, *Bioassay of 2,6-Toluenediamine Dihydrochloride for Possible Carcinogenicity*, Technical Report Series No. 200, NTP No. 80–20, NIH Publ. No. 80–1756, Bethesda, MD, 1980.
9. Waring, R. H. and Pheasant, A. E., Some phenolic metabolites of **2,4**-diaminotoluene in the rabbit, rat and guinea pig, *Xenobiotica*, 6, 257, 1976.

TRIALLYL ISOCYANURATE

Synonym. Cyanuric acid, N,N',N''-triallyl ester.

Properties. Odorless, colorless, oily liquid. Poorly soluble in water.

Applications. Used as a cross-linking agent in polymerization. A stabilizer for PVC.

Acute Toxicity. LD_{50} is 704 mg/kg BW in rats and 437 mg/kg BW in mice. Animals died in the first hours after administration. In 30 min after poisoning with high doses, survivors exhibited twitching of various muscle groups and of the extremities. Some animals experienced hind limb paralysis and impaired motor coordination. Mice tolerated a 300 mg/kg BW dose; none of rats died after oral administration of a 400 mg/kg BW dose.

Repeated Exposure revealed moderate cumulative properties. $^1/_5$ LD$_{50}$ caused five out of six rats to die within 2 weeks to 2 months. Daily administration of $^1/_{10}$ or $^1/_{20}$ LD$_{50}$ caused no mortality but only consistent retardation of BW gain in the treated animals.

Regulations. USFDA (1993) approved the use of T. as an accelerator of rubber articles intended for repeated use in producing, manufacturing, packing, processing, treating, packaging, transporting, or holding food (total not to exceed 1.5% by weight of the rubber product).

Reference:

See **BUTYL BENZYL ADIPATE, 299.**

TRIETHYLAMINE (TEA) (CAS No 121–44–8)

Synonym. *N,N'*-Diethylethanamine.

Properties. Colorless liquid with a penetrating, fishy, amine odor. Solubility in water is 19.7 g/l at 65°C or 71 g/l at 25°C.[02] Odor perception threshold is 4 mg/l,[1] 0.42 mg/l,[02] or <0.09 mg/l.[010] Taste perception threshold is 3 mg/l.

Applications. Used as a catalyst in polyurethane systems.

Acute Toxicity. LD$_{50}$ is 460 mg/kg BW in rats and 546 mg/kg BW in mice.[1] Sensitivity of rabbits is likely to be the same. Signs of intoxication include effect on the CNS: excitation, then inhibition, motor coordination disorder, and clonic convulsions.

Repeated Exposure. A part of animals exposed to $^1/_{10}$ LD$_{50}$ for 2.5 months died at the end of the experiment. The treatment led to retardation of BW gain and an increase in liver ascorbic acid content.

Long-term Toxicity. Rabbits were given 1 and 6 mg TEA/kg BW. In 3 to 4 months after the treatment onset, the higher dose caused liver function impairment.[1]

Reproductive Toxicity. Effect on fertility has been reported. TEA embryotoxic action was noted in 3-d-old chick embryos.[2]

Carcinogenicity. TEA could be metabolized into **diethylamine.** The exact mechanism for this deethylation is not known. Similar to other secondary amines, diethylamine might be nitrosated endogenously to form carcinogenic compound ***N*-nitrosodiethylamine.**[3] The latter should be regarded as if it were carcinogenic to man.

Chemobiokinetics were investigated in four volunteers. After oral administration, TEA is efficiently absorbed from the GI tract, rapidly distributed, and partially metabolized into **triethylamine-*N*-oxide.**[4] Formation of **diethylamine** from TEA also occurs. More than 90% of the dose was recovered in the urine as **triethylamine** and **triethylamine-*N*-oxide.** Exhalation of TEA is minimal.

Standards. Russia (1988). MAC and PML: 2 mg/l.

References:

1. Kagan, G. Z., Comparative evaluation of diethylamine and triethylamine in connection with sanitary protection of reservoirs, *Gig. Sanit.,* 9, 28, 1965 (in Russian).
2. Korhonen, A., Hemminki, K., and Vainio, H., Toxicity of rubber chemicals towards 3-day-old chicken embryos, *Scand. Work Environ. Health,* 9, 115, 1983.
3. Bellander, T., Osterdahl, B. G., Hagmar, L., et al., Excretion of *N*-mononitrosopiperasine in the urine in workers manufacturing piperasine, *Int. Arch. Occup. Environ. Health,* 60, 25, 1988.
4. Akesson, B., Vinge, E., and Skerfving, S., Pharmacokinetics of triethylamine and triethylamine-*N*-oxide in man, *Toxicol. Appl. Pharmacol.,* 100, 529, 1989.

TRIETHYLENEDIAMINE (CAS No 280–57–9)

Synonym. 1,4-Diazobicyclo(**2,2,2**)octane.

Properties. Yellowish, hygroscopic, crystalline powder with a specific, penetrating odor. Readily soluble in water. Taste and odor perception threshold appears to be 60 mg/l.[1]

Applications and **Exposure.** Used as a catalyst and a foaming agent in the production of polyurethane foams. Migration from the polyurethane coating into water (3-d exposure at the temperature of 37°C) was defined at the level of 0.5 mg/l.[2]

Acute Toxicity. LD$_{50}$ is reported to be 3.3 g/kg BW in rats, 2.25 g/kg BW in guinea pigs, and 1.1 g/kg BW in rabbits. No differences are evident in the sex- or species-sensitivity. Poisoning was accompanied mainly by NS impairment, with tension in the muscles and extremities, and clonic convulsions with subsequent death in the first 3 to 5 h after poisoning. The acute action threshold for the effect on STI appeared to be 1.55 g/kg BW. Gross pathological examination revealed distension of the stomach and small intestine, edema of the brain matter, and myocardial flaccidity.[1]

Repeated Exposure revealed slight cumulative properties. Administration of $^1/_5$ and $^1/_{10}$ LD$_{50}$ produced only mild signs of intoxication on day 3.

Long-term Toxicity. Rats were dosed by gavage with a 30 mg/kg BW oral dose for 6 months. The treatment affected the liver functions.[1]

Reproductive Toxicity. No selective embryotoxic effect is reported.[3] Gonadotoxic effect is found only when high doses, causing maternal toxicity, are administered.[4]

Standards. Russia (1988). MAC and PML: 6 mg/l.

References:

1. Troenkina, L. B., Hygienic substantiation of the MAC for triethylenediamine and monooxy-ethylpiperasine in water bodies, *Gig. Sanit.*, 5, 67, 1980 (in Russian).
2. See **EPICHLOROHYDRIN, #2.**
3. See **DIBUTYLTIN-S,S'-BIS(ISOOCTYL-MERCAPTOACETATE), #2,** 104.
4. Mazayev, V. T., Troenkina, L. B., and Gladun, V. I., The effect of aliphatic amines on reproduction in animals, *Gig. Sanit.*, 10, 66, 1986 (in Russian).

TRIETHYLENETETRAMINE (TETA) (CAS No 112–24–3)

Synonyms. *N,N*-Bis(2-aminoethyl)-**1,2**-ethanediamine; **1,8**-Diamino-**3,6**-diazaoctane; **3,6**-Diaza-octane-**1,8**-diamine; **1,4,7,10**-Tetraazadecane; Trientine; Trien.

Properties. Light-yellow, oily liquid. Readily soluble in water, ethanol, and vegetable oil.

Applications. T. is used as a curing agent for epoxy resins and as a thermo-setting resin.

Acute Toxicity. LD$_{50}$ is found to be 4.3 g/kg BW in rats, 1.6 g/kg BW in mice,[1,05] and 5.5 g/kg BW in rabbits. According to Dubinina and Fukalova (1979), LD$_{50}$ for rats and mice is 2.75 and 2 g/kg BW, respectively. Acute effect threshold appeared to be 30 mg/kg BW. Poisoning is accompanied by symptoms of CNS and GI tract damage.

Repeated Exposure revealed little evidence of cumulative properties. Rats were given $^1/_{10}$ and $^1/_{20}$ LD$_{50}$. The treatment caused decreased BW gain, reduction in the blood Hb level, and erythrocyte count, affected liver function, and lowered NS excitability. These changes were found to be reversible.[1]

Long-term Toxicity. In a 10-month study, mild changes in activity of a number of enzymes were noted in rats at a dose-level of 0.8 mg/kg. Animals recovered subsequently.[1]

Allergenic Effect. T. is reported to be a weak allergen.

Reproductive Toxicity. *Humans.* No malformations were observed among six infants of mothers treated with TETA for Wilson disease.[2] *Animals.* **Embryotoxic** effect of T. was noted in 3-d-old chick embryos.[3] *Teratogenic* action was observed on skin application in guinea pigs (Wayton, 1978). Rats received 0.17, 0.83, and 1.66% TETA in their diets throughout pregnancy. The treatment caused an increase in resorption rate at all dosages; fetal abnormalities (hemorrhage and edema) occurred at the two highest levels.[4]

Genotoxicity. Shows evident mutagenicity in *Salmonella* assay.[5]

Regulations. USFDA (1993) approved the use of T. (1) in resinous and polymeric coatings of a food-contact surface, (2) as a modifier for aminoresins for the uncoated or coated food-contact surface of paper and paperboard intended for use in producing, manufacturing, packing, transporting, or holding aqueous, fatty and dry food, and (3) as an accelerator for rubber articles intended for repeated use in contact with food up to 1.5% by weight of the rubber product.

Standards. Russia. PML in food: 0.02 mg/l.

References:

1. Stavreva, M. S., Toxicological and Hygienic Characteristics of Triethylenetetramine and Development of Recommendations for Using it as a Curing Agent for Anti-corrosion Epoxy Coatings for Food Industry, Author's abstract of thesis, Kiev, 1977, 23 (in Russian).
2. Walsche, J. M., Treatment of Wilson's disease with trientine (triethylenetetramine) dihydrochloride, *Lancet*, 1, 643, 1982.

3. See **TRIETHYLAMINE**, #2.
4. Keen, C. L., Cohen, N. L., Lonnerdal, B., et al., Teratogenesis and low copper status resulting from triethylenetetramine in rats, *Proc. Soc. Exp. Biol. Med.*, 173, 598, 1983.
5. Hulla, J. E., Rogers, S. J., and Warren, J. R., Mutagenicity of a series of polyamines, *Environ. Mut.*, 3, 332, 1981.

m- and *p*-XYLENEDIAMINE

Synonyms. ω'ω-Diamino-*m*-(*p*)-xylene.
Properties. *m*-X. is a liquid of low volatility. It is soluble in water and mixes with alcohol. *p*-X. is a white, crystalline substance.
Applications. Used as a curing agent in the production of epoxy resins.
Acute Toxicity. LD_{50} is 1.3 to 2 g/kg BW in rats and 0.47 to 1.4 g/kg BW in mice. Lethal doses affect the NS. Pronounced excitation is followed by apathy, unsteady gait, and respiratory disturbances. Gross pathological examination revealed visceral congestion.[05]
Repeated Exposure revealed marked cumulative properties (Grigorova et al., 1971).
Reproductive Toxicity. Acute *embryotoxic* effect was found to develop as a result of administration of $^1/_5$ LD_{50} to rats. $^1/_{100}$ LD_{50} produced no *teratogenic* effect.[05]
Allergenic Effect was observed on ingestion and skin application of *m*-X.
Reference:
See **BISPHENOL A**, #2, 171.

XYLENEDIAMINE CYANOETHYLATED

Properties. Liquid. Soluble in water and alcohol.
Applications. Used as a curing agent in the production of epoxy resins.
Acute Toxicity. LD_{50} is reported to be 3.96 g/kg BW in rats and 1.86 g/kg BW in mice.
Repeated Exposure failed to reveal cumulative properties.
Allergenic Effect was not found to develop either on inhalation or on skin application.
Reference:
See **BISPHENOL A**, #2, 171.

ZIRCONIUM COMPOUNDS

Applications. *Zr* salts are used as curing agents for silicones.
Acute Toxicity. LD_{50} of *Zr hydrocarbonate* (calculated as *Zr*) exceeds 10 g/kg BW.[03] LD_{50} is 1.6 g *Zr* acetate/kg BW, and that of *Zr sulfate* is 1.25 g/kg BW.[05]
Repeated Exposure. *Zr* compounds are low toxic chemicals. Of *Zr hydrocarbonate,* 20% in the diet of rats for 17 d and 5% in the diet of cats is tolerated by animals without signs of intoxication.[05]
Allergenic Effect of *Zr* compounds was studied by measuring the level of immunoglobulin-*M*-rosette-forming cells in response to injection of male sheep erythrocytes to the mouse spleen. The adjuvant effect on humoral immunity was more pronounced at doses of $^1/_{50}$ to $^1/_{100}$ LD_{50} than at doses of $^1/_5$ to $^1/_{10}$ LD_{50} (Horiba et al., 1986).
Genotoxicity. *Zr chloride* and *Zr oxide octahydrate* produced a doubling of the SCE frequency, and a three-point monotonic increase, with at least the highest dose at the $p < 0.001$ significance level.[1]
Carcinogenicity. Negative results for *Zr sulfate* are reported in mice.[09]
Regulations. USFDA (1993) approved the use of *Zr* compounds in resinous and polymeric coatings for a food-contact surface; *Zr oxide* may be used as a component of paper and paperboard for contact with dry food.
Reference:
See **FURFURAL**, #5.

Chapter 5

RUBBER INGREDIENTS

ALLYL GLYCIDYL ETHER (AGE) (CAS No 106–92–3)

Synonyms. Allyl **2,3**-epoxypropyl ether; **1**-Allyloxy-**2,3**-epoxypropane; [(2-Propenyloxy)methyl]-oxirane.

Properties. Liquid with a faint, specific odor. Water solubility is 172 g/l.

Acute Toxicity. LD_{50} is 433 mg/kg BW for mice and 700 mg/kg BW for rats. Death within 3 d.[1]

Repeated Exposure failed to reveal cumulative properties. K_{acc} is 6.2 (by Lim). The treatment led to alterations in the NS, liver, and kidneys.[1]

Genotoxicity. AGE is shown to be positive in *Salmonella* mutagenicity assay, creating reversion of the bacteria to histidine independence.[2]

References:
1. Shugaev, B. B. and Buckhalovsky, A. A., Toxicity of allyl glycidyl ether, *Gig. Sanit.*, 3, 92, 1983 (in Russian).
2. Wade, M. J., Mayer, J. W., and Hine, C. H., Mutagenic action of a series of epoxides in *Salmonella*, *Mutat. Res.*, 66, 367, 1979.

N-(2-BENZOTHIASOLYLTHIO)UREA

Synonym. *N*-Carbomoyl-**2**-benzothiasolesulphenamide.

Properties. Beige powder. Poorly soluble in water, soluble in alcohols and vegetable oils.

Applications. Used as a slow-acting vulcanization accelerator.

Acute Toxicity. In mice, LD_{50} is 5 g/kg BW; LD_0 appears to be 4 g/kg BW. Lethal doses do not affect behavior or BW gain. Gross pathological examination failed to reveal changes in morphology and weights of the visceral organs.[1]

Repeated Exposure. Rats tolerate ten administrations of 1 g/kg BW. The treatment caused liver and kidney functional impairment, exophthalmos, and retardation of BW gain. There were signs of thyroid function depression.[2]

Long-term Toxicity. Rabbits received 20 mg/kg for 6 months. Vascular permeability impairment and disturbed liver and kidney functions were revealed.[2]

References:
1. See **CYCLOHEXANONE, #3**, 77.
2. See *N*-(1,3-**DIMETHYLBUTYL**)-*N'*-**PHENYL**-*p*-**PHENYLENEDIAMINE**, 19.

N,N-BIS-(2-BENZOTHIAZOLYLTHIOMETHYL)UREA (CAS No 95–35–2)

Synonym. **1,3**-Bis(2-benzothiazolylthiomethyl)urea

Properties. A cream-colored powder, poorly soluble in water.

Applications. Used as a vulcanization accelerator for latex mixtures.

Acute Toxicity. LD_{50} was not attained in mice. The minimum lethal dose was 4 g/kg BW. High doses caused no changes in BW gain or in the relative weights of the visceral organs.

Long-term Toxicity. Rabbits received 20 mg/kg BW for 6 months. The treatment had no effect on behavior or general condition of the animals. Retardation of BW gain and early signs of acute hepatitis appeared by the end of the experiment. Histological examination revealed fatty dystrophy of parenchymatous cells in the liver and cloudy swelling of the renal tubular epithelium.

Reference:
See **CYCLOHEXANONE, #3**, 46.

230

1,1-BIS-(*tert*-BUTYLPEROXY)-3,3,5-TRIMETHYLCYCLOHEXANE(BBTC) (CAS No 6731–86–8)

Applications. BBTC is used, in particular, in the manufacture of rubber as well as in the hardening of unsaturated polyester resins and in the polymerization of styrene.

Short-term Toxicity. In a 13-week study, BBTC was added to MF-powdered basal diet and fed ad libitum to B6C3F$_1$ mice at dietary concentrations of 0.5 to 4%. The treated animals showed a tendency of anemia at 1% BBTC in the diet or more. Significant increase in the relative liver weight and decrease in the spleen weight was observed in a dose-dependent manner.

Swelling of hepatocytes was hystologically evident in animals fed 1% BBTC or more in the diet. Atrophy of the red and white pulp was noted in the spleen at 2 or 4% BBTC. Final BW was in excess of 90% of the control group values only at 0.5% dose-level (MTD).

Genotoxicity. BBTC did not cause mutagenic effect in *Salmonella* bioassay (Machigaki, 1987).

Carcinogenicity. B6C3F$_1$ mice received BBTC at dietary levels of 0.25 and 0.5% for 78 weeks. No differences were noted in mortality of treated and untreated animals. Spontaneous neoplasms were found in all groups, including the control group at the same rate. Thus, BBTC exerts no carcinogenic activity in B6B6C3F$_1$ mice.

Reference:

Mitsui, M., Furukawa, F., Sato, M., et al., Carcinogenicity study of 1, 1-bis(tert-butylperoxy)-3, 3,5-trimethylcyclohexane in B6C3F$_1$ mice, *Food Chem. Toxicol.*, 31, 929, 1993.

BUTYLAMINE (CAS No 109–73–9)

Synonym. 1-Aminobutane.

Properties. Colorless, oily liquid with a sour, ammonical odor. Mixes with water at all ratios. Soluble in alcohol. Odor perception threshold is 0.08 mg/l^{010} or 6 mg/l; taste perception threshold is 3.5 to 4 mg/l.[1] According to other data, odor perception threshold is 2.2 mg/l.[2]

Acute Toxicity. LD$_{50}$ in rats, mice, and guinea pigs is 430 to 450 mg/kg BW.[1] According to other data, LD$_{50}$ of *n*-B. is 370 to 380 mg/kg BW and that of **isobutylamine** is 230 mg/kg BW.[3] Poisoning produces weakness, ataxia, nasal discharges, dyspnea, salivation, and convulsions. ET$_{50}$ appeared to be 34 h. Gross pathological examination revealed local irritant effect, fatty infiltration, and necrotic lesions in the liver, and pulmonary edema.

Repeated Exposure failed to reveal cumulative properties. Rats and guinea pigs tolerated administration of $^1/_5$ LD$_{50}$. LD$_{50}$ at $^1/_5$ to $^1/_{20}$ reduced blood cholinesterase activity in rats. The 55 mg/kg BW dose caused an increase in the activity of glutaminate oxalate and glutaminate pyruvate transaminase and a reduction in blood prothrombin index. The dose of 9 mg/kg BW was ineffective in the short-term study in rats.[1]

Long-term Toxicity. Rats received $^1/_{1000}$ LD$_{50}$ for 6 months. The treatment produced no effect on conditioned reflex activity, blood cholinesterase, and liver diamino-oxidase activity or total serum cholesterol level. Rabbits were dosed by gavage with 0.15 and 8.5 mg/kg BW. Neither dose altered BW gain, prothrombin index, or histaminolytic function of the liver. Only serum transaminase activity was significantly increased at the highest dose level.[1]

Standards. Russia (1988). MAC and PML: 1 mg/l.

References:

1. Trubko, E. I., Studies on hygienic standard-setting for *N*-butylamines in water bodies, *Gig. Sanit.*, 11, 21, 1975 (in Russian).
2. Rudeiko, V. A., Kuklina, M. N., Romashov, P. G., et al., Hygienic rating of tertiary butylamine in water, *Gig. Sanit.*, 12, 70, 1985 (in Russian).
3. Cheever, R. L., Richards, D. E., and Plotnik, H. B., The acute oral toxicity of isomeric butylamines in the adult male and female rat, *Toxicol. Appl. Pharmacol.*, 63, 150, 1982.

p-CHLOROANILINE (PCA) (CAS No 106–47–8)

Synonyms. 1-Amino-4-chlorobenzene; 4-Chloroaniline; 4-Chlorobenzenamine.

Properties. Orthorhombic crystals (from alcohol or petroleum ether). Soluble in hot water, freely soluble in organic solvents.

Acute Toxicity. LD_{50} is reported to be 310 mg/kg BW in rats,[1] 100 mg/kg BW in mice, and 350 mg/kg BW in guinea pigs.[6]

Repeated Exposure led to cyanosis. MetHb formation develops with or without loss of Hb.

Short-term Toxicity. Hemopoietic system appeared to be the target of PCA toxicity. Exposure of Fisher 344 rats (5 to 80 mg/kg BW) and B6C3F$_1$ mice (7.5 to 120 mg/kg BW) to **PCA hydrochloride** given in water by gavage for 13 weeks failed to reveal treatment-related effect on organ weights (except spleen). Dose-related secondary anemia due to metHb formation and increased hematopoiesis in the liver and spleen as well as in the bone marrow (in rats but not in mice) were noted. Hemosiderin was found in the kidney, spleen, and liver.[2]

Long-term Toxicity. Aromatic amines produce reversible anemia.[3]

Genotoxicity. PCA produced gene mutations, SCE, and CA in Chinese hamster ovary cells *in vitro.*[4]

Carcinogenicity. B6C3F$_1$ mice and Fisher 344 rats were administered PCA in the diet and by gavage. Mice were given 2500 and 5000 ppm; rats received 250 and 500 ppm in their diet for 78 weeks. Retardation of BW gain was noted in the treated animals. PCA produced hemangiosarcomas in male and female mice.[5]

B6C3F$_1$ mice were given 3 to 30 mg PCA/kg, Fisher 344 rats were dosed with 2 to 18 mg PCA/kg by gavage in aqueous hydrochloric acid for 103 weeks. Hemangiosarcomas of the spleen and liver and hepatocellular adenomas and carcinomas in male mice have been reported.[6,7] Sarcomas of the spleen and splenic capsule were found in male rats in both studies. *Carcinogenicity classification.* IARC: Group 2B; NTP: E—N—E—E.

Chemobiokinetics. MetHb formation in erythrocytes results from conversion of heme iron from the ferrous to **ferric state.**[2]

References:

1. See **DIMETHYL MALEAT,** #1.
2. Chhabra, R. S., Thompson, M., Elwell, M. R., et al., Toxicity of *p*-chloroaniline in rats and mice, *Food Chem. Toxicol.,* 28, 717, 1990.
3. Linch, A. L., Biological monitoring for industrial exposure to cyanogenic aromatic nitro and amino compounds, *Am. Ind. Hyg. Assoc. J.,* 7, 426, 1974.
4. Anderson, B. E., Zeiger, E., Shelby, M. D., et al., Chromosome aberration and sister chromatid exchange test results with 42 chemicals, *Environ. Mol. Mutag.,* 16 (Suppl. 18), 55, 1990.
5. National Cancer Institute, *Bioassay of* **p**-*Chloroaniline for Possible Carcinogenicity,* NCI-CGTR-189, DHEW Publ. No. 79–1745, Bethesda, MD.
6. NTP, *Toxicology and Carcinogenesis Studies of* **p**-*Chloroaniline Hydrochloride in F344/N Rats and B6C3F₁ Mice (Gavage Studies),* Technical Report 351, NIH Publ. No. 89–2806, Research Triangle Park, NC.
7. Chhabra, R. S., Huff, J. E., Haseman, J. K., et al., Carcinogenicity of *p*-chloroaniline in rats and mice, *Food Chem. Toxicol.,* 29, 119, 1991.

N-CYCLOHEXYL-2-BENZOTHIAZOLE SULPHENAMIDE (CBS)
(CAS No 95–33–0)

Synonyms. Sulfenamide C; Santocure.

Properties. Odorless powder with a color ranging from beige to light-green. Poorly soluble in water, slightly soluble in alcohol.

Applications and **Exposure.** Used as a vulcanization accelerator in the rubber industry (for mixtures based on natural and synthetic rubber). Migration from butadiene-nitrile vulcanizates into water after the 3-d exposure at 20°C is up to 0.6 mg/l.[1]

Acute Toxicity. Mice tolerate a dose of 4 g/kg BW.[2] In rats 50% but none of mice died after two administrations of 5 g/kg BW. Gross pathological examination revealed acute circulatory disturbances, marked congestion, and parenchymal dystrophy of the visceral organs. There was also finte-drop adipose dystrophy of the liver.[3]

Short-term Toxicity. Rabbits were given 20 mg/kg BW for 3.5 months; initially every other day and then daily. The treatment produced no effect on BW gain, blood morphology, and relative weights of the visceral organs. Histological examination revealed changes in liver cell cytology and round cell infiltration along the intertrabecular triads.[2]

Allergenic Effect is shown to develop.[2]

Reproductive Toxicity. Wistar rats were given 0.001 to 0.5% CBS in the diet on day 0 to 20 of pregnancy. There was no evidence of *teratogenic effect,* death, or clinical signs of toxicity in animals at any dose-level.[4] According to other data,[5] CBS produces death and malformations in the developing chick embryo. *Embryotoxicity.* An embryonic mortality increased when the dose of 2 g CBS/kg BW was administered orally to female rats before the beginning of pregnancy on the first or third days of estrus or on day 4 and 11 of pregnancy.[6]

Carcinogenicity. There was no significant increase in tumor incidence in mice receiving CBS at a dietary level of 692 ppm for approximately 18 months.[7]

Standards. Russia. PML in food: 0.15 mg/l.

References:
1. See **ZINC PHENYLETHYLDITHIOCARBAMATE,** #1, 153.
2. See **CYCLOHEXANONE,** #3, 64 (in Russian).
3. See **CYCLOHEXANONE,** #3, 70.
4. Ema, M., Murai, T., Itami, T., et al., Evaluation of the teratogenic potential of the rubber accelerator N-cyclohexyl-2-benzothiazyl sulfenamide in rats, *J. Appl. Toxicol.,* 9, 187, 1989.
5. See **TRIETHYLAMINE,** #2.
6. Alexandrov, S. E., Effect of vulcanization accelerators on embryonic mortality in rats, *Bull. Exp. Biol. Med.,* 93, 107, 1982 (in Russian).
7. See 1-**BUTYL ALCOHOL,** #2.

4,4'-DIAMINODIPHENYL SULFIDE (CAS No 139–65–1)

Synonyms. Bis(4-aminophenyl)sulfide; Thioaniline; **4,4'**-Thiodianiline.

Properties. Needle-shaped crystals. Poorly soluble in water.

Applications. Used as a curing accelerator in the rubber industry.

Acute Toxicity. LD_{50} is found to be 9 g/kg BW in rats and 6.2 g/kg BW in mice. Signs of intoxication include reduced motor activity, body temperature, and reflex responses.[1]

Repeated Exposure revealed moderate cumulative properties. K_{acc} is 2 (by Lim).

Short-term and **Long-term Toxicity.** Male Fisher 344 rats received D. at dietary levels of 0.5 and 1% for 13 or 85 weeks. Changes observed consisted of liver necrosis and atrophy and hyperplasia of the renal tubular epithelium. The phospholipid composition in the liver, kidneys, and spleen was also changed.[2]

Genotoxicity. D. was shown to be negative in *Salmonella* mutagenicity assay with and without metabolic activation. It was not mutagenic for liver microsomes of rats that had previously received Aroclor 1254.[2] However, according to Endo et al.,[3] D. has a mutagenic potential.

Carcinogenicity. In the experiments described above, one case of hepatocellular carcinoma was reported.

References:
1. *Proc. Rostov-na-Donu Medical Institute,* 17, 54, 1974 (in Russian).
2. Benjamin, T., Evarts, R. P., Reddy, T. V., et al., Effect of **2,2'**-diaminodiphenylsulfide, a resin hardener, on rats, *J. Toxicol. Environ. Health,* 7, 69, 1981.
3. Endo T. et al., Mutagenicity and chemical structure of diaminoaromatic compounds, *Mutat. Res.,* 130, 361, 1984.

DI-2-BENZOTHIAZOLYL DISULFIDE (DBTD) (CAS No 120–78–5)

Synonyms. Altax; Benzothiazole disulfide; Dibenzothiazyl disulfide; Dibenzoyl disulfide; **2,2'**-Dithiobis(benzothiazole); Thiofide.

Properties. A greyish-yellow powder with a faint odor. Poorly soluble in water and alcohol. Perception thresholds for odor and taste are 45 and 50 mg/l, respectively.

Applications and **Exposure.** Widely used as a vulcanization accelerator for mixes based on diene-type natural and synthetic rubber and for butyl rubber. Migration from butadiene-nitrile vulcanizates into water after a 3-d exposure at 20°C was determined to be 0.42 mg/l.[1]

Acute Toxicity. LD_{50} is 4.3 g/kg BW in rats, 4.6 g/kg BW in mice, and 6.2 g/kg BW in rabbits.[2] However, according to Radeva,[3] Wistar rats did not die even after ingestion of a 8 g/kg BW dose.

Repeated Exposure revealed little evidence of cumulative properties. Rats received 400 and 800 mg/kg BW for 10 d. The treatment increased the amount of total and reduced glutathione in the blood and decreased serum alkaline phosphatase activity.[3]

Reproductive Toxicity. DBTD produced no teratogenic effect in rats.[4]

Carcinogenicity. Negative results are noted in mice.[09]

Standards. Russia (1988). MAC and PML: zero, PML in food: 0.15 mg/l.

References:

1. See **ZINC PHENYLETHYLDITHIOCARBAMATE,** #1, 153, 1987 (in Russian).
2. See **DI-2-BENZOTHIAZOLYL DISULFIDE,** #2.
3. Radeva, M., Study upon acute peroral toxicity of altax, *Khigiene i Zdraveopazvane,* 4, 326, 1980 (in Bulgarian).
4. See **4,4′-DIAMINODIPHENYL SULFIDE,** #4.

DIBENZOYL SULFIDE

Properties. White, crystalline powder with a pink tint and an unpleasant odor. Poorly soluble in water, readily soluble in alcohol.

Applications. Used as a plasticization accelerator of mixes based on natural and synthetic rubber.

Acute Toxicity. Exposure of rats and mice to 8 g/kg BW (in starch solution) caused 100% mortality. The animals exhibited decreased food consumption and progressive weakness; death occurred on days 1 to 17. Gross pathological examination revealed circulatory disturbances and slight fatty degeneration of the liver cells and renal tubular epithelium. The maximum tolerated dose was 4 g/kg BW.

Reference:

See **DIBENZOYL SULFIDE,** 55.

DIBUTYLAMINE (DBA) (CAS No 111–92–2)

Synonym. *N*-Butyl-1-butanamine.

Properties. Colorless, oily liquid with a fishy, amine odor. Solubility in water is 0.5%, readily soluble in alcohol. Odor perception threshold is 0.08 mg/l[010] or 2 mg/l; taste perception threshold is 3.5 to 4 mg/l.[1]

Applications and **Exposure.** Used in the production of rubber and synthetic fibers. It is found to be a degradation product in rubber materials that were made using dialkyldithiocarbamine acid derivatives as vulcanization accelerators. Reported migration levels of 0.2 to 0.5 mg/l usually depend on rubber and contact media compositions.[2] DBA was found at levels of up to 3.89 ppm in rubber nipples from baby pacifiers.[3]

Acute Toxicity. LD_{50} is reported to be 290 to 300 mg/kg BW in rats and mice and 230 mg/kg BW in guinea pigs. Poisoning produced a local irritating effect in the GI tract, fatty infiltration, and necrotic lesions in the liver.[1]

Repeated Exposure failed to reveal cumulative properties.[1] Rats and guinea pigs tolerate administration of $1/5$ LD_{50} for a month. $1/20$ LD_{50} increased oxygen consumption and decreased activity of liver deaminoxidase. Increased activity of serum transaminases and depressed liver histaminolytic activity were found in rabbits fed $1/8$ LD_{50}.[1]

Long-term Toxicity. Rats were fed dietary levels of 0.3 mg DBA/kg BW for 6 months. There were no treatment-related changes in conditioned reflex activity, blood cholinesterase and liver diaminoxidase activities, and total serum cholesterol. Rabbits tolerated doses of up to 7.6 mg/kg BW without changes in BW gain, prothrombin index, and liver histaminolytic function. However, the increased activity of serum transaminases was noted with the higher dose.[1]

Chemobiokinetics. DBA metabolism occurs by involving the microsome nonspecific oxidase system. Activation of peroxide oxidation of lipids takes place during the process of detoxication that may be the cause of dysfunction of the enzyme systems that are of fundamental importance in the mechanism of homeostasis maintenance.[4]

Standards. Russia (1988). MAC: 1 mg/l (organolept., odor).

References:

1. Trubko, E. I., Study on hygiene substantiation of the maximum allowable concentration for *N*-butylamines in water bodies, *Gig. Sanit.*, 11, 21, 1975 (in Russian).
2. See **SEBACIC ACID,** #2, 112.
3. Thompson, H. S., Billedeau, S. M., Miller, B. J., et al., Determination of *N*-nitrosoamines and *N*-nitrosoamine-precursors in rubber nipples from baby pacifiers by gas chromatography — thermal energy analysis, *J. Toxicol. Environ. Health*, 13, 615, 1984.
4. Sidorin, G. I., Lukovnikova, L. V., and Stroikov, Yu. N., Toxicity of some aliphatic amines, *Gig. Truda Prof. Zabol.*, 11, 50, 1984 (in Russian).

DIBUTYLNAPHTHALENE SULFOACID, SODIUM SALT (CAS No 25417–20–3)

Synonym. Nekal.

Properties. Yellowish-grey powder. Solubility in water is 150 g/l at 20°C. Soluble in alcohol. Gives water a specific odor of soap stock and a bitter, astringent taste. Perception threshold concentrations for odor vary from 0.16 mg/l (Ivanov and Khal'sov, 1966) to 13 mg/l,[1] that for taste from 0.08 mg/l (Ivanov and Khal'sov) to 1.3 mg/l.[1] The threshold concentration for foam formation is 0.5 mg/l.[1]

Applications. Used as an emulsifier in the production of synthetic rubber. A surfactant.

Acute Toxicity. LD_{50} is 1.25 g/kg BW in rats, 1.13 g/kg BW in mice, and 1.5 g/kg BW in guinea pigs. ET_{50} for rats is 13 h.[1]

Long-term Toxicity. Rats were exposed to 0.25 to 25 mg/kg BW doses for 6 months. The highest dose produced an increase in serum cholesterol level. Gross pathological examination revealed morphological changes in animals given 2.5 and 25 mg/kg BW doses.[1]

Reproductive Toxicity and **Genotoxicity** effects were not found.

Standards. Russia (1988). MAC and PML: 0.5 mg/l (organolept., taste).

Regulations. USFDA (1993) approved the use of D. as an accelerator in rubber articles intended for repeated use in producing, manufacturing, packing, processing, treating, packaging, transporting, or holding food (total not to exceed 1.5% by weight of the rubber product).

Reference:

Yegorova, N. A., Prediction of chronic toxicity of nekal by calculation and experimentation with the aim of setting its maximum allowable concentration in water, *Gig. Sanit.*, 4, 39, 1980 (in Russian).

DIETHANOLAMINE (DEA) (CAS No 111–42–2)

Synonyms. 2,2'-Aminodiethanol; Diethylolamine; **2,2'**-Dihydroxydiethylamine; Di(β-hydroxyethyl)amine; **2,2'**-Iminodiethanol.

Properties. The technical product is a glycerol-like, clear, colorless liquid with a faint odor of ammonia. The pure substance occurs in the form of colorless, deliquescent prisms. Readily mixes with water and alcohol. Odor perception threshold is reported to be 0.8 mg/l.[1] However, according to other data, it is 22 g/l.[02]

Applications. Used as a secondary vulcanization activator, in the manufacturing of paper and paperboard, etc.

Acute Toxicity. LD_{50} values are reported to be 0.78 g/kg[2] or 3.5 g/kg BW in rats, 3.3 g/kg BW in mice, and 2.2 g/kg BW in rabbits and guinea pigs. Poisoning caused irritation of the oral and GI tract mucosa and motor excitation. Death within 24 h from respiratory failure.

Gross pathological examination revealed hemorrhages and congestion in the visceral organs,[1] dilatation and degranulation of the endoplasmic reticulum, and mitochondrial swelling in the liver of mice given a lethal dose.[3] Sprague-Dawley rats exposed to 0.1 to 3.2 g neutralized DEA/kg BW displayed an increase in the relative liver and kidney weights. A single 0.4 g/kg BW dose caused renal tubular epithelium necrosis.[4]

Repeated Exposure. Rabbits and rats were exposed to $^1/_{10}$ LD_{50} for 2 months. The treatment produced changes in prothrombin-forming and detoxifying functions of the liver and in blood serum enzyme activity. BW gain and gas exchange were reduced. Histological examination revealed dystrophic, necrobiotic, and atrophic changes.[1] In the 14-d NTP study, no renal toxicity was reported in mice; however, renal degeneration including tubular epithelium necrosis was noted in rats. The NOAEL for this effect appeared to be 160 mg/kg BW.[5]

Regardless of the route of administration, DEA produced normochromic anemia in rats. Normocytic anemia without bone marrow depression or reticulocytosis was observed in male rats exposed to 4000 ppm DEA in drinking water for 7 weeks.[6] In the 14-d NTP study, microscopic examination of bone marrow and red blood cells failed to reveal important morphological changes; the LOAEL for this effect is close to 160 ppm DEA in drinking water.[5]

Short-term Toxicity. The mechanism of DEA toxicity is unknown but may be related to its high tissue accumulation and subsequent alteration of membrane phospholipid composition.[5] In the 13-week NTP study in Fisher 344 rats and B6C3F$_1$ mice, renal tubular epithelium necrosis was less severe than in the 14-d study. The LOAEL for this effect appeared to be 124 mg/kg BW in the more susceptible group of animals tested (female rats). The liver was predominantly affected in mice. Demyelination in the brain was reported to develop in rats but not in mice. Histological examination also revealed also cardiac myocyte degeneration and microscopic changes in the salivary glands in mice.[5]

Long-term Toxicity. Rats and rabbits were dosed by gavage with 1 to 100 mg/kg BW for 6 months. The highest dose caused retardation of BW gain, reduced oxygen consumption, prolonged prothrombin time, changes in the glycogen-forming and antitoxic functions of the liver and in blood serum enzyme activity, increased content of free *SH*-groups in the blood homogenate and liver tissue, and dystrophic and necrobiotic lesions to the visceral organs. The dose of 2 mg/kg BW was found to be ineffective for these changes.[1,6]

Reproductive Toxicity. *Gonadotoxicity.* In a 13-week NTP study in Fisher 344 rats, testicular degeneration appeared to be a direct toxic action of DEA. It was accompanied by a decrease in testis and epididymis weights; in animals given 2500 ppm DEA in drinking water, there were reduced sperm motility and count.[5]

Genotoxicity. DEA exhibited no mutagenic activity in *Salmonella* mutagenicity assay[7,8] but appeared to be a mutagen for *Dr. melanogaster*. It did not induce SCE and CA[9] and is shown to be negative in mouse lymphoma cells. DEA appeared to be positive in *in vitro* hepatocyte (isolated from rats, hamsters, or pigs) single strand-break assay.[10]

Chemobiokinetics. In rats, DEA is rapidly and nearly completely absorbed from the GI tract and distributed throughout the body. The highest concentrations of radiolabel are found in the liver and kidney, the target organs for the toxicity of DEA.[5] The major route of elimination is likely to be urinary excretion.

Standards. Russia (1988). MAC and PML: 0.8 mg/l (organolept., odor).

Regulations. USFDA (1993) approved the use of D. (1) as an adjuvant to control pulp content in the manufacture of paper and paperboard for contact with dry food prior to the sheet-forming operation, (2) in adhesives used as components of articles intended for use in packaging, transporting, or holding food, and (3) as a defoaming agent that may be used safely in the manufacturing of paper and paperboard intended for use in producing, manufacturing, packing, transporting, or holding food.

References:
1. See **ACRYLIC ACID**, #1, 105.
2. Smyth, H. F., Weil, C. S., West, Js., et al., An exploration of joint toxic action. II. Equitoxic versus equivolume mixtures, *Toxicol. Appl. Pharmacol.,* 17, 498, 1970.
3. Blum, K., Huizenga, C. G., Ryback, R. S., et al., Toxicity of diethanolamine in mice, *Toxicol. Appl. Pharmacol.,* 22, 175, 1972.
4. Korsrud, G. O., Grice, H. G., Goodman, T. K., et al., Sensitivity of several serum enzymes for the detection of thioacetamide-, dimethylnitrosamine- and diethanolamine-induced liver damage in rats, *Toxicol. Appl. Pharmacol.,* 26, 299, 1973.
5. NTP, *Toxicity Studies of Diethanolamine Administered Topically and in Drinking Water to F344 Rats and B6C3F$_1$ Mice,* Technical Report, NIH Publ. No 92–3342, U.S. Department of Health and Human Services, Research Triangle Park, NC, October 1992, p. 68.
6. See **CYCLOHEXANONE**, #3, 89.
7. Hartung, R., Rigas, L. K., and Cornish, H. H., Acute and chronic toxicity of diethanolamine, *Toxicol. Appl. Pharmacol.,* 17, 308, 1970.

8. See **ZINC PHENYLETHYLDITHIOCARBAMATE,** #3.
9. Dean, B. J., Brooks, T. M., Hodson-Walker, G., et al., Genetic toxicology testing of 41 industrial chemicals, *Mutat. Res.,* 153, 57, 1985.
10. Pool, B. L., Brenndler, S. Y., Liegible, U. M., et al., Employment of adult mammalian primary cells in toxicology: *in vivo* and *in vitro* genotoxic effects of environmentally significant *N*-nitrosodialkylamines in cells of the liver, lung, and kidney, *Environ. Mut.,* 15, 24, 1990.

DIETHYLAMINE (CAS No 109–89–7)

Synonym. Diethamine; *N*-Ethylethanamine.

Properties. A colorless, alkaline liquid with a sharp (musty, fishy, amine) odor. Readily miscible with water, soluble in alcohol. Odor and taste perception thresholds are 10 and 8 mg/l, respectively.[1] According to other data, odor perception threshold is 0.47 mg/l[02] or 0.02 mg/l.[010]

Applications and **Exposure.** Found to be a degradation product in rubber materials that were made using dialkyldithiocarbamine acid derivatives as vulcanization accelerators. Reported migration levels of 0.05 to 0.1 mg/l usually depend on rubber and contact media compositions.[2]

Acute Toxicity. LD_{50} was found to be 540 mg/kg BW in rats and 648 mg/kg BW in mice. No difference in species susceptibility was found. Administration led to CNS inhibition.[1] See **Dimethylamine.**

Repeated Exposure revealed no cumulative properties.

Short-term Toxicity. Mice were exposed to $^{1}/_{10}$ LD_{50} for 2.5 months. Some animals died; survivors developed reduced BW gain and increased content of ascorbic acid in the liver.

Long-term Toxicity. Rats and rabbits received 1 to 10 mg/kg BW for 6 months. The dose of 6 mg/kg BW affected the liver carbohydrate function.[1]

Reproductive Toxicity. See *Dimethylamine.*

Genotoxicity. D. is reported to be negative in the *Salmonella* mutagenicity assay.[3]

Carcinogenicity. Similar to other secondary amines, D. might be nitrosated endogenously to form a carcinogenic compound *N*-**nitrosodiethylamine** (Bellander et al., 1988). The latter should be regarded as if it were carcinogenic to man. See **Dimethylamine.**

Standards. Russia (1988). MAC and PML: 2 mg/l.

Regulations. USFDA (1993) approved the use of D. (1) in adhesives used as components of articles intended for use in packaging, transporting, or holding food and (2) as an activator of rubber articles intended for repeated use in producing, manufacturing, packing, processing, treating, packaging, transporting, or holding food (total not to exceed 5% by weight of the rubber product).

References:

1. Kagan, G. Z., Comparative evaluation of diethylamine and triethylamine in connection with sanitary protection of reservoirs, *Gig. Sanit.,* 9, 28, 1965 (in Russian).
2. See **SEBACIC ACID,** #2, 112.
3. See **ZINC PHENYLETHYLDITHIOCARBAMATE,** #3.

3-DIETHYLAMINOMETHYL-2-BENZOTHIAZOLETHIONE

Properties. White, crystalline powder with a bitter taste. Poorly soluble in water (decomposes), soluble in alcohol.

Applications. Used as a vulcanization accelerator.

Acute Toxicity. In mice, LD_{50} is 1.08 g/kg BW (technical grade product) and 1.57 g/kg BW (pure substance). High doses caused restlessness, disturbances of motor coordination, tail rigidity, clonic–tonic convulsions. Death occurs in the first 2 to 3 h.

Short-term Toxicity. Rats were dosed by gavage with 100 mg/kg BW dose for 4 months. There were no significant functional or morphological changes in the viscera other than local inflammation of the gastric and duodenal mucous membranes. Rabbits were administered 20 mg/kg BW (every other day for 2 months and daily for 1.5 months). The treatment produced an increase in the bilirubin index and pathomorphological changes in the liver.

Reference:

See **CYCLOHEXANONE,** #3, 35.

2-DIETHYLAMINOMETHYLTHIO BENZOTHIAZOLE

Properties. D. has an unpleasant odor and an extremely bitter taste. Hydrolyzes in water. Decomposes in acid and alkaline media with the formation of **2-mercaptobenzothiasole** (*Captax*)-q.v.

Applications. Used as a vulcanization accelerator in the rubber industry.

Acute Toxicity. Mice tolerate administration of 4 g/kg BW. However, a single oral dose of 4.5 g/kg BW resulted in the death of eight out of ten animals. Gross pathological examination revealed visceral congestion, point hemorrhages in the gastric mucosa, and mucosal edema (necrotic gastritis).

Repeated Exposure. There was neither mortality nor retardation of BW gain in mice gavaged with 2 g/kg BW in oil solution every other day over 2 weeks. A dose of 4 g/kg BW caused 50% mortality in mice.

Short-term Toxicity. Rats were dosed by gavage with 100 mg/kg BW for 4 months. The treatment produced a reduction in the temperature and growth of animals. The NS was unaffected, and there were no histological changes in the treated animals.

Standards. Russia (1988). PML: **n/m.**

Reference:

See **DIBENZOYL SULFIDE**, 26.

N,N-DIETHYL-2-BENZOTHIAZOLE SULFENAMIDE

Properties. Brown, oily liquid. Poorly soluble in water, soluble in alcohol. Its bitter taste restricts its use as a vulcanizor of rubber intended for contact with food.

Applications. Used as a slow-acting accelerator in the rubber industry.

Acute Toxicity. LD_{50} appeared to be 4 g/kg BW in rats, 4.96 g/kg BW in mice, and 6.2 g/kg BW in rabbits.[1] According to other data, rats and mice tolerated 5 g/kg BW dose administered in a starch solution. The same dose of the pure substance appeared to be the minimum lethal dose in mice and the LD_{80} in rats.[2] Treatment-related effects consisted of general inhibition, apathy, decreased food consumption, and hypothermia. Liver functions were affected (tests on hippuric acid and bromosulfophthalein elimination). Gross pathological examination revealed vascular disorders with pronounced visceral congestion, liver fatty dystrophy, and parenchymatous dystrophy of the myocardium.

Repeated Exposure demonstrated weak cumulative properties. None of mice died following three administrations of 5 g/kg BW dose in 3% starch solution. The same dosing produced minimum lethal effect in rats. Gross pathological examination revealed fatty and parenchymatous dystrophy of the central lobular liver cells. Reticuloendothelial cell infiltration around the blood vessels and bile ducts was also observed.[2] A 45-d treatment revealed some effect on the peripheral blood, liver, and CNS functions.[1]

Standards. Russia (1988). MAC and PML: 0.05 mg/l (organolept., taste).

References:

1. See **DI-2-BENZOTHIAZOLYL DISULFIDE**, #2.
2. See **CYCLOHEXANONE**, #3, 70.

DIETHYLDITHIOCARBAMATE, SODIUM SALT (DEDC) (CAS No 148–18–5)

Synonyms. Carbamate *EN;* Dithiocarb; Diethyldithiocarbamate trihydrate salt; Thiocarb.

Properties. White, crystalline powder. Soluble in water and alcohol.

Applications. Used as a vulcanization accelerator for rubber and vulcanizates based on them, for natural and synthetic latex.

Acute Toxicity. The LD_{50} values are 1.5 to 2.47 g/kg BW for mice and 1.5 to 3.32 g/kg BW for rats. High doses caused brief excitation, diarrhea, hind leg paralysis, and convulsions.

Repeated Exposure. Treatment with large doses produced anemia and leukopenia, CNS affection, and inhibition of oxidation enzyme activity, particularly in the brain and liver. Rabbits given 800 mg/kg developed leukopenia and decreased Hb level. Of animals, 30% died after five to eight administrations.

Carcinogenicity Classification. NTP: N—N—N—N.
Reference:
Korablev, M. V. and Lukienko, B. I., Effect of dithiocarbamine acid derivatives and structurally similar compounds on hypoxia, *Pharmacol. Toxicol.*, 30, 186, 1967 (in Russian).

DIETHYLPERFLUOROADIPATE

Synonym. Perfluoroadipic acid, diethyl ester.
Properties. Liquid. Almost insoluble in water, soluble in alcohol.
Acute Toxicity. There were no mortality or visible clinical signs of intoxication in rats given 5 to 8.4 g/kg BW doses.
Reference:
See *N,N'*-**BIS**(1,4-**DIMETHYLPENTYL**)-*p*-**PHENYLENEDIAMINE,** 142.

DIISOPROPYLBENZENE (CAS No 25321–09–9)

Properties. A mixture of isomers. A clear liquid with a sharp specific odor. Solubility in water is 10 mg/l. Odor perception threshold for *m*-D. is 0.1 mg/l; for *p*-D. it is 0.05 mg/l.
Acute Toxicity. LD_{50} of *m*-D. is 7.4 g/kg BW in rats and 2.1 g/kg BW in mice; LD_{50} of *p*-D. is 11.54 and 7.98 g/kg BW, respectively. High doses cause convulsions and narcosis.
Repeated Exposure. *p*-Isomer exhibits more pronounced cumulative properties (in comparison to *m*-D.). When $^1/_{10}$ LD_{50} was administered, K_{acc} appeared to be 1.56 (*p*-D.) and 3 (*m*-D.). The treatment caused no changes in the blood Hb level, erythrocyte count, or cholinesterase activity in the survivors.
Long-term Toxicity. In a 6-month study in rats, disturbance of CNS activity and changes in protein-forming function of the liver were noted at the higher doses tested.
Reproductive Toxicity. Following oral and inhalation exposure, disorder of the sexual cycle in rats and mice was noted (Yelisuiskaya, 1970).
Standards. Russia (1988). MAC and PML: 0.05 mg/l.
Reference:
Sologub, A. M. and Bogdanova, T. P., Experimental study on substantiation of maximum allowable concentration for *m*-diisopropylbenzene and *p*-diisopropyl benzene in water bodies, *Gig. Sanit.*, 9, 18, 1971 (in Russian).

DIMETHYLDIPROPYLENETRIAMINE

Properties. High-boiling liquid. Soluble in fats, organic solvents, and water.
Applications. Used as an inhibitor of the thermal polymerization of rubber and vulcanization stabilizer. Used in the synthesis of plasticizers and corrosion-resistant coatings.
Acute Toxicity. In rats, LD_{50} is 1.4 g/kg BW. Poisoning manifestations comprise excitation, with subsequent CNS inhibition and adynamia. Death occurs in 2 to 4 d.
Repeated Exposure. Rats received $^1/_3$ LD_{50} during three consecutive administrations. Disorganization in the intracellular structure of the liver, particularly that associated with detoxifying and energy biotransformation processes, was noted. D. penetrates via the skin to exert a pronounced toxic effect.
Reference:
See **DIBUTYLAMINE,** #4.

DIMETHYLDITHIOCARBAMATE, SODIUM SALT (CAS No 128–04–1)

Synonyms. Alcobam NM; Carbamate MN; Carbon S; Dibam; Dimethyldithiocarbamic acid, sodium salt.

Properties. Odorless, white, crystalline powder with a bitter taste. Readily soluble in water and alcohol.

Applications. Used as a vulcanization ultra-accelerator for mixes based on synthetic and natural rubbers (for example, butadiene rubber) and natural and synthetic latex. A fungicide.

Migration of **dithiocarbamates** from smoked sheet natural rubber vulcanizates for children's dummies was determinated to be up to 0.6 mg/l.[07]

Acute Toxicity. Rabbits received single doses of 0.5 to 1.5 g D./kg BW. The intake was accompanied by excitation with subsequent CNS inhibition, reduced tactile and pain sensitivity, and decreased food consumption. Hematological and ECG changes were noted. Animals given 1 and 1.5 g/kg BW died within 24 h. The lowest dose caused no mortality. Gross pathological examination revealed visceral congestion and pulmonary edema, hyperemia and focal hemorrhages in the gastric mucosa, and intestinal swelling. The relative weights of the lungs, heart, and stomach were increased. For mice and rats, the LD_{50} values were 1.5 and 1 g/kg BW, respectively.

Repeated Exposure failed to reveal cumulative properties. BW gain was increased in rabbits given 100 mg/kg (in an aqueous solution) for 15 weeks, as compared to the controls. Adynamia, changes in blood morphology, in the NS, and cardiovascular systems were noted. The fur of animals was unkempt and had turned yellow.

Reference:
See **BISPHENOL A, #2, 201.**

DIMORPHOLINOTHIURAM DISULFIDE

Synonym. *N,N′*-Di(oxydiethylene)thiuramdisulfide.

Properties. White-yellow, crystalline powder.

Applications. Used as a vulcanization accelerator.

Acute Toxicity. In rats, LD_{50} is 2.6 g/kg BW.

Short-term Toxicity. Rabbits were dosed by gavage with a 20 mg/kg BW dose every other day for the first two months and then daily for two further months. This treatment resulted in disorder of protein balance, fat (cholesterol) and elektrolyte (*Na* and *K*) metabolism, as well as in a reduction in alkaline phosphatase activity and in dystrophic changes and circulatory disturbances in the visceral organs.

Carcinogenicity. Nitroso-compounds are known to be powerful carcinogens.

Reference:
See **PHTHALIC ANHYDRIDE, #2, 108.**

N,N′-DIPHENYL-*N,N′*-DIETHYLTHIURAM DISULFIDE

Synonym. Thiuram *EF*.

Properties. White powder. Poorly soluble in water.

Applications. A vulcanization accelerator for rubber intended for use in a food-contact surface. In contrast to other dithiocarbamine accelerators, its use in a rubber in the course of vulcanization does not result in development of toxic substances.

Acute Toxicity. Rats tolerate the dose of 12 g/kg BW. Acute action threshold is 0.5 g/kg BW.

Repeated Exposure revealed no cumulative properties.

Long-term Toxicity. Rats received a 100 mg/kg BW dose for 6 months. The treatment caused anemia to develop. Normalization was observed to the tenth month. Doses of 50 and 100 mg/kg BW caused liver function deterioration. The NOAEL was identified to be 5 mg/kg BW.[1]

Reproductive Toxicity. *Gonadotoxic* effect was not observed.[1] *Teratogenicity.* SNK mice were given 5 and 50 mg/kg BW as oily emulsion during pregnancy. Both doses caused an increased rate of hemorrhages in the fetal organs and tissues and generalized edema. No morphofunctional changes in the fetuses were reported. *Embryotoxicity* effect was negligible.[2]

Genotoxicity. A weak mutagenic action on somatic and germ cells was observed in Wistar rats given 50 mg/kg BW for 2.5 months. The NOAEL for mutagenic and genotoxic effect appeared to be 5 mg/kg BW.[1]

240

Standards. Russia (1988). PML in food: 0.5 mg/kg; PML in drinking water: **n/m.**
References:
1. Khoroshilova, N. V., Toxicological characteristics of the new vulcanization accelerator thiuram EF, *Kauchuk i Resina,* 11, 50, 1979 (in Russian).
2. Salikova, L. S., Study of the embryotropic effect of thiuram, *Gig. Sanit.,* 3, 88, 1989 (in Russian).

1,3-DIPHENYLGUANIDINE (DPG) (CAS No 102–06–7)

Synonyms. *N,N'*-Diphenylguanidine; DPG accelerator; Melaniline; Vulkacid *DC.*

Properties. Odorless, crystalline powder, with a color ranging from white to light-yellow or lilac. Solubility in water is 0.08%, soluble in alcohol.

Applications and **Exposure.** A vulcanization accelerator widely used in the processing of acryl-butadiene, styrene-butadiene, chloroprene, siloxane, butyl rubber, etc. Predominantly used in combination with other accelerators. Migration from rubber into liquid media varies within the limits 0.02 to 1.3 mg/l. According to the data available, migration from butadiene-nitrile vulcanizates into water after 3-d exposure at 20°C was determined at the level of 2.9 mg/l.[1]

Acute Toxicity. The LD_{50} values are 190 to 850 mg/kg BW in rats, 258 to 625 mg/kg BW in mice, 507 mg/kg BW in guinea pigs, and 246 mg/kg BW in rabbits.[2–4] The acute action threshold appeared to be 25 mg/kg BW, identified by minimal signs of anemia and an increase in blood peroxidase activity. The NOEL seems to be 10 mg/kg BW. Sensitivity to DPG decreased with the age of animals. Poisoning was manifested by unsteady gait, side position, convulsions, and increased sensitivity to pain. There were also liver damage and an increase in its relative weight. Gross pathological examination revealed marked hyperemia and considerable morphological changes. Death occurred in 1 to 2 d.[4]

Repeated Exposure failed to reveal cumulative properties in rats. Moderate capacity for accumulation was found in mice. CD-1 mice were administered 0.06 to 16 mg DPG/kg BW for 8 weeks. There were no dose-related changes in BW gain, gross pathological and histological findings, or organ weights.[5]

Short-term Toxicity. Oral administration of 32 mg/kg BW to rats for 1 to 3 months produced signs of anemia, reduction in blood catalase and peroxydase activity, liver dysfunction, and pathomorphological changes. A stable reduction in the content of cholic acids in the bile when 5 g/kg BW was given daily to dogs was reported.[6]

Long-term Toxicity. Rabbits received 50 mg/kg BW by gavage every other day for 1.5 months and daily for the next 4 months. Toxic action was manifested by exhaustion, a decrease in erythrocyte count and Hb level, and dysbalance of blood proteins. Gross pathological examination revealed changes in the liver and kidney.[4] Administration of $1/100$ LD_{50} to rats for 10 months as solutions in milk and in water starch suspension caused a number of signs of intoxication, which were very marked in young animals: retardation of BW gain, hematological changes, increase in leukocyte and phagocyte activity, and reduction in the relative liver weight.[7]

Reproductive Toxicity. *Gonadotoxicity.* Reported data appear to be controversial. A testicular toxicity and fertility study was carried out in CD-1 mice.[7] Animals were treated by daily gavage during 8 weeks of the premating period. DPG did not exert any significant adverse effect on fertility at a dose-level up to 16 mg/kg BW. Following chronic exposure to DPG as a solution in 0.025% solution of acetic acid *ad libitum,* mice and hamsters exhibited changes in sperm morphology and a decline in the number of spermatozoa and in the weights of the testes. The shape of the seminal ducts was altered and the fertility index was reduced.[8] *Embryotoxicity.* In the above-described study, the treatment caused a reduced number of implants per pregnancy and fetal mortality in rats.[8] *Teratogenic* effect was not observed in mice.[9]

Genotoxicity. DPG is a direct acting mutagen; it is shown to be positive in *Salmonella* mutagenicity assay.[10]

Carcinogenicity. Rats were orally exposed to a 40 mg/kg BW dose once a week for 6 weeks; 6 months of follow-up revealed no tumors.

Chemobiokinetics. DPG is rapidly absorbed and distributed throughout body tissues: 30 min after administration of 100 mg DPG/kg BW, the substance was found in the blood; in an hour, it was discovered in all the visceral organs; after 24 h, it was found in the urine. DPG excretion with the urine ceased on day 6.

Standards. Russia (1988). MAC and PML in drinking water: 1 mg/l; PML in food: 0.15 mg/l.

Regulations. USFDA (1993) regulates the use of DPG as an accelerator for rubber articles intended for repeated use in contact with food up to 1.5% by weight of the rubber product.

References:

1. See **ZINC PHENYLETHYLDITHIOCARBAMATE,** #1, 153.
2. Bourne, H. G., Yee, H. T., and Seferian, S., The toxicity of rubber additives. Finding from a survey of 140 plants in Ohio, *Arch. Environ. Health,* 16, 700, 1968.
3. See **DIBUTYLTIN-*S,S*'-BIS(ISOOCTYLMERCAPTOACETATE),** #2, 60.
4. See **PHTHALIC ANHYDRIDE,** #2, 117.
5. Koeter, H. B. W., Regnier, J. F., and Marwijk, M. W., Effect of oral administration of **1,3**-diphenylguanidine on sperm morphology and male fertility in mice, *Toxicology,* 71, 173, 1992.
6. Burov, Yu. A., Toxic effects of diphenylguanidine, *Pharmacol. Toxicol.,* 27, 714, 1961 (in Russian).
7. Vlasyuk, M. G., Data on substantiation of permissible migration level for diphenylguanidine from rubber, *Gig. Sanit.,* 7, 35, 1978 (in Russian).
8. Bempong, M. A. and Hall, E. V., Reproductive Toxicity of **1,3**-diphenylguanidine: analysis of induced sperm abnormalities in mice and hamsters and reproductive consequence in mice, *J. Toxicol. Environ. Health,* 11, 869, 1983.
9. Yasuda, Y. and Tanimura, T., Effect of diphenylguanidine on development of mouse fetuses, *J. Environ. Pathol. Toxicol.,* 4, 451, 1980.
10. Bempong, M. A. and Mantley, R., Body fluid analysis of **1,3**-diphenylguanidine for mutagenicity as detected by Salmonella strains, *J. Environ. Pathol. Toxicol. Oncol.,* 6, 293, 1985.

4,4'-DIPHENYLMETHANE DIISOCYANATE (CAS No 101–68–8)

Synonym. 4,4'-Methylene di(phenylisocyanate).

Properties. Colorless or pale-yellow crystals with a faint odor.

Applications. Used in the production of polyurethane coatings and elastomers. Vulcanizing agent for rubber.

Acute Toxicity. In rats and mice, LD_{50} is 2.2 g/kg BW. Following administration of the lethal dose, animals died within 3 to 5 d. Gross pathological examination revealed changes in the blood-vascular and circulatory systems, tiny hemorrhages and areas of dystrophy in the myocardium.[1,2] At the same time, Wollrich and Roy (1969) considered a single administration of D. to be of low toxicity.

Repeated Exposure failed to reveal cumulative properties. K_{acc} is 12.8, when animals were exposed to a mixture of D. and polycyclic isocyanates.

Allergenic Effect was found to develop on *i/g* administration. D. caused intensive skin sensitivity in guinea pigs, similar to contact allergy.[3]

References:

1. *Problems of Industrial Hygiene and Industrial Pathology in Chemical Industry Workers,* Moscow, 1977, 63.
2. *Current Problems of Hygiene and Occupational Pathology in Some Sectors of Chemical Industry,* Moscow, 1978, 50 (in Russian).
3. Duprat, P., Gradiski, D., et Marignac, B., Pouvoir irritant et allergisant de deux isocyanates: toluene diisocyanate, diphenylmethane diisocyanate, *Eur. J. Toxicol. Environ. Hyg.,* 9, 43, 1976 (in French).

4,4'-DITHIODIMORPHOLINE (DTDM) (CAS No 103–34–4)

Synonyms. Dimorpholinodisulphide; *N,N*-Disulfidemorpholine; Sulphazane.

Properties. White, crystalline powder. Poorly soluble in water, soluble in alcohol. Undergoes decomposition during vulcanization and is converted into *N,N*-**tetrathiodimorpholine** (about 60%), which itself is a vulcanizing agent.[1] Exchange processes may occur between DTDM and Altax and Captax, with the formation of polysulfide compounds with a variable number of sulfur atoms.[2]

Applications and **Exposure.** Used as a vulcanization accelerator in the production of natural and synthetic rubber. Migration from butadiene-nitrile vulcanizates into water after a 3-d exposure at 20°C was determined to be up to 1 mg/l.[3]

Acute Toxicity. In mice, LD_{50} is 2.75 g/kg BW. In rats, reported LD_{50} values are in the range of 7 g/kg[4] to 4.3 g/kg BW.[5] The acute action threshold (for STI) appeared to be 0.5 g/kg BW. LD_{50} of **tetradithiodimorpholine** is 16.4 g/kg BW in rats.

Repeated Exposure. DTDM exhibited no cumulative properties.

Short-term Toxicity. Administration of $^1/_{10}$ and $^1/_{20}$ LD_{50} did not cause mortality in the animals tested. The CNS functions and parenchymatous organs were affected predominantly.[5]

Long-term Toxicity. Mice were dosed by gavage with 5 and 25 mg DTDM/kg BW for 10 months. The treatment caused no mortality or marked manifestations of intoxication. The higher dose produced changes in the STI and hepatic excretory function. By the end of the experiment there was inhibition of blood pyroracemic acid and cholinesterase activity and an increase in aminotransferase activity. A 5 mg/kg BW dose was found to be the NOAEL.[5] Administration of 7 mg/kg BW for 5 months caused a considerable reduction in liver weights. Blood formula and catalase and peroxidase activity were not affected. However, hemodynamic disorders in the visceral organs and desquamation of the renal tubular epithelium were found.[4] Rabbits were given 20 mg/kg BW for a 4-month period. The treatment caused disturbance of protein metabolism and increased blood sugar and cholesterol levels. Alkaline phosphatase activity was reduced.[6] Gross pathological examination revealed catarrhal inflammation of the stomach and small intestine and stimulation of the macrophage system.

Genotoxicity. DTDM is negative in *Salmonella* mutagenicity assay at concentrations of 0.001 to 10 mg/ml.[5]

Allergenic Effect is considered to be slight.

Standards. Russia. MAC: 0.5 mg/l.

Regulations. USFDA (1993) regulates the use of DTDM (1) in adhesives as a component of articles intended for use in packaging, transporting, or holding food and (2) as an accelerator for rubber articles intended for repeated use in contact with food up to 1.5% by weight of the rubber product.

References:

1. Blokh, G. A., *Organic Vulcanization Accelerators and Vulcanizing Agents for Elastomers,* Khimiya, Leningrad, 1978, 155.
2. Petrova, L. F. et al., Toxicological characteristics of dithiomorpholine, *Kauchuk i Rezina,* 12, 29, 1983 (in Russian).
3. See **ZINC PHENYLETHYLDITHIOCARBAMATE,** #1, 153.
4. Stankevich, V. V. and Shurupova, E. A., Hygienic evaluation of some rubber articles used in the food industry, *Gig. Sanit.,* 9, 24, 1976 (in Russian).
5. Sokolníkov, Z. A., The data to substantiate the permissible migration level of the vulcanization accelerator *N,N'*-dithiodimorpholine from rubber products, *Gig. Sanit.,* 12, 67, 1986 (in Russian).
6. See **DIBUTYL DIPHENATE,** #2, 88.

4,4′-DITHIODI(*N*-PHENYLMALEIMIDE)

Synonym. 4,4′-Dithiodiphenyldimaleimide.

Properties. Powder, poorly soluble in water.

Applications. Used as a vulcanizing agent for synthetic rubber.

Acute Toxicity. LD_{50} is 4.3 g/kg BW in mice. In rats, it was not attained.

Repeated Exposure revealed cumulative properties. In mice, K_{acc} is 4.5. A reduction in the content of free SH-groups in the blood serum was found in rats that received D. for 2 months.

Reference:

See **ADIPIC ACID,** #2, 261.

ETHYLAMINE (CAS No 75–04–7)

Synonyms. Aminoethane; Ethanamine; Monoethylamine.

Properties. Colorless liquid or gas with the odor of ammonia. Miscible with water and alcohol at all ratios. Odor perception threshold is 3.9 mg/l,[1] 4.3 mg/l,[02] or 0.27 mg/l (70 to 72% in water).[010] Taste perception threshold is 0.5 mg/l.

Applications. A stabilizer for rubber latexes.

Acute Toxicity. In rats, LD_{50} was reported to be 400 mg/kg[2] or 530 to 580 mg/kg BW (for rats and mice).[1] Rabbits appeared to be more susceptible. High doses affected the NS. Gross pathological examination revealed visceral congestion.

Repeated Exposure revealed marked cumulative properties. K_{acc} is 0.54. Oral administration of 90 mg/kg BW for two weeks caused death of eight rats out of ten.

Long-term Toxicity. In a 6-month study, rats were gavaged with 2.5 mg/kg BW dose. The treatment affected conditioned reflex activity and ascorbic acid content in the spleen and liver.[1]

Genotoxicity. ES was shown to be positive in *Dr. melanogaster* (Dubinin, 1970).

Standards. Russia. Recommended MAC and PML: 0.5 mg/l (organolept.,taste).

References:
1. See **BUTYLENE,** 99.
2. Smyth, H. F., Carpenter, C. P., Weil, C. S., et al., Range-finding toxicity data, List V, *Arch. Ind. Hyg. Occup. Med.*, 10, 61, 1954.

N-ETHYLANILINE (CAS No 578–54–1)

Synonyms. *N*-Ethylbenzenamine; *N*-Ethylphenylamine.

Properties. Very refracted, oily liquid. Poorly soluble in water, miscible with alcohol at all ratios.

Exposure. Rubber destruction product. E. could be detected at concentrations of 3 to 4 mg/l in the extracts of rubbers that are made using **zinc phenylethyldithiocarbamate** as a vulcanization accelerator. Migration from butadiene-nitrile vulcanizates into 0.3% lactic acid following 24-h exposure at 40°C was up to 0.52 mg/l,[1] and in water after 3-d exposure at 20°C, it was determined to be at the level of up to 0.8 mg/l.[2]

Acute Toxicity. LD_{50} is 1.26 g/kg BW in rats and 0.5 g/kg BW in mice. Poisoning is immediately followed by excitation, with subsequent inhibition. Dyspnea, cyanosis, and tonic convulsions have been noted to develop. Death occurs within 2 to 4 d. Gross pathological examination revealed congestion in the viscera and distension of the stomach and small intestine. Acute normochromic anemia developed in rats following a single oral administration of 145 mg/kg BW.[3]

Repeated Exposure revealed moderate cumulative properties. K_{acc} is 4.6 (by Kagan).

Short-term Toxicity. Rats were dosed by gavage with 1.5, 6, and 15 mg/kg BW for 4 months. By the end of the study, signs of intoxication appeared only at the high dose-level. Hypodynamia, loss of appetite, cyanosis, dyspnea, retardation of BW gain, and an increase of the relative weights of the liver and spleen were observed.[3]

Standards. Russia. MAC: 0.5 mg/l; recommended PML: 10 mg/l.

References:
1. Grushevskaya, N. Yu., Assessment of vulcanization accelerator of vulcacite-*P*-extra-*N* and some transformation product in sanitary and chemical investigation of rubber, *Gig. Sanit.*, 4, 59, 1987 (in Russian).
2. See **ZINC PHENYLETHYLDITHIOCARBAMATE,** #1.
3. See **BUTYL BENZYL ADIPATE,** 328.

ETHYLENE THIOUREA (ETU) (CAS No 96–45–7)

Synonyms. 2-Imidazolidinethione; 2-Imidazoline-2-thiol; 2-Mercaptoimidazoline.

Properties. Solubility in water is 20 g/l at 30°C and 90 g/l at 60°C. Half-life in water is 7 to 13 d.[1] Moderately soluble in methanol, ethanol, ethylene glycol.

Applications and **Exposure.** Used as an accelerator in synthetic (neoprene) rubber production. It is used also as a part of the curing system for polyacrylate rubber and as an intermediate for antioxidants. ETU was detected in food, namely in vegetables and fruits.

Acute Toxicity. LD_{50} is 780 mg/kg BW in rats and 3000 mg/kg BW in mice. Poisoning results in decreased serum SH-group content,[2] reduced hexenal-induced sleep, and aminopyridine methylase activity in liver microsomes.[3]

Repeated Exposure. In a 28-d study, rabbits were exposed to ETU in their drinking water at concentrations of 100 to 300 mg/l. Exposure to ETU decreased BW gain but did not significantly affect urinary sodium, potassium, glucose, or protein excretion and urinary osmolality. High doses resulted in ultrastructural alterations in the epithelial cells or renal proximal tubuli.[4]

Short-term Toxicity. After oral administration of 80 and 40 mg/kg BW to rats for 3 months pronounced cumulative effect was noted.[2] Like all urea derivatives it appears to exhibit strumogenic effect.[6]

Administration of 50 to 750 mg/kg in the diet of rats for 30 to 120 d resulted in decreased BW, thyroid/BW ratio and uptake of I^{131}. Thyroid hyperplasia at higher doses was also noted.[5] B6C3F$_1$ mice were fed ETU-containing diets for 13 weeks. Dose-related diffuse follicular cell hyperplasia of the thyroid and hepatocellular cytomegaly occurred at dose-level of 75 mg ETU/kg BW in both sexes. No effects were observed at lower doses. The NOAEL in this study was 38 mg/kg BW (NTP–92).

CD-1 mice were administered ETU in the diet at doses of 0.16 to 230 mg/kg BW. Follicular cell hyperplasia of the thyroid was observed at dose of about 20 mg/kg BW. The NOAEL in this study was shown to be 1.7 mg/kg BW in males and 2.4 mg/kg BW in females (G. P. O'Hara and L. J. DiDonato, 1985).[18] Beagle dogs received dietary concentrations of ETU for 13 weeks. The NOAEL of 0.39 mg/kg based on decreased Hb and erythrocyte count, packed cell volume, and increased cholesterol was determined at the dose of approximately 6 mg/kg (J. P. Briffaux, 1991).[18] In a 52-week feeding study in dogs, 1.8 mg/kg dose caused a reduction in BW gain, hypertrophy of the thyroid with colloid retention, a slight increase in thyroid weight, and pigment accumulation in the liver. The NOAEL in this study was 0.18 mg ETU/kg (J. P. Briffaux, 1992).[18]

Long-term Toxicity. Oral lifetime administration of up to 200 mg/kg BW caused upper cholesterolemia in rats and hamsters. The most affected organs are the liver (in hamsters) and thyroid (in rats).[7] In a 2-year oral study in rats, the NOAEL of 0.25 mg ETU/kg (lack of thyroid tumors) was reported.[19]

Reproductive Toxicity. *Teratogenicity.* ETU appears to be teratogenic itself, but not its metabolites. The initial target is the primitive neuroblast that undergoes necrosis, with further development of hydrocephalus. High doses caused teratogenic effects in hamsters, mice, guinea pigs, and rabbits. ETU is a specific neuroteratogen that induces communicating hydrocephalus *ex vacuo* at oral doses far lower than those causing any observable toxic symptom or 50% death (LD_{50}) in the rat dam.[8] Oral administration of 15 to 35 or 100 mg/kg BW,[9,10] or 200 mg/kg[11] during the organogenesis period in rats caused specific developmental abnormalities of the CNS and musculoskeletal, craniofacial systems, etc. Cats given 0.6 g/kg BW on days 16 to 35 of pregnancy exhibited maternal toxicity and a great number of developmental abnormalities (30% animals).[8] *Embryotoxicity.* Sprague-Dawley rats received 20 to 60 mg ETU/kg BW on day 11 of gestation. An exposure to the highest dose caused a number of neonatal deaths. Hydronephrosis was observed in this study.[12]

Genotoxicity. ETU induces DLM, micronuclei, or SCE in mice, and CA in rats treated *in vivo*. DNA synthesis in human fibroblasts was observed.[13]

Carcinogenicity. The primary effect of ETU in rats is on the thyroid gland. ETU-induced liver tumors were found in mice. In rats but not in hamsters fed with 60 mg/kg BW (males) and 200 mg/kg BW (females), a carcinogenic effect was shown.[7] Oral administration of 77 mg/kg BW to mice for 82 weeks caused thyroid and lung cancer.[14] After 80 weeks of feeding rats with the dose of 646 mg/kg BW, thyroid and lung tumors were observed in 100% of cases.[15] *Carcinogenicity classification.* IARC: Group 2B; NTP: CE—CE—CE—CE.

Chemobiokinetics. After oral administration of 100 mg/kg BW to rats, ETU was rapidly absorbed in the guts and penetrated all the visceral tissues and embryos. Accumulates only in the thyroid and lung alveols.[16] Not detected in the blood, probably due to intracellular transformation. After oral administration of 4 mg ^{14}C-ETU/kg BW, **imidasol** and **imidasoline** were found to be the main metabolites. Following *i/v* injection, **S-methylethyl thiourea** was detected as a metabolite in cats.[15,17] ETU is rapidly excreted with the urine.

Recommendations. JECFA (1993). ADI: 0 to 0.004 mg/kg BW.

Standards. USFDA prohibited use of ETU as a food additive.

References:

1. *Hazardous Substances in Industry,* Organic Compounds, Handbook, New data of 1974 to 1984, Levina, A. L. and Gadaskina, E. D., Eds., Khimiya, Leningrad, 1985, 209 (in Russian).
2. Antonovich, E. A., Toxicology of Dithiocarbamates and Hygienic Aspects of Their Use in Food Production, Author's abstract of thesis, L'vov, 1975, 32 (in Russian).
3. Lewerenz, H. J. and Plas, R., Effect of ethylene thiourea on hepatic microsomal enzymes in the rat, *Arch. Toxicol.,* (Suppl. 1), 189, 1978.
4. Kurttio, P., Savolainen, K., Naukkarinen, A., et al., Urinary excretion of ETU and kidney morphology in rats after continuous oral exposure to nabam or ethylenthiourea, *Arch. Toxicol.,* 65, 381, 1991.
5. Graham, S. L., Hansen, W. H., Davis, K. J., et al., Effect of 1-year administration of ETU upon the thyroid of the rat, *J. Agric. Food Chem.,* 21, 324, 1973.
6. Newsome, W. N., Residues of mancozeb, 2-imidazoline and ethylene thiourea in tomato and potato crops after field treatment with mancozeb, *Bull. Environ. Contam. Toxicol.,* 20, 678, 1978.
7. Gak, J. C., Graillot, C., et Turhaut, R., Difference in the sensitivity of the hamster and the rat to the effects of long-term administration of ethylene thiourea, *Eur. J. Toxicol. Appl. Pharmacol.,* 9, 303, 1976 (in French).
8. Khera, K. S., Ethylenethiourea: a review of teratogenicity and distribution studies and of assessment of reproduction risk, *Crit. Rev. Toxicol.,* 18, 129, 1987.
9. Saillenfait, A. M., Sabate, J. P., Langonne, I., et al., Difference in the developmental toxicity of ethylene thiourea and three-*N,N'*-substituted thiourea derivatives in rats, *Fundam. Appl. Toxicol.,* 17, 399, 1991.
10. Ruddick, J. A., Newsome, W. H., Nash, L., Correlation of teratogenicity and molecular structure: ethylene thiourea and relative compounds, *Teratology,* 13, 263, 1976.
11. Chernoff, N., Kavlock, R. J., Rogers, E. N., et al., Perinatal toxicity of maneb, ethylenethiourea and ethylenebisiso-thiocyanate sulfide in rodents, *J. Toxicol. Environ. Health,* 5, 821, 1979.
12. Daston, G. P., Rehnberg, B. F., Corver, B., et al., Functional teratogens of the rat kidney. II. Nitrofen and ethylene thiourea, *Fundam. Appl. Toxicol.,* 11, 401, 1988.
13. EPA GENOTOX Program, 1986.
14. See *N-***CYCLOHEXYL-2-BENZOTHIAZOLE SULPHENAMIDE,** #7.
15. Ulland, B. M., Weisburger, J. H., Weisburger, E. K., et al., Thyroid cancer in rats from ethylene thiourea intake, *J. Natl. Cancer Inst.,* 49, 583, 1972.
16. Kato, Y., Odanaka, Y., Teramoto, S., et al., Metabolic fate of ethylene thiourea in pregnant rats, *Bull. Environ. Contam. Toxicol.,* 16, 546, 1976.
17. Iverson, F., Khera, K. S., Hierlihy, S. L., et al., *In vivo* and *in vitro* metabolism of ethylene thiourea in the rat and the cat, *Toxicol. Appl. Pharmacol.,* 52, 16, 1980.
18. *Pesticide Residues in Food — 1993.* Toxicity Evaluations, JECFA FAO/WHO, Geneva, 20–29 September 1993, 167.
19. Graham, S. L., Davis, K. J., Hansen, W. H., and Graham, C. H., Effects of prolonged ethylenethiourea ingestion on the thyroid of the rat, *Food Cosmet. Toxicol.,* 13, 493, 1975.

HYDROQUINONE (HQ) (CAS No 123–31–9)

Synonyms. **1,4**-Benzenediol; Benzohydroquinone; Benzoquinol; **1,4**-Dihydroxybenzene; **1,4**-Dioxybenzene; Hydroquinol; Quinol.

Properties. Light-grey or light-brown crystals (colorless when pure). Solubility in water is 6 to 7% (20°C). Readily soluble in hot water, ether, and ethanol. Aqueous solutions with up to 3 g/l concentrations are odorless. A sweetish taste appears at a concentration of a few grams per liter. The threshold for effect on the color of water is 0.2 mg/l.[1]

Applications. Chemical intermediate in the rubber industry. Antioxidant and stabilizer for materials that polymerize in the presence of oxidizing agent. Polymerization inhibitor of vinyl monomers: styrene, methyl methacrylate, etc.

Acute Toxicity. The LD_{50} values were reported to be 340 mg/kg BW in mice and 720 mg/kg BW[1] or 370 to 390 mg/kg BW in rats. According to other data,[2] LD_{50} for rats is about 1000 mg/kg BW. Mice given three LD_{50} during 2.5 h died in a few hours. They developed convulsions.[1] The toxicity of HQ increased by a factor of two to three when it was given to fasting animals. After oral administration, the animals immediately developed clonic convulsions and motor coordination disorder. Methemoglobinemia occurred. Death within a few minutes.

Repeated Exposure to HQ revealed slight cumulative properties. The amount of microsomal albumen and cytochrome B_5 and P-450, the activity of cytochrome-c-reductase, and liver weights were unchanged after administration of 200 mg/kg to rats for 4 d.[3] However, the same dose given to female rats for 14 d is reported to be toxic, causing tremors and clonic seizures and death of some animals due to respiratory paralysis after four to five administrations. Doses of 50 and 100 mg/kg BW produced no toxic effect.[4]

Long-term Toxicity. *Humans.* Nineteen volunteers who received 300 to 500 mg HQ daily for 3 to 5 months for experimental purposes exhibited no signs of intoxication. *Animals.* Rats were dosed by gavage with HQ over a period of 6 months. The treatment with 50 mg/kg caused anemia and leukocytosis. Gross pathological examination revealed dystrophic changes in the small intestine, liver, kidneys, and myocardium of rats given 50 to 100 mg HQ/kg BW.[5] In a 2-year study in rats, there was no effect on BW gain (diet contains up to 1% HQ), no hematological or other pathological changes were reported. A 5% dose-level in the diet caused ~50% loss of BW within 9 weeks; the animals developed aplastic anemia, depletion of the bone marrow, liver cord-cell atrophy, superficial ulceration, and hemorrhages of the gastric mucosa.[2]

Reproductive Toxicity. Addition of 0.003 and 0.3% HQ to the diet of female rats for 10 d prior to mating did not affect reproduction.[6] HQ is unlikely to be a developmental toxicant. Pregnant rats were dosed by gavage with 30 to 300 mg/kg on days 6 to 15 of gestation. The NOEL for maternal and developmental toxicity was identified to be 100 mg/kg BW.[7] *Embryotoxicity.* HQ was administered by gavage to pregnant New Zealand white rabbits on days 6 to 18 of gestation in aqueous solution at dose-levels of 25, 75, and 150 mg/kg BW. Doses of 75 and 150 mg/kg affected food consumption and BW gain of dams during the treatment period. The NOAEL was found to be 25 mg/kg BW for maternal toxicity and 75 mg/kg BW for developmental effects.[8] Administration of 115 mg/kg in rat diet throughout the gestation period caused an increase of fetal resorptions.[9]

Gonadotoxicity. There was no adverse effect of HQ on the fertility of rats treated through two generations with 15 to 150 mg/kg.[10] The dose of 50 to 200 mg HQ/kg BW administered by gavage to female rats for 14 d affected the estrus cycle.[4]

Genotoxicity. HQ has demonstrated mutagenic activity in modified micronuclei assay.[11] It exhibited colchicine-like action, caused build-up of metaphases in the small intestine of mice and SCE in human lymphocyte culture.[12] *Carcinogenicity classification.* IARC: Group 3; NTP: SE—SE—NE—SE*.

Chemobiokinetics. HQ is readily absorbed from the GI tract and rapidly distributed throughout the body tissues, but the maximum accumulation is found in the liver and kidneys. HQ metabolism occurs via conjugation with glucuronic and sulfuric acids. The major metabolites are reported to be **HQ monosulfate** and **HQ monoglucuronide,** which are found in the urine.[3] HQ is rapidly excreted in the urine.[13] Excretion level does not depend on the period of exposure.

Standards. Russia (1988). MAC and PML: 0.2 mg/l (organolept., chromaticity). **EEC** (1990). SML: 0.6 mg/l.

Regulations. USFDA (1993) approved the use of HQ (1) in adhesives used as components of articles intended for use in packaging, transporting, or holding food, (2) as an inhibitor in cross-linked polyester resins to be used safely as articles or components of articles intended for repeated use in contact with food, (3) in resinous and polymeric coatings for polyolefin films to be used safely as a food-contact surface of articles intended for use in producing, manufacturing, packing, transporting, or holding food, and (4) as an inhibitor for the monomer in the uncoated or coated food-contact surface of paper and paperboard intended for use in producing, manufacturing, packing, transporting, or holding aqueous and fatty food.

References:
1. See **BUTYLENE,** 194.
2. Carlson, A. J. and Brewer, N. R., Toxicity studies on hydroquinone, *Proc. Soc. Exp. Biol.,* 84, 684, 1953.
3. Divincenzo, G. D., Hamilton, M. L., Reynolds, R. C., et al., Metabolic fate and disposition of ^{14}C-hydroquinone given orally to Sprague-Dawley rats, *Toxicology,* 33, 9, 1984.
4. Racz, G., Fuzi, J., Kemeny, G., et al., The effect of hydroquinone and plorizin on the sexual cycle of white rats, *Orvosi. Szemle.,* 5, 65, 1958 (in Hungarian).
5. Mozhayev, E. A. et al., *Pharmacol. Toxicol.,* 29, 238, 1984 (in Russian).
6. Ames, S. R., Ludwig, M. I., Swanson, W. J., et al., Effect of DPPD, methylene blue, BHT, and hydroquinone on reproductive process in the rat, *Proc. Exp. Biol. Med.,* 93, 39, 1956.
7. Krasavage, W. J., Blacker, A. M., English, J. C., et al., Hydroquinone: a developmental toxicity study in rats, *Fundam. Appl. Toxicol.,* 18, 370, 1992.
8. Murphy, S. J., Schroeder, R. E., Blacker, A. M., et al., A study of developmental toxicity of hydroquinone in the rabbit, *Fundam. Appl. Toxicol.,* 19, 214, 1992.
9. Telford, I. R., Woodruff, C. S., and Linford, R. H., Fetal resorption in the rats as influenced by certain antioxidants, *Am. J. Anat.,* 110, 29, 1962.
10. Blacker, A. M., Schroeder, R. E., English, J. C., et al., A two-generation reproduction study with hydroquinone in rats, *Teratology,* 43 (Abstr. P39), 429, 1991.
11. Robertson, M. L., Eastmond, D. A., and Smith, M. T., Two benzene metabolites, catechol and hydroquinone, produce a synergistic induction of micronuclei and toxicity in cultured human lymphocytes, *Mutat. Res.,* 249, 201, 1991.
12. Morimoto, K., Wolf, S., and Koizumi, A., Induction of sister chromatid exchanges in human lymphocytes by microsomal activation of benzene metabolites, *Mutat. Res.,* 119, 355, 1983.
13. *Data on Hygienic Assessment of Pesticides and Polymers,* Erisman Research Hygiene Institute, Moscow, 1977, 174 (in Russian).

MELAMINE CYANURATE (MC) (CAS No 37640–57–6)

Synonym. 2,4,6-Triamino-*s*-triasine, compound with s-triazinetroil.

Properties. White, crystalline substance. Poorly soluble in water and organic solvents.

Applications. Used as a thermostable filler for vulcanizate mixes.

Acute Toxicity. LD_{50} is 2.5 to 5.52 g/kg BW for rats and 3.46 g/kg BW for mice. Lethal doses cause adynamia and motor coordination disorder. Hematuria and epitaxis, pareses, and paralyses were found to develop. Death occurs with signs of respiratory distress. Gross pathological examination revealed severe circulatory disturbances, particularly in the brain.

Repeated Exposure revealed cumulative properties. K_{acc} is 1.23 (by Kagan), when rats are given $^1/_{10}$ LD_{50}.

Genotoxicity and **Reproductive Toxicity** effects were not observed after inhalation exposure to 1.15 mg MC/m³.

Reference:
Alexandryan, A. V., Toxicity and sanitary standardization of melamine cyanurate, *Gig. Truda Prof. Zabol.,* 1, 44, 1986 (in Russian).

MERCAPTANS

Synonyms. Thioalcohols; Thiols.

Properties. The lower mercaptans (C_3 to C_4) are highly volatile, colorless iiquids with a sharp, specific odor. Poorly soluble in water, readily soluble in alcohol. Taste and odor perception threshold is <0.02 mg/l. Odor threshold concentrations are 0.02 mg/l for **amylmercaptane, dodecylmercaptane,** and **isopropylmercaptane,** 0.00019 mg/l for **benzylmercaptane** and **ethylmercaptane,** 0.00005 mg/l for **allylmercaptane,** and 0.0001 mg/l for **tolylmercaptane.**

Occurs in rubber and cellophane, various coatings.

Methyl mercaptan (MM) (CAS No 74–93–1)

Synonyms. Methanethiol; Thiomethanol.

Properties. Gas with a sharp odor. Solubility in water is ~14 g/l at 25°C. Odor perception threshold is 0.00025 or 0.0011 mg/l.[010]

Chemobiokinetics. Metabolism of MM occurs through rapid methylation to form **dimethyl sulfide.** About 40% MM is excreted as CO_2 and 30% MM in the urine as **sulfates.** [14]C is detected in the methyl group of methionine, choline, and creatine.

Standards. Russia (1988). MAC and PML: 0.0002 mg/l (organolept., odor).

Reference:
See **DIPHENYLAMINE, #1,** V. 2, 375.

N-Butyl mercaptan (CAS No 109–79–5)

Synonyms. *n*-Butanethiol; *n*-Butyl thioalcohol; Thiobutyl alcohol.

Properties. Liquid. Solubility in water is 600 mg/l at 20°C. Odor perception threshold is reported to be 0.006 mg/l[1] or 0.000012 mg/l.[02]

Acute Toxicity. Some mice died from a 3 g/kg BW dose, some rats from 0.4 g/kg BW.[2] According to other data, LD_{50} for rats is 1.5 g/kg BW.

Reproductive Toxicity. Showed no teratogenic action following inhalation exposure of mice and rats.[3]

References:
1. See 6-**METHYL-2-VINYLPYRIDINE, #1.**
2. Blinova, E. A., Toxicity characteristics of ethylmercaptan on the basis of chronic studies, *Gig. Sanit.,* 1, 18, 1965 (in Russian).
3. Thomas, W. C., Seckar, J. A., Johnson, J. T., et al., Inhalation teratology studies of *n*-butyl mercaptan in rats and mice, *Fundam. Appl. Toxicol.,* 8, 170, 1987.

2-MERCAPTOBENZIMIDAZOL (2-MBI) (CAS No 583–39–1)

Synonyms. 2-Benzimidazolethiol; **1,3**-Dihydro-*2H*-benzimidazole-2-thione; Idazole *MB; o*-Phenylenethiourea; Thiobenzimidazole.

Properties. Yellow-white powder with a bitter taste. Poorly soluble in water, soluble in alcohol.

Applications. Used as a stabilizer in the production of synthetic rubber, polyethylene, and polypropylene.

Acute Toxicity. LD_{50} is 270 mg/kg BW in rats and 1250 mg/kg[1] or 105 mg/kg BW in mice.[2] Gross pathological examination revealed morphological changes in the lungs, cerebral vessels, and liver. The single lethal dose caused thyroid toxicity as measured by a 95% decrease in iodine uptake in rats.[3] Thyroid enlargement is associated with decreased plasma concentrations of circulating thyroxine and triiodothyronine and increased thyotropin levels in rats receiving a single oral dose.[4]

Repeated Exposure revealed marked cumulative properties. The treatment caused retardation of BW gain, impairment of liver function, a raised threshold of neuro-muscular excitation, and reduction in oxygen consumption. Gross pathological examination revealed congestion in the pulmonary vessels and accumulation of lymphocytes around the hepatic vessels.

Short-term Toxicity. Rabbits were dosed by gavage with 20 mg/kg for 4 months. The treatment caused retardation of BW gain, impairment of liver functions, hemorrhagic syndrome, and changes in hematological analyses.[2]

Immunotoxicity. 2-MBI is suspected to cause thymic involution.

Reproductive Toxicity. *Embryotoxic* effect was observed in rats.[5]

Genotoxic effect was demonstrated in rat fetal cells;[5] cytogenetic abnormalities and immunodeficiency were described in the progeny of rats.[6]

Carcinogenicity. 2-MBI is suspected of being capable of producing thyroid and liver tumors because it is structurally related to ETU.

Chemobiokinetics. 2-MBI is readily absorbed after oral administration. Its half-life in blood is approximately 83 h. Accumulates in the thyroid.[4]

References:
1. See **PHTHALIC ANHYDRIDE,** #2, 180.
2. See *N*-**BUTYLIDENEANILINE,** 72.
3. Searle, C. E., Lawson, A., and Hemmings, A. W., Antithyroid substances. I. The mercapto glyoxolines, *Biochem. J.,* 47, 77, 1950.
4. Janssen, F. W., Young, E. M., Kirkman, S. K., et al., Biotransformation of the immunomodulator (*p*-chlorophenyl)-**2,3**-dihydro-3-hydroxythiazolo [**3,2a**] benzimidazole-2-acetic acid and its relationship to thyroid toxicity, *Toxicol. Appl. Pharmacol.,* 59, 355, 1981.
5. Barilyak, I. R., Embryotoxic and mutagenic effects of **2**-mercaptoimidazole, *Physiol. Active Substances,* 6, 85, 1974 (in Russian).
6. Barilyak, I. R. and Melnik, E. K., Dynamics of the cytogenetic effect of some chemical preparations, *Rep. Ukrainian Acad. Sci.,* 1, 66, 1979 (in Ukrainian).

2-MERCAPTOBENZOTHIAZOLE (2-MBT) (CAS No 149–30–4)

Synonyms. 2-Benzothiazolethiol; Captax.

Properties. Pale monoclonic needles or leaflets, or easily dusting yellow powder with a characteristic odor and bitter taste. Poorly soluble in water, soluble in alcohol.

Applications and **Exposure.** The commonest vulcanization accelerator for vulcanizate mixes based on natural and synthetic rubbers. Migration from butadiene-nitrile vulcanizates into water after a 3-d exposure at 20°C was determined at the level of up to 1 mg/l.[1]

Acute Toxicity. In mice, rats, and rabbits, LD_{50} is 2.3 to 2.7 g/kg, or it is 3.8 g/kg BW in rats (Monsanto Company, unpublished report). All mice died from a 4 g/kg BW dose. Manifestations of toxic action include restlessness, excitation, impaired motor coordination, and convulsions. Death within 24 h.[2,3]

Repeated Exposure failed to reveal cumulative properties. The treatment of rabbits with 200 mg/kg, dogs with 50 to 100 mg/kg (for 6 d) and rats with 1 g/kg BW (for 10 d) did not affect the general condition of animals,[4] while a 45-d administration of 2-MBT altered hematological analyses and the functional state of the liver and CNS.

Short-term Toxicity. Rats were dosed with 50 mg/kg BW for 4 months. Only changes in conditioned reflex activity were marked; BW gain indices and blood morphology did not differ from those in the controls. Rabbits received 20 mg 2-MBT/kg BW every other day for 2 months and then daily for 1.5 months; dogs received 50 to 100 mg/kg BW for 3 to 4 months. No manifestations of toxic effect were noted in the animals.[4]

Reproductive Toxicity. *Embryotoxicity.* Severe embryo-lethality was noted in rats given 4.0 mg 2-MBT/kg BW on gestation day 8 or 10.[5] *Teratogenic* effect (following inhalation) was not found in rats.[6]

Carcinogenicity. Carcinogenic effect was found in rats gavaged with 750 mg/kg (males) and 375 mg/kg BW (females) doses but not in mice.[7] *Carcinogenicity classification.* NTP: SE*—SE—NE—EE.

Chemobiokinetics. Rats were given $^1/_{10}$ LD_{50}. In 1 h after administration, 0.015 mg 2-MBT/ml serum and 0.072 mg 2-MBT/ml urine were found. Accumulates primarily in the liver, brain, kidneys, and fatty tissues. Excretion is mainly via the kidneys.[8] Rats were given 0.51 mg 2-MBT/kg BW for a 14-d period prior to a single 0.503 mg/kg BW dose of ^{14}C-MBT. In 96 h, a small portion of radioactivity remained associated with erythrocytes, most of which was bonded to the membranes. Two metabolites were found in the urine, one of which was a **thioglucuronide derivative** of 2-MBT; the other was possibly a **sulfonic acid derivative.** Elimination was mainly with the urine and to a lesser extent in the feces (El Dareer et al., 1989).

Standards. Russia (1988). PML in food: 0.15 mg/l; MAC and PML in drinking water: 5 mg/l (organolept., odor).

Regulations. USFDA (1993) approved the use of 2-MBT (1) as a component of adhesives (as preservative only) to be used safely in contact with food, (2) as an accelerator for rubber articles intended for repeated use in contact with food (up to 1.5% by weight of the rubber product), and (3) in preparation of slimicides in the manufacturing of paper and paperboard that may be used, safely in contact with food.

References:
1. See **ZINC PHENYLETHYLDITHIOCARBAMATE,** #1, 153.
2. See **DI-2-BENZOTHIAZOLYL DISULFIDE,** #2.
3. See **PHTHALIC ANHYDRIDE,** #2, 28.
4. Litvinchuk, M. D., Studies on toxicity of 2-mercapto-benzothiazole, *Pharmacol. Toxicol.,* 4, 484, 1963 (in Russian).
5. *Problems of Epidemiology and Hygiene in the Lithuanian SSR,* Vilnius, 1969, 165 (in Russian).
6. Hardin, B. D., Bond, G. P., Sikov, M. R., et al., Testing of selected workplace chemicals for teratogenic potential, *Scand. J. Work Environ. Health,* 7 (Suppl. 4), 66, 1981.
7. See **DIMETHYL TEREPHTHALATE,** #6.
8. Datsenko, I. I. and Korneichuk, E. P., Kinetics of 2-mercaptobenzothiasole metabolism in the animals, *Gig. Sanit.* 1, 51, 1991 (in Russian).

2-MERCAPTOBENZOTHIAZOLE, COPPER SALT

Synonyms. Captax, copper salt; Copper, 2-benzothyazole-thiolate.

Properties. A yellow, fine-particled, easily dusting substance. Poorly soluble in water.

Applications. A vulcanization accelerator.

Acute Toxicity. LD_{50} was not attained in mice: administration of 10 g/kg BW caused only 27% animals to die on days 6 to 8 after treatment. High doses produce retardation of BW gain and reduction in relative liver weights.

Long-term Toxicity. Rabbits were dosed by gavage with 50 mg/kg BW every other day for the first 2 months and daily for a further 3 months. The treatment resulted in decrease of BW gain and changes in the hematological analyses and in the balance of serum protein fractions. Gross pathological examination revealed congestion in the visceral organs and changes in the suprarenals and myocardium.

Reference:
See **PHTHALIC ANHYDRIDE,** #2, 43.

2-MERCAPTOBENZOTHIAZOLE, DIPHENYLGUANIDINE SALT

Synonym. 2-MBT.DPG.

Properties. Light-grey, fine-particled powder. Readily soluble in alcohol.

Applications. A vulcanization ultra-accelerator.

Acute Toxicity. In mice, LD_{50} is 360 mg/kg BW. Poisoned animals looked immobile, with unkempt fur. Other manifestations of toxic action included impaired motor coordination, rigid neck and tail muscles, and labored breathing.

Long-term Toxicity. Rabbits were exposed to 50 mg/kg BW for 5.5 months. The treatment caused retardation of BW gain, leukocytosis, and dysbalance in serum protein composition. Gross pathological examination revealed lymphoid stroma infiltration, connective tissue growth along the interlobar septa, and focal, large-drop, fatty infiltration of the liver cell protoplasm; in the kidneys, there was a marked congestion, glomerular swelling, plasma in the Bowman capsules, and cloudy swelling of the renal tubular epithelium.

Reference:
See **PHTHALIC ANHYDRIDE,** #2, 51.

2-MERCAPTOTHIAZOLINE (2-MT) (CAS No 96–53–7)

Synonym. Thiazoline-2-thiol.

Properties. A colorless, odorless powder. Poorly soluble in cold water and alcohol, readily soluble in hot water.

Applications. A vulcanization accelerator for mixtures of natural and styrene-butadiene rubber.

Acute Toxicity. LD$_{50}$ is found to be 254 mg/kg BW in male mice.[1]

Short-term Toxicity. Rabbits were exposed to a 20 mg/kg BW dose every other day for 2 months and then daily for the same period of time. The treatment resulted in decrease of BW gain, reduced cholinesterase and alkaline phosphatase activity, and increased serum cholesterol and lipid levels. Gross pathological examination revealed dystrophic changes in the liver. Edema, congestion, and lymphoid elements were found in the myocardium. In the lungs, thickening of the alveolar septa with proliferation of cellular elements and focal desquamation of the epithelium were found. Kidney weights were increased.[1]

Reproductive Toxicity. 2-MT produced no *teratogenic* effect in rats.[2]

Regulations. USFDA (1993) approved the use of 2-MT as an accelerator of rubber articles intended for repeated use in producing, manufacturing, packing, processing, treating, packaging, transporting, or holding food (total not to exceed 1.5% by weight of the rubber product).

References:

1. See **PHTHALIC ANHYDRIDE**, #2, 57.
2. See **ETHYLENE THIOUREA**, #10.

5-METHYL-1,3-BIS(PIPERIDINOMETHYL)HEXAHYDRO-1,3,5-TRIAZINE-2-THIONE

Synonym. Bistriazine.

Properties. Poorly soluble in water, readily soluble in alcohol on heating.

Applications. A vulcanization accelerator.

Acute Toxicity. LD$_{50}$ is reported to be 482 mg/kg BW in rats, 984 mg/kg BW in mice, 710 mg/kg BW in rabbits, and 543 mg/kg BW in guinea pigs. Poisoning was accompanied by excitation, with subsequent CNS inhibition and convulsions. Death occurred from respiratory arrest.

Short-term Toxicity. Cumulative properties were observed in 4-month rat studies.

Long-term Toxicity. Rats and rabbits were exposed by gavage to oral doses of up to 9.6 mg/kg BW for 10 months. The treatment caused anemia and retardation of BW gain, a decline in total serum albumen, and a reduction in the ascorbic acid content in the suprarenals. Changes in the STI value were found at all doses tested. Gross pathological examination revealed necrotic foci in the spleen, parenchymatous dystrophy, and necrotic areas in the small intestine mucosa.

Reference:

Verbilov, A. A., Toxicological characteristics of new rubber vulcanization accelerators triazinethion and bistriazine, *Vrachebnoye Delo,* 6, 132, 1974 (in Russian).

5-METHYLHEXAHYDRO-1,3,5-TRIAZINE-2-THIONE

Synonym. Triazinethione.

Properties. Soluble in water; on heating, soluble in alcohol.

Applications. A vulcanization accelerator.

Acute Toxicity. Substance of very low if any acute toxicity. Administration of 15 g/kg BW to mice, rats, and guinea pigs three times at 3-h intervals caused no mortality or symptoms of intoxications. In 2 d, leukopenia and decrease in the concentration of SH-group in the blood serum were noted.[1]

Short-term Toxicity. No accumulation was found in a 4-month study when rats were given 0.5 g/kg BW dose.

Long-term Toxicity. Rats and rabbits were dosed by gavage for 10 months. A 50 mg/kg dose affected hippuric acid synthesis in the liver and caused prolongation of narcotic sleep and changes in the sugar curves as a result of galactose loading. Changes in oxidation-reduction enzyme activities and in SH-group content in the blood serum have been reported. Gross pathological examination revealed hemodynamic disturbances in the liver and areas of superficial necrosis in the stomach and small intestine, with desquamation of the upper mucosal layers.[1,2]

Standards. Russia. Recommended PML in drinking water: 10 mg/l.

References:
1. Verbilov, A. A., Toxicological characteristics of triazinethione — a new vulcanization accelerator for rubber, *Gig. Sanit.*, 7, 97, 1974 (in Russian).
2. See 5-**METHYL**-1,3-**BIS-(PIPERIDINOMETHYL)HEXAHYDRO**-1,3,5-**TRIAZINE**-2-**THIONE**.

N-METHYL-*N,4*-DINITROZOANILINE (CAS No 99–80–9)

Synonyms. Elastopar; *N,p*-Dinitroso-*N*-methylaniline; *N*-Methyl-*N,4*-di-nitrosobenzenamine; Nitrozan *K*.

Properties. Khaki-colored, amorphous powder. Poorly soluble in water; on heating, dissolves in alcohol.

Applications. M. is used as a modifier and activator of vulcanization in the mixtures of natural and synthetic rubber, and also of polyethylene.

Acute Toxicity. LD$_{50}$ is 0.9 g/kg BW in rats and 1.85 g/kg BW in mice. Poisoning is accompanied by general depression, adynamia, and diarrhea. The animals look unkempt. MetHb formation was found when doses of 0.2 to 0.9 g/kg BW were given. Survivors recovered in a week.[1]

Short-term Toxicity. Administration of 18 and 90 mg/kg BW caused death in the majority of the treated animals. Death occurred earlier with the smaller dose. Rabbits were given 200 mg/kg BW dose by gavage for 4 months. The treatment led to a reduction in alkaline phosphatase activity and an increase in aldolase activity. Gross pathological and histological examination revealed moderate congestion in the lungs, liver, kidneys, and myocardium.[2]

Long-term Toxicity. Chronic exposure affected liver and kidney functions in rats by the end of experiment (doses not indicated). Gross pathological examination revealed circulatory disturbances and parenchymatous dystrophy of the liver and myocardium and an increase in the relative weights of the brain, liver, and spleen at a 90 mg/kg BW dose-level. Observed effects appeared to be dose-dependent.[1]

Carcinogenicity. A number of **nitroso compounds** have been shown to have carcinogenic potential. M. is considered to be an equivocal tumorogenic agent by RTECS criteria.

References:
1. See **BISPHENOL A,** #2, 217.
2. See **PHTHALIC ANHYDRIDE,** #2, 187.

2-(MORPHOLINOTHIO)BENZOTHIAZOLE (CAS No 102–77–2)

Synonyms. 2-Benzothiazolyl *N*-morpholinosulfide; *N*-(Oxydiethylene)benzothiazole-2-sulphenamide; Santocure mor; Sulfenamide *M*.

Properties. Light-yellow powder. Poorly soluble in water, soluble in alcohol.

Applications. Used either alone or in mixtures with other accelerators for rubber and vulcanizates.

Acute Toxicity. LD$_{50}$ for mice is reported to be 1.87 g/kg BW.[1] According to other data,[2] rats and mice tolerate 5 g/kg BW. After administration, there was apathy and adynamia. Death occurred within 8 d.

Repeated Exposure. Three administrations of 5 g/kg BW caused no mortality in mice. Two doses of 0.5 g/kg BW led to reduced incorporation of radioactive iodine and to histological changes in the thyroid.

Short-term Toxicity. Rabbits were dosed by gavage with 20 mg/kg BW for 3.5 months. The treatment caused some liver impairment and bilirubin index increase, indicating cholagogical effect.[3] Histological examination revealed signs of fatty dystrophy of the renal tubular epithelium. Changes in the thyroid and liver dystrophy were found to develop.[2]

Reproductive Toxicity. M. administration has been shown to cause developmental abnormalities in rats.

Carcinogenicity. Nitroso-compounds of morpholine are powerful carcinogens. M. is reported to be neoplastic by RTECS criteria.

References:
1. See **PHTHALIC ANHYDRIDE,** #2, 54.
2. See **PHTHALIC ANHYDRIDE,** #2, 70.
3. Vorob'yeva, R. S. and Mesentseva, N. V., Experimental study on comparative toxicity of a new agent accelerating vulcanization, *Gig. Truda Prof. Zabol.,* 7, 28, 1962 (in Russian).

1-NAPHTHYLAMINE (CAS No 134–32–7)

Synonyms. 1-Aminonaphthalene; 1-Naphthaleneamine; Naphthalidam; Naphthalidine; α-Naphthylamine.

Properties. Grey needles, becoming red on exposure to air, or a reddish crystalline mass with an unpleasant odor. Solubility in water is 0.17%, readily soluble in alcohol.

Applications. A vulcanization accelerator for rubber.

Genotoxicity. No data are available on the genetic and related effects in humans (IARC Suppl. 6 to 406). 1-N. did not induce micronuclei in bone marrow cells of mice treated *in vitro;* it induced DNA strand breaks in mice but not in rats. N. increased the incidence of CA in cultured rodent cells, but the results for SCE, mutation, and DNA damages were inconclusive; no cell transformation was induced in Syrian hamster embryo cells (IARC 4–87).

Carcinogenicity. No carcinogenic effect was found following oral administration of 1-N. to hamsters or dogs.[1,2] In beagle dogs[1] the effect of highly purified 1-N. has been investigated. 1-N. was given orally, dissolved in corn oil, in gelatin capsules at a dose-level of 15 mg/kg BW. (This study has limitations due to a small number of animals tested.)

Chemobiokinetics. Metabolism of 1-N. varies in different animals. 1-N. is partly oxidized to 1-amino-2-naphthol, which is eliminated in the urine, but mainly as a glucuronide.[3] From a 5 mg dose, 0.2% was shown to be excreted in the urine. A part of 1-N. is excreted unchanged.

Regulations. USFDA (1993) listed 1-N. for use in adhesives as a component of articles intended for use in packaging, transporting, or holding food.

References:
1. Radomski, J. L., Deichmann, W. B., Altman, N. H., et al., Failure of pure 1-naphthylamine to induce bladder tumours in dogs, *Cancer Res.,* 40, 3537, 1980.
2. Purchase, I. F. H., Kalinowski, A. E., Ishmael, J., et al., Lifetime carcinogenicity study of 1-naphthylamine and 2-naphthylamine in dogs, *Br. J. Cancer,* 44, 892, 1981.
3. *Proc. 3rd Int. Cancer Cong.,* Moscow-Leningrad, 2, 314, 1983 (in Russian).

N-NITROSODIMETHYLAMINE (NDMA) (CAS No 62–75–9)

Synonyms. *N,N'*-Dimethylnitrosamine; *N*-Methyl-*N'*-nitroso methanamine.

Properties. Volatile, yellow, oily liquid of low viscosity. Soluble in water, lipids, alcohols, and other organic solvents.

Applications and **Exposure.** NDMA is used primarily as a research chemical; it occurs as an impurity; plasticizer for rubber and acrylonitrile polymers; an antioxidant. NDMA intake from a pack of cigarettes is about 0.001 mg. It is also the most commonly detected *N*-nitrosamine in food systems. Traces of nitrosamines in baby bottle rubber nipples and pacifiers could easily migrate to simulated saliva and milk. Out of 42 samples, 7 of these products contained greater than 0.03 ppm total volatile nitrosamines (mainly, NDMA and *N*-nitrosodi-*n*-butylamine).[1]

Acute Toxicity. The toxicity of nitroso-compounds is not of great interest because there is no relationship between toxic effect and carcinogenic activity.[2] LD_{50} is 40 to 50 mg/kg BW in rats,[2,3] 30 mg/kg BW in hamsters, and 20 mg/kg BW in dogs. Administration of lethal doses causes very pronounced liver lesions in dogs, rabbits, rats and guinea pigs. Gross pathological examination revealed liver centrolobular necrosis with subsequent hemorrhage. In rats, acute toxic effects occur mainly as cellular necrosis in the liver[4] and kidney.[5] Mink and also sheep and cattle are particularly sensitive to NDMA. A single dosage of 50 mg/kg, which is necrotizing for rat's liver, produced more acute liver damage in cats than in guinea pigs, rats, or monkeys.[6]

Repeated Exposure. Administration of 5 mg/kg BW daily produced the same liver damage in guinea pigs and rats as a single 50 mg/kg dose.[6] Mink received 2.5 to 5 mg NDMA/kg BW for

7 to 11 d. Administration led to extensive liver damage and hepatocyte necrosis. Cell proliferation was noted in the bile ducts. Ascites and hemorrhages in the GI tract were found to develop (Carter et al., 1969). Twelve administrations of 0.5 mg/kg doses to sheep caused death or the development of pronounced anoxia. Masticatory function was inhibited. Cats and lizards received 1 mg NDMA/kg BW for 30 d. The treatment revealed centrilobular necrosis and hemorrhages to be the main alteration detected in the liver of cats. Lizards appeared to be unaffected.[3]

Short-term Toxicity. Significant hepatotoxicity was observed in cattle given 0.1 mg/kg BW dose over a period of 1 to 6 months.

Long-term Toxicity. Dose-related mortality and hepatotoxicity were observed following oral exposure of CD-1 mice to 1 to 20 ppm NDMA.[7]

Reproductive Toxicity. A 30 mg/kg BW dose caused embryolethal effect, particularly when given on critical days of gestation.[8]

Immunotoxicity. Chronic exposure to NDMA resulted in a marked and persistent immunodepression of cellular and humoral responses in CD-1 mice. Immunodepression of immunoglobulin M antibody response to sheep erythrocytes was found to be time-related and dose-related; cellular immune response was markedly suppressed by 10 and 20 ppm doses. No signs of immunotoxicity were caused by 1 ppm NDMA.[7]

Genotoxicity. NDMA requires metabolic activation to exert toxic and carcinogenic effects. It was shown to have positive results in *Dr. melanogaster* and in *Salmonella* mutagenicity assay, but not in the DLM test or in cytogenicity studies.

NDMA is a classical alkylating agent.

Carcinogenicity. Although no direct NDMA-related cancer was reported in humans, nitrosoamines are considered to be potentially epigenetic compounds, due to chronic immunodepression.[6] Carcinogenic potential was revealed in all animals tested: mice, rats, hamsters, guinea pigs, rabbits, ducks, etc.[9-12] NDMA produced malignant neoplasms predominantly in the liver, kidneys, and respiratory tract. An increase in the lung adenoma response in strain A/J mice was observed after *i/p* and oral administration.[13] Doses of 2 to 50 mg/kg BW were shown to cause tumors, predominantly in the rat liver. Tomatis and Cefis[14] reported the formation of liver tumors in Syrian golden hamsters after administration of a total dose of 3.6 mg in 5 weeks and a single dose of 1.6 mg. The lowest dose tested in male RF mice (administered via drinking water and correspondent to a dose of 0.4 mg/kg BW) produced lung adenomas in 13 out of 17 animals and hemoangiocellular tumors in 2 out of 10 animals.[9] *Carcinogenicity classification.* IARC: Group 2A.

Chemobiokinetics. Is slowly absorbed from the stomach but rapidly from the upper part of the small intestine.[15] Oral doses below 40 mg/kg BW were completely metabolized by the liver and did not enter general circulation.[16] The liver and nasal mucosa are likely to be the principal sites of NDMA metabolism, and their injury effects may be related to NDMA metabolites. Oxidative *N*-dimethylation to form **formaldehyde** has been shown with liver microsomes of rats, mice, and hamsters.[17] Metabolic activation of NDMA is mediated by cytochrome P-450 and culminates in the formation of electrophilic intermediates, which may form covalent bonds with DNA or other macromolecules. This seems to be a mechanism of mutagenicity and carcinogenicity initiation.

Regulations. USFDA (1993) has set a 10 ppb limit on nitrosoamines in rubber nipples for baby bottles.

References:

1. Sen, N. P., Kushwaha, S. C., Seaman, S. W., et al., Nitrosamines in baby bottle nipples and pacifiers: occurrence, migration, and effect of infant formulas and fruit juices on *in vitro* formation of nitrosamines under simulated gastric conditions, *J. Agric. Food Chem.*, 33, 428, 1985.
2. Druckrey, H., Preussmann, R., Ivankovic, S., u. a., Organotropic action of 65 different *N*-nitrosocompounds on BD rats, *Z. Krebsforsch.*, 69, 103, 1967 (in German).
3. Maduagewu, E. N. and Anosa, V. O., Hepatotoxicity of dimethylnitrosamine in cats and lizards, *Toxicol. Lett.*, 9, 41, 1981.
4. Shank, R. C., Toxicology of *N*-nitrosocompounds, *Toxicol. Appl. Pharmacol.*, 31, 361, 1975.
5. Hard, G. C., MacKay, R. L., and Kockhar, O. S., Electron microscopic determination of the sequence of acute tubular and vascular injury induced in the rat kidney by carcinogenic dose of dimethylnitrosamine, *Lab. Invest.*, 50, 659, 1984.
6. Maduagewu, E. N. and Bassir, O., A comparative assessment of toxic effect of dimethylnitrosamine in six different species, *Toxicol. Appl. Pharmacol.*, 53, 211, 1980.

7. Desjardins, R., Fournier, M., Denizeau, F., et al., Immunodepression by chronic exposure to *N*-nitroso dimethylamine in mice, *J. Toxicol. Environ. Health,* 37, 351, 1992.
8. Napalkov, N. P. und Alexandrov, Y. A., On the effects of blastomogenic substances on the body during embryogenesis, *Z. Krebsforsch.,* 71, 32, 1968 (in German).
9. Magee, P. N. and Barness, J. M., Carcinogenic nitroso compounds, *Adv. Cancer Res.,* 10, 163, 1967.
10. Ishinishi, N., Tanaka, A., Hisanaga, A., et al., Comparative study on the carcinogenicity of *N*-nitrosodiethylamine, *N*-nitrosodimethylamino, *N*-nitroso-*n*-propylamine to the lung of Syrian golden hamster following intermittent instillation to the trachea, *Carcinogenesis,* 9, 947, 1988.
11. Clapp, N. K. and Toya, R. E., Effect of cumulative dose and dose rate on dimethylnitrosamine oncogenesis in RF mice, *J. Natl. Cancer Inst.,* 45, 495, 1970.
12. Terracini, B., Magee, P. N., and Barness, J. M., Hepatic pathology in rats of low dietary levels of dimethylnitrosamine, *Br. J. Cancer,* 21, 539, 1967.
13. Stoner, G. D., Greisiger, E. A., Schut, H. A. J., et al., A comparison of the lung adenoma response in strain A/J mice after intraperitoneal and oral administration of carcinogens, *Toxicol. Appl. Pharmacol.,* 72, 313, 1984.
14. Tomatis, L. and Cefis, F., The effect of multiple and single administration of dimethylnitrosamine to hamsters, *Tumori,* 53, 447, 1967.
15. Phillips, J. C., Heading, C. E., Lake, B. G., et al., Further studies on the effects of inhibitors on the metabolism and toxicity of dimethylnitrosamine, *Biochem. Soc. Trans.,* 3, 179, 1975.
16. Diaz Gomes, M. I., Swann, P. E., and Magee, P. N., The absorption and metabolism in rats of small oral doses of dimethyl nitrosamine, *Biochem. J.,* 164, 497, 1977.
17. Argus, M. F., Acros, J. C, Pastor, K. M., et al., Dimethylnitrosamine-demethylase: absence of increased enzyme catabolism and multiplicity of effector sites in repression, homoprotein involvement, *Chem. Biol. Interact.,* 13, 127, 1976.

N-NITROSODIPHENYLAMINE (NDPA) (CAS No 86–30–6)

Synonyms. Diphenylnitrosamine; 4-Nitroso-*N*-phenylbenzenamine; Vulcalent A.

Properties. Green or yellow plates or prisms with a bluish luster, or light-yellow, crystalline powder. Slightly soluble in water, soluble in acetone and ethanol. Sensitive to light and undergoes photolytic degradation. Gives water a specific, unpleasant odor (threshold concentration 0.1 mg/l).

Applications and **Exposure.** Used in mixes of natural and synthetic rubber and latex. Slows down the vulcanization and cross-linking of rubber mixes. Used in the production of organic dyestuffs. A stabilizer for polymers; an anti-scorching agent.

N-Nitrosamines were shown to migrate easily from rubber products to liquid infant formula, orange juice, and simulated human saliva.[1]

Acute Toxicity. All mice died in 24 h given doses of 4 to 5 g/kg BW. LD_{50} appeared to be 3.85 g/kg BW.[2] In rats, LD_{50} is 3 g/kg BW.[3] Toxicity increases when NDPA is given to animals in sunflower oil. In this case, LD_{50} is reported to be 1.68 g/kg BW in mice and 1.87 g/kg BW in rats. Signs of cyanosis are noted.[4]

Repeated Exposure revealed insignificant cumulative properties. A 40-d exposure to 70 or 200 mg/kg BW affected the peripheral blood, CNS, and enzyme activity in the blood and liver. Fisher 344 rats and $B6C3F_1$ mice were fed diets containing up to 46 g/kg for 7 or 11 weeks. Females did not survive the doses greater than 16 g/kg diet. Inflammatory lesions were found in the urinary bladder of mice (SHELL Chemical Co.).

Short-term Toxicity. Rabbits were dosed by gavage with 20 mg/kg for 4 months. The treatment caused retardation of BW gain and increase in serum aldolase activity. Gross pathological examination revealed parenchymatous dystrophy of the renal tubular epithelium, foci of peribronchial pneumonia, and sometimes, emphysema.[2]

Reproductive Toxicity. *Teratogenic* effect was found in 2-d-old chick embryos.[5] *Embryolethality* was observed in chickens and rats at a 100 mg/kg dose-level.

Genotoxicity. Causes genetic damage *in vitro* (in bacteria, yeast and cultured mammalian cells).[9] Mutagenic activity was shown in *Salmonella typhimurium* TA98 in the presence of S9 microsomal fraction and a co-mutagen.[6] NDPA did not induce unscheduled DNA synthesis (Lake et al., 1978).

Carcinogenicity. No carcinogenic effect was observed in B6C3F$_1$ mice given the diet containing 10 and 20 g NDPA/kg for 101 weeks. However, carcinogenic action was found in Fisher 344 rats treated with 1 or 4 g NDPA/kg diet for 100 weeks. The treatment produced transitional-cell carcinomas of the urinary bladder in animals of the high-dose group.[7,8] Carcinogenic potency is at least 100-fold lower than the potency of aliphatic nitrosamines. **Carcinogenicity classification.** IARC: Group 3; NTP: P—P—N—N.

Chemobiokinetics. NDPA is a powerful nitrosating agent readily donating NO to secondary amines.

Standards. Russia (1988). MAC and PML: 0.1 mg/l (organolept.).

Regulations. USFDA has set a 10 ppb limit on **nitrosoamines** in rubber nipples for baby bottles. FDA also established action levels of 5 ppb NDMA in malt beverages and 10 ppb in barley malt, which is expected to reduce or eliminate exposure from these sources.

References:

1. *Nitrosocompounds: Occurrence, Biological Effects and Relevance to Human Cancer,* IARC Sci. Publ. No. 57, O'Neill, I. K., von Borstel, R. C., Miller, C. T., et al., Eds., IARC, Lyon, 1984, 51.
2. See **PHTHALIC ANHYDRIDE, #2,** 157.
3. See *N*-**NITROSODIMETHYLAMINE, #2.**
4. Korolev, A. A., Shlepnina, T. G., Mikhailovsky, N. Ya., et al., Maximum allowable concentration of diphenylnitrosamine and nitroguanidine in water bodies, *Gig. Sanit.,* 1, 18, 1980 (in Russian).
5. See *N,N',N''*-**TRICHLOROMELAMINE, #2.**
6. *IARC Sci. Publ.,* 41, 695, 1982.
7. Cardy, R. H., Lijinsky, W., and Hildebrandt, P. K., Neoplastic and non-neoplastic urinary bladder lesions induced in F344 rats and B6C3F$_1$ hybrid mice by *N*-nitrosodiphenylamine, *Ecotoxicol. Environ. Saf.,* 3, 29, 1979.
8. National Cancer Institute, *Bioassay of N-Nitrosodiphenylamine for Possible Carcinogenicity,* Technical Report Series 164, DHEW Publ. No 79–1720, U.S. Government Printing Office, Washington, D.C., 1979.
9. McGregor, D., The genetic toxicology of *N*-Nitrosodiphenylamine, *Mut. Res.,* 317, 195, 1994.

N-OXYDIETHYLENE THIOCARBAMYL-*N*-OXYDIETHYLENE SULFENAMIDE (CAS No 13752–51–7)

Synonyms. Accelerator OTOS; *4*-[(Morpholinothiocarbonyl)thio]-morpholine; *N*-Oxydiethylenesulphenamide.

Applications. A vulcanization accelerator.

Repeated Exposure. Sprague-Dawley rats were exposed to 6.25 to 25 mg/kg for 8 weeks. The highest dose decreased BW gain and liver weight.

Long-term Toxicity. Male and female Sprague-Dawley rats were fed dietary levels of 20 to 60 mg/kg BW for 112 weeks. Retardation of BW gain was observed only with the highest dose. Gross pathological examination revealed hydronephrosis, capillary necrosis, hemorrhages in the kidneys and urinary bladder, and epithelial hyperplasia in the ureter. Similar results were reported with 25 to 500 mg/kg BW doses of *N*-**oxydiethylene**-2-**benzothiazole-sulfenamide** (a vulcanization accelerator).

Reproductive Toxicity. Did not affect the fertility of males.

Genotoxicity. Data available are insufficient for evaluating mutagenic hazard. *N*-O. is likely to be negative in DLM assay. However, the mutagenic effect is reported to occur in *in vitro* experiments.

Carcinogenicity. A significant increase in the incidence of benign and malignant tumors of the kidneys, urethra, and urinary bladder, pineal gland, suprarenals, and mammary gland was observed in the above-described chronic experiment.

Reference:

Hinderer, R. K., Lancas J. R., Knezevich A. L., et al., The effect of long-term dietary administration of the rubber accelerator, *N*-oxydiethylene thiocarbamide-*N*-oxydiethylene sulfenamide, to rats, *Toxicol. Appl. Pharmacol.,* 82, 521, 1968.

p-QUINONE DIOXIME (CAS No 105–11–3)

Synonyms. Dioxime *p*-benzoquinone; Dioxime **2,5**-cyclohexadiene-**1,4**-dione.

Properties. Brown, highly inflammable, fine crystalline powder. Solubility in water is 200 mg/l (20°C). Odor perception threshold is 1 g/l, taste perception threshold is 0.5 mg/l, and color change threshold is 0.1 mg/l.

Applications. A vulcanization agent in the synthetic rubber and acrylic resins production.

Acute Toxicity. The LD$_{50}$ values are reported to be 0.46 g/kg BW (Litton Bionetics) or 1.6 g/kg BW in rats, and 1.4 to 1.5 g/kg BW in mice. Poisoning is accompanied by apathy, with subsequent death within 5 h.[1]

Short-term Toxicity. Rabbits were dosed by gavage with 20 mg/kg BW (every other day for 2 months and daily for a further 2 months). The treatment decreased blood prothrombine time and increased the activity of serum aldolase and alkaline phosphatase. There were no changes in behavior, BW gain, or hematological analyses. Gross pathological examination revealed visceral congestion.[2]

Genotoxicity. Appears to be a direct-acting mutagen in *Salmonella*. Negative results were observed after oral administration to female rats in both the bone marrow micronucleus test and *in vitro* liver UDS test.[3]

Carcinogenicity. Positive results have been reported in female rats but not in mice or male rats.[09] Mitogenic activity might play a contributory role to the induction of bladder cancer in rats if it also acts as a mitogen in this tissue.[3] *Carcinogenicity classification.* IARC: Group 2B.

Standards. Russia (1988). MAC: 0.1 mg/l (organolept., color).

References:
1. See **PHTHALIC ANHYDRIDE,** #2, 152.
2. See **ACRYLONITRILE,** #7, 189.
3. Westmoreland, C., Gerge, E., York, M., and Gatehouse, D., *In vitro* genotoxicity studies with *p*-benzoquinone dioxime, *Environ. Mol. Mutag.,* 19, 71, 1992.

RESORCINOL (CAS No 108–46–3)

Synonyms. 1,3-Benzenediol; *m*-Dihydroxybenzene; *m*-Dioxybenzene; 3-Hydroquinone; 3-Hydroxyphenol; Resorcin.

Properties. Colorless crystals with a specific odor. Readily soluble in water (63.7%), acetone, and ethanol. R. produces a very slight effect on the organoleptic properties of water: odor perception threshold is 40 mg/l.

Applications. R. is used in the production of plasticizers and stabilizers, and in the manufacturing of rubber products. R. (alone or in combination with phenol) is used to make resins or resin intermediates by reaction with formaldehyde.

Acute Toxicity. LD$_{50}$ is 239 mg/kg BW in mice[1] and 300 mg/kg BW in rats.

Repeated Exposure failed to reveal cumulative properties. Repeated administrations caused a reduction in SH-group content in the blood serum.

Long-term Toxicity. Chronic exposure to high doses produced a change in the phagocytic activity of leukocytes, in cholinesterase activity in the brain and liver, in blood eosinophil count, and in the protein fractions balance in the blood serum (Nesmeyanova, 1953). The dose of 5 mg/kg BW reduced the number of SH-groups in the blood serum.

Reproductive Toxicity. Caused no sperm-head abnormalities in mice.[2]

Genotoxicity. R. has been shown to produce gene mutations in bacteria and CA in plant cells.[3] It caused no SCE in cultured Chinese hamster ovary cells V79, and was negative in the micronuclei test on mouse bone marrow.[2,4] *Carcinogenicity classification.* NTP: NE—NE—NE—NE.

Chemobiokinetics. R. is readily absorbed from the GI tract, rapidly metabolized, and excreted by rats. After ingestion of 112 mg/kg, the main part of the dose is excreted within 24 h in the urine and up to 3% in the feces. Accumulation was not observed.[5]

Standards. Russia. PML in drinking water: 5 mg/l.

Regulations. USFDA (1993) approved the use of R. in closures with sealing gaskets in containers intended for use in producing, manufacturing, packing, processing, preparing, treating, packaging, transporting, or holding food (0.24%), for use only as a reactive adjuvant substance employed in the production of gelatin-bonded cord compositions, for use in lining crown closures. The gelatin so used shall be technical grade or better.

References:

1. *Hygiene of Populated Areas,* Sci. Conf. Abstracts, Zdorov'ya, Kiev, 1967, 52 (in Russian).
2. Wild, D., King, M.-T., Eckhardt, K., et al., Mutagenic activity of aminophenols and diphenols, and relations with chemical structure, *Mutat. Res.,* 85, 456, 1981.
3. McCann, J., Choi, E., Yamasaki, E., et al., Detection of carcinogens as mutagens in the Salmonella/microsome test: assay of 300 chemicals, *Proc. Natl. Acad. Sci. U.S.A,* 72, 5135, 1975.
4. Darroudi, F. and Natarajan, A. T., Cytogenetic analysis of human peripheral blood lymphocytes (*in vitro*) treated with resorcinol, *Mutat. Res.,* 124, 179, 1983.
5. Kim, Y. C. and Matthews, H. B., Comparative metabolism and excretion of resorcinol in male and female Fisher 344 rats, *Fundam. Appl. Toxicol.,* 9, 409, 1987.

RESOTROPIN

Composition. A molecular combination of resorcinol and urotropin.

Properties. Fine, crystalline powder of a light-pink to light-grey color. Soluble in water and partially soluble in alcohol.

Applications. Used as a modifier of vulcanizate mixtures based on natural, styrene-butadiene, and isoprene rubber.

Acute Toxicity. In mice, the LD_{50} is 1.3 g/kg BW. A 0.5 g/kg BW dose caused clonic convulsions and paralysis of the hind legs in 30 to 40 min after administration. Death within 24 h.[1] LD_{50} of a **modifier RU-1** (a mechanical mixture of 98.5% R., 1% boric acid, and 0.5% alkamone) in mice is 570 mg/kg BW (Zhilova, 1970).

Short-term Toxicity. Rabbits exposed to 20 mg/kg BW for 4 months developed no changes in BW gain, relative weights of the visceral organs, and contents of blood albumen, aminoacids, and prothrombin. The treatment caused a reduction in alkaline phosphathase activity.[1] In rabbits given 5 mg modifier RU-1/kg BW for 3.5 months, manifestations of the toxic effect included a decrease in blood vitamin C content and an increase in cholinesterase activity and blood cholesterol level. Histological examination revealed inflammatory foci around the small bronchi in the lungs, lymphocyte and leukocyte infiltration of the large hepatic vessels, inflammatory changes of the mucosa, and superficial epithelial desquamation in the stomach and intestine,[1] as well as swelling of the renal tubular epithelium (Zhilova, 1970).

Reference:

See **PHTHALIC ANHYDRIDE,** #2, 190.

STEARIC ACID (SA) (CAS No 57–11–4)

Synonym. Octadecanoic acid.

Properties. Amorphous, white or yellowish powder, plate-like or flaky. Poorly soluble in water, soluble in ethanol.

Applications. A vulcanization activator, lubricant, and filler dispersant.

Acute Toxicity. Rats and mice tolerate administration of 5 g/kg BW without any signs of intoxication.

Repeated Exposure. The increased liver weights were the only finding in mice given 250 and 1000 mg/kg BW for 1.5 months.

Long-term Toxicity. In a 9-month gavage study, the doses of 20 and 100 mg/kg BW appeared to be harmless for exposed rats.

Allergenic Effect was not found in the guinea pigs studies.

Genotoxicity. SA appeared to be negative in the *Salmonella* mutagenisity assay (NTP–91).

Standards. Russia. Recommended PML: **n/m.**

Regulations. USFDA (1993) regulates the use of SA (1) in adhesives used as components of articles intended for use in packaging, transporting, or holding food and (2) in resinous and polymeric coatings in a food-contact surface of articles intended for use in producing, manufacturing, packing, transporting, or holding food.

Reference:
See **BUTYL STEARATE, #3.**

SULFUR (S) (CAS No 7704–34–9)

Properties. Yellow, grey-yellow, or greenish powder. Insoluble in water, poorly soluble in ethanol.

Applications. A vulcanizing agent in the rubber industry in the production of vulcanizates based on isoprene, butadiene, and other rubbers. Usually it is not a component of chloroprene rubber.

Acute Toxicity. *Humans.* S. does not cause acute poisoning, but some pharmacological preparations used in medicine could exhibit evident toxicity. The lethal oral dose of sedimented S. (*Sulfur praecipitatum, Lac sulfuricus —Sulfuric milk*) is about 12 g.

Repeated Exposure. S. accumulation is noted in the liver, kidneys, and spleen.

Short-term Toxicity. Hypochromic anemia and leukopenia were found in rats given 9.3 mg elemental S./kg BW for 4 months.

Chemobiokinetics. S. is transformed into **sulfides** and H_2S, which could be reabsorbed from the intestines. Sulfides are predominantly converted into **sulfates** and excreted via the urine.

Long-term administration of elemental S. increased its amount as well as the quantity of mineral sulfates, common oxidized and neutral S. in the urine, liver, kidney, and spleen.

Regulations. USFDA (1993) approved the use of S. (1) in adhesives used as components of articles intended for use in packaging, transporting, or holding food, (2) in closures with sealing gaskets that may be used safely on containers intended for use in producing, manufacturing, packing, processing, treating, packaging, transporting, or holding food (for use only as a vulcanizing agent in vulcanized natural or synthetic rubber gasket compositions and up to 4% by weight of the elastomer content of the rubber gasket composition), and (3) as a vulcanizing agent of rubber articles intended for repeated use in contact with foodstuffs (total not to exceed 1.5% by weight of the rubber product).

Reference:
Environmental Factors and Their Effect on the Health of Population, Issue 1, Zdorov'ya, Kiev, 1976, 75 (in Russian).

TETRAETHYLTHIURAM DISULFIDE (TETD) (CAS No 97–77–8)

Synonyms. Antabuse; Bis(diethylthiocarbamoyl) disulfide; Disulfiram; Ethyl dithiurame; Ethylthiram; Tetradine; Teturam; Thiuram E; Thiuram disulfide; Thiuranide.

Properties. White powder. Solubility in water is 2 mg/l (38°C); in ethanol it is 20 g/l. TETD does not affect the taste, odor, or color of water.

Applications and **Exposure.** A vulcanization accelerator in rubber production. TETD is used to obtain colored vulcanizates. Used to treat patients with alcoholism.

Acute Toxicity. LD_{50} is 3 to 3.4 g/kg BW in rats,[1,2] 3.7 g/kg[2] or 12 to 14 g/kg BW and above[1] in mice, and 4.7 g/kg BW in rabbits.[2] In guinea pigs, manifestations of toxic action appear at dose levels of 1 to 15 g/kg BW. TETD poisoning is accompanied by diarrhea and, in dogs, vomiting. Administration led to paralysis and, in rats, reduced thyroid activity. There was leukopenia in rabbits. Reduced oxygen consumption and prolonged chloroform narcosis time were observed in mice. Histological examination revealed inflammatory changes and hemorrhages in the GI tract mucosa, degenerative and necrotic foci in the liver, brain, spleen, suprarenals, and renal tubular epithelium. Toxicity of TETD is known to be significantly enhanced after the administration of alcohol: LD_{50} in rats is reduced by 75%.

Repeated Exposure failed to reveal cumulative properties.[2] Rats tolerate 30 administrations of 250 mg/kg BW, whereas rabbits died after six *i/g* administrations of 300 mg/kg BW and dogs from 250 to 500 mg/kg BW following three to seven administrations. Oral exposure for 1.5 months affected hematological analyses and liver and CNS functions. Pathological changes were found in the kidneys, liver, thyroid, spleen, and myocardium.

Long-term Toxicity. A dose of 0.1 g/kg BW given for a year caused no mortality in dogs and rabbits.

260

Reproductive Toxicity effect is reported.[3] TETD caused no teratogenic effect in rats (Alexandrov, 1974).

Genotoxicity. TETD did not cause CA in Chinese hamster ovary cells (NTP–91).

Carcinogenicity. Negative results have been reported in rats and mice (feeding study).[09] *Carcinogenicity classification.* NTP: N—N—N—N.

Regulations. USFDA (1993) approved the use of TETD (1) in adhesives used as components of articles intended for use in packaging, transporting, or holding food and (2) as an accelerator for rubber articles intended for repeated use in contact with food up to 1.5% by weight of the rubber product.

References:

1. Korablev, M. V., Pharmacology and Toxicology of Dithiocarbamic Acid Derivatives, Author's abstract of thesis, Kaunas, 1965, 28 (in Russian).
2. See **DI**-2-**BENZOTHIAZOLYL DISULFIDE, #2.**
3. Holck, H. G., Lish, P. M., Sjogren, D. W., et al., Effects of disulfiram on growth, longevity, and reproduction of the albino rat, *J. Pharmacol. Sci.,* 59, 1267, 1970.

TETRAHYDROPHTHALIC ACID, ANHYDRIDE (CAS No 85–43–8)

Synonym. 4-Cyclohexene-**1,2**-dicarboxylic acid, anhydride.

Properties. White crystals. Poorly soluble in water.

Applications. Used as an antiscorching agent in the rubber industry.

Acute Toxicity. In mice, LD_{50} is reported to be 2.7 g/kg BW.

Short-term Toxicity. Rabbits were given 20 mg/kg BW, initially every other day and then daily (for 2 months in each case). The treatment affected the protein-forming function of the liver, caused hypercholesteremia, a reduction in alkaline phosphatase, and aldolase activity, and leukocytosis. Gross pathological examination revealed stimulation of the reticuloendothelial system.

Reference:

See **PHTHALIC ANHYDRIDE, #2,** 164.

TETRAMETHYLTHIURAM DISULFIDE (TMTD) (CAS No 137–26–8)

Synonyms. Bis(dimethylthiocarbomoyl)disulfide; Thiuram D; Thiram.

Properties. White or yellow powder with an unpleasant odor, poorly soluble in water, soluble in hot alcohol. Does not affect the taste, odor, or color of water.

Applications and **Exposure.** The major use is as a vulcanization accelerator in the rubber industry. Vulcanizes polyethylene. Used as a pesticide and a fungicide on seeds. Human exposure may occur through food and water contaminated with pesticide residues. Migration from butadiene-nitrile vulcanizates into water after a 3-d exposure at 20°C was up to 0.6 mg/l.[1]

Acute Toxicity. LD_{50} is 0.4 to 5.4 g/kg BW in rats, 1.2 to 7.1 g/kg BW in mice, and 0.21 g/kg BW in rabbits[2,3](IARC 53–412). Manifestations of the toxic effect include apathy, paralysis of the extremities, cyanosis, decreased temperature, and slowed respiration. Death occurs within 2 d. Gross pathological examination revealed hemodynamic disturbances in the brain, parenchymatous organs and GI tract, degenerative and dystrophic changes in the liver, kidneys and heart, and hemorrhages and ulcerations of the gastric mucosa.[3,4] TMTD increased sensitivity to alcohol and prolonged hexenal sleep.

Repeated Exposure revealed cumulative properties. K_{acc} is 2.85 (by Kagan). Leukopenia is noted after 6 to 20 administrations of a 30 mg/kg dose.[2] Charles River albino mice received TMTD in the diet at doses of 54, 108, or 201 mg/kg BW in males and 62, 118, or 241 mg/kg BW in females for 4 weeks. The treatment caused retardation of BW gain and anemia in males at all dose-levels. Significant reduction in food intake was noted in both sexes at all dose-levels (Kehoe, 1989).[21]

Short-term Toxicity. TMTD is shown to be a polytropic toxicant, causing significant changes in the peripheral blood, liver, NS, and suprarenals. There are pathological changes in the stomach, pancreas, liver, spleen, and other organs.[5] In a 13-week study, CD-1 rats received 58 and 132 mg TMTD/kg BW. The treatment produced mild elevations of blood biochemical parameters, indicating renal or hepatic dysfunction.[6] Charles River rats received TMTD in the diet at doses of 2.5, 25, or 50 mg/kg over a period of 13 weeks. BW gain and food consumption were significantly reduced; changes in clinical chemistry and hematological parameters were observed in both sexes at the dose-levels of 25 and 50 mg/kg. The NOAEL of 2.5 mg/kg in rats and 3 mg/kg in dogs was determined in this study (Kehoe, 1988). The NOAEL of 2.2 to 2.3 mg/kg based on hematological changes noted in both sexes and even 0.84 mg/kg based on increased absolute liver weights and altered clinical chemistry were later established by this author.[21] Rats were dosed by gavage with 25 mg TMTD/kg BW for 2 to 12 weeks. In the course of treatment, animals exhibited a decline in pyridoxine and nicotinic acid utilization, disruption of copper metabolism and ceruloplasmin activity, and the occurrence of secondary changes in the metabolism of serotonine, catecholamines, etc.[7]

Long-term Toxicity. *Humans.* Thyroid disturbances were reported in a group of subjects exposed occupationally to TMTD.[8] *Animals.* CD-1 female rats were given 60 mg TMTD/kg BW for 80 weeks. There was a neurological syndrome with onset of ataxia in some animals. Some behavioral changes were observed. Gross pathological examination revealed chromatolysis of motor neurons.[6] In a 2-year study, Wistar rats were exposed to dietary levels of 3, 30, and 300 ppm for 104 weeks. The group of 300-ppm dosed rats had retarded growth and anemia.[9] In a 2-year study, rats were given TMTD in their diet. The NOAEL of 1.2 and 1.4 mg/kg in males and females, respectively, based on lower erythrocyte count, Hb and hematocrit levels, and degenerative changes of the sciatic nerve, was found in this study (Maita et al., 1991).[21]

Allergenic response to TMTD that could migrate from the rubber is suspected.[10]

Reproductive Toxicity. *Gonadotoxicity.* TMTD decreases fertility and induces damage to sperm morphology in rodents. Exposure to TMTD can cause sperm-head abnormalities in mice.[11] Oral administration of 132 mg/kg BW in the feed for 13 weeks decreased fertility in male rats. The dose of 96 mg/kg BW administered to females for 14 d prolonged the diestrus phase of the estrus cycle and caused a decrease in BW gain.[12] *Embryotoxicity.* TMTD caused embryolethality and embryotoxicity in rats and hamsters. It produced adverse effects on reproduction of CD-1 rats when given at a dose of 132 mg/kg BW for 13 weeks to males and 30 mg/kg BW or more for 2 weeks to females.[12] When CD-1 rats were treated during organogenesis or in the peri- and postnatal periods, toxic changes were found in the fetuses or offspring only at levels at which dams or adults experienced significant retardation of BW gain and food consumption.[6] Doses of 136 to 200 mg/kg BW administered by gavage during the organogenetic period (on gestation days 6 to 15) increased the mortality rate of embryos. Weights of surviving embryos were decreased (40 mg/kg BW). However, Swiss-Webster mice exhibited no significant developmental effects after exposure to up to 300 mg/kg BW doses administered by gavage on days 6 to 14 of gestation.[12] *Teratogenicity.* Pregnant NMRI mice were given daily oral doses of 10 to 30 mg TMTD per animal on days 5 to 15 or 6 to 17 of gestation. The treatment produced an increased number of resorptions during the intermediate and late stages of organogenesis. Fetal malformations were characterized by cleft palate, micrognathia, wavy ribs, and distorted bones.[13] Syrian hamsters were dosed with 250 mg/kg BW and more on day 7 or 8 of gestation. An increased rate of resorptions, decreased fetal weights, and an increased number of terata have been reported.[14] With respect to reproduction and postnatal development, the NOAEL was found to be greater than 8.9 and 14 mg/kg in male and female rats, respectively (York, 1991).[21]

Genotoxicity. Positive results were reported in *Salmonella* and in mammalian somatic cells.[15,16] TMTD induced unscheduled DNA synthesis and SCE in cultured human cells (IARC 53–413).

Carcinogenicity. No increase in tumor incidence at any sites was reported in a mouse study.[17] A dose-dependent reduction in the incidence of mononuclear cell leukemia was reported in Fisher 344 rats given 0.1 or 0.05% TMTD in the feed for 104 weeks.[18]

A high incidence of tumors of the nasal cavity was found in rats of both sexes given the combined but not separate treatment with TMTD (500 ppm) and sodium nitrite (2000 ppm) for 104 weeks. In addition, a 20% incidence of papilloma of the forestomach was noted in rats.[19] This experiment was not designed as a standard bioassay, since only one dose-level was used and only a limited number of rats (24 of each sex) was treated. Beagle dogs were treated with 0.4, 4, and 40 mg/kg BW for 104 weeks. The dose of 40 mg/kg BW caused severe toxic symptoms (nausea or vomiting, salivations, occasional clonic convulsion, ophthalmological changes). Doses of 4 mg/kg and 40 mg/kg BW produced liver failure and kidney damage.[18]

No neoplastic changes were found in mice given TMTD in their diet for 97 weeks (Trutter, 1992).[21] *Carcinogenicity classification.* IARC: Group 3.

Chemobiokinetics. TMTD is readily absorbed in the GI tract and rapidly distributed to the blood and all organs and tissues. It is likely to be reduced by glutathione to **dithiocarbamate,** which is oxidized or converted into the corresponding **metal complex.**[3] In 2 d following its ingestion, TMTD is found in the liver and spleen together with its metabolites: **amine salt dimethyldithiocarbamic acid** (DDCA) and **tetramethylthiourea,** and, in the lungs, **carbon disulfide** and the **amine salt** of DDCA. TMTD is known to form *N*-**nitrosodimethylamine** by reaction with nitrite in mildly acid solution. TMTD and the amine salt of DDCA are excreted from the body in the urine and feces, and as CS_2 via the lungs.[20]

Standards. FAO/WHO (1992). ADI: 0–0.01 mg/kg BW for humans. **EEC** (1990). Residue limits in food products: 3 mg/kg, but this value should be reviewed. **USEPA** (1989). Residue limits in some fruits and vegetables: 7 mg/kg. **Russia** (1988). MA and PML: 1 mg/l; residues in food not permitted.

Regulations. USFDA (1993) approved the use of TMTD (1) in adhesives as a component of articles intended for use in packaging, transporting, or holding food and (2) as an accelerator for rubber articles intended for repeated use in contact with food up to 1.5% by weight of the rubber product.

References:

1. See **ZINC PHENYLETHYLDITHIOCARBAMATE,** #1, 153.
2. See **TETRAETHYLTHIURAM DISULFIDE,** #1.
3. See **PHTHALIC ANHYDRIDE,** #2, 93.
4. See *N*-(1,3-**DIMETHYLBUTYL**)-*N'*-**PHENYL-**p**-PHENYLENEDIAMINE,** 5.
5. See **ETHYLENEIMINE,** #6, 83.
6. Lee, C. C. and Peters, P. J., Neurotoxicity and behavioral effects of thiram in rats, *Environ. Health Perspect.,* 17, 35, 1976.
7. Abramova, J. I. and Friedman, S. N., *Gig. Truda Prof. Zabol.,* 3, 45, 1973 (in Russian).
8. Kaskevich, L. M. and Bezugly, V. P., Clinical aspects of intoxication induced by tetramethyl-thiuram disulfide, *Vrachebnoye Delo,* 6, 128, 1973 (in Russian).
9. Maita, K., Tsuda, S., and Shirasu, Y., Chronic toxicity studies with thiram in Wistar rats and beagle dogs, *Fundam. Appl. Toxicol.,* 16, 667, 1991.
10. Rudzki, E., Ostaszewski, K., Grzywa, Z., et al., Sensitivity to some rubber additives, *Contact. Dermat.,* 2, 24, 1976.
11. Zdzienicka, M., Hryniewicz, M., and Pienkowska, M., Thiram-induced sperm-head abnormalities in mice, *Mutat. Res.,* 102, 261, 1982.
12. Short, R. D., Russel, J. Q., Minor, J. I., et al., Developmental toxicity of ferric dimethyldithio-carbamate and bis(dimethylthio-carbamoyl) disulfide in rats and mice, *Toxicol. Appl. Pharmacol.,* 35, 83, 1976.
13. Matthiaschk, G., Influence of L-cysteine on the teratogenicity of thiram in NMRI mice, *Arch. Toxicol.,* 30, 251, 1973.
14. Robens, J. F., Teratologic studies of carbaryl, diazinine, norea, disulfiran and thiram in small laboratory animals, *Toxicol. Appl. Pharmacol.,* 15, 152, 1969.
15. Paschin, Y. V. and Bakhitova, L. M., Mutagenic effects of thiuram in mammalian somatic cells, *Food Chem. Toxicol.,* 23, 373, 1985.
16. Hedenstedt, A., Rannug, U., Ramel, C., et al., Mutagenicity and metabolism studies on 12 thiuram and dithiocarbamate compounds used as accelerators in the Swedish rubber industry, *Mutat. Res.,* 68, 313, 1979.
17. Takahashi, M., Kokulo, T., Furukawa, F., et al., Inhibition of spontaneous leukemia in Fisher 344 rats by tetramethylthiuram disulfide, *Gann,* 74, 810, 1983.
18. Hasegawa, R., Takahashi, M., Furukawa, F., et al., Carcinogenicity study of tetramethylthiuram disulfide (thiram) in Fisher 344 rats, *Toxicology,* 51, 155, 1988.
19. Lijinsky, W., Induction of tumors of the nasal cavity in rats by concurrent feeding of thiram and sodium nitrite, *J. Toxicol. Environ. Health,* 13, 609, 1984.
20. *Pesticide Handbook,* Urozhai, Kiev, 1974, 222 (in Russian).
21. Pesticide Residues in Food — 1992, Toxicology Evaluation, Joint Meeting of the FAO/WHO Panel of Experts, Rome, 21–30 September 1992, 391.

TETRAMETHYLTHIURAM SULFIDE (CAS No 97–74–5)

Synonyms. Anhydrosulfide; Bis(dimethylthiocarbamoyl)sulfide; Dimethyldithiocarbamic acid; Tetramethylthiuram monosulfide; Thiuram *MM*.

Properties. Yellow, crystalline powder. Poorly soluble in water, soluble in alcohol.

Applications. A vulcanization accelerator for diene type rubber and rubber mixes. A fungicide.

Acute Toxicity. LD_{50} is 413 mg/kg BW in rats, and 820 to 1150 mg/kg BW in mice.[1,2] Lethal doses caused inhibition, diarrhea, paresis, hind limb paralysis, and convulsions. Hexenal sleep time was prolonged. The liver seemed to be predominantly affected. Dogs, cats, and rabbits tolerated the dose of 100 mg/kg BW. Guinea pigs were likely to be more sensitive.

Repeated Exposure revealed cumulative properties. Of animals, 70% died in 10 d from the doses of 10 to 40 mg/kg BW.[2,3] Severe leukopenia was observed in rabbits after four to six administrations of 60 to 80 mg/kg BW doses.

Short-term Toxicity. Exposure to 20 mg/kg BW for 3.5 months caused impaired liver function and changes in the lungs and blood formula. There was 50% rabbit mortality in this study.[1]

Carcinogenicity. Negative results were reported in mice.[09]

References:
1. See 2-(**MORPHOLINOTHIO)BENZOTHIAZOLE**, #3.
2. See **TETRAETHYLTHIURAM DISULFIDE**, #1.
3. See **TETRAMETHYLTHIURAM SULFIDE**, #3.

3,3′-THIOBIS(5-ISOPROPYLANISOL)

Properties. White, crystalline powder. Poorly soluble in water and alcohol.

Applications. An antiosone and antiscorching action agent in the rubber industry.

Acute Toxicity. Rats and mice tolerate administration of 6 g/kg BW. Poisoning is accompanied with signs of adynamia evident on day 3.

Reference:
See *N*-**BUTYLIDENEANILINE**, 26.

TITANIUM DIOXIDE (CAS No 13463–67–7)

Synonyms. Flamenco; Titan A; Titania; Titanium oxide; Titanium peroxide; Titanox; Unitane.

Properties. White, crystalline powder. Poorly soluble in water. Chemically inactive.

Applications. White pigment; a filler.

Short-term Toxicity. Feeding rodents with 0.6 to 0.9 g TiO_2/d caused no signs of intoxication.[1]

Long-term Toxicity. Male and female Fisher 344 rats were fed diets containing 1 to 5% TiO_2-coated mica for up to 130 weeks. There were no changes in survival, BW gain, hematological analysis, clinical chemistry parameters, or histopathology.[2] B6C3F₁ mice anf Fisher 344 rats were fed diets containing 2.5% or 5% TiO_2.[3] No difference in BW gain between experimental and control animals was observed in this study.

Genotoxicity. TiO_2 did not induce morphological transformation in Syrian hamster embryo cells or mutations in bacteria.[4,5] It was shown to be negative in mouse lymphoma assay.[6]

Carcinogenicity. TiO_2 did not produce a significant increase in the frequency of any type of tumor in any species tested in the above-described 103-week studies.[2,3] *Carcinogenicity classification.* IARC:Group 3; NTP: N—N—N—N.

Chemobiokinetics. Absorption from the GI tract after administration by gavage was found to be negligible.[7]

Regulations. USFDA (1993) regulates the use of TiO_2 (1) in adhesives as a component of articles intended for use in packaging, transporting, or holding food, (2) in cellophane for packaging food, (3) as a filler for rubber articles intended for repeated use, (4) in polysulfide polymer-polyepoxy resins in articles used as a surface intended for contact with dry food subject to the specified conditions, (5) in acrylate ester copolymer coating to be used safely as a food-contact surface of articles intended for packaging and holding food, including heating of prepared food, and (6) as a substance employed in the production of, or added to, textiles and textile fibers intended for use in contact with food. *Ti* **dioxide,** *Ti* **dioxide-barium sulfate,** and *Ti* **dioxide-magnesium silicate** may be used as colorants in the manufacturing of articles intended for use in contact with food. **EC Commission** (1962) approved the use of TiO_2 as a food color additive with the following specifications limiting impurities: antimony compounds, <100 mg/kg; zinc compounds, <50 mg/kg; soluble barium compounds, <5 mg/kg; and hydrochloric acid-soluble compounds, 3.4 g/kg.

References:

1. *Handbook on Toxicology of Metals,* Amsterdam, 1980, 627.
2. Bernard, B. K., Osheroff, M. R., Hofmann, A., et al., *J. Toxicol. Environ. Health,* 29, 417, 1990.
3. National Cancer Institute, *Bioassay of Titanium Dioxide for Possible Carcinogenicity,* Technical Report Series No. 97, Bethesda, MD, 1979.
4. Di Paolo, J. A. and Castro, B. C., Quantitative studies of *in vitro* morphological transformations of Syrian hamster cells by inorganic metal salts, *Cancer Res.,* 39, 1008, 1979.
5. Castro, B. C., Meyers, J., and Di Paolo, J. A., Enhancement of viral transformation for evaluation of the carcinogenic or mutagenic potential of metal salts, *Cancer Res.,* 39, 193, 1979.
6. See **BISPHENOL A, #6.**
7. Thomas, R. G. and Archuleta, R. F., Titanium retention in mice, *Toxicol. Lett.,* 6, 115, 1980.

TRICHLOROCYANURIC ACID (CAS No 87–90–1)

Synonyms. Trichlorocyan; Trichloroisocyanic acid; **1,3,5-**Trichloro-*s*-triazine-**2,4,6-(***1H,3H,5H***)**-trione.

Properties. White powder with a strong odor of chlorine (91.5% active chlorine).

Applications. An antiscorching agent.

Acute Toxicity. In mice, LD_{50} is reported to be 2.2 g/kg BW (in sunflower oil) and 0.75 g/kg BW (in vaseline). Gross pathological examination revealed changes in the GI tract.

N,N′,N″-TRICHLOROMELAMINE (CAS No 7673–09–8)

Synonyms. Chloromelamine; **2,4,6-**Tris(chloramino)-**1,3,5-**triazine.

Properties. Light-yellow powder with an odor of chlorine. Poorly soluble in water.

Applications. A vulcanization retarder.

Acute Toxicity. LD_{50} for mice is reported to be 490 mg/kg BW (Pesticide Index) or 3250 mg/kg BW. Gross pathological examination revealed point hemorrhages in the exterior wall of the stomach. It was partially liquified from the effect of high doses and easily torn. This effect is evidently associated with the splitting of chlorine and hydrochloride.[1]

Reproductive Toxicity. Embryotoxicity was shown in 2-d-old chick embryos.[2]

Immunotoxicity. Allergenic effect was not observed (Kostrodymova, 1976).

Genotoxicity. T. is likely to be mutagenic in *Dr. melanogaster*.[3] It does not interact with DNA and does not induce changes in its structure and function (Onoue et al., 1982).

Carcinogenicity. Addition of 0.03% and 0.3% T. to the diet caused lymphoid tissue tumors in mice (Hoshino and Tanooka, 1978).

Standards. Russia (1988). MAC: 1.4 mg/l (organolept., taste).

References:

1. See **PHTHALIC ANHYDRIDE, #2,** 167.
2. See *N,N′,N″*-**TRICHLOROMELAMINE, #2.**
3. Dubinin, V. I., *General Genetics,* Nauka, Moscow, 1970 (in Russian).

TRIETHANOLAMINE (CAS No 102–71–6)

Synonyms. **2,2′,2″-**Nitrilotriethanol; **2,2,2-**Trihydroxytriethylamine; Tri(2-hydroxyethyl)amine.

Properties. Viscous, glycerine-like, colorless, clear liquid with a faint odor of ammonia. Readily soluble in water and alcohol. Perception threshold for the effect on the organoleptic properties of water is 5 mg/l. According to D'yakov, odor perception threshold is 36 mg/l, taste perception threshold is 1.4 mg/l. Concentration of 1 g/l does not change the color of water. According to other data, T. organoleptic threshold in water is 160 mg/l.[01]

Applications. Vulcanization activator, particularly in the mixes based on styrene-butadiene rubber. T. is also used in the manufacturing of surface-active agents, solvents, and commercial cosmetics. It is the most widely used emulsifier.

Acute Toxicity. LD_{50} is reported to be 5.2 g/kg[1] or 8.4 g/kg BW[2] in rats, 5.4 g/kg[1] or 7.8 g/kg BW[2] in mice, and 5.3 g/kg BW in guinea pigs.[2] Administration of the high doses is accompanied by painful and tactile hyperesthesia and diarrhea on the second day. Mice look aroused and rats inhibited. Recovery in survivors occurs in 2 to 3 weeks. Gross pathological examination revealed congestion in the viscera and in the liver.

Repeated Exposure revealed cumulative properties. Retardation of BW gain, excitation, aggression, and changes in blood morphology were noted in rats given 520 mg T./kg BW for 2 months. Mortality reached 75% in this study.

Short-term Toxicity. Rats received 104 and 260 mg/kg BW for 4 months. From the beginning of the experiment, animals were aroused and aggressive. The treatment caused anemia, an increase in the level of residual blood nitrogen, and other changes which, however, had disappeared by the end of the experiment. The dose of 2 mg/kg was found to be the NOAEL.[2]

Long-term Toxicity. T. dissolved in distilled water at levels of 1 and 2% was given to male and female B6C3F$_1$ mice *ad libitum* in drinking water over a period of 82 weeks. There were no adverse effects on survival or organ weights.[3]

Reproductive Toxicity. Genotoxicity. T. was not shown to be mutagenic in *Bac. subtilis* and in DNA damage-inducible repair in an unscheduled DNA synthesis test.[4,5]

Carcinogenicity. Results of carcinogenicity studies are controversial. T. has been reported to produce an elevated incidence of malignant lymphoid tumors in female ICR-JCL mice fed diets containing 0.3 and 0.03% T. for their life-span over that of male mice on the same diet or control mice.[6] B6C3F$_1$ mice were given *ad libitum* T. dissolved in distilled water (1 and 2%). The dose-levels in females were reduced by half from week 69, because of associated nephro-toxicity.[3,7] This study provided no evidence of carcinogenic potential of T. in mice. Neoplasms developed in all groups of mice, including the control group. No dose-related increase of the incidence of any tumor was observed.

Regulations. USFDA (1993) approved the use of T. (1) in adhesives as a component of articles intended for use in packaging, transporting, or holding food; (2) in polyurethane resins to be used safely in food-contact articles intended for use in contact with dry food; (3) as an activator for rubber articles intended for repeated use in contact with food up to 5% by weight of the rubber product; (4) to adjust pH during the manufacturing of amino resins permitted for use as components of paper and paperboard intended for contact with dry food; (5) in resinous and polymeric coatings in a food-contact surface; (6) as a defoaming agent that may be used safely in the manufacturing of paper and paperboard; and (7) as a substance employed in the production of or added to textiles and textile fibers intended for use in contact with food.

References:

1. See **BUTYL BENZYL ADIPATE**, 323.
2. See **ACRYLIC ACID**, #1, 105.
3. Konishi, Y., Denda, A., Ushida, K., et al., Chronic toxicity and carcinogenicity studies of triethanolamine in B6C3F$_1$ mice, *Fundam. Appl. Toxicol.*, 18, 25, 1992.
4. *The Fifth Report of the Cosmetic Ingredient,* Review Expert Panel, 1983.
5. Inoue, K., Sunakawa, T., Okamoto, K., et al., Mutagenicity tests and *in vitro* transformation assays on trietanolamine, *Mutat. Res.*, 10, 305, 1982.
6. Hoshino, H. and Tanooka, H., Carcinogenicity of triethanolamine in mice and its mutagenicity after reaction with sodium nitrile in bacteria, *Cancer Res.*, 38, 18, 1978.
7. Mackawa, A., Onodera, H., Tanigawa, H., et al., Lack of carcinogenicity of triethanolamine in Fisher 344 rats, *J. Toxicol. Environ. Health*, 19, 345, 1986.

TRITHIOCYANURIC ACID (TCY) (CAS No 638–16–4)

Synonyms. 1,3,5-Triazine-2,4,6-trimercaptan; 2,4,6-Triazinetrithiol; **1,3,5**-Trimercaptotriazine.
Applications. A rubber curative.

Acute Toxicity. A substance of low oral and dermal toxicity and low potential for eye and skin irritation. LD_{50} for rats is 9.5 g/kg BW.

Repeated Exposure. Sprague-Dawley rats were exposed to 625, 2500, and 5000 mg/kg in the diet for 2 to 30 d. Some effect was noted on BW gain and survival at the higher levels of intake. Main effects concerned unusual lesions of the pinna and the distal portions of the tail. There were purplish discolorations of the ear margin and the tip of the tail in some animals (2.5 and 5 g/kg diet). These changes were apparently site-specific and have not been justified by histological examination. The NOAEL with regard to gross pathological lesions in a 30-d feeding study appeared to be 625 mg/kg in the diet.

Reference:

Koschier, F. J., Brown, D. R., and Friedman, M. A., Effect of dietary administration of the rubber curative trithiocyanuric acid to the rat, *Food Chem. Toxicol.,* 21, 495, 1983.

ZIRAM (CAS No 137–30–4)

Synonyms. Carbamate Z; Zinc dimethyldithiocarbamate; Methyl cymate; Methyl zineb; Methyl Ziram; Zimate.

Properties. White, crystalline powder. Solubility in water is 65 mg/l, soluble in ethanol (<2 g/100 ml at 25°C).

Applications and **Exposure.** A vulcanization accelerator for rubber; a pesticide; a foliar fumigant, mainly used on fruit and nuts. Migration from butadiene-nitrile vulcanizates into water after 3-d exposure at 20°C was up to 1 mg/l.[1]

Acute Toxicity. *Humans.* 0.5 l Z. solution of unknown concentration was fatal within a few hours of nonspecific pathology.[2] *Animals.* The LD_{50} values are reported to be 1.2 to 1.4 g/kg BW in rats, 0.34 to 0.8 g/kg BW in mice, and 0.1 to 0.2 mg/kg BW in rabbits and guinea pigs.[3–5] Administration of the lethal doses was accompanied by depression, followed by diarrhea, paresis, hind limb paralysis, and convulsions. Animals usually died on days 2 to 5.

Repeated Exposure revealed a marked cumulative effect (with administration of $^1/_{20}$ LD_{50}). K_{acc} is 1.2 (by Kagan). The dose of 100 mg/kg caused some of the rabbits to die after three to seven administrations. Marked leukopenia is reported in rabbits after administration of 70 mg/kg BW for a week.[4] A dose of 25 mg/kg BW appeared to be harmless in sheep. Administration of 5 and 25 mg/kg in the diet does not affect hematological analyses or the relative weights of the visceral organs in dogs. No morphological changes were found in the visceral organs and thyroid of rats that had received Z. with the feed (0.25% Z. in the diet) for a month.[6]

Short-term Toxicity. Rabbits were given 450 mg/kg BW for 3.5 months. Hematological analyses were affected. The treatment produced disorders of metabolism, blood protein balance, and liver function.[5]

Long-term Toxicity. Rats and rabbits were gavaged with 10 mg/kg BW for 6 months.[5] The treatment caused retardation of BW gain, a decline in cholinesterase activity, reduction in the content of SH-groups, anemia, and leukopenia. Gross pathological examination revealed changes in the viscera. In a 2-year feeding study in rats, epiphyseal abnormalities in the long bones of the hind legs were observed at the highest dose tested (2 g/kg in the diet).[7] Antonovich[8] has reported the LOAEL of 1 mg Z./kg BW in a 9-month oral study. According to other data, the safe dose is 12.5 mg Z./kg BW for rats and 5 mg Z./kg BW for dogs (JECFA, 1967).

Reproductive Toxicity. Z. caused an embryotoxic and teratogenic effect. It is capable of altering the period of gestation, reducing fertility, and causing fetal destruction and sterility.[9]

Embryotoxicity. The doses of 50 and 100 mg/kg BW administered by gastric intubation to pregnant CD rats reduced fetal weight. Doses of 25 mg/kg BW and above given on days 6 to 15 of gestation resulted in embryotoxic effects and maternal toxicity signs. The embryotoxic effect was also evident later, when administration of the substance had ceased.[9,10]

Teratogenicity. Newborn rats showed curvature of the tail and growth retardation.[9]

Genotoxicity. Mutagenic activity depends more on carbamate that on zinc. According to other data, the reason why the activity of zinc and other dithiocarbamates damages DNA is apparently inhibition of dismutase and catalase.[11,12] Their mutagenicity in the Ames test is increased in conditions favorable to the formation of anionic radicals of oxygen. These compounds show gonadotoxicity in short-term test systems.

Increased frequency of CA was seen in peripheral blood lymphocytes of workers who handled and packaged Z.[11] Z. was clastogenic in mammalian cells *in vitro* and *in vivo* and induced mutations in cultured rodent cells (IARC 53–431). No mutagenic effect was found in *Dr. melanogaster.*[12]

Carcinogenicity. B6C3F$_1$ mice were exposed to the dietary levels of 0.6 and 1.2 g Z. per kilogram BW over a period of 103 weeks. Increased incidence of benign lung tumors was noted in female mice. There was a dose-dependent increase of thyroid carcinomas in male Fisher 344/N rats given 0.3 and 0.6 g Z./kg BW for 103 weeks.[13] Twenty administrations of 75 mg/kg BW to mice for 2.5 months produced a weak carcinogenic effect (adenomata). The RNA content in the liver and lungs increased along with continuation of the period of observation.[14] *Carcinogenicity classification.* IARC: Group 3; NTP: P—N—N—E.

Chemobiokinetics. Z. is shown to accumulate in the visceral organs. Its metabolism in the body occurs to form much more toxic compounds: **tetramethylthiuram disulfide, dimethylamine salt of dithiocarbamic acid, carbon disulfide, dimethylamine.**[4] Metabolites were found in the blood, kidneys, liver, ovaries, spleen, and thyroid of female rats 24 h after oral administration of radiolabeled Z. When pregnant rats were given 55 mg/kg, a metabolic product, **dimethylamine salt of dimethyldithiocarbamic acid,** was found in the amniotic fluid, placenta, and fetal tissues.[9] Metabolites are predominantly removed in the urine and feces, and **dimethylamine** and **carbon disulfide** in the expired air. Unchanged Z. is excreted in the feces.[15]

Standards. FAO/WHO (1990). ADI of 0.02 mg/kg for humans was confirmed in 1980 (Codex Committee on Pesticide Residues). **U.S.** Residues in food: 7.0 mg/kg in fruits and vegetables (calculated as *zineb*). **Russia.** PML in food: 0.03 mg/l. No residues of Z. (pesticide) are permitted in food products.

Regulations. USFDA (1993) approved the use of Z. (1) in adhesives as a component of articles intended for use in packaging, transporting, or holding food and (2) as an accelerator of rubber articles intended for repeated use in producing, manufacturing, packing, processing, treating, packaging, transporting, or holding food (total not to exceed 1.5% by weight of the rubber product).

References:

1. See **ZINC PHENYLETHYLDITHIOCARBAMATE,** #1, 153.
2. Buklan, A. J., Acute ziram poisoning, *Forens. Med. Expertize,* 17, 51, 1974 (in Russian).
3. Hodge, H. C., Mayunard, E. A., Downs, W., et al., Acute and short-term oral toxicity tests of ferric dimethyldithiocarbamate (febram) and zinc dimethyldithiocarbamate (ziram), *J. Am. Pharm. Assoc.,* 41, 662, 1952.
4. See **TETRAETHYLTHIURAM DISULFIDE,** #1.
5. See α-**METHYLBENZYLPHENOLS,** mixture, #1, 78.
6. *Toxicology and Pharmacology of Pesticides and Other Chemical Compounds,* Zdorov'ya, Kiev, 1967, 163 (in Russian).
7. Enomoto, A., Harada, T., Maita, K., et al., Epiphyseal lesions of the femur and tibia in rats following oral chronic administration of zinc dimethyldithiocarbamate (ziram), *Toxicology,* 54, 45, 1989.
8. *Pesticides Handbook,* 2nd ed., Urozhai, Kiev, 1977, 187 (in Russian).
9. *Pesticides Handbook,* 2nd ed., Urozhai, Kiev, 1977, 181 (in Russian).
10. Giavini, E., Vismara, C., and Broccia, M. I., Pre- and postimplantation embryotoxic effects of zinc dimethyldithiocarbamate (ziram) in the rat, *Ecotoxicol. Environ. Saf.,* 7, 531, 1983.
11. Pilinskaya, M. A., Chromosomal aberrations in persons handling ziram under industrial conditions, *Genetics,* 6, 157, 1971 (in Russian).
12. *Toxicology and Pharmacology of Pesticides and Other Chemical Compounds,* Zdorov'ya, Kiev, 1987, 132 (in Russian).
13. NTP, *Carcinogenicity Bioassay of Ziram in F344/N Rats and B6C3F$_1$ Mice (Feed Study),* Technical Report No. 238, Research Triangle Park, NC, 1983.
14. *Pesticides Toxicology and the Clinical Features of Poisoning,* Proc. VNIIGINTOX, Kiev, 1968, 770 (in Russian).
15. Ismirova, N. and Marinov, V., Distribution and excretion of ^{35}S-ziram and metabolic products after 24 h following oral administration of the preparation to female rats, *Exp. Med. Morphol.,* 11, 152, 1972 (in Russian).

Chapter 6

SOLVENTS

ACETONE (CAS No 67–64–1)

Synonyms. Dimethylformaldehyde; Dimethyl ketone; β-Ketopropane; Methyl ketone; 2-Propanone; Pyroacetic acid.

Properties. Clear, colorless liquid with a characteristic odor. Mixes with water in all proportions. Odor threshold is 20 mg/l[02] or 40 to 70 mg/l; at these concentrations it does not affect the taste, color, or clarity of water. According to other data, taste threshold is 12 mg/l.[1]

Applications and **Exposure.** Used in the production of nitro- and acetylcellulose, epoxy and vinyl resins, and vulcanizates.

Acute Toxicity. *Humans.* Ingestion of a high dose resulted in a comatose condition with subsequent convulsions. Poisoning was accompanied by cyanosis of the skin, labored breathing, and increased blood pressure. Death was preceded by signs of cardiopulmonary insufficiency. At autopsy there were pulmonary and cerebral edema and hemorrhages in the renal tubular epithelium.[2] *Animals.* Dogs tolerate administration of 1 g/kg BW without any harmful effect. A dose of 4 g/kg BW causes stupor, and 8 g/kg BW causes death. In rats, LD_{50} is in the range of 8.5 to 10.7 g/kg BW; in mice, it is found to be up to 9.75 g/kg BW. Rabbits tolerate a dose of 1 g/kg, and LD_{50} varied from 3.8 to 5.3 g/kg BW.[3]

Repeated Exposure revealed slight cumulative ability. K_{acc} is 8.6 (by Cherkinsky). A. exhibits a low level of toxicity. Fisher 344/N rats and B6C3F₁ mice were given A. in drinking water for 14 d. At a concentration of 100,000 ppm, a decreased BW was noted but no histological changes were observed except centrilobular hepatocellular hypertrophy in mice (concentrations of 20,000 to 50,000 ppm).[4]

Short-term Toxicity. In a 13-week study, no hematological changes were found in mice given drinking water containing 20,000 to 50,000 ppm A. There was mild hepatocellular hypertrophy only in two out of ten female mice. An increased incidence and severity of nephropathy were the most prominent treatment-related findings.[4] In a 13-week study in Sprague-Dawley rats, no compound-related effects on BW, ophthalmic and urinalysis results, or mortality were reported.[5]

Long-term Toxicity. Rats were gavaged with 7 to 70 mg/kg BW for 6 months. The treatment did not affect general condition, behavior, or BW gain. The highest dose caused the greatest changes in the biochemical indices. Histological examination revealed hemo- and lymphodynamic disorders, dystrophic changes in the myocardium, and parenchymatous dystrophy of the liver and kidneys.[6]

Reproductive Toxicity. *Humans.* Female factory workers had undergone a long-term exposure to A. at a concentration below 200 mg/m³. An increased incidence of complications of pregnancy was reported.[7] *Animals. Gonadotoxicity.* Sperm motility was decreased and percentage of abnormal sperm was increased in rats given drinking water containing 50,000 ppm A. Lower concentrations produced signs of macrocytic anemia. Ketonemia during pregnancy may result in disturbance of CNS normal development and may cause some abnormalities both in humans and animals.[8] According to Mast et al.,[9] A. produced no **teratogenic** effect in rats and mice.

Genotoxicity. A. did not show mutagenic potential in the *Salmonella* test and did not induce SCE or CA in Chinese hamster ovary cells (NTP–88).

Chemobiokinetics. After absorption, A. is rapidly distributed throughout the tissues according to their water content and is excreted via the kidneys and lungs, and also in the sweat, unchanged or metabolized. Although A. has been considered to be a nonmetabolizable compound, it was shown to metabolize by three separate gluconeogenic pathways.[10] **Carbon dioxide** formed on oxidation of A. is excreted in the expired air.

Standards. Russia (1983). PML in food: 0.1 mg/l; PML in drinking water: 2 mg/l.

Regulations. USFDA (1993) approved the use of A. (1) as a component of adhesives for a food-contact surface, (2) as an ingredient of resinous and polymeric coatings for polyolefin films coming in contact with food, and (3) as an adjuvant in the preparation of slimicides in the manufacture of paper and paperboard that are in contact with food.

References:
 1. See **ACRYLONITRILE,** #7, 40.
 2. Alexandrova, V. V., Poisoning by liquid acetone, *Sudmedexpertiza,* 3, 57, 1980 (in Russian).
 3. See **ACETONE,** #3, 221.
 4. NTP, *Toxicity Studies of Acetone in F344/N Rats and B6C3F$_1$ Mice (Drinking Water Studies),* NTP Tox. 3, NIH Publ. No. 91–3122, 1991.
 5. Sonawane, B., de Rosa, C., Rubinstein, R., et al., Estimation of reference dose for oral exposure to acetone, 7th Annu. Meeting Am. Coll. Toxicol., November 16–19, 1986, 21.
 6. Omel'ynets, N. I., Mironets, N. V., and Martyzchenko, N. V., *Kosm. Biol. Aviokosm. Med.,* 12, 67, 1978 (in Russian).
 7. Nizyaeva, I. V., On hygienic assessment of acetone, *Gig. Truda Prof. Zabol.,* 6, 24, 1982 (in Russian).
 8. Dietz, D. D., Leininger, J. R., Rauckman, E. J., et al., Toxicity studies of acetone administered in the drinking water of rodents, *Fundam. Appl. Toxicol.,* 17, 347, 1991.
 9. Mast, T. G., Rommereim, R. L., Weigel, R. G., et al., Developmental toxicity study of acetone in mice and rats, *Teratology,* 39, 468, 1989.
10. Morris, J. B., Covanagh, D. G., Metabolism and deposition of propanol and acetone vapors in the upper respiratory tract of the hamster, *Fundam. Appl. Toxicol.,* 9, 34, 1987.

ALLYL ALCOHOL (AA) (CAS No 107–18–6)

Synonyms. 2-Propen-1-ol; 3-hydroxypropene.

Properties. Colorless, transparent liquid with a pungent mustard odor. Readily miscible with water and alcohol. Taste perception threshold is 0.1 mg/l; odor perception threshold is 0.33 to 0.66 mg/l (when heated to 60°C, 0.07 mg/l and 0.75 mg/l, respectively) (Karmazin). However, odor perception threshold of 14 mg/l is reported.[02] AA does not alter the transparency or color of water.

Acute Toxicity. Oral LD$_{50}$ values have been reported to be 140 mg/kg BW in rats, 75.5 mg/kg BW in mice, and 90 mg/kg BW in rabbits (Karmazin). Other median lethal doses indicated are 64, 96, and 71 mg/kg BW, respectively.[06] Poisoned animals displayed signs of CNS impairment (ataxia) and pulmonary edema. Histological examination showed necrotic and dystrophic changes in the liver, myocardium and kidneys, and edema of the connective tissue.

Repeated Exposure to AA produced a general toxic effect with specific liver cell damage. Reduced serum catalase activity was noted. Hematological analysis revealed leukocytosis; erythrocyte counts and Hb levels were not altered. Rats were dosed by gavage with 14 and 28 mg/kg BW for 10 d. Gross pathological examination revealed congestion in the liver, kidneys and spleen, and dystrophic and necrobiotic changes in the myocardium and liver. Severe swelling of the renal tubular epithelium was observed (Karmazin).

Long-term Toxicity. Rabbits were dosed with 2.5 mg AA/kg BW for 8 months. The treatment produced protein fraction dysbalance in the blood serum. Microscopic findings included necrotic foci in the liver and moderate congestion and parenchymatous dystrophy of the renal tubular epithelium (Al'meyev and Karmazin).

Reproductive Toxicity. AA produced no *teratogenic* effect in mice.[1] However, Slott and Hales found increased *embryolethality* at a dose of 0.1 mg AA and limb defects and other fetal malformations at a dose of 1 mg AA.[2] Both doses were injected directly into the amniotic cavity of day 13 rat embryos.

Genotoxicity. AA was found to be positive in the *Salmonella* mutagenicity assay.

Carcinogenicity. Negative results are reported in rats.[09]

Chemobiokinetics. AA undergoes transformation into **acrolein** under the action of alcohol dehydrogenase, with further conjugation to cellular glutathione to form an **aldehyde-glutathione** t, which is then gradually transformed into the corresponding **acid.** A reduction in the intra- ar content of SH-groups is of considerable importance in the pathogenesis of AA toxic effects.[3] The toxicity of AA (or its metabolite acrolein) is dependent on the concentration of glutathione.[5]

Standards. Russia (1988). MAC: 0.1 mg/l (organolept., taste).

References:
 1. Roschlau, G. and Rodenkirchen, H., Histological examination of the diaplacental action of carbon tetrachloride and allyl alcohol in mice embryos, *Exp. Pathol.,* 3, 255, 1969.

2. Slott, V. L. and Hales, B. F., Teratogenicity and embryolethality of acrolein and structurally related compounds in rats, *Teratology,* 32, 65, 1985.

3. Ohno, Y., Ormstad, K., Ross, D., et al., Mechanism of allyl alcohol toxicity and protective effects of low-molecular-weight thiols studied with isolated rat hepatocytes, *Toxicol. Appl. Pharmacol.,* 78, 169, 1985.

4. Atzori, L., Dore, M., and Congiu, L., Aspects of allyl alcohol toxicity, *Drug Metabol. Drug Interact.,* 7, 295, 1989.

AMYL ALCOHOL (AA) (CAS No 71–41–0)

Synonyms. *n*-Butylcarbinol; Pentanone; Pentasol; Pentyl alcohol.

Properties. A yellow liquid with an odor of fusel oil. Odor perception threshold is 0.12 mg/l^{010} or 0.7 mg/l; taste threshold concentration is 0.5 mg/l.[1] According to other data, odor threshold concentration is 1.3 mg/l, and the taste threshold slightly higher.[2]

Applications and **Exposure.** AA is used in the production of plastics, varnishes, paints, and synthetic detergents. Maximum daily intake in man is calculated to be 33 mg in the U.K. and 42 mg in the U.S. and in Europe.[3]

Acute Toxicity. In rats, LD_{50} is 3 to 4.25 g/kg BW.[2,4,5] CNS depression was the only observed effect of a single oral dose administration. Appearance and behavior of survivors did not differ from those of the controls.[2]

Repeated Exposure failed to reveal cumulative properties. Mice were given $^1/_5$ LD_{50}. An overall dose of 12 g/kg BW caused no animal mortality.[2]

Short-term Toxicity. Rats were administered AA, dissolved in corn oil, by oral intubation at dose-levels of 50 to 1000 mg/kg BW for 13 weeks. No reduction in BW gain or food and water consumption, and no changes in hematological indices or urinalyses, renal function, organ weights, or hystopathology were reported. Butterworth et al. considered the NOAEL to be 1000 mg/kg BW, which is about 2000 times the estimated maximum likely intake by man.[3]

Long-term Toxicity. In a 6-month study, rats received the dose of 95 mg/kg BW. There were changes in blood catalase and liver cholinesterase activity, as well as in the immune status of animals.[2] Liver necrosis was noted in rabbits following oral exposure to AA for up to 1 year.[6] Damage to the gastric mucosa was the only pathological finding observed in another study.[7]

Chemobiokinetics. AA is oxidized to **valeric acid** by aldehyde in rats.[8] Less than 0.03% of a single oral dose was excreted in the urine of rats within 8 h.[9]

Standards. Russia (1988). MAC and PML: 1.3 mg/l (organolept., odor). Recommended MAC: 0.1 mg/l.[7]

Regulations. USFDA (1993) has permitted AA for use as a synthetic flavoring substance and additive.

References:

1. *Proc. Med. Inst.,* Voronezh, 55, 56, 1966 (in Russian).

2. Korolev, A. A., Krasovsky, G. N., and Varshavskaya, S. P., Hygiene evaluation of amyl alcohol, primary and secondary octyl alcohols as regards their standardization in water bodies, *Gig. Sanit.,* 9, 88, 1970 (in Russian).

3. Butterworth, K. R., Gaunt, I. F., Heading, C. E., et al., Short-term toxicity of *n*-amyl alcoh rats, *Food Chem. Toxicol.,* 16, 203, 1978.

4. Jenner, P. M., Hagan, E. C., Taylor, J. M., et al., Food flavoring and compounds of re structure. I. Acute oral toxicity, *Food Cosmet. Toxicol.,* 2, 327, 1964.

5. See **ACRYLONITRILE,** #7, 31.

6. Straus, I. and Blocq, P., Etude experimentelle sur la cirrhose du foie, *Arch. Physiol.,* 10, 409, 1887 (in French).

7. *Integrated Problems of Hygiene and Health Protection in Siberian Regions,* Moscow, 1988, 48 (in Russian).

8. Haggard, H. W., Miller, D. P., and Greenberg, L. A., The amyl alcohols and their ketones: their metabolic fates and comparative toxicities, *J. Ind. Hyg. Toxicol.,* 27, 1, 1945.

9. Gaiollard, D. and Derache, R., Vitesse de la metabolisation de differents alcools chez la rat, *Cr. Seanc. Soc. Biol.,* 125, 1605, 1964 (in French).

BENZYL ACETATE (BA) (CAS No 140–11–4)

Synonyms. Acetic acid, phenylmethyl ester; Acetic acid, benzyl ester; α-Acetoxytoluene; Benzyl ethanoate.

Properties. Clear, colorless liquid with a powerful, specific odor (jasmine-like or pear-like). Solubility in water is 250 mg/l; readily soluble in alcohols.

Applications and **Exposure.** Is primarily a flavoring component, to a lesser degree a solvent of cellulose acetate, cellulose nitrate, and natural and synthetic resins. B. has been identified in several fruits and mushrooms. May be used safely as a synthetic flavoring substance for food.

Acute Toxicity. LD_{50} is reported to be 2500 to 2800 mg/kg in rats, 830 mg/kg in mice, and 220 mg/kg BW in guinea pigs and rabbits.[1] However, according to other data, LD_{50} is 3690 mg/kg in rats and 2640 mg/kg BW in rabbits.[2] Administration of the lethal doses results in CNS depression. Death occurs on days 1 to 5. Acute action threshold was identified to be 50 mg/kg BW for STI.[1] After oral administration of 4000 mg/kg BW, seven out of ten treated rats died within 2 h.[3]

Repeated Exposure failed to reveal evident cumulative properties. K_{acc} is 7.2 (by Lim). Rats were dosed with 50 to 200 mg/kg BW. The treatment caused transient changes in the NS function and blood formula. The NOAEL of 30 mg/kg BW was identified.[1] Fisher 344 rats were exposed by gavage to 62.5 to 1000 mg/kg BW in corn oil and to 250 to 4000 mg/kg BW during 14 consecutive days: 3 out of 10 rats receiving 1000 mg/kg died. Clinical signs included decreased BW, tremor, ataxia, and sluggishness. No histopathological data are available.[4]

Short-term Toxicity. In 13-week study, B6C3F$_1$ mice received the doses of 45 to 7200 mg BA/kg BW in their diet. The treatment caused retardation in BW gain; feed consumption was lower in females at the dose of 3600 mg/kg and above. No effects on hematology, clinical chemistry, or pancreatic enzyme parameters were noted.[6] Fisher 344 rats received 210 to 3360 mg BA/kg BW in their diet for 13 weeks. Tremor, ataxia, and urine stains were observed in animals exposed to 3360 mg/kg dose. Decreased serum cholesterol was noted in females dosed with 840 mg BA/kg and more.[6]

Long-term Toxicity. B6C3F$_1$ mice received 330 to 3000 mg BA/kg feed over the period of 103 weeks. Fisher 344 rats were given 3000 to 12,000 mg BA/kg feed for the same period of time. The treatment caused dose-related degeneration and atrophy of the olfactory epithelium, cystic hyperplasia of the nasal submucosal glands, and pigmentation of the nasalmucosae epithelium. The NOEL of 550 mg/kg BW was identified in this study.[6]

Allergenic Effect was not found on dermal applications.[1]

Genotoxicity. Mutagenic effect was seen in *Salmonella* assay, but no SCE or CA were observed in Chinese hamster ovary cells.[4,5]

Carcinogenicity. B6C3F$_1$ mice were dosed by oral intubation with 500 and 1000 mg/kg BW in corn oil for 103 weeks; there was an increased incidence of liver adenoma (not statistically significant) and of combined liver adenomas and carcinomas, forestomach tumors. Acinar-cell adenomas of the pancreas were found in Fisher 344 rats.[4] *Carcinogenicity classification.* IARC: Group 3; NTP: EE—NE—SE*—SE.

Chemobiokinetics. In the body, B. is hydrolyzed to **benzyl alcohol.** The benzyl radical is oxidized to **benzyl acid** and excreted primarily in the urine as **hippuric** and **benzylmercapturic acids.**[2,3]

Recommendations. JECFA (1993). ADI of 0 to 5 mg/kg BW is considered as total benzoic acid from all food additive sources.

References:

1. Fursova, T. N., Toxicologic and hygienic characteristics of benzyl acetate, *Gig. Sanit.,* 7, 17, 1985 (in Russian).
2. Graham, B. E. and Kuizenga, M. H., Toxicity studies of benzylacetate and related benzyl compounds, *J. Pharmacol. Exp. Ther.,* 84, 358, 1945.
3. von Ottingen, W. F., The aliphatic acids and their esters: toxicity and potential danger, *Arch. Ind. Health,* 21, 28, 1960.
4. Abdo, V. M., Huff, J. E., Haseman, J. K., et al., Benzylacetate carcinogenicity, metabolism, and disposition in Fischer 344 rats and B6C3F$_1$ mice, *Toxicology,* 37, 159, 1985.
5. Mortelmans, K., Haworth, S., Lawlor, T., et al., *Salmonella* mutagenicity tests. II. Results from testing of 270 chemicals, *Environ. Mutat.,* 8 (Suppl. 7), 1, 1986.

6. NTP, *Technical Report on the Toxicology and Carcinogenesis Studies of Benzyl Acetate in F344 Rats and B6C3F₁ Mice (Feed Studies),* NTP Technical Report No. 431, Board Draft, NIH, Research Triangle Park, NC, 1992.

BENZYL ALCOHOL (BA) (CAS No 100–51–5)

Synonyms. Benzenemethanol; Benzenecarbinol; α-Hydroxytoluene; Phenylcarbinol; Phenylmethanol; Phenylmethyl alcohol.

Properties. Colorless, transparent liquid with a faint aromatic odor and a sharp burning taste. Water solubility is 40 g/l; readily soluble in organic solvents.

Applications. Used in the production of cellulose acetate.

Acute Toxicity. LD_{50} is reported to be 2 g/kg BW in male rats, 1.66 g/kg BW in female rats, 1.36 g/kg BW in mice, 2.5 g/kg BW in guinea pigs, and 0.1 g/kg BW in rabbits. Administration of high doses causes NS inhibition. After oral administration of LD_{50}, 50% of rats died in 1 to 2 d, mice in 2 to 3 d.[1] ET_{50} is 90 min. Acute action threshold is 0.1 mg/kg.

Repeated Exposure revealed pronounced cumulative properties. K_{acc} is 2 to 3. Rats were administered BA at dose-levels of 20 to 200 mg/kg BW. Manifestations of toxic action included retardation of BW gain, an increase in STI, and behavioral changes. Some animals died following repeated intake of 100 and 200 mg BA/kg BW. Histological examination revealed hyperplasia in the spleen and peribronchial lymphatic follicules. Dose-dependence was noted. Calculated approximate NOEL appeared to be 100 mg/kg.[1,2]

Short-term Toxicity. Inhalation exposure demonstrated histological and histochemical changes in the visual analyzer.

Genotoxicity. BA is reported to be negative in the *Salmonella* mutagenicity assay, but it caused SCE in cultured mammalian cells.[3]

Carcinogenicity. BA caused no carcinogenic response in rats and mice.[3] *Carcinogenicity classification.* NTP: NE—NE—NE—NE.

Chemobiokinetics. BA undergoes *in vivo* oxidation to **benzoic acid** in man and rabbits.[4]

Recommendations. JECFA (1993). ADI: 0 to 5 mg/kg.

Regulations. USFDA (1993) approved the use of BA (1) in adhesives used as components of articles intended for use in packaging, transporting, or holding food and (2) in closures with sealing gaskets on containers intended for use in producing, manufacturing, packing, processing, preparing, treating, packaging, transporting, or holding food at levels not to exceed 1% by weight of the closure-sealing gasket composition.

References:

1. Rumyantsev, G. I., Novikov, S. M., Kozeeva, E. E., et al., Experimental studies of biological effects of benzyl alcohol, *Gig. Sanit.,* 7, 81, 1985 (in Russian).
2. Rumyantsev, G. I., Novikov, S. M., Fursova, T. N., et al., Experimental study of toxicity of phenylethyl alcohol and phenylethyl acetate, *Gig. Sanit.,* 10, 83, 1987 (in Russian).
3. See **BISPHENOL A,** #6.
4. See 1,3-**BUTANEDIOL,** #1, 15.

BUTYL ACETATE (CAS No 123–86–4)

Synonym. Acetic acid, butyl ester.

Properties. A liquid with an odor of ether. Water solubility is 1% or 6.8 g/l at 25°C.[02] Mixes with alcohol and vegetable oils at any ratio. Odor perception threshold is 1 mg/l or 0.17 mg/l,[02] or even 0.006 mg/l;[010] taste perception threshold is 0.3 mg/l. Does not affect color and transparency of water.

Acute Toxicity. LD_{50} is 14.1 g/kg BW in rats and 4.7 g/kg BW in mice (Sporn et al.). According to Bulbin, LD_{50} is 13.1 and 7.7 g/kg BW, respectively.[1] Gross pathological examination revealed congestion in the viscera and soft brain meninges. Histological examination showed kidney lesions.

Repeated Exposure failed to reveal cumulative properties. Rats were dosed with 0.8 and 1.6 g/kg BW for a month. The treatment caused adynamia, lymphocytosis, depression of leukocyte phagocytic activity, and increased vitamin C content in the suprarenals. Gross pathological examination revealed parenchymatous dystrophy in the cerebellum cells, especially evident at a higher dose (Bulbin).

Long-term Toxicity. In a 6-month study, no changes were found in the liver function and enzymatic activity of rats exposed to 0.5 mg B./kg BW.

Reproductive Toxicity. BA produced no teratogenic effect in rats, mice, and rabbits.[1,2]

Genotoxicity. BA is not shown to be mutagenic in *Dr. melanogaster* (NTP–82).

Standards. Russia (1988). PML in drinking water: 0.3 mg/l.

Regulations. USFDA (1993) regulates BA (1) as a component of adhesives intended for use in contact with food, (2) in cellophane to be used safely for packaging food (up to 0.1% by weight), and (3) as an ingredient of resinous and polymeric coatings for the food-contact surface of articles intended for use in producing, manufacturing, packing, transporting, or holding food.

References

1. Scheufler, H., Experimental testing of chemical agents for embryotoxicity, teratogenicity and mutagenicity. Ontogenic reactions of the laboratory mouse to these injections and their evaluation—a critical analysis method, *Biol. Rundsch.*, 14, 227, 1976.
2. Hackertt, P. L., Brown, M. G., Buschbom, R. L., et al., Teratogenic activity of ethylene and propylene oxide and *n*-butyl acetate, *Gov. Rep. Announce. Ind.*, Issue 26, NTIS/PB 83–258038, 1983.

1-BUTYL ALCOHOL (NBA) (CAS No 76–36–3)

Synonyms. 1-Butanol; *n*-Butanol; Butyric alcohol.

Properties. Clear, colorless liquid with a heavy unpleasant odor. Solubility in water is 9% at 15°C or 73,000 mg/l at 25°C;[02] mixes with ethyl, alcohol, ether, and other organic solvents at all ratios. In small concentrations, it gives an aromatic but not unpleasant odor to water. Odor perception threshold is 0.3 mg/l,[010] 1 mg/l,[1] or 7.1 mg/l;[02] according to other data, organoleptic threshold is 0.27 mg/l.[01]

Applications and **Exposure.** Used in production of coatings, nitrocellulose lacquers, and natural resins; used in rubber vulcanization and in production of foam plastics. Exposure of the general population is principally through the natural occurrence of NBA in foods and beverages and its use as a flavoring agent.

Acute Toxicity. *Humans.* Ingestion of a dose of more than 250 ml seems to be fatal, although there may be individual variations in sensitivity. NBA does not induce vomiting.[2] *Animals.* LD_{50} is reported to be 2.68 g/kg BW in mice,[3] 0.7 to 2.1 g/kg BW in rats, 3.5 g/kg BW in rabbits, and 1.2 g/kg BW in hamsters.[2] Its potency for intoxication is approximately six times that of ethanol.

Repeated Exposure revealed moderate cumulative properties. K_{acc} is 3.4. NBA is capable of both material and functional accumulation.[4] Chicks were given NBA starting with 15 mg and rising to 600 mg, for 2 months. Exposure caused slight anemia of the comb only; BW gain and development were unaffected.[1]

Reproductive Toxicity. Nelson et al. reported a teratogenic effect of NBA administered by inhalation to rats.[3]

Genotoxicity. NBA was shown to be negative in the *Salmonella* mutagenicity assay.[5] It did not produce CA in cultured human lymphocytes.[6]

Chemobiokinetics. Following ingestion, NBA is readily absorbed in the GI tract of experimental animals. NBA metabolism is found to occur through rapid and complete oxidation in the body, apparently via the formation of **butyric acid** and via the **aldehyde** to **carbon dioxide,** which is the major metabolite. A small amount (only 2 to 4%) of the alcoholic glucuronides appears in the urine.

Standards. USA (1981). Maximum permitted concentration in plant (in food additive modified hot extract): 50 mg/l. **Russia** (1988). MAC: 0.1 mg/l; PML in food: 0.5 mg/l.

Regulations. USFDA (1993) approved the use of NBA (1) as a component of adhesives used in the food-contact surface of articles; (2) in cellophane for food packaging; (3) as a component of paper and paperboard for contact with dry food; (4) as a solvent in polysulfide polymer-polyepoxy resins used as the dry food-contact surface in accordance with specified conditions; (5) as an ingredient of resinous and polymeric coatings for polyolefin films; (6) as a defoaming agent that may be used safely as a component of articles intended for contact with food; and (7) as a substance employed in the production of or added to textiles and textile fibers intended for use in contact with food.

References:
1. See **ACRYLONITRILE,** #1, 65.
2. IPCS, *Butanols—Four Isomers: 1-Butanol, 2-Butanol, tert-Butanol,* Isobutanol, Environmental Health Criteria 65, WHO, Geneva, 1987, 141.
3. Nelson, B. K., Brightwell, W. S., Khan, A., et al., Lack of selective developmental toxicity of three butanol isomers administered by inhalation to rats, *Fundam. Appl. Toxicol.,* 12, 469, 1989.
4. Rumyantsev, A. P., Lobanova, I. Ya., Tiunova, L. V., et al., Toxicology of Butyl Alcohol, *Chemical Industry: Series Toxicology, Sanitary Chemistry of Plastics,* 2, 24, 1979 (in Russian).
5. See **RESORCINOL,** #3.
6. Obe, G., Ristow, M. J., Herma, J., Chromosomal damage by alcohol *in vitro* and *in vivo, Adv. Exp. Med. Biol.,* 85a, 47, 1977.

2-BUTYL ALCOHOL (SBA) (CAS No 78–92–2)

Synonyms. *sec*-Butanol; Butylene hydrate; 2-Hydroxy butane; Methyl ethyl carbinol.

Properties. Colorless liquid with a characteristic sweet odor. Solubility in water is 125 g/l at 20°C, or according to other data, 200 g/l at 25°C.[02] Miscible with ethyl alcohol and ether. Odor perception threshold is 0.12 mg/l[010] or 19 mg/l.[02]

Applications and **Exposure.** A solvent. SBA occurs naturally in foods and beverages.

Acute Toxicity. *Humans.* Consumption is followed with abdominal pain, vomiting, and diarrhea. *Animals.* In rats, LD_{50} is 6.5 g/kg BW. Administration caused ataxia and narcosis. The potency of SBA for intoxication is approximately four times that of ethanol.[1] In rabbits, LD_{50} is 4.9 g/kg BW.[2]

Long-term Toxicity. In a two-generation rat reproduction study, SBA was administered via drinking water at concentrations of 0.3, 1, and 2% (through the first and second generations). The treatment with 2% concentration resulted in several changes, which represent mild toxicity and are reminiscent of stress lesions.[3]

Reproductive Toxicity. In the above-described studies, 0.3 and 1% concentrations of SBA in drinking water produced no effect on growth and reproduction of rats.[3] A concentration of 2% caused significant depression of growth of weaning rats with evidence of retarded skeletal maturation.

Chemobiokinetics. Approximately 97% of the dose is converted by alcohol dehydrogenase to **methyl ethyl ketone,** which is either excreted in the breath and urine or conjugated with a formation of glucuronide.[1] According to Dietz et al.,[4] the following metabolites are formed in the body: 2-**butanone,** 3-**hydroxy-2-butanone,** and 2,3-**butanediol.**

Standards. Russia (1988). MAC: 0.2 mg/l. U.K. MAFF (1978). Recommended residues of butan-2-ol in food should not exceed 30 mg/kg.

Regulations. USFDA (1993) approved the use of SBA as a component of the uncoated or coated food-contact surface of paper and paperboard that may be used safely in producing, manufacturing, packing, transporting, or holding dry food.

References:
1. See 1-**BUTYL ALCOHOL,** #2.
2. Munch, J. C., Aliphatic alcohols and alkyl esters: narcotic and lethal potencies to tadpoles and to rabbits, *Ind. Med.,* 41, 31, 1972.
3. 16th Annual Meeting of the Society of Toxicology, Abstracts of Papers, Toronto, March 27–30, 1977, p. 9.
4. Dietz, F. K., Rodriguez-Jiaxola, M., Traiger, G. J., et al., Pharmaco-kinetics of 2-butanol and its metabolites in the rat, *J. Pharmacokinet. Biopharm.,* 9, 553, 1981.

tert-BUTYL ALCOHOL (TBA) (CAS No 75–65–0)

Synonyms. *tert*-Butanol; **1,1**-Dimethylethanol; 2-Methyl-2-propanol; Trimethylcarbinol; Trimethylethanol.

Properties. Colorless liquid or white crystals with a camphor-like odor. Mixes with water and alcohol at any ratio. Odor perception threshold is 290 mg/l.[02]

Applications. Powerful solvent. Used in the manufacturing of polyolefins, copolymers of methacrylonitrile and methacrylic acid, and also in the production of drugs, perfumes, and lacquers.

Acute Toxicity. *Humans.* Ingestion is followed by abdominal pain, vomiting, and diarrhea. *Animals.* LD_{50} is 3.5 g/kg BW in rats and 3.6 g/kg BW in rabbits. Poisoning exerts a narcotic effect. The primary acute effects in animals are those of alcoholic intoxication. TBA potency for intoxication is approximately 1.5 times that of ethanol.[1,2]

Repeated Exposure. Rats were fed with 20 ml TBA/l liquid diet. The treatment entailed slight ataxia.[1]

Short-term Toxicity. In a 90-d study in B6C3F$_1$ mice and Fisher 344 rats, TBA was added to the diet at dose-levels of 0.25 to 4% (w/v).[3] Of TBA, 4% caused mortality in the treated animals. Decreased BW was noted at all doses in males and at a 4%-dose in females. Clinical signs of intoxication included ataxia in rats and mice. In females, hypoactivity was found to develop. Gross pathological examination revealed urinary tract calculi, renal pelvic and ureteral dilatation, and thickening of the urinary bladder mucosa. Histological examination revealed hyperplasia of the transitional epithelium. The NOAEL for the urinary tract lesions appeared to be about 800 mg/kg BW (male rats), 1570 mg/kg BW (male mice), or 1450 mg/kg BW (females of both species).

Reproductive Toxicity. Postnatal effects of TBA were found in the offspring exposed *in utero*, although Daniel and Evans[4] found no teratogenicity in rats and mice fed TBA up to 1% in the diet.

Genotoxicity. TBA seems to be a nongenotoxic compound. In the NTP study, it was shown to be negative in the *Salmonella* mutagenicity assay, mouse lymphoma cells, and *in vitro* cytogenetic assays.[1]

Chemobiokinetics. TBA is not a substrate for alcohol dehydrogenase and is slowly metabolized by mammals. Conjugation of the hydroxyl group with glucuronic acid is possible.[5] TBA is excreted in the urine as glucuronide, but also in the breath and urine as **acetone** or **carbon dioxide**.[1]

Standards. Russia (1988). MAC: 1 mg/l.

Regulations. USFDA (1993) approved the use of TBA as a defoaming agent in components of articles intended for use in producing, manufacturing, packing, transporting, or holding food.

References:

1. See 1-**BUTYL ALCOHOL,** #2.
2. See 2-**BUTYL ALCOHOL,** #2.
3. Lindamood, C., III, Farnel, D. R., Giles, H. D., et al., Subchronic toxicity studies of *tert*-butyl alcohol in rats and mice, *Fundam. Appl. Toxicol.,* 19, 91, 1992.
4. Daniel, M. A. and Evans, M. A., Quantitative comparison of maternal ethanol and maternal tertiary butanol diet on postnatal development, *J. Pharmacol. Exp. Ther.,* 222, 294, 1982.
5. Merritt, D. A. and Thomkins, G. M., Reversible oxidation of cyclic secondary alcohols by liver alcohol dehydrogenase, *J. Biol. Chem.,* 234, 2778, 1959.

CARBON TETRACHLORIDE (CCl_4) (CAS No 56–23–5)

Synonyms. Tetrachloromethane; Methane tetrachloride; Perchloromethane.

Properties. Colorless liquid. Water solubility is 800 mg/l at 20°C. Odor perception threshold is 0.52 mg/l.[02]

Applications and **Exposure.** A solvent, a cleaning agent, an intermediate in the production of chlorofluorocarbons. It has been detected in a variety of foodstuffs at levels from 0.1 to 20 g/kg and less frequently in drinking water. Sources of human exposure from the environment include CCl_4-contaminated air and water.

Acute Toxicity. *Humans.* No effects were reported after single oral doses of 2.5 to 15 ml CCl_4 (57 to 343 mg/kg BW), although changes may occur in the liver and kidney. Some individual adults suffer adverse effects (including death) from ingestion of as little as 1.5 mg/l (34 mg/kg BW), and doses of 0.18 and 0.92 ml may be fatal in children.[1] *Animals.* LD_{50} values for laboratory animals vary from 1 to 12.8 g/kg BW.[1] According to other data, LD_{50} is 6.2 g/kg BW in rats, 12 to 14 g/kg BW in mice, and 5.7 g/kg BW in guinea pigs and rabbits.[2] Single doses of about 4 g/kg result in lesions of the renal proximal tubules in rats and pulmonary Clara cells and endothelial cells in rats and mice. Adverse effects in the liver were shown after oral administration of 80 mg CCl_4/kg but not after 40 mg CCl_4/kg BW.[1,3]

Repeated Exposure. After oral administration of large doses, the most severe damage is observed in the liver (including hepatocellular necrosis), kidney, and lung. Lower doses of CCl_4 cause reversible changes.[1] Hepatotoxic effects were observed in CD-1 mice given 625 to 2500 mg CCl_4/kg BW for 14 d[4] and in rats given 20 mg CCl_4/kg BW and higher doses (in corn oil) for 9 d.[3] Increased liver lipid and triglyceride levels were reported in a 6-week study using doses of 40 and 76 mg/kg but not 22 mg/kg BW.[5] When young adult (8 to 9 weeks old) male Fisher 344 rats were given 40 mg CCl_4/kg BW by gavage for 10 consecutive days, a significant increase in the relative liver weights was noted. Histological examination revealed mild to moderate vacuolar degeneration and minimal to mild hepatocellular necrosis in these livers. When rats were dosed with 20 to 40 mg/kg BW, the serum levels of ALT and AST were elevated. However, no renal effects were observed.[6] Newborn rats seem to be less sensitive to liver damage by CCl_4 than 7-d-old rats.[7]

Short-term Toxicity. CCl_4 hepatotoxicity is a result of the parent compound metabolism to a highly reactive radical intermediate by the cytochrome P-450 mixed function system. CCl_4 increases lipid peroxidation, fatty infiltration, destruction of cytochrome P-450, and liver necrosis.[8] CD-1 mice were given the doses of 12 to 1200 mg CCl_4/kg BW in corn oil by gavage over a period of 3 months. The treatment resulted in increased serum enzyme levels, increased organ weights, and pathological changes.[4] In rats given 1 mg CCl_4/kg BW for 12 weeks, no adverse effects were found. Doses of 10 and 33 mg/kg BW resulted in enzyme release, centrilobular vacuolization, and necrosis in the liver.[3]

Long-term Toxicity. The same effects, namely, fatty infiltration, release of liver enzymes, inhibition of cellular enzyme activities, inflammation, and cellular necrosis, were observed following a long-term exposure. No adverse effects were observed in rats of both sexes fed dietary levels of 80 and 200 ppm CCl_4 until final sacrifice in 2 years. However, survival was below 50% at 21 months, and tissues were not examined microscopically in this study.[5]

Immunotoxicity. CCl_4 was administered *i/p* to female B6C3F$_1$ rats for 7 d. Doses of 0.5, 1, and 1.5 g/kg BW were found to produce a marked suppression of both humoral and cell-mediated immune functions.[9] No consistent alterations in the immune parameters examined in the Fisher 344 rat study were observed.[6] There was no difference in antibody response to SRBC in another group of rats dosed with 40 to 160 mg/kg. CCl_4 is not immunotoxic in the rat at the dosages that produce overt hepatotoxicity.

Reproductive Toxicity. There were no reproductive effects in rats fed the diet containing CCl_4 at concentrations of 80 and 200 ppm for 2 years.[5] CCl_4 *is not teratogenic.*[1]

Genotoxicity. No data are reported on genetic and related effects of CCl_4 humans (IARC). It is not genotoxic in the majority of mutagenicity bioassays. Mutagenic effects were not observed in a number of bacterial test systems or in cultured liver cells.[1] It caused cell transformation in Syrian hamster embryo cells,[10] but did not induce CA, unscheduled DNA synthesis, or DNA strand breaks in the cells of rodents treated *in vivo*. CCl_4 did not induce CA or SCE in rat cells *in vivo* (NTP–88; IARC 1–53 and 8–371).

Carcinogenicity. CCl_4 demonstrated carcinogenic potential through oral exposure, producing several types of tumors but mainly hepatic neoplasms. Doses of about 30 mg/kg BW or higher administered for 6 months or longer have been found to produce an increased frequency of hepatocellular tumors in mice, rats, and hamsters.[1] CCl_4 was found to be carcinogenic in the B6C3F$_1$ mice exposed to time-weighted average doses of 1250 and 2500 mg/kg BW for 78 weeks. The incidence of hepatocellular carcinoma was reported to be almost 100% in both sexes. The rate of carcinoma development was substantially lower (about 5%) in Osborne-Mandel rats exposed to 47 or 94 mg/kg and 80 or 159 mg/kg BW. Because there are doubts regarding the mechanism of tumorogenesis in the liver of this strain of mouse with agents that are known hepatotoxins (such as CCl_4), the appropriateness of a nonthreshold model for extrapolation is questionable.[11] Syrian golden hamsters were exposed to approximately 10 to 20 mg CCl_4/d for 43 weeks. Half of animals died in the course of treatment. Survivals developed liver cell carcinomas.[12] CCl_4 was not shown to be carcinogenic through inhalation exposure in animals. WHO considered it to be nongenotoxic carcinogen (1992). *Carcinogenicity classification.* USEPA: Group B2; IARC: Group 2B.

Chemobiokinetics. CCl_4 is absorbed readily from the GI tract and seems to be distributed in all major organs and tissues following absorption. It is converted into a **trichloromethyl** free radical,

278

which is the main metabolite and undergoes a variety of reactions, including hydrogen abstraction to form **chloroform,** dimerization to form **hexachloroethane,** and addition to cellular molecules. Further metabolism of the heme-bound trichloromethyl radical is postulated to result in the eventual formation of **carbonyl chloride (phosgene).**[13] CCl_4 and its volatile metabolites are excreted primarily in exhaled air and also in the urine and feces. The major part of the oral dose is excreted in 1 to 2 d.[1]

Standards. WHO (1992). Guideline value for drinking water: 0.002 mg/l. WHO recommends that no detectable residues (detection limit: 0.01 ppm) be allowed in food or feed but permits 50 mg/l in cookie cereals. **USEPA** (1991). MCL: 0.005 mg/l, MCLG: zero. **Russia** (1988). MAC: 0.006 mg/l.

Regulations. USFDA (1993) approved the use of CCl_4 (1) as a component of adhesives for articles intended for use in packaging, transporting, or holding food; (2) as a component of paper and paperboard that may be used safely in producing, manufacturing, packing, transporting, or holding dry food; (3) as an ingredient in resinous and polymeric coatings of food-contact surfaces; and (4) to prevent the transfer of inks employed in printing and decorating paper and paperboard used for food packaging.

References:

1. USEPA, *Final Draft Criteria Document for Carbon Tetrachloride,* TR-540–131A, Office of Drinking Water.
2. Chirkova, V. M., *Carbon Tetrachloride,* Issue No. 27, Soviet Toxicology Center, Moscow, 1983, 20 (in Russian).
3. Bruckner, J. V., Kim, H. J., Dallas, C. E., et al., Effect of dosing vehicles on the pharmacokinetics of orally administered carbon tetrachloride, Soc. Toxicol. Annual Meeting (Abstract), 1987.
4. Hayes, J. R., Condie, L. W., and Borcelleca, J. F. Acute, 14-day repeated dosing, and 90-day subchronic toxicity studies of carbon tetrachloride in CD-1 mice, *Fundam. Appl. Toxicol.,* 7, 454, 1986.
5. Alumot, E., Nachtomi, E., Mandel, E., et al., Tolerance and acceptable daily intake of chlorinated fumigants in the rat diet, *Food Cosmet. Toxicol.,* 14, 105, 1976.
6. Smelowicz, R. J., Simmons, J. E., Luebke, R. W., et al., Immunotoxicologic assessment of subacute exposure of rats to CCl_4 with comparison to hepatotoxicity and nephrotoxicity, *Fundam. Appl. Toxicol.,* 17, 186, 1991.
7. Dawkins, M. J. R., Carbon tetrachloride poisoning in the liver of newborn rat, *J. Pathol. Bacteriol.,* 85, 189, 1963.
8. Recknagel, R. O., A new direction in the study of carbon tetrachloride hepatotoxicity, *Life Sci.,* 33, 401, 1983.
9. Kaminski, N. E., Barnes, D. W., Jordan, S. D., et al., The role of metabolism in CCl_4-mediated immunosuppression, *in vitro* studies, *Toxicol. Appl. Pharmacol.,* 102, 9, 1990.
10. Amacher, E. D. and Zelljadt, I., The morphological transformations of Syrian hamster embryo cells by chemicals reportedly non-mutagenic to *Salmonella typhimurium, Carcinogenesis,* 4, 291, 1983.
11. National Cancer Institute, *Report on Carcinogenesis Bioassay of Chloroform,* Carcinogenesis Program, Division of Cancer Cause and Prevention, Bethesda, MD, 1976.
12. Della Porta, G., Terracini, B., and Shubik, P., Induction with carbon tetrachloride of liver cell carcinomas in hamsters, *J. Natl. Cancer Inst.,* 26, 855, 1961.
13. Shah, H., Hartman, S., and Weinhouse, S., Formation of carbonyl chloride in carbon tetrachloride metabolism by rat liver *in vitro, Cancer Res.,* 39, 3942, 1979.

CHLOROFORM (CAS No 67–66–3)

Synonyms. Trichloromethane.

Properties. Colorless, very volatile, sweet-tasting liquid with an unpleasant odor at 25°C. Solubility in water is 8 g/l, readily soluble in organic solvents. Odor perception threshold is 2.4 mg/l.[02]

Applications and **Exposure.** Used in the manufacturing of fluoropolymers, rubbers, and resins. An important extraction solvent for resins, gums, etc. A cosmetic ingredient. A by-product of drinking water chlorination.

Acute Toxicity. *Humans.* C. is a CNS depressant; it affects the liver and kidney functions in humans and animals. In humans, LD_0 is approximately 44 g.[1] A fatal dose may be as small as 211 mg/kg BW, with death due to respiratory or cardiac arrest.[2] *Animals.* The LD_{50} values are 1.25 g/kg BW in rats, 0.1 g/kg BW in mice, 0.82 g/kg BW in guinea pigs, 9.83 g/kg BW in rabbits, and 2.25 g/kg BW in dogs.[3]

Repeated Exposure. C. appears to be both hepatotoxic and neurotoxic in most animal species. Administration of $1/30$ LD_{50} for 1 month produced adipose dystrophy and cirrhotic and necrotic lesions in the liver (Miklashevsky et al., 1966).

Short-term Toxicity. Prolonged exposure to doses of more than 15 mg/kg can affect the kidney, liver, and thyroid. In a 14-d study, the NOAEL of 125 mg/kg BW was identified in mice (based on elevated serum enzyme levels).[4]

Long-term Toxicity. *Humans.* A number of epidemiology studies tend to support the finding of an increased risk of bladder, colon, and rectal cancer from exposure to chlorinated water. A positive correlation between C. levels in drinking water and mortality from stomach, large intestine, rectum, and bladder cancer was demonstrated by regression statistical analysis.[5] Although C. appears to be the single largest constituent in chlorinated water; these studies do not prove directly that C. is a human carcinogen, since chlorinated water contains many other chlorination by-products.[6] *Animals.* Administration of $1/50$ LD_{50} for 5 months produced adipose dystrophy and cirrhotic and necrotic lesions of the liver.[3] The LOAEL of 15 mg/kg BW was identified for liver fatty cysts in dogs exposed to C.[7] The calculated ADI and DWEL are 0.01 mg/kg BW and 0.5 mg/l, respectively. Sprague-Dawley rats were dosed by gavage with C. in a toothpaste-based vehicle at up to 60 mg/kg BW for 80 weeks. The treatment resulted in retardation of BW gain, a decrease in plasma cholinesterase activity, and a significant decrease in relative liver weights in female rats. The LOAEL of 60 mg/kg BW based on decreases in BW and plasma cholinesterase activity was identified in this study.[8] $B6C3F_1$ mice were exposed to 600 or 800 mg C./l (86 or 258 mg C./kg BW) via drinking water for 24 or 52 weeks.[9]

Manifestations of toxic action included decreased BW, focal areas of cellular necrosis in the kidneys and liver, and focal areas of hepatic lipid accumulation in the high-dosed group.

Reproductive Toxicity. No significant *teratogenicity* was found after oral administration of up to 126 mg/kg BW to pregnant rats or up to 50 mg/kg BW to pregnant rabbits. However, pronounced maternal toxicity was observed in animals given 50 mg/kg BW (both rats and rabbits).[10] The maternal NOAEL was 20 mg/kg BW in rats and 35 mg/kg BW in rabbits; the fetal NOAEL was 50 mg/kg BW in rats.

Genotoxicity. The available results are inconclusive. C. seems to be a weak mutagen. It was shown to cause genotoxic effect in a host-mediated assay in male mice,[11] in a sperm-head abnormality assay in mice,[12] and in *Dr. melanogaster.*

Carcinogenicity. C. induced hepatocellular carcinomas in mice when administered by gavage in oil-based vehicles (but not in drinking water) and renal tubular adenomas and adenocarcinomas in male rats, regardless of the carrier vehicle. Variability of the obtained result in carcinogenicity testing relates to the species, strain, and sex of the animals tested and also to the vehicle in which C. has been administered orally in each study.

C. may induce a tumor through a nongenotoxic mechanism. In the NCI study (1976), doses of 90 and 180 mg/kg BW were administered in corn oil by gavage to Osborne-Mendel rats and $B6C3F_1$ mice for 78 weeks. The treatment resulted in kidney epithelial tumors in male rats (8% in the low-dose, 24% in the high-dose group, 0% in the control). In another NCI study, male mice received 150 and 300 mg/kg BW, and female mice received doses of 250 and 500 mg/kg BW for 78 weeks. A statistically significant incidence of hepatocellular carcinomas in all the treated groups was found.[13] Roe et al.,[14] observed malignant and benign kidney tumors in male ICI mice given not less than 60 mg/kg dose by gavage in a toothpaste base or in arachis oil for 80 weeks. Jorgenson et al.,[15] found a dose-dependent increased incidence of renal tubular adenomas and adenocarcinomas in male rats (a concentration of up to 160 mg C./l in drinking water). Meanwhile, according to the data reported by the EPA Health Effects Laboratory (1985), 1800 ppm C. in drinking water caused no carcinogenicity promotion in Swiss mice. *Carcinogenicity classification.* IARC: Group 2B; USEPA: Group B2.

Chemobiokinetics. C. is readily and rapidly absorbed from the GI tract. More than 90% of ^{14}C-chloroform administered orally in olive oil to mice, rats, and monkeys (60 mg/kg BW) was absorbed in 48 h. Accumulation occurs in the fatty tissues and liver. Lesser amounts were found in the blood, brain, kidney, etc. Animals demonstrated a marked species difference in C. metabolism.[16]

Carbon dioxide seems to be the major metabolite in humans. According to other data, hepatotoxicity of C. is due, at least in part, to its metabolite **phosgene.**[17] Cytochrome P-450 in rat liver microsomes metabolizes C. to $COCl_2$ by a rate-determining oxidation of the carbon-hydrogen bond of C. to form presumably **trichloromethanol** ($HOCCl_3$). Formation of CO_2 may be a sequence of C. degradation to **methylene chloride** and then to **formaldehyde, formic acid,** and CO_2.[18] C. has been shown to produce other severe reactive metabolic intermediates in the GI tract. Almost the whole oral dose (0.1 to 1 g) administered to humans was excreted in the form of CO_2 or unchanged via the lungs.[19]

Guidelines. WHO (1992). Guideline value for drinking water: 0.2 mg/l.

Standards. USEPA (1991). MCL: 0.1 mg/l. **Russia** (1988). MAC: 0.06 mg/l.

Regulations. USFDA (1993) listed C. as a component of adhesives used as components of articles intended for use in packaging, transporting, or holding food.

References:

1. Gosselin, R. E., Hodge, H. C., Smith, R. P., et al., *Clinical Toxicology of Commercial Products, Acute Poisoning,* 4th ed., Williams & Wilkins, Baltimore, MD, 1976.
2. USEPA, *Health Assessment Document for Chloroform,* Final Report, EPA-600/8–84–004F, Office of Research and Development, Research Triangle Park, NC, 1985.
3. Cahier de Notes Documentaries, Inst. Nat. Rech. Secur., No. 26, Suppl., 1987, 87, (in French).
4. Munson, A. E., Sain, L. E., Sanders, V. M., et al., Toxicology of organic drinking water contaminants: trichloromethane, bromodichloromethane, dibromochloromethane and tribromomethane, *Environ. Health Perspect.,* 46, 117, 1982.
5. Hogan, M. D., Chi, P., Hoel, D. G., et al., Association between chloroform levels in finished drinking water supplies and various site-specific cancer mortality rates, *J. Environ. Pathol. Toxicol.,* 2, 873, 1979.
6. Wilkins, J. R., Reiches, N. A., and Kruse, C. W., Organic chemical contaminants in drinking water and cancer, *Am. J. Epidemiol.,* 110, 420, 1979.
7. Heywood, R., Sortwell, R. J., Noel, P. R. B., et al., Safety evaluation of toothpaste containing chloroform. III. Long-term study in beagle dogs, *J. Environ. Pathol. Toxicol.,* 2, 835, 1979.
8. Palmer, A. K., Street, A. E., Roe, F. J. C., et al., Safety evaluation of toothpaste containing chloroform. II. Long-term studies in rats, *J. Environ. Pathol. Toxicol.,* 2, 821, 1979.
9. Klaunig, J. E., Ruch, R. J., and Pereira, M. A., Carcinogenicity of chlorinated methane and ethane compounds administered in drinking water to mice, *Environ. Health Perspect.,* 69, 1986.
10. Thompson, D. J., Warner, S. D., and Robinson, V. B., Teratology studies on orally administered chloroform in the rat and rabbit, *Toxicol. Appl. Pharmacol.,* 29, 348, 1974.
11. Agustin, J. S. and Lim-Syliano, Mutagenic and clastogenic effects of cloroform, *Bull. Philos. Biochem. Soc.,* 1, 17, 1978.
12. Land, P. C., Owen, E. L., and Linde, H. W., Morphologic changes in mouse spermatogen after exposure to inhalation anesthetics during early spermatogenesis, *Anesthesiology,* 54, 53, 1981.
13. National Cancer Institute, *Carcinogenesis Bioassay of Chloroform,* NTIS PB-264018, Springfield, VA, 1976.
14. Roe, F. J. C., Palmer, A. K,, Worden, A. N., et al., Safety evaluation of toothpaste containing chloroform. I. Long-term studies in mice, *J. Environ. Pathol. Toxicol.,* 2, 799, 1979.
15. Jorgenson, T. A., Meierhenry, E. F., Rushbrook, C. J., et al., Carcinogenicity of chloroform in drinking water to male Osborne-Mendel rats and female B6C3F₁ mice, *Fundam. Appl. Toxicol.,* 5, 760, 1985.
16. Brown, D. M., Langley, P. F., Smith, D., et al., Metabolism of chloroform. I. The metabolism of ^{14}C-chloroform by different species, *Xenobiotica,* 4, 151, 1974; revised in USEPA, 1985.
17. Pohe, L. K., George, J. W., Martin, J. L., et al., Deuterium isotope effect in *in vivo* bioactivation of chloroform to phosgene, *Biochem. Pharmacol.,* 28, 561, 1979.
18. Rubinstein, D. and Kanics, L., The conversion of carbon tetrachloride and chloroform to carbon dioxide by rat liver homogenates, *Can. J. Biochem.,* 42, 1577, 1964; reviewed in USEPA, 1985.
19. Fry, B. J., Taylor, T., and Hathway, D. E., Pulmonary elimination of chloroform and its metabolite in man, *Arch. Int. Pharmacodyn.,* 196, 98, 1972.

1,2-DICHLOROETHANE (1,2-DCE) (CAS No 107–06–2)

Synonyms. Ethane dichloride; Ethylene dichloride (EDC); Glycol dichloride.

Properties. Colorless liquid with an odor reminiscent of chloroform. Water solubility is 8820 mg/l (20°C). Odor perception threshold is 2 to 3 mg/l; practical threshold is 5 mg/l.[1]

Applications and **Exposure.** Used mainly in the production of vinylidene chloride and other plastic intermediates, for cleaning, in paints, coatings, and adhesives. It also may be used as a solvent in PVC production. Concentrations found in drinking water in the U.S. reached 0.006 mg/l.

Acute Toxicity. *Humans.* Clinical symptoms of acute poisoning by ingestion include general weakness, nausea, dizziness, headache, vomiting of blood and bile, etc.,[2] and also unconsciousness, mental disorders, and cerebral and extrapyramidal disorders.[3] Death is most often attributed to circulatory and respiratory failure.[4] *Animals.* LD_{50} was reported to be 680 to 770 mg/kg BW in rats and 860 mg/kg BW in rabbits.[1,06] After administration, liver damage, myocardial edema, and damage to the coronary vessels were reported. Poisoned animals displayed CNS impairment and multiple hemorrhages. Gross pathological examination revealed dystrophic changes, mainly in the liver, but also in the kidney and other organs.

Short-term Toxicity. In a 13-week study, B6C3F$_1$ mice and Fisher 344/N rats were given drinking water containing **1,2**-DCE. There was an increase in the liver weights in mice, although histological lesions were not observed. The NOAELs were found to be 120 mg/kg BW (male rats) or 150 mg/kg BW (female rats); 780 mg/kg BW (male mice) or 2500 mg/kg BW (female mice), this based on mortality. Nine out of ten female mice exposed to **1,2**-DCE concentration of 8000 ppm in drinking water died before the end of the study.[5] Rats died after repeated oral administration of the dose of 300 mg/kg BW. This dose produced necrosis and fatty changes in the liver. No effects were observed in rats when the chemical was given orally at a 10 mg/kg BW dose for 90 d or at a 150 mg/kg BW dose for 2 weeks.[6]

Reproductive Toxicity. In a multigeneration reproduction study, male and female ICR Swiss mice were given the doses of 5 to 50 mg/kg via their drinking water. The treatment did not result in reproductive effects, as measured by fertility, gestation, viability, or lactation indices, pup survival and weight gain. No statistically significant dose-related developmental effects were observed as indicated by the incidence of fetal visceral or skeletal abnormalities.[7] No effect on the adult generations was reported after 25 weeks of dosing as measured by BW gain, fluid intake, or gross pathological examination. **1,2**-DCE is capable of crossing the placental barrier in pregnant rats.

Genotoxicity. **1,2**-DCE is weakly or not mutagenic in the *Salmonella* microsome assay system and in DNA polymerase-deficient *E. coli;*[8,9] it does not induce sex-linked recessive lethals in *Dr. melanogaster.*[10] A weak mutagenic effect was noted in a spot test in mice. Negative results were obtained in one DLM assay and two micronucleus assays in mice.[6]

Carcinogenicity. Epidemiological data have not established carcinogenicity of **1,2**-DCE, but in mice and rats, it was shown to increase the incidence of several types of tumors. It was not shown to be carcinogenic through inhalation exposure in animals. **1,2**-DCE induced circulatory system hemangiosarcomas in male Osborne-Mendel rats given oral doses of 97 or 195 mg/kg BW (males), and 149 or 299 mg/kg BW (females) over a period of 78 weeks.[11] *Carcinogenicity classification.* USEPA: Group B2; IARC: Group 2B; NTP: N—E—N—E.

Chemobiokinetics. **1,2**-DCE is readily absorbed from the GI tract. After oral administration, the adipose tissues, liver, and kidney seem to have the highest concentrations of **1,2**-DCE.[12] Dechlorination of **1,2**-DCE takes place in the presence of liver oxidase. It is metabolized into 2-**chloroethanol**.[13] Of the radioactivity of a single oral dose of 150 mg/kg, 96% was eliminated within 48 h after dosing.[9] The major part of the absorbed chemical is excreted rapidly via the urine, mainly as glutathione conjugates, and via the lungs, as carbon dioxide or the unchanged compound.

Guidelines. WHO (1992). Guideline values for drinking water: 0.03 mg/l. EEC (1992). Banned to certain uses owing to its effects on health and the environment.

Standards. USEPA (1991). MCL: 0.005 mg/l; MCLG/: zero. **Russia** (1988). MAC: ·0.02 mg/l.

References:

1. See **BISPHENOL A,** #16, 156.
2. McNally, W. D. and Fostvedt, G., Ethylene dichloride poisoning, *Ind. Med.,* 10, 373, 1941.

3. Akimov, G. A. et al., Neurological disorders in acute dichloroethane poisoning, *J. Neuropathol. Psychiatr.*, 78, 687, 1978 (in Russian).
4. Chesnokov, N. Y., Acute dichloroethane poisoning, *Vrachebnoye Delo,* 6, 127, 1976 (in Russian).
5. NTP, *Toxicity Studies of **1,2**-DCE in F344/N Rats, Sprague-Dawley Rats, Osborne Mendel Rats and B6C3F$_1$ Mice (Drinking Water and Gavage Studies),* NTP Tox. 4, NIH Publ. No 91–3123, 1991, 54.
6. IPCS, ***1,2**-Dichloroethane,* Health and Safety Guide, WHO, Geneva, 1991, 33.
7. Lane, R. W., Riddle, B. L., and Borcelleca, J. F., Effect of **1,2**-dichloroethane and **1,1,1**-trichloroethane in drinking water on reproduction and development in mice, *Toxicol. Appl. Pharmacol.*, 63, 409, 1982.
8. Brem, H., Stein, A., and Rozenkranz, C., The mutagenicity and DNA-modifying effect of haloalkanes, *Cancer Res.*, 34, 2576, 1974.
9. McCann, J., Simmon, V., Streitweisser, D., et al., Mutagenicity of chloroacetaldehyde, *Proc. Natl. Acad. Sci. U.S.A.*, 72, 3190, 1975.
10. Rapoport, I. A., The reactions of genic protein with 1,2-DCE, *Ser. Biol. Sci.*, 134, 745, 1960 (in Russian).
11. National Cancer Institute, *Bioassay of **1,2**-DCE for Possible Carcinogenicity,* NCI-CG-TR-55.
12. Reitz, R. H., Fox, T. R., Domoradzki, J. Y., et al., Pharmacokinetics and macromolecular interactions of ethylene dichloride, in *Ethylene Dichloride: A Potential Health Risk?,* Banbury report No. 5, Cold Spring Harbor Laboratory, Cold Spring Harbor, NY, 1980, 135.
13. Kokarovtzeva, M. G. and Kiseleva, N. I., Chloroethanol (ethylene chlorohydrin)—toxic metabolite of **1,2**-dichloroethane, *Pharmacol. Toxicol.*, 1, 118, 1978 (in Russian).

1,2-DICHLOROETHENE (CAS No 540–59–0)

Synonyms. Acetylene dichloride; **1,2**-Dichloroethylene; **1,2**-DCE.

Properties. Clear, colorless liquids. Water solubility is 6.3 mg/l (25°C, *trans*-**1,2**-DCE), and 3.5 mg/l (20°C, *cis*-**1,2**-DCE). Odor perception threshold is 0.26 mg/l (*trans*-**1,2**-DCE).[02]

Applications and **Exposure.** Solvent and chemical intermediate. Occurs as an impurity in commercial grade 1,1-**dichloroethylene** (vinylidene chloride). Due to their volatility and limited use, levels of either *cis*- or *trans*-**1,2**-DCE in food are expected to be negligible (USEPA, 1983).

Acute Toxicity. In rats, LD_{50} of isomer mixture is 770 mg/kg.[06] LD_{50} of *trans*-**1,2**-DCE is 1300 mg/kg BW.[1] Single doses of 400 and 1500 mg *cis*-**1,2**-DCE/kg BW but not of *trans*-**1,2**-DCE administered to rats produced significant elevations of liver alkaline phosphatase.[2]

Repeated Exposure. The doses of 21 and 210 mg *trans*-**1,2**-DCE/kg BW were administered by gavage to male CD-1 mice for 14 d. No changes in BW gain, the content of some serum enzymes, or blood urea nitrogen were noted. However, at the 210 mg/kg BW dose-level, fibrinogen level and prothrombin times were significantly decreased.[3]

Short-term Toxicity. CD-1 mice received *trans*-**1,2**-DCE for 90 d in their drinking water at dose-levels of 17, 175, and 387 mg/kg BW (males) or 23, 224, and 452 mg/kg BW (females). There were no changes in fluid consumption, BW gain, or gross pathology among the experimental animals. In male mice, significant increases in serum alkaline phosphatase were noted at the two highest dose-levels. In females, the thymus weight was significantly depressed at 224 and 452 mg/kg BW doses. The NOAEL of 17 mg/kg BW was identified based on normal serum chemistry values in male mice.[3] It was proposed (USEPA, WHO) that the value calculated for the *trans*-isomer be used for the *cis*-isomer also.

Immunotoxicity. No significant immunological effects were noted in mice exposed by gavage to 22 or 220 mg/kg BW for 14 d.[4] Administration of 175 and 387 mg/kg but not 17 mg *trans*-**1,2**-DCE/kg BW caused a significant decrease in antibody-forming cells of the spleen only in male mice.[5]

Genotoxicity. No mutagenic effect was noted in the *Salmonella* test, neither the *cis*- or *trans*-isomer of **1,2**-DCE-induced CA or SCE in the Chinese hamster lung fibroblast cell line.[6] ***Carcinogenicity classification.*** IARC: Group 3; USEPA: Group D.

Chemobiokinetics. Neutral, lipid-soluble substances are expected to be absorbed readily following oral or dermal exposure (USEPA, 1984). DCE are metabolized into epoxides, which can yield **dichloroacetaldehyde, dichloroethanol,** and **dichloroacetic acid.**[7,8]

Guidelines. WHO (1992). Guideline value for drinking water: 0.05 mg/l.

Standards. USEPA (1991). MCL and MCLG for *cis*-1,2- DCE: 0.07 mg/l; MCL and MCLG for *trans*-1,2-DCE: 0.1 mg/l.

Regulations. As outlined in the U.S. Vinyl Chloride Rule (52 Federal Register 25690) water supplies must test for *vinyl chloride* whenever the 1,2-DCE are found. The *vinyl chloride* MCL will adequately protect the public against any vinyl chloride that may be produced through the biodegradation of the 1,2-DCE.

References:

1. Freundt, J. J., Liebaldt, G. P., and Lieberwirth, E., Toxicity studies on *trans*-**1,2**-dichloroethylene, *Toxicology,* 7, 141, 1977.
2. Jenkins, L. J., Trabulus, M. J., Murphy, S. D. Biochemical effects of **1,1**-dichloroethylene in rats: comparison with carbon tetrachloride and **1,2**-dichloroethylene, *Toxicol. Appl. Pharmacol.,* 23, 501, 1972.
3. Barness, D. W., Sanders, V. M., White K. L., et al., Toxicology of *trans*-**1,2**-dichloroethylene in the mouse, *Drug Chem. Toxicol.,* 8, 373, 1985.
4. Munson, A. E., Saunders, V. M., Douglas, L. E., et al., *In vivo* assessment of immunotoxicity, *Environ. Health Perspect.,* 43, 41, 1982.
5. Shopp, G. M., Sanders, V. M., White, K. L., et al., Humoral and cell-mediated immune status of mice exposed to *trans*-**1,2**-dichloroethylene, *Drug Chem. Toxicol.,* 8, 393, 1985.
6. Sawada, M., Sofuni, T., and Ishidate, M. Cytogenetic studies on **1,1**-dichloroethylene and its two isomers in mammalian cells *in vitro* and *in vivo*, *Mutat. Res.,* 187, 157, 1987.
7. Henschler, D., Metabolism and mutagenicity of halogenated olefins: a comparison of structure and activity, *Environ. Health Perspect.,* 21, 61, 1977.
8. Costa, A. K., The chlorinated ethylenes: Their hepatic metabolism and carcinogenicity, *Diss. Abstr. Int. [B],* 44, 1797, 1983.

DICHLOROMETHANE (DCM) (CAS No 75–09–2)

Synonyms. Methane dichloride; Methylene chloride.

Properties. DCM occurs as a colorless liquid at room temperature. Water solubility is 20 g/l, miscible with alcohols and oils at all ratios. Odor perception threshold is 7.5 or 9.1 mg/l;[02] taste perception threshold is 15 mg/l. According to other data, organoleptic threshold is 5.6 mg/l.[01]

Applications and **Exposure.** Used in the production of cellulose esters, resins, and rubber. A substituent of formaldehyde in plastics production. Used also in the manufacturing of polyurethanes. DCM has been used as an extraction solvent in food processing.

Acute Toxicity. A dose of 7.5 g/kg BW is very close to LD_{100}.[1] LD_{50} in rats is approximately 2.1 to 3 g/kg BW.[2] In mice, it is 5.5 g/kg BW. Doses of about 2.1 to 2.15 g/kg BW are lethal for rabbits;[3] doses of 3 to 5 g/kg BW are lethal for dogs. Other data[4] suggested that LD_{50} is 1.25 g/kg in rats, 1 g/kg in mice, and 2 g/kg BW in rabbits. Poisoned animals exhibit prolonged excitation, with subsequent ataxia and CNS inhibition. Animals experienced clonic convulsions. Death occurred from respiratory failure. Gross pathological examination revealed no abnormalities other than cerebral hyperemia and liver dystrophy (in some animals).

Repeated Exposure. Mice tolerated ten administrations of 750 mg/kg. Gross pathological examination revealed dystrophic changes in the liver.

Short-term Toxicity. Wistar rats were exposed to 15 mg/kg BW in drinking water (assuming daily drinking water consumption rates of 12 ml/100 g) for 13 weeks. No treatment-related effects were observed.[5] Guinea pigs and rats were dosed by gavage with 0.4 and 377 mg/kg BW for 5 and 6 months, respectively. At the higher dose-level, only ascorbic acid content in the suprarenals was found to be altered in guinea pigs, but no other functional or structural changes were reported.[6]

Long-term Toxicity. In a 2-year study in Fisher 344 rats given doses of 5 to 250 mg/kg in their drinking water, the NOAEL of 6.5 mg/kg BW was identified based on the absence of effects on BW gain, hematological parameters, and histopathological changes in the liver.[7] Doses 5 to 250 mg/kg BW were administered in deionized water to Fisher 344 rats for 104 weeks. An additional group received 250 mg/kg BW for 78 weeks followed by a 26-week recovery period. Doses of 125 and 250 mg/kg affected BW and water and food consumption and produced histomorphological hepatic changes (cellular alterations and fatty change). Under the experimental conditions of this study, the NOAEL appeared to be 5 mg/kg BW.[7] The dose of 250 mg/kg BW given in drinking water to B6C3F$_1$ mice for 104 weeks produced hepatocellular alterations (increased fat content in the liver). In this study, the NOAEL was identified to be 185 mg/kg BW.[8]

Reproductive Toxicity. On inhalation exposure, no adverse effects were reported. Exposure did not induce visceral malformations in fetal mice or rats. In rats, postnatal behavioral development was affected by prenatal exposure to DCM.[9] Single doses caused maternal toxicity.[10]

Genotoxicity. No data were available on the genetic and related effects of DCM in humans (IARC). DCM did not induce unscheduled DNA synthesis in human cells *in vitro*. DCM was positive in the *Salmonella* mutagenicity assay[11] and has been shown to transform rat embryo cells and to enhance viral transformation of Syrian hamster embryo cells.[12] It did not induce CA in bone marrow cells of rats or micronuclei in mice treated *in vivo*. CA, but neither mutations nor DNA damage in rodent cells, was observed *in vitro* (IARC, Suppl. 7–195). DCM failed to increase the frequency of either SCE or CA in mouse bone marrow cells following *i/p* exposures to 100 to 2000 mg/kg BW.[13]

Carcinogenicity. Carcinogenicity studies on DCM have yielded inconsistent and contradictory results. *Humans.* An epidemiological study failed to show a positive correlation between DCM exposure and increased cancer incidence.[14,15] Good human epidemiology of workers exposed for many years to DCM (inhalation) did not justify the validity for humans of results obtained in B6C3F$_1$ mice.[16] *Animals.* In a 2-year study, Fisher 344 rats were exposed to 50 and 250 mg/kg BW in their drinking water. Hepatological changes detected in the target dose groups (both sexes) included an increased incidence of foci/areas of cellular alteration. Fatty liver changes were noted at 125 and 250 mg/kg BW dose-levels after 78 and 104 weeks of treatment. DCM did not induce carcinogenic effect.[7] USEPA (1985) considered the dose of 250 mg/kg BW to be a borderline for carcinogenicity in this study. *Carcinogenicity classification.* NTP: SE—CE*—CE*—CE*.

Chemobiokinetics. DCM is expected to be completely absorbed when ingested in mice and rats.[17,18] It is primarily distributed to the liver and fat.[19] The main metabolites are **carbon monoxide,** formed via the cytochrome P-450 system, and CO_2, formed via GSH metabolism of either the parent compound or carbon monoxide. ^{14}C-DCM metabolites did not bind to the RNA or DNA of rat hepatocytes, although radioactivity was associated with lipid and protein.[20] Syrian hamsters metabolize DCM more slowly than mice. According to other data, species such as hamsters and humans having much lower rates of DCM metabolism via DCM-protein cross-links may not generate toxicologically significant concentrations of **formaldehyde** and **DNA-protein cross-links.**[21] Excretion occurs primarily via the lungs.

Standards. WHO (1992). Guideline value for drinking water: 0.02 mg/l. **USEPA** (1991). Proposed MCL: 0.005 mg/l, MCLG: 0. **Russia** (1988). MAC and PML: 5 mg/l (organolept.).

Regulations. FAO/WHO (1983) withdrew the previously allocated temporary ADI of 0 to 0.5 mg/kg and recommended that the use of DCM as an extraction solvent be limited in order to ensure that its residues in food are as low as practicable. **USFDA** (1993) approved the use of DCM in adhesives used as components of articles intended for use in packaging, transporting, or holding food.

References:
1. See **BUTYLENE,** 41.
2. See **ACETONITRILE,** #4.
3. See **BENZYL ACETATE,** #3.
4. *Methylene Chloride,* Environ. Health Criteria No. 32, WHO, Moscow, 1988, 55 (in Russian).
5. Bornmann, G. und Loeser, A., Zur Frage einer chronish-toxischen Wirkung von Dichlorome-than, *Z. Lebensm.-Untersuch. Forschung.,* 136, 14, 1967 (in German).
6. *Hygiene Problems in the Production and Use of Polymeric Materials,* Erisman Research Hygiene Institute, Moscow, 1969, 41 (in Russian).
7. Serota, D. G., Thakur, A. K., and Ulland, B. M. A two-year drinking-water study of dichloro-methane in rodents. I. Rats, *Food Chem. Toxicol.,* 24, 951, 1986.

8. Serota, D. G., Thakur, A. K., Ulland, B. M., et al., A two-year drinking-water study of dichloromethane. II. Mice, *Food Chem. Toxicol.,* 24, 959, 1986.
9. Hatch, G. G., Conclin, P. M., Christensen, C. I., et al., Chemical enhancement of viral transformation in Syrian hamster embryo cells by gaseous and volatile chlorinated methanes and ethanes, *Cancer Res.,* 43, 1945, 1983.
10. See **DICHLOROMETHANE,** #10.
11. Green, T., Proven, W. M., Collinge, D. C., et al., Macromolecular interactions of inhaled methylene chloride in rats and mice, *Toxicol. Appl. Pharmacol.,* 93, 1, 1988.
12. Price, P. J., Hassett, C. M., and Mansfield, J. I., Transforming activities of trichloroethylene and proposed industrial alternatives, *In Vitro,* 14, 290, 1978.
13. Westbrook-Collins, B., Allen, J. W., Shariet, Y., et al., Further evidence that dichloromethane does not induce chromosome damage, *J. Appl. Toxicol.,* 10, 79, 1990.
14. Friedlander, B. R., Hearne, F. T., Hall, S. Epidemiologic investigation of employees chronically exposed to methylene chloride, *J. Occup. Med.,* 20, 657, 1978.
15. Ott, M. G., Skory, L. K., Holder, B. B., et al., Health evaluation of employees occupationally exposed to methylene chloride. General study design and environmental considerations, *Scand. J. Health,* 9 (Suppl. 1), 1, 1983.
16. Hearne, F. T., Health risk assessment, *Reg. Toxicol. Pharmacol.,* 17, 219, 1993.
17. Angelo, M. J., Pritchard, A. B., Hawkin, D. R., et al., The pharmacokinetics of dichloromethane. II. Disposition in Fischer 344 rats following intravenous and oral administration, *Food Chem. Toxicol.,* 24, 975, 1986.
18. Angelo, M. J., Pritchard, A. B., Hawkin, D. R., et al., The pharmacokinetics of dichloromethane. I. Disposition in B6C3F$_1$ mice, *Food Chem. Toxicol.,* 24, 965, 1986.
19. McKenna, M. J. and Zempel, J. A., The dose-dependent metabolism of [^{14}C]dichloromethane chloride following oral administration to rats, *Food Cosmet. Toxicol.,* 19, 73, 1981.
20. Cunningham, M. L., Gandolfi, A. J., Brendel, K. M., et al., Covalent binding of halogenated volatile solvents to subcellular macromolecules in hepatocytes, *Life Sci.,* 29, 1207, 1981.
21. Casanova, M., Deyo, D. F., Heck, H. d'A., Dichloromethane (methylene chloride): metabolism to formaldehyde and formation of DNA-protein cross-links in B6C3F$_1$ mice and Syrian golden hamsters, *Toxicol. Appl. Pharmacol.,* 114, 162, 1992.

DIETHYLENE GLYCOL, DIVINYL ETHER (DEGDE) (CAS No 764–99–8)

Synonym. Diethyleneglycol bis(2-vinyloxyethyl) ether.

Properties. Colorless, transparent liquid with a specific odor. Readily soluble in water. Odor perception threshold is 1.99 mg/l; taste perception threshold is slightly higher. It does not alter the transparency or color of water and forms no foam.

Acute Toxicity. LD$_{50}$ is 6.39 g/kg BW in rats and 2.57 g/kg BW in mice. On autopsy, there were congestion in the visceral organs and dystrophic changes in the renal tubular epithelium, liver, and brain.

Repeated Exposure revealed slight cumulative properties. The liver, kidneys, and NS are the target organs for DEGDE toxicity. Severity and nature of the toxic effect are similar to that of 2-methyl-**1,3**-dioxolane (q.v.). Morphological changes are not observed.

Chemobiokinetics. DEGDE metabolism occurs via hydrolysis to form **diethylene glycol,** which is then excreted with the urine.

Standards. Russia (1988). MAC and PML: 1 mg/l (organolept., odor).

Reference:

Buzina, L. Z. and Rudi, F. A., Substantiation of MAC for glycol vinyl ethers in water bodies, *Gig. Sanit.,* 3, 12, 1977 (in Russian).

DIETHYLENE GLYCOL, MONOBUTYL ETHER (DEGBE) (CAS No 101–76–2)

Synonyms. 2-(2-Butoxyethoxy) ethanol; Butylcarbitol.

Properties. Liquid with a distinct odor of ether. Soluble in water, oils, and organic solvents. Threshold concentration for the change in organoleptic properties of water is 0.8 mg/l.[01]

Applications. Used as a cellulose nitrate solvent, in the synthesis of resins and plasticizers, and in the production of lacquers.

Acute Toxicity. LD_{50} is 4.5 g/kg BW in rats and 6 g/kg BW in mice.

Repeated Exposure revealed moderate cumulative properties.[1] K_{acc} is 4.4 (by Lim).

Allergenic Effect is not observed on oral administration of **DEG monomethyl ester.**[2]

Reproductive Toxicity. The doses of 250 to 1000 mg DEGBE/kg BW were given over a 60-d period prior to mating to male rats, and from the 14th day prior to mating until day 13 or the weaning of the offspring to females. No adverse effect on fertility, embryos, fetuses or neonates was noted.[3]

Genotoxicity. Shows no mutagenic activity in a set of *in vitro* tests and in experiments on *Dr. melanogaster.*[4]

Regulations. USFDA (1993) approved the use of DEMBE (1) as a component of adhesives intended for use in contact with food and (2) as a component of the uncoated or coated food-contact surface of paper and paperboard that may be used safely in producing, manufacturing, packing, transporting, or holding dry food.

References:

1. Krotov, Yu. A., Lykova, A. S., Skachkov, M. A., et al., Sanitary and hygienic characteristics of diethylene glycol ethers (Carbitols), with special reference to air pollution control, *Gig. Sanit.,* 2, 14, 1981 (in Russian).
2. Pastushenko, T. V., Golka, N. V., Kondratyuk, V., et al., Study of the skin irritant and sensitizing effect of diethylene glycol monomethyl ether, *Gig. Sanit.,* 10, 81, 1985 (in Russian).
3. Nolen, G. A., Gibson, W. B., Benedict, J. H., et al., Fertility and teratogenic studies of diethylene glycol monobutyl ether in rats and rabbits, *Fundam. Appl. Toxicol.,* 5, 1137, 1985.
4. Thompson, E. D., Coppinger, W. J., Valencia, R., et al., Mutagenicity testing of diethylene glycol monobutyl ether, *Environ. Health Perspect.,* 57, 105, 1984.

DIETHYLENE GLYCOL, MONOETHYL ETHER (DGEE) (CAS No 111–90–0)

Synonyms. Carbitol; Carbitol cellosolve; Dowanol; **2-(2-**Ethoxyethoxy) ethanol; Ethylcarbitol.

Properties. DGEE occurs as a colorless or slightly yellowish, very hygroscopic liquid having a characteristic ether odor. Possesses the properties of both alcohol and ether. Readily soluble in water and alcohol. Odor perception threshold is <0.21 mg/l[1010] or 8.75 mg/l; practical threshold is 24.3 mg/l. Heating increases odor intensity by 4.5 times. Chlorination of aqueous solutions does not lead to odor intensification or to the appearance of additional odors.[1]

Applications. A solvent for nitrocellulose lacquers and for resins used in cosmetics.

Acute Toxicity. LD_{50} is reported to be 6.3 to 7.8 g/kg BW in rats, 3.9 to 7.2 g/kg BW in mice, and 3 g/kg BW in guinea pigs. Poisoned animals displayed CNS functional disorder with paresis and paralysis of the extremities. High doses affected the urinary tract. Mice and guinea pigs die on days 1 to 2, rats on days 1 to 3.

Repeated Exposure. DGEE exhibited pronounced[1] or moderate[2] cumulative properties. Rats were dosed by gavage with 30 to 750 mg/kg for a month. The treatment caused marked signs of intoxication: retardation of BW gain, damage of hemopoiesis, and oxidation-reduction processes and changes in the visceral organs.

Long-term Toxicity. Mice were administered the doses of up to 75 mg/kg BW for 6 months. Blood analyses were changed in the 0.75 mg/kg BW and higher dose groups. The high-dosed animals developed reduced glucose concentration in blood. Pathological changes at the mid and high doses included increased urinary levels of chlorides and proteins and a reduced amount of creatinine. Gross pathological examination revealed morphological changes in the visceral organs and a reduction in their relative weights (at the dose of 75 mg/kg BW only).[1]

Reproductive Toxicity. *Gonadotoxicity.* The dose of 30 mg/kg BW appears to be ineffective in rats. ***Embryotoxic*** effect was not found to develop following administration of 75 and 750 mg/kg BW doses. ***Teratogenic*** effect was not observed in rats after inhalation and dermal application.[3]

Genotoxicity. DGEE was positive in the *Salmonella* mutagenicity assay (1 ml/plate), but not in the micronuclear test.[4]

Chemobiokinetics. In rabbits, excretion occurs in the form of glucuronides.

Standards. Russia. PML: 0.3 mg/l.

Regulations. USFDA (1993) approved the use of DGEE (1) as a component of adhesives intended for use in contact with food and (2) as a component of the uncoated or coated food-contact surface of paper and paperboard that may be used safely in producing, manufacturing, packing, transporting, or holding dry food.

References:
1. Kondratyuk, V. A., Sergeta, V. N., Pis'ko, G. T., et al., Experimental derivation of a maximum allowable concentration for the monoethyl ether of diethylene glycol in water bodies, *Gig. Sanit.*, 4, 74, 1981 (in Russian).
2. See **DIETHYLENE GLYCOL, MONOBUTYL ETHER,** #1.
3. Nelson, B. K., Setzer, J. V., Brightwell, W. S., et al., Comparative inhalation teratogenicity of four industrial glycol ether solvents in rats, *Teratology*, 25, 64A, 1982.
4. Berte, F., Bianchi, A., Gregotti, C., et al., *In vivo* and *in vitro* toxicity of carbitol, *Bull. Chim. Farm.*, 125, 401, 1986.

DIETHYLENE GLYCOL, MONOVINYL ETHER (DEGVE) (CAS No 929–37–3)

Synonym. Vinylcarbitol.

Properties. Colorless, transparent liquid with a specific odor. Readily soluble in water. Odor perception threshold is 1.83 mg/l; taste perception threshold is slightly higher. DEGVE does not alter transparency or color of water and forms no foam.

Acute Toxicity. LD_{50} is reported to be 4.93 g/kg BW in rats and 4.45 g/kg BW in mice.

Repeated Exposure. See **DEGDVE.**

Standards. Russia (1988). MAC and PML: 1 mg/l (organolept., odor).

Reference:
See **DIETHYLENE GLYCOL, DIVINYL ETHER.**

DIETHYL ETHER (DE) (CAS No 60–29–7)

Synonyms. Ethyl ester; Sulfuric ester.

Properties. DE occurs as a colorless, highly volatile liquid. Solubility in water is 77.7 g/l at 20°C or 56 g/l at 25°C. Readily soluble in alcohol. Odor perception threshold is 0.3 mg/l or 0.75 mg/l.[02]

Acute Toxicity. In mice, LD_{50} is 1.76 g/kg BW. Poisoning is accompanied by narcosis (at DE blood concentration of 100 to 140 mg%), adynamia, and motor coordination disorder.

Repeated Exposure resulted in the rapid development of habituation.

Long-term Toxicity. In a 6-month study, rats were given orally 0.2, 5, and 50 mg/kg BW; guinea pigs received 0.2 and 5 mg/kg BW. At higher doses, effect on the serum protein fractions ratio and conditioned reflex activity was found to develop. Gross pathological examination revealed catarrhal, desquamatous gastritis and parenchymatous dystrophy in some animals.

Genotoxicity. There are no data available on the genetic and related effects of DE in humans. DE does not induce SCE in cultured Chinese hamster ovary cells and is not mutagenic to fungi and bacteria (IARC 11–285).

Standards. Russia. PML in drinking water: 0.3 mg/l (organolept., odor).

Reference:
See **ACRYLIC ACID,** #1, 148.

DIGLYCIDYL RESORCINOL ETHER (DGRE) (CAS No 101–90–6)

Synonyms. *m*-Bis(2,3-epoxypropoxy)benzene; Bisglycidyloxybenzene; Resorcinol glycidyl'ether.

Properties. A straw-yellow liquid with a phenolic odor. Miscible with acetone, chloroform, and methyl alcohol.

Applications. DGRE is used as a liquid epoxy resin and as a reactive diluent in the manufacturing of other epoxy resins; a curing agent for polybisulfide rubber.

Acute Toxicity. No mortality was noted in rats and mice exposed to air that was saturated with DGRE (IARC 11–125). In mice, rabbits, and rats, the LD_{50} values have been found to be 980, 1240, and 2570 mg/kg BW, respectively.[1]

Repeated Exposure. The doses of 750, 1500, and 3000 mg/kg BW were administered to Fisher 344/N rats and B6C3F$_1$ mice in corn oil by gavage for 14 consecutive days: mortality rate was increased in these animals.[2]

Short-term Toxicity. Rats and mice were fed 50, 100, and 200 mg/kg BW for 13 weeks. An increase in mortality was marked in this study as well as hyperkeratosis, basal cell hyperplasia, and squamous cell papillomas of the forestomach.[2]

Long-term Toxicity. In a 2-year oral study, Fisher 344 rats exhibited an increase in the incidence of bronchopneumonia (NTP–85).

Genotoxicity. DGRE has been found to be positive in *Salmonella* mutagenicity assay with and without metabolic activation and in the mouse lymphoma assay. It caused CA and SCE in Chinese hamster ovary cells.[3]

Carcinogenicity. Neoplasms were not found in the strain of mouse lung adenoma model (*i/p* injections of 0.75 g/kg once a week for 16 weeks).[1] Fisher 344 rats were exposed to 12 and 50 mg/kg BW; B6C3F$_1$ mice received 50 and 100 mg/kg BW for 103 weeks. The incidence of neoplastic and nonneoplastic changes of the forestomach was increased.[2] *Carcinogenicity classification.* NTP: P*—P*—P*—P*.

References:

1. Hine, C. H., Guzman, R. J., Coursey, M. M., et al., An investigation of the oncogenic activity of two representative epoxy resins, *Cancer Res.*, 18, 20, 1958.
2. Murthy, A. S. K., McConnell, E. E., Huff, J. E., et al., Forestomach neoplasms in Fisher 344/N rats and B6C3F$_1$ mice exposed to diglycidyl resorcinol ether—an epoxy resin, *Food Chem. Toxicol.*, 28, 723, 1990.
3. See **BISPHENOL A, #6**.

DIMETHYLACETAMIDE (DMAA) (CAS No 127–19–5)

Synonym. Acetic acid, *N,N'*-dimethylamide.

Properties. A colorless liquid with a faint odor. Readily soluble in water, which does not affect its organoleptic properties.

Applications. Used in the synthesis of plastics and semipermeable polyamide membranes for reverse-osmotic water desalination.

Acute Toxicity. LD_{50} values range from 4.2 to 4.85 g/kg BW in mice, and from 4.3 to 5.2 g/kg BW in rats.[1,2] Manifestations of the acute toxic effect included CNS inhibition, convulsions, and paresis in mice.

Repeated Exposure. DMAA is classified as having moderate cumulative properties.[1,2] Rats were dosed with 2 ml/kg BW for 10 d. The treatment caused 75% animal mortality within 6 to 10 d. Gross pathological examination revealed extensive hemorrhages in the stomach and lungs. Rats tolerate 0.5 ml/kg dose (1:5 dilution), ingested for a month, without any signs of intoxication. The 90 and 450 mg/kg BW doses, however, were found to induce functional changes in the CNS and affected the enzyme-forming function of the liver. A reversible BW loss and anemia were noted. Gross pathological examination revealed hepatic and testicular lesions. A 16 mg/kg BW dose appeared to be ineffective.

Long-term Toxicity. The 6-month treatment affected the CNS functions, liver, and blood analysis in rats.[1]

Reproductive Toxicity. *Embryotoxic* effect was observed at a dose level of 20 mg/kg BW in rats. *Teratogenic* response was noted in rats but not in rabbits.[3]

Genotoxicity. DMAA is reported to be negative in the *Salmonella* mutagenicity assay.[4]

Standards. Russia (1988). MAC and PML: 0.2 mg/l.

References:

1. Bogdanov, M. V., Korolev, A. A., Kinzirsky, A. S., et al., Experimental substantiation of MAC for dimethylacetamide in water bodies, *Gig. Sanit.*, 6, 76, 1980 (in Russian).
2. Kreibig, T. U. A., *Arzneimittel-Forsch.*, 19, 1073, 1969 (in German).

3. Johannsen, F. R., Levinskas, G. J., and Schardein, J. L., Teratogenic response of dimethylace-tamide in rats, *Fundam. Appl. Toxicol.,* 9, 550, 1987.
4. See **ZINC PHENYLETHYLDITHIOCARBAMATE,** #3.

N,N'-DIMETHYLFORMAMIDE (DMF) (CAS No 68–12–2)

Synonyms. *N,N'*-Dimethylmethanamide; *N*-Formyldimethylamine.

Properties. DMF contains a significant percentage of monomethylformamide. Clear, colorless liquid. Mixes with water and alcohol at all ratios. Soluble in acetone and chloroform. Odor perception threshold is 50 mg/l.[1]

Applications and **Exposure.** Used as a solvent for acrylic fibers and polyurethane and for many vinyl-based polymers and copolymers intended for use as surface coatings. Human exposure occurs primarily through inhalation and dermal absorption.

Acute Toxicity. *Humans.* Ethanol intolerance is one of the earliest manifestations of excessive exposure to DMF, followed by nausea, vomiting, abdominal pain, and the release of liver cytotoxic enzymes in the plasma.[2] *Animals.* All rats die after administration of 8 g/kg BW, and mice after administration of 5 g/kg BW. According to other data, a single administration of 2.25 g/kg BW to rats causes death due to liver necrosis; LD_{50} is found to be 3 to 7 g/kg BW.[3] LD_{50} is reported to be 3.9 to 6.4 g/kg BW in mice, more than 5 g/kg BW in rabbits,[4,5] and 3 to 4 g/kg BW in guinea pigs (IARC 47–178). Animals developed general depression, anesthesia, loss of appetite and BW, tremors, convulsions, hemorrhage from the nose and mouth, liver injury, and coma immediately preceding death.[6] DMF is more toxic in younger than in older rats, with oral LD_{50} of less than 1 g/kg BW in newborn, 1.4 g/kg BW in 14-d-old, 4 g/kg BW in young adult, and 6.8 g/kg BW in adult animals.[7]

Repeated Exposure. DMF causes dose-related liver injuries in most species tested, including humans.[2] Doses of 620 or 1240 mg DMF/kg BW were given in the diet to mice over a period of 30 d. The treated animals displayed anorexia and loss of BW.[8] In Mongolian gerbils given DMF in drinking water with concentration of 10 g/l for 30 d, no changes in BW, liver, or kidney were reported.[9] Rats received DMF in their drinking water at concentrations of 102 or 497 mg/l for 49 d. Animals exhibited no behavioral changes. Dose-related deviations in cerebral and glial cell enzyme activities were noted.[10]

Short-term Toxicity. Consumption of drinking water containing 17 to 34 g/l DMF for 80 d caused increased mortality due to liver necrosis.[6] Administration of 160 to 1850 mg DMF/kg diet given for 119 d produced a dose-related increase in the liver weights; there were no histological or biochemical changes. The NOEL of 246 to 326 mg/kg was identified in this study.[11] In a 90-d oral study, slight anemia, leukocytosis, and hypercholesterolemia were observed at dose-levels of 60 and 300 mg/kg. The NOEL is reported to be 12 mg/kg BW.[3]

Reproductive Toxicity. DMF is unlikely to be **gonadotoxic** to rats; it produces no effect on fertility. Doses of 0.2 to 2000 mg/kg produce no sperm abnormalities in mice.[12] *Embryotoxicity.* Administration of 193 mg/kg BW by gavage to pregnant mice on days 6 to 15 of pregnancy resulted in fetal weight decrease and malformations.[13] A dose of 580 mg/kg BW caused embryolethality. In rabbits exposed by gavage to the doses of 47 to 68 mg/kg BW on days 6 to 18 of pregnancy, a decreased number of implantations and three cases of hydrocephalus were reported.[14] Rabbits appear to be more sensitive than rats and mice in dermal testing.[15] *Teratogenicity.* Administration of $^1/_{20}$ LD_{50} during pregnancy (460 mg/kg BW over the whole period) caused no changes in the body of mothers but disrupted embryogenesis, leading to abnormalities or fetal death.[6] However, according to other data, oral administration to pregnant animals led to an increased rate of malformations in the absence of overt maternal toxicity.[14] The dose of 200 mg/kg BW caused malformations in fetuses. The NOAEL appeared to be 182 mg/kg BW (mice) and 166 mg/kg BW (rats). According to Thiersch,[6] the treatment with 0.5 to 2 ml/kg BW had no teratogenic effect in experimental rats.

Genotoxicity. *Humans.* An increased frequency of CA was found in peripheral lymphocytes of industrial workers exposed to DMF.[17] *Animals.* DMF was generally found to be inactive, both *in vitro* and *in vivo,* in an extensive set of short-term tests for genetic and related effects.[7,18] It did not cause mutation in mouse lymphoma cells (NTP–86) or CA and SCE in Chinese hamster ovary cells (NTP–85).

Carcinogenicity. *Humans.* An excess risk for testicular germ-cell tumors was identified among workers exposed to a solvent mixture containing 80% DMF.[18] *Animals.* No adequate carcinogenicity studies have been reported. In a 107-week study, BD rats were given the doses of 75 and 150 mg/kg BW until the total dose of 38 g/kg BW had been given. The treatment did not produce any tumorigenic effect.[19] However, IARC considered this study to be inadequate for evaluation (IARC 47–186). *Carcinogenicity classification.* IARC: Group 2B.

Chemobiokinetics. Following ingestion, DMF is readily absorbed in the GI tract of experimental animals.[20] Metabolism occurs predominantly in the liver. The main product of biotransformations is *N*-**methylformamide**[21] or *N*-**hydroxymethyl-***N'***-methylformamide.** According to other data, DMF altered the hepatic microsomal monooxygenase system and glutathione metabolism.[22] **Monomethylformamide** and **formamide** are found in the urine.

Standards. **USA** (1983). Maximum permissible concentration (MPC) of 10 mg/kg applies to certain specified color additives that may be used in food, drugs, or cosmetics. **Russia** (1988). MAC and PML: 10 mg/l.

Regulations. **USFDA** (1993) approved the use of DMF (1) in adhesives as a component of articles intended for use in packaging, transporting, or holding food and (2) as an adjuvant in the preparation of slimicides in the manufacturing of paper and paperboard that may be used safely in contact with food.

References:
1. See **ACRYLONITRILE,** #1, 177.
2. Scailteur, J. P. and Lauwerys, R. R., Dimethylformamide (DMF) hepatotoxicity, *Toxicology,* 43, 231, 1987.
3. Kennedy, G. L. and Sherman, H. Acute and subacute toxicity of dimethylformamide and dimethylacetamide following various routes of administration, *Drug Chem. Toxicol.,* 9, 147, 1986.
4. *Toxicology of New Industrial Chemical Substances,* Medgiz, Moscow, 1, 1961, 58 (in Russian).
5. Sheveleva, G. A., A Study of the Influence of Formaldehyde and Dimethylformamide on the Maternal Body, Fetal Development and Offspring, Author's abstract of thesis, Moscow, 1971, 25 (in Russian).
6. Kennedy, G. L., Biological effects of acetamide, formamide, and their monomethyl and dimethyl derivatives, *CRC Crit. Rev. Toxicol.,* 17, 129, 1986.
7. See **ACETONITRILE,** #2.
8. Aucair, M. and Hameau, N., *Comp. Rend. Soc. Biol.,* 158, 245, 1964.
9. Llewellyn, G. C., Hastings, W. S., and Kimbrough, T. D., The effects of dimethylformamide on female Mongolian gerbils meriones unguicklatus, *Bull. Environ. Contam. Toxicol.,* 11, 467, 1974.
10. Savolainen, H., Dose-dependent effects of peroral dimethylformamide administration on rat brain, *Acta Neuropathol.,* 53, 249, 1981.
11. Becci, P. J., Voss, K. A., Johnson, W. D., et al., Subchronic feeding study of *N,N'*-dimethylformamide in rats and mice, *J. Am. Coll. Toxicol.,* 2, 371, 1983.
12. Antoine, J. L., Arany, J., Leonard, A., et al., Lack of mutagenic activity of dimethylformamide, *Toxicology,* 26, 207, 1983.
13. IPCS, *Dimethylformamide,* Environmental Health Criteria 114, WHO, Geneva, 1991, 70.
14. von Merkle, J. und Zeller, H., Studies on acetamides and formamides for embryotoxic and teratogenic activities in the rabbit, *Arzneimittel-Forsch.,* 30, 1557, 1980 (in German).
15. Hellwig, J., von Merkle, J., Klimisch, H. J., et al., Studies on the prenatal toxicity of *N,N'*-dimethylformamide in mice, rats and rabbits, *Food Chem. Toxicol.,* 29, 193, 1991.
16. Thiersch, J. B., *Malformations Congenitales des Mammiferes,* Paris, 1971, 95 (in French).
17. Ducatman, A. M., Conwill, D. E., and Crawl, J., Germ cell tumours of the testicle among aircraft repairment workers, *J. Urol.,* 136, 834, 1986.
18. Serres, F. J. and Ashby, J., Eds., *Evaluation of Short-term Tests for Carcinogens,* (Progress in Mutation Research, Vol. I) Elsevier, Amsterdam, 1981, 827.
19. Druckrey, H., Preussmann, R., Ivankovich, S. U. A., Organotropic carcinogenic effects of 65 different *N*-Nitroso-compounds on BD rats, *Z. Krebsforsch.,* 69, 103, 1967 (in German).
20. Massmann, W., Toxicological investigation of dimethylformamide, *Br. J. Ind. Med.,* 13, 51, 1956.

21. Kimmerle, G. und Eben, A., Metabolism studies of *N,N'*-dimethylformamide. I. Studies in rats and dogs, *Int. Arch. Arbeitsmed.,* 34, 109, 1975 (in German).
22. Imazu, K., Fujishiro, K., and Inoue, N., Effects of dimethylformamide on hepatic microsomal monooxygenase system and glutathione metabolism in rats, *Toxicology,* 72, 41, 1992.

p-DIOXANE (CAS No 123–91–1)

Synonyms. Diethylene dioxide; Diethylene ether; Diethylene oxide; **1,4**-Dioxacyclohexane; Dioxyethylene ether; Tetrahydro-**1,4**- dioxin.

Properties. Volatile, colorless liquid with a mild ethereal odor. Miscible with water, alcohol, and majority of organic solvents. Odor perception threshold is 0.8 mg/l[010] or 1.24 mg/l; taste perception threshold is 0.3 mg/l.[1]

Applications. Solvent for cellulose acetate, ethyl cellulose, benzyl cellulose, lacquers, plastics, resins, and polyvinyl polymers.

Acute Toxicity. A substance of low toxicity. The liver is shown to be a target organ.[2]

Reproductive Toxicity. D. produced no *teratogenic effect* in rats gavaged with up to 1 ml D./ kg BW on days 6 to 15 of gestation.[3] Maternal and fetal weights were reduced at the highest dose.

Genotoxicity. Does not exhibit mutagenic activity and does not react with DNA.

Carcinogenicity. There is sufficient evidence for carcinogenicity in experimental animals.[4] The group of 26 rats received 1% D. in their drinking water for 63 to 73 weeks. Liver cancer was found to develop in six animals; renal cancer was observed in one rat.[5]

Regulations. USFDA (1993) regulates D. under FD&CA as an indirect food additive. It is listed for use in adhesives as a component of articles intended for use in packaging, transporting, or holding food.

References:

1. See **ACRYLIC ACID,** #1, 137.
2. *Bull. IRPTC,* 6, 17, 1983.
3. Giavini, E., Vismara, C., and Brocera, M. L., Teratogenesis study of dioxane in rats, *Toxicol. Lett.,* 26, 85, 1985.
4. Stott, W. T., Quast, J. E., and Watanabe, P. G., Differentiation of the mechanism of oncogenicity of **1,4**-dioxane and **1,3**-hexachlorobutadiene in the rat, *Toxicol. Appl. Pharmacol.,* 60, 287, 1981.
5. Argus, M. F. et al., *J. Natl. Cancer Inst.,* 35, 949, 1965.

ETHANOLAMINE (CAS No 141–43–5)

Synonyms. 2-Aminoetanol; Colamine; β-Hydroxyethylamine; MEA inhibitor; Monoethanolamine.

Properties. EA occurs as a colorless, oily liquid having a faint odor of ammonia. Miscible with water and alcohol at any ratio. Odor perception threshold is 625 mg/l[1] or even 20 g/l;[02] taste perception threshold 700 mg/l. A 15 mg/l concentration of EA in water causes a burning sensation in the mouth.[1]

Acute Toxicity. LD_{50} is 2.05 g/kg BW in rats, 1.47 g/kg BW in mice, 1 g/kg BW in rabbits, and 0.82 g/kg BW in guinea pigs.[2] Poisoned animals displayed labored breathing, motor excitation, and convulsions.

Repeated Exposure. Cumulative properties are not very pronounced. K_{acc} is 5.5. Rats and mice tolerated a total dose of 4 g/kg BW administered for 2 months (Gurfein et al.).

Short-term Toxicity. When inhaled, EA caused CNS excitation, probably due to acetylcholine inhibition.

Reproductive Toxicity. *Embryotoxicity.* Pregnant Long Evans rats were exposed to 50, 300, and 500 mg/kg BW oral doses of EA during the dose-critical period of organogenesis. This treatment caused a dose-dependent increase in intrauterine death.[3] *Teratogenicity.* Malformations and intrauterine growth retardation were more frequent in male than in female offspring at all dose-levels.[3]

Genotoxicity. EA demonstrated direct cytotoxicity without cell transformation in Chinese hamster embryonic cells *in vitro*.[4] It was negative in the *Salmonella* mutagenicity assay (Hedenstedt, 1973).

Chemobiokinetics. EA metabolism occurs via partial deamination in the body to form **ethylene glycol,** which is then partially oxidized to **oxalic acid.**

Standards. Russia (1988). MAC and PML: 0.5 mg/l.

Regulations. USFDA (1993) approved the use of EA (1) in adhesives as a component of articles intended for use in packaging, transporting, or holding food; (2) as an adjuvant in the preparation of slimicides in the manufacturing of paper and paperboard that may be used safely in contact with food; and (3) as a defoaming agent that may be used safely in the manufacturing of paper and paperboard intended for use in producing, manufacturing, packing, transporting, or holding food.

References:

1. Rodionova, L. F., Experimental data on substantiation of maximum allowable concentration for ethanolamine in water bodies, *Gig. Sanit.,* 2, 9, 1964 (in Russian).
2. Sidorov, K. K. and Timofievskaya, L. A., Data on substantiation of MAC for monoethanolamine in the workplace air, *Gig. Truda Prof. Zabol.,* 9, 55, 1979 (in Russian).
3. Mankes, R. F., Studies on the embryotoxic effect of etanolamine embryopathy in pups contiguous with male siblings *in utero, Teratogen. Carginogen. Mutagen.,* 6, 403, 1986.
4. Inone, K., Sunakawa, T., Okoto, K., et al., Mutagenicity tests and *in vitro* transformation assays on triethanolamine, *Mutat. Res.,* 101, 305, 1982.

ETHYL ACETATE (EA) (CAS No 141-78-6)

Synonyms. Acetic acid, ethyl ester; 'Acetic ether'; Acetoethyl ether.

Properties. EA occurs as a clear, colorless liquid with a characteristic odor. Solubility in water is 8%, miscible with alcohol at all ratios. Unstable in water. Odor perception threshold is 6.3 mg/l[010] or 20 mg/l; taste perception threshold is 10 mg/l.[1] However, according to Saratikov,[2] odor perception threshold is 2 mg/l (20°C); according to other data, it is found to be 2.6 mg/l.[02]

Applications. Used in the production of materials based on cellulose ethers, alkyd and PVA resins.

Acute Toxicity. LD_{50} is found to be 6.1 g/kg BW in rats, 4.1 g/kg BW in mice, 5.5 g/kg BW in guinea pigs, and 7.65 g/kg BW in rabbits.[2] Manifestations of the toxic effect comprised a reduction in motor activity and excitability and a narcotic state subsequently supervening. Gross pathological examination revealed visceral congestion and tiny point hemorrhages.

Repeated Exposure failed to reveal cumulative properties. Rats tolerated administration of 1 g/kg BW dose for a month without any sign of intoxication or retardation of BW gain.

Long-term Toxicity. Rats were dosed by gavage with 1 and 10 mg/kg BW for 6 months. The treatment produced changes in the NS and excretory liver function.[2]

Reproductive Toxicity. Gonadotoxic effect was observed[3] when EA has been inhaled at concentrations of 100 and 200 mg/m[3].

Chemobiokinetics. Following ingestion, EA is very rapidly hydrolyzed to form **ethyl alcohol.**[1]

Regulations. USFDA (1993) approved the use of EA (1) in cellophane for packaging food and (2) in resinous and polymeric coatings for polyolefin films to be used safely as a food-contact surface of articles intended for use in producing, manufacturing, packing, transporting, or holding food.

References:

1. Gallacher, E. J. and Loomis, T. A., Metabolism of ethylacetate in the rat: hydrolysis to ethyl alcohol *in vitro* and *in vivo, Toxicol. Appl. Pharmacol.,* 34, 309, 1975.
2. Saratikov, A. S., Trofimovich, E. M., Burova, A. V., et al., Substantiation of the maximum allowable concentration for ethylacetate in water, *Gig. Sanit.,* 4, 66, 1983 (in Russian).
3. Proc. All-Union Constit. Toxicol. Conf., Abstracts, November 25–27, 1980, Moscow, 1980, 126 (in Russian).

ETHYLENE CHLOROHYDRIN (CAS No 107-07-3)

Synonyms. β-Chloroethyl alcohol; 2-Chloroethanol; 2-Ethylene glycol chlorohydrin.

Properties. Colorless, volatile liquid with an odor of ethyl alcohol. Miscible with water at all ratios. Odor perception threshold is 50 mg/l.

Applications. Used in epoxy resin manufacturing.

Acute Toxicity. LD_{50} is 71 mg/kg BW in rats and 91 mg/kg BW in mice. CNS functions are altered by high doses.[1]

Repeated Exposure failed to reveal cumulative properties. Rats tolerated administration of $^1/_5$ LD_{50} for 20 d.

Long-term Toxicity. In a 6-month study, rats administered 5 mg/kg BW dose displayed hypercholesterolemia, a disorder of some liver functions, and changes in ascorbic acid content in the liver and blood. The treatment caused nephrotoxic effect and impairment of pancreatic excretion. Gross pathological examination revealed necrotic areas in the GI tract mucosa, parenchymatous hydropic dystrophy of the liver, and renal dilatation.[1]

Reproductive Toxicity. E. produced teratogenic effect in rats but not in mice and rabbits.[2,3]

Genotoxicity. E. produced CA in bone marrow cells. The 0.05 mg/kg BW dose appeared to be ineffective.

Standards. Russia (1988). MAC and PML: 0.1 mg/l.

References:

1. Semenova, V. N., Kazanina, S. S., Fedyanina, V. N., et al., Data on substantiation of maximum allowable concentration for ethylene chlorohydrin in water bodies, *Gig. Sanit.*, 8, 13, 1978 (in Russian).
2. LaBorde, J. B., Kimmel, C. A., Jones-Price, C., et al., Teratogenic evaluation of ethylene chlorohydrin in mice and rabbits, *Toxicologist*, 2, 71, 1982.
3. Courtney, K. D., Andrews, J. E., and Grady, M., Teratogenic evaluation of ethylene chlorohydrin, *J. Environ. Sci. Health*, B17, 381, 1982.

ETHYLENE GLYCOL (EG) (CAS No 107–21–1)

Synonyms. 1,2-Dihydroxyethane; **1,2**-Ethanediol; Ethylene alcohol; Ethylene dihydrate; Glycol; Glycol alcohol; 2-Hydroxyethanol.

Properties. Clear, viscous, odorless, sweet liquid. Readily miscible with water and alcohol. Odor perception threshold is 1320 mg/l; taste perception threshold is 450 mg/l.[1] However, according to other data,[01] organoleptic threshold is 127 mg/l.

Applications. A solvent. EG is also used in production of resinous products, especially polyester fibers and resins, and polyethylene terephthalate.

Acute Toxicity. *Humans.* Doses of 1 to 2 g/kg BW (100 ml) are lethal in humans.[2] *Animals.* A single administration of 1 ml/kg BW did not cause intoxication.[3] LD_{50} is reported to be 13 g/kg BW in rats, 8.05 g/kg BW in mice, 5 g/kg BW in rabbits, and 11 g/kg BW in guinea pigs. Poisoning is accompanied by a short period of stimulation followed by depression, ataxia, refusal of food, and labored breathing and vomiting. Gross pathological examination revealed hemorrhages in the GI tract walls.[1] EG is known to be a vascular and protoplasmic poison, causing vascular edema and necrosis. In acute exposure, it appears to act as a typical narcotic; it causes erythrocyte hemolysis and disrupts the oxidation-reduction processes.[4,5]

Repeated Exposure failed to reveal a cumulative effect in rats given $^1/_5$ LD_{50} for 20 d. K_{acc} is 8.3 (by Lim).[5] Rats were given 1% EG solution instead of drinking water *ad libitum* for 2 weeks. Hematological and biochemical analyses showed anemia and an increase in hepatic microsomal cytochrome P-450.[6]

Long-term Toxicity. In a 6-month study, monkeys were dosed with 17 to 28 mg EG/kg BW. The treatment resulted in protein precipitation and hydrolytic degeneration of the proximal part of the renal tubules. Doses of 33 to 137 mg/kg BW caused oxalate precipitation in the proximal sections of the tubules, with epithelial necrosis. Hyaline was found in the distal sections of the Henle's loops and the convoluted tubules. The structure of the tubules was destroyed. In a 6-month study, an increase in the urea and indican content in the blood serum, a reduction in prothrombin time, and weakening of hepatic secretory function (bromosulfophthalein test) were found to develop.[1]

Allergenic Effect was noted on skin application.[7]

Reproductive Toxicity. Effect on reproduction was found at the toxic dose-level. *Embryotoxicity.* In a long-term study, mice received 0.25 to 1% EG in their drinking water. The treatment caused a slight decrease in the number of litters (in the second generation) per fertile pair and of live pups per litter, and in live pups a slight decrease in weights in the 1% dose group.[8] Schuler et al.[9] noted the number of dead pups per litter at birth to be elevated and postnatal survival to be reduced in a Chernoff-Kavlock assay design in which pregnant Swiss mice were dosed by gavage with 11 mg undiluted EG/kg BW on days 7 to 14 of gestation. Sprague-Dawley rats and CD-1 mice were exposed by gavage to 750 to 5000 mg/kg BW doses on days 6 to 15 of pregnancy. The treatment reduced BW gain of dams. There was a dose-dependent increase in postimplantation fetal mortality and reduction in fetal BW in rats (doses of 2500 to 5000 mg/kg BW) and mice (750 mg/kg BW and above).[10] According to other data, doses of 14 to 1400 mg/kg BW did not produce embryotoxic or teratogenic effect, but functions and morphology of the liver were affected.[11] No indication of developmental toxicity was found in artificially inseminated New Zealand white rabbits administered EG by gavage on gestation days 6 through 19 at doses of 100 to 2000 mg/kg BW.[12] The NOAEL for developmental toxicity was at least 2 g/kg BW in this study. Such NOAEL for Swiss mice administered EG by gavage was established at 150 mg/kg level.[13] *Teratogenic* effect was noted at dose-levels of 5000 mg/kg BW in rats and 3000 mg/kg BW in mice.[14] CD-1 mice received 1% aqueous solution of EG. Developmental abnormalities of the bone system comprised facial bone malformations, reduced size of skull bones, fusion of the ribs, and malformations of the bones of the thorax and vertebral column.[8] EG produced a *gonadotoxic* effect in rats. Doses of up to 5 mg/kg caused a consistent 50 to 66% reduction in cytochrome-*C*-oxidase activity. There was a reduction in alkaline phosphatase activity in the epididymis tissues when 5 mg/kg BW was given twice.[14]

Genotoxicity. EG was not shown to produce CA and SCE in Chinese hamster ovary cells (NTP–85).

Carcinogenicity. Fisher 344 rats and CD-1 mice were fed diets yielding approximate EG dosages of 0.04, 0.4, and 1 g/kg BW. The high dose caused retardation of BW gain, an increase in the blood urea nitrogen, in water intake, and in mortality rate in males. Urinary **calcium oxalate** crystals were noted in high-dosed rats. Histopathological examination revealed tubular cell hyperplasia, tubular dilatation, and parathyroid hyperplasia.[15] There was no evidence of carcinogenic effect in rodents. *Carcinogenicity classification.* NTP: XX—XX—NE—NE.

Chemobiokinetics. EG is metabolized to form **oxalic acid** in an amount depending on the administered dose and species of experimental animals.[16] In response to EG exposure, a number of oxalates are found in different body tissues. Administration of 15 ml/kg BW to monkeys resulted in precipitation of oxalates in the renal tubules and in necrosis of the renal tubular epithelium.[3] Doses of 1 and 2 mg/kg BW given by *i/v* injection affected the process of **glycolate** metabolism, and the latter is the principal factor determining the toxicity of EG. The compensatory increase of glycolate excretion in the urine is in its turn the reason for a less pronounced dose-dependence of EG clearance from the blood.[17] The main end-product of EG metabolism in rabbits is expired CO_2 (60% of the dose over 3 d). Of EG, 10% is excreted unchanged in the urine and 0.1% as **oxalic acid.** In addition to EG, **glycolic (hydroxyacetic) acid** and small amounts of **oxalic acid** are excreted in the urine.[2]

Standards. Russia (1988). MAC and PML: 1 mg/l.

Regulations. USFDA (1993) approved the use of EG (1) in adhesives as a component of articles intended for use in packaging, transporting, or holding food; (2) in resinous and polymeric coatings for a food-contact surface; (3) in resinous and polymeric coatings for polyolefin films intended for use in producing, manufacturing, packing, transporting, or holding food; (4) in cross-linked polyester resins for repeated use as articles or components of articles coming in contact with food; (5) in polyurethane resins for contact with dry food; and (6) as slimicides in the manufacturing of paper and paperboard that may be used safely in contact with food.

References:
1. Plugin, V. P., Ethylene glycol and diethylene glycol as an object in the sanitary protection of reservoirs, *Gig. Sanit.*, 3, 16, 1968 (in Russian).
2. Balazs, T., Jackson, B., and Hite, M., Nephrotoxicity of ethylene glycol, cephalosporins and diuretics, *Monogr. Appl. Toxicol.*, 1, 487, 1982.
3. McChesney, E. W., Golberg, L., Parekh, C. K., et al., Min BH: reappraisal of the toxicology of ethylene glycol. II. Metabolism studies in laboratory animals, *Food Cosmet. Toxicol.*, 9, 21, 1971.
4. See 1,4-**BUTANEDIOL,** #2, 102.

5. Filatova, V. S., Smirnova, E. S., Gronsberg, E. Sh., et al., Data on hygienic norm-setting of ethylene glycol in workplace air, *Gig. Truda Prof. Zabol.*, 6, 28, 1982 (in Russian).

6. Imazu, K., Fujishiro, K., Inoue, N., et al., Effects of ethylene glycol on drug metabolizing enzymes in rat liver, *Sanagyo Ika Daigaku Zasshi*, 13 (Abstr.), 13, 1991 (in Japanese).

7. See **DIBUTYLTIN-S,S'-BIS(ISOOCTYLMERCAPTOACETATE),** #2, 102.

8. Lamb, J. C., Maronpot, R. R., Gulati, D. K., et al., Reproductive and developmental toxicity of ethylene glycol in the mouse, *Toxicol. Appl. Pharmacol.*, 8, 100, 1985.

9. Schuler, R. L., Hardin, B. D., Niemeier, R. W. et al., Results of testing fifteen glycol ethers in a short-term *in vivo* reproductive toxicity assay, *Environ. Health Perspect.*, 57, 141, 1984.

10. Price, C. G., Kimmel, C. A., Rochelle, W., et al., The developmental toxicity of ethylene glycol in rats and mice, *Toxicol. Appl. Pharmacol.*, 81, 113, 1985.

11. Bariliak, I. R. and Kozachuk, S. Yu., Investigations on cytotoxicity of monoatomic alcohols in bone marrow, *Cytol. Genet.*, 22, 49, 1988 (in Russian).

12. Tyl, R. W., Price, C. J., Marr, M. C., et al., Developmental toxicity evaluation of ethylene glycol by gavage in New Zealand White rabbits, *Fundam. Appl. Toxicol.*, 20, 402, 1993.

13. Tyl, R. W., Ficher, L. C., Kubena, M. F., et al., Determination of a developmental toxicity NOAEL for EG by gavage in Swiss mice, *Teratology*, 39, 487, 1989.

14. Byshovets, T. F., Barilyak, I. R., Korkach, V. I., et al., Gonadotoxic effect of glycols, *Gig. Sanit.*, 9, 84, 1987 (in Russian).

15. De Pass, L. R., Garman, R. H., Woodside, M. D., et al., Chronic toxicity and oncogenicity studies of ethylene glycol in rats and mice, *Fundam. Appl. Toxicol.*, 7, 547, 1986.

16. See **ANILINE,** #11, 259.

17. Marshall, T. C., *J. Toxicol. Environ. Health*, 10, 397, 1982.

ETHYLENE GLYCOL, ACETACETHYL ETHER (EGAEE)

Synonym. Acetatethylcellosolve.

Properties. Colorless, transparent liquid. Soluble in water and ethanol. Does not affect pH, transparency, and color of water, does not form foam and film on the water surface. Odor perception threshold is 6.9 mg/l.

Applications. A solvent for cellulose ethers and resins. Used in the manufacturing of plastics and protective coatings.

Acute Toxicity. LD$_{50}$ is 2.7 g/kg BW in rats and 1.9 g/kg BW in rabbits. Death occurs in 2 to 4 d.

Repeated Exposure revealed moderate cumulative properties.

Standards. Russia. Recommended MAC: 0.14 mg/l.

Reference:

Yatsina, O. V., Plaksienko, N. F., Pys'ko, G. T., et al., Hygienic regulation of cellosolves in water bodies, *Gig. Sanit.*, 10, 78, 1988 (in Russian).

ETHYLENE GLYCOL, DIETHYL ETHER (EGDEE) (CAS No 629–14–1)

Synonyms. Diethyl cellosolve; Ethyl glume.

Properties. Liquid.

Applications. Used in the manufacturing of protective coatings.

Acute Toxicity. LD$_{50}$ is found to be 4.4 g/kg BW in rats and 2.44 g/kg BW in guinea pigs. The principal effect exerted by EG ethers in animals at acute exposure to high doses is damage to the kidneys. On microscopic examination, tubular degeneration, along with almost complete necrosis of the cortical tubules, is observed. Additional changes are hematuria, narcosis, and GI tract irritation. Animals exhibited inactivity, weakness, dyspnea, marked testicular toxicity: degeneration of germinal epithelium and testicular atrophy.

Reproductive Toxicity. EGDEE exhibits adverse developmental effects. Pregnant CD-1 outbred Albino Swiss mice and New Zealand White rabbits were dosed by gavage with EGDEE dissolved in distilled water during major organogenesis. Doses tested were 50 to 1000 mg/kg BW (mice) and 25 to 100 mg/kg BW (rabbits). No maternal mortality was observed in mice; at the high dose-level, decrease in fetal BW and malformation incidence (mainly, exencephaly and fused ribs) were found. The NOAEL for developmental toxicity appeared to be 50 mg/l (in mice) or 25 mg/kg BW (in rabbits).

Genotoxicity. EGDEE is shown to be positive in the *Salmonella* mutagenicity assay (NTP-91).

Reference:

George, J. D., Price, C. J., Marr, M. C., et al., The developmental toxicity of ethylene glycol diethyl ether in mice and rabbits, *Fundam. Appl. Toxicol.*, 19, 15, 1992.

ETHYLENE GLYCOL, MONOBUTYL ETHER (EGBE) (CAS No 111–76–2)

Synonyms. 2-Butoxyethanol; Butyl cellosolve; Butyl glycol.

Properties. Colorless, transparent liquid having a faint, specific odor. Miscible with water at 1:1, or, according to other data, at any ratio at 25°C;[02] soluble in oil and ethanol. Does not affect transparency and color of water, does not form foam and film on the water surface. Odor perception threshold is 9.3 mg/l.[1]

Applications. Solvent for cellulose ethers, resins, and lacquers. Used in the manufacturing of plastics, surface coatings, and cleaners.

Acute Toxicity. LD_{50} is 0.5 ml/kg BW in humans, 775 mg/kg BW in rats, and 320 mg/kg BW in rabbits.[2,3] Grant et al.[4] have found less severe signs of hematotoxicity in 4-week-old Fisher 344 rats given 500 and 1000 mg/kg BW by gavage than was reported in the more recent studies of Ghanayem et al.,[5] where 9- to 13-week-old animals have been tested. Histological changes in the liver and kidney were also observed. Two hours after administration of EGBE, the relative weight of the spleen was more than doubled. A greater sensitivity of older animals to the lethal doses of EGBE seems to be a relevant explanation of this variation.[5]

Repeated Exposure. Administration of $^1/_{10}$ LD_{50} to rats caused retardation of BW gain, decrease in glucose and cholesterol concentration in blood serum. The dose-dependence was noted.[2] Male Sprague-Dawley rats were exposed to 2000 to 6000 ppm and female rats to 1600 to 4800 ppm of EGBE in drinking water for 21 d. BW of females was affected at either dose, and it was also decreased in males in the high-dose group.[6] Gross pathological examination of rats exposed to 0.5 g EGBE/kg and 1 g EGBE/kg BW revealed thymus atrophy, hyperplasia of the spleen and bone marrow, lymphocytopenia, and reticulocytosis.[4]

Immunotoxicity. In the above-described studies, thymus weights were reduced in all the treated animals.[6] A decrease in specific antibody production was observed at the low dose.

Reproductive Toxicity. In a continuous breeding reproduction study, Swiss CD-1 mice were given doses of 0.7 to 2.1 g/kg BW in drinking water for 7 d prior to and during a 98-d cohabitation period. Effects on reproduction were only evident in the females and occurred at doses that elicited general toxicity.[7] Cardiovascular developmental effect was not observed in Fisher 344 rats exposed by gavage to 200 and 300 mg EGBE/kg BW during the 3-d period of organogenesis.[8]

Chemobiokinetics. In 2 d following oral administration, EGBE was discovered in the stomach, liver, spleen, kidneys, and other visceral organs (independently of the dose). EGBE metabolism occurs through oxidation to form **butoxyacetic acid** and conjugation with glucuronic acid and sulfates. Passes predominantly with the urine and expired CO_2.[5]

Regulations. USFDA (1993) approved the use of EGBE (1) in adhesives as a component of articles intended for use in packaging, transporting, or holding food and (2) as a solvent in polysulfide polymer-polyepoxy resins in articles intended for packaging, transporting, or otherwise containing dry food in accordance with specified conditions.

Standards. Russia. Recommended MAC: 0.12 mg/l.

References:

1. See **HYDROQUINONE**, #13, 236.
2. See **ETHYLENE GLYCOL, ACETACETHYL ETHER.**
3. Lomova, G. V. and Klimova, E. I., *Gig. Truda Prof. Zabol.*, 2, 38, 1974 (in Russian).
4. Grant, D., Slush, S., Jones, H. B., et al., Acute toxicity and recovery in the hemopoietic system of rats after treatment with ethylene glycol monobutyl ether, *Toxicol. Appl. Pharmacol.*, 77, 187, 1985.

5. Ghanayem, B. I., Burka, L. T., and Matthews, H. B., Metabolic basis of ethylene glycol monobutyl ether induced toxicity: role of alcohol and aldehyde dehydrogenases, *J. Pharmacol. Exp. Ther.,* 242, 222, 1987.
6. Exon, J. H., Hather, G. G., Bussiere, J. L., et al., Effect of subchronic exposure of rats to 2-methoxyethanol or 2-butoxyethanol: thymic athrophy and Immunotoxicity, *Fundam. Appl. Toxicol.,* 16, 830, 1991.
7. Heindel, J. J., Gulati, D. K., Russell, V. S., et al., Assessment of ethylene glycol monobutyl ether and ethylene glucol monophenyl ether reproductive toxicity using continuous breeding protocol in Swiss CD-1 mice, *Fundam. Appl. Toxicol.,* 15, 683, 1990.
8. Sleet, R. B., Price, C. J., Marr, M. C., et al., Cardiovascular development in F-344 rats following phase-specific exposure to 2-butoxyethanol, Abstract, *Teratology,* 43 (Abstr.), 466, 1991.

ETHYLENE GLYCOL, MONOETHYL ETHER (EGEE) (CAS No 110–80–5)

Synonyms. 2-Ethoxyethanol; Ethyl cellosolve; Ethylglycol; Oxitol.

Properties. Colorless, transparent liquid with a faint sweetish odor. Miscible with water and many organic solvents. Does not affect transparency and color of water, does not form foam and film on the water surface. Odor perception threshold is reported to be 25 or 9 mg/l,[1] or 190 mg/l;[02] taste perception threshold is 10 mg/l.[2]

Applications. Used in the production of nitro- and acetylcellulose, natural and synthetic resins. Used in the manufacturing of plastics, lacquers, and protective coatings.

Acute Toxicity. See **EGDEE.** LD_{50} is reported to be 3 to 5 g/kg BW in male rats, 2.3 to 5.4 g/kg BW in female rats, 1.28 to 3.1 g/kg BW in rabbits, and 1.4 g/kg BW in guinea pigs.[3,4]

Repeated Exposure revealed hematological, biochemical, and morphological changes.[4]

Short-term Toxicity. A reduction in blood Hb level and changes in the hematocrit, liver, kidneys, and ovaries were observed in rats and dogs administered the doses of 45 and 750 mg/kg BW over a period of 13 weeks.[1]

Long-term Toxicity. In the NTP-88 study, male and female rats were dosed by gavage with 0.5, 1, and 2 g/kg BW.[5] High mortality of animals was noted at the 2 g/kg dose-level; males exhibited testicular lesions.

Reproductive Toxicity. *Gonadotoxicity.* EGEE is reported to produce testicular atrophy, degenerative changes in the germinal epithelium, pathological changes in the sperm-head, and infertility.[6] Testicular effects produced by EGEE may be caused by its active metabolite, **ethoxyacetic acid** (EAA). Testicular atrophy was observed in male mice given oral doses of 0.5 to 4 g/kg BW for 5 weeks. The dose of 0.5 g/kg appeared to be ineffective.[7] Male rats were exposed to 0.25 to 1 g/kg for 11 d. Decreased testes weight, spermatocyte depletion, and degeneration were noted. The NOEL for gonadotoxic effect appeared to be 0.25 mg/kg BW.[8] ***Teratogenicity.*** Adverse maternal and developmental effects following EGEE administration were found to develop. An increase in the incidence of abnormal skeletal development was noted in the fetuses of rats exposed to 93 to 186 mg/kg BW on days 1 to 21 of gestation period. The NOAEL for these defects appeared to be 46.5 mg/kg.[1] ***Embryotoxic effects*** of EGEE are reported. Administration of 500 to 1000 mg/kg BW caused decreased liver and testicular weights in young rats on day 11 of administration. A 250 mg/kg BW dose caused no changes.[6]

Genotoxicity. EGEE is not mutagenic in *Salmonella* but induced SCE and CA in Chinese hamster ovary cells in the NTP study; it is not mutagenic in *Dr. melanogaster.*[9]

Carcinogenicity. The final report of NTP-88 carcinogenicity studies has not been published.

Chemobiokinetics. EGEE is partly broken down in the body. Administration by gastric intubation resulted in two major urinary metabolites in rats, **EAA** and *N*-**ethoxyacetyl glycine**.[10] EAA has been detected in the urine of workers exposed to EGEE and EGEEA. In rats, metabolism proceeded mainly through oxidation via alcohol dehydrogenase to EAA, with some subsequent conjugation of the acid metabolite with glycine. Excretion of the unchanged substance is via the urine.

Regulations. USFDA (1993) approved the use of EGEE in adhesives as a component of articles intended for use in packaging, transporting, or holding food.

Standards. Russia. Recommended MAC: 0.63 mg/l.

References:

1. Stenger, E. G., Aeppli, L., Muller, D. U. A., Zur Toxicologie des Athylenglyckol-Monoathy-lathers, *Arzneimittel- Forsch.*, 21, 880, 1971 (in German).
2. See **ACROLEIN,** 70.
3. Occupational Exposure to Ethylene Glycol Monomethyl Ether and Ethylene Glycol Monoethyl Ether and Their Acetates Criteria for a Recommended Standard, U.S. Department of Health and Human Services, NIOSH, Washington, D.C., 1991, 296.
4. See **ETHYLENE GLYCOL, ACETACETHYL ETHER.**
5. Melnick, R. L., Toxicity of ethylene glycol and ethylene glycol monoethyl ether in Fisher 344 rats and B6C3F₁ mice, *Environ. Health Perspect.*, 57, 147, 1984.
6. Hardin, B. D., Reproductive toxicity of the glycol ethers, *Toxicology,* 27, 259, 1983.
7. Nagano, K., Nakayama, E., Koyano, M., et al., Mouse testicular atrophy induced by ethylene glycol monoalkyl ethers, *Jpn. J. Ind. Health,* 21, 29, 1979.
8. Foster, P. M., Creasy, D. M., Foster, J. R., et al., Testicular toxicity of ethyleneglycol mono-methyl and monoethyl ethers in rats, *Toxicol. Appl. Pharmacol.,* 69, 385, 1983.
9. McGregor, D. B., Genotoxicity of glycol ethers, *Environ. Health Perspect.,* 57, 97, 1984.
10. Cheever, K. L., Plotnick, H. B., Richards, D. E., et al., Metabolism and excretion of 2-ethoxy-ethanol in the adult male rat, *Environ. Health Perspect.,* 57, 241, 1984.

ETHYLENE GLYCOL, MONOETHYL ETHER, ACETATE (EGEEA)
(CAS No 111–15–9)

Synonyms. 2-Ethoxyethyl acetate; Cellosolve acetate.

Properties. Colorless liquid with a mild, nonresidual, sweet, musty odor. Readily soluble in water. Odor threshold concentration is 0.056 mg/l.[010]

Applications. Used in the manufacturing of surface coatings (especially those based on epoxy resins).

Acute Toxicity. LD$_{50}$ is 3.9 to 5 g/kg BW in male rats, 2.9 g/kg BW in females, and 1.91 g/kg BW in guinea pigs.[1] See **EGDEE.**

Reproductive Toxicity. Embryolethality, visceral and skeletal abnormalities, and reduced fetal weights were found in the offspring of rats treated with EGEEA.[2] Male mice were given the doses of 0.5 to 4 g/kg BW for 5 weeks. The NOAEL for testicular atrophy was reported to be 0.5 g/kg BW.[3] Male rats were exposed orally to 726 mg/kg BW for 11 d. The LOAEL for testicular atrophy and spermatocyte depletion appeared to be 726 mg/kg BW.[4]

Chemobiokinetics. Ethoxyacetic acid seems to be the major metabolite of EGEEA. It has been detected in the urine of workers exposed to EGEEA.

Regulations. USFDA (1993) approved the use of EGEEA in adhesives as a component of articles intended for use in packaging, transporting, or holding food.

References:

1. See **ETHYLENE GLYCOL, MONOETHYL ETHER,** #3.
2. See **ETHYLENE GLYCOL, MONOETHYL ETHER,** #6.
3. See **ETHYLENE GLYCOL, MONOETHYL ETHER,** #7.
4. See **ETHYLENE GLYCOL, MONOETHYL ETHER,** #8.

ETHYLENE GLYCOL, MONOMETHYL ETHER (EGME) (CAS No 109–86–4)

Synonyms. 2-Methoxyethanol; Methyl cellosolve; Methyl glycol; Methyl oxitol; Monomethyl ethylene, glycol ether; Monomethyl glycol.

Properties. Colorless, transparent, volatile liquid with a mild, nonresidual odor. Soluble in water and ethanol. Does not affect transparency and color of water, does not form foam and film on the water surface. Odor perception threshold is 14.3 or <0.09 mg/l.[010] According to other data, organoleptic threshold is 100 mg/l.[01]

Applications. Used in the production of nitro- and acetylcellulose, natural and artificial resins. Used in the manufacturing of protective epoxy resin coatings for metals. A surfactant.

Acute Toxicity. LD_{50} is found to be 2.46 to 3.25 g/kg BW in male rats, 3.4 g/kg BW in female rats, 0.89 to 1.425 g/kg BW in rabbits, and 0.95 g/kg BW in guinea pigs.[1,2] High doses caused anuria. Death occurred in rabbits in 2 to 4 d after poisoning. In man, EGME affects hematology and CNS.

Repeated Exposure. Gross pathological examination in rats exposed to 100 and 150 mg/kg BW for 3 weeks revealed reduced weights of the visceral organs, normochromic and normocytic anemia, hemorrhagic changes in the bone marrow, and ovarian atrophy.[3] Inflammatory changes were observed in the bladder mucosa.[4]

Long-term Toxicity. Chronic exposure causes hemodynamic injuries, dystrophic changes in the brain, liver, kidneys, myocardium, etc.

Immunotoxicity. Rats given doses >100 mg EGME/kg BW displayed significant thymic depression.[5] A dose-related increase in natural killer cell cytotoxic activities and decrease in specific antibody production in rats were reported.[6] Mice, however, appeared to be insensitive to the immunosuppressive effects of EGME at the doses producing such effects in rats.[7] House[8] did not observe changes in the immunological functions or host-resistance in B6C3F$_1$ female mice dosed by gavage ten times over 2 weeks with 25 to 100 mg EGME or 2-methoxyacetic acid/kg BW. Nevertheless, thymic involution was revealed at 50 to 100 mg/kg doses.

2-Methoxyacetic acid, the main metabolite of EGME has been shown to cause an immunodepressive effect when administered by gavage to young adult Fisher 344 rats. Thymic involution has been reported.[9]

Reproductive Toxicity. EGME produces a dose-related *embryotoxic* and *teratogenic effect* in mice, hamsters, and guinea pigs. Fisher 344 rats were treated on days 6 to 15 of gestation by dosed feed or gavage (doses 12.5 to 100 mg/kg BW). This exposure caused only a small decrease in maternal BW gain during treatment. Litter weights and postnatal survival were decreased (100 mg/kg BW group); the percentage of resorption was increased in 50 and 100 mg/kg BW-dosed groups. The number of live pups was decreased in 25 to 100 mg/kg BW-dosed groups.[10] Out of 13 pregnancies, 3 ended in death in female monkeys given oral doses of EGEE on gestation days 20 to 45. The LOAEL was considered to be 12 mg/kg BW in this study.[11] Bifurcated or split cervical vertebrae were found in the offspring of female mice treated on days 7 to 14 of gestation. The LOAEL of 31 mg/kg BW was established.[12] The NOAEL for induction of malformations after a single administration on gestation day 11 was 100 mg/kg BW.[4] *Gonadotoxicity.* EGME causes testicular atrophy, degenerative changes in the germinal epithelium, pathological changes in the sperm-head, and infertility.[13] EGME was found to deplete the spermatocytes of rats and mice given a single oral dose of 0.5 to 1.5 g/kg BW. It produced morphological abnormalities in rat spermatozoa that had been exposed as spermatocytes.[14] In a 5-week study, male mice were exposed to oral doses of 0.5 to 4 g/kg BW. The NOAEL of 0.5 mg/kg BW for testicular atrophy was identified.[15] The ineffective dose for lesions and degeneration in primary spermatocytes and spermatids in male rats dosed for 11 d appeared to be 0.05 g/kg BW.[16]

Genotoxicity. EGME has been shown to increase the rate of DLM in rats. It was negative in the *Salmonella* mutagenicity assay[17] but produced damage in mouse sperm-head morphology following inhalation exposure. Equivocal results were obtained in *Dr. melanogaster* (NTP-82).

Chemobiokinetics. *Humans.* Methoxyacetic acid was found in the urine of seven male volunteers exposed to 5 ppm EGME,[18] but it was not found in the human urine in acute poisoning cases. *Animals.* EGME is metabolized via alcohol dehydrogenase to **methoxyacetaldehyde** and via aldehyde dehydrogenase to **methoxyacetic acid,** which seems to be a major oxidative metabolite, and urine appears to be a major route of excretion.[9,19] EGME is removed mainly in the urine; it does not accumulate in the testis.[16]

Regulations. USFDA (1993) approved the use of EGME in adhesives as a component of articles intended for use in packaging, transporting, or holding food.

Standards. Russia (1990). Tentative MAC: 0.6 mg/l.

References:
1. See **ETHYLENE GLYCOL, ACETACETHYL ETHER.**
2. See **ETHYLENE GLYCOL, MONOETHYL ETHER,** #3.
3. Grant, D., Sulsh, S., Jones, H. B., et al., Acute toxicity and recovery in the hemopoietic system of rats after treatment with ethylene glycol monomethyl and monobutylethers, *Toxicol. Appl. Pharmacol.*, 77, 187, 1985.

300

4. Horton, V. L., Sleet, R. B., John-Greene, J. A., et al., Developmental phase-specific effects of ethylene glycol monomethyl ether in CD-1 mice, *Toxicol. Appl. Pharmacol.,* 80, 108, 1985.
5. Henningsen, G. H., Sendelbach, L. E., Braun, A. G., et al., *Society of Toxicology, 228 Annual Meeting,* Abstract, Atlanta, GA, 1989.
6. **ETHYLENE GLYCOL, MONOBUTYL ETHER,** #6.
7. Smialowicz, R. J., Riddle, M. M., Williams, W. C., et al., Differences between rats and mice in the immunosuppressive activity of **2**-methoxyethanol and **2**-methoxyacetic acid, *Toxicology,* 74, 57, 1992.
8. House, R. V., Lauer, L. D., Murray, M. J., et al., Immunological studies in B6C3F$_1$ mice following exposure to ethylene glycol monomethyl ether and its principal metabolite methoxyacetic acid, *Toxicol. Appl. Pharmacol.,* 77, 358, 1985.
9. Smialowicz, R. J., Riddle, M. M., Lueb, R. W., et al., Immunotoxicity of **2**-methoxyethanol following oral administration in F344 rats, *Toxicol. Appl. Pharmacol.,* 109, 494, 1991.
10. Morrissey, R. E., Harris, M. W., and Schwetz, B. A., Developmental toxicity screen: results of rat studies with diethylhexyl phthalate and ethylene glycol monomethyl ether, *Teratogen. Carcinogen. Mutagen.,* 9, 119, 1989.
11. Scott, W. J., Fradkin, R., Wittfoht, W., et al., Teratologic potential of **2**-ethoxyethanol and transplacental distribution of its metabolite, **2**-methoxyacetic acid, in nonhuman primates, *Teratology,* 39, 363, 1989.
12. Nagano, K., Nakayama, E., Oobayashi, H., et al., Embryotoxic effects of ethylene glycol monomethyl ether in mice, *Toxicology,* 20, 335, 1981.
13. Hardin, B. D. and Lyon, J. P., Summary and overview: NIOSH symposium on toxic effects of glycol ethers, *Environ. Health Perspect.,* 57, 273, 1984.
14. Anderson, D., Brinkworth, M. H., Jenkinson, P. C., et al., Effect of ethylene glycol monomethyl ether on spermatogenesis, dominant lethality, and F_1 abnormalities in the rat and mouse after treatment of F_0 males, *Teratogen. Carcinogen. Mutagen.,* 7, 141, 1987.
15. See **ETHYLENE GLYCOL, MONOETHYL ETHER,** #7.
16. See **ETHYLENE GLYCOL, MONOETHYL ETHER,** #8.
17. See **ETHYLENE GLYCOL, MONOETHYL ETHER,** #9.
18. Groeseneken, D., Veulemans, H., Masschelein, R., et al., Experimental human exposure to ethylene glycol monomethyl ether, *Int. Arch. Occup. Environ. Health,* 61, 243, 1989.
19. Miller, R. R., Metabolism and disposition of glycol ethers, *Drug Metab. Rev.,* 18, 1, 1987.

ETHYLENE GLYCOL, MONOPHENYL ETHER (EGPE) (CAS No 122–99–6)

Synonyms. β-Hydroxyethyl phenyl ether; **1**-Hydroxy-**2**-phenoxyethane; **2**-Phenoxyethanol; Phenoxethol; Phenylcellosolve.

Properties. Colorless, oily liquid with a faint aromatic odor and burning taste. Solubility in water is 26.7 g/l. Soluble in alcohols and ethers.

Applications. Primarily used in latexes, paints, cosmetics, and in the manufacturing of protective coatings.

Acute Toxicity. In rats, LD$_{50}$ is reported to be 1.26 or 1.4 g/kg BW (Rowe and Wolf, 1982). See **EGDEE.**

Repeated Exposure. Female New Zealand white rabbits were dosed by gavage with 100 to 1000 mg/kg BW for up to ten consecutive days. The treatment resulted in dose-related intravascular hemolitic anemia (decreased erythrocyte count, Hb level, packed cell volume, regenerative erythroid response in the bone marrow, etc.) and renal tubule damage.[1]

Reproductive Toxicity. Swiss CD-1 mice received 0.4 to 4 g/kg BW administered via feed for 7 d prior to and during a 98-d cohabitation period. Effects on reproduction were only evident in the female and occurred at doses that elicited general toxicity.[2] EGPE appeared to be neither embryo/fetotoxic nor teratogenic when applied dermally in rabbits.

Chemobiokinetics. Phenoxyacetic acid was identified as a major blood metabolite.[1]

Regulations. USFDA (1993) approved the use of EGPE in adhesives as a component of articles intended for use in packaging, transporting, or holding food.

References:

1. Breslin, W. G., Phillips, J. E., Lomax, L. G., et al., Hemolitic activity of ethylene glycol phenyl ether in rabbits, *Fundam. Appl. Toxicol.,* 17, 466, 1991.
2. Heindel, J. J., Gulati, D. K., Russell, V. S. et al., Assessment of ethylene glycol monobutyl ether and ethylene glycol monophenyl ether reproductive toxicity using continuous breeding protocol in Swiss CD-1 mice, *Fundam. Appl. Toxicol.,* 15, 683, 1990.

FORMAMIDE (CAS No 75–12–7)

Synonyms. Formic acid, amide; Methanamide.

Properties. Colorless and odorless liquid. Readily soluble in water and lower alcohols. At a concentration of 25 mg/l does not give water any foreign odor, taste, or coloration. After chlorination or heating of water solutions of F. up to 60°C, organoleptic properties of water are not affected.[1]

Acute Toxicity. LD_{50} is 5.7 to 6.1 g/kg BW in rats, and 2.1 to 3.15 g/kg BW in mice (administered as water solutions).[1,2] The toxic symptoms include CNS affection (motor coordination disorder, decreased excitability and muscle tonus in a part of animals, respiratory disorders). Conjunctivitis developed. In 2 d after poisoning, animals displayed hind limb paresis and lowered pain sensitivity. Death occurred in 3 to 4 d. Histological examination revealed vascular disturbances and signs of parenchymatous dystrophy in the liver. Marked disturbances in the blood and lymph circulation occurred. Gross pathological examination revealed congestion and moderate edema, dystrophic changes in the spinal cord, and parenchymatous dystrophy of the renal tubular epithelium.[2]

Repeated Exposure. When mice were dosed by gavage with 0.3 g/kg BW, cumulative properties were pronounced. K_{acc} is 1.4 (by Kagan).

Long-term Toxicity. In a 6-month study, male rats were administered F. by gavage as water solutions. A dose of 5 mg/kg BW caused retardation in bromosulfophthalein excretion, an increase in protein content in the urine, and shortening of STI.[1]

Reproductive Toxicity. Gonadotoxicity. Necrosis of the spermatogenic epithelium and a reduction in the spermatozoa count were reported following oral administration of 5 to 8 g/kg BW. Long-term inhalation exposure to 6 mg/m³ did not affect fertility of rats.[2] *Embryotoxicity.* Pregnant rats were exposed by gavage to 2 to 3 g F./kg BW. Administration on day 10 of pregnancy increased the rate of resorptions, and, subsequently, embryo weights and size. *Teratogenic effect* was found in rabbits.[3]

Regulations. USFDA (1993) approved the use of F. in adhesives as a component of articles intended for use in packaging, transporting, or holding food.

References:

1. Saratikov, A. S., Trofimovich, E. M., Novozheeva, T. P., et al., Hygienic regulation of formamide in water bodies, *Gig. Sanit.,* 3, 72, 1987 (in Russian).
2. *Toxicology of New Industrial Chemical Substances,* Vol. 9, Meditsina, Moscow, 1967, 86 (in Russian).
3. Merkle, J. and Zeller, H., Studies on acetamides and formamides for embryotoxic and teratogenic activities in the rabbit, *Arzneimittel-Forsch.,* 30, 1557, 1980 (in German).

FURFURYL ALCOHOL (FA) (CAS No 98–00–0)

Synonyms. 2-Furancarbinol; 2-Furanmethanol; Furfural alcohol; Furfuralcohol; 2-Hydroxymethylfuran.

Properties. FA occurs as a colorless or slightly yellowish liquid, having a faint burning odor and bitter taste. Miscible with water, very soluble in alcohol and ether.

Applications. Used in the production of urea and furan resins in compositions of corrosion-resistant polymers. A solvent.

Acute Toxicity. LD_{50} is reported to be 275 mg/kg BW in rats, 160 mg/kg BW in mice, and 600 mg/kg BW in rabbits (Ephraim). Poisoning produced CNS depression.

Reproductive Toxicity. Inhalation exposure caused no specific gonadotoxic or embryotoxic effect.[1]

302

Allergenic Effect. Weak evidence.[2]

Genotoxicity. FA is shown to produce SCE in Chinese hamster ovary cells (NTP–85).

Standards. Russia. Tentative MAC: 0.7 mg/l.

Regulations. USFDA (1993) regulates the use of FA in adhesives used as components of articles intended for use in packaging, transporting, or holding food.

References:
1. Gadalina, I. D. and Malysheva, M. V., Substantiation of the MAC for furyl alcohol in the industrialized zone air, *Gig. Truda Prof. Zabol.*, 9, 52, 1981 (in Russian).
2. See **FURAN,** #2.

GLYCERIN MONOALLYL ETHER (CAS No 25136–53–2)

Synonyms. Glycerol allyl ether; (2-Propenyloxy)propanediol.

Properties. Liquid.

Acute Toxicity. In mice, LD_{50} is 1.75 g/kg BW. All animals died within 24 h.

Repeated Exposure failed to reveal cumulative properties.

Reference:

Mel'nikova, L. V., Toxicity and hazard of several chemicals, *Gig. Sanit.*, 1, 94, 1981 (in Russian).

HEPTYL ALCOHOL (CAS No 111–70–6)

Synonyms. Enanthic alcohol; 1-Heptanol; Hydroxyheptane.

Properties. A colorless liquid with a strong odor. Solubility in water is 0.9 g/l. Mixes with alcohol at any ratio. Odor perception threshold is 2 mg/l; taste perception threshold is 0.5 mg/l.[1]

Applications. Used in the production of phenolformaldehyde resins, plasticizers, and emulsifiers.

Acute Toxicity. LD_{50} is found to be 6 g/kg[2] or 1.5 g/kg BW in mice,[3] 0.5 g/kg BW in rats, and 0.75 g/kg BW in rabbits. Clinical signs of intoxication comprised CNS affection and respiratory failure.

Repeated Exposure revealed no cumulative properties in rats exposed to 300 mg/kg BW for 1 month.

Long-term Toxicity. In a 6-month study, gross pathological examination revealed edema and dystrophic changes in the liver and kidneys of rats.[1]

Standards. Russia (1988). MAC and PML: 0.05 mg/l.

References:
1. See **BUTYLENE,** p. 152.
2. See **TETRAHYDROFURAN,** #10, 1963, p. 51 (in Russian).
3. Voskoboinikova, V. B., Substantiation of maximum allowable concentration for floatoreagent IM-68 and hexyl, heptyl and octyl alcohols, it is composed of, *Gig. Sanit.*, 3, 16, 1966 (in Russian).

HEPTYLCYCLOPENTANONE (CAS No 137–03–1)

Properties. Colorless liquid with a specific odor. Soluble in water and alcohol.

Applications. Used in the production of vinyl resins, nitrocellulose, and synthetic rubber.

Acute Toxicity. LD_{50} is 10.55 g/kg BW in rats and 6.8 g/kg BW in mice. For other data, see **Hexyl methyl ketone.**

n-HEXANE (CAS No 110–54–3)

Synonyms. Hexane; Skellysolve *B*.

Properties. Colorless, easily evaporated liquid with a specific odor. Solubility in water is 1.0 mg/l, soluble in ethanol. Odor perception threshold is 0.0064 mg/l.[02]

Applications and **Exposure.** A component of paints, a solvent in the manufacturing of polyethylene, polypropylene, etc.

Acute Toxicity. The most common toxic response is the development of central and peripheral neuropathy.

Short-term Toxicity studies revealed a neuropathic action.

Reproductive Toxicity. Inhalation exposure of pregnant Fisher 344 rats to 1000 ppm on days 8 to 16 of gestation resulted in depression of postnatal growth of pups up to 3 weeks after birth.[1] Pregnant CD-1 mice exposed by gavage to doses up to 9.9 g/kg BW on days 6 to 15 of gestation gave birth to litters that had no signs of an increased teratogenic effect.[2]

Genotoxicity. Doses of 500 to 2000 mg/kg administered *i/p* failed to increase the incidence of SCE in an *in vivo* mouse bone marrow cytogenetic assay.[3] The number of CA was slightly increased.

Chemobiokinetics. Accumulation in the tissues depends on lipid content in these tissues. *n*-H. is oxidized in the liver. Excretion occurs via the lungs and kidneys.

Regulations. USFDA (1993) approved the use of F. (1) as a component of adhesives to be used safely in contact with food, (2) as a defoaming agent that may be used safely as a component of food-contact articles, and (3) in resinous and polymeric coatings for polyolefin films to be used safely as a food-contact surface of articles intended for use in producing, manufacturing, packing, transporting, or holding food.

References:
1. Bus, J. S., White, E. L., Tyl, R. W., et al., Perinatal toxicity and metabolism of *n*-hexane in Fisher 344 rats after inhalation exposure during gestation, *Toxicol. Appl. Pharmacol.*, 51, 295, 1979.
2. Marks, T. A., Fisher, P. W., and Staples, R. E., Influence of *n*-hexane on embryo and fetal development of mice, *Drug Chem. Toxicol.*, 3, 393, 1980.
3. NTP, *Toxicity Studies of n-Hexane in B6C3F₁ Mice (Inhalation Studies)*, NTP Tox. 2, NIH Publ. No 91–3121, 1991, 32.

HEXYL ALCOHOL (HA) (CAS No 111–27–3)

Synonym. 1-Hexanol.

Properties. A liquid. Solubility in water is 5.9 g/l. Concentrations of 21 and 37 mg/l in the water give a rating of 1 for odor and taste, respectively. Odor disappears only after 3 d.[1] According to other data, odor threshold concentration is 0.01 mg/l.[010]

Acute Toxicity. In mice, LD_{50} is reported to be 1.95 g/kg BW.

Repeated Exposure failed to reveal cumulative properties.

Long-term Toxicity. The NOAEL must be not lower than that established experimentally for the more toxic heptyl alcohol.[1,2]

Reproductive Toxicity. HA produced no adverse reproductive effects in rats exposed to vapor at a concentration of 14,000 mg/m.[3]

Standards. Russia (1988). MAC and MPL: 0.01 mg/l (for *normal, secondary,* and *tert*-HA).

References:
1. See **HEPTYL ALCOHOL,** #3.
2. See **ACRYLONITRILE,** #1, 65.
3. Nelson, B. K., Brightwell, W. S., Khan, A., et al., Teratological evaluation of 1-pentanol, 1-hexanol and 2-ethyl-1-hexanol administered by inhalation to rats, *Teratology*, 37, 479, 1988.

HEXYL METHYL KETONE (CAS No 111–13–7)

Synonym. 2-Octanone.

Properties. A colorless liquid with a specific odor. Poorly soluble in water, mixes with alcohol at any ratio.

Applications. Used in the production of vinyl resins, nitrocellulose, and synthetic rubber.

Acute Toxicity. The LD_{50} values are reported to be 3.1 g/kg BW in mice and 9.2 g/kg BW in rats. Poisoned animals displayed lethargy. Death occurred within 2 to 4 d.

Repeated Exposure failed to reveal cumulative properties.

Reference:
Occupational Hygiene and Pathology and Industrial Toxicology, Vol. II, Sumgait, 1977, 145 (in Russian).

ISOBUTYL ACETATE (CAS No 110–19–01)

Synonyms. Acetic acid, isobutyl ester; Acetic acid, 2-methylpropyl ester.

Properties. Clear, colorless liquid. Solubility in water is 6.7 or 5.9 g/l at 25°C; mixes with alcohol at any ratio. Odor perception threshold is 1 or 0.15 mg/l,[02] taste perception threshold is 0.5 mg/l. According to other data, odor threshold concentration is 0.35 mg/l.[010]

Acute Toxicity. LD_{50} is 15 g/kg BW in rats, 6.68 g/kg BW in mice, 6.66 g/kg BW in guinea pigs, and 3.7 g/kg BW in rabbits.

Long-term Toxicity. The NOAEL appeared to be 5 mg/kg BW.

Standards. Russia. PML in drinking water: 0.5 mg/l (organolept., taste).

Reference:
See **PHTHALIC ANHYDRIDE,** #1, 56.

ISOBUTYL ALCOHOL (CAS No 78–83–1)

Synonyms. Isobutanol; Isopropylcarbinol; Fermentation butyl alcohol; 1-Hydroxymethylpropane; 2-Methyl-1-propanol.

Properties. Colorless refractive liquid having a sharp, alcoholic odor similar to that of amyl alcohol. Solubility in water is 8.7 to 9.5%; mixes with alcohols and ethers at any ratio. Odor perception threshold is 0.1 to 10 mg/l.[1] According to other data, organoleptic perception threshold is 0.36 mg/l.[01]

Applications and **Exposure.** Used in the production of nitrocellulose lacquers and resins; I. is also used in the manufacturing of isobutyl esters (plasticizers) and in the vulcanization of rubber. Its natural occurrence in food could entail exposure of the population.

Acute Toxicity. LD_{50} is 3.1 g/kg BW in rats and 3.5 g/kg BW in mice.[2] There are no species differences in susceptibility. The threshold narcotic dose for rabbits is 1.4 g/kg and the LD_{50} is 3 g/kg BW.[1] Poisoning produced signs of alcohol intoxication.

Repeated Exposure revealed weak cumulative properties.[2] Rats tolerate administration of 0.6 and 0.3 g/kg BW for a month.

Short-term Toxicity. Rats were given 1 mol/l solution of I. as their sole drinking liquid for 4 months. No adverse effects on the liver were recorded. Exposure to a 2 mol/l solution for 2 months produced a reduction in fat, glycogen, and RNA contents, and in the overall size of the hepatocytes.[3]

Long-term Toxicity. The threshold concentration seems to be more than 100 mg/l.[1]

Immunotoxicity. A positive specific microprecipitation reaction (up to 40%) was noted in 1, 3, and 7 d following inhalation exposure.[2]

Reproductive Toxicity. Rats received the dose of 0.05 mg/kg BW and higher with their drinking water. The treatment affected the course of pregnancy in rats.[4]

Carcinogenicity. In a lifetime study, rats were dosed orally twice a week with 0.2 ml/kg BW. The treatment resulted in toxic liver damage and hyperplasia of blood-forming tissues. An increase in tumor incidence was noted.[5] Methodological inadequacies in the study led to questions of its significance. *Carcinogenicity classification.* IARC: Group 3.

Chemobiokinetics. I. is metabolized by alcohol dehydrogenase to **isobutyric acid** via the aldehyde and may be involved in the tricarboxylic acid cycle. The urinary metabolites are **acetaldehyde, acetic acid, isobutyraldehyde, isovaleric acid,** and unmetabolized I.[6] Only a negligible amount is eliminated unchanged or as the glucuronide in the urine.

Standards. Russia (1983). MAC: 0.15 mg/l, PML in food: 0.5 mg/l.

Regulations. USFDA (1993) approved the use of I. (1) in adhesives as a component of articles intended for use in packaging, transporting, or holding food; (2) as a substance employed in the production of or added to textiles and textile fibers intended for use in contact with foods; and (3) as a defoaming agent that may be used safely as a component of articles and in the manufacturing of paper and paperboard intended for use in contact with food.

References:
1. See **ACRYLONITRILE,** #1, 65.
2. Kushneva, V. S., Koloskova, G. A., and Koltunova I. G., Experimental data for hygienic standardization of isobutyl alcohol in the air of the work environment, *Gig. Truda Prof. Zabol.,* 1, 46, 1983 (in Russian).
3. Hilbom, M. E., Franssila, K., and Forsander, O. A., Effects of chronic ingestion of some lower aliphatic alcohols in rats, *Res. Commun. Chem. Pathol. Pharmacol.,* 9, 177, 1974.
4. Nadeyenko, V. G. et al., Embryotoxic effect of isobutanol, *Gig. Sanit.,* 2, 6, 1980 (in Russian).
5. Gibel, W., Lohs, K. H., und Wildner C. P., Experimental research on the carcinogenic effect of solvents, using propanol-1, 2-methylpropanol-1, and 3- methylbutanol-1 as examples, *Arch. Geschwulstforsch.,* 45, 19, 1975 (in German).
6. Saito, M., Studies on the metabolism of lower alcohols, *Nichidai Igaka Zasshi,* 34, 569, 1975 (in Japanese).

METHYL ACETATE (CAS No 72–20–9)

Synonym. Acetic acid, methyl ether.

Properties. Clear, colorless liquid with a unique odor. Solubility in water is 319 g/l at 20°C[1] or 220 g/l at 25°C;[02] mixes with alcohol at any ratio. Odor and taste perception threshold is 5 to 10 mg/l. According to other data, odor perception threshold is 3 mg/l.[02]

Applications. Used in the manufacturing of lacquers and films; a diluent and a solvent for nitrocellulose, acetylcellulose, and many resins.

Acute Toxicity. LD_{50} is reported to be 2.9 g/kg BW in rats, 2.4 g/kg BW in mice and rabbits, and 3.6 g/kg BW in guinea pigs.

Repeated Exposure revealed a cumulative effect. K_{acc} was found to be 3.65 (by Cherkinsky) when $1/5$ LD_{50} was administered to animals. A dose of 250 mg/kg produced an increase in blood cholinesterase activity and in the number of segmented nucleus neutrophils (Meleshchenko et al.)

Standards. Russia (1988). MAC and PML: 0.1 mg/l.

Regulations. USFDA (1993) approved the use of F. (1) as a component of the uncoated or coated food-contact surface of paper and paperboard and (2) in adhesives used as components of articles intended for use in packaging, transporting, or holding food.

Reference:
See **ACRYLONITRILE,** #7, 38.

METHYL ALCOHOL (MA) (CAS No 67–56–1)

Synonyms. Methanol; Wood alcohol; Wood spirit.

Properties. Clear, colorless, volatile liquid with a faint alcoholic odor when pure; crude product may exhibit a repulsive, pungent odor. Mixes with water and ethyl alcohol at any ratio. Odor perception threshold is 4.26 mg/l[010] or 30 to 50 mg/l,[1] or, according to other data,[02] 740 mg/l. At these concentrations, no taste is detected in water.

Acute Toxicity. In rats, LD_{50} is 10.6 g/kg BW (average over a year, 120 animals). Interspecies variations are negligible.[2] Very young and old animals are more sensitive to MA. In rats with BW of up to 50 g, LD_{50} is 7.4 ml/kg BW; at BW of 80 to 160 g, 13 ml/kg BW; and at BW of 300 to 470 g, 8.8 ml/kg BW.[3] Edema of the optic disc and of other areas of the CNS developed on a single administration of 2 g/kg BW.[4] Administration of 3.5 ml 50% solution to male and female rats caused changes in the CNS, GI tract, and parenchymatous and immunocompetent organs. A lesser dose resulted in mild disturbances of blood circulation. A greater dose produced gross pathological changes in the CNS and lungs.[5] Mild CNS depression, tremor, ataxia, and recumbency were reported in minipigs YU given a single oral dose by gavage.[6] Animals did not develop optic nerve lesions.

Repeated Exposure. Administration of 10 to 500 mg/kg BW to rats caused focal parenchymatous degeneration in the liver, an increase in hepatocyte size, and a change in the activity of certain microsomal enzymes.[7] In a month, a dose of 1.5 mg/kg had already affected conditioned reflex activity in rats (Guseva, 1969).

Long-term Toxicity. In a 5.5-month study, rabbits were dosed by gavage with 2.5 ml MA/kg BW. Animals displayed retardation of BW gain and changes in the optic nerve axons and in the cerebral cortex cells (Guseva, 1969). Rats received 3.25 ml MA/kg BW for 6 months. The treatment caused a decline in the heart contraction rate and in skin temperature by the end of the experiment. Myocardial hypoxia developed.[8]

Reproductive Toxicity was observed on administration of $^1/_2$ to $^1/_{500}$ LD$_{50}$. No dose-dependence was noted.[9]

Embryotoxicity. Female mice received a 10% MA aqueous solution for 2 months; rats were given a 1% MA aqueous solution for 6 months. The treatment caused an increase in neonatal mortality compared with the control.[8] Nelson et al.,[10] found a *teratogenic effect* at high inhalation levels in rats.

Chemobiokinetics. MA metabolism in the body occurs via oxidation in the direction of the **methanol/formaldehyde/formic acid/carbon dioxide.** Its oxidation and elimination from the body occur slowly.[1] A single administration of 2 g/kg led to development of metabolic adiposis and the accumulation of **formic acid** in the blood of male Macaco monkeys.[4] High doses of ingested MA are retained in the body for up to 7 to 8 d. In rats, MA is removed unchanged and as **carbon dioxide** in the expired air. Of MA, 3% is excreted in the urine, and a further 3% as **salts** or **esters of formic acid.**[11]

Standards. Russia (1986). MAC and PML: 3.0 mg/l, PML in food: 0.2 mg/l.

Regulations. USFDA (1993) approved MA for safe use (1) as a solvent in polyester resins, (2) in resinous and polymeric coatings for a food-contact surface, (3) as a component of paper and paper board for contact with dry food, (4) in the poly(**2,6**-dimethyl-**1,4**-phenylene)oxide resins as a component of an article intended for use in contact with food subject to the specified provisions, (5) in polyurethane resins for use in contact with dry food, (6) as a component of adhesives for a food-contact surface, (7) as a defoaming agent that may be used safely as a component of food-contact articles, (8) as a substance employed in the production of or added to textiles and textile fibers intended for use in contact with food, and (9) in cross-linked polyester resins to be used safely as articles or components of articles intended for repeated use in contact with food.

References:

1. See **ACETONE,** #3, 209.
2. Trifonov, Yu. A., Tutdyev, A. A., Tiunov, L. A., et al., Correlation of seasonal death of methanol-intoxicated rats with energetic activity of liver mitochondria, *Gig. Sanit.,* 8, 82, 1987 (in Russian).
3. See **ACETONITRILE,** #2.
4. Martin-Amat, J., Tephly, T. R., McMartin, K. E. et al., Methyl alcohol poisoning. II. Development of a model for ocular toxicity of methyl alcohol poisoning using the rhesus monkey, *Arch. Ophthalmol.,* 95, 1847, 1977.
5. *Reports 2nd All-Union Congr. Forens. Medicine,* Irkutsk-Moscow, 1987, 220 (in Russian).
6. Dorman, D. C., Dye, J. A., Nassise, M. P., et al., Acute methanol toxicity in minipigs, *Fundam. Appl. Toxicol.,* 20, 341, 1993.
7. Skirko, B. K., Ivanitsky, A. M., Pilenitsina, R. A., et al., Study on the toxic action of methanol in experiments with complete and protein deficient nutrition, *Probl. Nutr.,* 5, 70, 1976 (in Russian).
8. Rudnev, M. I. and Nozdrachev, S. I., Effect of methanol on some body functions, *Vrachebnoye Delo,* 6, 125, 1975 (in Russian).
9. Sprinchak, G. K., in All-Union Constitutive Toxicol Conf., Abstracts, Moscow, November 25–27, 1980, 131 (in Russian).
10. Nelson, B. K., Brightwell, W. S., MacKenzie, D. R., et al., Teratological assessment of methanol and ethanol at high inhalation levels in rats, *Fundam. Appl. Toxicol.,* 5, 727, 1985.
11. See **ANILINE,** #11, 167.

METHYL CYCLOHEXYL KETONE

Synonyms. Acetocyclohexane; Hexahydrobenzophenone.

Properties. Colorless liquid with a specific odor. Readily soluble in water.

Applications. Used in the production of vinyl resins, nitrocellulose, and synthetic rubber.

Acute Toxicity. LD_{50} is 2.5 g/kg BW in rats and 3.6 g/kg BW in mice. For other data see **Hexyl methyl ketone.**

METHYL ETHYL KETONE (MEK) (CAS No 78–93–3)

Synonyms. 2-Butanone; 3-Butanone; Ethylmethyl ketone; Methylacetone.

Properties. Colorless liquid with an acetone odor. Water solubility is 26.8%, miscible with alcohol at all ratios. Odor perception threshold is reported to be 1 to 2 mg/l[1,010] or 8.4 mg/l.[02]

Applications. Used as a solvent in the manufacturing of plastics, resins, and lacquers.

Acute Toxicity. LD_{50} is found to be 2.74 g/kg BW in rats, 4.05 g/kg BW in mice, and 13 g/kg BW in rabbits (Union Carbide Data Sheet). Following single oral administration, MEK inhibited the NS and had a mild toxic effect on the liver.[2]

Reproductive Toxicity. *Humans.* An increase in the incidence of children born with CNS defects following maternal exposure was reported.[3] *Animals.* Pregnant Swiss mice were relatively insensitive to the toxic effects of MEK at inhaled concentrations up to 3000 ppm. However, the offspring of mice exhibited significant signs of developmental toxicity at the 3000 ppm exposure level. Neither maternal nor developmental toxicity was observed at 1000 ppm of MEK or below.[4]

Genotoxicity. MEK is unlikely to be mutagenic in mammalian or bacterial cell systems, but caused aneuploidy in yeast.[5]

Carcinogenicity. Epidemiological studies revealed no carcinogenic effect.

Chemobiokinetics. Induction of cytochrome P-450 systems by MEK and/or its metabolites has been reported.[6] MEK undergoes reduction to **secondary alcohol.** It is passed with the urine in the form of glucuronide. Of the administered dose, 30 to 33% is released unaltered through the lungs.[2]

Standards. Russia (1988). MAC and PML: 1 mg/l (organolept., odor).

Regulations. USFDA (1993) regulates MEK (1) as a component of adhesives to be used safely in a food-contact surface, (2) as a cellophane ingredient for packaging food (residue limit 0.1%), and (3) as an ingredient for resinous and polymeric coatings for polyolefin films to be used safely as a food-contact surface of articles intended for use in producing, manufacturing, packing, transporting, or holding food.

References:
1. See **ACRYLONITRILE,** #1, 76.
2. *Bull. IRPTC,* 6, 18, 1983.
3. Holmberg, P. C. and Nurminen, M., Congenital defects of the central nervous system and occupational factors during pregnancy: a case-referent study, *Am. J. Ind. Med.,* 1, 167, 1980.
4. Schwetz, B. A., Mast, T. J., Weigel, R. J., et al., Developmental toxicity of inhaled methyl ethyl ketone in Swiss mice, *Fundam. Appl. Toxicol.,* 16, 742, 1991.
5. *Residue Reviews,* Gunther, F. A. and Gunther, J. D., Eds., Springer Verlag, New York, 97, 212, 1986.
6. Abdel-Rahman, M. S., Hetland, L. B., and Couri, D., Toxicity and metabolism of methyl *n*-butyl ketone, *Am. Ind. Hyg. Assoc. J.,* 37, 95, 1976.

N-METHYLFORMAMIDE (CAS No 123–39–7)

Synonym. Formic acid, *N*-methylamide.

Properties. Liquid. Soluble in water and alcohol.

Acute Toxicity. Ingestion of 50 ml by man results in acute poisoning. Manifestations of the toxic effect include liver damage, jaundice, and dyspepsia.[1] LD_{50} is reported to be 1.58 to 2.6 g/kg BW in mice and 4 g/kg BW in rats.

Reproductive Toxicity. *Teratogenic* effect is found in rats, mice, and rabbits.[2,3] The dose of 100 mg/kg administered on day 7 of pregnancy caused 90% mortality. Severe maternal toxicity and subsequent **embryolethality** were noted in pregnant rats and rabbits exposed to dermal application of 1 to 1.5 g N-M./kg BW at the period of organogenesis.[4]

Carcinogenicity. A 400 mg/kg BW dose caused some animals to die and initiated development of malignant neoplasms in surviving progeny.

References:

1. Vasil'eva, V. N. and Sukharevskaya, G. M., Justification of safe discharge of industrial sewage containing urea, *Gig. Truda Prof. Zabol.,* 12, 53, 1966 (in Russian).
2. See *N,N'*-**DIMETHYLFORMAMIDE,** #16.
3. See **FORMAMIDE,** #3.
4. Stula, E. F. and Krauss, W. C., Embryotoxicity in rats and rabbits from cutaneous application of amide-type solvents and substituted ureas, *Toxicol. Appl. Pharmacol.,* 41, 35, 1977.

METHYL ISOBUTYL KETONE (MIBK) (CAS No 108–10–1)

Synonyms. Hexone; Isopropyl acetone; **4**-Methyl-**2**-pentanone.

Properties. Clear liquid with a sweet, sharp odor. Solubility in water is 17 g/l at 20°C or 18 g/l at 25°C. Odor perception threshold is 1.3 mg/l[02] or 0.1 mg/l.[010]

Applications and **Exposure.** MIBK is used in the synthesis of cellulose and polyurethane lacquers and paint solvents. It is permitted as a flavoring agent and is used in food-contact materials. It occurs naturally in food. Levels of detection in certain foods is of the mg/kg range.

Acute Toxicity. LD_{50} is 4.56 g/kg BW in rats and 1.6 to 3.2 g/kg BW in guinea pigs.[1]

Short-term Toxicity. In a 90-d study on mice, rats, dogs, and monkeys, only male rats developed hyaline droplets in the proximal tubules of the kidney. In a 90-d gavage study in rats, a NOAEL of 50 mg/kg BW was identified.[2] Inhalation exposure affects the CNS and liver.

Reproductive Toxicity. Some maternal toxicity and retardation of ossification were observed in rats and mice exposed to 3000 ppm MIBK during the organogenesis period.[3]

Genotoxicity. MIBK appeared to be nonmutagenic in different test systems, *in vivo* and *in vitro*.

Chemobiokinetics. After administration, MIBK is widely distributed throughout the body and readily metabolized to water-soluble excretory products and can induce metabolic activation in the liver. Excretion of metabolites occurs mainly via the urine.[2]

Regulations. In the **EC countries** and in the **U.S.,** MIBK is allowed as a component of food-packaging materials.[2] A limit of 5 mg/l is suggested in beverages (EEC). **USFDA** (1993) approved the use of MIBK in adhesives as a component of articles intended for use in packaging, transporting, or holding food.

References:

1. See **VINYLTOLUENES,** #2, 1, 535.
2. IPCS, *Methyl Isobutyl Ketone,* Health and Safety Guide, WHO, Geneva, 1991, 28.
3. Tyl, R. W., France, K. A., Fisher, L. C., et al., Developmental toxicity evaluation of inhaled methyl isobutyl ketone in Fisher 344 rats and CD-1 mice, *Fundam. Appl. Toxicol.,* 8, 310, 1987.

MONOCHLOROBENZENE (MCB) (CAS No 108–90–7)

Synonyms. Benzene chloride; Chlorobenzene; Phenyl chloride.

Properties. A colorless, very refractive liquid with a faint, not unpleasant odor. Water solubility is 500 mg/l at 20°C or 110 mg/l at 25°C.[02] Freely soluble in alcohol, ether, and benzene. Taste and odor perception threshold is 0.05 mg/l,[02] or 0.01 to 0.12 mg/l,[1,01] or 0.001 to 0.003 mg/l (WHO, 1984).

Applications and **Exposure.** MCB is used in diisocyanate manufacturing, in silicone resin production, and also in the production of perchlorovinyl resins and to obtain lacquers, enamels, and glues. A solvent of nitrocellulose. Concentrations of 1 to 5 μg/l are found in water.

Acute Toxicity. *Humans.* A 2-year-old male who swallowed 5 to 10 ml of stain remover that consisted almost entirely of MCB became unconscious, did not respond to skin stimuli, showed muscle spasms, and became cyanotic. The child made a full recovery.[2] *Animals.* Rats tolerated an oral dose of 1 g/kg BW, but 4 g/kg BW was lethal to all. The LD_{50} for rats was reported to range from 2.4 to 3.3 g/kg BW. Guinea pigs tolerated an oral dose of 1.6 g/kg, but 2.8 g/kg BW was fatal. LD_{50} is 5.06 g/kg BW. Rats and mice are more sensitive than guinea pigs. LD_{50} is 1.45 to 2.3 g/kg BW in mice and 2.25 to 2.8 g/kg BW in rabbits. Following administration of lethal doses, ataxia, labored breathing, prostration, or lethargy were observed. Death occurred in the first 3 d from paralysis of the respiratory centers.[3,4] The principal morphological effects are hepatic and renal necrosis.

Repeated Exposure. Administration of $^1/_5$ LD$_{50}$ revealed marked cumulative properties. K$_{acc}$ is 1.25 (by Cherkinsky). The treated animals displayed asthenia, adynamia, and anorexia.[1] Neutrophilosis and a reduction in NS excitability were noted in rats that received $^1/_{10}$ LD$_{50}$ (Shamilov). In a 14-d study, the dose of 1 g/kg BW appeared to be lethal to rats; 0.5 g/kg BW did not affect survival and produced no clinical signs of intoxication.[4]

Short-term Toxicity. Male and female B6C3F$_1$ mice and Fisher 344 rats were given MCB at doses of 60 to 750 mg/kg BW by gavage for 13 weeks. Clinical signs of toxicity were not observed. A reduction in BW gain occurred at 250 mg/kg and higher doses. A marked increase in liver weights occurred in a dose-related manner. Histological examinations revealed toxic lesions in the liver, kidney, spleen, bone marrow, and thymus of MCB-exposed rats and mice. Porphyrinuria was detected at the higher doses. MCB is reported to cause renal necrosis at 250 mg/kg BW.[4,5] TDI was calculated to be 0.086 mg/kg BW from the NOAEL of 60 mg/kg BW.

Long-term Toxicity. *Humans.* Inhalation of MCB in the workplace for up to 2 years resulted in CNS disturbances. *Animals.* Long-term oral exposure to high doses affected mainly the liver, kidney, and hematopoietic system. Rabbits and guinea pigs received oral doses of 0.1 and 1 mg/kg BW for 11 to 14.5 months. The treatment caused behavioral changes, retardation of BW gain, and altered hemogram. Histological examination revealed visceral congestion, a focal atrophic gastritis with mucosal fibrosis, parenchymatous hepatitis, and dystrophy of the renal tubular epithelium (Obukhov). Rats were dosed by gavage with 0.1 mg/kg BW for 9 months. Manifestations of toxic action included CNS and hemopoiesis inhibition, anemia, reticulocytosis, thrombocytosis, and eosinophilia.[1] There were no MCB-related toxic effects in a 2-year study when rats and mice were given the doses of 0.03 to 0.12 mg/kg BW in corn oil for 103 weeks.[5]

Reproductive Toxicity. In a two-generation reproduction study, levels of 50 to 450 ppm had no adverse effects on reproductive performance or fertility of male and female rats.[6]

Immunotoxicity. B6C3F$_1$ mice exposed to the mixture of 25 common groundwater contaminants containing 0.2 mg MCB/l, for 14 or 90 d, showed some immune function changes which could be related to rapidly proliferating cells, including suppression of hematopoietic system cells and antigen-induced antibody-forming cells. There was no effect on T- and B-cell numbers in any group. Altered resistance to challenge with an infectious agent also occurred in mice given the highest concentration.[7] There was an increase in leukocyte phagocytic activity and in serum γ-globulin at a dose of 0.1 mg/kg BW in rats.[1]

Genotoxicity. The weight of evidence indicates that MCB is not mutagenic, although it does bind to DNA *in vitro* and *in vivo*.[8] The *in vivo* level of binding, however, is low. MCB has been found to be nonmutagenic in the Ames test; it is unable to induce unscheduled DNA synthesis in primary cultures of hepatocytes but is able in transformed adult rat liver epithelial cells *in vitro*.

Carcinogenicity. MCB caused a slight increase in the frequency of neoplastic liver nodules at the highest tested dose of 120 mg/kg BW,[4,5] providing some but not clear evidence of carcinogenicity in male rats. Carcinogenic effect was not observed in female Fisher 344/N rats or male and female B6C3F$_1$ mice. *Carcinogenicity classification.* USEPA: Group D; NTP: E—N—N—N.

Chemobiokinetics. Major metabolites of MCB are *p*-**chlorophenyl mercapturic acid,** 4-**chlorocatechol,** and *p*-**chlorophenol.**[9,10] The main route of excretion is in the urine. The toxicity of chlorinated benzenes could be explained by formation of **mercapturic acid,** for the synthesis of which sulfur-containing amino acids are used. Subsequently, the greater the number of halogen atoms in the benzene molecule, the lesser the mercapturic acid formed in the body and the lower the toxicity.[11]

Guidelines. WHO (1992). Guideline value for drinking water is identified at the level of 0.3 mg/kg. However, this value far exceeds the lowest reported taste and odor perception threshold in water. The levels of 0.01 to 0.12 mg/l are likely to give rise to consumer complaints of foreign odor and taste.

Standards. USEPA (1991). MCL and MCLG: 0.1 mg/l. **Russia** (1988). MAC and PML in drinking water: 0.02 mg/l.

Regulations. USFDA (1993) listed MCB as a component of adhesives intended for use in packaging, transporting, or holding food.

References:

1. See **DICHLOROBENZENES,** #1.
2. Reich, H., Puran (monochlorobenzene) poisoning in 2-year-old child, *Samml. von Vergiftungsfallen,* 5, 193, 1934 (in German).

3. Sanotsky, I. V. and Ulanova, I. P., *Criteria of Safety in Assessing the Danger of Chemical Compounds,* Meditsina, Moscow, 1975, 328 (in Russian).

4. Kluwe, W. M., Dill, G., Persing, R., et al., Toxic responses to acute, subchronic and chronic administrations of monochlorobenzene to rodents, *J. Toxicol. Environ. Health.,* 15, 745, 1985.

5. NTP, *Toxicology and Carcinogenicity Studies of Chlorobenzene in Fisher 344/N Rats and B6C3F₁ Mice,* NTP Technical Report Series No. 261, NIH Publ. No 86–2517, 1985.

6. Nair, R. S., Barter, J. A., and Schroeder, R. E., A two generation reproduction study with monochlorobenzene vapour in rats, *Fundam. Appl. Toxicol.,* 9, 678, 1987.

7. Germolec, D. R., Young, R. S. H., Ackermann, M. P., et al., Toxicology study of a chemical mixture of 25 ground water contaminants. II. Immunosuppression in B6C3F₁ mice, *Fundam. Appl. Toxicol.,* 13, 377, 1989.

8. Grilli, S., Arfrllini, G., Colacci, A., et al., *In vivo* and *in vitro* covalent binding of chlorobenzene to nucleic acids, *Jpn. J. Cancer Res.,* 76, 745, 1985.

9. Lindsay-Smith, J. R., Shaw, B. A. J., and Foulkes, D. M., Mechanisms of mammalian hydroxylation: some novel metabolites of chlorobenzene, *Xenobiotica,* 2, 215, 1972.

10. Parke, D. V. and Williams, R. T., The metabolism of halogenobenzenes, (a) Meta-dichlorobenzene. (b) Further observations on the metabolism of chlorobenzene, *Biochem. J.,* 59, 415, 1955.

11. See *p*-**TOLUENE SULFONAMIDE.**

NONYL ALCOHOL (NA) (CAS No 143–08–8)

Synonym. 1-Nonnanol.

Properties. Colorless to yellowish-colored liquid with a strong odor reminiscent of toilet soap. Poorly soluble in water (75 mg/l), readily soluble in alcohol. Odor and taste perception threshold is 0.1 mg/l. Does not affect the color and clarity of water.

Applications. Used in the rubber and dyestuff industry, in the production of plasticizers, emulsifiers, etc.

Acute Toxicity. LD_{50} is reported to be 12 to 19 g/kg BW in rats and 20 g/kg BW in mice (Yegorov and Andrianov). Signs of intoxication include increased reflex excitability, tail rigidity, convulsive twitching, motor coordination disorder, and respiratory distress.[1]

Long-term Toxicity. Doses of up to 0.05 mg/kg BW were ineffective for behavior, BW gain, hematological analysis, and vitamin C content in the visceral organs of rats and rabbits. Gross pathological examination revealed signs of parenchymatous dystrophy in the viscera (Kostovetsky, 1966).

Reproductive Toxicity. Inhalation exposure to NA produced no teratogenic effect in rats (Nelson et al., 1989).

Standards. Russia. MAC and PML: 0.01 mg/l.

Reference:
See **BUTYLENE,** 152.

OCTYL ALCOHOL (OA) (CAS No 111–87–5)

Synonyms. Caprylic alcohol; Octanol.

Properties. Colorless liquid with a penetrating odor. Solubility in water is 0.568 g/l; odor perception threshold is 0.05 mg/l (for primary and secondary OA).[1] According to other data, it appeared to be 0.13 to 0.2 mg/l.[2]

Acute Toxicity. LD_{50} of *primary* OA is reported to be 20 g/kg BW in rats and 15 g/kg BW in mice. That of *secondary* OA is 3.7 g/kg BW in rats, 3.1 g/kg BW in mice, 9.5 g/kg BW in guinea pigs, and 9.3 g/kg BW in rabbits. Intoxication is followed by narcosis. The survivors do not differ from the controls in appearance and behavior.[1]

Repeated Exposure revealed no cumulative properties. Rats tolerate administration of four LD_{50}s given as daily doses of $^1/_5$ LD_{50}.

Long-term Toxicity. Rats were dosed by gavage with 75 mg/kg BW of *sec*-OA. The treatment affected blood catalase and liver cholinesterase activity in rats. On chronic inhalation exposure (0.9

to 540 mg/m^3) of rats, OA caused NS inhibition and impairment of the oxidation-reduction processes and liver functions, as well as disorders in the blood acid-alkali balance.[3]

Immunotoxicity. A 75 mg/kg BW dose of *sec*-OA affected the immunobiological indices in rats.

Reproductive Toxicity. Inhalation during pregnancy caused no *embryotoxic* or *teratogenic effect*.[3]

Standards. Russia (1988). MAC and PML: 0.05 mg/l (organolept.).

Regulations. USFDA (1993) approved the use of OA (1) in adhesives as a component of articles intended for use in packaging, transporting, or holding food; (2) in resinous and polymeric coatings in a food-contact surface of articles intended for use in producing, manufacturing, packing, transporting, or holding food; and (3) as a lubricant (0.1%) in cellophane to be used safely for packaging food (2-ethylhexyl alcohol).

References:

1. Korolev, A. A., Krasovsky, G. N., and Varshavskaya, S. P., Hygienic evaluation of amyl alcohol, primary and secondary octyl alcohols in regard to their standardization in water bodies, *Gig. Sanit.,* 9, 88, 1970, (in Russian).
2. See **ACRYLONITRILE,** #7, 138.
3. Krashenina, G. I. and Kosiborod, N. R., Substantiation of the maximum allowable concentration for octyl alcohol in the air, *Gig. Sanit.,* 6, 82, 1986 (in Russian).

OXALIC ACID, DIETHYL ETHER (CAS No 95–92–1)

Synonyms. Diethyl ethanedioate; Diethyl oxalate; Ethyl oxalate.

Properties. Clear, colorless liquid. Poorly soluble in water; mixes with alcohol and hydrolyzes in water in presence of alkalis.

Applications. Used in the synthesis of cellulose ethers.

Acute Toxicity. LD$_{50}$ is 1.5 g/kg BW in rats and 2 g/kg BW in mice. Species susceptibility is not evident. Lethal doses caused CNS inhibition. Narcotic sleep was found to develop. Death occurred in weeks.

Repeated Exposure revealed pronounced accumulation. K$_{acc}$ appeared to be 1.1 (by Lim).

Reference:

See **TRIETHYLENE GLYCOL DI(2-ETHYLHEXOATE).**

1-PROPANOL (NPA) (CAS No 71–23–8)

Synonyms. Ethyl carbinol; 1-Hydroxypropane; *n*-Propanol; Propyl alcohol.

Properties. Colorless liquid with a characteristic odor of alcohol. Miscible with water and ethyl alcohol at all ratios. Odor perception threshold is in the range of 23 to 40 mg/l, but according to other data, it is <0.033 mg/l.[010] Taste perception threshold is 12 mg/l.[1,2]

Applications and **Exposure.** Used in the production of synthetic polymers, such as polyvinylbutyral, cellulose esters, lacquers, and PVC adhesives. Exposure of human beings may occur through the ingestion of food or beverages containing NPA.

Acute Toxicity. NPA exhibits low toxicity, except in very young rats. (LD$_{50}$ is found to be 560 to 660 mg/kg BW.)[3] LD$_{50}$ is 1.87 to 6.5 g/kg BW in rats, 6.8 g/kg BW in mice, and 2.82 g/kg BW in rabbits.[4] The principal toxic effect appeared to be depression of the CNS. NPA is more neurotoxic than ethanol. In rabbits, LD$_{50}$ for narcosis is 1440 mg/kg BW.

Repeated Exposure. Four daily doses of 2.16 g/kg BW given to rats did not produce death or gross pathological changes in their livers.[4] Rats were exposed by feeding to the doses of 45 mg/kg BW for 2 months.[1] The treatment produced changes in serum enzyme activity and signs of dystrophy in the liver and brain.

Reproductive Toxicity. NPA produced a teratogenic effect in rats at high inhalation concentrations.[5]

Genotoxicity. NPA did not produce SCE or micronuclei in mammalian cells *in vitro*.[6]

Carcinogenicity. Wistar rats were exposed to the oral doses of 240 mg/kg BW or to subcutaneous doses of 48 mg/kg BW throughout their life-span. A significant increase in the incidence of liver

312

sarcoma was found in the group dosed subcutaneously.[7] According to IARC, because of technical limitations, this study seems to be inadequate.

Chemobiokinetics. NPA is rapidly absorbed and distributed following ingestion. It is oxidized to **propionic acid** and then to CO_2. Traces of glucuronide conjugates are found in rabbits. NPA is excreted in the urine or in expired air.[8]

Standards. Russia (1988). MAC: 0.25 mg/l (organolept.), PML in food: 0.1 mg/l.

Regulations. USFDA (1993) approved the use of NPA in cellophane to be used safely for packaging food (residue limit 0.1%).

References:

1. *Hygiene of Populated Areas,* Proc. Sci. Conf., Kiev, 1969, 40 (in Russian).
2. *Data on Hygiene Assessment of Pesticides and Polymers*, Erisman Research Hygiene Institute, Moscow, 1977, 281 (in Russian).
3. Purchase, I. F. H., Studies in kaffircorn malting and brewing. XXII. The acute toxicity of some fusel oils found in Bantu beer, *S. A. Med. J.,* 43, 795, 1969.
4. Taylor, J. M., Jenner, P. M., and Jones, W. I., A comparison of the toxicity of some allyl, propenyl, and propyl compounds in the rat, *Toxicol. Appl. Pharmacol.,* 6, 378, 1964.
5. Nelson, B. K., Brightwell, W. S., MacKensie-Taylor, D. R., et al., Teratogenicity of *n*-propanol and isopropanol administered at high inhalation concentrations to rats, *Food Chem. Toxicol.,* 26, 247, 1988.
6. IPCS, *2-Propanol,* Environmental Health Criteria No. 103, WHO, Geneva, 1990, 132.
7. Gibel, W., Lohs, K., and Wildner, G. P., Experimental study on the cancerogenic activity of propanol-**1, 2**-methylpropanol-**1** and **3**-methylbutanol-**1**, *Arch. Geschwulstforsch.,* 45, 19, 1975 (in German).
8. See **ANILINE,** #11, 167; See also 2-**Propanol.**

2-PROPANOL (IP) (CAS No 67–63–0)

Synonyms. Dimethylcarbinol; Isopropanol (IP); Isopropyl alcohol.

Properties. Colorless liquid. Mixes with water and alcohol at any ratio. Odor perception threshold is 1.13,[1] 3.2,[010] or 160 mg/l.[02]

Applications. Used in the production of coatings such as phenolic varnishes and nitrocellulose lacquers.

Acute Toxicity. *Humans.* Ingestion caused an alcoholic intoxication and narcosis. *Animals.* LD_{50} is 5 to 5.5 g/kg BW in rats and 3.6 to 4.5 g/kg BW in male mice. Poisoning is accompanied by a narcotic effect.[1,2] Death occurred within 24 h in the lethargy condition because of respiratory failure. Gross pathological examination revealed congestion, edema, and hemorrhages into the interstitial tissues of the parenchymatous organs, which displayed inflammatory and dystrophic changes.[3]

Repeated Exposure. *Humans.* Eight volunteers received daily doses of 6.4 mg/kg BW for 6 weeks. This intake did not affect blood and urinalysis, as well as liver and kidney functions.[4] *Animals.* Moderate cumulative properties have been noted. K_{acc} has been found to be 4.9 and 4 (by Lim) for mice and rats, respectively. Rats were dosed by gavage with 115 mg/kg BW in the feed for 2 months. Changes in serum enzyme activity and dystrophic changes in the liver and brain have been reported.

Long-term Toxicity. In a two-generation reproduction study, IP was administered to rats via drinking water at 2% concentration (through the first and second generation). The treatment resulted in several changes that represent mild toxicity and are reminiscent of stress lesions.[5] Rats received drinking water containing IP for 27 weeks. At the end of the exposure period, all exposed females showed growth retardation.[2]

Reproductive Toxicity. In a three-generation study in rats, average daily intake of 1.29 to 1.47 g/kg BW did not produce adverse effects on growth and reproductive functions of animals.[6] Doses of 1.8 g/kg BW were administered to rats for 3 months before mating. *Embryolethality* in rats was increased twofold. Rats given 252 and 1008 mg/kg BW for 2 months exhibited an extended duration of the estral cycle.[7]

Timed-pregnant Sprague-Dawley rats were dosed orally with aqueous IP solutions at dose levels of 400, 800, and 1200 mg IP/kg BW on gestation days 6 through 15 at a dosing volume of 5 ml/kg.

New Zealand white rabbits were dosed orally at dose levels of 120, 240, and 480 mg IP/kg BW on gestation days 6 through 18 at a dosing volume of 2 ml/kg. No evidence was found of increased *teratogenicity* at any doses tested in rats and rabbits. The NOAELs for both maternal and developmental toxicity were 40 mg/kg BW in rats, and 240 and 480 mg/kg BW respectively in rabbits.[8] IP given orally to timed-pregnant rats at the doses as high as 1200 mg/kg BW caused no biologically significant findings in the behavioral tests, no changes in organ weights, and no pathological findings in offsprings that could be attributed to IP exposure.[13]

An **allergenic effect** was not observed on the skin application test.[3]

Genotoxicity. IP is found to be negative in a variety of genotoxicity tests, but inhalation exposure produced a mutagenic effect in rats (cytogenic analysis of bone marrow cells).[9]

Chemobiokinetics. IP is absorbed rapidly and distributed throughout the body, partially as **acetone.** It undergoes oxidation in the body with the formation of **acetic acid.** Following ingestion of an unknown amount of rubbing alcohol, IP as well as its metabolite acetone were found in the spinal fluid of two persons at levels similar to those in the serum.[10] Excretion of both substances is limited and does not exceed 4% of the dose in rats, rabbits, and dog.[11,12] The major route of excretion is via the lungs.

Standards. Russia (1988). MAC: 0.25 mg/l, PML in food: 0.1 mg/l.

Regulations. USFDA (1993) approved the use of IP (1) in adhesives as a component of food-contact articles, (2) in cellophane to be used safely for packaging food (residue limit 0.1%), (3) as a defoaming agent that may be used safely as a component of articles and in the manufacturing of paper and paperboard intended for use in contact with food, and (4) as a substance employed in the production of or added to textiles and textile fibers intended for use in contact with food.

References:
1. Galeta, S. G., Hygienic Substantiation of the MAC for Isopropyl Alcohol in Water Bodies, Author's abstract of thesis, Medical Institute, Donetsk, 1967, 14 (in Russian).
2. Lehman, A. J. and Chase, H. F., The acute and chronic toxicity of isopropyl alcohol, *J. Lab. Clin. Med.,* 29, 561, 1944.
3. Guseinov, V. G., Toxicologic and hygienic characteristics of isopropyl alcohol, *Gig. Truda Prof. Zabol.,* 7, 60, 1985.
4. Wills, J. H., Jameson, E. M., and Coulston, F., Effects on man of daily ingestion of small doses of isopropyl alcohol, *Toxicol. Appl. Pharmacol.,* 15, 560, 1969.
5. Gallo, M. A., Oser, B. L., Cox, G. E., et al., Studies on the long-term toxicity of 2-butanol, in 16th Annual Meeting Soc. Toxicol., Toronto, Abstracts of Papers, March 27–30, 1977, p. 9.
6. Lehman, A. J., Schwerma, H., and Rickards, E., Acquired tolerance in dogs, rate of disappearance from the blood stream in various species, and effects on successive generation of rats, *J. Pharmacol. Exp. Ther.,* 85, 61, 1945.
7. Antonova, V. I. and Salmina, Z. A., Substantiation of the Maximum Allowable Concentration for isopropyl alcohol as regards to its effect on gonads and offspring, *Gig. Sanit.,* 1, 8, 1978 (in Russian).
8. Tyl, R. W., Masten, L. W., Marr, M. C., et al., Developmental toxicity evaluation of isopropanol administered by gavage in rats and rabbits, *Fund. Appl. Toxicol.,* 22, 139, 1994.
9. Aristov, V. N., Red'kin, Yu. V., Bruskin, Z. Z., et al., Experimental data on the mutagenous effects of toluene, isopropanol, and sulfur dioxide, *Gig. Truda Prof. Zabol.,* 7, 33, 1981 (in Russian).
10. Natowicz, M., Donahue, J., Gorman, L., et al., Pharmacokinetic analysis of a case of isopropanol intoxication, *Clin. Med.,* 31, 326, 1985.
11. Rietbrock, N. and Abshagen, U., Pharmacokinetics and metabolism of aliphatic alcohols, *Arzneimittel-Forsch.,* 21, 1309, 1971 (in German).
12. See 1-**PROPANOL,** #6.
13. Bates, H. K., McKee, R. H., Bieler, G. S., et al., Developmental neurotoxicity evaluation of orally administered isopropanol in rats, *Fund. Appl. Toxicol.,* 22, 152, 1994.

TETRACHLOROETHYLENE (PCE) (CAS No 127–18–4)

Synonyms. Ethylene tetrachloride; Perclene; Perchloroethylene; Tetlen; Tetracap; **1,1,2,2**-Tetrachloroethene; Tetraleno.

Properties. Colorless, highly volatile, nonflammable liquid with a sweetish taste. Odor reminiscent of ether or chloroform. Water solubility is 150 mg/l at 25°C. Odor perception threshold is 0.17 mg/l[02] or 0.3 mg/l (USEPA, 1987).

Applications and **Exposure.** Used as a solvent for resins, lacquers, and paints. Primarily used in the dry-cleaning industry. Trace amounts are found in food and ground and surface water but seem to be minimal due to PCE high volatility.

Acute Toxicity. LD_{50} is reported to be 3.8 g/kg BW in male and 3 g/kg BW in female rats.[1] In mice, LD_{50} is 8.4 to 10.3 g/kg BW[2] or 6.4 to 8 g/kg BW.[3] The principal manifestations in humans and animals of acute exposure to PCE are CNS depression, ataxia, and respiratory and cardiac arrest. Gross pathological examination revealed fatty infiltration of the liver and heart and changes in the respiratory and circulatory systems in dogs given 0.3 to 0.4 g/kg BW.[2]

Repeated Exposure. Swiss-Cox mice received calculated doses of 14 to 1400 mg/kg BW in corn oil for 6 weeks.[4] The treatment caused histopathological lesions, including impairment of hepatic triglyceride levels, DNA content, and serum enzyme activity. In mice dosed with up to 100 mg PCE/kg BW, there was an increase in liver triglyceride concentration and in liver relative weights. Histological changes were noted in mice given 0.1 g/kg BW for 11 d[5] but not in rats exposed to a 0.016 mg/kg BW dose.[2]

Short-term Toxicity effects included damage to the liver and kidney. Sprague-Dawley rats received doses of 14 to 1400 mg PCE/kg BW via their drinking water for 3 months.[1] The exposure revealed retardation of BW gain in high dose-level groups of animals. Rats were given 10 mg PCE/kg BW. The treatment caused an 18 to 38% increase in the excretion of 17-ketosteroids, with subsequent normalization in the 3rd to 4th month of the experiment.[6]

Genotoxicity. T. has not been shown to exhibit DNA binding in short-term studies. It induced single-strand DNA breaks in the mouse but did not cause CA in rat bone marrow or human lymphocytes.[2,7]

Reproductive Toxicity. No fetal toxicity or teratogenicity was found in pregnant mice and rats exposed to concentrations of 300 ppm.[8]

Carcinogenicity. Daily administration of high oral doses (by gavage) increased the incidence of hepatocellular carcinomas in both sexes of $B6C3F_1$ mice but not in Osborne-Mendel rats.[7] Due to a high level of mortality among the rats, these data are not considered to be adequate. Inhalation exposure to PCE resulted in an increased incidence of renal tubular cell adenomas and adenocarcinomas[9] that could be considered a result of formation of a highly reactive metabolite and cell damage produced by renal accumulation of α-2μ globulin, both in male rats only.[10] *Carcinogenicity classification.* USEPA: Group B2; IARC: Group 2B; NTP: IS*—IS*—P*—P*.

Chemobiokinetics. PCE is readily absorbed in the GI tract. The main metabolites are **tetrachloroethene oxide** and **trichlorometabolites. Trichloroethylene** and **trichloroacetic acid** (25% of excreta) are found in the urine. PCE is not shown to exhibit direct interaction with hepatic DNA. PCE is metabolized to trichloroacetic acid in mice and Fisher 344 rats and in humans. Only $B6C3F_1$ mice exhibit peroxisome proliferation and carcinogenicity in the liver.[10] PCE is removed predominantly via exhalation. Following oral administration of 1 mg tetrachloro-(^{14}C)ethylene per kilogram to Sprague-Dawley rats, about 70% of radioactivity is excreted in expired air as PCE, and 26% as CO_2 and nonvolatile metabolites in the urine and feces.[11] **Oxalic acid** is considered to be an important metabolite. Disposition of PCE is a saturable, primarily dose-dependent process in rats.

Guidelines. WHO (1992). Guideline value for drinking water: 0.04 mg/l.

Regulations. USFDA (1993) approved the use of PCE in adhesives as a component of articles intended for use in packaging, transporting, or holding food.

References:
1. Hayes, J. R., Condie, L. W., and Borcelleca, J. F., The subchronic toxicity of tetrachloroethylene (perchloroethylene) administered in the drinking water of rats, *Fundam. Appl. Toxicol.,* 7, 119, 1986.
2. *Tetrachloroethylene,* Environmental Health Criteria Series 31, WHO, Geneva, 1984, 48.
3. *Problems of Medical Chemistry,* Ufa, 1980, 22 (in Russian).
4. Buben, J. A. and O'Flaherty, E. J., Delineation of the role of metabolism in the hepatotoxicity of trichlorothylene and perchlorothylene: a dose-effect study, *Toxicol. Appl. Pharmacol.,* 78, 105, 1985.

5. Schumann, A. M., Quast, J. F., and Watanabe, P. G., The pharmacokinetics and macromolecular interaction of perchloroethylene in mice and rats as related to oncogenicity, *Toxicol. Appl. Pharmacol.*, 55, 207, 1980.

6. *Endocrine System and Toxic Environmental Factors*, Leningrad, 1980, 139 (in Russian).

7. National Cancer Institute, Bioassay of Tetrachloroethylene for Possible Carcinogenicity, Department of Health, Education and Welfare, NIH 77–813, Washington, D.C., 1977.

8. See **DICHLOROMETHANE**, #10.

9. NTP, *Toxicology and Carcinogenesis Studies of Tetrachloroethylene (Perchloroethylene) in F344/N Rats and B6C3F₁ Mice (Inhalation Studies)*, NIH TR 311, U.S. Department of Health and Human Services, Research Triangle Park, NC, 1986.

10. Green, T., Chloroethylenes: a mechanistic approach to human risk evaluation, *Annu. Rev. Pharmacol. Toxicol.*, 30, 73, 1990.

11. Pegg, D. G., Zempel, J. A., Braun, W. H., et al., Disposition of tetrachloro(^{14}C)ethylene following oral and inhalation exposure in rats, *Toxicol. Appl. Pharmacol.*, 51, 465, 1979.

TETRAHYDROFURFURYL ALCOHOL (TA) (CAS No 97–99–4)

Properties. Colorless liquid with a pronounced odor of diethyl ether. Miscible with water and ethyl alcohol at any ratio. Odor perception threshold is 8.6 mg/l. TA does not alter the taste of water at a concentration of 5 g/l.[1,2]

Acute Toxicity. LD_{50} values are 2.5 g/kg BW in rats, 2.3 g/kg BW in mice, and 3 g/kg BW in guinea pigs.[1] Acute poisoning resulted in a narcotic effect.

Repeated Exposure revealed cumulative properties.[1] Mice were dosed by gavage with 100 mg/kg BW for 1.5 months. The treatment caused retardation of BW gain, transient hind limb paralysis, leukocytosis, and anemia. Some of the animals died. A 40 mg/kg BW dose appeared to be ineffective.

Long-term Toxicity. Similar changes were found in the 6-month study in mice exposed to an oral dose of 10 mg/kg BW.[1] The NOAEL was identified to be 5 mg/kg BW.[2]

Allergenic Effect is not observed on chronic oral exposure to a dose of 20 mg/kg BW.[3]

Standards. Russia (1990). PML in drinking water: 3 mg/l (organolept., odor).

Regulations. USFDA (1993) approved the use of TA (1) as a component of adhesives to be used safely in a food-contact surface and (2) as a defoaming agent in the manufacturing of paper and paperboard intended for use in producing, manufacturing, packing, transporting, or holding food.

References:

1. See **ACROLEIN**, #1, 91.

2. *Proc. Sanitary - Hygiene Med. Inst.*, Leningrad, 106, 51, 1974 (in Russian).

3. Ilichkina, A. G., Method of investigation of allergenic properties of chemicals in the course of their standard-setting in water bodies, *Gig. Sanit.*, 12, 42, 1979 (in Russian).

TOLUENE (CAS No 108–88–3)

Synonyms. Methylbenzene; Methacide; Phenylmethane; Toluol.

Properties. Clear, colorless liquid with a characteristic odor. Solubility in water is 470 mg/l at 16°C or 540 mg/l at 25°C, miscible with alcohol at all ratios. Odor perception threshold is 0.17 mg/l[10] or 0.042 mg/l.[02] The reported taste perception thresholds vary from 0.4 to 0.12 mg/l.

Applications and **Exposure.** T. is found to migrate from synthetic coating materials commonly used to protect drinking water storage tanks.[1] There is limited information concerning the oral intake of T., and this intake is likely to be much lower compared to the intake via air.

Acute Toxicity. In rats, oral LD_{50} varies from 2.6 to 7.5 g/kg BW.[2,3] Administration of the high doses led to an increase in the concentration of mediators in the brain, with subsequent effects on behavioral responses.

Repeated Exposure failed to reveal cumulative properties. Exposure to T. may cause adverse effects on the NS, kidneys, and liver. Several daily doses of 2 to 4 g/kg BW caused mortality in mice. Administration of 216 to 433 mg/kg BW for 25 d produced no signs of intoxication.[4]

Short-term Toxicity. In a NTP-69 study, rats were administered T. in corn oil at dose-levels of 312 to 5000 mg/kg BW for 13 weeks. Liver-to-brain ratio was significantly increased in males receiving the dose of 625 mg/kg BW. A NOAEL of 312 mg/kg, adjusted to 223 mg/kg BW for exposure of 5 days per week, was established in this study.

Long-term Toxicity. Rats tolerated doses of 118 to 590 mg/kg BW given to them for a half a year. Rabbits received 0.25 mg/kg and 1 mg T./kg BW for 9.5 months, and 10 mg T./kg BW for 5 months. In both studies, there were no changes in general condition, in blood and biochemical analyses, and in the morphological and histological structure of the visceral organs.[4]

Reproductive Toxicity. *Humans.* Shepard summarized incidences of embryotoxic and teratogenic effects.[5] *Animals.* ***Embryotoxic*** and ***fetotoxic effect*** but not teratogenic effect has been observed at high dose-levels in mice and rats but not in rabbits. In one of two oral studies, embryolethality occurred at dose-levels of more than 260 mg/kg BW with a ***teratogenic effect*** (increased incidence of cleft palate) at the highest dose-level (870 mg/kg BW) only. Another oral study in mice was limited to behavioral parameters, with an effect seen at 400 mg/l in drinking water but not at 80 mg/l.[6]

Genotoxicity. T. was not found to be genotoxic. *Humans.* No changes were found in the peripheral blood lymphocytes of workers exposed daily to T.[7] T. also does not cause SCE in human lymphocyte culture.[8] *Animals.* There is no cytogenic effect in the bone marrow cells of mice on ten administrations of 0.0001 to 0.1 LD_{50}. When given to males for 5 weeks, it has no effect on the incidence of DLM in mice. Positive results were found in a micronuclear test in random-bred SNK mice given 8 to 1000 mg T./kg in sunflower oil by gavage. The safe concentration of T. in water for a genotoxic effect is 20 mg/l.[9] T. did not cause CA and SCE in Chinese hamster ovary cells (NTP–85).

Carcinogenicity. In a 2-year study, T. was administered to Sprague-Dawley rats in olive oil by stomach tube. An increase in the total number of animals with malignant tumors (types unspecified) was found.[10] The incomplete reporting of tumor pathology in this study was indicated. Available studies are inadequate for evaluation (IARC 47–106). *Carcinogenicity classification.* IARC: Group 3; NTP: NE—NE—NE—NE.

Chemobiokinetics. T. appears to be absorbed completely from the GI tract after oral intake. After absorption, it is rapidly distributed in the body and accumulates preferentially in the highly vascularized organs and adipose tissue, successively followed by adrenals, kidneys, liver, and brain. T. metabolism occurs through oxidation of the methyl group to **benzyl alcohol** by the microsomal mixed-function oxidase system in the liver. Another biotransformation route includes subsequent oxidation, sulfonation, and conjugation with glutatione. Of the dose, 80% is excreted as **hippuric acid** in the urine of man and rabbit. In the lung, part of the resorbed amount of T. is excreted unchanged.[11]

Guidelines. WHO (1992). Guideline value for drinking water: 0.7 mg/l. The levels of 0.024 to 0.17 mg/l are likely to give rise to consumer complaints of foreign taste and odor.

Standards. USEPA (1991). MCL and MCLG: 1 mg/l. **Russia** (1988). MAC: 0.5 mg/l (organolept., odor).

Regulations. USFDA (1993) approved the use of T. (1) in adhesives as a component of articles intended for use in packaging, transporting, or holding food; (2) in resinous and polymeric coatings for polyolefin films to be used safely as a food-contact surface; (3) as a component of the uncoated or coated food-contact surface of paper and paperboard that may be used safely in producing, manufacturing, packing, transporting, or holding dry food; (4) in semirigid and rigid acrylic and modified acrylic plastics to be used safely as articles intended for use in contact with food; (5) as a solvent in polysulfide polymer-polyepoxy resins used as the surface contacting dry food; and (6) in cellophane to be used safely for packaging food (residue limit of 0.1%).

References:
1. Bruchet, A., Shipert, E., and Alban, K., Investigation of organic coating material used in drinking water distribution systems, *J. Fr. d'Hydrol.,* 19, 101, 1988 (in French).
2. van der Heijeden, C. A., Mulder, H. C. M., de Vrijer, F., et al., Integrated Criteria Document: Toluene Effects, Appendix to report No. 75847310, Natl. Inst. Public Health Environ. Protection, Bilthoven, The Netherlands, 1988.
3. See α-**METHYL STYRENE,** #1.
4. See **ACRYLONITRILE,** #1, 109.

5. Shepard, T. H., *Catalog of Teratogenic Agents,* The Johns Hopkins University Press, Baltimore, MD, 1992, 388.
6. IPSC, *Toluene,* Environmental Health Criteria, No. 52, WHO, Geneva, 1985.
7. Maki-Paakkanen, J., Husgafvel-Pursiainen, K., Kolliomaki, P. L., et al., Toluene-exposed workers and chromosomal aberrations, *J. Toxicol. Environ. Health,* 6, 775, 1980).
8. Gerner-Schmidt, P. and Freidrich, U., The mutagenic effect of benzene, toluene, and xylene studied by the SCE technique, *Mutat. Res.,* 85, 313, 1978.
9. See **NITROBENZENE,** #12.
10. Maltoni, C., Conti, B., and Cotti, G., Experimental studies on benzene carcinogenicity at the Bologna Institute of Oncology: current results and ongoing research, *Am. J. Ind. Med.,* 7, 415, 1985.
11. See **ANILINE,** #11, 25.

TRIBUTYLAMINE (CAS No 102–82–9)

Synonym. *N,N'*-Dibutyl-1-butanamine.

Properties. Hygroscopic, colorless, oily liquid with a penetrating odor. Solubility in water is 0.1%, readily soluble in alcohols and fats. Odor and taste perception threshold is 0.9 mg/l.[1]

Applications. A principal solvent in vulcanization processes.

Acute Toxicity. LD$_{50}$ is reported to be 115 mg/kg BW in mice, 455 mg/kg BW in rats, 350 mg/kg BW in guinea pigs, and 615 mg/kg BW in rabbits. Poisoned animals displayed a short period of excitation, with subsequent inhibition and adynamia. Death occurred within 2 to 4 d. Histological examination revealed dystrophic changes in the liver, necrosis, white pulp hyperplasia in the spleen, and myocardial edema.

Repeated Exposure failed to reveal cumulative properties. Mice tolerated administration of $^1/_5$ LD$_{50}$ for 20 d.[1] A 3-d administration of $^1/_3$ LD$_{50}$ caused enzyme disorganization of the intracellular structures of the liver cells, especially of those associated with detoxifying and energy biotransformation processes.[2]

Long-term Toxicity. Rabbits were dosed by gavage with 0.6 and 6 mg/kg BW for 6 months. The treatment did not affect BW gain, but in the higher dose group, changes were noted in the blood clotting time system. There was an increase in glutaminate oxalate transaminase activity and a decline in blood histamine level, as well as in liver deaminoxidase activity. Histological examination revealed splenic hyperplasia at the higher dose, liver fatty dystrophy, and intestinal mucosal edema.[1]

Chemobiokinetics. See *Dibutylamine.*

Standards. Russia (1988). MAC and PML: 0.9 mg/l (organolept., odor, taste).

References:

1. Le Din Min, Hygienic Assessment of Liquid Effluents from the Manufacture of Normal Butylamines and Protection of Reservoirs, Author's abstract of thesis, Sanitary–Hygiene Medical Institute, Leningrad, 1977, 24 (in Russian).
2. See **DIBUTYLAMINE,** #4.

1,1,1-TRICHLOROETHANE (TCA) (CAS No 71–55–6)

Synonyms. Chlorothene; Methylchloroform; TCEN.

Properties. Colorless, nonflammable liquid. Water solubility is 44 mg/l at 25°C. Absorbs some water. Soluble in acetone, methanol, and ether.

Acute Toxicity. *Humans.* Nonlethal acute intoxication occurred after oral ingestion of a liquid ounce of TCA (0.6 g/kg BW).[1] Such an intake is accompanied by nausea, vomiting, and diarrhea. The lethal dose in man is likely to be 5 ml. *Animals.* LD$_{50}$ for several species of animals ranges from 5.7 to 14.3 g/kg BW.[2] It is 14.3 g/kg BW in rats and 8.6 g/kg BW in guinea pigs.[3] A single oral dose of about 1.4 g/kg BW depressed some hepatic microsomal metabolic indices in rats (including cytochrome P-450 and epoxide hydratase).[4]

Repeated Exposure. Rats were gavaged with 0.5 g TCA/kg BW for 9 d. There was relatively little evidence of toxicity. Doses of 5 and 10 g/kg BW caused a transient hyperexcitability and protracted narcosis.[5]

Short-term Toxicity. Doses of 0.5 to 5 g/kg BW were administered to rats by gavage for up to 12 weeks. Doses of 2.5 to 5 g/kg BW reduced BW gain and produced CNS effects. Approximately 35% of these rats died during the first 7 weeks of the experiment. The dose of 0.5 g/kg BW turned out to be ineffective.[5]

Long-term Toxicity. Principal noncarcinogenic effects include CNS inhibition, increased liver weights, and cardiovascular changes. Rats were given 0.75 and 1.5 g TCA/kg BW in corn oil by gavage for 78 weeks. Similarly, mice received 2.8 and 5.6 g/kg BW for 78 weeks. Diminished BW gain and decreased survival time were reported in both rats and mice. Selected dose-levels were very high in this study, and only 3% of animals survived to the end of the experiment.[6]

Reproductive Toxicity. There were no dose-dependent effects on fertility, gestation, and viability indices in mice exposed to TCA at dose-levels up to 1 g/kg BW in drinking water.[3] Neither maternal toxicity nor significant effect on the morphological development of CD rats were found in the study where animals received 3, 10, and 30 ppm TCA in their drinking water for 14 d prior to cohabitation and for up to 13 d during the cohabitation period and during pregnancy.[7]

Genotoxicity. TCA appeared to be mutagenic to various strains of *Salmonella* with metabolic activation.[3] However, positive and negative results in *Salmonella* mutagenicity assays provided minimal evidence of DNA binding in test systems.

Carcinogenicity. Rats and mice were gavaged with TCA in corn oil at doses of 375 or 750 mg/kg BW (rats) and 1500 or 3000 mg/kg BW (mice) for 103 weeks. No treatment-related tumors were observed in male rats. The study was inadequate for evaluation of female rats because the high dose was toxic and there was a large number of accidental deaths.[8] However, in mice there was an increased incidence of hepatocellular carcinoma. *Carcinogenicity classification.* USEPA: Group D; IARC: Group 3; NTP: IS—IS—IS—IS.

Chemobiokinetics. TCA is metabolized to a very limited extent (no more than 6% of the dose). The metabolites include **trichloroethanol, TCA-glucuronide,** and **trichloroacetic acid,** which are excreted primarily in the urine. Approximately 1% of TCA, however, is excreted unchanged by the lungs.[9] According to D'Urk et al.,[10] acetylene is found to be a new metabolite of T. in Sprague-Dawley rats following inhalation at the concentration of 2000 ppm.

Guidelines. WHO (1992). Provisional guideline value for drinking water: 2 mg/l.

Standards. USEPA (1991). MCL and MCLG: 0.2 mg/l.

Regulations. USFDA (1993) regulates TCA (1) as a component of adhesives to be used safely in a food-contact surface and (2) as a cross-linking agent in polysulfide polymer-polyepoxy resins used as the surface contacting dry food.

References:
1. Stewart, R. D. and Andrews, J. T., Acute intoxication with methyl chloroform vapor, *JAMA,* 195, 705, 1966.
2. Torkelson, T. R., Oyen, F., McCollister, D., et al., Toxicity of **1,1,1**-trichloroethane as determined on laboratory animals and human subjects, *Am. Ind. Hyg. Assoc. J.,* 19, 353, 1958.
3. Lane, R. W., Riddle, B. L., and Borzelleca, J. F., Effect of **1,2**-dichloroethane and **1,1,1**-trichloroethane in drinking water on reproduction and development in mice, *Toxicol. Appl. Pharmacol.,* 63, 409, 1982.
4. Vainio, H., Parkki, M. A., and Marniemi, J. A., Effects of aliphatic chlorohydrocarbons on drug-metabolizing enzymes in rat liver *in vivo, Xenobiotica,* 6, 599, 1976.
5. Bruckner, J. V., Muralidhara, S., Mackenzie, W. F. et al., Acute and subacute oral toxicity studies of **1,1,1**-trichloroethane in rats, *Toxicologist,* 5, 100, 1985.
6. National Cancer Institute, Bioassay of **1,1,1**-Trichloroethane for Possible Carcinogenicity, Tech. Rep. Ser. No. 3, January, 1977.
7. George, J. D., Price, C. J., Marr, M. C., et al., Developmental toxicity of **1,1,1**-trichloroethane in CD rats, *Fundam. Appl. Toxicol.,* 13, 641, 1989.
8. NTP, *Carcinogenesis Bioassay of 1,1,1-Trichlorethane in F344/N Rats and B6C3F₁ Mice,* NTP, 1983.
9. Monster, A. C., Boersma, G., and Steenweg, M., Kinetics of **1,1,1**-trichloroethane in volunteers; influence of exposure concentration and work load, *Int. Arch. Occup. Environ. Health,* 42, 293, 1979.
10. D'Urk, H., Poyer, J., Lee, K. C., et al., Acetylene, a mammalian metabolite of 1,1,1-trichloroethane, *Biochem. J.,* 286, 353, 1992.

1,1,2-TRICHLOROETHANE (TCA) (CAS No 79–00–5)

Synonym. Vinyl trichloride.

Properties. Clear, colorless liquid with a sweet odor. Water solubility is 4.5 g/l at 20°C. Soluble in ethanol and diethyl ether.

Applications and **Exposure.** An intermediate in the production of vinylidene chloride; a solvent for natural resins, chlorinated rubbers, adhesives, and coatings laid down on films.

Acute Toxicity. In CD-1 mice, LD_{50} is found to be 378 mg/kg BW (males) and 491 mg/kg BW (females). Signs of toxicity included sedation, gastric irritation, lung hemorrhage, and liver and kidney damage.[1]

Repeated Exposure. In a 14-d study, doses of 3.8 and 38 mg/kg BW caused no toxic effects in CD-1 mice.[2]

Short-term Toxicity. In a 90-d study, CD-1 mice were given 0.02 to 2 g/l in their drinking water. A decrease in BW gain and liver glutathione level in males was observed. Decreased hematocrit and Hb blood levels were found in females.[1,2]

Long-term Toxicity. In rabbits, an increase in excretion of 17-ketosteroids was noted on months 1 to 3 of the treatment with a dose of 100 mg/kg BW. Normalization was observed up to the fifth month of the experiment.[3]

Immunotoxicity. In the above-described studies,[1,2] antibody-forming function and phagocytic activity of the peritoneal macrophages was altered.

Reproductive Toxicity. The oral dose that killed 10% of pregnant mice caused no developmental toxicity in the offspring of survivors.[4]

Genotoxicity. No mutagenic effect in bacteria was noted.

Carcinogenicity. $B6C3F_1$ mice were given doses of 150 and 300 mg/kg BW by gavage in corn oil for 8 weeks, then doses of 200 and 400 mg/kg BW for 70 weeks; Osborne-Mendel rats received 35 and 70 mg/kg BW for 20 weeks, then 50 and 100 mg/kg BW for 58 weeks. Hepatocellular neoplasms and adrenal phaeochromocytomas were observed in mice. No increase in tumor incidence was found in rats.[5] *Carcinogenicity classification.* USEPA: Group C; NTP: N—N—P*—P*.

Chemobiokinetics. In rats, T. is metabolized by hepatic cytochrome P-450 to **chloroacetic acid**[6] and to **inorganic chloride.**

Standards. USEPA (1991). MCL: 0.005 mg/l; MCLG: 0.003 mg/l.

Regulations. USFDA (1993) regulates TCA as a component of adhesives to be used safely in a food-contact surface.

References:

1. White, K. L., Sanders, V. M., Barness, D. W., et al., Toxicology of **1,1,2**-trichloroethane in the mouse, *Drug Chem. Toxicol.*, 8, 333, 1985.
2. Sanders, V. M., White, K. L., Shopp, G. M., et al., Humoral and cell-mediated immune status of mice exposed to **1,1,2**-trichloroethane, *Drug Chem. Toxicol.*, 8, 357, 1985.
3. See **TETRACHLOROETHYLENE, #6.**
4. Seidenberg, J. M., Anderson, D. G., and Becker, R. A., Validation of an *in vivo* development toxicity screen in the mouse, *Teratogen. Carcinogen. Mutagen.*, 6, 361, 1986.
5. National Cancer Institute, *Bioassay of 1,1,2-Trichloroethane for Possible Carcinogenicity*, Tech. Rep. Ser. 74, DHEW Publ. No 78–1324, U.S. Department of Health, Education and Welfare, Washington, D.C., 1978.
6. Ivanetich, K. M. and van den Honert, L. N., Chloroethanes: their metabolism by hepatic cytochrome P-450 *in vitro*, *Carcinogenesis*, 2, 697, 1981.

TRICHLOROETHYLENE (TCE) (CAS No 79–01–6)

Synonyms. Acetylene trichloride; Chlorylene; Ethylene trichloride; Ethynyl trichloride; Trichloran; Trichloroethene; Triclene; Trilene.

Properties. Colorless, volatile, noncombustible liquid with an odor of chloroform. Water solubility is 1 g/l at 20°C. Miscible with alcohols and dissolves oils. Organoleptic perception threshold is 0.31 mg/l (WHO, 1987). According to other data, odor perception threshold is 0.5 mg/l; practical perception threshold is 1 mg/l. Can be tasted at concentrations 50 to 100 times greater.[1]

Acute Toxicity. *Humans.* TCE seems to be a classical CNS depressant. Fatal hepatic failure has been observed following the use of TCE as an anesthetic. Oral exposure to 15 to 25 ml TCE resulted in vomiting and abdominal pain, followed by transient unconsciousness.[2] *Animals.* LD_{50} is 4.92 g/kg BW in rats.[06] In mice, it is reported to be 2.4 g/kg (females) and 2.44 g/kg BW (males).[3] Predominantly affects CNS activity and liver functions. Single oral administration of 0.5 to 1.5 g/kg BW causes an increase in liver weight and a decline in DNA concentration in the liver of rats and mice.

Repeated Exposure. In a 14-d study, male CD-1 mice were given $^1/_{10}$ and $^1/_{100}$ LD_{50} (24 and 240 mg/kg BW, respectively). An increase in liver weights was observed at the higher dose-level.[3] The NOAEL of 100 mg/kg BW is reported[4] for minor effects on relative liver weights in a 6-week reproduction study in rats. Female Swiss mice were dosed with 0.5, 1, and 2 mg T./kg BW for 4 weeks. Increase in liver size and stimulation of proliferation of sinusoid cells, and degeneration and necrotization of hepatocytes were found in all dosed animals. Kidneys were affected at the 2 mg/kg dose-level.[4]

Short-term Toxicity. CD-1 mice of both sexes received TCE in their drinking water at concentrations of 0.1 to 5 mg/l for 4 to 6 months. A decrease in BW gain was noted in the high-dose group. An increase in liver weights was accompanied by increased nonprotein sulfhydryl levels in males; an increase in kidney weights was accompanied by an increase in protein and ketone levels in the urine.[3] Fisher 344 rats and $B6C3F_1$ mice were administered TCE in corn oil by gavage for 13 weeks. Survival in mice was greatly decreased at 3 g TCE/kg and 6 g TCE/kg BW dose-levels. Histological examination revealed changes in the renal tubular epithelium at 1 g/kg (rats) and 3 g/kg BW (mice) dose-levels.[5]

Long-term Toxicity. Oral administration to mice induced hepatic peroxysome proliferation; however, no such effect was observed in rats.[6] Mice received 0.1 to 5 mg TCE/l in their drinking water for 4 to 6 months. The treatment caused an increase in liver and kidney weights and in the concentration of ketones and albumen in the urine of mice.[3] The NOAEL of 0.5 mg/kg BW is reported for rats and rabbits.[1] However, in more recent studies, the LOAEL of 500 mg/kg BW in rats and 1000 mg/kg BW in mice was determined for signs of toxic nephrosis.[5]

Immunotoxicity. Exposure to 24 and 240 mg/kg BW for 14 d resulted in depression of the cellular but not of humoral immunity in mice.[7]

Reproductive Toxicity. *Gonadotoxicity.* Testicular and epididymal weights were found to be decreased, but no histological changes were noted in Fisher 344 rats fed the diet containing 75 to 300 mg TCE/kg.[8] Reduced sperm motility was observed in CD-1 mice given 750 mg/kg in continuous-breeding fertility study.[9] *Embryotoxicity.* TCE passes through the placenta and penetrates the fetal blood.[10] Inhalation exposure of rats to relatively high concentrations (1800 ppm) caused an embryotoxic effect, manifested as retarded development. $B6C3F_1$ mice were gavaged from day 1 to 5, 6 to 10, or 11 to 15 of gestation with TCE in corn oil at dose-levels of $^1/_{10}$ or $^1/_{100}$ LD_{50}. No maternal or reproductive effects have been found at either dose-level. TCE was not found to cause any effect on reproduction at a dose-level up to $^1/_{10}$ LD_{50}.[11] *Teratogenicity.* TCE is not found to be teratogenic when administered in corn oil at doses of 0.5 and 1 g/kg BW[12] and when inhaled by pregnant rats and mice.

Genotoxicity. No data are available on the genetic and related effects of TCE in humans, but TCE produced SCE and unscheduled DNA synthesis in human lymphocytes *in vitro*. TCE was shown to be mutagenic in a number of bacterial strains. It induced micronuclei and somatic mutations (in the spot test), sperm anomalies and DNA strand breaks in the kidney and liver but not in the lung of mice treated *in vivo*. TCE does not produce DLM. It induced transformation of mouse and rat cells but not of Syrian hamster cells. No SCE were observed in Chinese hamster cells *in vitro* or unscheduled DNA synthesis in rat hepatocytes (IARC Suppl. 7–365).

Carcinogenicity. *Humans.* There is no epidemiological evidence of TCE carcinogenicity: epidemiological data are inadequate to refute or demonstrate a human carcinogenic potential (USEPA, 1985). *Animals.* TCE carcinogenic potential was evident in mice and rats of both sexes, exposed by inhalation and orally.

Epichlorohydrin-free TCE was reported to be carcinogenic in $B6C3F_1$ mice when administered in corn oil at 1 g/kg BW for 103 weeks.[13] A hepatocellular carcinogenic response was found in $B6C3F_1$ mice (average daily doses, 1169 and 2339 mg/kg for males and 869 and 1739 mg/kg for females) but not in Osborne-Mendel rats exposed to 549 and 1097 mg technical grade TCE/kg.[14] Henschler et al. concluded that TCE containing ECH and epoxybutane causes tumors in test animals, but purified TCE was not carcinogenic in ICR/HA mice. However, USEPA concluded that Henschler's study used mice that are known to be less responsive to hepatocellular carcinomas than the mice in several other studies. *Carcinogenicity Classification.* USEPA: Group B2; IARC: Group 3; NTP: N—N—P*—P.

Chemobiokinetics. TCE is readily absorbed in the GI tract. The principal products of TCE metabolism measured in the urine are **trichloroacetaldehyde, trichloroetahnol, trichloroacetic acid,** and conjugated derivatives (glucuronides) of TCE.[15] In mice, metabolism is rapid and gives rise to relatively large amounts of the acid, which could induce liver peroxisome proliferation and cancer. In rats and humans, the rate of oxidation is limited, and subsequently, cancer risk to humans should be decreased.[16] TCE and its metabolites are excreted in exhaled air, urine, sweat, feces, and saliva.

Guidelines. WHO (1992): Provisional guideline value for drinking water 0.07 mg/l. However, the lowest reported odor threshold for TCE is 0.3 mg/l.

Standards. USEPA (1991). MCL: 0.005 mg/l, MCLG: zero. **Russia** (1988): MAC and PML in drinking water: 0.06 mg/l.

Regulations. USFDA (1993) approved the use of TCE in adhesives used as components of articles intended for use in packaging, transporting, or holding food.

References:

1. See **CYCLOHEXANOL,** 308.
2. Stephens, C. A., Poisoning by accidental drinking of trichloroethylene, *Br. Med. J.,* 2, 218, 1945.
3. Tucker, A. N., Sanders, V. M, Barness, D. W., et al., Toxicology of trichloroethylene in the mouse, *Toxicol. Appl. Pharmacol.,* 62, 351, 1982.
4. Goel, S. K., Rao, G. S., Pandya, K. P., et al., Trichloroethylene toxicity in mice; a biochemical, hematological and pathological assessment, *Ind. J. Exp. Biol.,* 30, 402, 1992.
5. *Carcinogenesis Studies of Trichloroethylene (without Epichlorohydrin) in F344/N Rats and B6C3F₁ Mice (Gavage Studies),* NTP TR 90–1799, 1990.
6. Elcombe, C. R., Rose, M. S., and Pratt, I. S., Biochemical, histological and ultrastructural changes in rat and mouse liver following the administration of trichloroethylene: possible relevance of species differences in hepatocarcinogenicity, *Toxicol. Appl. Pharmacol.,* 79, 365, 1985.
7. Sanders, V. M., Tucker, A. N., White. K. L., et al., Humoral and cell-mediated immune status in mice exposed to trichloroethylene in the drinking water, *Toxicol. Appl. Pharmacol.,* 61, 358, 1982.
8. *Trichloroethylene: Reproduction and Fertility Assessment in F344 Rats When Administered in the Feed, Final Report,* NTP–86–085, Research Triangle Park, NC, 1986.
9. *Trichloroethylene: Reproduction and Fertility Assessment in CD-1 Mice When Administered in the Feed, Final Report,* NTP–86–068, Research Triangle Park, NC, 1985.
10. Laham, S. L., Studies on placental transfer, Trichloroethylene, *Ind. Med. Surg.,* 39, 46, 1970.
11. Cosby, N. C. and Dukelow, W. R., Toxicology of maternally ingested trichloroethylene (TCE) on embryonal and fetal development in mice and TCE metabolites on *in vitro* fertilization, *Fundam. Appl. Toxicol.,* 19, 268, 1992.
12. *Carcinogenesis Bioassay for Trichloroethylene,* NTP TR No 82–1799 (Draft), 1982.
13. Henschler, D., Elsasser, W., Romen, W., et al., Carcinogenicity study of trichloroethylene, with and without epoxide stabilizers, in mice, *J. Cancer Res. Clin. Oncol.,* 107, 149, 1984.
14. National Cancer Institute, *Carcinogenesis Bioassay of Trichloroethylene,* U.S. Department of Health, Education, and Welfare, Public Health Service, Washington, D.C., 1976.
15. Ikeda, M., Imamura, T., and Ohtsvji, H., Urinary excretion of total trichlorocompounds, trichloroethanol, trichloroacetic acid, as a measure of exposure to trichloroethylene and tetrachloroethylene, *Br. J. Ind. Med.,* 29, 46, 1970.
16. Green, T., Chloroethylenes: a mechanistic approach to human risk evaluation, *Annu. Rev. Pharmacol. Toxicol.,* 30, 73, 1990.

TURPENTINE (CAS No 8006–64–2)

Synonym. Oil of turpentine.
Composition. A complex mixture, mainly of $C_{10}H_{16}$ terpene hydrocarbons.

Properties. Clear and colorless or green-yellowish sticky masses. T. exhibits a unique, pleasant odor and a sharp, abrasive taste. Almost insoluble in water. Soluble in alcohol and ether. Odor perception threshold is 0.2 mg/l.

Acute Toxicity. In rats, LD_{50} is reported to be 5.76 g/kg BW. The urinary system is predominantly affected.

Standards. Russia (1988). MAC and PML: 0.2 mg/l (organolept., odor).

Regulations. USFDA (1993) approved the use of T. in adhesives used as components of articles intended for use in packaging, transporting, or holding food.

XYLENES (CAS No (mixed) 1330–20–7)

Synonyms. Dimethylbenzenes; Xylols.

Properties. Colorless, highly combustible liquid. Water solubility of X. isomers is up to 160 to 198 mg/l at 25°C. Odor threshold concentration is reported to be 0.02 to 2.2 mg/l.[010] The data on the taste threshold value are limited to the concentration range of 0.3 to 1 mg/l; these concentrations produced a detectable taste and odor.

Applications and **Exposure.** Used in the production of synthetic rubbers and polyether fibers. Used as an intermediate in the manufacturing of plastics. Leaching from synthetic coating materials commonly used to protect drinking water storage tanks is reported.[1] X. are found in the drinking water in the U.S. and Switzerland. Air seems to be the major source of exposure.

Acute Toxicity. *Humans.* Accidental human poisonings are followed by adverse effects on the kidneys and liver. *Animals.* In rats, LD_{50} is found to be 3.6 to 5.8 g/kg (3.57 g *o*-X./kg BW; 4.99 g *m*-X./kg BW; 3.91 g *p*-X./kg BW).[2]

Short-term Toxicity. The LOAEL was identified to be 200 mg/kg BW. Ultrastructural changes were observed in the liver.[3,4]

Long-term Toxicity. In an NTP–86 study, rats were administered 250 and 500 mg X./kg BW in corn oil by gavage for 103 weeks.[5] The dose of 250 mg/kg BW was considered to be the NOAEL, since the mean BW of low-dose and vehicle-control male rats and those of dosed and vehicle-control female rats were comparable. Rabbits were given the dose of 48 mg X./kg BW in starch emulsion for 5.5 months. Eosinophilia and lymphopenia were observed. Gross pathological examination revealed adipose dystrophy of the liver and kidneys, and pyelitis. There were no clinical signs of intoxication.[6]

Reproductive Toxicity. Maternal toxicity with concurrent embryotoxicity and teratogenicity was found in mice. The NOAEL of 255 mg/kg BW was established in this oral study.[7] *o*- and *p*-Isomers appear to be more hazardous to the offspring than is the *m*-Isomer.[8] Embryotoxic effect was found in rats and mice, but no teratogenic effect was observed in rats.[9,10] An insignificant amount of X. passes through the placenta.

Genotoxicity. *Humans.* SCE were not found in peripheral lymphocytes of exposed workers. *Animals.* Negative results were shown in the *Salmonella* test and in mammalian somatic cells (both *in vitro* and *in vivo*) (IARC 47–146). X. did not produce CA in bone marrow cells in mice.[11]

Carcinogenicity. *Humans.* Available information on X. has been reviewed by the USEPA and was found to be inadequate for determining potential carcinogenicity in humans.[12] *Animals.* No increase in the incidence of tumors was observed in either mice or rats following the administration of a technical-grade X. There was no evidence of carcinogenicity in a 2-year oral study.[5] *Carcinogenicity classification.* IARC: Group 3; USEPA: Group D.

Chemobiokinetics. Data on the absorption after ingestion are not available. Adipose tissue is likely to be the site of storage. The general pathways of *m*-X. metabolism involve initial side chain and aromatic hydroxylation catalyzed by cytochrome P-450.[13] X. is converted in the body primarily to **methyl benzoic acid** that is excreted in the urine as **methyl hippuric acid.**

Guidelines. WHO (1992). Guideline value for drinking water, 0.5 mg/l. The levels of 0.02 to 1.8 mg/l are likely to give rise to consumer complaints of foreign taste and odor.

Standards. USEPA (1991). MCL and MCLG: 10 mg/l. **Russia** (1988). MAC and PML in drinking water: 0.05 mg/l (organolept., odor).

Regulations. USFDA (1993) approved the use of X. (1) as components of the uncoated or coated food-contact surface of paper and paperboard that may be used safely in producing, manufacturing, packing, transporting, or holding dry food; (2) as plasticizers for rubber articles intended for repeated use in contact with food up to 30% by weight of the rubber product; (3) in adhesives used as components of articles intended for use in packaging, transporting, or holding food; (4) in semirigid and rigid acrylic and modified acrylic plastics for single and repeated use in food-contact articles; and (5) as a solvent in polysulfide polymer-polyepoxy resins in articles intended for packaging, transporting, holding, or otherwise containing dry food in accordance with specified conditions.

References:

1. Bruchet, A., Shipert, E., et Alban, K., Investigation of organic coating material used in drinking water distribution systems, *J. Fr. d'Hydrol.,* 19, 101, 1988 (in French).
2. Jori, A. et al., *Ecotoxicol. Environ. Saf.,* 11, 44, 1986.
3. Bowers, D. E., Cannon, M. S., and Jones, D. H., Ultrastructural changes in the liver of young and aging rats exposed to methylated benzenes, *Am. J. Vet. Res.,* 43, 679, 1982.
4. Janssen, P., van der Heijden, C. A., and Knaap, A. G. A. C., Short summary and evaluation of toxicological data on xylene, Document from Toxicol. Advisory Centre, Natl. Inst. Public Health Environ. Protection, The Netherlands, 1989.
5. NTP, Toxicology and Carcinogenesis Studies of Xylenes (mixed) (60% **m**-Xylene, 14% **p**-Xylene, 9% **o**-Xylene, 17% Ethylbenzene) in F344/N Rats and B6C3F$_1$ Mice (Gavage Studies), NTP Tech. Rep. Ser. No. 327, 1986.
6. See **ACRYLIC ACID,** #1, 100, 1960.
7. Marks, T. A., Ledoux, T. A., and Moore, J. A., Teratogenicity of a commercial xylene mixture in the mouse, *J. Toxicol. Environ. Health,* 9, 97, 1982.
8. Hood, R. D. and Ottley, M. S., Developmental effects associated with exposure to xylenes: a review, *Drug Chem. Toxicol.,* 8, 281, 1985.
9. Ungvary, G., Tatrai, D., Hudak, A., et al., Studies on embryotoxic effects of *o*-, *m*- and *p*-xylenes, *Toxicology,* 18, 81, 1980.
10. Brown-Woodman, P. D. C., Webster, W. S., Picker, K., et al., Embryotoxicity of xylene and toluene: an *in vitro* study, *Ind. Health,* 29, 139, 1991.
11. See **NITROBENZENE,** #12.
12. *U.S. Fed. Reg.,* Vol. 56, No. 20, 3543, 1992.
13. Sedivec, V. and Flek, J., The adsorption, metabolism and excretion of xylenes in man, *Int. Arch. Occup. Environ. Health,* 37, 205, 1976.

Chapter 7

OTHER ADDITIVES AND INGREDIENTS

ACETAMIDE (CAS No 60–35–5)

Synonyms. Acetic acid, amide; Acetimidic acid; Ethanamide; Methanecarboxamide.

Properties. Colorless, odorless needles. Solubility in water is 97%; solubility in alcohol is 25% at 20°C.

Acute Toxicity. Rats and mice tolerate 7.5 and 8 g/kg BW, respectively. In rats, LD_{50} is reported to be 30 g/kg BW. Poisoning is accompanied by lethargy and respiratory distress in rats and excitation (convulsion) with subsequent CNS inhibition in mice (Garrett and Dangherti, 1951).

Repeated Exposure. Rats were given 400 mg/kg BW for 36 d. Decreased growth without other signs of intoxication or pathological lesions was observed.[1]

Reproductive Toxicity. *Embryotoxic* effect is observed in rabbits. *Teratogenic* effect is found in rats, mice, and rabbits.[2]

Genotoxicity. A. was shown to be positive in *in vivo* micronucleous test (Fritzenschaft et al., 1993).

Carcinogenicity. Wistar 1-month-old male rats were fed a diet containing 1.25, 2.5, or 5% A. for 4 to 12 months. This treatment caused development of liver tumors (trabecular carcinomas and adenocarcinomas with lung metastases).[3] An increased incidence of tumors (hepatomas) was noted in the study,[4] where Wistar rats were exposed to 2.5% A. in the diet for 1 year.

References:
1. Caujolle, F., Chanh, P. H., Dat-Xuong, N., U. A., Toxicity studies on acetamide and its *N*-methyl- and *N*-ethyl derivatives, *Arzneimittel-Forsch.*, 20, 9042, 1970 (in German).
2. See *N*-**METHYLFORMAMIDE**, #3.
3. Jackson, B. and Dessau, F., Liver tumor in rats fed acetamide, *Lab. Invest.*, 10, 909, 1961.
4. Weisburger, J. H., Yamamoto, R. S., Glass, R. M., et al., Prevention by arginine glutamate of the carcinogenicity of acetamide in rats, *Toxicol. Appl. Pharmacol.*, 14, 163, 1969.

ADIPONITRILE (ADN) (CAS No 111–69–3)

Synonyms. Adipic acid, dinitrile; **1,4**-Dicyanobutane; Hexanedinitrile.

Properties. Light, odorless, oily liquid. Poorly soluble in water, soluble in alcohol. Odor perception threshold is 31 mg/l; practical threshold is 63 mg/l.[1]

Acute Toxicity. *Humans.* Ingestion of several milliliters of ADN resulted in vomiting, chest tightness, headache, weakness, cyanosis, increased heart rate, etc. A man recovered within 4 h following sodium thiosulfate treatment and then suffered a recurrent episode (Ghiringhelley). *Animals.* LD_{50} is reported to be 500 or 105 mg/kg in rats, 48 mg/kg in mice, 20 mg/kg in rabbits, and 50 mg/kg BW in guinea pigs.[1,2] LD_{50} of technical grade ADN, containing 1% of ammonia, is 1.0 g/kg BW in rats.[3] According to Daniel and Kennedy,[4] LD_{50} is 138 mg/kg BW for fasted rats and 301 mg/kg BW for nonfasted rats. Poisoning is accompanied by excitation, convulsions, and respiratory arrest.

Repeated exposure revealed little evidence of cumulative properties. Mice were given 10 mg A./kg BW. The treatment caused retardation of BW gain and reduction in their capacity to work. Of animals, 25% died after 12 administrations. Gross pathological examination revealed dystrophic changes in the viscera. A 250 mg/kg dose given to rats for 10 d produced temporary retardation of BW gain; albumen appeared in the urine.[3]

Long-term Toxicity. A 2-year feeding study in rats resulted in adrenal degeneration in females at feeding levels of 0.5, 5, and 50 ppm. Other parameters were found to be unaltered.[5] In a 6-month study in rabbits, the main manifestations of toxic action included impairment of enzyme indices reflecting the state of the oxidation-reduction systems and changes in the content of blood SH-groups and blood nucleic acids.[1]

Reproductive Toxicity. Sprague-Dawley rats were administered the doses of 20, 40, and 80 mg/kg BW on days 6 to 19 of gestation. Maternal toxicity effect was observed at 40 and 60 mg/kg dose-levels. The highest dose caused a slight fetotoxicity. Teratogenic effect has not been observed.[6]

Chemobiokinetics. ADN is likely to break down in the body to form CN^-. Its toxicity is possibly determined by the action of intact molecules. According to Ghiringhelley,[1] 83.5% of administered dinitrile is broken down in the body to form **hydrogen cyanide.**

Standards. Russia (1986). MAC and PML: 0.1 mg/l.

References:

1. See **ANILINE,** #1, 85.
2. Dieke, S. H., Allen, G. S., and Richter, C. P., The acute toxicity of thioureas and related compounds to wild and domestic Norway rats, *J. Pharmacol.,* 90, 260, 1947.
3. Plokhova, E. I. and Rubakina, A. P., Toxicity characteristics of the dinitrile of adipic acid, *Gig. Truda Prof. Zabol.,* 9, 56, 1965 (in Russian).
4. Daniel, O. L. and Kennedy, G. L., The effect of fasting on the acute oral toxicity of nine chemicals in rats, *J. Appl. Toxicol.,* 4, 320, 1984.
5. Svirbely, J. L. and Floyd, E. P., Toxicologic studies of acrylonitrile, adiponitrile, and BB-oxy-dipropionitrile. III. Chronic Studies, U.S. Department of Health, Education, and Welfare, Occupational Health and Safety, Robert A. Taft, Eng. Center J-4614, 1964.
6. Johannsen, F. R., Levinskas, G. J., Berteau, P. E., et al., Evaluation of the teratogenic potential of three aliphatic nitriles in the rat, *Fundam. Appl. Toxicol.,* 7, 33, 1986.

ALKYLARYLSULFONATE

Synonyms. D-40; Witconate.

Composition. Synthetic anionic detergent, containing 40% sodium alkylsulfonate, approximately 2% moisture, 1% unsulfonated oil, and the balance sodium sulfate.

Properties. Threshold concentration is 0.3 to 0.7 mg/l for odor and 0.4 to 0.6 mg/l for taste. Classified as an anionic surfactant.

Applications. Used as an emulsifier.

Acute Toxicity. LD_{50} is 2.3 to 1.4 g/kg in rats, 1.5 to 2 g/kg in mice, and 1.13 g/kg BW in hamsters.

Short-term Toxicity. Six volunteers received 100 mg A./d with their food over a period of 4 months. Ingestion did not result in any toxic effect.

Reference:

See **ACRYLONITRILE,** #7, 96.

ALKYLBENZENESULFONATE, SODIUM SALT (CAS No 6841–30–3)

Properties. At 20°C, a light-yellow, paste-like liquid that forms layers; at 70°C, it is homogeneous. Smells of soap. Soluble in water. Odor perception threshold is 23.8 mg/l; taste perception threshold is 17.3 mg/l. Practical thresholds are 54.4 and 35.3 mg/l, respectively. There are no chlorinated phenol odors on chlorination. Foam formation threshold is 0.4 mg/l.[1]

Applications. Used as a foam-forming surfactant, a wetting agent.

Acute Toxicity. LD_{50} is 3 g/kg in rats, 2.1 g/kg in mice, and 1.9 g/kg BW in guinea pigs.

Repeated exposure revealed evident cumulative properties.[1] Rats were given 300 mg A./kg BW for a month. The treatment reduced leukocyte count, blood Hb level and cholesterol concentrations, cholinesterase, alkaline and acid phosphatase activity, and increased SH-group concentration in the blood serum of rats. This dose was considered to be the LOAEL.

Long-term Toxicity. Volunteers consumed 100 mg daily doses for 6 months without any toxic effect observed. In a 6-month study, guinea pigs were given drinking water with a concentration of 2 g/l; rats received 500 mg/l for 2 years. No harmful effect was noted in both studies.[2]

Reproductive Toxicity. Gonadotoxic effect was not observed at dose-levels of 12 to 300 mg/kg administered for 3 months.

Carcinogenicity. Negative results for linear A. are reported in rats.[09]

Standards. Russia (1988). MAC: 0.4 mg/l (organolept., foam).

Regulations. USFDA (1993) approved the use of A. as emulsifiers and/or surface-active agents in the manufacturing of nonfood articles or components of articles.

References:
1. Kondratyuk, V. A., Pastushenko. T. V., Gun'ko, L. M., et al., Substantiation of the hygienic standard for sodium alkylsulfate in water, *Gig. Sanit.*, 6, 81, 1983 (in Russian).
2. See **ACRYLONITRILE, #7, 26.**

ALKYLSULFATES, SODIUM SALT (AS) (Primary)

Properties. Water-soluble liquid containing 27 to 32% of the pure substance. The threshold concentration of Primary AS from sperm whale fat is 0.3 mg/l for odor and 0.4 mg/l for foam formation. The threshold concentration of Primary AS from 'second unsaponifiables' is 0.6 mg/l for odor and 0.5 mg/l for foam formation.[1]

Applications. A base for detergents, used also as an ingredient of coatings.

Acute Toxicity. In rats, LD_{50} is 1.7 g/kg BW. Manifestations of toxic action include circulatory disturbances, reduction in the RNA content in the liver, myocardium and lung, and DNA content in the liver. Increased glycogen content in all organs has been noted.[2,3] For Primary AS from sperm whale fat, LD_{50} is 3.82 g/kg in rats, 3.44 g/kg in guinea pigs, and 2.8 g/kg BW in mice. For Primary AS from 'secondary unsaponifiables', LD_{50} is 7 g/kg in rats, 3.75 g/kg in mice, and 6.85 g/kg BW in guinea pigs.[1]

Long-term Toxicity. In a 6-month study, a 100 mg/kg BW dose caused changes in enzyme activity at various points in the metabolic chain.

Allergenic Effect. The LOAEL is identified to be 10 mg/kg BW (on skin application).

Standards. Russia (1988). MAC and PML: 0.5 mg/l (organolept., odor). In the case of Primary AS from sperm whale fat, MAC, and PML in drinking water 0.3 mg/l (organolept., foam).

Regulations. USFDA (1993) approved the use of AS (1) as emulsifiers and/or surface-active agents in the manufacturing of articles or components of articles at levels not to exceed 2% PVC and/or vinyl chloride copolymers and (2) as emulsifiers for vinylidene chloride polymers and copolymers at levels not to exceed a total of 2.6% by weight of the coating solids.

References:
1. Rusakov, N. V., Allergenic effects of different chemicals by oral administration to the body, *Gig. Sanit.*, 2, 13, 1984 (in Russian).
2. Mozhayev, E. V., *Pollution of Reservoirs by Surfactants, Health and Safety Aspects*, Meditsina, Moscow, 1976, 94 (in Russian).
3. Voloshchenko, O. I. and Medyanik, L. A., *The Safety and Toxicology of Household Chemicals*, Zdorov'ya, Kiev, 1983, 142 (in Russian).

ALKYLSULFATES, SODIUM SALT (Secondary)

Properties. Light-yellow to amber-colored liquid, readily soluble in water. Threshold perception concentration is 0.6 mg/l for odor and 0.5 mg/l for foam formation.[1]

Applications. Base for liquid detergents; used as a foaming agent in the production of foam concrete, as a dispersant and an emulsifier.

Acute Toxicity. In rats, LD_{50} is 1.65 to 5.63 g/kg BW.[1]

Allergenic Effect. When given *i/g*, the threshold concentration for such an effect was shown to be 0.1 mg/l.[2]

Regulations. USFDA (1993) approved the use of AS as antistatic and antifogging agents in food-packaging materials at levels not to exceed 3% by weight of polystyrene- or rubber-modified polystyrene.

References:
1. See **ALKYLSULFATES, SODIUM SALT, #2.**
2. See **ALKYLSULFATES, SODIUM SALT, #1.**

6-AMINOHEXYLAMINOMETHYLTRIETHOXYSILANE (CAS No 15129–36–9)

Synonyms. *N*-[(Triethoxysilyl)methyl]-**1,6**-hexanediamine.

Properties. Clear, light-yellow liquid. Soluble in water and alcohol.

Applications. Used as a water-repelling and finishing agent.

Acute Toxicity. After administration of 0.5 to 3 g/kg BW, mice became lethargic but rapidly recovered. General inhibition and adynamia persisted for 2 h in some animals. A certain number of mice died within 7 d. Gross pathological and microscopic examinations revealed pronounced congestion in the visceral organs and fatty dystrophy of the liver cells, myocardium, and renal tubular epithelium.

Reference:

See *N,N'*-**DIBUTYL-***N,N'***-DINITROSO-***p***-PHENYLENE DIAMINE,** 118.

ρ-AMINOPROPYLTRIETHOXYSILANE (CAS No 919–30–2)

Properties. Clear, light-yellow liquid.

Applications. Used as a water repellent or finishing agent.

Acute Toxicity. Mice and rats received A. in oil solution. The treatment caused a transient inhibition. LD_{50} is found to be 4 g/kg BW. No interspecies differences were observed in this study.

Reference:

See 6-**AMINOHEXYLAMINOMETHYLTRIETHOXYSILANE.**

AMMONIA (CAS No 7664–41–7)

Properties. The term 'ammonia' refers to nonionized (NH_3) and ionized (NH_4^+) species. Under room temperature and atmospheric pressure, A. occurs as a colorless gas with a pungent, repulsive smell. The gas dissolves readily in water, forming the ammonium cation. In solution it forms, and is in equilibrium with, ammonium ions. Water solubility is 421 g/l at 20°C. Ammonium solutions are alkaline and react with acids to form ammonium salts. Odor perception threshold at alkaline pH is approximately 1.5 mg/l[1] or 0.037 mg/l.[010] According to other data, the taste and odor perception threshold is likely to be 35 mg/l.[2]

Applications and **Exposure.** In the U.S., 10% A. is used in production of fibers, rubber, and plastics. Ammonium is a natural component of foods. Minor amounts of ammonium compounds are added to food as acid regulators, stabilizers, flavoring agents, and fermentation aids. Water contamination can also arise from cement pipe linings. The levels detected in ground and surface water are not higher than 0.2 mg/l. Human exposure from environmental sources is negligible in comparison with endogenous synthesis. Daily intake via food and drinking water seems to be about 18 mg.[2]

Acute Toxicity. In rats, LD_{50} is found to be 350 mg/kg; in cats, it is 750 mg/kg BW.[3] LD_{50} equal to 4070 to 5020 mg/kg is reported for rats.[4] Guinea pigs orally exposed to watery solutions of different **ammonium salts (nitrate, acetate, bromide, chloride, sulfate)** at dose-levels of 200 to 510 mg/kg died of acute lung edema. NS dysfunction (changes in respiration rhythms and depth, weakness and difficulties of locomotion, hyperexcitability to tactile, acoustic and pain stimuli) was shown to develop.[2] Rabbits were treated by instillation with 0.8 and 1 g nonacidic **A. carbonate/** kg (283 and 354 mg NH_4/kg BW, respectively). Hyperemia of the renal cortex was noted.[2]

Repeated Exposure. A. causes acidotic effect. It is believed that signs of intoxication might be noted at an exposure level above 0.2 g/kg BW.

Short-term Toxicity. Male Sprague-Dawley rats received A. via drinking water at a dose of 510 mg NH_4Cl/l. Urinalysis and histological examination failed to reveal changes in the kidneys. Consumption of drinking water with a concentration of 10 g NH_4Cl/l for 2.5 months caused no signs of renal damage.[2,5] Weaned rats were administered orally 0.1 to 5 g/kg BW for 30, 60, and 90 d. Only young animals reacted to the 5 g/kg dose by reduced food consumption and elevated intake of water. No retardation of BW gain or histological changes were noted to develop, and only a mild fatty degeneration of the liver in the adult animals was found after a 90-d exposure.

Long-term Toxicity. A. itself is not of direct importance for health at concentrations below 0.5 mg/l, nor at higher concentrations.

Ammonium hydroxide at a dose of 100 mg/kg BW was administered to rabbits by gavage initially every second day, then daily for a maximum of 17 months. The treatment resulted in enlargement of the adrenals. After several weeks of treatment, blood pressure was increased.[2,5] Sprague-Dawley rats received **ammonium chloride** in their drinking water at a dose of 478 mg/kg BW (males for 330 d, females for 300 d). Significant decrease of the nonlipid bone mass (the upper thigh) and of the calcium content was observed. The blood pH value and the carbonic acid content in the plasma decreased. Reduced BW gain with lower fat accumulation was also observed.[5]

Reproductive Toxicity. Administration of 100 to 200 mg/kg BW doses of different A. compounds caused enlargement of the ovaries and uterus. Breast hypertrophy with milk secretion, follicle ripening, and formation of the corpus luteum was observed in unpuberal female rabbits. Administration of a 200 mg/kg BW dose to pregnant rats in their drinking water inhibited the fetal growth but did not produce any teratogenic effect.[6]

Genotoxicity. A. was shown to be negative in the *Salmonella* mutagenicity assay. CA were observed in Chinese hamster fibroblasts without metabolic activation. A. did not induce micronuclei in the bone marrow cells of mice after *i/p* administration.[7,8]

Carcinogenicity. Ammonium hydroxide was administered to Swiss mice and C3H mice in their drinking water for their lifetime (the calculated doses were 140 to 270 mg/kg BW). No increase in tumor incidence was found.[5] Negative results were reported for **ammonium chloride** in female mice.[09]

Chemobiokinetics. A. is produced by animal and microbial metabolism; it is a product of the bacterial degradation of amino and nucleic acids. A. is absorbed from the GI tract, but not through the skin, and is rapidly distributed throughout the body. In the liver, A. is incorporated in **urea** as a part of the urea cycle; furthermore, urea in the liver is absorbed into blood circulation and transferred to the kidneys in order to be excreted in the urine. A. seems to be a key metabolite in the mammalian body and is formed through deamination of amino acids in the liver and GI tract by the enzymatic breakdown of food products and with the help of microorganisms.[1]

Standards. WHO (1991) did not consider it necessary to recommend the guideline value for drinking water. Concentration of 1.5 mg/l is likely to give rise to consumer complaints of foreign odor and taste. **EEC** (1992). MAC: 0.5 mg NH_4/l. **JECFA** (1982). No ADI allocated in food. **Russia** (1988). MAC: 2 mg N/l. **Netherlands** (1986). Limit value: 10 mg N/l. **Czechoslovakia** (1975). MAC in drinking water: 0.5 mg/l.

Regulations. USFDA (1993) approved the use of different **ammonium salts (borate, citrate, persulfate, polyacrylate, potassium hydrogen phosphate, sulfamate, thiocyanate, thiosulfate)** in adhesives as components of articles intended for use in packaging, transporting, or holding food. **Ammonium benzoate** may be used as a preservative only; **ammonium bifluoride** and **ammonium silico-fluoride** are allowed to be used only as a bonding agent for aluminum foil, stabilizer, or preservative (up to 1% by weight of the finished adhesive). In the **EEC**, paints, varnishes, printing inks, adhesives, and similar products that contain A. in solution at concentrations greater than 35% are considered toxic and corrosive and, at concentrations of 10 to 35%, as harmful and irritant, and must be packaged and labeled accordingly.[5] The maximum concentration of A. in finished cosmetic products must not exceed 6% calculated as NH_3. If the concentration exceeds 2%, the label must read: **CONTAINS AMMONIA.**

References:

1. *Hazardous Substances Data Bank,* On-line: January 1990.
2. *Ammonia,* Environmental Health Criteria 54, WHO, Geneva, 1986.
3. Institute Natl. de Recherche et du Securite (INRS), Ammoniae et solutions aqueuses, Fiche toxicologique 16, Cahiers de notes documentaires Issue 128, 1987 p. 461 (in French).
4. Ishidate, M., Sofuni, T., Yoshikawa, K, et al., Primary mutagenicity screening of food additives currently used in Japan, *Food Cosmet. Toxicol.,* 22, 623, 1984.
5. USEPA, *Health Effect Assessment for Ammonia,* EPA/600/8–88/017, Cincinnati, OH, 1987.
6. *Ammonia,* IPCS Health Safety Guide No 37, WHO, Geneva, 1990, 30.
7. Reichert, J. and Lochtmann, S., Occurrence of nitrite in water distribution system, *JMF Wasser-Abwasser.,* 125, 442, 1984 (in German).
8. Hayashi, M., Kishi, M., Sofuni, T., et al., Micronucleus tests in mice on 39 food additives and 8 miscellaneous chemicals, *Food Cosmet. Toxicol.,* 26, 487, 1988.

ASBESTOS
CHRYSOTILE (CAS No 12001–29–5); AMOSITE (12172–73–5); ANTHOPHYLLITE (17068–78–9); CROCIDOLITE (12001–28–4)

Composition. A. are fibrous silicate minerals of the serpentine and amphibole mineral groups. The different forms of A. consist of silica (40 to 60%) and also of oxides of iron, magnesium, and other metals. The length of A. fibers in drinking water is 0.0005 to 0.002 mm, the diameter is 0.00003 to 0.0001 mm, and the weight of a single fiber is 10^{-12} to 10^{-16} g.

Applications and **Exposure.** Used as a filler for A.-cement and A.-filled polymer materials. A. gets to water by means of dissolution of A.-containing sheet and pipe in the distribution system. Exfoliation of A. fibers from A.-containing pipes is related to the corrosiveness of the water supply, increasing in aggressive waters. A. may enter foodstuffs from the water and impure talc used in their production (chewing gums, rice sticks, etc.).

Long-term Toxicity. *Humans.* In the studies of populations in Duluth, Canadian cities, Connecticut, Florida, and Utah,[1–3] all of which were ecological in nature and in which population mobility could not be adequately assessed, there was no consistent evidence of an association between cancer mortality or incidence and ingestion of A. concentrations in water and in digestive cancers combined. All GI tract tumors and tumors of the esophagus and colon in both sexes were observed in an ecological study in the San Francisco Bay area; the reanalysis of the data, taking potential confounding factors into consideration, has undermined the significance of these results.[1,2] The mobility of the population in this area is high and coincidental geographical associations were not apparent when San Francisco and all other Bay area counties were considered separately. In an analytical epidemiological study that was inherently more sensitive than the ecological studies mentioned above, there was no consistent evidence of a cancer risk due to ingestion of A. in drinking water in Puget Sound, where levels ranged up to 200 MFL (million fibers per liter).[2,3] *Animals.* Information on the transmigration of ingested A. through the GI tract to other tissues is also contradictory.[4,5] Although it is not possible to conclude with certainty that ingested fibers do not cross the intact GI tract, available data indicate that penetration, if it occurs at all, is extremely limited.

Reproductive Toxicity. CD-1 mice received approximately 4 to 400 mg chrysotile per kilogram on days 1 to 15 of pregnancy. The treatment did not affect survival of the progeny. *In vitro* administration interfered with implantation upon transfer of exposed blastocysts to recipient females but did not result in a decrease in postimplantation survival. A. was not considered as teratogen in these studies.[6]

Genotoxicity. In cultured human cells, conflicting results were reported for the induction of CA and negative results for the induction of SCE by chrysotile and crocidolite; amosite and crocidolite did not induce DNA strand breaks. A single oral administration of chrysotile did not increase the frequency of micronuclei in mice or CA in bone marrow cells of rhesus monkeys treated *in vivo*.[7] Amosite, anthophyllite, chrysotile, and crocidolite induced transformations of Syrian hamster embryo cells. In cultured rodent cells, amosite, anthophyllite, chrysotile, and crocidolite induced CA, and amosite, chrysotile, and crocodolite induced SCE. Chrysotile did not induce unscheduled DNA synthesis in rat hepatocytes (IARC 2–17).

Carcinogenicity. *Humans.* A. is a known human carcinogen by the inhalation route. Although well studied, there has been little convincing evidence of the carcinogenicity of ingested A. in epidemiological studies of populations with drinking water supplies containing high concentrations of A. (WHO, 1991). On the basis of available epidemiological data then, for the majority of cancer sites and for all ages and sexes, no excess risk is present, even for high levels of A. in the drinking water.[8] In 1984, the U.S. National Research Council concluded that "the association of asbestos with an increased risk of malignancies other than lung cancer and mesothelioma has not been confirmed in animal studies and has not been observed consistently in human studies." *Animals.* In extensive studies in animal species, A. has not induced consistent increases in the incidence of tumors of the GI tract.[4,9] A. was administered to Syrian golden hamsters as 1% amosite, or short-range, or intermediate-range chrysotile in the diet over their lifetime, and also to Fisher 344 rats as 1% tremolite or amosite in the diet over their lifetime. No treatment-related effects in animals were observed; increase in the incidence of benign epithelial neoplasms in the GI tract of male rats (treated with chrysotile) was not statistically significant. Moreover, no increase in tumor incidence was observed in Fisher 344 rats ingesting short-range chrysotile (98% shorter than 0.01 mm), which was composed of fiber sizes more similar to those found in drinking water.[10,11] The weight of evidence shows that ingested A. is not hazardous to health (WHO, 1991). *Carcinogenicity classification* (inhalation exposure). IARC: Group 1; USEPA: Group A.

Chemobiokinetics. The fate of absorbed A. fibers is imperfectly understood at present. There was evidence that particles of A., entering the GI tract with drinking water or food, penetrate its walls and enter the bloodstream. On being disseminated throughout the parenchymatous organs, they may stimulate DNA biosynthesis and exert a powerful carcinogenic effect. Now there is considerable disagreement concerning whether or not A. fibers ingested in drinking water can migrate from the GI lumen through the walls of the GI tract in sufficient numbers to cause adverse local or systemic effects.

Standards. USEPA (1991). MCL and MCLG: 7 MFL (million fibers per liter).

Regulations. WHO (1991). Ingested A. is not hazardous to health and because of it, a health-based guideline value for A. in drinking-water is not recommended. A.-cement pipes should not be used in areas where the water supply is aggressive.[8] **USFDA** has taken no action to date with regard to A. in food because there is no evidence that the ingestion of small amounts of A. found in food poses any human health risk. FDA regulates A. as (1) a component of adhesives intended for use in contact with food, (2) in cross-linked polyester resins for use in articles or components of articles intended for repeated use in contact with food, and (3) as a filler in rubber articles intended for repeated use in producing, manufacturing, packing, processing, treating, packaging, transporting, or holding food.

References:

1. Wigle, D. T., Cancer mortality in relation to asbestos in municipal water supplies, *Arch. Environ. Health,* 32, 185, 1977.
2. Harrington J. M., Craun, G. F., Meigs, J. W., et al., An investigation of the use of asbestos cement pipe for public water supply and the incidence of gastrointestinal cancer in Connecticut 1935–1973, *Am. J. Epidemiol.,* 107, 96, 1978.
3. IARC Monographs, Suppl. 7: *Overall Evaluation of Carcinogenicity,* An Updating of IARC Monographs Volumes 1 to 42, Lyon, 1987.
4. Toft, P., Meek, M. E., Wigle, D. T., et al., Asbestos in drinking water, *CRC Crit. Rev. Environ. Control,* 14, 151, 1984.
5. DHHS Working Group, Report on cancer risk associated with ingestion of asbestos, *Environ. Health. Perspect.,* 72, 253, 1987.
6. Schneider, V. and Maurer, R. R., Asbestos and embryonic development, *Teratology,* 15, 273, 1977.
7. Montizaan, G. K., Knaap, A. G., and van der Heijden, C. A., Asbestos: toxicology and risk assessment for the general population in the Netherlands, *Food Chem. Toxicol.,* 27, 53, 1989.
8. Macrae, K. D., Asbestos in drinking water and cancer, *J. R. Coll. Physicians,* London, 22, 7, 1988.
9. NTP, *Toxicology and Carcinogenesis Studies of Chrysotile Asbestos in Fisher-344 rats,* NIH Publ. No 86–2551, Tech. Rep. No. 295, Public Health Service, U.S. Department of Health and Human Services, Washington, D.C., 1985.
10. McConnell, E. E., Shefner, A. M., Rust, J. H., et al., Chronic effects of dietary exposure to amosite and chrysotile asbestos in Syrian golden hamsters, *Environ. Health Perspect.,* 53, 11, 1983.

1,1-AZOBISFORMAMIDE (CAS No 123–77–3)

Synonyms. *ADA;* Azobiscarbonamide; Azodicarbonic acid, diamide; Azoform *A;* Porofor *ChKħZ 21.*

Properties. Fine, yellow-orange crystalline powder. Poorly soluble in cold water, more readily soluble in hot water. Insoluble in ethanol. Does not react with plasticizers and other components of plastics.

Applications. High-temperature blowing (pore-formation) agent for PVC, polyolefins, polyamides, polyepoxys, polysiloxanes, polyacrilonitrile-butadiene-styrene copolymers, etc. A vulcanizing agent and vulcanization accelerator.

Acute Toxicity. No animals died following administration of 1.5 to 2 g/kg BW doses. Poisoning was accompanied by BW loss and changes in blood morphology. In Wistar rats, LD_{50} is 6.4 g/kg BW.[1] According to Joiner, however, this dose is shown to be harmless for rats and mice.[05]

Repeated exposure revealed cumulative properties: two out of five rats died after four administrations of 1.5 g/kg BW. Gross pathological examination revealed atony of the stomach and intestine. The liver and kidneys looked unusually dark, and the brain was hyperemic. However, according to other data, administration of 1 g/kg BW to rats for 8 weeks was found to be harmless for animals.[05]

Long-term Toxicity. Because ADA is readily reduced to **biurea** during the baking process, feeding studies were performed in which biurea was added to the bread diet at levels of 100 to 1000 times the normal use levels (7.5 ppm in bread dough). The studies elicited no adverse response for the 2-year investigation period.[2] Dogs and rats received 5 and 10% biurea in their diet for a year.[2] No adverse effect was reported in rats. In dogs, evidence of renal pathology was noted after about 4 months. Histological examination revealed a deposition of calculi in the kidney, ureter, and bladder of the majority of animals.

Chemobiokinetics. Of the dose given by gavage to Fisher 344 rats, 30% was found to be absorbed in 72 h after administration.[3] Upon inhalation, ADA is readily converted to **biurea** under physiological conditions, and biurea was the only ^{14}C-labeled compound present in excreta.

Standards. Russia. PML in drinking water: 0.2 mg/l.

Regulations. USFDA (1993) approved the use of A. in closures with sealing gaskets on containers intended for use in producing, manufacturing, packing, processing, preparing, treating, packaging, transporting, or holding food not to exceed 1.2% by weight of the closure-sealing gasket composition.

References:

1. Ferris, B. G., Peters, J. M., Burgess, W. A., et al., Apparent effect of an azodicarbonamide on the lungs, A preliminary report, *J. Occup. Med.*, 19, 424, 1977.
2. Oser, B. L., Oser, M., and Morgareige, K., Studies of the safety of azodicarbonamide as a flour maturing agent, *Toxicol. Appl. Pharmacol.*, 7, 445, 1965.
3. Mewhinney, J. A., Ayres, P. H., Bechtold, W. E., et al., The fate of inhaled A. in rats, *Fundam. Appl. Toxicol.*, 8, 372, 1987.

BENZALDEHYDE (CAS No 100–52–7)

Synonyms. Benzenecarbonal; Benzenecarboxaldehyde; Benzoic aldehyde; Phenylmethanal.

Properties. Clear, colorless liquid with the odor of bitter almonds. Solubility in water is 3 g/l. Mixes with alcohol at all ratios. Odor perception threshold is 0.002 mg/l. Taste perception threshold is 0.003 mg/l.[1]

Applications and **Exposure.** Used in the production of phenolaldehyde and other resins. Is found to be a degradation product of **dibenzoyl peroxide** and to migrate as such from rubber articles into contact media.

Acute Toxicity. Rats tolerate administration of the maximum oral dose of 0.5 g/kg BW.

Repeated Exposure. Rats received 0.6 g/kg BW over a period of 5 weeks. The treatment resulted in adiposis of hepatocytes, liver focal necrosis, periportal cell infiltration, and pneumonia.[2] *Carcinogenicity classification.* NTP: NE—NE—SE*—SE*.

Chemobiokinetics. In rabbits and dogs, B. is removed from the body with the urine as **hippuric(N-benzoylaminoacetic) acid.**

Standards. Russia (1988). MAC and PML: 0.003 mg/l (organolept., taste).

References:

1. Klein, L. et al., *Aspects of River Pollution,* London, 1957.
2. Herrman, H. U. A., *Arzneimittel- Forsch.,* 9, 1244, 1966 (in German).

BENZO[*a*]PYRENE (BaP) (CAS No 50–32–8)

Synonyms. Benz[*a*]pyrene; **3,4**-Benzo[*a*]pyrene; **1,2**-Benzopyrene; **3,4**-Benzopyrene.

Properties. Pale-yellow needles. Water solubility is 0.0012 mg/l at 25°C, soluble in alcohol.

Occurrence and **Exposure.** Polycyclic aromatic hydrocarbons (PAHs) have no industrial use. BaP is only a member of a class of more than 100 compounds belonging to the family of PAHs. BaP may occur in plastics as a contaminant of starting material and additives. BaP is usually found in drinking water in combination with other PAHs. The typical level of BaP in U.S. drinking water is estimated to be 0.00000055 mg/l, and daily intake of total PAH is reported to be 0.000027 mg, 0.000001 mg of which is BaP.[1]

Acute Toxicity. The oral LD_{50} for various PAH is reported to range between 490 and 18,000 mg/kg BW.[2] Effect induced in animals following acute exposure to PAHs include inflammation, hyperplasia, hyperkeratosis, and ulceration of the skin, pneumonia, damage to the hematopoietic and lymphoid systems, immunosuppression, adrenal necrosis, ovotoxicity, and antispermatogenic effects.[2]

Reproductive Toxicity. BaP may cross the placenta of mice and pigs. Pregnant CD-1 mice were given oral doses of 10 to 160 mg/kg BW on days 7 through 16 of gestation. No toxic effects were shown in the dams. In the offspring, total sterility was noted in 97% animals in groups administered 40 and 160 mg/kg.[3]

According to Schardein,[4] BaP produced no teratogenic effects in rats and mice.

Immunotoxicity. Pregnant mice were exposed to a single dose of 150 mg BaP/kg BW *i/p*. The offspring was severely immunosuppressed. This effect may have led to the subsequent widespread development of tumors in these animals.

Genotoxicity. The diol-epoxide metabolites of BaP are considerably more mutagenic than the parent compound. Mutations have been induced in cultured human lymphoblastoma cells.[5] Induction of SCE chromatid exchanges in Chinese hamsters following *i/p* administration of BaP has also been reported.[6]

Carcinogenicity. PAH health effect of primary concern is carcinogenicity. Adequate data were identified only for BaP that is one of the most potent carcinogens: primary tumors have been produced in mice, rats, hamsters, guinea pigs, rabbits, ducks, and monkeys following intragastric, subcutaneous, dermal, or intratracheal administration, both in the site of administration and in other tissues. *Humans.* There is no information from epidemiological studies regarding the carcinogenicity of BaP alone to humans. *Animals.* BaP was given to CC-57 mice in a solution (triethylene glycol) at a dose of 0.001 to 10 mg, ten times, at weekly intervals. As a result, the animals developed malignant neoplasms of the rumen, predominantly squamous cell carcinoma with keratinization, more rarely without keratinization, and benign papillomata.[7] The percentage of animals with tumors declined as the dose was reduced. The LOAEL was identified as 0.01 mg; when this was given orally, rumen papillomata developed in 7.7% mice. The NOAEL appeared to be 0.001 mg, which caused only proliferative changes similar to those in the control group. This dose showed no effect on descendants. Administration of BaP in the diet (0.001 to 0.25 mg/kg food) for a period of 98 to 197 d[8] resulted in an increased incidence of forestomach tumors in CFW mice (papillomas and squamous cell carcinomas). Gastric tumors were found in more than 70% of mice fed 50 to 250 ppm BaP for 4 to 6 months. There were no gastric tumors in 287 control mice, while 178 out of 454 mice fed various levels of BaP developed gastric tumors. *Carcinogenicity classification.* USEPA: Group B2; IARC: Group 2A

Chemobiokinetics. BaP is absorbed principally through the GI tract and lungs. Absorbed BaP is rapidly distributed to the organs and tissues, and may be stored in the mammary and adipose tissue.[9] Metabolism of BaP occurs primarily in the liver. It is initially converted by mixed function oxidases resulting in the formation of **diol epoxides.** These metabolites are subsequently detoxified in a series of conjugation reactions. BaP is metabolized with the formation of 27 metabolites, which are excreted through the bile and subsequently in the feces.[10]

Recommendations. There are insufficient data available to derive drinking-water guidelines for individual PAH but only for BaP, one of the most potent among PAH compounds tested to date. **JECFA** (1983). ADI: 0 to 5 mg/kg BW.

Guidelines. WHO (1992). Guideline value for drinking water: 0.0007 mg/l.

334

Standards. USEPA (1991). Proposed MAC is 0.0002 mg/l for **BaP, antracene, benz[a]anthracene, benz[b]fluoranthene, benzo[g,h,i] perylene, benzo[k]fluoranthene, chrysene, dibenz[a,h]anthracene, fluorene, indeno[1,2,3,c,d]pyrene, phenanthrene** and **pyrene**. **Russia** (1988). MAC: 0.000005 mg/l. **France** (1990). MAC: 0.00001 mg/l.

References:
1. Santodonato, J., Howard, P., and Basu, D., Health and ecological assessment of polynuclear aromatic hydrocarbons, *J. Environ. Pathol. Toxicol.,* 5, 1, 1981.
2. Montizaan, G. K., Kramers, P. G. N., Janus, J. A., et al., Integrated Criteria Document, Polynuclear Aromatic Hydrocarbons (PAH): Effects of 10 Selected Compounds, Natl. Inst. Publ. Health Environ. Protection, The Netherlands, 1989.
3. MacKenzie, K. M. and Angevine, D. M., Infertility in mice exposed *in utero* to BaP, *Biol. Reprod.,* 24, 183, 1981.
4. See **BISPHENOL A,** #7, 843.
5. Danheiser, S. L., Liber, H. L., and Thilly, W. G., Long-term, low-dose BaP-induced mutation in human lymphoblasts competent in xenobiotic metabolism, *Mutat. Res.,* 210, 142, 1989.
6. Raszinsky, K., Basler, A., and Rohrborn, G., Mutagenicity of polycyclic hydrocarbons. V. Induction of sister chromatid exchanges *in vivo, Mutat. Res.,* 66, 65, 1979.
7. Yanysheva, N. Ya., Chernichenko, I. A., Balenko, N. V., et al., *Carcinogenic Substances and Their Environmental Safety Standards,* Zdorov'ya, Kiev, 1977, 136 (in Russian).
8. Neal, J. and Rigdon, R. H., Gastric tumors in mice fed benzo[a]pyrene: a quantitative study, *Tech. Rep. Biol. Med.,* 25, 553, 1967.
9. Weyand, E. N. and Bevan, D. R., Species differences in disposition of benzo[a]pyrene, *Drug Metabol. Dispos.,* 15, 442, 1987.
10. See **ANILINE,** #12, 186.

BUTYRIC ACID (BAc) (CAS No 107–92–5), BUTYRIC ANHYDRIDE (BAn)

Synonyms. Butanoic acid; Ethylacetic acid.

Properties. BAc is a colorless, oily liquid with an unpleasant odor (the odor of rancid oil in diluted solution). Mixes with water and alcohol.[1] BAn is a colorless liquid with a suffocating odor. Solubility in water is 40 g/l. Soluble in alcohol. BAc odor threshold concentration is 0.001 mg/l.[010]

Applications. Used in the production of acetobutyrate cellulose, etrol, etc.

Acute Toxicity. LD_{100} of BAc is 2.5 g/kg BW for rats and mice. Rats tolerate 0.5 g/kg BW. Immediately after administration of the lethal doses, animals assume side position, wheezing and convulsions are observed. Gross pathological examination revealed marked visceral congestion and extensive necrosis of the gastric mucosa. With the small doses, dystrophic changes in the kidneys, catarrhal phenomena in the gastric mucosa, and other morphological changes were noted. Rats tolerate 5 g BAn/kg BW; mice tolerate 0.5 g BAn/kg BW. LD_{50} for mice is 3.5 g BAn/kg. Poisoning is accompanied by the symptoms of CNS inhibition and adynamia. Death occurred on days 10 to 15. Gross pathological examination failed to reveal visible changes in the viscera. Histological examination showed mild dystrophy of the liver and kidneys.

Long-term Toxicity. ADI for BAc appeared to be 1.25 mg/kg BW.

Regulations. USFDA (1993) regulates BAn for use (1) as a component of adhesives to be safely used in food-contact surface, and (2) as a component of the uncoated or coated food-contact surface of paper and paperboard that may be used safely in contact with dry food.

Reference:
Toxicology of New Industrial Chemical Substances, Medgiz, Moscow, 1962, p. 32 (in Russian).

CALCIUM SULFATE (CAS No 7778–18–9)

Properties. Colorless crystals. Gypsum hemihydrate ($CaSO_4 \cdot 5H_2O$), after mixing with water is converted into dihydrate, which sets into a firm mass.

Applications. A filler. Used in construction and medicine.

Toxicity. Oral toxicity has not been studied.

Regulations. USFDA (1992) regulates the use of *Ca* **sulfate** as a colorant in the manufacturing of articles intended for use in contact with aqueous and fatty food.

CARBON BLACK (CB) (CAS No 1333–86–4)

Synonym. Amorphous carbon.

Composition and **Properties.** A quasi-graphitic form of carbon of small particle size formed during the incomplete combustion or thermal decomposition of hydrocarbons contained in natural or industrial gases (gas black, furnace black, channel black) and in oil and coal treatment products (lamp black or activated charcoal, carboraffin, medicoal, norit, ultracarbon).

Applications. Used as a filler for most vulcanizates, as well as for polyolefins, etc. It improves the mechanical properties of vulcanizates and is a dye and antioxidant when exposed to heat, light, and, in particular, UV-radiation.

Acute Toxicity. In rats, LD_{50} is 5 g/kg BW.[1]

Genotoxicity. No data are available on genetic and related effects in humans. Extracts of various commercial CB were mutagenic to *Salmonella* in the presence of an exogenous metabolic system (IARC 1–53, IARC 20–371).

Carcinogenicity. Different sources of CB are carriers of resin containing polycyclic aromatic hydrocarbons, including benzo[a]pyrene, pyrene, antracene and fluorene. Oral administration to mice did not produce GI tract tumors seen after administration of solvent (benzene) extracts (IARC 33–35). No increase in the development of colon tumors occurred in mice or rats fed CB in the diet.[2] Biological activity of the carcinogens contained in CB may be demonstrated by a solvent capable of eluting them from CB. The extractability of carcinogenic hydrocarbons is proportional to the particle size of CB. Highly carcinogenic appear to be benzene extracts of CB *PM-15* (95% tumors), *PGM-33* (92.5%), and *PM-50* (92.5%); CB *PM-70* produced a moderate effect (37.5%), and CB *PM-100* had a slight effect (2.5%). Of benzo[a]pyrene, 25 and 31 mg/kg are found in CB *TGM-33* and *TM-15*, respectively, while *TG-10* contains 32.8 mg/kg.[3] *Carcinogenicity classification.* IARC: carbon black–Group 3, carbon black extract–Group 2B.

Regulations. USFDA (1993) approved the use of CB (channel process or furnace combustion process; total CB not to exceed 50% by weight of the rubber product; furnace combustion black content not to exceed 10% by weight of the rubber products intended for use in contact with milk or edible oils): (1) as a filler for food-contact rubber articles intended for repeated use in producing, manufacturing, packing, processing, treating, packaging, transporting, or holding food; (2) in polysulfide polymer-polyepoxy resins to be used safely for packaging, transporting, holding, or otherwise containing dry food in accordance with specified conditions; (3) in phenolic resins (channel process) to be used as a food-contact surface of molded articles intended for repeated use in contact with nonacid food (pH above 5); (4) as a colorant in the manufacturing of articles intended for use in contact with food (channel process, prepared by the impringement process from stripped natural gas); and (5) in adhesives (channel process) used as components of articles intended for use in packaging, transporting, or holding food. **Russia.** Ministry of Health allows CB *PM-15* to be used temporarily in vulcanizates intended for food-contact purposes; CB *PM-70* and *DG-100* are recommended for use in materials coming in contact with food and pharmacology products.

References:

1. *Bull. IRPTC*, 6, 18, 1983.
2. Pence, B. C. and Buddingh, F., The effect of carbon black ingestion on **1,2**-dimethyl hydrazine-induced colon carcinogenesis in rats and mice, *Toxicol. Lett.*, 25, 273, 1985.
3. Pylev, L. N. and Iankova, G. D., Studies on possible absorption of benzo[a]pyrene and its content in the several domestic sorts of carbon black, *Gig. Truda Prof. Zabol.*, 4, 52, 1974 (in Russian).

CASTOR OIL (CO) (CAS No 8001-79-4)

Synonyms. Neoloid; Oil of Palma Christi; Phorboyl; Ricinus oil.

Properties. Pale-yellow or almost colorless, clear, viscous liquid with a faint odor and a characteristic taste. Soluble in 95% ethanol.

Applications and **Exposure.** Is used as a drying oil for paints, varnishes, plastics, and resins. It is also used in numerous cosmetics. Apart from its laxative effect, CO is likely to be used without harm.

Acute Toxicity. CO is considered minimally toxic when administered orally to humans; the estimated LD is 1 to 2 pt undiluted oil.[1]

Repeated Exposure. Rhesus monkeys were administered 1 ml CO/kg BW by gavage for 4 d. Mild morphological changes in the small intestine characterized by lipid droplets along the mucosal epithelium and in the underlying lamina propria were observed.[2] This was considered a possible indication that CO had reduced lipid metabolism in the intestinal epithelium.

Short-term Toxicity. Dietary concentrations as high as 10% did not affect survival or BW gain in Fisher 344 rats or B6C3F$_1$ mice in 13-week studies. No hematological changes and histopathological lesions in the liver or other visceral organs were observed in the treated animals.[3]

Allergenic Effect. Several cases of sensitization to CO in cosmetics were reported including an allergic reaction to make-up remover.[4] A lipstick containing CO may cause contact dermatitis.[5] A hypersensitivity reaction has been associated with ingestion of CO.[6]

Reproductive Toxicity. *Humans. Teratogenicity.* Mauhoub et al.[7] observed a 3-month-old infant with ectrodactyly, vertebral defects, and growth retardation. The mother had taken one castor oil seed during each of the first 3 months of pregnancy. *Animals. Gonadotoxicity.* There were no significant changes in male reproductive characteristics (sperm count and motility) or in the length of the estrus cycles of rats or mice given diets containing CO for 13 weeks.[1]

Genotoxicity. CO was negative in the *Salmonella* mutagenicity studies; it did not induce SCE or CA in Chinese hamster ovary cells or micronuclei in the peripheral blood erythrocytes of mice evaluated at the end of the 13-week studies.[1]

Chemobiokinetics. Low doses of CO are readily absorbed, but as the oral dose increases, absorption decreases and laxation occurs. At doses of 4 g/kg BW in adults, absorption seems to be complete. JECFA considered this dose to be the NOAEL.

Standards. JECFA. ADI for man: 0 to 0.7 mg/kg BW.

Regulations. USFDA (1993) regulates use of CO as (1) a component of the uncoated or coated food-contact surface of paper and paperboard intended for use in producing, manufacturing, packing, transporting, or holding aqueous and fatty food; (2) in adhesives used as components of articles intended for use in packaging, transporting, or holding food; (3) as a plasticizer in rubber articles intended for repeated use in contact with food in an amount up to 30% by weight of the rubber product; (4) in cellophane for packaging food; (5) in resinous and polymeric coatings in a food-contact surface; (6) in acrylic and modified acrylic plastics; (7) in cross-linked polyester resins to be used safely as articles or components of articles intended for repeated use in contact with food; (8) in closures with sealing gaskets (up to 2% by weight) for food containers; and (9) as a defoaming agent that may be used safely in the manufacturing of paper and paperboard and as components of articles intended for use in producing, manufacturing, packing, transporting, or holding food.

References:

1. Gosselin, R. E., Hodge, H. C., Smith, R. P., Gleason, M. N., Eds. *Clinical Toxicology of Commercial Products,* 4th ed., Williams and Wilkins, Baltimore, MD, 1976, 152.
2. Diener, R. M. and Sparano, B. M., Effects of various treatments on the histochemical profile of the gastrointestinal mucosa in rhesus monkeys, *Toxicol. Appl. Pharmacol.,* 13, 412, 1968.
3. Irwin, R., *Toxicity Studies of Castor Oil in F344/N Rats and B6C3F$_1$ Mice (Dosed Feed Studies),* NTP Tox. 12, NIH Publ. No 92–3131, 1992, 30.
4. Brandle, I., Boujnah-Khouadja, A., and Fousserau, J., Allergy to castor oil, *Contact Dermat.,* 9, 424, 1983.
5. Say, S., Lipstick dermatitis caused by castor oil, *Contact Dermatit.,* 9, 75, 1983.
6. McGuire, T., Rothenberg, M. B., and Tiler, D. C., Profound shock following intervention for chronic untreated stool retention, *Clin. Pediatr.,* 23, 459, 1983.

7. Mauhoub, M. E., Khalifa, M. M., Jaswal, O. B., et al., Ricin syndrome, a possible new syndrome associated with ingestion of castor-oil seed in early pregnancy: a case report, *Ann. Trop. Paediatr.*, 3, 57, 1983.

CHLORAL (CAS No 75–87–6)

Synonyms. Trichloroacetaldehyde; Trichloroethanol.

CHLORAL HYDRATE (CH) (CAS No 302–17–0)

Synonyms. Trichloroacetaldehyde monohydrate; **1,1,1**-Trichloro-**2,2**-dihydroxyethane; **2,2,2**-Trichloro-**1,1**-ethanediol.

Properties. Colorless, oily liquid with a sharp, penetrating, unpleasant odor. Soluble in water, readily soluble in alcohol. Combines readily with water and becomes completely converted to **CH** (colorless crystals with specific odor). CH solubility in water is 470 g/100 ml. Taste and odor perception threshold is 16 mg C./l; practical threshold is 32 mg C./l. Heating and chlorination of such solutions did not affect odor intensity. The color of solutions remained unaffected even by high concentrations.[1]

Applications and **Exposure.** Used in the production of rigid polyurethane foams and other materials intended for contact with food. It is formed as a by-product of chlorination when chlorine reacts with humic acids. CH has been widely used as a sedative, an anesthetic, or a hypnotic drug in humans at oral doses of up to 14 mg/kg BW; it is still used in pediatric medicine and dentistry. CH has been found in drinking water at concentrations up to 0.02 mg/l.

Acute Toxicity. LD_{50} of CH is 725 mg/kg in rats, 850 mg/kg in mice, 1400 mg/kg in rabbits, and 940 mg/kg BW in guinea pigs. According to Sanders et al.[2] in Sprague-Dawley rats, the LD_{50} values are 1442 mg CH/kg BW (males) and 1265 mg CH/kg BW (females). Administration of lethal doses led to a brief period of excitation, with marked activity and subsequent CNS inhibition and motor coordination disorder. Animals assumed a side position, became comatose, and died within a few hours.[1] Gross pathological examination revealed hemodynamic disturbances and parenchymatous dystrophy of the liver and kidneys.[3]

Repeated Exposure. CH has no marked cumulative properties. Rats tolerate administration of $\frac{1}{5}$ and $\frac{1}{10}$ LD_{50}, though the total dose was equal to three LD_{50}.[2] In a 14-d gavage study, male mice were exposed to the doses of 0.1 and 0.01 LD_{50} of CH. Poisoning caused an increase in liver weights and a decrease in spleen weights at the highest dose-level.[2]

Short-term Toxicity. In a 90-d drinking water study in mice (the same doses of CH as in the 14-d study), there was a dose-related hepatomegaly in male but not in female mice, accompanied by significant changes in serum chemistry and hepatic microsomal parameters. (The latter was observed in both sexes.) No other significant toxicological effects were found.[2] The LOAEL was identified to be 16 mg/kg BW for liver enlargement. Mild liver toxicity was noted in rats receiving CH in their drinking water over a period of 90 d.[4]

Long-term Toxicity. B6C3F₁ mice were exposed via their drinking water to 1 g CH/l (166 mg/kg BW) for 104 weeks. The liver was found to be a primary target organ. An increase in liver weights and hepatocellular necrosis were noted.[5] In a 7-month study in rats, CH was found to exhibit a polytropic action. Primary target tissues are the liver and cardiovascular system. Conditioned reflex activity was affected. Diminished bromosulfophthalein retention by the liver and increased activity of serum transaminases were noted. The NOAEL of 0.01 mg/kg BW was identified in this study.[1] USEPA (1990) considered RfD of 0.0016 mg/kg.

Reproductive Toxicity. No increase of the adverse effect was found in the offspring of 71 women who took CH as a drug in their 4 lunar months.[6]

Immunotoxicity. In the above-described studies,[2] no alterations were found in either humoral or cell-mediated immunity. However, female mice treated for 90 d demonstrated a significant inhibition of humoral immunity function.[7]

Genotoxicity. CH was shown to be mutagenic in *Salmonella* and in *Dr. melanogaster* but not to bind to DNA. It caused SCE and disruption of chromosomal segregation in cell division,[8] and produced aneuploidy in human lymphocytes (Vagnarelli et al., 1990) and in cultured hamster cells (Degrassi and Tanzorella, 1988).

Carcinogenicity. The exposure of male B6C3F$_1$ mice to up to 1 g CH/l in their drinking water[6] resulted in an increased incidence of hepatocellular carcinomas (46% of survivors), hepatocellular adenomas (29%), and combined tumors (71%). A single oral dose of 10 mg CH/kg BW administered to 15-d-old male mice resulted in a significant increase in the rate of liver tumors (including trabecular carcinomas after 48 to 92 weeks of observation).[9]

Standards. WHO (1992). Guideline value (Chloral) for drinking water: 0.011 mg/l. **Russia** (1988). MAC and PML: 0.2 mg/l.

Regulations. USFDA (1993) regulates CH (1) as an ingredient in paper and paperboard intended for contact with dry food and (2) in adhesives used as components of articles intended for use in packaging, transporting, or holding food.

References:

1. Kryatov, I. A., Hygienic evaluation of sodium *p*-chlorobenzenesulphonate and chloral as water pollutants, *Gig. Sanit.,* 3, 14, 1970 (in Russian).
2. Sanders, V. M., Kauffmann, B. M., White, K. L., et al., Toxicology of chloral hydrate in the mouse, *Environ. Health Perspect.,* 44, 137, 1982.
3. See **BISPHENOL A,** #2, 175 (in Russian).
4. Daniel, F. B., Robinson, M., Stober, J. A., et al., Ninety-day toxicity study of chloral hydrate in the Sprague-Dawley rat, *Drug Chem. Toxicol.,* 15, 15, 1992.
5. Daniel, F. B., De Angelo, A. B., Stober, J. A., et al., Hepatocarcinogenicity of chloral hydrate, 2-chloroacetaldehyde, and dichloracetic acid in the male B6C3F$_1$ mouse, *Fundam. Appl. Toxicol.,* 19, 159, 1992.
6. Heinonen, O. P., Slone, D., and Shapiro, S., *Birth Defects and Drugs in Pregnancy,* Publishing Sciences Group, Littleton, MA, 1977.
7. Kaufmann, B. M., White, K. L., Sanders, V. M., et al., Humoral and cell-mediated immune status in mice exposed to chloral hydrate, *Environ. Health Perspect.,* 44, 147, 1982.
8. Gu, W. Z., Sele, B., Jalbert, P., et al., Induction of sister chromatid exchanges by trichloroethylene and its metabolites, *Toxicol. Eur. Res.,* 3, 63, 1981.
9. Rijhainghani, K. S., Abrahams, C., Swerdlow, M. A., et al., Induction of neoplastic lesions in the liver of C$_{57}$BL × C3HF$_1$ mice by chloral hydrate, *Cancer Detect. Prev.,* 9, 279, 1986.

CHLORITE, SODIUM SALT (CAS No 7758–19–2)

Synonym. Chlorous acid, sodium salt.

Properties. White, slightly hygroscopic crystals or flakes. Water solubility is 390 g/l at 17°C.

Applications. Used to bleach wood pulp in paper processing, in the production of waxes and varnishes. It is the only chlorite salt produced commercially in significant quantities.

Acute Toxicity. LD$_{50}$ is reported to be 105 mg/kg BW in rats[1] and 493 mg/kg BW in quail.[2]

Repeated Exposure. No adverse effect on thyroid function was noted in monkeys given C. in their drinking water at doses of up to 60 mg/kg BW for 30 to 60 d.[3]

Short-term Toxicity. Chlorite affects red blood cells, resulting in metHb formation in cats and monkeys. *Animals.* C. was shown to produce hemolitic anemia in several animal species at concentrations of 100 mg/l or higher in drinking water. In the 30- to 60-d studies in Sprague-Dawley rats, there was no induction of methemoglobinemia at C. drinking water concentration of up to 500 mg/l. After 90 d of treatment, red blood cell glutathione levels were 40% below the controls in the 100 mg/l group, with at least a 20% reduction in the rats receiving 50 mg/l. The NOAEL of 1 mg/kg was identified in this study.[4] *Humans.* No similar signs were noted in humans at much lower doses.

Long-term Toxicity. Rats received 1 to 1000 mg C. in their drinking water for 2 years. Animals consuming C. at concentrations of 100 and 1000 mg/l (9.3 and 81 mg/kg) exhibited treatment-related renal pathology.[5]

Reproductive Toxicity. There was a minimal adverse effect on reproduction in rats and mice given C. in their drinking water at concentrations of 100 mg/l or higher. Female A/J mice given C.

in drinking water (up to 60 mg/kg BW) throughout gestation and then for 28 d of lactation exhibited no change in litter size, weights at birth, or neonatal survival throughout lactation, but pups had a significant reduction in BW gain.[6] No adverse effects were noted in Long Evans rats receiving 1 to 100 mg C./l in their drinking water.[7] A concentration of 100 mg/l in drinking water given to maternal mice (14 mg/kg BW) through gestation and lactation caused a decrease in BW in pups at weaning. The LOAEL for developmental effects appeared to be 14 mg/kg BW.[8]

Genotoxicity. C. was shown to be positive in bacterial tests; it produced CA in cultured mammalian cells. In mice treated *in vivo,* there were inconclusive results with regard to the induction of micronuclei, while a single study showed no CA or abnormal sperm morphology (IARC 52–196). According to Meier et al.,[9] C. caused CA neither in the micronucleus test nor in the cytogenic assay in mouse bone marrow cells following gavage dosing.

Carcinogenicity. Male and female B6C3F$_1$ rats received 0.025% and 0.05% C. in their drinking water for 80 weeks; Fisher 344 rats were given 0.03 and 0.06% C. in their drinking water for 85 weeks. The treatment resulted in a marginal increase of the rate of lung tumors in male mice. A significant increase in tumor incidence was found in rats.[10,11] No increase in tumor incidence was noted in mice given 250 mg C./l in drinking water over a period of 85 weeks[12] and in rats in a 2-year drinking water study.[5] *Carcinogenicity classification.* IARC: Group 3.

Chemobiokinetics. Radiolabel, derived from C., given to rats by gavage (1.15 mg/kg), was absorbed and excreted as **chloride** (32%) and **chlorite** (6%).[13]

Guidelines. WHO (1992). Provisional guideline value for drinking water: 0.2 mg/l.

Regulations. USFDA (1993) approved the use of C. (1) at levels from 125 to 250 ppm as a slimicide in the manufacturing of paper and paperboard that contact food and (2) in adhesives used as components of articles intended for use in packaging, transporting, or holding food.

References:

1. Musil, J. R., Knotek, Z., Chalupa, J., et al., Toxicologic aspects of chlorine dioxide application for the treatment of water containing phenols, *Technol. Water,* 8, 327, 1964.
2. Fletcher, D., Acute oral toxicity study with sodium chlorite in bobwhite quail, Industrial Bio-Test Laboratory, Report to Olin Corp. (IBT No J2119), 1973.
3. Bercz, J. P., Jones, L., Garner, L., et al., Subchronic toxicity of chlorine dioxide and related compounds in drinking water in the non-human primates, *Environ. Health Perspect.,* 46, 47, 1982.
4. Heffernan, W. P., Guion, C., and Bull, R. J., Oxidative damage to the erythrocyte induced by sodium chlorite *in vivo, J. Environ. Pathol. Toxicol.,* 2, 1487, 1979.
5. Haag, H. B., The effect on rats of chronic administration of sodium chlorite and chlorine dioxine in the drinking water. Report to the Mathieson Alkali Works from the Medical College of Virginia, 1949, 5.
6. Moore, C. S. and Calabrese, E. J., The effects of chlorine dioxide and sodium chlorite on erythrocytes of A/J and C57L/J mice, *Environ. Pathol. Toxicol.,* 4, 513, 1980.
7. Carlton, B. D., Habash, D. L., Basaran, A. H., et al., Sodium chlorite administration to Long Evans rats: reproductive and endocrine effects, *Environ. Res.,* 42, 238, 1987.
8. Moore, G. S. and Calabrese, E. J., Toxicological effects of chlorite in the mouse, *Environ. Health Perspect.,* 46, 31, 1982.
9. Meier, J. R., Bull, R. J., Stober, J. A., et al., Evaluation of chemicals used for drinking water disinfection for production of chromosomal damage and sperm-head abnormalities in mice, *Environ. Mut.,* 7, 201, 1985.
10. Yokose, Y., Uchida, K., Nakae, D., et al., Studies of carcinogenicity of sodium chlorite in B6C3F$_1$ mice, *Environ. Health Perspect.,* 76, 205, 1987.
11. Shimoyama, T., Hiasa, Y., Kitahori Y., et al., Absence of carcinogenic effect in rats, *J. Nara. Med. Assoc.,* 36, 710, 1985.
12. Kurokawa, Y., Takayama, S., Konishi, Y., et al., Long-term *in vivo* carcinogenicity tests of potassium bromate, sodium hypochlorite and sodium chlorite conducted in Japan, *Environ. Health Perspect.,* 69, 221, 1986.
13. Abdel-Rahman, M. S., Couri, D., and Bull, R. J., Metabolism and pharmacokinetics of alternative drinking water disinfectants, *Environ. Health Perspect.,* 46, 19, 1982.

CYCLOHEXANE (CAS No 110–82–7)

Synonyms. Hexahydrobenzene; Hexamethylene.

Properties. Colorless liquid with a penetrating odor. Solubility in water is 55 mg/l;[02] miscible with alcohol. Odor perception threshold is 1 g/l,[1] or 0.3 mg/l,[2] or 0.011 mg/l.[02]

Applications. Used in the production of polyamides and synthetic fibers.

Acute Toxicity. In mice, LD_{50} is 4.7 g/kg BW. Poisoned animals displayed CNS inhibition, with subsequent excitation and clonic convulsions. Death within 3 to 5 min, possibly as a result of respiratory failure.[3] Rats are less susceptible, though they show age-dependent susceptibility to C.: at BW up to 50 g, LD_{50} is 8 ml/kg BW, at 80 to 160 g, it is 39 ml/kg BW, and at 300 to 470 g, it is 16.5 ml/kg BW.[4]

Short-term Toxicity. Rats received 200 and 400 mg C./kg BW. The treatment did not affect general condition, behavior, blood morphology, or liver function. A 400 mg/kg dose lowered catalase and cholinesterase activity.

Genotoxicity. C. is found to be positive in the *Salmonella* mutagenicity assay.

Chemobiokinetics. C. is metabolized to form metabolites **cyclohexanol** (38%) and **(q)**-*trans*-1,2-**cyclohexanediol** (7.0%), which are excreted in rabbits as glucuronides in the urine. Of administered C., 30% are excreted unchanged via the lungs and 9% as CO_2.[6]

Standards. Russia (1988). MAC and PML: 0.1 mg/l.

Regulations. USFDA (1993) regulates C. (1) as a component of adhesives used in articles intended for use in packaging, transporting, or holding food; (2) as a component of the uncoated or coated food-contact surface of paper and paperboard that may be used safely in producing, manufacturing, packing, transporting, or holding dry food; and (3) as a defoaming agent that may be safely used as a component of food-contact articles.

References:

1. See **ADIPIC ACID,** #4, 30.
2. Ogata, M. and Meyake, *J. Water Res.,* 7, 1493, 1973.
3. See **PHTHALIC ANHYDRIDE,** #1, 61.
4. See **ACETONITRILE,** #2.
5. See **RESORCINOL,** #3.
6. See **ANILINE,** #11, 266.

CYCLOHEXANONE OXIME (CO)

Genotoxicity. CO is found to be positive in *Salmonella* mutagenicity assay (NTP–92).

Chemobiokinetics. In male F344 rats, **CO** is rapidly eliminated, primarily as metabolites in urea. Metabolites arise by hydrolysis of the oxime to cyclohexanone which is then reduced to cyclohexanol and eliminated as the glucuronide or further oxidized to the diol which is then conjugated and eliminated (NTP–92).

CYCLOHEXYLAMINE (CAS No 108–91–8)

Synonyms. Aminocyclohexane; Cyclohexanamine; Hexahydroaniline.

Properties. Colorless liquid with a penetrating odor and a bitter taste. Completely soluble in water and alcohol. Odor perception threshold is 25 mg/l.[02]

Applications. Used in the production of silicon elastomers, plasticizers, rubber chemicals, and dyestuffs; an emulsifying agent.

Acute Toxicity. In rats, LD_{50} is 600 mg/kg BW (5% solution).

Short-term Toxicity. C. hydrochloride was added to the diet of mice and Wistar and DA rats for 13 weeks. The 400 mg/kg BW dose caused a decrease of BW gain and food consumption only in rats.[1] Retardation of BW gain and other changes were noted in rats, guinea pigs, and rabbits given the 100 mg/kg BW dose for 3 months.[05]

Reproductive Toxicity. *Teratogenicity.* No fetal changes were recorded in rats gavaged with the dose of 36 mg C./kg BW on days 7 through 13 of gestation[2] and in mice fed 100 mg C./kg BW on days 6 through 11 of gestation.[3] *Gonadotoxicity.* A limited toxic effect seems to be testicular atrophy (decreased weight and histology) in rats fed dietary levels of 2000 ppm or more; in mice, there was no evidence of testicular damage even at a dietary concentration of 3000 ppm.[4]

Carcinogenicity. Negative results for **C. hydrochloride** and **C. sulfate** have been reported in rats and mice.[09]

Chemobiokinetics. Possible differences in metabolism in mice and rats (see above) have been reported.

References:

1. Gaunt, I. F., Hardy, J., Grasso, P., et al., Long-term toxicity of cyclohexylamine hydrochloride in the rat, *Food Cosmet. Toxicol.,* 14, 55, 1976.
2. Tanaka, S., Nakaura, S., Kawashima, K., et al., Teratogenicity of food additives. II. Effect of cyclohexylamine and cyclohexylamine sulfate on fetal development in rats, *Shokuhin Eiseigaku Zasshi,* 14, 542, 1973 (in Japanese).
3. Takano, K., and Suzuki, M., Cyclohexylamine, a chromosome-aberration producing substance: no teratogenicity in the mouse, *Cong. Anomalies,* 11, 51, 1971 (in Japanese).
4. Roberts, A., Renwick, A. G., Ford, G., et al., The metabolism and testicular toxicity of cyclo-hexylamine in rats and mice during chronic dietary administration, *Toxicol. Appl. Pharmacol.,* 98, 216, 1989.

2,6-DICHLORO-*p*-PHENYLENEDIAMINE (CAS No 609–20–1)

Synonym. 1,4-Diamino-2,6-dichlorobenzene.

Applications. Used in the manufacturing of polyamide fibers and polyurethane.

Acute Toxicity. In rats, LD_{50} is reported to be 0.7 g/kg BW.

Genotoxicity. Is found to cause CA in *in vitro* studies (NTP–92).

Carcinogenicity. No potential for carcinogenicity was observed in male rats given D. at dose-levels of 1000 or 2000 ppm in their feed. Meanwhile, there was an increase in tumor incidence in females fed 2000 or 6000 ppm doses. In mice (dose-levels of 1000 or 3000 ppm), hepatocellular adenomas (in males), hepatocarcinomas, and adenomas were reported. *Carcinogenicity classification:* NTP: N—N—P—P.

Reference:

NTP Technical Report TR-036, Litton Bionetics, Inc., 1980.

ETHYLBENZENE (EB) (CAS No 100–41–4)

Synonyms. Ethylbenzol; Phenylethane.

Properties. Colorless liquid at room temperature with an aromatic odor. Solubility in water is 0.017% at 25°C. Readily soluble in alcohol and in most organic solvents. Odor perception threshold is 0.029 mg/l.[02] Taste perception threshold values vary from 0.072 to 0.2 mg/l.[01]

Applications and **Exposure.** EB is used primarily in the production of styrene, of cellulose acetate silk, and in styrene-butadiene rubber manufacturing. EB may migrate from polystyrene packaging materials into food. Concentrations of 0.0025 to 0.021 mg/l in milk beverage and soup have been reported.[1] EB is distributed widely in the environment because it is often used in foils and as a solvent.

Acute Toxicity. LD_{50} for Wistar rats of both sexes is 3.5 g/kg BW.[2] The maximum tolerated dose for rats is 2 g/kg BW. The dose of 1 g/kg affects NS function.[3] Poisoned animals displayed primary excitation with subsequent severe CNS depression, impaired motor coordination, and later, a progressive decline in body temperature, respiratory and cardiac activity, convulsions, and death from respiratory center paralysis. Histological examination revealed signs of venous congestion with point hemorrhages[2] and moderate changes in the kidneys. Administration of 1 g/kg BW to rabbits caused leukocytosis and thrombocytopenia (Faustov and Volchkova, 1958).

Repeated Exposure failed to reveal the evidence of accumulation. The treatment caused changes in hematological analyses (neutrophilia and relative lymphopenia) and in the rate of motor conduction along the sciatic nerve. Histological examination revealed alterations in the structure of the visceral organs (Mitran et al., 1986).

Long-term Toxicity. In rats given 400 mg EB/kg BW for 6 months, effects on the liver and kidneys were shown. The NOAEL of 136 mg/kg BW was reported.[3] Rabbits received 200 mg/kg BW for the same period of time. The treatment caused retardation of BW gain and marked changes in CNS function and in blood morphology. Gross pathological examination revealed changes in the visceral organs. In rabbits given water containing 2 mg EB/l, there were no such changes.[4]

Immunotoxicity. Simultaneous immunization and poisoning of rats caused some inhibition of the plasmocytic reaction and a stable rise in antibody titer. A decline in the blood serum bactericidal capacity was noted.[5]

Reproductive Toxicity. A mild teratogenic and embryotoxic effect was observed in rats following inhalation exposure to 2.4 g/m^3 on days 7 to 15 of pregnancy (Tatrai et al., 1982). EB is shown to cross the placenta.

Genotoxicity. EB did not show mutagenic activity in bacteria, yeasts, insects, mammalian cells *in vitro,* and mammals. No induction of SCE or CA were found in Chinese hamster ovary cells, but a weak positive response was reported for SCE induction in human lymphocytes cultured with S9.[1,4]

Carcinogenicity. Positive results are noted in rats.[09]

Chemobiokinetics. EB, in a liquid form, is easily absorbed by humans via the skin and via the GI tract. Distribution and excretion are rapid processes. In humans, it is stored in the fat. EB is likely to be metabolized almost completely into **mandelic acid** and **fenylglyoxalic acid,** both compounds being excreted in the urine. Metabolism in experimental animals differs from that in humans in the formation of **benzoic acid** as the major metabolite, along with **mandelic acid.** Urinary excretion of metabolites is almost complete within 24 h.[1]

Guidelines. WHO (1992). Guideline value for drinking water: 0.3 mg/l. The levels of 0.0024 to 0.2 mg/l are likely to give rise to consumer complaints of foreign odor and taste.

Standards. Russia (1988). MAC and PML in drinking water: 0.01 mg/l (organolept., taste). **USEPA** (1991). MCL and MCLG: 0.7 mg/l.

References:

1. ECETOC Joint Assessment of Commodity Chemicals, *Ethylbenzene,* CAS No 100–41–4, No. 7: Report from the European Chemical Industry Ecology and Toxicology Centre, August 1, 1986.
2. See **CYCLOHEXANOL,** 62, 1962.
3. See α-**METHYL STYRENE,** #1.
4. Norppa, H. and Vainio, H., Induction of SCE by styrene analogues in cultured human lymphocytes, *Mutat. Res.,* 116, 379, 1983.
5. USEPA, Office of Drinking Water Health Advisories, *Rev. Environ. Contam. Toxicol.,* 106, 123, 1988.

2-ETHYLHEXANOL (2-EH) (CAS No 104–76–7)

Synonyms. Ethylhexanol; **2**-Ethylhexyl alcohol.

Properties. Colorless liquid with a musty odor. Odor threshold concentration is 0.075 mg/l.[010]

Applications and **Exposure.** Used in the production of plastics, in the manufacturing of plasticizers, **di(2-ethylhexyl) phthalate** and **di(2-ethylhexyl) adipate.** 2-EH is the major metabolite of DEHP. Occurs naturally in food.

Acute Toxicity. LD_{50} is reported to be about 2 g/kg BW in rats, mice, and guinea pigs and about 1.2 g/kg BW in rabbits.[06]

Repeated Exposure. $B6C3F_1$ mice were given by gavage 100 to 1500 mg 2-EH/kg BW for 11 d. The treated animals demonstrated ataxia and lethargy. A dose of 100 mg/kg appeared to be ineffective in mice.[5]

Short-term Toxicity. Wistar rats were given 1335 mg 2-EH/kg for 7 d. Treatment caused an increase in relative liver weight and changes in activity of some enzymes.[6] $B6C3F_1$ mice received 25 to 500 mg 2-EH/kg BW by gavage over a period of 3 months. Toxic effect was observed at 250 and 500 mg/kg dose levels.[5]

Long-term Toxicity. The doses of 200 mg 2-EH/kg BW in mice and 50 mg 2-EH/kg BW in rats were found to be ineffective.[5]

Reproductive Toxicity. *Teratogenicity.* Dermal application on days 6 to 15 of gestation at dose levels of 0.5 to 3 ml/kg BW caused no developmental toxicity in Fisher 344 rats at and below levels that produced maternal toxicity.[1] The NOAEL was identified to be at least 2520 mg/kg BW. Pregnant Wistar rats received repeated oral doses of 130 to 1300 mg 2-EH/kg BW in an aqueous emulsion on days 6 to 15 of gestation. The treatment with the highest dose led to marked maternal toxicity. Skeletal abnormalities were noted in the offspring. The NOAEL for these effects appeared to be 130 mg/kg BW in this study.[2] No signs of developmental toxicity were observed in timed pregnant Swiss mice after oral administration of 17 to 190 mg/kg BW doses on gestation days 0 through 17 (by microencapsulation and incorporation in the diet).[3]

Genotoxicity. 2-EH has been shown to be positive in the *Salmonella* mutagenicity assay. Meanwhile, JECFA (1993) does not consider 2-EH to be genotoxic.

Carcinogenicity. 2-EH was not shown carcinogenic activity when administered by gavage to rats for 24 months or to mice for 18 months.[5]

Chemobiokinetics. Following oral administration of up to 300 mg/kg, 2-EH is shown to be readily absorbed in the GI tract of rats and completely eliminated in the urine within 22 h; 2-**ethylhexanoic acid** is the major metabolite.[4]

Recommendations. JECFA (1993). ADI: 0 to 0.5 mg/kg BW.

References:
1. Tyl, R. W., Fisher, L. C., Kubena, M. F., et al., The developmental toxicity of 2-ethylhexanol applied dermally to pregnant Fisher 344 rats, *Fundam. Appl. Toxicol.,* 19, 176, 1992.
2. USEPA, Office of Toxic Substances, *Chem. Prog. Bull.,* 12, 20, 1990.
3. Price, C. J., Tyl, R. W., Marr, M. C., et al., Developmental toxicity evaluation of DEHP metabolites in Swiss mice, *Tertatology,* 43, 457, 1991.
4. Albro, P. W., The metabolism of 2-ethylhexanol, *Xenobiotica,* 5, 625, 1975.
5. *Toxicological Evaluation of Certain Food Additives and Contaminants,* WHO Food Additives Ser. 32, WHO/IPCS, Geneva, 1993, 35.
6. Lake, B. G., Gangolli, S. D., Grasso, P., et al., Studies on the hepatic effects of orally administered di-(ethylhexyl)phthalate in the rat, *Toxicol. Appl. Pharmacol.,* 32, 355, 1975.

FERROCENE (CAS No 102–54–5)

Synonyms. Bis(cyclopentadienyl)iron; Dicyclopentadienyl iron.

Properties. Orange crystals with a camphoric odor. A relatively volatile, organometallic compound. Poorly soluble in water, soluble in alcohol.

Applications. Used as a chemical intermediate, a catalyst in the manufacturing of plastics. A photosensitizer of plastics.

Acute Toxicity. LD_{50} is reported to be 1190 to 2260 mg/kg BW in rats and 380 to 930 mg/kg BW in mice. Gross pathological examination showed enteritis and deposition of administered substance in the fatty tissue of the liver.[1]

Repeated Exposure. No mortality or clinical signs of intoxication were observed in Fisher 344 rats and $B6C3F_1$ mice exposed to 2.5 to 40 mg F./m^3 for 2 weeks.[2]

Short-term Toxicity. Mice were given $\frac{1}{20}$, $\frac{1}{10}$, or $\frac{1}{5}$ LD_{50} for 3.5 months. Retardation of BW gain and changes in the functional condition of the CNS were observed. A decline in Hb blood level and impaired liver function were found in rats administered $\frac{1}{20}$ LD_{50} for 3 months. Liver weights were increased and F. accumulation was noted in the liver, spleen, lungs, and suprarenals. Gross pathological examination revealed exudative glomerulonephritis and changes in the thyroid.[1]

Allergenic effect is not observed.

Genotoxicity. F. appeared to be positive in the sex-linked recessive lethal test in *Dr. melanogaster*[3] and in the *Salmonella* mutagenicity assay with metabolic activation. A long-term administration of $\frac{1}{10}$ LD_{50} to rats caused a mutagenic effect in the bone marrow cells. One-twentieth LD_{50} appeared to be ineffective in this study.[4] F. produced a doubling of the SCE frequency and a three-point monotonic increase with at least the highest dose.[5]

Chemobiokinetics. F. is rapidly metabolized. Carbon portion of the molecule is nearly completely eliminated, primarily in the urine.

Carcinogenicity. F. has structural similarities to other metallocenes that have been shown to be carcinogenic.

References:
1. See **ADIPIC ACID,** #2, 275.
2. Sun, J. D., Dahl, A. R., Gillett, N. A., et al., Two-week repeated inhalation exposure of F344 rats and B6C3F$_1$ mice to ferrocene, *Fundam. Appl. Toxicol.,* 17, 150, 1991.
3. Zimmering, S., Mason, J. M., Valencia, R., et al., Chemical mutagenesis testing in Drosophila. II. Result of 20 coded compounds tested for the National Toxicology Program, *Environ. Mut.,* 7, 81, 1985.
4. See **ISOPROPYLBENZENE HYDROPEROXIDE,** #6, 319.
5. See **FURFURAL,** #5.

FERROCENE DERIVATIVE

Synonym. *FEP-2.*

Properties. Dark-brown, oily liquid with a penetrating odor. Poorly soluble in water. Contains diethylferrocene as an impurity.

Applications. Organometallic photosensitizer that is effective in plastics intended for agricultural use.

Acute Toxicity. In female rats, LD$_{50}$ is 17.1 g/kg BW for the technical grade product. Male mice and rats of both sexes tolerate the administration of 20 g/kg BW of the distilled FEP-2 and technical grade product. No changes in behavior or in BW gain were observed. However, gross pathological and histological examination revealed exudative glomerulonephritis in these animals.

Repeated exposure failed to reveal marked cumulative properties. K$_{acc}$ is 6. The treatment caused a reduction in Hb blood level and an increase in the relative weights of the thyroid, liver, and kidney. Histological examination revealed deposition of dark-brown drops in the viscera.

Short-term Toxicity. Mice were dosed by gavage with 200 and 1000 mg/kg BW, rats with 500 and 1000 mg/kg BW as an oil suspension for 4 months. Animals looked unkempt, with tousled fur. A 1000 mg/kg dose caused 50% mortality in mice of both sexes; a dose of 200 mg/kg caused 23% mortality.

Long-term Toxicity. In a 13-month study, rats and mice were dosed by gavage with 50 mg FEP/kg BW. The treatment produced a pronounced toxic effect. The NOAEL of 5 mg/kg BW was identified in this study.

Reproductive Toxicity. *Embryotoxic* and *teratogenic* effects were not found on administration of 1 g/kg BW to pregnant rats.

Genotoxicity. Metaphase analysis of bone marrow cells revealed a cytogenic effect in animals treated with 50 mg/kg BW in a long-term study. The NOEL for this effect is 20 mg/kg BW.

Standards. Russia. PML in food: 0.5 mg/l; Recommended PML in drinking water: **n/m.**

Reference:
See **BUTYL STEARATE,** #1, 34.

FORMIC ACID (FA) (CAS No 64–18–6)

Synonyms. Aminic acid; Formylic acid; Hydrogen carboxylic acid; Methanoic acid.

Properties. White, crystalline powder or colorless, highly caustic liquid with a pungent odor. Miscible with water, alcohol, and ether.[1] Odor perception threshold is 1700 mg/l.[02]

Applications and **Exposure.** Used as a plasticizer for vinyl resins and a coagulant for latex. Thermal degradation of polyethylene during manufacturing may result in the release of FA. FA is present in a free acid state in a number of plants. Human exposure occurs due to consumption of food and water containing FA.

Acute Toxicity. *Humans.* Ingesting by man is accompanied by severe intoxication (sometimes, hematuria, anuria, uremia, circulatory failure, or pneumonia) and death.[2] *Animals.* LD$_{50}$ is 1.1 to 1.85 g/kg BW in rats, 0.7 to 1.1 g/kg BW in mice, and 4 g/kg BW in dogs.[3]

Repeated Exposure. Young rats received FA in the diet or drinking water at levels of 0.5 to 1% for 6 weeks. The treatment caused a reduction in BW at the higher dose-levels. Hypochromic anemia and mild lymphocytosis developed in rats receiving FA in the diet.[4]

Reproductive Toxicity. *Embryotoxicity.* Survival of the offspring of female rats exposed to 1% FA in their drinking water for up to 7 months was reduced to 50 to 67%.[5] *Gonadotoxicity.* In a 13-week inhalation study, there were no effects on sperm motility, density, and testicular or epididymal weights in rats and mice. No changes were found to develop in the estrus cycle.[1]

Genotoxicity. FA is not mutagenic by itself but could give positive results being tested in concentrations that produced nonphysiological *pH* levels. It is not shown to be mutagenic in the *Salmonella* assay, with or without S9;[6] it was not reported to induce SCE in Chinese hamster V79 cells treated with a maximum dose of 2 mM,[7] but some increase in SCE in cultured human lymphocytes was found at a concentration of 10 mM.[8]

Chemobiokinetics. Following absorption, FA is oxidized to CO_2 and H_2O, predominantly in the liver, partly metabolized in the tissues, and partly excreted unchanged in the urine.[9] FA is an inhibitor of the mitochondrial cytochrome oxidase causing histotoxic hypoxia. The most significant acid load results from the hypoxic metabolism. Urinary acidification is affected by FA.[10]

Regulations. USFDA (1993) listed FA as a constituent of paper and paperboard used for food packaging at levels not to exceed GMP. It is considered to be *GRAS.*

Recommendations. 124 mg/l is proposed as water standard.[2]

References:

1. NTP, *Toxicity Studies of Formic Acid Administered by Inhalation to Fisher 344 Rats and B6C3F₁ Mice*, NIH Publ. 92–3342, July 1992, 41.
2. Sittig, M., Ed., *Handbook of Toxic and Hazardous Chemicals and Carcinogens*, 2nd ed., Noyes Publishers, Park Ridge, NJ, 1985, 465.
3. *Dangerous Properties of Industrial Materials,* Vol. 3, 7th ed., Van Nostrand Reinhold, New York, 1989, 1766.
4. Patty's *Industrial Hygiene and Toxicology,* 3rd revised ed., John Wiley & Sons, New York, 1981, 4096.
5. Tracor Jitco, Inc., Scientific Literature Reviews on Generally Recognized as Safe Food Ingredients—Formic Acid and Derivatives, Prepared by Tracor Jitco, Inc., for FDA, NTIS Publ. PB-228 558, U.S. Department of Commerce, Washington, D.C., 1974.
6. See **BISPHENOL A,**
7. Basler, A., Hyde, Wvd., and Scheutwinkel-Reich, M., Formaldehyde-induced sister-chromatid exchanges *in vitro* and the influence of the exogenous metabolizing systems S9 mix and primary rat hepatocytes, *Arch. Toxicol.,* 58, 10, 1985.
8. Siri, P., Jarventaus, H., and Norppa, H., Sister chromatid exchanges induced by vinyl esters and respective carboxylic acids in cultured human lymphocytes, *Mutat. Res.,* 279, 75, 1992.
9. Clay, R. L., Morphy, E. C., and Watkins, W. D., Experimental ethanol toxicity in the primate: analysis of metabolic acidosis, *Toxicol. Appl. Pharmacol.,* 34, 49, 1975.
10. Liesivuori, J. and Savolainen, H., Methanol and formic acid toxicity: biochemical mechanisms, *Pharmacol. Toxicol.,* 69, 157, 1991.

GLUTARALDEHYDE (GA) (CAS No 111–30–8)

Synonyms. 1,3-Diformylpropane; Glutaral; Glutaric dialdehyde; Glutarol; **1,5-**Pentanedial; **1,5-**Pentanedione; Sonacide; Verucasep.

Properties. Oily liquid. Soluble in water and ethanol at all ratios.

Applications. Used as a replacement for formaldehyde, in the manufacturing of paper and paperboard intended for food packaging; a bactericide.

Acute Toxicity. LD_{50} of the 50% aquatic solution is reported to be 1.3 ml/kg BW in rats and 1.59 ml/kg BW in rabbits; that of 25% aquatic solution is 1.87 and 8 ml/kg BW, and that of 5% aquatic solution is 3.3 and 16 ml/kg BW, respectively. In mice, LD_{50} is found to be 0.1 g/kg BW.[1] According to Ohno et al.,[2] LD_{50} in the young rats (5- to 6-weeks-old) is 283 mg/kg BW; LD_{50} in the old animals (57- to 60-weeks-old) is 141 mg/kg BW. Signs of intoxication include piloerection, red periocular and perinasal encrustations, sluggish movement, rapid breathing, and diarrhea. At necropsy, no gross pathological changes were noted in survivors; animals that died displayed distention, congestion and hemorrhagia of the stomach and small intestine, and lesions in the kidneys, liver, and lungs.[1]

Short-term Toxicity. Gross pathological examination failed to reveal changes in the visceral organs of rats given 0.5, 2.5, and 5% GA in the diet for 3 months. This study affirmed the GRAS status of GA when it is used in the food industry to cross-link edible collagen sausage casings.[3] Rats received 0.25% GA in their drinking water for 11 weeks. Animals exhibited no evidence of damage in the NS and CNS. There were no data indicating neurotoxic action of GA.[4]

Immunotoxicity. Contact hypersensitivity has been reported in mice and guinea pigs, which resulted from dermal application of 0.3 to 3.3% GA for 5 to 14 d.[5]

Reproductive Toxicity. According to Ballantyne,[1] there were no *embryotoxic* effects in the offspring of mice treated by gavage with up to 30 mg GA/kg BW on days 7 to 12 of gestation. In the recent Ema et al. studies,[6] rats were dosed by gastric intubation with 25 to 100 mg/kg BW on days 6 to 15 of pregnancy. The highest dose caused maternal toxicity and decreased fetal weights. Nevertheless, no *teratogenicity* or postimplantation loss was noted, even at a dose that induced maternal toxicity.

Genotoxicity. Exhibits mutagenic activity that could be explained as a result of oxidative damage to the DNA; GA is markedly cytotoxic.[7] GA is shown to be negative in *E. coli* and in the DLM test (30 to 60 mg/kg oral doses).[8] NTP results ranged from no activity to weakly positive in the *Salmonella* reversion assay.[8] GA appeared to be a potent mutagen in the mouse lymphoma cell line.[9]

Chemobiokinetics. Fisher 344 rats and New Zealand rabbits were exposed *i/v* to 1,5-^{14}C-GA. GA metabolism occurs through oxidation by rat liver mitochondria and in the kidney. It is excreted predominantly as CO_2. Urinary excretion of radioactivity was found to be in the range of 8 to 12% in rats and 15 to 28% in rabbits.[1]

Regulations. USFDA (1993) allowed the use of GA (1) as an antimicrobial agent in pigment and filler slurries used in the manufacturing of paper and paperboard at levels not to exceed 300 ppm by weight of the slurry solids and (2) in adhesives used as components of articles intended for use in packaging, transporting, or holding food.

References:

1. Ballantyne, B., Review of Toxicological Studies and Human Health Effects—Glutaraldehyde, Union Carbide Corp., Danbury, CT, 1986.
2. Ohno, K., Yasuhara, K., Kawasaki, Y., et al., Comparative study of glutaraldehyde acute toxicity in the old and young rats, *Bull. Natl. Inst. Hyg.* Sci., 109, 92, 1991 (in Japanese).
3. Devro, Inc., Glutaraldehyde oral toxicity tests indicate no related lesions, *Food Chem. News.*, 26, 42, 1984.
4. Spencer, P. C., Bischoff, M. C., and Schaumburg, H. H., On the specific molecular configuration of neurotoxic aliphatic hexacarbon compounds causing central peripheral distal axonopathy, *Toxicol. Appl. Pharmacol.*, 44, 17, 1978.
5. Stern, M. C., Holsapple, M. P., McClay, J. A., et al., Contact hypersensitivity response to glutaraldehyde in guinea pigs and mice, *Toxicol. Ind. Health*, 5, 31, 1989.
6. Ema, M., Itami, T., and Kawasaki, H., Teratological assessment of glutaraldehyde in rats by gastric intubation, *Toxicol. Lett.*, 63, 147, 1992.
7. Tamada, M., Sasaki, S., Kadono, Y., et al., Mutagenicity of glutaraldehyde in mice, *Bobkin Bobai*, 6, 62, 1978.
8. Beauchamp, R. O., St Clair, M. B. G., Fenell, T. R., et al., A critical review of the toxicology of glutaraldehyde, *Crit. Rev. Toxicol.*, 22, 143, 1992.
9. Hawort, S., Lawlor, T., Mortelmans, K., et al., Salmonella mutagenicity tests results for 250 chemicals, *Environ. Mut.*, Suppl. 1, 3, 1983.

GLYCIDE (CAS No 556–52–5)

Synonyms. Allyl alcohol oxide; **2,3**-Epoxy-1-propane; Glycidol; Glycidyl alcohol; **3**-Hydroxy-**1,2**-epoxypropane; **3**-Hydroxypropylene oxide; Oxiranemethanol.

Properties. Colorless, oily liquid. Miscible with water and alcohol at any ratio.

Applications. Used in the manufacturing of epoxy resins.

Acute Toxicity. In rats, LD_{50} is 850 mg/kg BW. Poisoning is accompanied by excitation, with subsequent CNS inhibition causing tremor and twitching of the facial muscles. In mice, LD_{50} is 430 mg/kg BW. Death occurs within 3 d.[1]

Repeated Exposure exhibits moderate cumulative properties.

Reproductive Toxicity. *Teratogenicity* effect was not found in rats given 200 mg/kg BW dose, although the litter of one female was of significantly lower weights.[2]

Genotoxicity. G. is shown to be positive in the *Salmonella* mutagenicity assay.[3] Produced a doubling of the SCE frequency, and a three-point monotonic increase, with at least the highest dose at the $p < 0.001$ significance level.[4]

Chemobiokinetics. G. is readily absorbed and distributed after gavage administration. G.-derived radioactivity was excreted primarily in the urine, to a lesser extent in the expired air as CO_2, and in the feces (NTP–92). *Carcinogenicity classification.* NTP: CE*—CE*—CE—CE*.

Chemobiokinetics. G. is readily absorbed and distributed after gavage administration.

References:
1. *Toxicology of New Industrial Chemical Substances,* Meditsina, Moscow, 15, 1979, 97 (in Russian).
2. See **EPICHLOROHYDRIN, #13.**
3. Wade, M. J., Mayer, J. W., and Hine, C. H., Mutagenic action of series of epoxides in *Salmonella, Mutat. Res.,* 66, 367, 1979.
4. See **FURFURAL, #5.**

HEXAMETHYLENETETRAMINE (HMT) (CAS No 100–97–0)

Synonyms. Formamine; Hexamine; Methenamine; Urotropine.

Properties. Odorless, colorless, hygroscopic crystals with a caustic, sweet, but subsequently bitter taste. Solubility in water is 81% (12°C), solubility in alcohol is 3.2% at 12°C. Decomposes in a slightly acidic solution, forming **ammonia** and **formaldehyde.** Aqueous solutions are clear, colorless, and odorless. Taste perception threshold is 60 mg/l.[1]

Applications and **Exposure.** A component of stabilizers, plasticizers, and catalysts used in the production of amine resins and of foam plastics. A vulcanization accelerator for rubber. HMT is used in the food industry as an antimicrobial additive (in fish products and in cheese-making).

Acute Toxicity. Rats and mice tolerate doses of 2 to 15 g/kg BW. Administration of high doses caused no pathological changes in the visceral organs or species differences in susceptibility to HMT.[1] However, according to other data,[06] LD_{50} in mice appears to be 512 mg/kg BW.

Repeated Exposure. *Humans.* No harmful reactions or complications were observed in patients receiving HMT as an antiseptic at dose-levels of 4 to 6 g/d for weeks. However, doses of 8 g/d for 3 to 4 weeks produced bladder irritation, painful and frequent micturition, albuminuria, and hematuria.[2] *Animals.* Mice tolerated a 10-d treatment with 5 g/kg BW dose. Capacity to work was not altered.[1]

Short-term Toxicity. Rats were gavaged with 400 mg/d for 90 or 333 d.[3] The yellow fur discoloration observed in animals was shown to be a consequence of a reaction between formaldehyde in the urine and kynurenine in the rat hair.[4]

Long-term Toxicity. Cats were fed up to 1250 mg/kg BW daily doses for 2 years. No effect was observed on food consumption and BW gain. There were also no histological changes in animal tissues.[5] No adverse effects and histopathology were reported in rats given the doses of 200 and 400 mg/d for 1 year.[3] Rats were dosed by gavage with 10 mg/kg BW for 7 months. The treatment caused liver dysfunction and behavioral effects. A dose of 0.1 mg/kg was considered to be ineffective in this study. Bearing in mind that HMT may be hydrolyzed with the formation of **formaldehyde,** MAC for HMT was recommended to be set at the same level as that of formaldehyde.[1]

Reproductive Toxicity. *Embryotoxicity.* Pregnant rats were exposed to 5 and 50 mg HMT/kg BW via their drinking water. In a five-generation study lasting 42 months, no alterations attributable to HMT were demonstrated in animals or fetuses and placentas at each dose-level or at every period of examination.[6] There were no effects on body growth, survival, reproduction, and viability of the offspring in rats fed a diet containing 400 to 1600 mg/kg for 2 years.[5] Addition of 1% HMT to drinking water did not cause any embryotoxic effect in Wistar rats (after mating, the treatment of females continued during pregnancy and lactation). No adverse effects were also observed in a three-generation study in rats. (There were 1 to 2% concentrations of HMT in drinking water.)[7] *Teratogenicity.* No effects on reproduction or litter numbers were reported in dogs that were fed the doses from 125 to 1875 mg HMT/kg BW for 32 months. However, two-thirds of litters had malformations.[5] In another dog study (0.06 and 0.125% HMT in the diet from day 4 to 56 after mating) the reproductive function of animals was not affected.[8] *Gonadotoxicity.* No differences in fertility were detected in two rat generations. (The exposed rats consumed about 100 mg HMT/kg BW.)[9]

Genotoxicity. HMT appeared to be a mutagen for *Dr. melanogaster* (Rapoport, 1946).

Carcinogenicity. Carcinomas and adenocarcinomas were reported to occur in female rats after 2 years of feeding with 1% HMT in the diet. However, a further study at the same dose-levels showed no differences in tumor incidence between the treated and control animals.[4] Tumor incidence in rats fed a diet containing 0.16% HMT from weaning to death was not higher than that observed in control rats.[9]

Chemobiokinetics. Toxicity of ingested HMT depends on the dose and the rate at which it is hydrolyzed in the presence of protein in the acid contents of the stomach. Intensive breakdown of HMT is shown to proceed also in the kidneys and urinary bladder. The most toxic metabolite is likely to be **formaldehyde.**

HMT is absorbed rapidly in the GI tract in man and appears in the urine in a few minutes.

Standards. Russia (1988). MAC: 0.5 mg/l.

Regulations. USFDA (1993) approved the use of F. (1) as a component of adhesives to be used safely in contact with food; (2) as an ingredient of paper and paperboard for contact with dry food; (3) as an accelerator for rubber articles intended for repeated use in contact with food up to 1.5% by weight of the rubber product; (4) as a polymerization cross-linked agent for protein, including casein in the manufacturing of paper and paperboard; (5) in urea-formaldehyde resins in molded articles intended for use in contact with food only as a polymerization control agent; (6) in closures with sealing gaskets that may be used safely on containers intended for use in producing, manufacturing, packing, processing, treating, packaging, transporting, or holding food (1%); and (7) in phenolic resins to be used safely as a food-contact surface of molded articles intended for repeated use in contact with nonacid food (pH above 5);

References:

1. See **ANILINE,** #1, 140.
2. Goodman, L. S. and Gilman, A., *The Pharmacological Basis of Therapeutics,* 2nd ed., McMillan, New York, 1955, 1087.
3. Brendel, R., Untersuchungen an ratten zur ventraglichkeit von hexamethylenetetramin, *Arzneimittel.-Forsch.,* 14, 51, 1964 (in German).
4. Kewitz, H. und Welsh, F., Ein gelber farbstoff aus formaldehyd und Kynurenin bei hexaminbehandelten ratten, Naunyn-Schmiedeberg, *Arch. Exp. Pathol. Pharmacol.,* 254, 101, 1966 (in German).
5. Restani, P. and Corrado, L. G., Oral toxicity of formaldehyde and its derivatives, *Crit. Rev. Toxicol.,* 21, 315, 1991.
6. Malorny, G., Rietbrock, N., und Schassan, H.-H., Uber den stoffwechsel des trimethylaminoxyds, *Naunyn-Schmiedebergs Arch. Exp. Pathol. Pharmacol.,* 246, 62, 1963 (in German).
7. Della Porta, G., Cabral, J. R., and Parmiani, G., Studio della tossicita' tranceplacentare e di cancerogenesi in ratti trattati con esametilentetramina, *Tumori,* 56, 325, 1970 (in Italian).
8. Hurni, H. and Ohder, H., Reproduction study with formaldehyde and hexamethylenetetramine in beagle dogs, *Food Cosmet. Toxicol.,* 11, 459, 1973.
9. Natvig, H., Andersen, J., and Rasmussen, E. W., A contribution to the toxicological evaluation of hexamethylene tetramine, *Food Cosmet. Toxicol.,* 9, 491, 1971.

1-HYDROXYETHYLFERROCENE (CAS No 1277–49–2)

Synonym. α-Methylferrocenemethanol.

Applications. Used as a light-sensitizer of plastics.

Acute Toxicity. The LD_{50} values are 380 to 930 mg/kg BW for mice and 1190 to 2660 mg/kg BW for rats. Gross pathological and microscopic examinations revealed enteritis and deposition of administered substances in the fat tissue and liver, gastric mucosal edema and increased thyroid and suprarenal activity.

Short-term Toxicity. Experimental animals were given suspensions and emulsions of 1-H. in sunflower oil for 3 months: $\frac{1}{20}$, $\frac{1}{10}$, and $\frac{1}{5}$ LD$_{50}$ (mice), and $\frac{1}{20}$ LD$_{50}$ (rats). Gross pathological examination revealed increased liver weights, deposition of 1-H. in the liver, spleen, lungs, and suprarenals, exudative glomerulonephritis, and changes in the thyroid.

Allergenic effect has not been noted.

Reference:
See **ADIPIC ACID,** #2, 275.

HYDROXYLAMINE (CAS No 7803–49–8)

Synonym. Oxammonium.

Properties. Unstable, large, white flakes or needles. Mixes with water at all ratios to form a weak base, **hydroxylamine hydrate.**

Applications. Used in the production of polycaprolactam.

Acute Toxicity. The lethal doses of **H. chloride** are 10 to 75 mg/kg BW for rabbits and 200 to 300 mg/kg BW for dogs. Poisoning is accompanied by CNS inhibition, convulsions, and paralysis. Death occurs from respiratory arrest.

Long-term Toxicity. A four- to fivefold enlargement of the spleen is observed in rats administered 330 to 380 mg/kg BW for 5.5 months. The thyroid shrank to half its size. Erythrocyte hemolysis and methemoglobinemia in mice R1R and blood clotting system disturbance[2] were also noted.

Allergenic Effect. H. has been found to increase sensitivity to allergens.

Reproductive Toxicity. *Embryotoxic* effect has been observed only at high doses. *Teratogenic* effect was noted in New Zealand white rabbits[3] but not in rats.[4]

Genotoxicity. H. and its derivatives are shown to cause CA in animal and plant cells. Such an effect seems to be the result of a direct action on DNA as a sequence of a radical formation, or it is caused indirectly.[5] H. is reported to increase slightly the rate of SCE and chromatid aberrations without recombination.[6]

References:
1. Arnol'dova, K. A. and Speransky, N. N., Some aspects of hydroxylamine chloride action on animals, *Gig. Truda Prof. Zabol.,* 12, 39, 1963 (in Russian).
2. Gross, P., Biological activity of hydroxylamine: a review, *CRC Crit. Rev. Toxicol.,* 14, 87, 1985.
3. Desess, J. M., Demonstration of the embryotoxic effects of hydroxylamine on the New-Zealand white rabbit, *Anat. Rec.,* 196, 45a, 1980.
4. Chaube, S. and Murphy, M. L., The effect of hydroxyurea and related compounds on the rat fetus, *Cancer Res.,* 26, 1448, 1966.
5. Auerbach, Ch., *Mutation Research [Problems of Mutagenesis],* Mir, Moscow, 1978, 315 (Russian, translation from English).
6. Peticone, P., Becchetti, A., Frediani, A., et al., *Atti. Assoc. Genet. Ital.,* 22, 219, 1977 (in Italian).

KAOLIN (Clay) (CAS No 1332–58–7)

Synonyms. Altowhite; Bentone; Bolus alba; China clay; Porcelain clay; White bole.

Composition and **Properties.** A product of the erosion of feldspars, mica, granites, and other strata. Consists mainly of mineral kaolinite, $Al_2O_3 \cdot 2SiO_2 \cdot 2H_2O$. A white or whitish powder, consisting of tiny hydrophilic particles with a lamellar structure. Finely-dispersed plastic rock.

K.-modified is produced by treating kaolin with a reaction product of isopropyl titanate and oleic acid in which 1 mol of isopropyl titanate reacts with 1 to 2 mol of oleic acid. The reaction product will not exceed 8% of K.-modified.

Applications. Filler.

Toxicity. Oral exposure had not been studied. It seems to be nontoxic.

Regulations. USFDA (1993) approves the use of K.-modified that meets CFR provisions (1) in olefin polymers as articles or components of articles intended for use in contact with food and (2) as a pigment, colorant, or opacifier in olefin polymers at levels not to exceed 40% by weight of the olefin polymers. K. is considered to be GRAS.

NAPHTHALENE (CAS No 91–20–3)

Synonyms. Naphthaline; Naphthene; Tar camphor; "White resin"; White tar.

Properties. Colorless, lamellar crystals with a characteristic odor and a sweetish, astringent taste. Readily soluble in organic solvents, poorly soluble in water (34.4 mg/l at 25°C). Odor perception threshold is 0.01 or 0.021 mg/l;[02] taste perception threshold is higher. Odor disappears in 3 to 4 d. The half-life in water is 24 to 48 h.[1]

Acute Toxicity. The LD_{50} values are reported to be 1250 mg/kg in rats, 580 mg/kg in mice, and 1200 mg/kg BW in guinea pigs. Death within 3 d. Poisoning is accompanied by apathy and adynamia. Doses of 400 and 600 mg/kg BW affected the renal tubular epithelium; a 200 mg/kg BW dose caused selective damage to the bronchial epithelium in mice (O'Brien Kym et al., 1985).

Repeated exposure revealed the capacity for functional accumulation. A total dose of 354 mg/kg BW given to mice for 8 d caused a 50% mortality (Stillwell et al., 1982). Administration of 27 to 267 mg/kg BW doses for 14 d caused 5 to 10% mortality and retardation of BW gain at the highest dose-level. There was a decrease in thymus weights in males and in spleen weights in females.[2] Rats were dosed with $\frac{1}{5}$, $\frac{1}{25}$, and $\frac{1}{125}$ LD_{50} for a month. At the end of the treatment they developed decreased BW gain, anemia and reticulocytosis, and a reduction in cholinesterase activity in the blood urea and creatinine levels.[1]

Short-term Toxicity. Administration of 5.5 to 133 mg/kg BW doses for 90 d did not cause mortality, retardation of BW gain, and any effect on serum enzyme or electrolyte level in rats.[2]

Long-term Toxicity. Changes of the glomerulonephritis type were observed in rats given 0.15 and 1.5 mg/kg BW.

Immunotoxicity effect was not found following repeated or short-term exposure.[2]

Reproductive Toxicity. Tolerated dose administered during pregnancy caused high maternal mortality. An embryotoxicity effect was slight.[3] The threshold dose for this effect is 0.75 mg/kg BW.

Genotoxicity. N. produces CA in the somatic cells. (Threshold dose is 0.015 mg/kg BW.) It is mutagenic for microorganisms.[4]

An **allergenic effect** is not observed after *i/g* exposure.

Carcinogenicity Classification. USEPA: Group D; NTP: XX—XX—NE—SE.

Chemobiokinetics. Following ingestion, N. accumulates in the liver, GI tract, heart, and lungs (in pigs and cattle). Its metabolism occurs with the help of the microsomal monooxygenase system. N. forms corresponding **epoxides,** which are subsequently converted into **dihydrodiols,** which themselves are converted into **phenols** and **catechols.** The latter are conjugated with glucuronides or sulfates. In mice, a large part of N. is metabolized to a glutathione adduct, which is then converted by hydrolysis, deamination, and decarboxylation into corresponding **mercaptolactic** and **mercaptoacetic acids. Methylthioesters** are also formed (Stillwell et al., 1982).

Standards. Russia (1988): MAC and PML 0.01 mg/l (organolept.).

Regulations. USFDA (1993) regulates the use of monosulfonated N. in adhesives used as components of articles intended for use in packaging, transporting, or holding food.

References:

1. Matorova, N. I., Data on the substantiation of maximum allowable concentration for naphthalin and naphthalene in water bodies, *Gig. Sanit.,* 11, 78, 1982 (in Russian).
2. Shopp, G. M., White, K. L., Holsapple, M. P., et al., Naphthalene toxicity in CD-1 mice: general toxicology and immunotoxicology, *Fundam. Appl. Toxicol.,* 4, 406, 1984.
3. See **DIMETHYL PHTHALATE,** #11.
4. See **RESORCINOL,** #3.

OP-7

Synonym. Monoalkylphenols, polyethylene glycol esters, mixture.

Properties. OP-7 is a viscous, oily liquid with a color ranging from light-yellow to brown, and with a bitter taste and specific odor. Water solubility is 5 g/l at 25°C. Odor perception threshold is

0.45 mg/l, taste perception threshold is more than 0.9 mg/l, and foam formation threshold concentration is 1 to 2 mg/l.[1]

Applications. Used in the production of synthetic rubber and plastic synthetic coatings. A component of sealing pastes for hermetically sealing food cans.

Acute Toxicity. In rats, the LD_{50} is reported to be 3 or 7.9 g/kg BW.[2] Death occurs within 1 to 2 d.[3]

Long-term Toxicity. Rats that received OP-7 in their drinking water at a concentration of 2 g/l developed mild fatty dystrophy (Oyeva, 1961).

Standards. Russia (1988). MAC and PML: 0.4 mg/l.

References:

1. See **ACRYLONITRILE,** #7, 139.
2. Il'in, I. E., Investigation of toxic products of transformation formed in the course of water chlorination, *Gig. Sanit.,* 2, 11, 1980 (in Russian).
3. See **ALKYLSULFATES, SODIUM SALT,** #2.

OP-10

Synonym. Polyethylene glycol, alkylphenyl ester.

Composition and **Properties.** A product of condensation of **alkylphenol** with ten molecules of **ethylene oxide.** Brown, oily, viscous liquid with a specific odor. Readily soluble in water (5 g/l). When shaken with water, forms a stable foam; when shaken with mineral and vegetable oils, forms stable emulsions. A concentration of 1.8 mg/l gives water an odor rating of 1 mg/l; a concentration of 3.6 mg/l corresponds to a rating of 2. It can be tasted at a concentration of more than 3.6 mg/l. Foam formation occurs at the concentration of 0.09 (Goyeva, 1961) or 1 mg/l.

Applications. Used in the production of synthetic rubber and plastics, synthetic anticorrosion coatings, and as an emulsifier in the production of latex articles for medical and food purposes.

Acute Toxicity. LD_{50} is reported to be 5 g/kg BW for mice and 3.5 to 5.48 g/kg BW for rats.[1] The main sign of poisoning seems to be CNS inhibition. Gross pathological examination revealed congestion in the visceral organs.

Repeated Exposure showed moderate cumulative properties. K_{acc} is 5 (by Lim).

Long-term Toxicity. Rats were dosed by gavage with 0.02 mg/kg BW. Adynamia was found to develop in a month after onset of the treatment. To the third month, a decline in the STI value and a reduction in the lung, kidney, and spleen weights were noted.[1]

Standards. Russia (1988). MAC and PML: 1.5 mg/l (organolept., odor).

Reference:

Mel'nikova, V. V. and Zhilenko, V. N., On toxicity of alkylphenyl ether of polyethylene glycol, *Kauchuk i resina,* 6, 56, 1980 (in Russian).

PETROLATUM

Composition and **Properties.** P. is obtained from **paraffin oils** by deparaffinization. P. is a mixture of high molecular weight **heavy hydrocarbons** with residual **oil** and **ceresin.** It does not alter the chromaticity of water, even at a concentration of up to 100 mg/l. Benzo[a]pyrene is not found in the samples of P.

Applications. Used for waterproofing.

Repeated Exposure failed to reveal cumulative properties. Mice that received 1 g/kg BW for 10 d exhibited neither mortality nor behavioral changes.[1]

Long-term Toxicity. Administration of a 10 mg/kg BW dose caused BW gain decrease and anemia. Gross pathological examination revealed a number of visceral changes. Rats given 2-d water extracts of P. did not exhibit signs of intoxication.[1] Sprague-Dawley rats were fed diets containing 10% ground wax (petrolatum) for 2 years (PAH content up to 0.64 mg/kg). There were no effects on survival rates and BW gain. Histological examination revealed no other wax-associated toxic effects.[2]

Carcinogenicity. No increase in tumor incidence was noted in a 2-year study in Sprague-Dawley rats.[2]

Regulations. USFDA (1993) regulates P. (1) as a component of adhesives, (2) as an ingredient of resinous and polymeric coatings for food-contact surfaces, (3) as a plasticizer for rubber articles intended for repeated use in contact with food up to 30% by weight of the rubber product, (4) as a defoaming agent that may be used safely as a component of articles intended for use in contact with food, (5) as a component of paper and paperboard intended for contact with dry food, (6) in the manufacturing of closures with sealing gaskets for food containers, and (7) as a substance employed in the production of or added to textiles and textile fibers intended for use in contact with food. P. may be used safely (8) as a component of nonfood articles in contact with food (specifications in the U.S. Pharmacopoeia XX for white P.). It is used or intended for use as a protective coating of the surfaces of metal or wood tanks used in the fermentation process (GMP requirements) and may contain any antioxidant permitted in food.

References:
1. See **DICHLOROMETHANE,** #6, 139.
2. The 39th Technical Report of JECFA, WHO, Geneva, 1992, 25.

PETROLEUM WAX (PW)

Composition. PW is a mixture of **solid hydrocarbons,** paraffinic in nature, derived from petroleum and refined. Reinforced wax consists of PW to which certain optional substances required for its production or for imparting desired properties have been added.[1]

Applications. PW is used as a component of nonfood articles in contact with food. Reinforced wax is used in producing, manufacturing, packing, processing, transporting, or holding food.

Toxicity and **Carcinogenicity.** Hydrocarbon waxes consumed in the diet are not absorbed or metabolized in significant amounts. Paraffin and microcrystalline waxes are nontoxic and noncarcinogenic.[2]

Regulations. JECFA (1992). ADI is not specified for the uses indicated in the specifications (chewing gum base, protective coating, defoaming agent, and surface-finishing agent).[2] **USFDA** (1993) listed PW for use (1) in adhesives as a component of articles intended for use in packaging, transporting, or holding food; (2) in resinous and polymeric coatings for polyolefin films to be used safely as a food-contact surface; (3) as a component of the uncoated or coated food-contact surface of paper and paperboard intended for use in contact with aqueous and fatty food; (4) as a defoaming agent that may be used safely as a component of articles or in the manufacturing of paper and paperboard intended for use in contact with food; (5) in cellophane to be used safely for packaging food; (6) in cross-linked polyester resins to be used safely as articles or components of articles intended for repeated use in contact with food; and (7) as a plasticizer in rubber articles intended for repeated use in producing, manufacturing, packing, processing, treating, packaging, transporting, or holding food (total not to exceed 30% by weight of the rubber products).

PW may contain any oxidant permitted in food, a total of more than 1% by weight of residues of the following polymers: homopolymers and/or copolymers derived from one or more of the mixed n-alkyl $(C_{12}-C_{18})$ methacrylate esters, where the C_{12} and C_{14} alkyl groups are derived from concut oil and the C_{16} and C_{18} groups are derived from tallow; **2**-hydroxy-**4**-n-octoxybenzophenone at a level not to exceed 0.01% by weight of the PW; poly(alkylacrylate) (CAS No 27029–57–8) as a processing aid in the manufacture of PW.

Synthetic PW may be used safely in application and under the same conditions where naturally derived PW is permitted.

Reinforced wax (RW) may include (1) GRAS substances, (2) substances subjected to CFR sanctions for use in RW, and (3) copolymers of isobutylene modified with isoprene, PW (Types I and II), polyethylene, rosins, and rosin derivatives, synthetic wax polymer (not to exceed 5% by weight of the PW).

References:
1. 21 Code of Federal Regulations, #178.3710, 1992.
2. See **PETROLATUM,** #2.

1,2-PHENYLENEDIAMINE (*o*-PDA) (CAS No 95–54–5)

Synonyms. *o*-Aminoaniline; *o*-Benzenediamine; *o*-Diaminobenzene; *o*-Phenylenediamine.

Properties. Light-yellow crystals. Solubility in water is 3% at 20°C. Readily soluble in alcohol. Gives water an odor of rotten hay. Odor perception threshold is 436 mg/l at 20°C (rating 1) and 364.5 mg/l at 60°C. Chlorination does not affect the perception threshold. Gives water a color ranging from light-yellow to dark-brown. (Chlorination decreases the threshold from 0.05 to 0.01 mg/l.) Does not cause suspensions, films, or foam to form in water.[1]

Applications and **Exposure.** Used in the paint and dyestuffs industry and in the production of coatings.

Acute Toxicity. In rats, mice, and guinea pigs, LD_{50} is 660, 470, and 360 mg/kg BW, respectively. CNS is predominantly affected. An excitation is followed by depression, motor coordination disorder, and impaired respiratory rhythm. Animals experience bloody nasal discharge, paralysis, and convulsions. Death occurs within the first 24 h.[1,2]

Repeated Exposure revealed moderate cumulative properties. K_{acc} is 1.48 (by Lim). Administration of $\frac{1}{10}$ to $\frac{1}{50}$ LD_{50} defined the LD_{50}/LOEL ratio to be 250.

Long-term Toxicity. Administration of *o*-PDA to rats caused erythrocyte depletion and increased activity of alkaline phosphatase, transaminase, and aldolase.[1,2]

Allergenic Effect was not found on *i/g* administration[2] but was noted on dermal application.[3]

Reproductive Toxicity. *Gonadotoxicity.* *o*-PDA produces a consistent reduction in the number and mobility of spermatozoa. Administration of 0.96 and 9.6 mg/kg BW to rats led to an increase in the number of pathological forms and tubules with epithelial desquamation and the 12th stage of meiosis.[2] ***Embryotoxicity.*** Dyban et al. (1970) revealed a marked distention and engorgement of the large blood vessels in embryos and of the medium-sized vessels of the liver, thoracic, and abdominal cavities (15% of fetuses). The NOEL for gonado- and embryotoxic effect is reported to be 1.8 mg/kg BW. A ***teratogenic*** effect is observed in chickens on introduction of 0.5 ml into the yolk sac of 4-d-old embryos. Observed malformations included ophthalmic defects, cleft palate, and skeletal abnormalities (Karnofsky and Lacon, 1962).

Genotoxicity. Contradictory data have been reported. A single administration of $\frac{1}{5}$ or $\frac{1}{50}$ LD_{50} of *o*-PDA does not increase the number of CA in bone marrow cells.[1,2] Sbrana and Loprieno[4] found an increase in the number of CA in murine bone marrow cells and in cultured human lymphocytes.

Carcinogenicity. Wild et al.[5] revealed a carcinogenic effect of *o*-PDA and consider it to be potentially genetically harmful to humans.

Standards. Russia (1988). MAC: 0.01 mg/l (organolept.,color).

References:
1. Galushka, A. I., Manenko, A. K., Gzhegotsky, M. I., et al., Hygienic substantiation of maximum admissible concentration for *o*-phenylenediamine and methylcyanocarbamate dimer in water bodies, *Gig. Sanit.,* 6, 78, 1985 (in Russian).
2. Galushka, A. I., Kogut, O. N., Dodoleva, I. K., et al., Toxicity of dimer methylcyanocarbamate, *o*-phenylenediamine, trilane, *Gig. Sanit.,* 1, 73, 1986 (in Russian).
3. Kurliandsky, B. A., Alexeeva, O. G., Livke, T. N., et al., Complex comparative assessment of sensitivity effect of phenylenediamine isomers for MAC-setting at workplace air, *Gig. Truda Prof. Zabol.,* 8, 46, 1987 (in Russian).
4. Sbrana, J. and Loprieno, U., The cytogenetic effect of *o*-phenylenediamine in mammalian and human cells, *Mutat. Res.,* 147, 318, 1985.
5. Wild, D., King, M. T., and Eckhart, K., Cytogenetic effect of *o*-phenylenediamine in the mouse, Chinese hamster, and guinea pigs and derivatives, evaluated by the micronuclei test, *Arch. Toxicol.,* 43, 249, 1980.

SURFACTANTS

Properties. White or yellowish powders, pastes, or liquids with an aromatic odor. Readily soluble in water with foam formation.

The Effect of Some Surfactants (Perception Thresholds) on the Organoleptic Indices of Water (mg/l)[1]

| | Odor | | | |
Surfactants	Rating 1	Rating 2	Taste	Foam formation
Nonionogenic				
OP-7	0.45	0.9	>0.9	0.1
OP-10	1.8	3.6	>3.6	0.09
Proxanol 186	6	10	6,400	0.09
Proxamine 385	14	26	5,400	0.09
Schistose alkylphenol	0.8	1.7	100	0.1
Sintamide 5	180	300	11,000	0.17
Sintanol DS-10	3.0	7.8	2,000	0.08
Sintanol MC-10	24	64	170	0.09
Sintanol DT-7	9	20	400	0.1
Anion active				
Alkylsulfate based on secondary nonsaponifiable alcohols	0.6	1.0	140	0.5
Alkylsulfonate		200	500	0.5
Azolate B	0.07	0.11	0.1	0.5
Chlorine sulfonol	75	100	300	0.5
DNS	110	160	125	0.6
DS-RAS	80	150	110	0.5
Schistose sulfonol	150	230	280	0.7
Secondary alkylsulfate	0.6	1.0	72	0.5
Sulfonol NP-3	70	150	60	0.4
Sulfonol NP-1		200	500	0.5

Acute Toxicity. LD_{50} is reported to be 1.1 to 7 g/kg BW in rats, 0.85 to 5 g/kg BW in mice,[1,2] and 0.85 to 6.85 g/kg BW in guinea pigs. According to other data, in the case of anionic surfactants, LD_{50} is 1 to 10.3 g/kg BW, and in the case of nonionogenic surfactants, it is 3.5 to 9.65 g/kg BW.[3]

Genotoxicity. S. exhibited slight mutagenic effect.

Chemobiokinetics. I/g administration with water at concentrations of 100 to 1000 mg/l for a month resulted in an enhanced lipid synthesis in the aorta as was evident from tests using labeled cholesterol and ^{14}C-acetate.[1,2]

Standards. Russia (1988). MAC and PML: 0.1 mg/l (organolept.) for **schistose amylphenol, oxanol L-7** and **KSH-9, OP-7** and **OP-10, proxamine 385** and **186, sintamide-5,** and sintanols **VN-7, VT-15, DS-10, DT-7,** and **MC-10.** MAC and PML 0.5 mg/l (organolept.) for **alkylbenzenesulfonates** (ABS), **alkylsulfonates, alkylsulfates,** and the **disodium salt** of **monoalkylsulfosuccinic acid.**

References:

1. See **ALKYLSULFATES, SODIUM SALT,** #2.
2. Mozhayev, E. A., Yurasova, O. I., Charyev, O. G., et al., The effect of surfactants on the lipid metabolism in white rats, *Gig. Sanit.*, 2, 85, 1986 (in Russian).
3. See **ACRYLONITRILE,** #7, 128.

SYNTHETIC NAPHTHENIC ACIDS (SNA) (CAS No 13308-24-5)

Composition. A mixture of **naphthenic** (cyclo-alkanecarbonic) **acids** obtained in oxidation treatment of a **naphthene concentrate** of **petroleum oils** from Azerbaijan.

Properties. An oily liquid of a light-yellow to brown color with a penetrating odor. Readily soluble in water. Threshold for organoleptic effect is 0.3 mg/l.[1]

Applications. SNA are used in the production of stabilizers and plasticizers.

Acute Toxicity. LD_{50} is 1.75 to 6.42 g/kg BW in rats and 1.77 to 7.17 g/kg BW in mice. Administration of 20 and 50% oil solutions resulted in a mild stimulation with subsequent CNS inhibition and adynamia. Animals assumed a side position. Death occurred in 2 d. Gross pathological examination failed to reveal pathological changes, but relative liver weights were increased.[2,3]

Repeated Exposure revealed cumulative properties: 50% of animals died from the overall dose of 2.16 g/kg BW having been daily exposed to $^{1}/_{10}$ LD_{50}.[4] K_{acc} (by Lim) is reported to be 0.33.[2]

However, according to other data, accumulation does not take place. Some of the animals died only when $\frac{1}{5}$ LD_{50} was administered over a period of 1.5 months but not when $\frac{1}{10}$ LD_{50} was given.[3] Repeated exposure did not affect STI and blood morphology. Consistent BW loss was noted. Rats were dosed with 200 mg SNA/kg given as a 10% oil solution for 1 month. The treatment caused retardation of BW gain but did not affect biochemical and hematological indices.

Long-term Toxicity. The NOAEL of 5 mg/kg BW was identified.[1]

Allergenic Effect was not found on dermal application.

Standards. Russia (1988). MAC: 0.3 mg/l (organolept.).

References:

1. See **ACETONE,** #3, 160.
2. *Problems of Occupational Hygiene, Industrial Toxicology and Occupational Pathology,* Sumgait, 9, 1974, 37 (in Russian).
3. Uzhdaviny, E. R. and Glukharev, Yu. A., Toxic properties of synthetic naphthenic and alkylbenzenoic acids, *Gig. Truda Prof. Zabol.,* 9, 48, 1984 (in Russian).
4. *Toxicologic Evaluation of Some New Plasticizers, Additives and Oil Coolants,* Azerbaijan. State Publisher, Baku, 1979, 33 (in Russian).

TALC (powder) (CAS No 14807–96–6)

Synonyms. Agalite; Asbestine; Soapstone; Steatite; Supreme; Talcum.

Properties. Fine, crystalline powder with a color ranging from white/pale-green to brown. The particles are mainly lamellar in shape.

Applications and **Exposure.** Talc is a component of many plastics as a stabilizer, reinforcer, and filler used at up to 70% (w/w). Synthetic rubber includes ground talc as a filler on compounding formulations.

Acute Toxicity. LD_{50} is not attained.

Short-term Toxicity. Wistar rats were administered 100 mg talc/d for 101 d. No significant depression of mean life-span was reported.[1]

Reproductive Toxicity. *Teratogenicity* effect was not observed in rats and mice given 1.6, in hamsters given 1.2, and in rabbits given 0.9 g/kg BW (IARC 42-205).

Genotoxicity. No data are available on the genetic and related effects in humans. Talc did not cause CA in human WI38 cells treated with talc at concentrations of 2 to 200 µg/ml *in vitro.*[2] Talc did not induce DLM or CA in bone marrow cells of rats treated with doses of 30 to 5000 mg/kg BW *in vivo* (Litton Bionetics, 1974), and is not mutagenic to yeast or to bacteria in host-mediated assay (IARC 42–185).

Carcinogenicity. *Humans.* Talc particles were found in stomach tumors from Japanese men, possibly due to ingestion of talc-treated rice.[2] *Animals.* Wistar rats were given 50 mg/kg in the diet or standard diet for life (average survival, 649 d). No increase in tumor incidence was found in this study.[3] In the diet of Wistar-derived rats, 100 mg Italian talc/d was introduced for 5 months and caused no tumors in the treated animals.[1] IARC considered that sufficient evidence is available for the carcinogenicity of talc containing asbestiform fibers (under inhalation exposure but not in the diet). *Carcinogenicity classification.* IARC: Group 1 (containing asbestiform fibers), Group 3 (not containing asbestiform fibers).

Regulations. JECFA (1989). ADI is not specified.

References:

1. *Inhaled Particles,* Vol. IV, Part 2, Walton W. H., and Govern, B., Eds., Pergamon Press, Oxford, 1977, 647.
2. Merliss, R. R., Talc-treated rice and Japanese stomach cancer, *Science,* 173, 1141, 1971.
3. Gibel, W., Lohs, K., Horn, K.-H., U. A., Experimental study of the carcinogenic activity of asbestos fibres, *Arch. Geschwulsforsch.,* 46, 437, 1976 (in German).

TALL OIL (TO) (CAS No 8002–26–4)

Synonyms. Acintol *C;* Liquid rosin; Talleol; Tallol; The sap of the pine tree.

Composition. TO consists of **fatty acids** (saturated and unsaturated) and **resin acids** (primary, secondary, neutral and hydroxy acids). Contains a small amount of **malodorous substances.**

356

Properties. An oily, dark-brown liquid with a soapy odor and taste. Odor perception threshold is 0.16 mg/l; taste perception threshold is 1.25 mg/l.

Applications. TO is used in the manufacturing of plastics and synthetic coatings, predominantly alkyd resins and rubber.

Acute Toxicity. LD_{50} in mice is 7.32 g/kg BW; 12.5 g/kg BW appears to be LD_{100}.[1] According to Oshchenkova et al. (1985), LD_{50} is 780 mg/kg BW in rats.

Repeated Exposure. Rats were administered 900 mg/kg BW for 10 d. The treatment caused a slight decrease in BW gain and some changes of leukocyte phagocytic activity.

Long-term Toxicity. In a 6-month study, rabbits were dosed by gavage with 0.03 and rats were dosed with 1.5 mg/kg BW. There were no toxic manifestations in the treated animals.[1,2]

Genotoxicity effect was not found.

Standards. Russia. Recommended MAC and PML: 0.2 mg/l (organolept., odor).

Regulations. USFDA (1993) considered TO as GRAS. It is listed (1) in adhesives used as components of articles intended for use in packaging, transporting, or holding food, (2) as a constituent of cotton and cotton fabrics used for dry food packaging at levels not to exceed GMP, (3) in resinous and polymeric coatings in a food-contact surface of articles intended for use in producing, manufacturing, packing, transporting, or holding food, and (4) as a defoaming agent that may be used safely in the manufacturing of paper and paperboard intended for use in producing, manufacturing, packing, transporting, or holding food.

Reference:

Chen Nai Tun, Experimental data on substantiating the maximum allowable concentration of tall oil in water bodies, *Gig. Sanit.*, 5, 9, 1962 (in Russian).

TRIMELLITIC ANHYDRIDE (TA) (CAS No 552–30–7)

Synonyms. Anhydrotrimellitic acid; **1,2,4**-Benzenetricarboxylic acid, **1,2**-anhydride; **1,3**-Dihydro-**1,3**-dioxo-**5**-isobenzofurancarboxylic acid; **1,3**-Dioxo-**5**-phthalancarboxylic acid; Trimellitic acid, **1,2**-anhydride.

Properties. White crystals in the form of flakes. Readily hydrolyzed in water to trimellitic acid. Poorly soluble in water. React with alcohols. Soluble in organic solvents.

Applications. Used in the production of polyester resins, polyimide compounds, adhesives, and dyes. Used as a plasticizer in materials intended to store and cover food and in the synthesis of various anticorrosive surface coatings and pharmaceutical products.

Acute Toxicity. LD_{50} is 1.9 to 5.6 g/kg BW in rats and 1.25 g/kg BW in mice.[1,2] Poisoned mice experienced irritation of the gastrointestinal mucosa, accompanied by hyperemia and hemorrhage, and sometimes by perforations.

Repeated Exposure revealed moderate cumulative properties. Hb level was lowered and CNS activity was affected.[1]

Allergenic Effect was observed due to formation of conjugates between protein groups and TA. TA is a potent respiratory sensitizer. Dermal sensitization reactions were observed. TA-specific antibody can be transferred from mother to fetus in rats and guinea pigs.[3]

Reproductive Toxicity effect is not found in mice.

Genotoxicity. TA is found to be positive in the *Salmonella* mutagenicity assay.

Chemobiokinetics. Following inhalation exposure, the major retention is found to be in the lymph nodes associated with the lung. TA reacts with the free amino groups on proteins to form conjugates.[4]

Regulations. USFDA (1993) approved the use of TA in resinous and polymeric coatings for use only as a cross-linking agent at the level not to exceed 15% by weight of the resin in contact with food under all conditions of use, except that resins intended for use with food containing more than 8% alcohol must contact such food only under the specified conditions.

References:

1. See **ADIPIC ACID,** #2, 178.
2. IPCS, *Trimellitic Anhydride,* Health and Safety Guide No. 71, WHO, Geneva, 1992, 30.
3. Leach, C. L., Hatoum, N. S., Zeiss, C. R., et al., Immunologic tolerance in rats during 13 weeks of inhalation exposure to trimellitic anhydride, *Fundam. Appl. Toxicol.*, 12, 519, 1989.

4. Thrasher, J. D., Madison R., Broughton, A., et al., Building-related illness and antibodies to albumin conjugates of formaldehyde, toluene diisocyanate, and trimellitic anhydride, *Am. J. Ind. Med.,* 15, 187, 1989.

TRIMETHYLAMINE (CAS No 75–50–3)

Synonym. *N,N'*-Dimethylmethanamine.

Properties. Gas with a pungent, fishy, ammoniacal odor. Readily soluble in water (410 g/l at 19°C) and alcohol. Odor perception threshold appears to be 0.04 mg/l, practical threshold is 0.16 mg/l. At 60°C, practical odor threshold is 0.05 mg/l. However, odor perception threshold is also reported to be 0.0002 mg/l.[02]

Applications. A component of ion-exchange resins.

Acute Toxicity. *Humans.* Intake of 15 mg T. hydrochloride/kg BW induces nausea and ichthyohydrosis (Calvert, 1973). *Animals.* LD$_{50}$ is 535 mg/kg BW in rats, 460 mg/kg BW in mice, 240 mg/kg BW in rabbits, and 315 mg/kg BW in guinea pigs. Gross pathological examination of rats that received a lethal dose revealed lesions in the GI tract, including mucosal defects and liver fatty dystrophy, as well as tiny necrotic foci and vacuolization in the renal tubular epithelium.[1]

Repeated Exposure revealed no cumulative properties: administration of $\frac{1}{10}$ LD$_{50}$ for a month caused no animal mortality.

Long-term Toxicity. Rabbits were dosed by gavage with 2.4 mg/kg for 6 months. The treatment produced anemia, reduction in prothrombin time and decrease in monoaminooxidase activity, as well as an increase in the enzyme capacity of serum aminotransferases.[1]

Allergenic Effect. A decline in the agglutinin titer has been noted in rabbits immunized with typhoid vaccine.

Chemobiokinetics. *N*-oxidation is the major route of metabolism, while *N*-demethylation is negligible and only significant at the higher dose-levels.[2]

Standards. Russia (1988). MAC: 0.05 mg/l (organolept., odor).

References:

1. Trubko, E. I. and Teplyakova, E. V., Hygienic norm-setting for trimethylamine in water bodies, *Gig. Sanit.,* 8, 79, 1981 (in Russian).
2. Al-Waiz, M., Mitchell, S. C., Idle, J. R., et al., The relative importance of *N*-oxidation and *N*-demethylation in the metabolism of trimethylamine in man, *Toxicology,* 43, 117, 1987.

XYLENOLS (CAS No 1300–71–6)

Synonyms. Dimethylphenols; DMP.

Properties. Colorless liquid. Solubility in water is 100 mg/l at 21°C. Taste and odor perception thresholds are different in various isomers and vary from 0.12 to 0.25 mg/l. Threshold concentration in chlorinated water is much lower: 0.001 to 0.002 mg/l. Heating accentuates the odor in the presence of 2,5-DMP and 2,6-DMP.

Acute Toxicity. LD$_{50}$ is reported to be 300 to 400 mg/kg BW (2,5-DMP and 2,6-DMP) and 610 to 730 mg/kg BW (3,5-DMP and 3,4-DMP) in rats, 380 to 480 mg/kg BW in mice, and 700 to 1300 mg/kg BW in rabbits (Maazik, 1986). According to Uzhdavini et al.,[1] when 2,4-DMP is administered in oil solution, LD$_{50}$ is 3200 mg/kg BW in rats and 810 mg/kg BW in mice. On administration of 3,5-DMP, LD$_{50}$ appeared to be 2250 and 836 mg/kg BW, respectively. Lethal doses result in ataxia and rapid onset of clonic convulsions. Death occurs within 24 h. Guinea pigs seem to be much less sensitive: administration of a 1.2 g/kg dose provoked only mild symptoms of intoxication.

Repeated Exposure revealed weak cumulative properties only in the case of 3,4-DMP (in rats given $\frac{1}{5}$ LD$_{50}$).

Short-term Toxicity. Rats were dosed by gavage with $\frac{1}{10}$ LD$_{50}$ (3,4-DMP and 2,6-DMP) for 10 weeks. The treatment caused retardation of BW gain. Gross pathological examination revealed liver damage. Phenol was not detected in the urine.

Long-term Toxicity. Rats were exposed by gavage to oral doses of 14 mg 3,4-DMP/kg BW for 8 months. Manifestations of toxic action included a reduction in BW gain, erythrocyte count, and

Hb level in the blood. Content of SH-groups in the blood serum and blood pressure were decreased at a dose-level of 6 mg 2,6-DMP/kg. Gross pathological findings proved changes in the liver, kidney, spleen, and heart.

Standards. Russia (1988): MAC and PML: 0.12 mg **2,5-DMP** and **2,6-DMP/l**), and 0.25 mg **3,4-**DMP and **3,5-DMP/l**.

Regulations. USFDA (1993) regulates DMP for use as a component of resinous and polymeric coatings coming in contact with food.

Reference:

Uzhdavini, E. R., Mamayeva, A. A., and Gilev, V. G., Toxic properties of **2,6-** and **3,5-**dimethylphenols, *Gig. Truda Prof. Zabol.,* 10, 52, 1979 (in Russian).

INDEX

A